# ANNUAL REVIEW OF NEUROSCIENCE

# ANNUAL REVIEW OF NEUROSCIENCE

## VOLUME 19, 1996

W. MAXWELL COWAN, *Editor*
Howard Hughes Medical Institute

ERIC M. SHOOTER, *Associate Editor*
Stanford University School of Medicine

CHARLES F. STEVENS, *Associate Editor*
Salk Institute for Biological Studies

RICHARD F. THOMPSON, *Associate Editor*
University of Southern California

ANNUAL REVIEWS INC.    4139 EL CAMINO WAY    P.O. 10139    PALO ALTO, CALIFORNIA 94303-0139

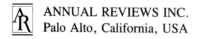

ANNUAL REVIEWS INC.
Palo Alto, California, USA

*International Standard Serial Number: 0147-006X*
*International Standard Book Number: 0-8243-2419-6*

Annual Review and publication titles are registered trademarks of Annual Reviews Inc.

♾ The paper used in this publication meets the minimum requirements of
American National Standard for Information Sciences—Permanence of Paper
for Printed Library Materials, ANSI Z39.48-1984.

Annual Reviews Inc. and the Editors of its publications assume no responsibility for the
statements expressed by the contributors to this *Review*.

Typesetting by Kachina Typesetting Inc., Tempe, Arizona; John Olson, President;
Toni Starr, Typesetting Coordinator; and by the Annual Reviews Inc. Editorial Staff

PRINTED AND BOUND IN THE UNITED STATES OF AMERICA

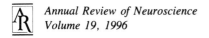

*Annual Review of Neuroscience*
*Volume 19, 1996*

# CONTENTS

# SOME RELATED ARTICLES IN OTHER *ANNUAL REVIEWS*

From the *Annual Review of Biochemistry*, Volume 65 (1996):

*Molecular Biology of Mammalian Amino Acid Transporters*, M. S. Malandro and M. S. Kilberg

*Connexins, Connexons, and Intracellular Communication*, D. A. Goodenough, J. A. Goliger and D. L. Paul

*Biochemistry and Structural Biology of Transcription Factor IID (TFIID)*, S. K. Burley and R. G. Roeder

*Hematopoietic Receptor Complexes*, J. A. Wells and A. M. de Vos

From the *Annual Review of Cell and Developmental Biology*, Volume 11 (1995):

*Receptor-Mediated Protein Sorting to the Vacuole in Yeast: Roles for Protein Kinase, Lipid Kinase, and GTP-Binding Proteins*, J. H. Stack, B. Horazdovsky, and S. D. Emr

*Nitric Oxide: A Neural Messenger*, S. R. Jaffrey and S. H. Snyder

*Integrins: Emerging Paradigms of Signal Transduction*, R. M. Nelson, A. Venot, M. P. Bevilacqua, R. J. Linhardt, and I. Stamenkovic

From the *Annual Review of Genetics*, Volume 29 (1995):

*Membrane Protein Assembly: Genetic, Evolutionary, and Medical Perspectives*, C. Manoil and B. Traxler

*Yeast Transcriptional Regulatory Mechanisms*, K. Struhl

*Trinucleotide Repeat Expansion and Human Diseases*, C. T. Ashley, Jr. and S. T. Warren

*Signal Pathways that Establish the Dorsal-Ventral Pattern of the Drosophila Embryo*, D. Morisato and K. V. Anderson

*Molecular Genetic Aspects of Human Mitochondrial Disorders*, N.-G. Larsson and D. A. Clayton

*Inherited Hearing Defects in Mice*, K. P. Steel

From the *Annual Review of Medicine*, Volume 47 (1996):

*The Cardiac Ion Channels: Relevance to Management of Arrhythmias*, D. M. Roden and A. L. George, Jr.

*Nicotine Addiction and Treatment*, J. E. Rose

*Differential Diagnosis and Management of Cushing's Syndrome*, C. Tsigos and G. P. Chrousos

From the *Annual Review of Pharmacology and Toxicology,* Volume 36 (1996):

From the *Annual Review of Physiology,* Volume 58 (1996):

From the *Annual Review of Psychology*, Volume 47 (1996):

ANNUAL REVIEWS INC. is a nonprofit scientific publisher established to promote the advancement of the sciences. Beginning in 1932 with the *Annual Review of Biochemistry*, the Company has pursued as its principal function the publication of high-quality, reasonably priced *Annual Review* volumes. The volumes are organized by Editors and Editorial Committees who invite qualified authors to contribute critical articles reviewing significant developments within each major discipline. The Editor-in-Chief invites those interested in serving as future Editorial Committee members to communicate directly with him. Annual Reviews Inc. is administered by a Board of Directors, whose members serve without compensation.

For the convenience of readers, a detachable order form/envelope is bound into the back of this volume.

*Annu. Rev. Neurosci. 1996. 19:1–26*

# HUMAN IMMUNODEFICIENCY VIRUS AND THE BRAIN

## *Jonathan D. Glass*[1,2] *and Richard T. Johnson*[1,3,4]

Departments of [1]Neurology, [2]Pathology (Neuropathology), [3]Neuroscience, and [4]Molecular Biology and Genetics, Johns Hopkins University School of Medicine, Baltimore, Maryland 21287

KEY WORDS: AIDS, neurology, neuropathology, dementia, neurotoxicity

## ABSTRACT

Human immunodeficiency virus (HIV) infects the nervous system in the majority of patients, causing a variety of neurological syndromes throughout the course of the disease. This review focuses on the effects of HIV in the central nervous system, with an emphasis on HIV-associated dementia. HIV-associated dementia occurs in a subset of patients with AIDS; it is unclear why these patients and not all patients develop the disease. Several factors are likely to be involved in the pathogenesis of HIV-associated dementia, including neurotoxins released from the virus and/or infected macrophages and microglia, immunologic dysregulation of macrophage function, and specific genetic strains of HIV. These factors, and their possible interactions, are discussed.

## INTRODUCTION

Over the past 15 years, the human immunodeficiency virus (HIV) has emerged globally as one of the most influential infectious agents of all time. The impact of this virus can be seen in almost every facet of society, from international politics to intramural school sports. Universal precautions have become the rule, even with the most casual of human contacts. In the field of medicine, HIV and the disease it causes, acquired immune deficiency syndrome (AIDS), have introduced us to new and unusual types of disorders and promoted an explosion of interest in the fields of infectious disease and immunology.

As of late 1994, about 15 million people are infected with HIV worldwide, with greater than 1 million infected in the United States. More than 3 million people have died of AIDS. Although an encouraging decline in the rate of new infections has occurred in certain subgroups of patients (e.g. homosexual men), the overall rate of new infections continues to increase. This trend is particularly alarming in Southeast Asia, where the seroconversion rate in intravenous

1

drug users in Bangkok has run as high as 5% per month (Anderson et al 1991). In Africa, AIDS has already devastated heterosexual populations in many urban areas, and in the United States and Europe, AIDS continues to rise in intravenous drug users, their sexual partners, and their progeny. Considering that virtually all persons infected with HIV will eventually develop AIDS, these increasing infection rates portend enormous economic burdens, societal disruption, and human suffering worldwide (Mann 1993).

This review focuses on current information regarding the pathogenesis of HIV-associated neurological disease. The clinical information presented refers to neurological disease in adults and is not necessarily generalizable to the neurological syndromes seen in HIV-infected children (Belman et al 1988). We emphasize HIV-associated dementia; however, many of the concepts concerned with damage to the nervous system during HIV infection are clearly relevant to HIV-associated myelopathy and peripheral neuropathy. Two species of HIV, types 1 and 2, have been identified, but data are sparse on the neurological pathogenicity or virulence of HIV-2. Unless otherwise specified, all data presented in this review are for HIV-1.

## NEUROLOGICAL SYNDROMES ASSOCIATED WITH HIV

The remarkable array of neurological syndromes that occur throughout the course of HIV infection affect the central nervous system, the peripheral nervous system, and muscle. These disorders can be categorized either as related to opportunistic infections of the nervous system or as direct or indirect effects of the virus itself. Some disorders are manifested early and some late during the infectious process, and the pathological changes include inflammatory, demyelinating, and degenerative changes. This spectrum of diseases associated with a single virus infection is unique in the field of virology (Figure 1).

Although rare, neurological disease can occur at the time of initial infection with HIV (even before seroconversion) and during the period when there is chronic lymphadenopathy without the profound immunosuppression of AIDS. These early manifestations include aseptic meningitis or encephalitis, acute and chronic inflammatory demyelinating polyneuropathies, mononeuritis multiplex associated with peripheral nerve vasculitis, and HIV-associated polymyositis (McArthur 1987). These disorders are thought to have an autoimmune pathogenesis; patients with demyelinating polyneuropathies and polymyositis typically respond to immunosuppressive or immunomodulatory therapies in a manner similar to that of HIV-negative patients with similar clinical syndromes.

Neurological morbidity and mortality occur mainly after the onset of profound immunodeficiency (see Figure 1). The progressive dementia occurring

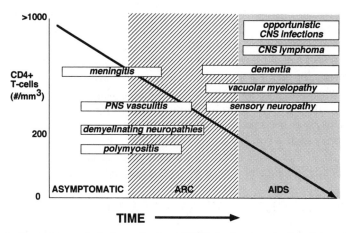

*Figure 1*  The major neurological complications of HIV infection plotted against their general time of occurrence after infection and blood CD4 count. Most of the neurological morbidity and mortality occurs during AIDS (CD4 < 200/mm$^3$). (ARC = AIDS-related complex.) [Modified from Johnson et al (1988).]

during AIDS was originally described by Navia et al (1986b) and named the AIDS-dementia complex. This disorder has been given a variety of other names (McArthur & Harrison 1994) and is referred to in this article as HIV-associated dementia. HIV-associated dementia occurs in approximately 20% of patients and results in the loss of independence and a shortened life expectancy (McArthur et al 1993). HIV-associated dementia is characterized by cognitive impairment, i.e. mental slowness, forgetfulness, and poor concentration; behavioral changes, including apathy, lethargy, and diminished emotional responses; and motor symptoms characterized by the loss of fine motor control, unsteady gait, and tremor. The progression of this disorder is variable, but it often culminates in mutism, incontinence, and generalized spasticity. Because of the prominence of withdrawal and motor symptoms, the syndrome has been referred to as a subcortical form of dementia.

HIV-associated myelopathy occurs in approximately 20% of patients with late-stage AIDS and is clinically characterized by progressive spasticity and loss of proprioception, predominantly in the lower extremities. Pathologically these findings correlate with vacuolar changes that are most prominent in the posterior and lateral columns of the thoracic spinal cord (termed vacuolar myelopathy) and are morphologically similar to those seen in vitamin B12 deficiency (Petito et al 1985). Autopsy studies have shown that vacuolar myelopathy occurs in up to 50% of patients who died with AIDS; only the more severe cases show symptoms during life (Dal Pan et al 1994).

The peripheral neuropathy seen in AIDS is distinct from the inflammatory neuropathies seen earlier in the infection. Patients typically develop a distal, symmetric axonal polyneuropathy, which frequently is complicated by painful dysesthesias causing significant morbidity (de la Monte et al 1988). Approximately 40% of patients develop this polyneuropathy prior to death, but as with vacuolar myelopathy, autopsy studies show that a much higher percentage of patients who died with AIDS show morphological abnormalities in peripheral nerves (Griffin et al 1995a,b). Even though HIV-associated dementia, myelopathy, and neuropathy often occur in the same patient, it is unclear whether they share a combined or even a parallel pathogenesis. The pathological features of these three disorders are quite distinct, and the specific role of HIV infection in the pathogenesis of each is still unknown.

## NEUROPATHOLOGY OF HIV INFECTION

The neuropathological effects of HIV infection can be separated into abnormalities related to opportunistic processes (Table 1) and those either directly or indirectly related to the presence of HIV in the brain [for reviews, see Rhodes (1987), Gray et al (1988), Sharer (1992)]. Direct infection of the brain with HIV is associated with abnormalities collectively termed HIV encephalitis. HIV encephalitis is quite different from other viral encephalitides (e.g. herpes simplex, measles); inflammatory infiltrates of lymphocytes are virtually absent, and tissue destruction is usually minimal. The lesion is characterized by the presence of HIV-infected macrophages and microglia that are usually found in perivascular regions or coalesced into inflammatory microglial nodules. Infected macrophages sometimes form giant syncytia, called multinucleated giant cells,

**Table 1**  Frequency of pathological diagnoses of the major opportunistic processes in brain[a]

| | | |
|---|---|---|
| Toxoplasmosis | 7% | 1984–89 = 10.3% |
| | ($n$ = 22) | 1990–94 = 4.8%[b] |
| Cytomegalovirus (excludes | 19% | Disseminated = 38% |
| inclusions only without | ($n$ = 60) | Encephalitis = 25% |
| tissue disruption) | | Ventriculitis = 13% |
| | | Polyradiculitis = 8% |
| Progressive multifocal | 4% | |
| leukoencephalopathy | ($n$ = 13) | |
| Cryptococcus | 8% | |
| | ($n$ = 27) | |
| Primary CNS lymphoma | 11% | |
| | ($n$ = 34) | |

[a] Data are from the Johns Hopkins AIDS Neuropathology brain bank; 347 patients dying with AIDS or HIV infection between 1984 and 1994.
[b] No cases in 1993 or 1994.

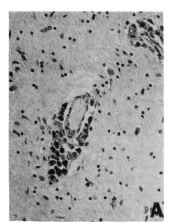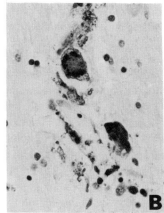

*Figure 2*  Immunocytochemical localization of HIV in the brain, using a monoclonal antibody to the gp41 transmembrane protein. (*A*) The typical perivascular location of HIV infected macrophages and (*B*) two multinucleated giant cells, as well as mononuclear cells, around a blood vessel and stained with the anti-HIV antibody. (Magnification: *A* = 300 times, *B* = 570 times.)

that either are associated with microglial nodules or may occur in isolation from regions of inflammation (Figure 2). Rarely, HIV encephalitis causes tissue cavitation and necrosis; this severe form is more common in cases of pediatric AIDS (Kure et al 1991, Brustle et al 1992). HIV encephalitis should not be confused with HIV-associated dementia, which is a designation reserved for the clinical syndrome. The relationship between HIV encephalitis and HIV-associated dementia is still an open question and is addressed below.

White matter abnormalities have received considerable attention because of the frequency of changes seen on magnetic resonance images and the subcortical nature of HIV-associated dementia (Navia et al 1986b). Several abnormalities of white matter have been recognized; the most common are HIV leukoencephalopathy (Budka 1991) and diffuse myelin pallor (Kleihues et al 1985, Navia et al 1986a, Rosenblum 1990). HIV leukoencephalopathy appears to be a variant of HIV encephalitis that is centered within the white matter and that includes myelin breakdown, macrophage infiltration, and the presence of HIV—sometimes with multinucleated giant cells. The more common finding of diffuse myelin pallor is an abnormality of the histochemical staining characteristics of myelin in the deep white matter, with relative sparing of subcortical U-fibers. Diffuse myelin pallor cannot be considered a demyelinating or destructive lesion because specific immunolabeling shows normal amounts of each of the major myelin proteins (Power et al 1993). Interestingly, the presence of diffuse myelin pallor correlates with the presence of dementia

(Navia et al 1986a, Glass et al 1993), so the etiology of this histological phenomenon may provide clues to the pathogenesis of HIV-associated dementia. It has been suggested that diffuse myelin pallor is due to increased amounts of interstitial water and reflects a breakdown in the blood-brain barrier. Abnormalities of barrier function in AIDS have been demonstrated by several groups (Smith et al 1990, Petito & Cash 1992, Power et al 1993), who have suggested that a leaky blood-brain barrier may allow toxic serum proteins access to cortical and subcortical structures.

Astrocytosis is almost always found in the brains of patients dying with AIDS, and it may be one of the earliest recognizable abnormalities in patients with dementia (McArthur et al 1989). A pathological entity termed diffuse poliodystrophy is identified by the presence of reactive astrocytosis and microglial activation in the cerebral cortex and subcortical gray structures, sometimes associated with a spongiform change (Ciardi et al 1990). In culture models, astrocytes have been shown to be essential for HIV-induced neuronal toxicity (see below); consequently their increased presence in the brains of patients with AIDS may not only be reactive but may play a primary pathogenetic role in neurological disease.

Gross changes in regional brain anatomy in patients infected with HIV-1 have been studied both by magnetic resonance imaging and by measurements of pathological specimens. There is a general consensus that patients with AIDS develop brain atrophy manifested by shrinkage of the cerebral cortex and deep gray structures, and by enlargement of the cerebral ventricles. Several studies have correlated these changes with the clinical syndrome of HIV-associated dementia. Using magnetic resonance images, Dal Pan et al (1992) and Aylward et al (1993) showed a relative decrease in the size of the caudate nucleus in patients with HIV-associated dementia and an indirect association between the severity of dementia and the size of the caudate. No differences in cortical volumes were detected. Data from pathological specimens have demonstrated both thinning of the cerebral cortex (Wiley et al 1991) and overall cerebral atrophy in patients dying with AIDS; however, cerebral atrophy may be a generalized finding in AIDS that is not necessarily specific for HIV-associated dementia (Oster et al 1993). We have recently completed a comparison of cortical and subcortical brain volumes in autopsy specimens from patients prospectively characterized for the diagnosis of HIV-associated dementia (P Subbiah & JD Glass, unpublished data). We found that in comparison to seronegative controls, patients who died with AIDS showed smaller volumes of brain structures independent of the diagnosis of HIV-associated dementia.

Routine neuropathological examination of the cerebral cortex, even in patients with the most severe form of HIV-associated dementia, rarely shows a recognizable degree of neuronal loss, as might be seen in patients dying with Alzheimer's disease. Several groups, however, using various methods to assess neuronal densities, have concluded that neuronal loss occurs in AIDS (Ketzler et al 1990; Everall et al 1991, 1993; Wiley et al 1991). Counting neurons in the cerebral cortex is a daunting task, requiring methods designed to avoid overgeneralization and systematic bias (Mayhew 1992). Whether the reduction in neuronal density (or actual loss of neurons) is a finding restricted to patients with HIV-associated dementia has not been thoroughly studied; the available data thus far indicate that as with cortical atrophy, neuronal loss may be a more generalized finding in AIDS patients and not necessarily confined to patients with dementia (Seilhean et al 1993, Everall et al 1994).

Microscopic abnormalities of neuronal processes have been beautifully demonstrated with Golgi preparations and laser confocal microscopy in brains of patients with AIDS and in experimental models of HIV neurotoxicity (Wiley et al 1991, Masliah et al 1992, Hill et al 1993, Toggas et al 1994). These studies show simplification of dendritic arborization and loss of presynaptic terminals, which may be related to the presence of HIV in these regions. The relationship of these changes to the clinical syndrome of HIV-associated dementia has not been investigated.

### Clinical-Pathological Correlations

As alluded to above, there is a general lack of information regarding the presence or specificity of any individual pathological finding with the clinical syndrome of HIV-associated dementia. The few published studies have shown only a moderate correlation between HIV-associated dementia and the presence at autopsy of multinucleated giant cells, diffuse myelin pallor, and/or HIV encephalitis (Navia et al 1986a, Glass et al 1993, Wiley & Achim 1994). In fact, there are many well-documented examples of nondemented patients with severe HIV-related neuropathology and, more intriguingly, of patients with severe HIV-associated dementia but with none of these pathological changes (Navia et al 1986a, Vazeux et al 1992, Glass et al 1993). This lack of a tight correlation between clinical symptomatology and structural abnormalities has encouraged theories of pathogenesis invoking indirect mechanisms of neuronal dysfunction, which may show subtle structural correlates not easily recognized with conventional neuropathological techniques. As of this writing, the pathological substrate of HIV-associated dementia remains an open question.

## ROLE OF THE VIRUS

HIV-1 and HIV-2 are retroviruses belonging to the lentivirus family. They contain two single-stranded RNA molecules that upon infection of the host cell, are synthesized into a double-stranded DNA provirus by the viral enzyme reverse transcriptase (RNA-dependent DNA polymerase). This RNA-to-DNA

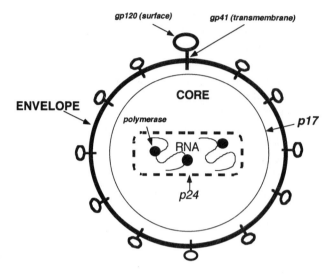

*Figure 3* Simplified diagram depicting the major structural components of HIV (see text).

transcription is done with a remarkable lack of fidelity, which in part, explains the hypermutability of HIV (Preston et al 1988). There are three major genes coding for structural proteins: *gag,* which encodes the core proteins including p24, the most abundant protein; *pol,* which encodes reverse transcriptase and the integrase; and *env,* which encodes the surface envelope proteins, including the gp160 complex, which is composed of the transmembrane glycoprotein gp41 and the surface glycoprotein gp120. The latter external protein contains multiple conformational loops, including the V3 loop that provides both the attachment site for the major cellular HIV receptor CD4 and an adjacent site that determines macrophage tropism (Chesebro et al 1992). Small spliced genes—including *tat, rev,* and *nef*—code for proteins that regulate virus replication. A simplified structural diagram depicting the major components involved in infectivity, antigenicity, and neurotoxicity (discussed below) is presented in Figure 3 [for complete reviews of viral structure and regulation, see Narayan & Clements (1989) and Levy (1994)].

When HIV was isolated in 1983 from CD4-positive lymphocytes (T-helper and -effector cells), a simple scenario for the pathogenesis of AIDS was proposed. In this schema, growth was limited to the CD4-positive lymphocytes; their destruction led to immunodeficiency; and opportunistic infections and tumors led to death. The dementia seen in AIDS patients was attributed to cytomegalovirus or an as yet unidentified opportunistic agent. A role for HIV itself in the pathogenesis of the neurological disorders associated with AIDS generally was not considered until about 1985, with the realization that HIV is a neurotropic virus that infects the nervous system frequently (Ho et al 1985, Levy et al 1985, Resnick et al 1985, Shaw et al 1985). Subsequent studies of cerebrospinal fluids of recently infected asymptomatic gay men showed that brain infection occurred early during the incubation period in the majority of patients (McArthur et al 1988). A patient who died 15 days after accidental intravenous injection of HIV showed infected cells within the brain (Davis et al 1992). Further experience with this disease has discounted the importance of opportunistic infections and focused on a primary role for HIV in neurological disease and especially in the pathogenesis of HIV-associated dementia.

Autopsy studies of patients who died with AIDS have demonstrated that HIV is present in the brain in a majority of cases (Sinclair & Scaravilli 1992, Achim et al 1994; JD Glass & RT Johnson, unpublished data), but how HIV enters the nervous system is still unclear. Early studies of HIV localization in the brain as well as in vitro studies demonstrated that endothelial cells may be infected (Wiley et al 1986, Moses et al 1993), suggesting that free virus may enter the nervous system by this route. The majority of data, however, do not support HIV infection of a significant number of brain endothelial cells. A more likely scenario is that HIV enters the nervous system as a passenger within infected blood monocytes, i.e. the Trojan horse hypothesis (Williams & Blakemore 1990). These monocytes may enter the brain in response to chemoattractant signals released by damage from opportunistic pathogens, or they may home specifically to brain because they are infected with HIV. Studies with both HIV and simian immunodeficiency virus (SIV; see below) have shown exclusively macrophage-tropic strains in the brain (Cheng-Mayer et al 1989, Epstein et al 1991, Sharma et al 1992, Power et al 1994). Peripheral inoculation of macrophage-tropic but not of lymphocyte-tropic strains of SIV are associated with subsequent brain infection (Sharma et al 1992). Why macrophage-tropic strains specifically enter and replicate in the nervous system is unknown, but this selectivity may reflect the expression of cellular adhesion molecules and their ligands on infected monocytes and brain endothelium (Sasseville et al 1992).

The identification of the infected cells within the nervous system has been controversial. Although cells of astrocyte or neuronal origin can be infected in vitro, in vivo it is generally agreed that cells of macrophage origin are the predominant cells productively infected. These include parenchymal microglia and macrophages and the perivascular macrophages, which are located in the space between the endothelial basement membrane and the astrocyte foot process and may be difficult to distinguish spatially from endothelial cells. Two recent reports demonstrated astrocytic infection in pediatric brains infected with AIDS (Saito et al 1994, Tornatore et al 1994), and there has been one report suggesting HIV infection of neurons (Nuovo et al 1994). Our own studies using immunocytochemistry combined with in situ polymerase chain reaction (PCR) on tissue sections demonstrated that the predominant infected cells are macrophages and microglia, with evidence of viral DNA in occasional cells staining for glial fibrillary acid protein (K Takahashi & SL Wesselingh, unpublished data). The rarity of infection in cells of neuroectodermal origin, however, raises questions as to their biological significance.

The issue of how much virus is present and whether the viral load or burden is correlated with HIV-associated dementia has been approached in several ways. The conclusions from these studies differ, and there is no general consensus about whether the amount of virus in the brain determines either the presence or the severity of clinical symptomatology. Studies of blood and cerebrospinal fluid have focused on the presence of HIV p24 antigen (Goudsmit et al 1986, Portegies et al 1989, Royal et al 1994, Singer et al 1994), which can be measured reliably in acid-hydrolyzed specimens. At this time, however, p24 antigen levels are not routinely measured in clinical practice, which reflects the variability of test results and the lack of a clear predictive value for specific disease processes. Research studies that have addressed this issue in HIV-associated dementia have produced conflicting results. Singer et al (1994) found a direct correlation between the mean p24 antigen levels in spinal fluid and the presence of HIV-associated dementia, whereas Portegies et al (1989) concluded that p24 antigen expression in the spinal fluid was a poor predictor of neurological disease. Royal et al (1994) compared three groups of patients based on neuropsychological status and also found a correlation between detectable levels of p24 in both blood and cerebrospinal fluid and the presence of HIV-associated dementia, suggesting a higher viral load in these patients. Only 50% of demented patients had detectable p24 in their spinal fluids, a finding that is likely an underestimate of the number of patients with HIV in the brain. Spinal fluid analysis frequently underestimates the severity of abnormalities in parenchymal brain diseases, and it remains unclear whether looking for HIV in spinal fluid, even by more sensitive PCR-based techniques (Tourtellotte et al 1993), will answer the question of viral load in HIV- associated dementia.

Pathologically based studies of viral burden in HIV-associated dementia have generally suffered from either a lack of prospectively collected clinical information or the use of semiquantitative or nonquantitative measures of viral load, or both. Wiley and colleagues (Achim et al 1994, Wiley & Achim 1994, Wiley et al 1991) have correlated viral load as determined by several methods—including immunocytochemistry, gp41 ELISA, and PCR—with the presence of pathologically determined HIV encephalitis. Although they found that these measures correlate with HIV-related pathological changes, these data do not address the question of viral load in clinically determined HIV- associated dementia. In a more recent study, these investigators showed that immunocytochemical staining for HIV gp41 correlated with HIV-associated dementia as determined by retrospective chart review (Wiley & Achim 1994). We also addressed this question using gp41 immunocytochemistry (JD Glass, unpublished data) in a prospectively characterized population and found that most patients, demented or nondemented, showed immunoreactivity for gp41 in the brain and that the abundance of staining was not necessarily correlated with either the presence or the severity of dementia. We found that some patients with severe HIV-associated dementia show little or no evidence of productive HIV infection in the brain, an observation that was previously reported in a pediatric population (Vazeux et al 1992).

Immunocytochemistry, even at its best, is only useful as a semiquantitative measure, a problem that is amplified when addressing questions in postmortem human tissue. Quantifying HIV presents a further complication, since the virus is not evenly distributed throughout the brain. The answer to the question of viral load in pathological specimens is likely to come from quantitative PCR techniques (Bell et al 1993; RT Johnson, JD Glass, JC McArthur & B Chesebro, unpublished data).

Given that HIV enters the brain in the majority of cases (neuroinvasive) and infects cells (neurotropic), the major question is what determines its ability to cause neurological disease (neurovirulence). Several groups have looked at the possibility that the genetic strain of HIV in the brain, rather than the viral load, may determine whether a patient develops HIV-associated dementia (Epstein et al 1991; Li et al 1991, 1992; Pang et al 1991; Keys et al 1993; Power et al 1994). There is precedent for this mechanism of clinical disease in experimental and naturally occurring viral diseases in animals, including neurotropic retroviral diseases (Georgsson et al 1989, Narayan & Clements 1989, Masuda et al 1992, Sharma et al 1992). In a recent study, Power et al (1994) cloned and sequenced, directly from the brains of 22 patients dying with AIDS, a 430-bp region of the HIV envelope that included the V3 loop of gp120. These patients were prospectively evaluated for the diagnosis of HIV-associated dementia by serial neurological and psychometrical examinations and followed to the time of death. Analysis of these brain-derived sequences showed that virus from both demented and nondemented patients had the macrophage-tropic sequences within the V3 loop. However, there were differences at amino acid positions 305 and 329 in the 14 demented patients when compared to the 8 nondemented patients dying with AIDS. The group with HIV-associated dementia also showed a significantly greater number of unique amino acid residues in this region. These findings lay the foundation for an intriguing mechanism of pathogenicity, that of viral strain, which may explain the puzzling dichotomy of AIDS patients dying with or without HIV-associated dementia. The issue of possible interactions between the virus and host factors in the pathogenesis of HIV-associated dementia is virtually unexplored territory.

## MEDIATORS OF NEUROTOXICITY

The issue of neurotoxicity in HIV infection has been extensively studied by in vitro methods in a variety of model culture systems. Using these models, investigators have attempted to separate the ability of HIV or HIV-infected macrophages and microglia to cause direct injury to the nervous system by the interaction of HIV-infected cells with other cell types to produce neurotoxic substances.

### Neurotoxins from HIV-Infected Macrophages

Whether infected macrophages and microglia secrete toxins that play a role in neurological dysfunction has been a topic of intense study and continued debate (Giulian et al 1990, Dhawan et al 1992, Pulliam et al 1991, Bernton et al 1992). In 1990, Giulian and colleagues (1990) showed that HIV-infected monocytoid cell lines (U937 and THP-1) produced diffusible toxic substances able to kill neurons from embryonic chick ciliary ganglion and embryonic rat spinal cord. Characterization of this neurotoxin showed it to be >2 kDa in size, both heat and cold stable, and resistant to protease inactivation. The neurotoxic activity was prevented by $N$-methyl-D-aspartate (NMDA) receptor antagonists, suggesting a role for NMDA receptors in toxicity. Pulliam et al (1991) also identified neurotoxic activity in HIV-infected monocytes by using a system of cultured aggregates from human fetal brains. Cytopathic effects were seen in all cell types but were specific for HIV-infected monocytes and could not be reproduced using lipopolysaccharide-activated macrophages.

Bernton et al (1992) and Tardieu et al (1992) also investigated the possibility of direct neurotoxicity of HIV-1–infected monocytes and could not demonstrate that toxic soluble factors were released from infected monocytes. Bernton et al made the important observation that contamination of culture systems with *Mycoplasma* species creates a picture of neurotoxicity unrelated to the presence of HIV and that similar neurotoxic effects are induced by tumor necrosis factor–$\alpha$ (TNF-$\alpha$) and lipopolysaccharide. Tardieu et al concluded that the cytopathic effect of HIV-infected monocytes occurs only by direct cell contact and not by the release of a soluble neurotoxin.

At the time of this writing, an adequate resolution to the apparent inconsistencies in the data from different laboratories is lacking. There is, however, good reason to believe that HIV-infected macrophages and microglia indeed play a significant role in the pathogenesis of HIV-related neurological disease. Because macrophages and microglia are the cells infected with HIV, they must, by definition, be included in any scheme that involves HIV in the pathogenesis of HIV-associated dementia.

## Toxicity of HIV Proteins

Many investigators have shown that HIV proteins or peptides are toxic to neurons and glia in model systems both in vitro and in vivo. The most prominent neurotoxic viral protein is gp120, the envelope glycoprotein that acts as the ligand for the CD4 receptor during cellular infection. On the basis of data from culture and animal models, it is hypothesized that gp120 can be released from the surface of free virus or from the cytoplasmic membrane of HIV-infected cells and produce its effects at a distance (Brenneman et al 1988; Dreyer et al 1990; Lipton et al 1991, 1994a; Hill et al 1993). Experimental evidence shows that gp120 may act to cause neuronal injury either directly via a non–CD4-mediated mechanism (Harouse et al 1991) or indirectly through neurotoxic mediators released by gp120-stimulated macrophages or astrocytes (Lipton 1994a,b). In cultures of rat retinal ganglion cells, exposure to picomolar amounts of gp120 caused a marked rise in intracellular free calcium, presumably through the activation of dihydropyridine-sensitive voltage-gated calcium channels (Dreyer et al 1990). This effect was not blocked by addition of anti-CD4 antibodies, a finding that suggests that binding to CD4 receptor is not necessary for neuronotoxicity. Several studies have demonstrated that gp120-mediated calcium entry and toxicity involves activation of neuronal NMDA receptors (Lipton 1994a, Lipton & Rosenberg 1994) and that this excitotoxic mechanism may involve generation of nitric oxide intermediates (Dawson et al 1993).

The effects of gp120 can be alleviated in cultured rat hippocampal neurons and in rat pups by analogues of the neuronal growth factor vasoactive intestinal peptide (VIP), suggesting that gp120 may compete for activation of a receptor that is essential for neuronal survival (Brenneman et al 1988, Hill et al 1993). Additional potential effects of gp120 include inhibition of myelin formation (Kimura-Kuroda et al 1994) and alteration of ion transport in astrocytes (Benos et al 1994a,b).

## Cytokines

Indirect mechanisms of gp120 neurotoxicity may involve its binding to CD4 receptors on brain macrophages and microglia and/or astrocytes, inciting the release of neurotoxins or mediators of neurotoxicity (Morganti-Kossmann et al 1992, Banati et al 1993, Lipton 1994b). For example, cytokines, which are small soluble proteins that normally function as regulators of the immune response, have been the focus of intense study because of the growing body of evidence suggesting that abnormal production, release, or relative amounts of cytokines in the brain play a role in AIDS-related neurological diseases. All cell types in the nervous system are capable of producing one or more cytokines, but the majority of data demonstrate that in the context of disease, brain macrophages and microglia and astrocytes are the most important cytokine producers (Merrill & Chen 1991, Benveniste 1994). In response to the addition of purified gp120, cultured human monocytes release the cytokines TNF-α, interleukin-1β and -6 (IL-1β, IL-6), and monocyte and granulocyte-macrophage colony stimulating factor (M-CSF and GM-CSF) (Clouse et al 1991). All of these cytokines (and possibly others) have actions that are known to fit well into theoretical schemes for pathogenesis of AIDS-related neurological disease.

TNF-α is a proinflammatory cytokine that can up-regulate the production of HIV-1 in infected cells (Griffin et al 1991, Mellors et al 1991, Tadmori et al 1991, Wilt et al 1995), induce the proliferation of astrocytes (Barna et al 1990), and influence the secretion of other cytokines by astrocytes (Benveniste 1994). TNF-α is toxic to cultured oligodendrocytes and myelin (Robbins et al 1987, Selmaj & Raine 1988) and may alter the function of cultured neurons (Shibata & Blatteis 1991, Soliven & Albert 1992). Both astrocytes and macrophages secrete TNF-α in vitro (Chung & Benveniste 1990, Lieberman et al 1990, Wesselingh et al 1990), but in the brains of patients dying with AIDS, macrophages and microglia are the dominant cells producing TNF-α (Tyor et al 1992; SL Wesselingh, unpublished data). Clinically, elevated levels of TNF-α have been found in the sera of AIDS patients, and several observations indicate that the presence of TNF-α in the nervous system is related to HIV-associated dementia. Increased amounts of TNF-α were found in cerebrospinal fluid samples from adults and children with dementia (Mintz et al 1989, Grimaldi et al 1991). In a prospective clinical-pathological study of AIDS in adults, the levels of mRNA for TNF-α in brains of demented patients were significantly higher than those found in either nondemented AIDS patients or

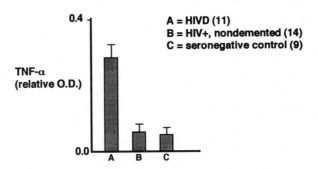

*Figure 4*  Relative quantities of TNF-α mRNA (determined by Southern blot) from the frontal subcortical white matter of seronegative controls, patients with HIV-associated dementia, and nondemented patients dying with AIDS. There is a significant increase in TNF-α mRNA in demented patients as compared to either of the other group [as described in Wesselingh et al (1993)].

seronegative controls (Glass et al 1993, Wesselingh et al 1993) (Figure 4), a finding that suggests a role for TNF-α in the pathogenesis of AIDS-related neurological disease.

Other proinflammatory cytokines of particular interest for HIV-related neurological disorders are IL-1b, IL-6, TGF-β, and M- and GM-CSF. Data from in vitro culture models and from human studies suggest that each of these cytokines may be elevated in response to HIV infection (Gallo et al 1989, 1994; Vitkovic et al 1991; Cupp et al 1993; Tyor et al 1992; Genis et al 1992; Wahl et al 1991; Benveniste 1994); however, these findings have not been universal. Wesselingh et al (1993), quantitating mRNA levels from postmortem brain samples found elevated amounts of mRNA for TNF-α, but reduced amounts of mRNAs for IL-1b and IL-6 in patients with HIV-associated dementia. This finding suggests that macrophages and microglia may be activated to induce differentially the transcription of specific cytokines, possibly through the action of nuclear regulators such as nuclear factor κB (NfκB) (Lieberman et al 1990, Griffin et al 1991).

Differential expression and production of cytokines from macrophages or lymphocytes may alter the normal balance of proinflammatory and antiinflammatory effects of the cytokine system, creating a milieu where dysregulation of otherwise normal cytokine responses produces a vicious cycle of cellular activation and cytotoxicity. Shearer and colleagues have recently proposed such a mechanism for T cells, where an imbalance between TH-1 and TH-2 cell products imposes a systemic environment conducive to HIV replication and progressive immunosuppression (Clerici & Shearer 1993, Mosmann 1994). A similar mechanism may occur in the nervous system during AIDS, where TH-2–type cytokines may be deficient, allowing for unchecked activation of macrophages. Evidence for this comes from Wesselingh et al (1994), who found increased levels of TNF-α and decreased levels of IL-4 and IL-10 in brain, spinal cord, and peripheral nerve tissues from patients with dementia, myelopathy, and neuropathy, respectively.

## Arachidonic Acid Metabolites and Platelet-Activating Factor

Cellular membrane phospholipids are metabolized through the production of arachidonic acid and subsequently via the cyclooxygenase pathway to the formation of prostaglandins and thromboxanes or via the lipoxygenase pathway to the formation to leukotrienes. These substances affect neuronal function and interact in the regulation of cytokine production and may play a role as neurotoxins or mediators of neurotoxicity in a variety of neurological diseases (Egg et al 1980, Merrill et al 1983). Macrophages are highly enriched in lipids containing arachidonic acid and are phagocytes of cellular debris; thus, they are major producers of arachidonic acid and its metabolites. Given the central role of macrophages and microglia and their products in HIV-related neuro-logical disease, these macrophage products are of interest. Griffin et al (1994) found elevated cerebrospinal fluid levels of prostaglandins $E_2$ and $F_{2a}$ and thromboxane $B_2$ in cerebrospinal fluids of patients with HIV-associated dementia compared to those in HIV-infected patients without dementia and seronegative controls. Genis et al (1992) showed elevated amounts of arachi-donic acid metabolites in cocultures of astrocytes and HIV-infected monocytes, and Wahl and colleagues (1989) demonstrated that HIV gp120 can increase production of these molecules in monocyte cultures. A direct neurotoxic effect of these metabolites has not been demonstrated; however, these molecules may induce the production of neurotoxic mediators such as TNF-$\alpha$ (Genis et al 1992), serving to fan the fire of the neurotoxic cascade.

Platelet activating factor (PAF), another molecule of interest, is produced in HIV-infected monocyte and astrocyte cocultures (Genis et al 1992) and is elevated in the spinal fluids of patients with AIDS and HIV-associated demen-tia, as well as in patients with systemic neoplastic and metabolic diseases (Gelbard et al 1994). This molecule is directly neuronotoxic to human fetal cortical cultures and rat retinal cultures through an NMDA-mediated process, a mechanism that has also been demonstrated for HIV gp120-mediated neuro-toxicity (Dreyer et al 1990, Lipton et al 1991).

## Quinolinic Acid

Quinolinic acid is a downstream metabolite of L-tryptophan, which is produced via activation of the kynurenine pathway in systemic tissues and possibly in brain. The idea that quinolinic acid is a neurotoxin has been championed by Heyes and colleagues, who have shown its elevation in the serum and cere-brospinal fluid in human brain diseases (Heyes et al 1989, 1991, 1993) and in animal models of neurological disease (Heyes et al 1990, Heyes & Nowak 1990). Quinolinic acid may enter the brain by crossing the blood-brain barrier, or it could be produced locally by macrophages and microglia or astrocytes in response to interferon-$\gamma$ or other cytokines (Heyes et al 1991). Quinolinic acid acts as an excitotoxin at the NMDA receptor, as do several other proposed neurotoxins, causing elevations of intracellular calcium and possibly cell death. Elevated levels of quinolinic acid are found in a variety of inflammatory brain diseases lacking the characteristic pathological changes associated with AIDS; this suggests that quinolinic acid may be a marker and not a mediator of disease.

Other products of activated macrophages are elevated in cerebrospinal fluid, including neopterin and $B_2$-microglobulin (Brew et al 1989, 1990; McArthur et al 1992). There is a moderate correlation with the levels of these markers and dementia, possibly reflecting that macrophage activation is maximal in demented patients or is involved in the disease process.

# ANIMAL MODELS

As in any human disease, animal models are important tools for exploring disease mechanisms and testing potential therapies. Unfortunately, HIV is as human-specific as its name implies and, with very few exceptions, infects only the human host. Nevertheless, immunodeficiency disorders induced in animals by other retroviruses have provided valuable information about pathogenesis that may be applied to HIV. The most useful virus for the understanding of neurological disorders is the simian immunodeficiency virus (SIV), which in its natural hosts, African monkeys, causes no disease but, when transmitted to Asian macaques, causes an immunodeficiency disease similar to human AIDS. Simian AIDS is accompanied by cognitive and motor neurological abnormalities and neuropathological changes that resemble those caused by HIV in humans (Sharer et al 1991, Hill et al 1993, Rausch et al 1994). The data that best link SIV and HIV-induced neurological disease come from a group of 15 rhesus monkeys trained on a battery of cognitive and motor tasks and later inoculated with SIV (Murray et al 1992, Rausch et al 1994). Eight of 10 inoculated monkeys developed productive infection, of which 3 developed cognitive changes and 7 showed abnormalities in motor skills. All of the infected animals showed elevated levels of quinolinic acid in plasma and cerebrospinal fluid, which did not accompany the onset of cognitive impairment in the affected animals. The extensive neuropathologic analysis mirrored the findings in human AIDS, showing extensive astrocytosis in all animals, multinucleated giant cell formation in only one animal, and no consistent correlation between clinical impairment and anatomical location or pathological extent of lesions. One animal with the earliest and most pronounced clinical impairments showed minimal neuropathologic changes and no SIV immunoreactivity in the brain (Rausch et al 1994), simulating the paradoxes of clinical-pathological correlation and brain viral burden in HIV-associated dementia.

Even though there are many similarities between the disorders caused by SIV in monkeys and HIV in humans, it can be reasonably argued that the differences between the viruses make this model less than perfect. Consequently, several groups are pursuing direct HIV infection of nonhuman primates or xenografting of HIV-infected human cells into rodent hosts. Successful infection with HIV has been demonstrated in chimpanzees and gibbon apes (Watanabe et al 1991), but these animals have not yet developed disease. Pigtailed macaques (*Macaca nemestrina*) can also be infected with HIV and develop immune deficiency (Agy et al 1992), but neurological disease has not been reported. HIV-2 was shown to cause an AIDS-like condition in baboons (Barnett et al 1994), but there was no report of neurological abnormalities.

Xenografting techniques have demonstrated infection of the nervous systems of rodents with HIV and are likely to prove valuable tools for studying the pathogenesis and therapy of HIV-related neurological disease. Mice with severe combined immune deficiency (*scid* mice) will accept grafts of human monocytes into the brain that can later be infected by direct inoculation of HIV (Tyor et al 1993). These grafted and infected mice develop some of the neuropathological hallmarks of human HIV infection. Human fetal brain and retinal explants grafted into the anterior eye chamber of immunosuppressed rats have been infected with HIV, and these grafts exhibited some of the pathological abnormalities attributed to HIV in in vitro culture systems (Cvetkovich et al 1992).

Transgenic mice expressing HIV or specific HIVproteins are another potentially useful model of HIV-related neurological disease. Toggas et al (1994) inserted a truncated envelope transgene encoding HIV gp120 into the glial fibrillary acidic protein (GFAP) gene and created transgenic mice expressing and secreting gp120 from astrocytes. These mice showed neuropathological changes in the cortex and hippocampus, including astrogliosis, microglial activation, and synaptic and dendritic abnormalities. Some degree of caution must be expressed in interpreting this as a model for the human condition, since the presence of gp120 protein could not be demonstrated in these mouse brains and since astrocytes are not the primary target for HIV in humans. Even with these limitations, this and other transgenic models can be powerful tools for the study of the effects of HIV in the brain.

Other animal models that are under study and deserve mention are feline immunodeficiency virus (FIV) (Phillips et al 1994) and murine AIDS (MAIDS) (Sei et al 1992). These viruses cause immunological abnormalities in their respective hosts, but accompanying neurological disease needs to be defined.

## SUMMARY AND HYPOTHESES

The initial systemic infection with HIV causes a disruption of normal immune regulation and abnormal immunoglobulin responses, and these changes probably lead to autoimmune diseases such as inflammatory demyelinating neuropathies. The virus in brain is localized predominantly in macrophages and microglia, but during the incubation period this usually represents an asymptomatic infection, and in general there is an absence of severe disease until after CD4 cells decline in the peripheral blood and immunodeficiency is evident. This systemic failure of immune responses may allow the greatest viral replication in the brain, but there is little evidence of this. This failure could also alter the control of cytokines and their release by infected cells, and this systemic change may be a factor in the development of neurological

**Table 2**  List of the major issues in the pathogenesis of HIV-associated dementia[a]

| Known | Questions |
| --- | --- |
| 1. HIV infects the nervous system early. | 1. Does HIV enter as a free virus or is it carried within infected cells? |
| 2. In the CNS, HIV primarily infects cells of macrophage origin. | 2. How does infection of nonneural cells lead to neurological disease? |
| 3. Only a subset of patients get HIV-associated dementia. | 3. How do the patients with dementia differ from those without dementia? |
| 4. Immunodeficiency is necessary but not sufficient for development of HIV-associated dementia. | 4. Does immunodeficiency allow for dysregulation of immune effector cells and molecules systemically or selectively in the CNS? |
| 5. Neurons are reduced in number or morphologically abnormal in AIDS, but they are not infected with HIV. | 5. Do neuronal changes correlate with HIV-associated dementia? |
| 6. Neuropathological abnormalities and quantities of HIV in the brain may not correlate with clinical disease. | 6. Is clinical disease dependent on<br>a. virus load?<br>b. virus strain?<br>c. neurotoxic viral proteins?<br>c. neurotoxic viral proteins?<br>d. virally induced secretion of cytokines or toxins? |

[a] Both accepted facts and current related question are presented.

disease. Factors released in the systemic circulation may disrupt the blood-brain barrier or factors may be generated by macrophages in the brain, as has been suggested by the studies of TNF-$\alpha$.

With each new discovery about the effects of HIV or HIV-infected cells in the nervous system, new models are generated to include interactions and effects on neurons, neurites, glia, or other cells. Our current model of HIV-induced neurological dysfunction still contains many question marks, but revolves around primary infection of brain macrophages and microglia and immune dysregulation (Table 2). It must be reiterated that two major features of HIV-associated dementia are (a) not all patients are affected and (b) infection of the nervous system is unlikely the sole determinant of clinical disease. HIV infects the brain in most HIV-infected persons, and this infection occurs early, but most patients remain neurologically asymptomatic until they develop AIDS. There is no direct correlation, however, between the presence of dementia and systemic CD4 counts; many patients with long and profound immunodeficiency do not develop dementia.

This pattern of clinical disease in HIV infection is clearly different from that of other viral brain infections, where clinical signs tend to correlate with the selective vulnerability of specific neural cells to infection. For example, infection of meningeal and ependymal cells leads to the clinical syndrome of viral meningitis; infection of motor neurons by poliovirus leads to poliomyelitis; and infection of oligodendrocytes by JC virus leads to progressive multifocal leukoencephalopathy. Furthermore, in experimental infections of the nervous system, the quantity of virus in the brain shows a reasonable correlation with disease onset, severity, and outcome. In the case of HIV-associated dementia, none of these generalizations hold true. Infection is largely limited to macrophages and microglia for which we know of no specific clinical correlate. The relationship between disease or disease severity and viral load is unconvincing, and the presence of the neurovirulent strains do not correlate with an increased amount of virus or changes in host cells to explain the nature of their neurovirulence. How neurovirulence is expressed is unknown; there are several indirect mechanisms that need further study, including the roles of viral proteins, cytokines, and other neurotoxins.

A reasonable argument can be made for HIV being involved only indirectly in the pathogenesis of neurological disease, through its ability to create an environment of immune dysregulation. Support for this hypothesis comes from studies of AIDS-related vacuolar myelopathy and sensory neuropathy. In these disorders, little or no virus is found in the affected tissues even though there is macrophage activation and infiltration, and cytokine production (Petito et al 1994, Yoshioka et al 1994; SL Wesselingh, unpublished). During AIDS, macrophages may be activated by the presence of infection or injury, but they may not be susceptible to deactivation, creating a vicious cycle of further macrophage activation, cytokine production, and release of neuro- toxins.

ACKNOWLEDGMENTS

We acknowledge the valuable input and discussion by Drs. JC McArthur, SL Wesselingh, JW Griffin, and DE Griffin. We thank Dr. Pamela Talalay for reviewing the manuscript and Mr. Rod Graham for manuscript preparation. This work was supported by grants NS26643 and NS01577 from the National Institute of Neurological Diseases and Stroke.

## Literature Cited

Achim CL, Wang R, Miners DK, Wiley CA. 1994. Brain viral burden in HIV infection. *J. Neuropathol. Exp. Neurol.* 53:284–94

Agy MB, Frumkin LR, Corey L, Coombs RW, Wolinsky SM, et al. 1992. Infection of *Macaca nemestrina* by human immunodeficiency virus type-1. *Science* 257:103–6

Anderson RM, May RM, Boily MC, Garnett GP, Rowley JT. 1991. The spread of HIV-1 in Africa: sexual contact patterns and the predicted demographic impact of AIDS. *Nature* 352:581–88

Aylward EH, Henderer JD, McArthur JC, Brettschneider PD, Harris GJ, et al. 1993. Reduced basal ganglia volume in HIV-1–associated dementia: results from quantitative neuroimaging. *Neurology* 43:2099–104

Banati RB, Gehrmann J, Schubert P, Kreutzberg GW. 1993. Cytotoxicity of microglia. *Glia* 7:111–18

Barna BP, Estes ML, Jacobs BS, Hudson S, Ransohoff RM. 1990. Human astrocytes proliferate in response to tumor necrosis factor alpha. *J. Neuroimmunol.* 30:239–43

Barnett SW, Murthy KK, Herndier BG, Levy JA. 1994. An AIDS-like condition induced in baboons by HIV-2. *Science* 266:642–46

Bell JE, Busuttil A, Ironside JW, Rebus S, Donaldson YK, et al. 1993. Human immunodeficiency virus and the brain: investigation of virus load and neuropathologic changes in pre-AIDS subjects. *J. Infect. Dis.* 168:818–24

Belman AL, Diamond G, Dickson D, Horoupian D, Llena J, et al. 1988. Pediatric acquired immunodeficiency syndrome: neurological syndromes. *Am. J. Dis. Child.* 142:29–35

Benos DJ, Hahn BH, Bubien JK, Ghosh SK, Mashburn NA, et al. 1994a. Envelope glycoprotein gp120 of human immunodeficiency virus type 1 alters ion transport in astrocytes: implications for AIDS dementia complex. *Proc. Natl. Acad. Sci. USA* 91:494–98

Benos DJ, McPherson S, Hahn BH, Chaikin MA, Benveniste EN. 1994b. Cytokines and HIV envelope glycoprotein gp120 stimulate Na$^+$/H$^+$ exchange in astrocytes. *J. Biol. Chem.* 269:13811–16

Benveniste EN. 1994. Cytokine circuits in brain: implications for AIDS dementia complex. In *HIV, AIDS, and the Brain*, ed. RW Price, SW Perry, pp. 71–88. New York: Raven

Bernton EW, Bryant HU, Decoster MA, Orenstein JM, Ribas JL, et al. 1992. No direct neuronotoxicity by HIV-1 virions or culture fluids from HIV-1–infected T cells or mono-

cytes. *AIDS Res. Hum. Retroviruses* 8:495–503

Brenneman DE, Westbrook GL, Fitzgerald SP, Ennist DL, Elkins KL, et al. 1988. Neuronal cell killing by the envelope protein of HIV and its prevention by vasoactive intestinal peptide. *Nature* 353:639–42

Brew BJ, Bhalla RB, Fleisher M, Paul M, Khan A, et al. 1989. Cerebrospinal fluid beta2 microglobulin in patients infected with human immunodeficiency virus. *Neurology* 39:830–34

Brew BJ, Bhalla RB, Paul M, Gallardo H, McArthur JC, et al. 1990. Cerebrospinal fluid neopterin in human immunodeficiency virus type 1 infection. *Ann. Neurol.* 28:556–60

Brustle O, Spiegel H, Leib SL, Finn T, Stein H, et al. 1992. Distribution of human immunodeficiency virus (HIV) in the CNS of children with severe HIV encephalomyelopathy. *Acta Neuropathol.* 84:24–31

Budka H. 1991. Neuropathology of human immunodeficiency virus infection. *Brain Pathol.* 1:163–75

Cheng-Mayer C, Weiss C, Seto D, Levy JA. 1989. Isolates of human immunodeficiency virus type 1 from the brain may constitute a special group of the AIDS virus. *Proc. Natl. Acad. Sci. USA* 86:8575–79

Chesebro B, Wehrly K, Nishio J, Perryman S. 1992. Macrophage-tropic human immunodeficiency virus isolates from different patients exhibit unusual V3 envelope sequence homogeneity in comparison with T-cell–tropic isolates: definition of critical amino acids involved in cell tropism. *J. Virol.* 66:6547–54

Chung IY, Benveniste EN. 1990. Tumor necrosis factor-alpha production by astrocytes: induction by lipopolysaccharide, IFN-gamma and IL-1beta. *J. Immunol.* 144:2999–3007

Ciardi A, Sinclair E, Scaravilli F, Harcourt-Webster NJ, Lucas S. 1990. The involvement of the cerebral cortex in human immunodeficiency virus encephalopathy: a morphological and immunohistochemical study. *Acta Neuropathol.* 81:51–59

Clerici M, Shearer GM. 1993. A T$_H$2 switch is a critical step in the etiology of HIV infection. *Immunol. Today* 14:107–11

Clouse KA, Cosentino LM, Weih KA, Pyle SW, Robbins PB, et al. 1991. The HIV-1 gp120 envelope protein has the intrinsic capacity to stimulate monokine secretion. *J. Immunol.* 147:2892–901

Cupp C, Taylor JP, Khalili K, Amini S. 1993. Evidence for stimulation of the transforming growth factor beta 1 promoter by HIV-1 Tat in cells derived from CNS. *Oncogene* 8: 2231–36

Cvetkovich TA, Lazar E, Blumberg BM, Saito Y, Eskin TA, et al. 1992. Human immunodeficiency virus type 1 infection of neural xenografts. *Proc. Natl. Acad. Sci. USA* 89: 5162–66

Dal Pan GJ, Glass JD, McArthur JC. 1994. Clinicopathological correlations on HIV-1–associated vacuolar myelopathy: an autopsy-based case-control study. *Neurology* 44: 2159–64

Dal Pan GJ, McArthur JH, Aylward E, Selnes OA, Nance-Sproson TE, et al. 1992. Patterns of cerebral atrophy in HIV-1 infected individuals: results of a quantitative MRI analysis. *Neurology* 42:2125–30

Davis LE, Hjelle BL, Miller VE, Palmer DL, Llewellyn AL, et al. 1992. Early viral brain invasion in iatrogenic human immunodeficiency virus infection. *Neurology* 42:1736–39

Dawson VL, Dawson TM, Uhl GR, Snyder SH. 1993. Human immunodeficiency virus type 1 coat protein neurotoxicity mediated by nitric oxide in primary cortical cultures. *Proc. Natl. Acad. Sci. USA* 90:3256–59

de la Monte SM, Gabuzda DH, Ho DD, Brown RH, Hedley-Whyte ET, et al. 1988. Peripheral neuropathy in the acquired immunodeficiency syndrome. *Ann. Neurol.* 23:485–92

Dhawan S, Toro LA, Jones E, Meltzer MS. 1992. Interactions between HIV-infected monocytes and the extracellular matrix: HIV-infected monocytes secrete neutral metalloproteases that degrade basement membrane protein matrices. *J. Leukoc. Biol.* 52: 244–48

Dreyer EB, Kaiser PK, Offermann JT, Lipton SA. 1990. HIV-1 coat protein neurotoxicity prevented by calcium channel antagonists. *Science* 248:364–67

Egg D, Herold M, Rumpf E, Gunther R. 1980. Prostaglandin F$_{2alpha}$ levels in human cerebrospinal fluid in normal and pathological conditions. *J. Neurol.* 222:239–48

Epstein LG, Kuiken C, Blumberg BM, Hartman S, Sharer LR, et al. 1991. HIV-1 V3 domain variation in brain and spleen of children with AIDS: tissue-specific evolution within host-determined quasispecies. *Virology* 180:583–90

Everall I, Luthert P, Lantos P. 1993. A review of neuronal damage in human immunodeficiency virus infection: its assessment, possible mechanism and relationship to dementia. *J. Neuropathol. Exp. Neurol.* 52:561–66

Everall IP, Glass JD, McArthur JC, Spargo E, Lantos P. 1994. Neuronal loss in the superior frontal and temporal gyri does not correlate with the degree of HIV-associated dementia. *Acta Neuropathol.* 88:538–44

Everall IP, Luthbert PJ, Lantos PL. 1991. Neuronal loss in the frontal cortex in HIV infection. *Lancet* 337:1119–21

Gallo P, de Rossi A, Sivieri S, Chieco-Bianchi L, Tavolato B. 1994. M-CSF production by HIV-1–infected monocytes and its intrathecal synthesis: implications for neurological HIV-1–related disease. *J. Neuroimmunol.* 51:193–98

Gallo P, Frei K, Rordorf C, Lazdins J, Tavolato B, et al. 1989. Human immunodeficiency virus type 1 (HIV-1) infection of the central nervous system: an evaluation of cytokines in cerebrospinal fluid. *J. Neuroimmunol.* 23: 109–16

Gelbard HA, Nottet HSLM, Swindells S, Jett M, Dzenko KA, et al. 1994. Platelet-activating factor: a candidate human immunodeficiency virus type 1–induced neurotoxin. *J. Virol.* 68:4628–35

Genis P, Jett M, Bernton EW, Boyle T, Gelbard HA, et al. 1992. Cytokines and arachidonic metabolites produced during human immunodeficiency virus (HIV)–infected macrophage-astroglia interactions: implications for the neuropathogenesis of HIV disease. *J. Exp. Med.* 176:1703–18

Georgsson G, Houwers DJ, Palsson PA, Petursson G. 1989. Expression of viral antigens in the central nervous system of visna-infected sheep: an immunohistochemical study on experimental visna induced by virus strains of increased neurovirulence. *Acta Neuropathol.* 77:299–306

Giulian D, Vaca K, Noonan CA. 1990. Secretion of neurotoxins by mononuclear phagocytes infected with HIV-1. *Science* 250: 1593–96

Glass JD, Wesselingh SL, Selnes OA, McArthur JC. 1993. Clinical-neuropathological correlation in HIV-associated dementia. *Neurology* 43:2230–37

Goudsmit J, Paul DA, Lange JMA, Speelman H, Van der Noordaa J, et al. 1986. Expression of human immunodeficiency virus antigen (HIV-Ag) in serum and cerebrospinal fluid during acute and chronic infection. *Lancet* 1:177–80

Gray F, Gherardi R, Scaravilli F. 1988. The neuropathology of the acquired immune deficiency syndrome (AIDS): a review. *Brain* 111:245–66

Griffin DE, Wesselingh SL, McArthur JC. 1994. Elevated central nervous system prostaglandins in human immunodeficiency virus–associated dementia. *Ann. Neurol.* 35: 592–97

Griffin GE, Leung K, Folks TM, Kunkel S, Nabel GJ. 1991. Induction of NF-xB during monocyte differentiation is associated with activation of HIV-gene expression. *Res. Virol.* 142:233–38

Griffin JW, Crawford TO, Tyor WR, Glass JD,

Price DL, et al. 1995a. Sensory neuropathy in AIDS. I. Neuropathology. *Brain*. In press

Griffin JW, Tyor WR, Glass JD, Crawford TO, Griffin DE, et al. 1995b. Sensory neuropathy in AIDS. II. Immunopathology. *Brain*. In press

Grimaldi LME, Martino GV, Franciotta DM, Brustia R, Castagna A, et al. 1991. Elevated alpha-tumor necrosis factor levels in spinal fluid from HIV-1–infected patients with central nervous system involvement. *Ann. Neurol.* 29:21–25

Harouse JM, Bhat S, Spitalnik SL, Laughlin M, Stefano K, et al. 1991. Inhibition of entry of HIV-1 in neural cell lines by antibodies against galactosyl ceramide. *Science* 253: 320–22

Heyes MP, Brew BJ, Martin A, Price RW, Salazar AM, et al. 1991. Quinolinic acid in cerebrospinal fluid and serum in HIV-1 infection: relationship to clinical and neurological status. *Ann. Neurol.* 29:202–9

Heyes MP, Gravell M, London WT, Eckhaus M, Vickers JH, et al. 1990. Sustained increases in cerebrospinal fluid quinolinic acid concentrations of rhesus macaques (*Macaca mulatta*) naturally infected with simian retrovirus type D. *Brain Res.* 531:148–58

Heyes MP, Nowak TSJ. 1990. Delayed increases in regional brain quinolinic acid follow transient ischemia in the gerbil. *J. Cereb. Blood Flow Metab.* 10:660–67

Heyes MP, Rubinow D, Lane C, Markey SP. 1989. Cerebrospinal fluid quinolinic acid concentrations are increased in acquired immune deficiency syndrome. *Ann. Neurol.* 26: 275–77

Heyes MP, Saito K, Major EO, Milstien S, Markey SP, et al. 1993. A mechanism of quinolinic acid formation by brain in inflammatory neurological disease: attenuation of synthesis from L-tryptophan by 6-chlorotryptophan and 4-chloro-3-hydroxyanthranilate. *Brain* 116:1425–50

Hill JM, Mervis RF, Avidor R, Moody TW, Brenneman DE. 1993. HIV envelope protein-induced neuronal damage and retardation of behavioral development in rat neonates. *Brain Res.* 603:222–33

Ho DD, Rota TR, Schooley RT, Kaplan JC, Allan JD, et al. 1985. Isolation of HTLV-III from cerebrospinal fluid and neural tissues of patients with neurologic syndromes related to the acquired immunodeficiency syndrome. *N. Engl. J. Med.* 313:1493–97

Johnson RT, McArthur JC, Narayan O. 1988. The neurobiology of human immunodeficiency virus infection. *FASEB J.* 2:2970–81

Ketzler S, Weis S, Haug H, Budka H. 1990. Loss of neurons in the frontal cortex in AIDS brains. *Acta Neuropathol.* 80:92–94

Keys B, Karis J, Fadeel B, Valentin A, Norkrans G, et al. 1993. V3 sequence of paired HIV-1 isolates from blood and cerebrospinal fluid cluster according to host and show variation related to the clinical stage of disease. *Virology* 196:475–83

Kimura-Kuroda J, Nagashima K, Yasui K. 1994. Inhibition of myelin formation by HIV-1 gp120 in rat cerebral cortex culture. *Arch. Virol.* 137:81–99

Kleihues P, Lang W, Burger PC, Budka H, Vogt M, et al. 1985. Progressive diffuse leukoencephalopathy in patients with acquired immune deficiency syndrome (AIDS). *Acta Neuropathol.* 68:333–39

Kure K, Llena JF, Lyman WD, Soeiro R, Weidenheim KM, et al. 1991. Human immunodeficiency virus-1 infection of the nervous system: an autopsy study of 268 adult, pediatric, and fetal brains. *Hum. Pathol.* 22:700–10

Levy JA. 1994. *HIV and the Pathogenesis of AIDS*. Washington, DC: ASM

Levy JA, Shimabukuro J, Hollander H, Mills J, Kaminsky L. 1985. Isolation of AIDS-associated retroviruses from cerebrospinal fluid and brain of patients with neurological symptoms. *Lancet* 2:586–88

Li Y, Hui H, Burgess CJ, Price RW, Sharp PM, et al. 1992. Complete nucleotide sequence, genome organization, and biological properties of human immunodeficiency virus type 1 in vivo: evidence for limited defectiveness and complementation. *J. Virol.* 66:6587–600

Li Y, Kappes JC, Conway JA, Price RW, Shaw GM, et al. 1991. Molecular characterization of human immunodeficiency virus type 1 cloned directly from uncultured human brain tissue: identification of replication-competent and -defective viral genomes. *J. Virol.* 65:3973–85

Lieberman AP, Pitha PM, Shin ML. 1990. Protein kinase regulates tumor necrosis factor mRNA stability in virus-stimulated astrocytes. *J. Exp. Med.* 172:989–92

Lipton SA. 1994a. HIV displays its coat of arms. *Nature* 367:113–14

Lipton SA. 1994b. HIV coat protein gp120 induces soluble neurotoxins in culture medium. *Neurosci. Res. Commun.* 15:31–37

Lipton SA, Rosenberg PA. 1994. Excitatory amino acids as a final common pathway for neurologic disorders. *N. Engl. J. Med.* 330: 613–21

Lipton SA, Sucher NJ, Kaiser PK, Dreyer EB. 1991. Synergistic effects of HIV coat protein and NMDA receptor-mediated neurotoxicity. *Neuron* 7:111–18

Mann JM. 1993. AIDS in the 1990s: a global analysis. *The Pharos* 56:2–5

Masliah E, Ge N, Morey M, DeTeresa R, Terry RD, et al. 1992. Cortical dendritic pathology in human immunodeficiency virus encephalitis. *Lab. Invest.* 66:285–91

Masuda M, Remington MP, Hoffman PN,

Ruscetti SK. 1992. Molecular characterization of a neuropathologic and nonerythroleukemogenic variant of Friend murine leukemia virus PVC-211. *J. Virol.* 66:2798–806

Mayhew TM. 1992. A review of recent advances in stereology for quantifying neural structure. *J. Neurocytol.* 21:313–28

McArthur JC. 1987. Neurologic manifestations of AIDS. *Medicine* 66:407–37

McArthur JC, Becker PS, Parisi J, Trapp BD, Selnes OA, et al. 1989. Neuropathological changes in early HIV dementia. *Ann. Neurol.* 26:681–84

McArthur JC, Cohen BA, Farzedegan H, Cornblath DR, Selnes OA, et al. 1988. Cerebrospinal fluid abnormalities in homosexual men with and without neuropsychiatric findings. *Ann. Neurol.* 23:S34–S37 (Suppl.)

McArthur JC, Harrison MJG. 1994. HIV-associated dementia. In *Current Neurology*, 14: 275-320. St. Louis: Mosby-Year Book

McArthur JC, Hoover DR, Bacellar H, Miller EN, Cohen BA, et al. 1993. Dementia in AIDS patients: incidence and risk factors. *Neurology* 43:2245–52

McArthur JC, Nance-Sproson TE, Griffin DE, Hoover D, Selnes OA, et al. 1992. The diagnostic utility of elevation in cerebrospinal fluid beta$^2$-microglobulin in HIV-1 dementia. *Neurology* 42:1707–12

Mellors JW, Griffith BP, Ortiz MA, Landry ML, Ryan JL. 1991. Tumor necrosis factor-alpha/cachectin enhances human immunodeficiency virus type 1 replication in primary macrophages. *J. Infect. Dis.* 163:78–82

Merrill JE, Chen ISY. 1991. HIV-1, macrophages, glial cells, and cytokines in AIDS nervous system disease. *FASEB J.* 5:2391–97

Merrill JE, Gerner RH, Myers LW, Ellison GW. 1983. Regulation of natural killer cell cytotoxicity by prostaglandin E in the peripheral blood and cerebrospinal fluid of patients with multiple sclerosis and other neurological diseases. *J. Neuroimmunol.* 4:223–37

Mintz M, Rapaport R, Oleske JM, Connor EM, Koenigsberger MR, et al. 1989. Elevated serum levels of tumor necrosis factor are associated with progressive encephalopathy in children with acquired immunodeficiency syndrome. *Am. J. Dis. Child.* 143:771–74

Morganti-Kossmann MC, Kossmann T, Wahl SM. 1992. Cytokines and neuropathology. *Trends Physiol. Sci.* 13:286–91

Moses AV, Bloom FE, Pauza CD, Nelson JA. 1993. Human immunodeficiency virus infection of human brain capillary endothelial cells occurs via a CD4/galactosylceramide-independent mechanism. *Proc. Natl. Acad. Sci. USA* 90:10474–78

Mosmann TR. 1994. Cytokine patterns during the progression to AIDS. *Science* 265:193–94

Murray EA, Rausch DM, Lendvay J, Sharer LR, Eiden LE. 1992. Cognitive and motor impairments associated with SIV infection in rhesus monkeys. *Science* 255:1246–49

Narayan O, Clements JE. 1989. Biology and pathogenesis of lentiviruses [Review]. *J. Gen. Virol.* 70:1617–39

Navia BA, Cho E-S, Petito CK, Price RW. 1986a. The AIDS dementia complex. II. Neuropathology. *Ann. Neurol.* 19:525–35

Navia BA, Jordan BD, Price RW. 1986b. The AIDS dementia complex. I. Clinical features. *Ann. Neurol.* 19:517–24

Nuovo GJ, Gallery F, MacConnell P, Braun A. 1994. In situ detection of polymerase chain reaction–amplified HIV-1 nucleic acids and tumor necrosis factor-alpha RNA in the central nervous system. *Am. J. Pathol.* 144:659–66

Oster S, Christoffersen P, Gundersen HJG, Nielsen JO, Pakkenberg B, et al. 1993. Cerebral atrophy in AIDS: a stereological study. *Acta Neuropathol.* 85:617–22

Pang S, Vinters HV, Akashi T, O'Brien WA, Chen ISY. 1991. HIV-1 *env* sequence variation in brain tissue of patients with AIDS-related neurologic disease. *J. AIDS* 4: 1082–92

Petito CK, Cash KS. 1992. Blood-brain barrier abnormalities in the acquired immunodeficiency syndrome: immunohistochemical localization of serum proteins in postmortem brain. *Ann. Neurol.* 32:658–66

Petito CK, Navia BA, Cho ES, Jordan BD, George DC, et al. 1985. Vacuolar myelopathy pathologically resembling subacute combined degeneration in patients with the acquired immunodeficiency syndrome. *N. Engl. J. Med.* 312:874–79

Petito CK, Vecchio D, Chen Y-T. 1994. HIV antigen and DNA in AIDS spinal cords correlate with macrophage infiltration but not with vacuolar myelopathy. *J. Neuropathol. Exp. Neurol.* 53:86–94

Phillips TR, Prospero-Garcia O, Puaoi DL, Lerner DL, Fox HS, et al. 1994. Neurological abnormalities associated with feline immunodeficiency virus infection. *J. Gen. Virol.* 75:979–87

Portegies P, Epstein LG, Hung STA, de Gans J, Goudsmit J. 1989. Human immunodeficiency virus type 1 antigen in cerebrospinal fluid. *Arch. Neurol.* 46:261–64

Power C, McArthur JC, Johnson RT, Griffin DE, Glass JD, et al. 1994. Demented and nondemented patients with AIDS differ in brain-derived human immunodeficiency virus type 1 envelope sequences. *J. Virol.* 68: 4643–49

Power CP, Kong PA, Crawford TO, Wesselingh S, Glass JD, et al. 1993. Cerebral white matter changes in acquired immunodeficiency syndrome dementia: alterations in the blood-brain barrier. *Ann. Neurol.* 34:339–50

Preston BD, Poiesz BJ, Loeb LA. 1988. Fidelity of HIV-1 reverse transcriptase. *Science* 242:1168–71

Pulliam L, Herndier BG, Tang NM, McGrath MS. 1991. Human immunodeficiency virus-infected macrophages produce soluble factors that cause histological and neurochemical alterations in cultured human brains. *J. Clin. Invest.* 87:503–12

Rausch DM, Heyes MP, Murray EA, Lendvay J, Sharer LR, et al. 1994. Cytopathologic and neurochemical correlates of progression to motor/cognitive impairment in SIV-infected rhesus monkeys. *J. Neuropathol. Exp. Neurol.* 53:165–75

Resnick L, DiMarzo-Veronese F, Schupback J, Tourtellotte WW, Ho DD, et al. 1985. Intra-blood-brain-barrier synthesis of HTLV-III specific IgG in patients with neurologic symptoms associated with AIDS or AIDS-related complex. *N. Engl. J. Med.* 313:1498–504

Rhodes R. 1987. Histopathology of the central nervous system in the acquired immunodeficiency syndrome. *Hum. Pathol.* 18:636–43

Robbins DS, Shirazi Y, Drysdale BE, Lieberman A, Shin HS, et al. 1987. Production of cytotoxic factor for oligodendrocytes by stimlulated astrocytes. *J. Immunol.* 139:2593–97

Rosenblum MK. 1990. Infection of the central nervous system by the human immunodeficiency virus type 1: morphology and relation to syndromes of progressive encephalopathy and myelopathy in patients with AIDS. *Pathol. Annu.* 25:117–69

Royal W III, Selnes OA, Concha M, Nance-Sproson TE, McArthur JC. 1994. Cerebrospinal fluid human immunodeficiency virus type 1 (HIV-1) p24 antigen levels in HIV-1-related dementia. *Ann. Neurol.* 36:32–39

Saito Y, Sharer LR, Epstein LG, Michaels J, Mintz M, et al. 1994. Overexpression of *nef* as a marker for restricted HIV-1 infection of astrocytes in postmortem pediatric central nervous tissues. *Neurology* 44:474–81

Sasseville VG, Newman WA, Lackner AA, Smith MO, Lausen NC, et al. 1992. Elevated vascular cell adhesion molecule-1 in AIDS encephalitis induced by simian immunodeficiency virus. *Am. J. Pathol.* 141:1021–30

Sei Y, Arora PK, Skolnick P, Paul IA. 1992. Spatial learning impairment in a murine model of AIDS. *FASEB J.* 6:3008–13

Seilhean D, Duyckaerts C, Vazeux R, Bolgert F, Brunet P, et al. 1993. HIV-1-associated cognitive/motor complex: absence of neuronal loss in the cerebral neocortex. *Neurology* 43:1492–99

Selmaj KW, Raine CS. 1988. Tumor necrosis factor mediates myelin and oligodendrocyte damage in vitro. *Ann. Neurol.* 23:339–46

Sharer LR. 1992. Pathology of HIV-1 infection of the central nervous system. A review. *J. Neuropathol. Exp. Neurol.* 51:3–11

Sharer LR, Michaels J, Murphey-Corb M, Hu F-S, Kuebler DJ, et al. 1991. Serial pathogenesis of SIV brain infection. *J. Med. Primatol.* 20:211–17

Sharma DP, Zink MC, Anderson M, Adams R, Clements JE, et al. 1992. Derivation of neurotropic simian immunodeficiency virus from exclusively lymphocytetropic parental virus: pathogenesis of infection in macaques. *J. Virol.* 66:3550–56

Shaw GM, Harper ME, Hahn BH, Epstein LG, Gajdusek DC, et al. 1985. HTLV-III infection in brains of children and adults with AIDS encephalopathy. *Science* 227:177–82

Shibata M, Blatteis CM. 1991. Differential effects of cytokines on thermosensitive neurons in guinea pig preoptic area slices. *Am. J. Physiol.* 261:R1096–103

Sinclair E, Scaravilli F. 1992. Detection of HIV proviral DNA in cortex and white matter of AIDS brains by non-isotopic polymerase chain reaction: correlation with diffuse poliodystrophy. *AIDS* 6:925–32

Singer EJ, Syndulko K, Fahy-Chandon BN, Shapshak P, Resnick L, et al. 1994. Cerebrospinal fluid p24 antigen levels and intrathecal immunoglobulin G synthesis are associated with cognitive disease severity in HIV-1. *AIDS* 8:197–204

Smith TW, DeGirolami U, Henin D, Bolgert F, Hauw J-J. 1990. Human immunodeficiency virus (HIV) leukoencephalopathy and the microcirculation. *J. Neuropathol. Exp. Neurol.* 49:357–70

Soliven B, Albert J. 1992. Tumor necrosis factor modulates $Ca^{2+}$ currents in cultured sympathetic neurons. *J. Neurosci.* 12:2665–71

Tadmori W, Mondal D, Tadmori I, Prakash O. 1991. Transactivation of human immunodeficiency virus type 1 long terminal repeats by cell surface tumor necrosis factor-$\alpha$. *J. Virol.* 65:6425–29

Tardieu M, Hery C, Peudenier S, Boespflug O, Montagnier L. 1992. Human immunodeficiency virus type 1-infected monocytic cells can destroy human neural cells after cell-to-cell adhesion. *Ann. Neurol.* 32:11–17

Toggas SM, Masliah E, Rockenstein EM, Rall GF, Abraham CR, et al. 1994. Central nervous system damage produced by expression of the HIV-1 coat protein gp120 in transgenic mice. *Nature* 367:188–93

Tornatore C, Chandra R, Berger JR, Major EO. 1994. HIV-1 infection of subcortical astrocytes in the pediatric central nervous system. *Neurology* 44:481–87

Tourtellotte WW, Schmid P, Conrad A, Syndulko K, Singer EJ, et al. 1993. Quantifying HIV-1 proviral DNA via the polymerase chain reaction (PCR) in cerebrospinal fluid (CSF) and blood of seropositive individuals with and without neurologic abnormalities. *Neurology* 43:A265 (Abstr.)

Tyor WR, Glass JD, Becker PS, Griffin JW, Bezman J, et al. 1992. Cytokine expression in the brain during AIDS. *Ann. Neurol.* 31: 349–60

Tyor WR, Power C, Gendelman HE, Markham RB. 1993. A model of human immunodeficiency virus encephalitis in *scid* mice. *Proc. Natl. Acad. Sci. USA* 90:8658–62

Vazeux R, Lacroix-Ciaudo C, Blanche S, Cumont M-C, Henin D, et al. 1992. Low levels of human immunodeficiency virus replication in the brain tissue of children with severe acquired immunodeficiency syndrome encephalopathy. *Am. J. Pathol.* 140:137–44

Vitkovic L, Wood GP, Major EO, Fauci AS. 1991. Human astrocytes stimulate HIV-1 expression in a chronically infected promonocyte clone via interleukin-6. *AIDS Res. Hum. Retroviruses* 7:723–27

Wahl LM, Corcoran ML, Pyle SW, Arthur LO, Harel-Bellan A, et al. 1989. Human immunodeficiency virus glycoprotein (gp120) induction of monocyte arachidonic acid metabolites and interleukin 1. *Immunology* 86:621–25

Wahl SM, Allen JB, McCartney-Francis N, Morganti-Kossmann MC, Kossmann T, et al. 1991. Macrophage- and astrocyte-derived transforming growth factor beta as a mediator of central nervous system dysfunction in acquired immune deficiency syndrome. *J. Exp. Med.* 173:981–91

Watanabe M, Ringler DJ, Fultz PN, MacKey JJ, Boyson JE, et al. 1991. A chimpanzee-passaged human immunodeficiency virus isolate is cytopathic for chimpanzee cells but does not induce disease. *J. Virol.* 65:3344–48

Wesselingh S, Gough NM, Finlay-Jones JJ, McDonald PJ. 1990. Detection of cytokine mRNA in astrocyte cultures using the polymerase chain reaction. *Lymphokine Res.* 9:2

Wesselingh SL, Glass J, McArthur JC, Griffin JW, Griffin DE. 1994. Cytokine dysregulation in HIV-associated neurological disease. *Adv. Neuroimmunol.* 4:199–206

Wesselingh SL, Power C, Glass JD, Tyor WR, McArthur JC, et al. 1993. Intracerebral cytokine mRNA expression in acquired immunodeficiency syndrome dementia. *Ann. Neurol.* 33:576–82

Wiley CA, Achim C. 1994. Human immunodeficiency virus encephalitis is the pathological correlate of dementia in acquired immunodeficiency syndrome. *Ann. Neurol.* 36:673–76

Wiley CA, Masliah E, Morey M, Lemere C, DeTeresa R, et al. 1991. Neocortical damage during HIV infection. *Ann. Neurol.* 29:651–57

Wiley CA, Schrier RD, Nelson JA, Lampert PW, Oldstone MBA. 1986. Cellular localization of human immunodeficiency virus infection within the brains of acquired immune deficiency syndrome patients. *Proc. Natl. Acad. Sci. USA* 83:7089–93

Williams AE, Blakemore WF. 1990. Monocyte-mediated entry of pathogens into the central nervous system. *Neuropathol. Appl. Neurobiol.* 16:377–92

Wilt SG, Milward E, Zhou JM, Nagasato K, Patton H, et al. 1995. In vitro evidence for a dual role of TNF-α in HIV-1 encephalopathy. *Ann. Neurol.* 37:381–94

Yoshioka M, Shapshak P, Srivastava AK, Stewart RV, Nelson SJ, et al. 1994. Expression of HIV-1 and interleukin-6 in lumbosacral dorsal root ganglia of patients with AIDS. *Neurology* 44:1120–30

Annu. Rev. Neurosci. 1996. 19:27–52

# RNA EDITING

## Larry Simpson

Howard Hughes Medical Institute and Departments of Biology and Medical
Microbiology and Immunology, University of California, Los Angeles, California
90024

## Ronald B. Emeson

Departments of Pharmacology and Molecular Physiology and Biophysics,
Vanderbilt University School of Medicine, Nashville, Tennessee 37232-6600

KEY WORDS:    glutamate receptor, dsRAD, adenosine deaminase, gRNAs, AMPA

### ABSTRACT

RNA editing is a term describing a variety of novel mechanisms for the modification of nucleotide sequences of RNA transcripts in different organisms. These
editing events include (*a*) the U-insertion and -deletion type of editing found in
the mitochondrion of kinetoplastid protozoa, (*b*) the C-insertion editing found
in the mitochondrion of *Physarum*, (*c*) the C-to-U substitution editing of the
mammalian apoB mRNA, (*d*) a similar C-to-U substitution editing of mRNAs
in higher plant mitochondria and chloroplasts and in tRNAs of marsupials and
rats, (*e*) a diverse nucleotide substitution editing of tRNAs in *Acanthomoeba*
mitochondria, and (*f*) the A-to-I type of editing found in the mammalian glutamate receptor subunits. These diverse phenomena involve several different enzymatic mechanisms. In several cases, duplex RNAs with internal or external
guide sequences help determine the site specificity of editing. The A-to-I editing
observed in RNAs encoding non-NMDA glutamate receptor subunits may be
due to the actions of a double-stranded RNA-specific adenosine deaminase that
is widespread in higher organisms. Although the function of many RNA editing
events is unclear, the biological importance of RNA editing in other systems
may prove as significant as the nucleotide modifications regulating the cation
selectivity and electrophysiological profiles elaborated by non-NMDA glutamate
receptors in the mammalian brain.

## INTRODUCTION

The term RNA editing is used to designate a variety of RNA modification
phenomena, some of which may be unrelated both mechanistically and evolutionarily. The term was first used to describe a novel genetic phenomenon
occurring in the mitochondria of kinetoplastid protozoa (Simpson 1972), in

27

which uridine (U) residues are inserted and occasionally deleted at specific sites, usually within coding regions of mRNA transcripts, creating open reading frames (Benne et al 1986; Van der Spek et al 1988; Feagin et al 1988a,b; Shaw et al 1988, 1989; Simpson & Shaw 1989; Bhat et al 1990; Koslowsky et al 1990). Soon thereafter, several other examples of RNA sequence modifications from diverse organisms were reported: a single cytidine (C)–to-U substitution in the apolipoprotein B (apo B) transcript in mammals (Powell et al 1987; Chen et al 1987, 1990; Wu et al 1990; Boström et al 1990), multiple C-to-U substitutions in transcripts of the mitochondrial and chloroplastic DNAs in higher plants (Gualberto et al 1989, Covello & Gray 1989, Hiesel et al 1989, Gray & Covello 1993, Schuster et al 1993, Araya et al 1994), and an insertion of guanosine (G) residues within coding sequences of negative strand RNA viruses (Thomas et al 1988, Cattaneo et al 1989, Vidal et al 1990, Paterson & Lamb 1990).

The initial broad definition of RNA editing as "any modification of the sequence of an mRNA molecule within coding regions, except for splicing of introns" (Simpson & Shaw 1989) rapidly proved inadequate to describe the full panoply of novel genetic phenomena that were appearing in the literature. Specific nucleotide changes in mitochondrial tRNAs in *Acantho-moeba castellani* (Lonergan & Gray 1993a,b) and marsupials (Janke & Pääbo 1993), specific insertion of C residues in multiple mitochondrial transcripts in *Physarum polycephalum* (Mahendran et al 1991), a single U-to-C change in the mRNA for the Wilms' tumor suppressor gene (Sharma et al 1994), and a specific A-to-G substitution in a glutamate receptor mRNA in mammals (Sommer et al 1991, Cha et al 1994, Egebjerg et al 1994, Puchalski et al 1994) were described. Several of these modifications were found to be regulated and to have significant phenotypic consequences. The glutamate receptor editing event is of particular interest to the field of neuroscience, and in this review, we survey the various types of RNA editing that have been reported and compare and contrast these phenomena with a more detailed consideration of recent advances in our knowledge of the glutamate receptor type of editing.

## U-ADDITION AND -DELETION RNA EDITING IN TRYPANOSOME MITOCHONDRIA

The U-addition and -deletion type of editing that occurs in transcripts of structural genes in mitochondrial DNA from kinetoplastid protozoa is the prototype for RNA editing. The extent of the editing varies from a few U's added at a few sites (5′ editing) to hundreds of U's added at hundreds of sites (pan editing), depending on the gene and the species (Benne et al 1986; Van der Spek et al 1988; Shaw et al 1988, 1989; Feagin et al 1988a,b; Simpson &

Shaw 1989; Bhat et al 1990; Koslowsky et al 1990). Deletions also occur but at a lower frequency than additions. In *Leishmania tarentolae,* the transcripts of 12 out of the 20 genes encoded in the mitochondrial maxicircle molecule are edited to varying extents (Shaw et al 1988, 1989; Feagin et al 1988b; Thiemann et al 1994). Those genes whose transcripts are edited are termed cryptogenes (Simpson & Shaw 1989). In the case of pan-edited cryptogenes (Bhat et al 1990; Koslowsky et al 1990; Feagin et al 1988a; Read et al 1992, 1994b; Souza et al 1992, 1993; Thiemann et al 1994), the editing is so extensive that the genomic sequences have no recognizable open reading frames at all. This type of RNA editing is mediated by small 3'-oligo-uridylated RNA molecules termed guide RNAs (gRNAs), which are complementary to short regions of the edited transcripts, if G-U base pairs are allowed (Blum et al 1990, Blum & Simpson 1990). The editing information is contained in the form of extra A and G residues, which can form base pairs with the inserted U residues. Each gRNA mediates the editing of a block of the preedited mRNA, and adjacent editing blocks overlap with each other (Figure 1). Editing proceeds from 3' to 5', with the polarity determined by the creation of upstream gRNA anchor sequences by downstream editing (Maslov & Simpson 1992, Sugisaki & Takanami, 1993, Corell et al 1993). Some of the gRNA genes are located in the maxicircle genome (Blum et al 1990), but the majority are in the minicircular component of the kinetoplastid DNA (kDNA) (Sturm & Simpson 1990, 1991; Pollard et al 1990; Pollard & Hajduk 1991; Corell et al 1993; Riley et al 1994). There is a single tubular mitochondrion per cell with a single kDNA network, and each network contains approximately 10,000 minicircles ranging from 0.5 to 2.5 kb and 20–50 maxicircle molecules ranging in size from 20–36 kb, depending on the species (Simpson 1972, 1987). The number of minicircle-encoded gRNA genes varies from 60–80 in *L. tarentolae* to over 900 in *Trypanosoma brucei.* The latter species contains multiple redundant gRNAs that encode the same editing information but differ in sequence, as a result of allowing G-U base pairing (Corell et al 1993).

The precise mechanism of this type of editing remains an open question, but several models have been proposed (Figure 2). In the enzyme cascade model (Blum et al 1990) there is an initial cleavage at a mismatched base, followed by the 3' addition of a U residue from UTP or from the 3'-oligo[U] tail of the gRNA (Sollner-Webb 1991), which can base-pair with a guide nucleotide in the gRNA, and re-ligation of the mRNA. In the transesterification model (Cech 1991, Blum et al 1991), the U residues are transferred from the 3' end of the gRNA to editing sites by two successive transesterifications, such as occur in RNA splicing. Several enzyme activities have been identified that are candidates for an involvement in editing: a 3' terminal uridylyl transferase (TUTase) (Bakalara et al 1989), an RNA ligase (White & Borst 1987, Bakalara et al 1989), an RNA helicase (Missel & Göringer 1994), an endonuclease that

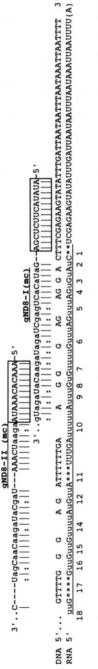

*Figure 1* Two overlapping gRNAs mediating the editing of blocks I and II of ND8 mRNA in *L. tarentolae*. The duplex anchor sequences are boxed. G-U base pairs are indicated by "·" and A-U and G-C base pairs by "|". Editing sites are numbered 3′ to 5′. Deletions are indicated by "*". The 3′-oligo[U] tails of the gRNAs are not shown. [Reprinted from Simpson & Thiemann (1995) with permission.]

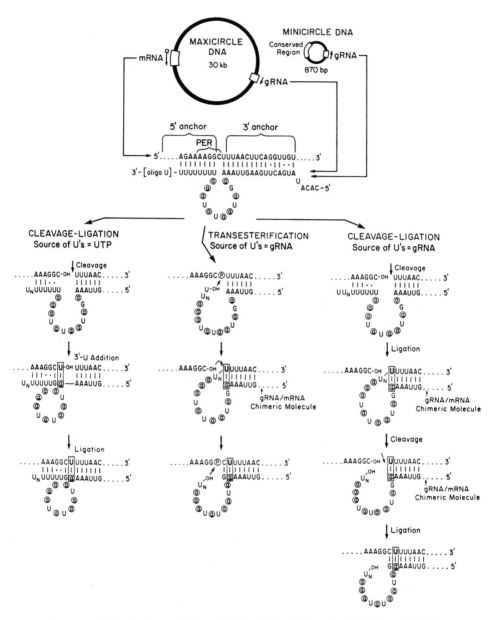

*Figure 2* Models for the mechanism of RNA editing in kinetoplastid mitochondria. PER = preedited region. Only U insertions are shown. The guiding a and g nucleotides in the gRNA are shown in lower case and circled. The gRNA shown is gCYb-I. [Reprinted from Simpson & Thiemann (1995) with permission.]

cleaves within the preedited region (PER) of the mRNA (Simpson et al 1992, Harris et al 1992), and a grNA-mRNA chimeric-forming activity (Harris & Hajduk 1992, Koslowsky et al 1992, Blum & Simpson 1992).

Two in vitro systems have also been reported that show RNA editing–like activities: (*a*) a U-deletion activity in *T. brucei* mitochondrial extracts, in which exogenous cognate gRNA determines by basepairing the number of U's deleted at the first editing site of the A6 (=MURF4) mRNA (Seiwert & Stuart 1994) (U additions were not observed in this system); (*b*) a U-addition activity in *L. tarentolae* mitochondrial extracts that is limited precisely to the mRNA preedited region (Frech et al 1995). The latter activity sediments as a ribonucleoprotein complex at 25S (the G complex) (Peris et al 1994). In addition, several smaller RNP complexes have been identified in *L. tarentolae* that contain gRNA and TUTase activity, the so-called T complexes (Peris et al 1994). RNP complexes and a small set of proteins that can be UV cross-linked to gRNA have been identified in *T. brucei* mitochondrial extracts (Göringer et al 1994, Köller et al 1994, Read et al 1994a). Detailed analysis of these in vitro systems should produce a clearer understanding of this type of RNA editing.

A great deal has been learned recently about the evolution of the trypanosome U-addition and -deletion type of editing. Analysis of edited genes in several kinetoplastid species that were located on a phylogenetic tree by rRNA sequence analysis (Fernandes et al 1993, Landweber & Gilbert 1994, Maslov et al 1994) showed that this type of editing was probably present in an ancestor of the entire kinetoplastid lineage, which represents one of the earliest extant eukaryotic cells possessing a mitochondrion. It is still unclear, however, whether U-editing was present in the eubacterial ancestor of the kinetoplastid mitochondrion or even earlier or whether it first appeared in the kinetoplastid mitochondrion (Fernandes et al 1993; Landweber & Gilbert 1994; Maslov & Simpson 1994; Simpson & Maslov 1994a,b).

## C-ADDITION EDITING IN *PHYSARUM* MITOCHONDRIA

A unique and complex type of editing occurs with transcripts of the 60-kb mitochondrial genome of *P. polycephalum,* another primitive eukaryotic cell. The protein coding genes contain −1 frameshifts located approximately every 25 nucleotides (nt), which are overcome at the RNA level by the insertion of single C residues (Mahendran et al 1991, 1994). A similar C-insertion phenomenon occurs posttranscriptionally in the mitochondrial rRNAs and tRNAs, both in conserved and variable regions, and in duplex regions and single-stranded loops. In addition to C insertions, a low level of U, G, and A single nucleotide insertions, dinucleotide insertions, and even C-to-U substitutions were observed (Gott et al 1993). No polarity appears to be involved, which is unlike the trypanosome editing, and no evidence for the existence of gRNAs has been obtained (Mahendran et al 1991, 1994). There is a clear bias for C insertion to occur after purine-pyrimidine dinucleotides. However, the mechanism and site specificity of this type of editing is unknown, and the suggestion has been made that multiple types of editing with multiple mechanisms may be occurring in this system.

## C-TO-U SUBSTITUTIONAL EDITING: MAMMALIAN APO B mRNA AND PLANT MITOCHONDRIAL TRANSCRIPTS

In the mammalian small intestine, a single C residue within the glutamine codon 2153 (CAA) of the apo B transcript is changed to a U residue, resulting in a UAA stop codon and a truncated protein with different physiological functions (Chen et al 1987, Powell et al 1987, Driscoll et al 1989). A breakthrough in the analysis of this phenomenon was the development in 1989 of an in vitro editing system using cytoplasmic (S-100) cellular extracts (Driscoll et al 1989). Smith et al (1991) demonstrated the presence of 11S and 27S ribonucleoprotein complexes involved in in vitro editing. Recently, a subunit of the editing complex was isolated by a functional cloning method and identified as a cytidine deaminase (Teng et al 1993, Hadjiagapiou et al 1994, Johnson et al 1993, Driscoll & Zhang 1994, Yamanaka et al 1994). A 43-kd and a 60-kd protein were shown to represent additional components of the editing complex by UV cross-linking to substrate RNA (Harris et al 1993, Navaratnam et al 1993).

The site specificity of this editing event is conferred by a conserved 26-nt downstream mooring sequence and upstream enhancer sequences (Backus & Smith 1991, 1992; Shah et al 1991; Backus et al 1994). The 3' sequence block can be transposed elsewhere and shown to confer specific C-to-U editing in vitro (Backus & Smith 1991).

A superficially similar type of substitution editing occurs in higher plant mitochondria but is much more extensive and more varied (Araya et al 1994, Gray & Covello 1993). Multiple sites in transcripts of structural and in one case rRNA genes are modified by specific C-to-U (Gualberto et al 1989; Covello & Gray 1989; Gray & Covello 1993; Araya et al 1992, 1994; Schuster et al 1993; Hiesel et al 1989; Graves et al 1990) and, in a few cases, U-to-C substitutions (Schuster et al 1990a) and even C deletions (Gualberto et al 1991). Most substitutions are at the first or second positions of codons, leading to restoration of conserved amino acids. In addition, editing restores the universal genetic code to plant mitochondria by changing CGG, which was proposed to encode tryptophan in plants (Fox & Leaver 1981), back to UGG, which encodes arginine. Partially edited transcripts are rare and usually involve silent positions; there is no apparent polarity of editing as in trypanosome U editing (Wissinger et al 1991, Schuster et al 1990b, Yang & Mulligan 1991). In addition, antisense gRNAs such as found in trypanosome editing could not be detected. Neither the enzymatic mechanism nor the control of site specificity is understood, although an in vitro editing system has been reported (Araya et al 1992).

This type of editing occurs in a wide variety of flowering plants, both monocots and dicots, but not in the bryophyte, *Marcantia polymorpha* (Oda et al 1992). Fully edited nuclear homologues of edited mitochondrial genes occur in some species, which presumably evolved by a transfer of edited mRNAs or cDNAs into the nucleus and subsequent integration (Schuster & Brennicke 1994, Nugent & Palmer 1991).

Transcripts from approximately half of the structural genes of the chloroplast genome from monocots and dicots are also edited by C-to-U substitutions, which create conserved amino acids (Hoch et al 1991; Kudla et al 1992; Maier et al 1992a,b; Freyer et al 1993; Zeltz et al 1993; Bock et al 1994; Neckermann et al 1994; Ruf et al 1994). Similarities with mitochondrial editing suggest that these organelles may share common components of the editing machinery.

## tRNA NUCLEOTIDE SUBSTITUTION EDITING

Modification of specific nucleotides and transglycosylation reactions have been previously described in tRNAs, but the term editing is not usually used to describe these phenomena. Recently, single-nucleotide substitution editing of tRNAs has been reported to occur in several systems. In several mitochondrial tRNAs in the lower eukaryotic protist *A. castellani*, single-nucleotide conversions (U-to-A, U-to-G, C-to-A, and A-to-G) localized to the first three nucleotides in the 5′ half of the acceptor stem restore correct base pairing (Lonergan & Gray 1993a,b). The opposite strand appears to serve as an internal guide sequence for these editing events. Janke & Pääbo (1993) reported a different type of tRNA editing in marsupial mitochondria. A GCC (glycine) anticodon of an aspartic acid tRNA is changed to a GUC anticodon by a C-to-U substitution, thereby restoring the correct coding property. And in the rat, a nuclear-encoded tRNA is edited at two sites upstream of the anticodon by C-to-U substitutions (Maréchal-Drouard et al 1993). Finally, in *Physarum* mitochondria, several encoded RNAs contain single C insertions that restore G-C base pairs in several duplex regions, and a single U insertion that restores a conserved motif (Mahendran et al 1994). Nothing is known about the mechanism or mechanisms for these diverse examples of tRNA editing.

## G-ADDITION AND A-TO-I SUBSTITUTION EDITING IN NEGATIVE STRAND RNA VIRUSES

In several paramyxoviruses—including simian virus 5, measles virus, Sendai virus, and mumps virus—a unique P gene gives rise to two mRNAs (Thomas et al 1988, Cattaneo et al 1989, Vidal et al 1990, Paterson & Lamb 1990, Curran & Kolakofsky 1990, Pelet et al 1991, Curran et al 1991). One mRNA is a faithful copy of the DNA, and the other contains either one or two Gs inserted within a run of five to six Gs. G deletions have also been reported to occur (Jacques et al 1994). The resulting frameshifts allow ribosomal access to a second downstream reading frame, resulting in an alternate P protein with a different C-terminal sequence. The insertion and deletion of G residues differs from the other types of RNA editing in that it occurs during transcription and is thought to be due to a stuttering of the RNA polymerase involving a pause and a slippage (Vidal et al 1990).

Another type of editing of negative strand RNA viruses also has been reported. In RNA from defective measles viruses isolated from the brains of patients suffering from subacute sclerosing panencephalitis, clusters of multiple U-to-C transitions in the positive strand (or A-to-G transitions in the minus strand) were observed, a phenomenon known as biased hypermutation (Cattaneo et al 1988, 1989). The mechanism is thought to involve a double-stranded RNA (dsRNA)–specific adenosine deaminase, which will be discussed in more detail in the section on glutamate receptor editing. The modification is thought to occur inadvertently during transcription or replication by formation of duplex RNA structures.

## C-TO-U SUBSTITUTION EDITING IN HEPATITIS δ VIRUS RNA

The hepatitis δ virus is a subviral pathogen of humans and contains a circular RNA genome that is 1679-nt long. A single U residue at nucleotide 1012 in the genomic RNA strand is converted to a C residue in a posttranscriptional event (Luo et al 1990, Zheng et al 1992). This editing leads to the modification of a stop codon and the extension of a reading frame to generate the large form of the δ antigen, a protein required for completion of the viral life cycle. This substitution editing requires a duplex RNA region with some sequence specificity and uses cellular factors (Casey et al 1992, 1993; Smith et al 1992; Gottlieb et al 1994). The apparent utilization of an internal guide sequence is reminiscent of gRNA-mediated editing in trypanosomes and acceptor stem editing of tRNAs in *Acanthomoeba* mitochondria.

## A-TO-I EDITING OF MAMMALIAN GLUTAMATE RECEPTOR mRNAs

L-glutamate is the principal excitatory neurotransmitter in the brain and activates cation-selective receptor channels involved in fast synaptic neurotransmission and the induction of long-term cellular changes associated with memory acquisition and learning (Collingridge & Lester 1989, Collingridge & Singer 1990, Ito 1989). Pharmacological and physiological studies have provided evidence for the existence of distinct subtypes of ion-transporting glutamate receptors (GluRs), named according to the agonists $N$-methyl-D-aspartate (NMDA), $\alpha$-amino-3-hydroxy-5-methylisoxazole-4-propionate (AMPA), and kainic acid. AMPA and kainate receptors demonstrate a rapid kinetic response, conducting mainly monovalent cations ($Na^+$ and $K^+$), whereas the NMDA channel is characterized by a slower response time, the requirement for glycine as a coagonist, and a high $Ca^{2+}$ permeability that is controlled in a voltage-dependent manner by extracellular magnesium concentrations. Many of the important physiological and pathological functions of glutamate receptors have been attributed to the $Ca^{2+}$ permeability of these integral ion channels (Dubinsky & Rothman 1991, Pizzi et al 1991, Randall & Thayer 1992), yet until recently only NMDA receptor channels were shown to be permeable to both monovalent and $Ca^{2+}$ ions (Dubinsky & Rothman 1991, Pizzi et al 1991, Randall & Thayer 1992). The influx of $Ca^{2+}$ into the postsynaptic neuron is believed to underlie the activity-dependent changes in synaptic strength critical for memory formation (Christie & Abraham 1992, Collingridge & Singer 1990) and the $Ca^{2+}$-dependent cellular processes leading to neurodegeneration (Collingridge & Lester 1989).

Molecular cloning of cDNAs encoding glutamate receptors of the AMPA subtype have revealed that such ionotropic receptors are generated by the assembly of GluR-A, -B, -C, and -D subunits into homo- and heteromeric channels (Boulter et al 1989, Keinanen et al 1990, Nakanishi et al 1990, Sakimura et al 1990). Messenger RNAs encoding AMPA receptor subunits are expressed widely, yet the mRNAs for each individual subunit share overlapping but nonidentical patterns of expression throughout the central nervous system (Sommer et al 1990). These cDNAs predict the synthesis of protein subunits, approximately 900 amino acids in length, with three membrane-spanning domains (Figure 3) representing the major determinants of transmembrane topology and channel architecture (Hollman et al 1994). A putative channel-lining hydrophobic domain (TMII) is thought not to span the membrane itself, but to lie in proximity to the intracellular face of the plasma membrane or to loop into the membrane without traversing it (Hollman et al 1994). Although all four AMPA receptor subunits share significant amino acid sequence identity within the membrane-spanning regions of each receptor subunit, there is significantly decreased sequence conservation in the amino- and carboxyl-terminal regions of the proteins (Boulter et al 1989, Keinanen et al 1990, Nakanishi et al 1990). Alternative splicing of GluR-A, -B, -C, and -D transcripts results in the production of two protein isoforms, termed flip and flop, which contain distinct 38–amino acid domains immediately preceding the fourth transmembrane region for each receptor subunit (Figure 3). The two alternatively spliced molecules display differing expression patterns in the mature and developing brain (Monyer et al 1991) and demonstrate distinct functional properties differing in both the kinetics and amplitude of agonist-induced responses (Sommer et al 1990).

Electrophysiological studies with recombinant AMPA receptor subunits have revealed that differing heteromeric subunit combinations alter both the current-voltage (I-V) relationship and the ion transport properties through the gated pore (Dingledine et al 1992, Hollmann et al 1991, Verdoorn et al 1991). The homomeric GluR-B channel and heteromeric assemblies containing this subunit show near linear I-V relations and a negligible $Ca^{2+}$ permeability. Most importantly, a double rectifying I-V relationship and a substantial $Ca^{2+}$ conductance can be observed only in the absence of the GluR-B subunit, demonstrating the functional dominance of the GluR-B subunit in heteromeric AMPA receptors. Amino acid sequence comparisons of AMPA receptor subunits have revealed that TMII is identical in each of the four GluR polypeptides, except that GluR-B contains a positively charged arginine residue at a position in which the other subunits contain a neutral glutamine moiety. Site-directed mutational analyses have demonstrated that the disparate channel properties demonstrated by various heteromeric AMPA complexes are independent of the subunits themselves but change after the positively charged arginine is introduced into the TMII segment (Verdoorn et al, 1991); therefore, this single arginine residue is responsible for dictating the linear I-V relationship and low divalent cation permeability demonstrated by GluR-B (Dingledine et al 1992, Hume et al 1991, Mishina et al 1991, Verdoorn et al 1991). Current properties through non-NMDA channels in cultured hippocampal neurons (Iino et al 1990) and in hippocampal slices (Jonas & Sakmann 1992) indicate that most naturally occurring AMPA receptors contain the GluR-B subunit. However, Bergmann glial cells of the rat cerebellum (Burnashev et al 1992, Keinanen et al 1990, Monyer et al 1991) and nonpyramidal neurons of rat neocortex (Jonas et al 1994) demonstrate low or negligible GluR-B mRNA expression and are permeable to calcium ions both in vivo and in vitro.

Sequence analyses of genomic DNA encoding the GluR-B subunit have indicated the presence of a glutamine codon (CAG) within TMII, even though

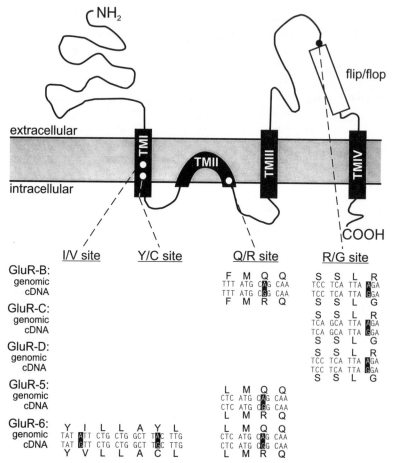

*Figure 3* A summary of the editing events for mRNAs encoding non-NMDA glutamate receptor subunits. A schematic representation of the proposed topology for GluR-A is presented indicating the relative positions of editing sites within non-NMDA receptor subunits; the toplogy of subunits other than GluR-A (Hollman et al 1994) have not yet been determined. The genomic, cDNA, and amino acid sequences surrounding the editing sites are presented and modified nucleosides are presented in inverse lettering.

an arginine codon (CGG) is found in the cDNA encoding the GluR-B subunit at this position (Q-R site; Figure 3) (Sommer et al 1991). Multiple genes and alternative splicing have been excluded as possible sources of the arginine codon. Therefore it has been proposed that the RNA species encoding the GluR-B subunit is edited posttranscriptionally such that the A residue of the glutamine codon is converted to the G residue present in the arginine triplet. Quantitative PCR analyses of adult rat and mouse brain RNA have demonstrated that virtually all GluR-B transcripts encode this critical arginine residue within TMII, whereas GluR-A, -C, and -D transcripts encode only a glutamine-containing form (Sommer et al 1991). RNA editing is also responsible for an A-to-G conversion allowing for the introduction of a glycine codon (GGA) in cDNAs encoding the GluR-B, -C, and -D subunits in place of a genomically-encoded arginine triplet (AGA) (Figure 3) (Lomeli et al 1994). This editing site (R-G site) occurs immediately before the alternatively spliced flip-flop regions in the extracellular loop between the second and third membrane-spanning domains and serves to increase the rate of recovery from receptor desensitization (Lomeli et al 1994). The extent of R-G editing is low during embryonic stages (E14) but increases dramatically during postnatal development.

In addition to RNA modifications regulating the electrophysiological function of heteromeric AMPA receptors, subunits of the kainate receptor complex (GluR-5 and GluR-6) also are subject to RNA editing processes similar to those observed for GluR-B mRNAs (Kohler et al 1993, Sommer et al 1991). GluR-5 and GluR-6 transcripts are present in both arginine- and glutamine-containing forms in adult rat brain demonstrating 40 and 80% RNA editing, respectively (Figure 3). Two additional sites in the TMI region of GluR-6 also are targeted by RNA editing and result in an altered $Ca^{2+}$ permeability for GluR-6 homomeric channels (Kohler et al 1993). These two additional editing sites both represent adenosine-to-guanosine alterations resulting in the coding of valine and cysteine in GluR-6 mRNAs rather than the isoleucine and tyrosine moieties predicted from the genomic DNA sequence (Figure 3). In contrast with AMPA receptor channels, the presence of a glutamine residue in TMII determines channels with low $Ca^{2+}$ permeability, whereas an arginine in this position confers an increased $Ca^{2+}$ flux if TMI is fully edited (Kohler et al 1993). If the sites in TMI are in the nonedited form, $Ca^{2+}$ permeability is less dependent upon the presence of glutamine or arginine in TMII.

Recent studies of regulatory sequences mediating the posttranscriptional processing of GluR-B mRNAs have indicated that the intron distal to the exon encoding TMII (intron 11) harbors determinants essential for accurate and efficient RNA editing at the Q-R site (Figure 4) (Egebjerg et al 1994, Higuchi et al 1993). Deletional analyses of GluR-B minigene constructs, transfected into a rat pheochromocytoma (PC12) or rat neuronal (N2a) cell line, have demonstrated that sequence elements critical for RNA editing reside within the proximal 380 nt of intron 11. DNA sequence analyses of this intronic region have shown the presence of a 10-nt sequence having exact complementarity to the exonic sequence immediately surrounding the Q-R site (Figure 4). It has been proposed that this editing-site complementary sequence (ECS) may be involved in the formation of a duplex RNA structure that is essential for RNA editing (Egebjerg et al 1994, Higuchi et al 1993). Additionally, an

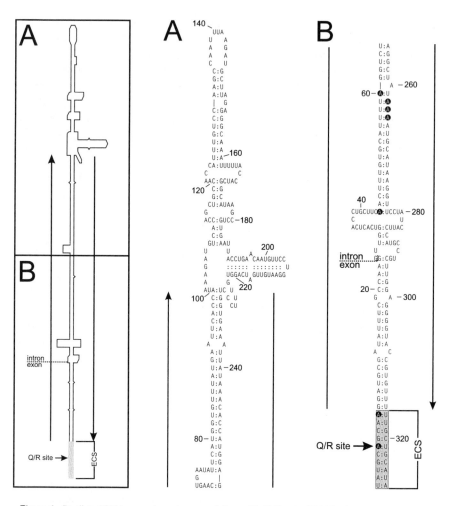

*Figure 4* Predicted RNA secondary structure of the rat GluR-B pre-mRNA in the proximal region of intron 11. The positions of edited nucleotides are represented by closed circles, the intron-exon boundary is indicated, and the imperfect inverted repeat is designated by arrows. Base-pairing interactions between the region surrounding the Q-R site and the ECS are indicated by a gray rectangle. Nucleotide positions are numbered relative to the Q-R site.

imperfect inverted repeat sequence upstream from the ECS was identified that also contributes to the formation of a region of dsRNA within the GluR-B primary RNA transcript. Mutations disrupting the base-pairing within the inverted repeat or between the ECS and the Q-R site significantly decreased or abolished the editing of transfected GluR-B transcripts (Higuchi et al 1993).

Compensatory mutations designed to maintain the RNA secondary structure generally restored editing efficiency, indicating that secondary structure rather than the primary nucleotide sequence is critical for the editing process. In addition to editing at the Q-R site, additional intronic G residues were identified in place of gene-specified A residues at multiple sites (Figure 4); the extent of editing at these intronic positions was significantly less than that observed at the Q-R site. Due to intronic sequence requirements, the editing of GluR-B transcripts must represent a nuclear posttranscriptional processing event occurring prior to or coincident with RNA splicing (Egebjerg et al 1994, Higuchi et al 1993). Previous studies have suggested that the C-to-U editing of apo B transcripts, which was discussed above, is also a nuclear RNA processing event based upon the subcellular distribution of modified apo B transcripts (Lau et al 1991).

Similar RNA duplex structures were identified in the regions surrounding the R-G sites for pre-mRNAs encoding the GluR-B, -C, and -D subunits (Lomeli et al 1994). In GluR-B, this dsRNA domain results from base-pairing interactions between the 3'-end of exon 13 and the proximal 60-nt of intron 13. As previously seen in analyses of the Q-R site, mutations that disrupted the proposed R-G RNA duplex abolished editing, whereas compensatory mutations significantly increased editing efficiency. In addition to the proposed region of dsRNA, distal intronic elements were necessary for maximal editing efficiency, although not absolutely required for editing at the R-G site (Lomeli et al 1994).

## Catalytic Mechanisms

The editing of RNA transcripts encoding non-NMDA glutamate receptor subunits results from the creation of a G-like nucleotide from a genomically encoded A residue (Sommer et al 1991). This nucleotide directs the incorporation of C by reverse transcriptase and produces a codon that alters the coding potential of the resultant mRNA transcripts. Alterations in base-pairing potential from a genomically encoded A residue to a G moiety could occur by three general biochemical mechanisms, including nucleotide excision and replacement, base exchange through transglycosylation, or direct modification of the base (Figure 5). Unlike the C-to-U substitution observed in apo B transcripts, an A-to-G conversion would require multiple catalytic steps (Bass 1993). Following precedents found in other biochemical systems, the adenosine C6 amino group could be replaced with a ketone by the deamination of the A residue to inosine (I) by using an adenosine deaminase–like activity (Merkler et al 1990, Polson et al 1991). Amination of C2 could then be achieved by oxidation of I to produce xanthosine by an enzyme similar to IMP dehydrogenase (Hedstrom & Wang 1990), and finally the C2 ketone of xanthosine

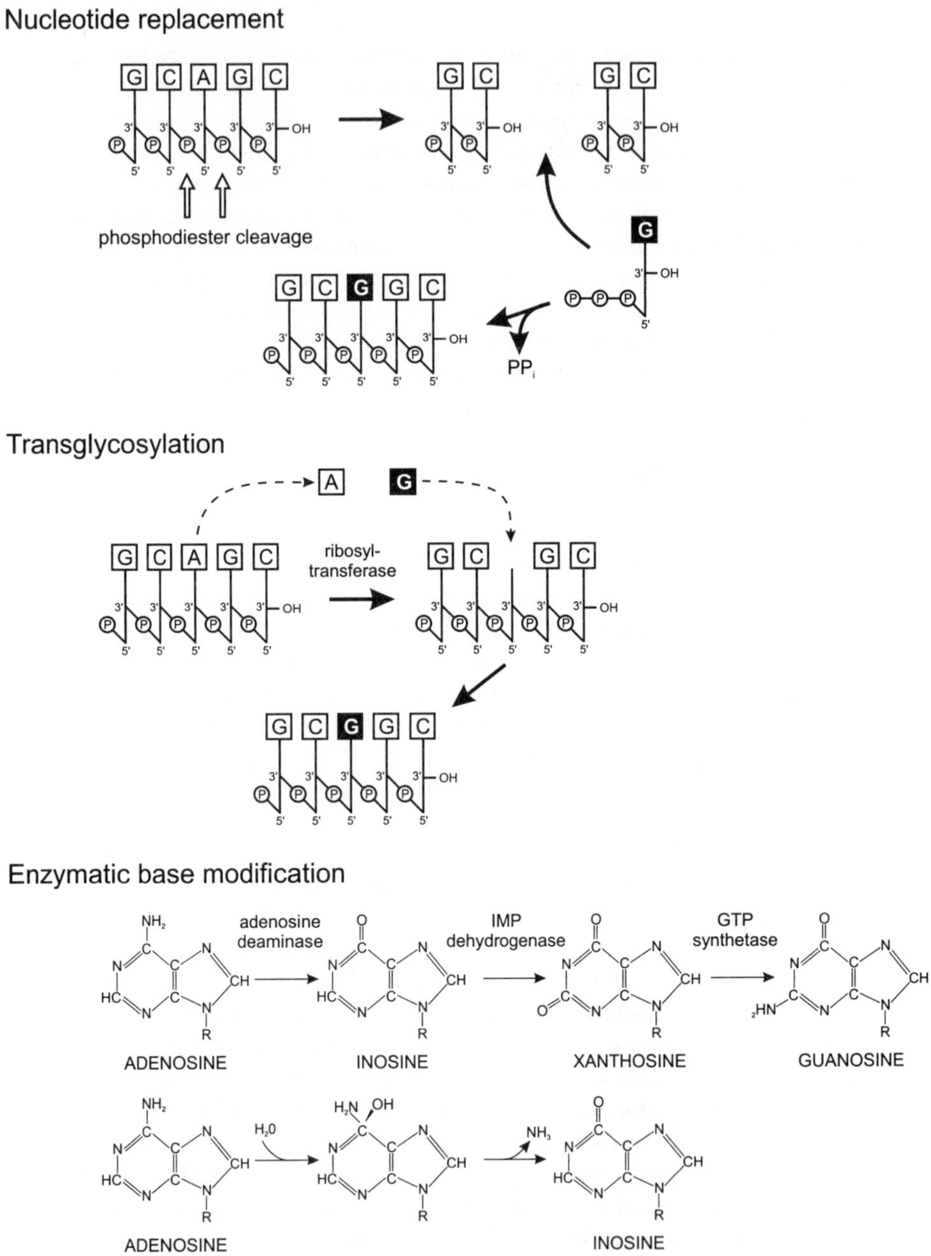

## Nucleotide replacement

phosphodiester cleavage

PP$_i$

## Transglycosylation

ribosyl-
transferase

## Enzymatic base modification

adenosine
deaminase

IMP
dehydrogenase

GTP
synthetase

NH$_2$

CH

HC

N

R

ADENOSINE

INOSINE

XANTHOSINE

GUANOSINE

$_2$HN

H$_2$O

H$_2$N OH

NH$_3$

ADENOSINE

INOSINE

*Figure 5* Molecular mechanisms for the A-to-G conversion in mRNAs encoding non-NMDA glutamate receptor subunits. A schematic diagram is presented for three distinct models by which an adenosine residue may be converted to a guanosine-like nucleoside; ribose, R. [Adapted from Bass (1993).]

could be aminated by an enzyme such as GTP synthetase (Lewis & Villafranca 1989).

The simplest possible mechanism of GluR-B RNA editing would be the creation of an I residue by hydrolytic deamination of the specific A residue at position 6 of the purine ring (Bass 1993, Chan 1993, Sommer et al 1991). Similar site-specific deaminations at position 4 of the pyrimidine ring have been identified for the production of apo B48 (Hodges et al 1991) and have been proposed to explain the multiple C-to-U transitions in plant mitochondrial RNAs (Covello & Gray 1989, Gray & Covello 1993). Reaction products from in vitro editing analyses have demonstrated that the RNA processing of apo B transcripts results from a nucleotide-specific deamination converting C to U (Greeve et al 1991, Johnson et al 1993, Navaratnam et al 1993). Similarly, the editing of GluR-B transcripts could involve a site-specific deamination converting A to I (Sommer et al 1991). I, like G, preferentially base pairs with C, suggesting that the actual RNA modification involved in the editing of GluR-B transcripts at the Q-R site may result from a CIG (arginine) codon in the mature mRNA rather than the CGG (arginine) codon inferred from the isolated cDNA sequence. The observation that editing of GluR-B transcripts at both the Q-R and R-G sites requires the presence of an RNA duplex has suggested that an adenosine deaminase–like activity with a dsRNA specificity may be the operant enzyme involved in the posttranscriptional processing of glutamate receptor RNAs (Egebjerg et al 1994; Higuchi et al 1993; Kim et al 1994a,b; Kim & Nishikura 1993; Lomeli et al 1994; Polson & Bass 1994; Sommer et al 1991). The identification of additional edited residues within the intronic region of the proposed RNA duplex (Figure 4), distal to the Q-R site, lends further support to this hypothesis (Higuchi et al 1993).

A dsRNA unwinding activity, originally described in *Xenopus laevis* embryos, was identified based upon its ability to alter the mobility of synthetic dsRNA transcripts on nondenaturing gels subsequent to embryo microinjection (Bass & Weintraub 1987, 1988; Rebagliatti & Melton 1987). Further analyses revealed that this phenomenon resulted from the conversion of A residues within the RNA duplex to I residues by hydrolytic deamination (Bass & Weintraub 1988, Polson et al 1991, Wagner et al 1989). The conversion of A to I by this dsRNA adenosine deaminase (dsRAD or DRADA) results in the production of I-U mismatches rather than Watson-Crick A-U base pairs, thereby destabilizing the RNA duplex. Since its initial discovery in *X. laevis,* dsRAD has been shown to be expressed in multiple mammalian tissues and cultured cells as well as diverse organisms throughout the animal kingdom (Bass 1993; Kim et al 1994a,b; Wagner & Nishikura 1988).

Although the physiological role of this RNA-modifying activity has not been determined, it has been implicated in the modification of the transactivation response (TAR) element of human immunodeficiency virus 1 RNA (Sharmeen et al 1991), in matrix protein transcripts of negative strand RNA viruses, as discussed above (Bass et al 1989a,b; Polson & Bass 1994), as well as in the modification of RNAs encoding subunits of glutamate-gated ion channels (Egebjerg et al 1994; Higuchi et al 1993; Kim et al 1994a,b; Kim & Nishikura 1993; Lomeli et al 1994; Polson & Bass 1994; Sommer et al 1991). In vitro studies of dsRAD have indicated that this enzymatic activity is specific for A residues within an RNA duplex, although it can modify dsRNAs of many different sequences (Bass & Weintraub 1988, Wagner et al 1989) and can bind to double-stranded substrates that do not contain A (Hough & Bass 1994). Up to 50% of the A residues in both strands of a synthetic dsRNA substrate can be converted to I residues (Bass & Weintraub 1988, Nishikura et al 1991). Because dsRAD can modify such a large percentage of the A residues in longer dsRNA substrates, it has generally been considered to be nonspecific with regard to A residue preference. More recent studies of A selectivity by dsRAD have indicated a 5′-neighbor preference (A = U > C > G) and a high selectivity for specific A residues within a short RNA duplex (Polson & Bass 1994). These characteristics have been used to explain how dsRAD may mediate non-NMDA receptor RNA editing without the need for additional factors to more accurately specify target A residues (Polson & Bass 1994).

A full investigation of the biochemical mechanisms mediating GluR-B RNA editing requires an identification of the exact chemical nature of the modified nucleotide product, a determination of whether this nucleotide is produced while maintaining the RNA phosphodiester backbone, and a determination of whether the new base results from chemical modification or transglycosylation. To elucidate the molecular mechanisms underlying this subunit-specific RNA editing event, an in vitro system for the editing of GluR-B RNA was developed using synthetic RNA substrates and nuclear extracts prepared from HeLa cells (Rueter et al 1995, Melcher et al 1995, Yang et al 1995). Results from these studies further implicated dsRAD in the posttranscriptional processing of GluR-B transcripts. Thin-layer chromatographic analyses of radiolabeled RNA substrates revealed that GluR-B editing resulted from the conversion of A to I by enzymatic base modification (Rueter et al 1995, Melcher et al 1995, Yang et al 1995). Analyses of mutant GluR-B RNA transcripts indicated that this enzymatic activity required an RNA duplex, while competition analyses verified that single-stranded RNA, single-stranded DNA, and double-stranded DNA could not serve as competitive inhibitors of the reaction. Biochemical fractionation studies using a double-stranded RNA affinity column have suggested that GluR-B editing and dsRNA-deamination activities are distinct, but they may have dsRAD, or a similar deaminase, as a common component (Yang et al 1995). Future studies will be required to determine if the recently purified (Hough & Bass 1994; Kim et al 1994a,b; O'Connell & Keller 1994) and cloned (Kim et al 1994a,b; O'Connell et al 1995) dsRAD is responsible for non-

NMDA receptor subunit RNA editing and to determine if a single enzymatic activity is responsible for the posttranscriptional modification of mRNAs encoding multiple glutamate receptor subunits.

## A-TO-I EDITING OF *DROSOPHILA* SODIUM CHANNEL mRNA

The para locus of *Drosophila* encodes the predominant voltage-gated sodium channel of both larval and adult central and peripheral nervous systems (Loughney et al 1989, Hong & Ganetzky 1994). Extensive analyses of para cDNAs have revealed three putative RNA editing sites within the para coding sequence (R Reenan, personal communication). The most thoroughly characterized of these is a site in homology domain III involving multiple editing events within a single exon. The frequency of editing at this site is approximately 75% in cDNA analyses from whole adult fly. However, analyses of tissue-specific splice forms of para transcripts indicates that this posttranscriptional modification may be tissue specific. The most frequent editing event involves three A-to-G (I) conversions resulting in a Q-to-R amino acid substitution like that observed for GluR-B, a silent change, and an N-to-D amino acid substitution. Like GluR-B, this region is predicted to form an RNA duplex, but the precise sequences necessary for editing remain to be determined. The functional significance of these editing events on para sodium channel function is currently being investigated.

## CONCLUSIONS

Studies on the various types of RNA editing reveal that nucleotide sequences can be modified at the RNA level in a variety of ways and that these modifications are frequently regulated and are biologically significant. The trypanosome U-addition and -deletion editing, the apo B mRNA C-to-U substitution editing, and the glutamate receptor mRNA A-to-I editing are perhaps the best understood phenomena, and they appear to have little in common other than the name and the fact that RNA molecules are involved. In several cases, duplex RNAs with internal or external guide sequences appear to be involved in determining the particular editing sites, but this may not be a general phenomenon. Little is yet known about the mechanisms and site specificity selection of the plant mitochondrial, plant chloroplast, and *Physarum* mitochondrial types of editing.

The A-to-I conversion, exemplified by the glutamate receptor editing and produced by the actions of a dsRAD-like activity, is likely to be quite widespread, owing to the ubiquity of this activity in higher organisms. The biological importance of RNA editing in other systems may prove as significant as the posttranscriptional modifications governing the cation selectivity and electrophysiological profiles elaborated by non-NMDA glutamate receptors in the mammalian brain.

## Literature Cited

Araya A, Bégu D, Litvak S. 1994. RNA editing in plants. *Physiol. Plant Pathol.* 91:543–50

Araya A, Domec C, Bégu D, Litvak S. 1992. An in vitro system for the editing of ATP synthase subunit 9 mRNA using wheat mitochondrial extracts. *Proc. Natl. Acad. Sci. USA* 89:1040–44

Backus JW, Schock D, Smith HC. 1994. Only cytidines 5′ of the apolipoprotein B mRNA mooring sequence are edited. *Biochim. Biophys. Acta Gene Struct. Expr.* 1219:1–14

Backus JW, Smith HC. 1991. Apolipoprotein B mRNA sequences 3′ of the editing site are necessary and sufficient for editing and editosome assembly. *Nucleic Acids Res.* 19: 6781–86

Backus JW, Smith HC. 1992. Three distinct RNA sequence elements are required for efficient apolipoprotein B (apoB) RNA editing in vitro. *Nucleic Acids Res.* 20:6007–14

Bakalara N, Simpson AM, Simpson L. 1989. The *Leishmania* kinetoplast-mitochondrion contains terminal uridylyltransferase and RNA ligase activities. *J. Biol. Chem.* 264: 18679–86

Bass B. 1993. In *The RNA World,* ed. RF Gesteland, JF Atkins, pp. 383–428. Cold Spring Harbor: Cold Spring Harbor Lab.

Bass B, Weintraub H. 1988. An unwinding activity that covalently modifies its double-stranded RNA substrate. *Cell* 55:1089–98

Bass B, Weintraub H, Cattaneo R, Billeter M. 1989. Biased hypermutation of viral RNA genomes could be due to unwinding/modification of double-stranded RNA. *Cell* 56:331

Benne R, Van den Burg J, Brakenhoff J, Sloof P, Van Boom J, Tromp M. 1986. Major transcript of the frameshifted coxII gene from trypanosome mitochondria contains four nucleotides that are not encoded in the DNA. *Cell* 46:819–26

Bhat GJ, Koslowsky DJ, Feagin JE, Smiley BL, Stuart K. 1990. An extensively edited mitochondrial transcript in kinetoplastids encodes a protein homologous to ATPase subunit 6. *Cell* 61:885–94

Blum B, Bakalara N, Simpson L. 1990. A model for RNA editing in kinetoplastid mitochondria: "guide" RNA molecules transcribed from maxicircle DNA provide the edited information. *Cell* 60:189–98

Blum B, Simpson L. 1990. Guide RNAs in kinetoplastid mitochondria have a nonencoded 3′ oligo-(U) tail involved in recognition of the pre-edited region. *Cell* 62:391–97

Blum B, Simpson L. 1992. Formation of gRNA/mRNA chimeric molecules in vitro, the initial step of RNA editing, is dependent on an anchor sequence. *Proc. Natl. Acad. Sci. USA* 89:11944–48

Blum B, Sturm NR, Simpson AM, Simpson L. 1991. Chimeric gRNA-mRNA molecules with oligo(U) tails covalently linked at sites of RNA editing suggest that U addition occurs by transesterification. *Cell* 65:543–50

Bock R, Kössel H, Maliga P. 1994. Introduction of a heterologous editing site into the tobacco plastid genome: the lack of RNA editing leads to a mutant phenotype. *EMBO J.* 13:4623–28

Boström K, Garcia Z, Poksay KS, Johnson DF, Lusis AJ, Innerarity TL. 1990. Apolipoprotein B mRNA editing. Direct determination of the edited base and occurrence in non–apolipoprotein B-producing cell lines. *J. Biol. Chem.* 265:22446–52

Boulter J, Hollmann M, O'Shea-Greenfield A, Hartley M, Deneris E, et al. 1989. Molecular cloning and functional expression of glutamate receptor subunit genes. *Neuron* 3:589–96

Burnashev N, Khodorova A, Jonas P, Helm PJ, Wisden W, et al. 1992. Calcium-permeable AMPA-kainate receptors in fusiform cerebellar glial cells. *Science* 256:1566–70

Casey JL, Bergmann KF, Brown TL, Gerin JL. 1992. Structural requirements for RNA editing in hepatitis δ virus: evidence for a uridine-to-cytidine editing mechanism. *Proc. Natl. Acad. Sci. USA* 89:7149–53

Casey JL, Bergmann KF, Brown TL, Gerin JL. 1993. Determinants of RNA editing in hepatitis delta virus. *Prog. Clin. Biol. Res.* 382:5–11

Cattaneo R, Kaelin K, Baczko K, Billeter M. 1989. Measles virus editing provides an additional cysteine-rich protein. *Cell* 56:759–64

Cattaneo R, Schmid A, Eschle D, Baczko K, Meulen V, Billeter M. 1988. Biased hypermutation and other genetic changes in defective measles viruses in human brain infections. *Cell* 55:255–65

Cech TR. 1991. RNA editing: world's smallest introns. *Cell* 64:667–69

Cha JH, Kinsman SL, Johnston MV. 1994. RNA editing of a human glutamate receptor subunit. *Mol. Brain Res.* 22:323–28

Chan L. 1993. RNA editing: exploring one mode with apolipoprotein B mRNA. *Bioessays* 15:33–41

Chen S-H, Habib G, Yang C, Gu Z, Lee B, et al. 1987. Apolipoprotein B-48 is the product of a messenger RNA with an organ-specific in-frame stop codon. *Science* 238:363–66

Chen S-H, Li X, Liao WSL, Wu JH, Chan L. 1990. RNA editing of apolipoprotein B mRNA. Sequence specificity determined by in vitro coupled transcription editing. *J. Biol. Chem.* 265:6811–16

Christie BR, Abraham WC. 1992. NMDA-dependent heterosynaptic long-term depression in the dentate gyrus of anaesthetized rats. *Synapse* 10:1–6

Collingridge GL, Lester RA. 1989. Excitatory amino acid receptors in the vertebrate central nervous system. *Pharmacol. Rev.* 41:143–210

Collingridge GL, Singer W. 1990. Excitatory amino acid receptors and synaptic plasticity. *Trends Pharmacol. Sci.* 11:290–96

Corell RA, Feagin JE, Riley GR, Strickland T, Guderian JA, et al. 1993. *Trypanosoma brucei* minicircles encode multiple guide RNAs which can direct editing of extensively overlapping sequences. *Nucleic Acids Res.* 21:4313–20

Covello PS, Gray MW. 1989. RNA editing in plant mitochondria. *Nature* 341:662–66

Curran J, Boeck R, Kolakofsky D. 1991. The Sendai virus P gene expresses both an essential protein and an inhibitor of RNA synthesis by shuffling modules via mRNA editing. *EMBO J.* 10:3079–85

Curran J, Kolakofsky D. 1990. Sendai virus P gene produces multiple proteins from overlapping open reading frames. *Enzyme* 44:244–49

Dingledine R, Hume R, Heinemann S. 1992. Structural determinants of barium permeation and rectification in non-NMDA glutamate receptor channels. *J. Neurosci.* 12:4080–87

Driscoll D, Wynne J, Wallis S, Scott J. 1989. An in vitro system for the editing of apolipoprotein B mRNA. *Cell* 58:519–25

Driscoll DM, Zhang Q. 1994. Expression and characterization of p27, the catalytic subunit of the apolipoprotein B mRNA editing enzyme. *J. Biol. Chem.* 269:19843–47

Dubinsky JM, Rothman SM. 1991. Intracellular calcium concentrations during "chemical hypoxia" and excitotoxic neuronal injury. *J. Neurosci.* 11:2545–51

Egebjerg J, Kukekov V, Heinemann SF. 1994. Intron sequence directs RNA editing of the glutamate receptor subunit GluR2 coding sequence. *Proc. Natl. Acad. Sci. USA* 91:10270–74

Feagin JE, Abraham J, Stuart K. 1988a. Extensive editing of the cytochrome c oxidase III transcript in *Trypanosoma brucei*. *Cell* 53:413–22

Feagin JE, Shaw JM, Simpson L, Stuart K. 1988b. Creation of AUG initiation codons by addition of uridines within cytochrome b transcripts of kinetoplastids. *Proc. Natl. Acad. Sci. USA* 85:539–43

Fernandes AP, Nelson K, Beverley SM. 1993. Evolution of nuclear ribosomal RNAs in kinetoplastid protozoa: perspectives on the age and origins of parasitism. *Proc. Natl. Acad. Sci. USA* 90:11608–12

Fox TD, Leaver CJ. 1981. The *Zea mays* mitochondrial gene coding cytochrome oxidase subunit II has an intervening sequence and does not contain TGA codons. *Cell* 26:315–23

Frech G C, Bakalara N, Simpson L, Simpson AM. 1995. In vitro RNA editing-like activity in a mitochondrial extract from Leishmania tarentolae. *EMBO J.* 14:178–87

Freyer R, Hoch B, Neckermann K, Maier RM, Kössel H. 1993. RNA editing in maize chloroplasts is a processing step independent of splicing and cleavage to monocistronic mRNAs. *Plant J.* 4:621–29

Göringer HU, Koslowsky DJ, Morales TH, Stuart K. 1994. The formation of mitochondrial ribonucleoprotein complexes involving guide RNA molecules in *Trypanosoma brucei*. *Proc. Natl. Acad. Sci. USA* 91:1776–80

Gott JM, Visomirski LM, Hunter JL. 1993. Substitutional and insertional RNA editing of the cytochrome c oxidase subunit 1 mRNA of *Physarum polycephalum*. *J. Biol. Chem.* 268:25483–86

Gottlieb PA, Prasad Y, Smith JB, Williams AP, Dinter-Gottlieb G. 1994. Evidence that alternate foldings of the hepatitis δ RNA confer varying rates of self-cleavage. *Biochemistry* 33:2802–8

Graves PV, Bégu D, Velours J, Neau E, Belloc F, Litvak S, Araya A. 1990. Direct protein sequencing of wheat mitochondrial ATP synthase subunit 9 confirms RNA editing in plants. *J. Mol. Biol.* 214:1–6

Gray MW, Covello PS. 1993. RNA editing in plant mitochondria and chloroplasts. *FASEB J.* 7:64–71

Greeve J, Navaratnam N, Scott J. 1991. Characterization of the apolipoprotein B mRNA editing enzyme:no similarity to the proposed mechanism of RNA editing in kinetoplastid protozoa. *Nucleic Acids Res.* 19:3569–76

Gualberto JM, Bonnard G, Lamattina L, Grienenberger JM. 1991. Expression of the wheat mitochondrial nad3-rps12 transcription unit: correlation between editing and mRNA maturation. *Plant Cell* 3:1109–20

Gualberto JM, Lamattina L, Bonnard G, Weil JH, Grienenberger JM. 1989. RNA editing in wheat mitochondria results in the conservation of protein sequences. *Nature* 341: 660–66

Hadjiagapiou C, Giannoni F, Funahashi T, Skarosi SF, Davidson NO. 1994. Molecular cloning of a human small intestinal apolipoprotein B mRNA editing protein. *Nucleic Acids Res.* 22:1874–79

Harris M, Decker C, Sollner-Webb B, Hajduk S. 1992. Specific cleavage of pre-edited mRNAs in trypanosome mitochondrial extracts. *Mol. Cell Biol.* 12:2591–98

Harris ME, Hajduk SL. 1992. Kinetoplastid RNA editing: in vitro formation of cytochrome b gRNA-mRNA chimeras from synthetic substrate RNAs. *Cell* 68:1091–99

Harris SG, Sabio I, Mayer E, Steinberg MF, Backus JW, et al. 1993. Extract-specific heterogeneity in high-order complexes containing apolipoprotein B mRNA editing activity and RNA-binding proteins. *J. Biol. Chem.* 268:7382–92

Hedstrom L, Wang C. 1990. Mycophenolic acid and thiazole adenine dinucleotide inhibition of *Tritrichomonas foetus* inosine 5′ monophosphate dehydrogenase: implications on enzyme mechanism. *Biochemistry* 29:849–54

Hiesel R, Wissinger B, Schuster W, Brennicke A. 1989. RNA editing in plant mitochondria. *Science* 246:1632–34

Higuchi M, Single FN, Kohler M, Sommer B, Sprengel R, Seeburg PH. 1993. RNA editing of AMPA receptor subunit GluR-B: a base-paired intron-exon structure determines position and efficiency. *Cell* 75: 1361–70

Hoch B, Maier RM, Appel K, Igloi GL, Kössel H. 1991. Editing of a chloroplast mRNA by creation of an initiation codon. *Nature* 353: 178–80

Hodges PE, Navaratnam N, Greeve JC, Scott J. 1991. Site-specific creation of uridine from cytidine in apolipoprotein B mRNA editing. *Nucleic Acids Res.* 19:1197–201

Hollmann M, Hartley M, Heinemann S. 1991. Ca²⁺ permeability of KA-AMPA-gated glutamate receptor channels depends on subunit composition. *Science* 252:851–53

Hollmann M, Maron C, Heinemann S. 1994. N-glycosylation site tagging suggests a three transmembrane domain topology for glutamate receptor GluR1. *Neuron* 13:1331–43

Hong C, Ganetzky B. 1994. Spatial and temporal expression patterns of two sodium channel genes in *Drosophila. J. Neurosci.* 14:5160–69

Hough RF, Bass BL. 1994. Purification of the *Xenopus laevis* double-stranded RNA adenosine deaminase. *J. Biol. Chem.* 269:9933–39

Hume RI, Dingledine R, Heinemann SF. 1991. Identification of a site in glutamate receptor subunits that controls calcium permeability. *Science* 253:1028–31

Iino M, Ozawa S, Tsuzuki K. 1990. Permeation of calcium through excitatory amino acid receptor channels in cultured rat hippocampal neurones. *J. Physiol.* 424:151–65

Ito M. 1989. Long-term depression. *Annu. Rev. Neurosci.* 12:85–102

Jacques J-P, Hausmann S, Kolakofsky D. 1994. Paramyxovirus mRNA editing leads to G deletions as well as insertions. *EMBO J.* 13: 5496–503

Janke A, Pääbo S. 1993. Editing of a tRNA anticodon in marsupial mitochondria changes its codon recognition. *Nucleic Acids Res.* 21:1523–25

Johnson DF, Poksay KS, Innerarity TL. 1993. The mechanism for Apo-B mRNA editing is deamination. *Biochem. Biophys. Res. Commun.* 195:1204–10

Jonas P, Racca C, Sakmann B, Seeburg PH, Monyer H. 1994. Difference in Ca²⁺ permeability of AMPA-type glutamate receptor channels in neocortical neurons caused by differential GluR-B subunit expression. *Neuron* 12:1281–89

Jonas P, Sakmann B. 1992. Glutamate receptor channels in isolated patches from CA1 and CA3 pyramidal cells of rat hippocampal slices. *J. Physiol.* 455:143–71

Keinanen K, Wisden W, Sommer B, Werner P, Herb A, et al. 1990. A family of AMPA-selective glutamate receptors. *Science* 249: 556–60

Kim U, Garner TL, Sanford T, Speicher D, Murray JM, Nishikura K. 1994a. Purification and characterization of double-stranded RNA adenosine deaminase from bovine nuclear extracts. *J. Biol. Chem.* 269:13480–89

Kim U, Nishikura K. 1993. Double-stranded RNA adenosine deaminase as a potential mammalian RNA editing factor. *Semin. Cell Biol.* 4:285–93

Kim U, Wang Y, Sanford T, Zeng Y, Nishikura K. 1994b. Molecular cloning of cDNA for double-stranded RNA adenosine deaminase, a candidate enzyme for nuclear RNA editing. *Proc. Natl. Acad. Sci. USA* 91:11457–61

Kohler M, Burnashev N, Sakmann B, Seeburg PH. 1993. Determinants of Ca²⁺ permeability in both TM1 and TM2 of high affinity kainate receptor channels:Diversity by RNA editing. *Neuron* 10:491–500

Koslowsky DJ, Bhat GJ, Perrollaz AL, Feagin JE, Stuart K. 1990. The MURF3 gene of *T. brucei* contains multiple domains of extensive editing and is homologous to a subunit of NADH dehydrogenase. *Cell* 62:901–11

Koslowsky DJ, Göringer HU, Morales TH, Stuart K. 1992. In vitro guide RNA/mRNA chimaera formation in *Trypanosoma brucei* RNA editing. *Nature* 356:807–9

Köller J, Nörskau G, Paul AS, Stuart K, Göringer HU. 1994. Different *Trypanosoma brucei* guide RNA molecules associate with an identical complement of mitochondrial proteins in vitro. *Nucleic Acids Res.* 22:1988–95

Kudla J, Igloi GL, Metzlaff M, Hagemann R, Kössel H. 1992. RNA editing in tobacco chloroplasts leads to the formation of a translatable psbL mRNA by a C to U substitution within the initiation codon. *EMBO J.* 11:1099–103

Landweber LF, Gilbert W. 1994. Phylogenetic analysis of RNA editing: a primitive genetic phenomenon. *Proc. Natl. Acad. Sci. USA* 91:918–21

Lau PP, Xiong WJ, Zhu HJ, Chen SH, Chan L. 1991. Apolipoprotein B mRNA editing is an intranuclear event that occurs posttranscriptionally coincident with splicing and polyadenylation. *J. Biol. Chem.* 266:20550–54

Lewis DA, Villafranca JJ. 1989. Investigation of the mechanism of CTP synthetase using rapid quench and isotope partitioning methods. *Biochemistry* 28:8454–59

Lomeli H, Mosbacher J, Melcher T, Hoger T, Geiger JRP, et al. 1994. Control of kinetic properties of AMPA receptor channels by nuclear RNA editing. *Science* 266:1709–13

Lonergan KM, Gray MW. 1993a. Predicted editing of additional transfer RNAs in *Acanthamoeba castellanii* mitochondria. *Nucleic Acids Res.* 21:4402

Lonergan KM, Gray MW. 1993b. Editing of transfer RNAs in *Acanthamoeba castellanii* mitochondria. *Science* 259:812–16

Loughney K, Kreber R, Ganetzky B. 1989. Molecular analysis of the para locus, a sodium channel gene in *Drosophila. Cell* 58:1143–54

Luo GX, Chao M, Hsieh SY, Sureau C, Nishikura K, Taylor J. 1990. A specific base transition occurs on replicating hepatitis delta virus RNA. *J. Virol.* 64:1021–27

Mahendran R, Spottswood MS, Ghate A, Ling M-I, Jeng K, Miller DL. 1994. Editing of the mitochondrial small subunit rRNA in *Physarum polycephalum. EMBO J.* 13:232–40

Mahendran R, Spottswood MR, Miller DL. 1991. RNA editing by cytidine insertion in mitochondria of *Physarum polycephalum. Nature* 349:434–38

Maier RM, Hoch B, Zeltz P, Kössel H. 1992a. Internal editing of the maize chloroplast ndhA transcript restores codons for conserved amino acids. *Plant Cell* 4:609–16

Maier RM, Neckermann K, Hoch B, Akhmedov NB, Kössel H. 1992b. Identification of editing positions in the ndhB transcript from maize chloroplasts reveals sequence similarities between editing sites of chloroplasts and plant mitochondria. *Nucleic Acids Res.* 20:6189–94

Maréchal-Drouard L, Ramamonjisoa D, Cosset A, Weil JH, Dietrich A. 1993. Editing corrects mispairing in the acceptor stem of bean and potato mitochondrial phenylalanine transfer RNAs. *Nucleic Acids Res.* 21:4909–14

Maslov DA, Avila HA, Lake JA, Simpson L. 1994. Evolution of RNA editing in kinetoplastid protozoa. *Nature* 365:345–48

Maslov DA, Simpson L. 1992. The polarity of editing within a multiple gRNA-mediated domain is due to formation of anchors for upstream gRNAs by downstream editing. *Cell* 70:459–67

Maslov DA, Simpson L. 1994. RNA editing and mitochondrial genomic organization in the cryptobiid kinetoplastid protozoan, *Trypanoplasma borreli. Mol. Cell Biol.* 14:8174–82

Melcher T, Maas S, Higuchi M, Keller W, Seeburg PH. 1995. Editing of α-amino-3-hydroxy-5-methyl isoxazole-4-propionic acid receptor GluR-B pre-mRNA in vitro reveals site selective adenosine to inosine conversion. *J. Biol. Chem.* 270:8566–70

Merkler DJ, Brenowitz M, Schramm VL. 1990. The rate constant describing slow-onset inhibition of yeast AMP deaminase by coformycin analogues is independent of inhibitor structure. *Biochemistry* 29:8358–64

Mishina M, Sakimura K, Mori H, Kushiya E, Harabayashi M, Uchino S, Nagahari K. 1991. A single amino acid residue determines the Ca$^{2+}$ permeability of AMPA-selective glutamate receptor channels. *Biochem. Biophys. Res. Commun.* 180:813–21

Missel A, Göringer HU. 1994. *Trypanosoma brucei* mitochondria contain RNA helicase activity. *Nucleic Acids Res.* 22:4050–56

Monyer H, Seeburg PH, Wisden W.1991. Glutamate-operated channels: developmentally early and mature forms arise by alternative splicing. *Neuron* 6:799–810

Nakanishi N, Shneider NA, Axel R. 1990. A family of glutamate receptor genes: evidence for the formation of heteromultimeric receptors with distinct channel properties. *Toxicon* 28:1333–46

Navaratnam N, Morrison JR, Bhattacharya S, Patel D, Funahashi T, et al. 1993. The p27 catalytic subunit of the apolipoprotein B mRNA editing enzyme is a cytidine deaminase. *J. Biol. Chem.* 268:20709–12

Navaratnam N, Shah R, Patel D, Fay V, Scott J. 1993. Apolipoprotein B mRNA editing is associated with UV cross-linking of proteins to the editing site. *Proc. Natl. Acad. Sci. USA* 90:222–26

Neckermann K, Zeltz P, Igloi GL, Kössel H, Maier RM. 1994. The role of RNA editing in conservation of start codons in chloroplast genomes. *Gene* 146:177–82

Nishikura K, Yoo C, Kim U, Murray JM, Estes PA, et al. 1991. Substrate specificity of the dsRNA unwinding/modifying activity. *EMBO J.* 10:3523–32

Nugent JM, Palmer JD. 1991. RNA-mediated transfer of the gene coxII from the mitochondrion to the nucleus during flowering plant evolution. *Cell* 66:473–81

O'Connell MA, Keller W. 1994. Purification and properties of double-stranded RNA-specific adenosine deaminase from calf thymus. *Proc. Natl. Acad. Sci. USA* 91: 10596–600

O'Connell MA, Krause S, Higuchi M, Hsuan JJ, Totty NF, et al. 1995. Cloning of cDNAs encoding mammalian double-stranded RNA-specific adenosine deaminase. *Mol. Cell Biol.* 15:1389–97

Oda K, Yamato K, Ohta E, Nakamura Y, Takemura M, et al. 1992. Gene organization deduced from the complete sequence of liverwort *Marchantia polymorpha* mitochondrial DNA. A primitive form of plant mitochondrial genome. *J. Mol. Biol.* 223:1–7

Paterson RG, Lamb RA. 1990. RNA editing by G-nucleotide insertion in mumps virus P-gene mRNA transcripts. *J. Virol.* 64:4137–45

Pelet T, Curran J, Kolakofsky D. 1991. The P gene of bovine parainfluenza virus 3 expresses all three reading frames from a single mRNA editing site. *EMBO J.* 10:443–48

Peris M, Frech GC, Simpson AM, Bringaud F, Byrne E, et al. 1994. Characterization of two classes of ribonucleoprotein complexes possibly involved in RNA editing from *Leishmania tarentolae* mitochondria. *EMBO J.* 13:1664–72

Pizzi M, Ribola M, Valerio A, Memo M, Spano P. 1991. Various Ca$^{2+}$ entry blockers prevent glutamate-induced neurotoxicity. *Eur. J. Pharmacol.* 209:169–73

Pollard VW, Hajduk SL. 1991. *Trypanosoma equiperdum* minicircles encode three distinct primary transcripts which exhibit guide RNA characteristics. *Mol. Cell Biol.* 11:1668–75

Pollard VW, Rohrer SP, Michelotti EF, Hancock K, Hajduk SL. 1990. Organization of minicircle genes for guide RNAs in *Trypanosoma brucei*. *Cell* 63:783–90

Polson AG, Bass BL. 1994. Preferential selection of adenosines for modification by double-stranded RNA adenosine deaminase. *EMBO J.* 13:5701–11

Polson AG, Crain PF, Pomerantz SC, McCloskey JA, Bass BL. 1991. The mechanism of adenosine to inosine conversion by the double-stranded RNA unwinding/modifying activity: a high-performance liquid chromatography-mass spectrometry analysis. *Biochemistry* 30:11507–14

Powell LM, Wallis SC, Pease RJ, Edwards YH, Knott TJ, Scott J. 1987. A novel form of tissue-specific RNA processing produces apolipoprotein-B48 in intestine. *Cell* 50: 831–40

Puchalski RB, Louis J-C, Brose N, Traynelis SF, Egebjerg J, et al. 1994. Selective RNA editing and subunit assembly of native glutamate receptors. *Neuron* 13:131–47

Randall RD, Thayer SA. 1992. Glutamate-induced calcium transient triggers delayed calcium overload and neurotoxicity in rat hippocampal neurons. *J. Neurosci.* 12:1882–95

Read LK, Göringer HU, Stuart K. 1994a. Assembly of mitochondrial ribonucleoprotein complexes involves specific guide RNA (gRNA)-binding proteins and gRNA domains but does not require preedited mRNA. *Mol. Cell Biol.* 14:2629–39

Read LK, Myler PJ, Stuart K. 1992. Extensive editing of both processed and preprocessed maxicircle CR6 transcripts in *Trypanosoma brucei*. *J. Biol. Chem.* 267:1123–28

Read LK, Wilson KD, Myler PJ, Stuart K. 1994b. Editing of *Trypanosoma brucei* maxicircle CR5 mRNA generates variable carboxy terminal predicted protein sequences. *Nucleic Acids Res.* 22:1489–95

Rebagliatti MR, Melton DA. 1987. Antisense RNA injections in fertilized frog eggs reveal an RNA duplex unwinding activity. *Cell* 48: 599–605

Riley GR, Corell RA, Stuart K. 1994. Multiple guide RNAs for identical editing of *Trypanosoma brucei* apocytochrome b mRNA have an unusual minicircle location and are developmentally regulated. *J. Biol. Chem.* 269:6101–8

Rueter SM, Burns CM, Coode SA, Mookherjee P, Emeson RB. 1995. Glutamate receptor RNA editing in vitro by enzymatic conversion of adenosine to inosine. *Science.* 267: 1491–94

Ruf S, Zeltz P, Kössel H. 1994. Complete RNA editing of unspliced and dicistronic transcripts of the intron-containing reading frame IRF170 from maize chloroplasts. *Proc. Natl. Acad. Sci. USA* 91:2295–99

Sakimura K, Bujo H, Kushiya E, Araki K, Yamazaki M, et al. 1990. Functional expression from cloned cDNAs of glutamate receptor species responsive to kainate and quisqualate. *FEBS Lett.* 272:73–80

Schuster W, Brennicke A. 1994. The plant mitochondrial genome: physical structure, information content, RNA editing, and gene migration to the nucleus. *Annu. Rev. Plant Physiol. Plant Mol. Biol.* 45:61–78

Schuster W, Hiesel R, Brennicke A. 1993. RNA editing in plant mitochondria. *Semin. Cell Biol.* 4:279–84

Schuster W, Hiesel R, Wissinger B, Brennicke A. 1990a. RNA editing in the cytochrome b locus of the higher plant *Oenothera berteriana* includes a U-to-C transition. *Mol. Cell Biol.* 10:2428–31

Schuster W, Wissinger B, Unseld M, Brennicke A. 1990b. Transcripts of the NADH-dehydrogenase subunit 3 gene are differentially edited in *Oenothera* mitochondria. *EMBO J.* 9:263–69

Seiwert SD, Stuart K. 1994. RNA editing: transfer of genetic information from gRNA to precursor mRNA in vitro. *Science* 266: 114–17

Shah RR, Knott TJ, Legros JE, Navaratnam N, Greeve JC, Scott J. 1991. Sequence requirements for the editing of apolipoprotein B mRNA. *J. Biol. Chem.* 266:16301–4

Sharma PM, Bowman M, Madden SL, Rauscher FJ, Sukumar S. 1994. RNA editing in the Wilms' tumor susceptibility gene, *WT1*. *Genes Dev.* 8:720–31

Sharmeen L, Bass B, Sonenberg N, Weintraub H, Groudine M. 1991. Tat-dependent adenosine-to-inosine modification of wild-type transactivation response RNA. *Proc. Natl. Acad. Sci. USA* 88:8096–100

Shaw J, Campbell D, Simpson L. 1989. Internal frameshifts within the mitochondrial genes for cytochrome oxidase subunit II and maxicircle unidentified reading frame 3 in *Leishmania tarentolae* are corrected by RNA editing: evidence for translation of the edited cytochrome oxidase subunit II mRNA. *Proc. Natl. Acad. Sci. USA* 86:6220–24

Shaw J, Feagin JE, Stuart K, Simpson L. 1988. Editing of mitochondrial mRNAs by uridine addition and deletion generates conserved amino acid sequences and AUG initiation codons. *Cell* 53:401–11

Simpson AM, Bakalara N, Simpson L. 1992. A ribonuclease activity is activated by heparin or by digestion with proteinase K in mitochondrial extracts of *Leishmania tarentolae*. *J. Biol. Chem.* 267:6782–88

Simpson L. 1972. The kinetoplast of the hemoflagellates. *Int. Rev. Cytol.* 32:139–207

Simpson L. 1987. The mitochondrial genome of kinetoplastid protozoa: genomic organization, transcription, replication and evolution. *Annu. Rev. Microbiol.* 41:363–82

Simpson L, Maslov DA. 1994a. RNA editing and the evolution of parasites. *Science* 264: 1870–71

Simpson L, Maslov DA. 1994b. Ancient origin of RNA editing in kinetoplastid protozoa. *Curr. Opin. Genet. Dev.* 4:887–94

Simpson L, Shaw J. 1989. RNA editing and the mitochondrial cryptogenes of kinetoplastid protozoa. *Cell* 57:355–66

Simpson L, Thiemann OH. 1995. mRNA editing. In *Eukaryotic mRNA Processing,* ed. AR Krainer. Oxford: IRL. In press

Smith HC, Kuo S-R, Backus JW, Harris SG, Sparks CE, Sparks JD. 1991. In vitro apolipoprotein B mRNA editing: identification of a 27S editing complex. *Proc. Natl. Acad. Sci. USA* 88:1489–93

Smith JB, Gottlieb PA, Dinter-Gottlieb G. 1992. A sequence element necessary for self-cleavage of the antigenomic hepatitis delta RNA in 20 M formamide. *Biochemistry* 31: 9629–35

Sollner-Webb B. 1991. RNA editing. *Curr. Opin. Cell Biol.* 3:1056–61

Sommer B, Keinanen K, Verdoorn TA, Wisden W, Burnashev N, et al. 1990. Flip and flop: a cell-specific functional switch in glutamate-operated channels of the CNS. *Science* 249:1580–85

Sommer B, Köhler M, Sprengel R, Seeburg PH. 1991. RNA editing in brain controls a determinant of ion flow in glutamate-gated channels. *Cell* 67:11–19

Souza AE, Myler PJ, Stuart K. 1992. Maxicircle CR1 transcripts of *Trypanosoma brucei* are edited, developmentally regulated, and encode a putative iron-sulfur protein homologous to an NADH dehydrogenase subunit. *Mol. Cell Biol.* 12:2100–7

Souza AE, Shu H-H, Read LK, Myler PJ, Stuart KD. 1993. Extensive editing of CR2 maxicircle transcripts of *Trypanosoma brucei* predicts a protein with homology to a subunit of NADH dehydrogenase. *Mol. Cell Biol.* 13:6832–40

Sturm NR, Simpson L. 1990. Kinetoplast DNA minicircles encode guide RNAs for editing of cytochrome oxidase subunit III mRNA. *Cell* 61:879–84

Sturm NR, Simpson L. 1991. *Leishmania tarentolae* minicircles of different sequence classes encode single guide RNAs located in the variable region approximately 150 bp from the conserved region. *Nucleic Acids Res.* 19:6277–81

Sugisaki H, Takanami M. 1993. The 5'-terminal region of the apocytochrome b transcript in *Crithidia fasciculata* is successively edited by two guide RNAs in the 3' to 5' direction. *J. Biol. Chem.* 268:887–91

Teng B, Burant CF, Davidson NO. 1993. Molecular cloning of an apolipoprotein B messenger RNA editing protein. *Science* 260: 1816–19

Thiemann OH, Maslov DA, Simpson L. 1994. Disruption of RNA editing in *Leishmania tarentolae* by the loss of minicircle-encoded guide RNA genes. *EMBO J.* 13:5689–700

Thomas SM, Lamb RA, Paterson RG. 1988. Two mRNAs that differ by two nontemplated nucleotides encode the amino coterminal proteins P and V of the Paramyxovirus SV5. *Cell* 54:891–903

Van der Spek H, Van den Burg J, Croiset A, Van den Broek M, Sloof P, Benne R. 1988. Transcripts from the frameshifted MURF3 gene from *Crithidia fasciculata* are edited by U insertion at multiple sites. *EMBO J.* 7: 2509–14

Verdoorn T, Burnashev N, Monyer H, Seeburg P, Sakmann B. 1991. Structural determinants of ion flow through recombinant glutamate receptor channels. *Science* 252:1715–18

Vidal S, Curran J, Kolakofsky D. 1990. A stuttering model for paramyxovirus P mRNA editing. *EMBO J.* 9:2017–22

Wagner RW, Nishikura K. 1988. Cell cycle expression of RNA duplex unwindase activity in mammalian cells. *Mol. Cell Biol.* 8: 770–77

Wagner RW, Smith JE, Cooperman BS, Nishikura K. 1989. A double-stranded RNA unwinding activity introduces structural alterations by means of adenosine to inosine conversions in mammalian cells and *Xenopus* eggs. *Proc. Natl. Acad. Sci. USA* 86: 2647–51

White T, Borst P. 1987. RNA end-labeling and RNA ligase activities can produce a circular ribosomal RNA in whole cell extracts from trypanosomes. *Nucleic Acids Res.* 15:3275–90

Wissinger B, Schuster W, Brennicke A. 1991. Trans splicing in *Oenothera* mitochondria: nad1 mRNAs are edited in exon and transsplicing group II intron sequences. *Cell* 65: 473–82

Wu JH, Semenkovich CF, Chen S-H, Li W-H, Chan L. 1990. Apolipoprotein B mRNA editing. Validation of a sensitive assay and developmental biology of RNA editing in the rat. *J. Biol. Chem.* 265:12312–16

Yamanaka S, Poksay KS, Balestra ME, Zeng G-Q, Innerarity TL. 1994. Cloning and mutagenesis of the rabbit apoB mRNA editing protein. A zinc motif is essential for catalytic activity, and noncatalytic auxiliary factor(s) of the editing complex are widely distributed. *J. Biol. Chem.* 269:21725–34

Yang AJ, Mulligan RM. 1991. RNA editing intermediates of cox2 transcripts in maize mitochondria. *Mol. Cell Biol.* 11:4278–81

Yang J-H, Sklar P, Axel R, Maniatis T. 1995. Editing of glutamate receptor subunit B premRNA in vitro by site-specific deamination of adenosine. *Nature* 374:77–86

Zeltz P, Hess WR, Neckermann K, Börner T, Kössel H. 1993. Editing of the chloroplast rpoB transcript is independent of chloroplast translation and shows different patterns in barley and maize. *EMBO J.* 12:4291–96

Zheng H, Fu T-B, Lazinski D, Taylor J. 1992. Editing on the genomic RNA of human hepatitis delta virus. *J. Virol.* 66:4693–97

Annu. Rev. Neurosci. 1996. 19:53–77

# APOLIPOPROTEIN E AND ALZHEIMER'S DISEASE

*Warren J. Strittmatter and Allen D. Roses*

Departments of Medicine (Neurology) and Neurobiology, Joseph and Kathleen Bryan Alzheimer's Disease Research Center, Box 2900, Duke University Medical Center, Durham, North Carolina 27710

KEY WORDS:   neurofibrillary tangles, neuritic plaques, tau, microtubules

## ABSTRACT

The apolipoprotein E locus (*APOE*) is associated with variations in the age of onset and risk of Alzheimer's disease. The *APOE4* allele increases the probability of disease at an earlier age. In contrast, the *APOE3* and *APOE2* alleles decrease the probability of disease and increase the age of onset. Therefore the metabolism of apolipoprotein E is relevant to Alzheimer's disease. Isoform-specific interactions of apolipoprotein E with other molecules may determine the rate of disease expression through molecular pathways that appear unique to the disease. In addition, some isoform-specific interactions of apolipoprotein E have been demonstrated with the defining pathological lesions of Alzheimer's disease, the neurofibrillary tangle and neuritic plaque. Several hypotheses of disease pathogenesis are now based on the relevance of apolipoprotein E.

## Alzheimer's Disease: Clinical Phenotype

The Alzheimer's diseases occur in the later decades of life and are characterized by progressive dementia, which is the diffuse deterioration of mental function. Thought and memory processes are primarily affected; affective and behavioral changes may follow. In 1984, the work group of the National Institutes of Communicative Diseases and Stroke and the Alzheimer's Disease and Related Disorders Association published criteria for the clinical diagnosis of Alzheimer's disease (AD) that were based on features of a patient's history, physical and neurological examination, and laboratory investigations, including brain imaging (McKhann et al 1984). These diagnostic criteria require the exclusion of other causes of memory loss and impaired cognitive function, such as multiple infarcts, intracranial mass lesions, infections, and toxic and metabolic disorders. Fulfilling these clinical criteria permits the diagnosis of possible or probable AD. The patient with AD experiences gradually increasing forgetfulness, decreasing attention span, and alterations in mood, often with

53

frustration and agitation, resulting in increasing difficulties in meeting the demands of daily living. These problems progress until the patient ultimately cannot attend to his or her simplest needs and becomes bedridden, totally dependent on caregivers. The interval between initial diagnosis and death varies considerably, usually between 3 and 15 years.

According to the defined diagnostic criteria, a diagnosis of definite AD can be made only by microscopic examination of brain tissue, either at biopsy or, more commonly, autopsy. The neuropathological criteria, also established by a consensus group (Khachaturian 1985), require the presence of neuritic plaques and neurofibrillary tangles (NFTs), at specified densities. The diagnosis of definite AD has therefore been defined by phenotypic neuropathological findings. The relationship between the plaques and tangles and the mechanism causing Alzheimer's dementia is unknown and controversial. Discovering the relationship between AD neuropathology and the pathogenesis of the disease is a central issue in developing and testing hypotheses to uncover the molecular and cellular mechanisms that ultimately result in the dementia.

## Neuritic Plaques as a Phenotype of Alzheimer's Disease

Neuritic plaques are extracellular structures with complex and incompletely characterized molecular and cellular constituents. Plaques contain $A\beta$, a peptide of 39–43 amino acids that is produced by proteolytic cleavage of the amyloid precursor protein (APP) in its normal metabolism [for review see Selkoe (1994)]. $\beta$-Pleated sheet fibrils of $A\beta$ interact with Congo red dye or thioflavin silver stains to produce the defining amyloid appearance. $A\beta$ peptide aggregates in these structures. Other proteins found in the neuritic plaque include apolipoprotein E (Namba et al 1991, Strittmatter et al 1993a, Wisniewski & Frangione 1992), APP, $\alpha_1$-antichymotrypsin, IgG, several complement proteins, amyloid P, and glycosaminoglycans and Sp40,40 (Ghiso et al 1993). The complete molecular composition of the fibrillar structures in the plaque, the mechanism of assembly, and their role in the disease are unknown. With the availability of antibodies against $A\beta$ peptide, the operative criteria for diagnosis have been subtly expanded in many laboratories to include $A\beta$-containing plaques without amyloid. Neuritic plaques may or may not contain amyloid deposits.

Some patients may clinically display probable AD but lack sufficient plaques for a diagnosis of definite AD. A variant "tangle-only" AD has been described with the clinical phenotype of AD and the neuropathology of NFTs, without neuritic plaques. Other patients with progressive dementia have pathological features such as Lewy or Pick bodies and are given other neuropathological diagnoses. Overlap cases with both Lewy bodies and amyloid plaques have also been described (Hansen et al 1990). Furthermore, many nondemented people are found at autopsy to have sufficient plaques for the diagnosis of AD. The classification of disease(s) with similar clinical manifestations but with a variety of neuropathological phenotypes is a source of much of the current discussion and confusion in the literature.

## Neurofibrillary Tangles as a Phenotype of Alzheimer's Disease

NFTs are dense bundles of long unbranched filaments in the cytoplasm of some neurons. These filamentous structures are paired helical filaments. Each filament is 10 nm in width, and they are helically twisted about each other, with a periodic full twist every 160 nm. NFTs may be so dense that they distort the neuronal cell body and displace the nucleus. Paired helical filaments may also be found in neurites undergoing degeneration. The filaments consist primarily, and probably exclusively, of the microtubule-associated protein (MAP) tau [for review see Goedert et al (1991)].

Tau normally binds and stabilizes microtubules and promotes the assembly of microtubules by polymerizing tubulin. Microtubules are necessary for neurite extension and maintenance and for the transport of materials along the axon and dendrites in both orthograde and retrograde directions. In AD, tau becomes abnormally phosphorylated and self-assembles into the pathological paired helical filaments, thereby forming NFTs (Biernat et al 1992, Crowther et al 1994, Goedert et al 1992). The topology and abundance of NFTs in patients' brains that meet neuritic plaque criteria for AD relate better to the severity of dementia than do the neuritic plaques (Arriagada et al 1992). NFTs are also in neurons in other neurodegenerative diseases. Rather than examining this phenotypic phenomenon as a model for neuronal degeneration, many researchers discount NFTs as nonspecific. The NFT may represent a phenotype common to several neurodegenerative diseases and, rather than being nonspecific, may provide clues to a common pathogenesis.

NFTs are intracellular and are thus limited by the surviving number of viable neurons. Extracellular "ghost" tangles are the remnants of some of these dead cells. Thus NFT counts can be limited by a ceiling effect. As the number of neurons containing NFTs declines, the proportion of tangle-bearing neurons may become stable. The formation of NFTs and the death of neurons occur over time and represent an ongoing progression. At any point in disease progression, however, the absolute numbers of tangle-bearing neurons may be the same. Whether NFTs themselves cause neuronal death or are simply phenotypic manifestations of dying cells is currently debatable.

## Genetic Heterogeneity of the Alzheimer Diseases

Mutations at several different loci result in the same clinical phenotype as AD. There are currently two identified genes and at least two additional genetic

**Table 1**   Genetic classification of the Alzheimer diseases

| Type | Chromo-some | Gene | Age | % |
|------|-------------|------|-----|---|
| *AD1*: Early-onset familial autosomal dominant inheritance | 21 | APP | mean = 50s, 42–68+ | <20 families |
| *AD2*: Late-onset familial and sporadic APOE associated | 19 | APOE | >55, mean ~70 | 60–90 |
| *AD3*: Early-onset familial autosomal dominant inheritance | 14 | UNK[a] | mean = 40s, 32–60+ | <2 |
| *AD4*: Early-onset familial autosomal dominant inheritance | UNK | UNK | mean = 50s | Volga-German founder |
| AD+: Late-onset familial and sporadic not *APOE* associated | UNK | UNK | >75 | 10–30 |
| AD++: Other forms of AD | | | | |

[a] UNK, unknown.

loci that are linked to the clinical complex of symptoms, signs, and neuropathological criteria by which AD is defined (Bird et al 1988, Goate et al 1991, Schellenberg et al 1992, St. George-Hyslop et al 1992, Strittmatter et al 1993). Locus heterogeneity is the term used when different genes cause the same disease. Table 1 presents the current genetic classification of the ADs, which is based on the order in which the nomenclature was assigned. *AD1* is the APP gene on chromosome 21 (Goate et al 1991). Mutations of APP are rare causes of AD, with less than 20 families in the world whose disease resulted from one of several mutations of APP. For most families, the disease results from valine-to-isoleucine mutations at codon 717. Other family-specific mutations occur at codon 717, and a few unique forms occur at codons 692 and 670–671 (Chartier et al 1991, Hendriks et al 1992, Lannfelt et al 1994a, Murrell et al 1991). Thus there is also allelic heterogeneity at the *AD1* locus. *AD1* is inherited as an autosomal dominant trait and has age of onset of 40–60 years. The age of onset of two *AD1* families appears to also be affected by the genotype of the apolipoprotein E (*APOE,* gene; apoE, protein) alleles located on chromosome 19 (St. George-Hyslop et al 1994).

The *APOE* locus was designated *AD2* and is a susceptibility gene for late-onset AD (Saunders et al 1993a,b; Strittmatter et al 1993a). Variations in the *APOE* genotype are associated with differences in the distribution of age of onset of AD (Corder et al 1993, 1994) (Figure 1). Although each allele is inherited from a parent, the effect on age of onset involves both alleles: the inherited genotype (Roses et al 1994). Unlike autosomal dominant or recessive traits, a specific mutation of one or both alleles does not necessarily cause disease. There are three common alleles of *APOE. APOE*-ε3 (*APOE3*) is the most common allele, representing approximately 78% of all chromosomes;

Proportion Unaffected

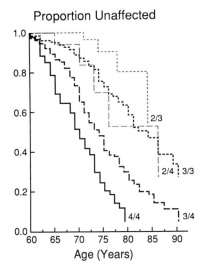

*Figure 1*  Risk of remaining unaffected by AD as a function of *APOE* genotype. The data for this figure is derived from Corder et al (1994) and include 115 late-onset AD subjects and 243 age-matched controls, as well as 150 affected and 197 unaffected members of 66 AD families from the United States. The age of onset of subjects with each indicated genotype was plotted. Onset curves were estimated by Kaplan-Meier product limit distributions. No patients with onset before age 60 years were used in this analysis. The proportion of each genotype will vary as a function of the ethnic or racial origin of the population (Table 2). Epidemiologically based distributions will be needed before risks can be estimated in the general population.

*APOE4*, 15%; and *APOE2*, 7%. The proportion of different *APOE* alleles varies between racial and ethnic groups, particularly with regard to their relative proportions of *APOE2* and *APOE4* (Crews et al 1993, Mayeux et al 1993, Noguchi et al 1993, Okuizumi et al 1994, Sakoda et al 1994, Ueki et al 1993, Yoshizawa et al 1994). In AD the mean age of onset can vary from less than 70 years for *APOE4/4* individuals who constitute approximately 2% of the age-matched population, to a mean of more than 90 years for *APOE2/3* individuals (~10%) (Roses et al 1994) (Figure 1, Table 2).

Strittmatter et al (1993a) first reported the association of *APOE4* with late-onset familial AD. Saunders et al (1993b) extended the *APOE4* association to 176 autopsy-verified sporadic AD patients (Table 2) and found that although the US white control population allele frequency of *APOE4* was 0.16, that of the AD patients was 0.40 (Table 2). These data have been confirmed and extended to AD groups in populations around the world. In Japan, for example, multiple investigators have confirmed that the *APOE4* allele frequency is lower (<0.09) in the control population; the *APOE4* frequency increased to >0.28 in AD patients (Noguchi et al 1993, Sakoda et al 1994, Tsuda et al 1994, Ueki

**Table 2a**    Apolipoprotein E genotypes and Alzheimer disease[a]

| Genotype APOE | Alzheimer USA | % | Controls CEPH | % | Alzheimer Japan | % | Controls Japan | % |
|---|---|---|---|---|---|---|---|---|
| 4/4 | 29/176 | 17 | 2/91 | 2 | 15/196 | 8 | 2/134 | 1 |
| 3/4 | 76/176 | 43 | 19/91 | 21 | 77/196 | 39 | 21/134 | 16 |
| 3/3 | 58/176 | 33 | 52/91 | 57 | 95/196 | 48 | 107/134 | 80 |
| 2/4 | 7/176 | 4 | 7/91 | 8 | 1/196 | 0.5 | 0/134 | 0 |
| 2/3 | 6/176 | 3 | 10/91 | 11 | 7/196 | 4 | 2/134 | 1 |
| 2/2 | 0 | 0 | 1/91 | 1 | 1/196 | 0.5 | 2/134 | 2 |

**Table 2b**    Allele frequency and allele bearer frequency in Alzheimer disease and controls

|  | ALZ/USA | Control/USA | ALZ/Japan | Control/Japan |
|---|---|---|---|---|
| Allele frequency[b] | | | | |
| APOE4 | 0.40 | 0.16 | 0.28 | 0.09 |
| APOE3 | 0.56 | 0.74 | 0.70 | 0.88 |
| APOE2 | 0.04 | 0.10 | 0.02 | 0.02 |
| Allele bearer frequency[c] | | | | |
| 4/4 + 3/4 + 2/4 | 0.64 | 0.31 | 0.47 | 0.17 |
| 3/3 + 2/3 + 2/2 | 0.36 | 0.69 | 0.53 | 0.83 |

[a] The Alzheimer disease series used for the USA was 176 autopsy-confirmed sporadic definite Alzheimer disease patients. Published by Saunders et al (1993b). The controls were the grandparent generation from the CEPH series of reference families and are similar to other control groups (Saunders et al (1993a). The Japanese Alzheimer disease and control series were collected by Professor Shoji Tsuji, from the Brain Research Institute, University of Niigata, Japan, and are representative of several published Japanese series (Noguchi et al 1993, Okuizumi et al 1994, Sakoda et al 1994, Ueki et al 1993, Yoshizawa et al 1994).
[b] Proportion of chromosomes of each allele.
[c] Proportion of individuals who carry one or two APOE4 alleles.

et al 1993). Similar racial variations in *APOE4* allele frequency in the African-American and Hispanic control populations have been confirmed, as has the association of *APOE4* with AD (Mayeux et al 1993). Thus there are ethnic and racial differences in the proportion of each genotype in various populations, which changes the apparent epidemiology of AD. Therefore, in Japan, which has a lower proportion of *APOE4/4* individuals and a shift towards

*APOE3/4* individuals, it might be expected that the average age of onset would be older by 6–8 years. Because the allele frequency of *APOE4* is lower, as is the proportion of the Japanese population with one or two *APOE4* alleles, the prevalence of AD would be expected to be lower. Both the average age of onset and the lower prevalence of AD in Japan, especially in the 60- to 70-year-old range, have been reported (Noguchi et al 1994, Okuizumi et al 1994, Sakado et al 1994).

Depending on the age criteria used for AD case ascertainment, the allele frequency of *APOE4* can vary. For example, there is a single report in which no association was made between *APOE4* and 29 Swedish sporadic AD patients, but the minimum age of the patients ascertained in that series was 75 years (Lannfelt et al 1994b). Thus ascertainment bias selected a small series of AD patients who were older than the mean age of onset of AD and well past the major effect of *APOE4* on the age-of-onset distribution (Figure 1). Excluding this small series, in less than two years after the initial reports of *APOE4* association with AD, more than 60 confirmations of clinical sporadic, clinical familial, and autopsy-confirmed series were published (Amouyel et al 1993, Anwar et al 1993, Ben Shlomo et al 1993, Borgaonkar et al 1993, Czech et al 1993, Group 1993, Houlden et al 1993, Lucotte et al 1993, Mayeux et al 1993, Noguchi et al 1993, Payami et al 1993, Poirier et al 1993, Rebeck et al 1993, Ueki et al 1993).

Corder et al (1993) demonstrated that there was a dose effect of the inheritance of *APOE4* on the distribution of age of onset in familial AD. With each *APOE4* allele inherited, the risk of developing AD increases and the distribution of the age of onset decreases. In a series of AD families, sporadic AD patients, and case controls, Corder et al (1994) showed that the inheritance of an *APOE2* allele decreased the risk and increased the mean age of onset. Figure 1 illustrates the distribution of age of onset of the five common *APOE* genotypes for individuals 60 years or older (Roses et al 1994). It was subsequently documented that the *APOE4* association is present before 60 years of age as well (Okuizumi et al 1994, van Duijn et al 1994). The initial reports specifically excluded AD patients below age 60 years from the analyses to avoid confounding interpretations with the undefined forms of early-onset AD (Saunders et al 1993a,b; Strittmatter et al 1993a). The allele frequency for AD patients in the 50- to 59-year-old group is similar to that of the 60- to 69-year-old group, although there are fewer AD patients in the younger decade and may be more *APOE4/4* homozygous patients.

Figure 1 illustrates the differences in the rate of disease (phenotypic) expression as a function of both inherited *APOE* alleles (Roses 1994, Roses et al 1994). The mean age of onset for individuals inheriting the *APOE4/4* genotype is less than 70 years (inclusion of AD cases less than 60 years old would magnify this effect), whereas the mean age of onset for *APOE2/3* individuals is greater than 90 years. *APOE2/2* represents <0.05% of the population, and insufficient data were available for the onset distribution curve. Thus the *APOE* genotype difference can account for more than two decades of difference in the rate of disease expression. It should also be noted that if the best common genotype, *APOE2/3* (approximately 10% of the US population), is extended to age 150 years, everyone would have developed AD by age 140 years. Thus the effect of *APOE* alleles on the expression of AD can be viewed as increasing the rate of a universal process to express disease early enough to become evident in the general population. A specific mutation of *APOE* is not causative, but *APOE4* > *APOE3* > *APOE2* appears to increase the rate of neuronal pathology and AD. Age has been considered the major risk factor for AD. With an increase in the average expected life span and more people reaching at-risk ages, the prevalence of AD increases. Within each population, the effect of age on AD is determined by the relative distributions of *APOE* alleles.

Table 2 lists the relative proportion of *APOE* genotypes in a large group of autopsy-verified AD patients from the United States and of aged controls, clearly illustrating that while 64% of AD patients inherited at least one *APOE4* allele, 36% of AD patients inherited no *APOE4* allele (Saunders et al 1993b). A representative series from Japan in which the allele frequency of *APOE4* in the control population is lower is also illustrated (Okuizumi et al 1994). The decreased proportion of AD patients who are *APOE4/4* relative to *APOE3/4* and *APOE3/3* in Japan would predict an older mean age of onset, while the lower allele frequency of *APOE4* would predict a lower disease prevalence, especially in the 50- to 70-year-old population. In fact, AD is less common in Japan and has an older mean age of onset (~78 years in Japan compared to ~70 years in the United States).

*AD3* is an as yet undefined locus on chromosome 14q2.3 (Schellenberg et al 1992, St. George-Hyslop et al 1992). *AD3* is inherited as an autosomal dominant trait and is responsible for more than 90% of the autosomal dominant early-onset AD. The mean age of onset appears to be slightly younger than that observed in APP mutation families. In preliminary reports there appears to be no additional effect of *APOE* genotypes on the age of onset of *AD3* (Van Broeckhoven et al 1994). The gene responsible for AD3 has not yet been identified and is the focus of positional cloning strategies successfully used for many other inherited diseases. The identity of the *AD3* gene will clearly increase the information concerning mechanisms of pathogenesis leading to AD. Any hypothesis would be speculative until the relevant gene is identified.

*AD4* is also an unknown locus and remains to be linked (Schellenberg et al 1992, Yu et al 1994). The *AD4* families have been clinically well documented and characterized and appear to represent a founder effect. The families originate from two nearby German-speaking villages in the Volga River region of

Ukraine (Bird et al 1988, 1989). The *AD4* locus has been formally excluded from chromosomes 21 and 14 and suggested for exclusion from chromosome 19, so a fourth locus is expected (Schellenberg et al 1992, Yu et al 1994). The mean age of onset is in the 50- to 60-year range.

Additional AD loci are expected to emanate from the linkage methods currently being applied in late-onset AD families. The effect of the *APOE* alleles on the expression of AD appears to be the strongest single susceptibility locus, but it is also clear that other loci may be operative. This is particularly evident in some of the oldest and largest late-onset pedigrees (mean age of onset >75 years), where there is little contribution from *APOE4* association and little variation in onset as a consequence of the *APOE* genotype. Until another susceptibility locus is identified, discussion of its role in the process leading to dementia is speculative. Other loci should be identified by complex analyses that incorporate the effect of *APOE* alleles into the linkage calculations.

## Application of Linkage Methods to Early-Onset and Late-Onset AD

The three identified forms of early-onset AD—AD1, AD3, and AD4—have been investigated using standard positional cloning methodologies that have been applied successfully to many other inherited diseases. The collection of large families, each of which can exhibit significant linkage or exclude linkage, led to the appreciation of locus heterogeneity of AD (Hardy & Duff 1993; Pericak-Vance et al 1988, 1991; Schellenberg et al 1991; St. George-Hyslop et al 1990). For late-onset AD the use of standard likelihood methods of analyses was confounded by an unclear mode of inheritance (Pericak-Vance et al 1988, 1991). In addition the allele frequency of the disease-associated allele (*APOE4*) was much higher than that usually encountered in autosomal dominant diseases. Familial aggregation was initially recognized in late-onset AD, but whether this represented chance ascertainment of a common disease in large families or a true genetic effort was unclear. Although the mode of inheritance could theoretically be autosomal dominant with variable penetrance, the data were confounded by the late age of onset, competing causes of death, and diagnostic phenocopies. Thus the genetic contribution to late-onset AD was not widely accepted during a period of major progress in the linkage of many other diseases (Lander & Schork 1994).

The strategy applied to late-onset AD was to investigate whether the familial aggregation could be explained by chance alone or whether it could be related to a genetic locus. By using classical linkage studies, the chromosome 21 and 14 loci were excluded in the late-onset AD pedigrees (Pericak-Vance et al 1988). Absence of linkage would prove nothing, but linkage of late-onset AD would support a genetic contribution. The families were then used to test for allele sharing among relatives using nonparametric methods of analysis (Pericak-Vance et al 1991).

The advantage of the application of excessive allele sharing methods to these families was that no assumptions regarding the mode of inheritance were required (Weeks & Lange 1988). By using the same highly polymorphic genetic markers that were used in classical linkage studies, a region of a chromosome containing multiple markers with excessive allele sharing was sought. In the first application of this methodology to complex disease, Pericak-Vance et al (1991) found that several adjacent probes (1990 genetic map) from the chromosome 19q13.1–13.3 region showed statistically significant excessive allele sharing. Because the marker data were generated from large families, multipoint analyses could be performed using standard linkage techniques by assuming an autosomal dominant inheritance with variable penetrance. The linkage to chromosome 19q13.1–13.3 was confirmed using standard multipoint lod score methods. When affecteds-only were used in the analysis, the multipoint lod score was significant. It is fair to say in retrospect that these analyses were not widely accepted in the field, so only our research group actively sought an AD gene on chromosome 19 (Lander & Schork 1994).

There were several very exciting candidate genes in the 19q region, but no sequence differences were demonstrated. During the course of a set of independent biochemical experiments studying the binding of proteins by the Aβ fragment of APP, a single protein among many proteins in the preparation exhibited functionally irreversible binding. After purification and partial amino-terminus sequencing, the protein was identified as apoE (Strittmatter et al 1993a). The *APOE* locus had been known to be located in the middle of the relevant region of chromosome 19q, but there had been no prior suggestion that it could be genetically relevant to AD.

The relationship of *APOE* to AD was then tested. There are three easily measured alleles of *APOE*, which permitassociation analyses. The most common isoform of apoE in the general population, apoE3, is secreted as a 299–amino acid protein containing a single cysteine residue at position 112 [see Weisgraber (1994) for review]. The other two common isoforms, apoE2 and apoE4, differ at one of two positions (residues 112 and 158) from apoE3 by cysteine-arginine interchanges: ApoE2 contains a cysteine at position 158, and apoE4 contains an arginine at 112.

In the initial association analysis, one affected patient (nonproband) was selected from each of 30 AD families and found to have an allele frequency of 0.50 compared to 0.16 in age-matched controls (Strittmatter et al 1993a). A similar allele frequency of 0.50 was found when all affecteds in the families were analyzed. Saunders et al (1993a,b) then examined several other AD series, including (*a*) 176 autopsy-confirmed sporadic AD patients, (*b*) a consecutive series of 80 probable or possible AD patients at a Memory Disorders Clinic and their spouses, (c) the first affected twin in a series of 62 affected twin pairs that had been independently ascertained, and (d) one affected from each of 16 early-onset pedigress (either APP mutations or chromosome 14 linked). The *APOE4* allele frequency significantly increased (~0.40) in each of the first three groups but did not significantly increase in the early-onset pedigrees. Thus the *APOE4* allele association was strongly associated with familial and sporadic late-onset AD. However, another gene could still be linkage disequilibrium with *APOE* and associated with late-onset AD. Linkage disequilibrium would imply that another undetermined gene was located on chromosome 19q adjacent to *APOE* so that over many generations the likelihood of recombination between that putative gene and *APOE* would be extremely small. Thus any data linking or associating *APOE* could actually be due to an adjacent gene.

Three sets of independent experiments were performed to test the specificity of the *APOE4* association with AD. The first involved direct testing of nearby highly polymorphic markers for linkage disequilibrium. The second series of experiments were epidemiological analyses of the relationship of the *APOE* alleles to the distribution of age of onset of AD. The third set of experiments investigated the phenotypic involvement of apoE in the neuropathology of AD, an independent report of which appeared in the literature during our early studies (Namba et al 1991).

Highly polymorphic genes flanking *APOE,* including the *APOC2* locus, which is located less than 40 kb away, did not show any association with AD. There was no evidence for linkage disequilibrium (Chartier et al 1994, Liddell et al 1994, Mayeux et al 1993, Tsuda et al 1994, Yu et al 1994). Although Schellenberg et al (1987) had previously reported an association between a single *APOC2* restriction fragment length polymorphism and AD, closely related Volga-German founder-effect families were included in this early analysis, which probably accounted for their results. An association with *APOC2* has not been confirmed in a subsequent unrelated series of of studies of AD families (Chartier et al 1994, Liddell et al 1994, Mayeux et al 1993, Tsuda et al 1994, Yu et al 1994).

The effect of the inheritance of different *APOE* alleles on the age of onset distribution provided strong evidence that *APOE* was the relevant susceptibility gene (Corder et al 1993, 1994). The fact that one allele, *APOE4,* increased the risk and lowered the mean age of onset, while another allele of the same gene, *APOE2,* decreased the risk and increased the mean age of onset further supported *APOE* as a major susceptibility gene for AD. The biological effect on the rate of disease expression in the absence of evidence for an association of any adjacent marker reduces the possibility of another gene in linkage disequilibrium.

The phenotypic evidence that apoE protein participates in the characteristic neuritic plaques and NFTs of AD (reviewed below) also implicates apoE in AD pathology (Namba et al 1991, Strittmatter et al 1993a, Wisniewski & Frangione 1992). The localization of apoE in neurons in AD, in other neurode-generative diseases, and in controls raises new possibilities regarding the metabolic roles of apoE in neurons (Han et al 1994a,b; Strittmatter et al 1993a). Localization of apoE in neuronal cytoplasm and peroxisomes created new possibilities for normal apoE-dependent metabolic interactions, in addition to the possible isoform-specific role of apoE in AD (Han et al 1994a).

## Specificity of the APOE Association for AD-Type Dementias

At present there is a diagnostic paradox in the field of AD. On one side, the frequency of neuritic or amyloid plaques is used as the diagnostic criteria; on the other, there may be clinically identical dementias that are influenced by the *APOE* genotype but do not meet AD diagnostic criteria. Which is AD? For example, some demented individuals lack plaques but have Lewy bodies, a different neuropathological entity (Armstrong et al 1991; Hansen et al 1990, 1993; Masliah et al 1990, 1993; Wisniewski et al 1991). Lewy bodies are composed of aggregated neurofilament proteins. Some patients with Lewy bodies also have AD–like changes, meaning they also have some plaques that may be insufficient in number to be labeled AD. When the two groups are examined separately, the preliminary consensus is that Lewy body disease (without plaques) is not associated with an increase in the *APOE4* allele frequency, but Lewy body variant (with plaques) has an increased *APOE4* allele frequency (Benjamin et al 1994, Galasko et al 1994, Hansen et al 1994, Pickering et al 1994). Because the density of amyloid plaques is related to the *APOE4* allele and because apoE4 binds Aβ more avidly than apoE3 (Rebeck et al 1993, Schmechel et al 1993, Strittmatter et al 1993), there may be no difference between Lewy body disease and the Lewy body variant. Defining the difference by whether or not plaques are present may simply select the *APOE4* allele carriers preferentially (Strittmatter et al 1993a).

There is now significant confusion regarding the clinical classifications of AD and other clinically identical dementias. Dementias with other phenotypes may have similar pathogeneses, perhaps modified by other susceptibility poly-morphisms. Some rearranging of classification is expected as more genetic data become available, which is similar to the experience in other areas of medicine. Rsearchers now know that mutations at the same locus can lead to quite distinct phenotypes and that identical phenotypes can exhibit locus het-erogeneity. Phenotypic neuropathology variations are the bases of differential diagnosis of the dementias. At present, there are insufficient data to clarify the diagnostic puzzles with high degrees of certainty.

One example may illustrate the dilemma. Less than ten autopsies of AD patients from APP717 mutation families have been published. One patient, who had a clinical course with onset at age 49 years similar to that of her sister, had insufficient Aβ immunostained plaques to meet diagnostic criteria for AD (Strittmatter et al 1994). Yet the patient had the APP717 mutation and a typical disease progression. The patient had early-onset AD despite not meeting a criterion based on amyloid plaque numbers. Perhaps in the near future there will be an opportunity to reevaluate the phenotypic definitions of the ADs with the incorporation of genetic information.

## Genetic Strategies for Common Late-Onset Diseases

Late-onset AD was the first common disease to have a major susceptibility gene identified by using positional gene identification strategies. The strategy made use of a statistical method based explicitly on identity by state (IBS) sharing, called the Affected Pedigree Member (APM) analysis (Weeks & Lange 1988). Using this relatively rapid and computationally simple method, Pericak-Vance et al (1991) were able to screen the genome until they identified a series of polymorphic probes for which there was excessive allele sharing, which yielded a positive APM test statistic. The power of using the APM method with large AD families was that the initial screen was relatively rapid (even in 1990, when the density of markers was considerably less), but multipoint lod score analyses could subsequently be used to confirm the suggested APM linkage region of the genome. Many current investigations use the affected sib pair methods for genomic screens of complex diseases (Lander & Schork 1994). When only sib pairs have been ascertained and collected, the ability to rapidly check nearby markers for confirming linkage by another method is not readily available.

There were, however, problems with the interpretation of the lod score analysis for a susceptibility gene. These methods had previously been applied to diseases where the frequency of the disease-causing allele was relatively rare, usually less than 0.001. With a relatively common allele, such as APOE4 with an allele frequency of 0.16, the analysis was confounded by the inability to distinguish which of the alleles was disease causing (Lander & Schork 1994). With the benefit of hindsight, investigators now know that all alleles contribute to the rate of susceptibility to AD and that none are disease causing. The lod score analysis was unusual. When only affected individuals were included in the analysis, the multipoint lod score was significant at 4.22 (Pericak-Vance et al 1991). With the addition of more subjects in the families, none of whom were affected but some of whom could be expected to become affected, the lod score was reduced to 2.54. In autosomal dominant late-onset diseases such as Huntington's disease, addition of data from subjects not yet affected would usually lead to a slight increase in the lod score. The high allele frequency in unaffected individuals of the relevant susceptibility gene, including homozygosity, was a factor usually not observed in diseases with rare alleles unless inbreeding had occurred (Lander & Schork 1994). For practical purposes, confirmation of the APM method by affecteds-only lod score analysis identified the chromosome 19q13.1–13.3 region. In a recent review of the dissection of complex traits, Lander & Schork (1994) state that the linkage was dismissed by many observers and only explained after *APOE4,* the major susceptibility gene, was identified using disease association and population genetic epidemiological analyses. The practical implications of using large family aggregations for disease research, even though they are very difficult to collect, may still be missed by many observers.

One might anticipate that the identification of other susceptibility genes for common and heterogenous diseases will face similar problems. Most of the strategies currently being used involve a genomic screen with several hundred affected sib pairs but without the secondary resource of large families for multipoint lod score analyses. The problems of determining what constitutes a statistical linkage are confounded by the second confirmation step. One can hope that the markers used for linkage are close to other markers that are in linkage disequilibrium or that the probe itself is the relevant gene so that a biological association with an allele can be established. Collecting large families may be expensive, time-consuming, and require considerable clinical expertise, but the possibility of using affecteds-only multipoint lod score analyses across a chromosomal region of interest may make such efforts worthwhile, if not necessary. Large family ascertainment is frequently thought to be impossible, but one of the major contributions of Pericak-Vance et al (1991) was to demonstrate that pedigrees with aggregations of common diseases collected over years could be subjected to genetic analyses. A full genomic screen may be technically possible in a few weeks, but the resource to be sceened is the critical element of complex disease research.

There was another practical advantage in the application of the APM to AD that could be viewed as a signal-to-noise advantage. Positive APM linkage statistics were also found for markers from chromosome 21 and chromosome 14, even though there was only a single APP mutation family and two chromosome 14–linked families in the analysis. It was relatively easy to identify which families gave the positive data. This advantage is lost when families are not available for secondary analyses.

Unfortunately most successful positional cloning efforts have ended in the discovery of a new disease-causing gene, the structure, function, and metabolism of which is unknown initially. Considerable effort must then be applied to its characterization. The *APOE* gene and its protein product are, however, well known from two decades of study in lipid metabolism and heart disease. What was unknown was the genetic association with late-onset AD and the critical involvement of apoE in neuronal metabolism.

## Isoform-Specific Interactions of ApoE

The single amino acid differences among the common isoforms of apoE result in significant differences in the biochemistry and cellular metabolism of these proteins and in marked differences in the risk of AD. The single cysteine in apoE3 permits disulfide bond formation with other molecules, including itself (Weisgraber & Shinto 1991). ApoE3 forms a disulfide-linked heterodimer with apolipoprotein A-II and a homodimer with another apoE3 molecule. The apoE4 isoform lacks a cysteine and cannot form disulfide complexes.

The apoE isoforms differ in their interactions with the low-density lipoprotein (LDL) receptor and in binding cholesterol-containing lipid particles (Gregg et al 1983, Weisgraber et al 1982). ApoE3 and apoE4 bind the LDL receptor with high affinity, whereas apoE2 binds with very low affinity (approximately 1% of apoE3). Severe atherosclerosis is common in some individuals homozygous for apoE2 and is thought to be due to decreased receptor-mediated apoE endocytosis. ApoE binds other cell surface receptors, including the LDL receptor related protein, the very low density lipoprotein (VLDL) receptor, and a scavenger receptor, but the interactions of the various isoforms of apoE with these receptors have not been extensively investigated. The ability of apoE to bind phospholipid particles is isoform specific; apoE2 and apoE3 preferentially associate with high-density lipoprotein (HDL) particles, and apoE4 preferentially associates with VLDL particles.

The molecular mechanisms of these isoform-specific differences have been studied. ApoE contains two functionally important domains, one that binds the LDL receptor and the other that binds lipoprotein particles (VLDL or HDL). Thrombin cleaves apoE at residues 191 and 215, yielding a 22-kDa amino-terminal fragment and a 10-kDa carboxyl-terminal fragment. The region of apoE that binds the LDL receptor is contained in the amino-terminal domain (residues 1–191) (Bradley et al 1982); the major lipid-binding region of apoE resides in the carboxyl-terminal domain, with residues carboxyl-terminal to 244 playing a major role (Weisgraber 1990). The cysteine-arginine interchange at position 112, which distinguishes apoE3 from apoE4, is not contained in the major lipid-binding region of apoE, yet it influences the distribution of these isoforms among the various lipoprotein particle classes. ApoE3 and apoE2 preferentially bind HDLs, and apoE4 binds the triglyceride-rich LDL particles, both VLDL and intermediate-density lipoproteins (IDL). A domain-domain interaction has been suggested to account for the distribution effect. The determination of the three-dimensional structure of the receptor-binding

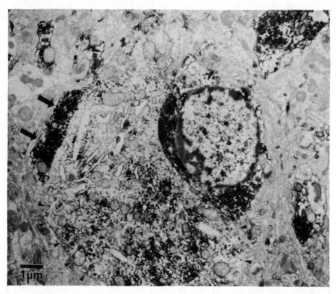

*Figure 2* Localization of apoE to the neuronal cytoplasm. Electron microscopic analysis of immunoreacted vibratome section from surgical specimen of lateral temporal lobe from a 30-year-old patient who had a temporal lobectomy for epilepsy. Cortical tissue several centimeters from the epileptic focus was examined using apoE immunoelectron microscopy (Han et al 1994). In this study visualization of apoE immunoreactivity involved the use of several different apoE antibodies and a secondary reaction using biotinylated horse antimouse IgG (diluted 1:50) followed by washes and incubation for 30 min in avidin-biotin-peroxidase complex (Vector Laboratories). ApoE immunoreactivity is stained black. The densely apoE (very black staining) immunoreactive cell in right center with relatively condensed nuclear chromatin is assumed to represent a satellite glial cell due to its close apposition to a large, less strongly immunoreactive region (arrowheads indicate border) filling much of the photograph. This region is presumed to represent proximal dendrite or perinuclear soma of a large cortical neuron given the size and shape of this process and the intensity of immunoreaction. On the left side of this lightly immunoreactive cell is a strongly immunoreactive process (*arrows*) that is also apposed to the presumed neuron. The overall appearance suggests the combined profile of satellite glial cells and a neuronal cell body.

domain of apoE has helped model isoform-specific functional properties, including the differences in LDL receptor binding and lipid particle binding.

## Intraneuronal Localization of Apolipoprotein E

In situ hybridization studies on brain have indicated that apoE mRNA is expressed in astrocytes and microglia and not in neurons. However, recent immunohistochemical studies on AD brain have shown apoE in both tangle-bearing and tangle-free nerve cells, indicating that apoE may play a role in neuronal metabolism and in neuronal degeneration or regeneration in the central nervous system (Han et al 1994a,b; Strittmatter et al 1993a). In a series of 24 patients with AD, immunoreactive apoE was demonstrated in hippocampal neurons, independent of *APOE* genotype (Han et al 1994b). Immunoreactive apoE was also observed in hippocampal granule cells of 7 out of 7 Parkinson's disease patients, several patients with other neurodegenerative diseases, and 2 out of 6 aged controls without known brain disease (Han et al 1994b). Only a small proportion of the apoE immunoreactive neurons in AD patients also contained NFTs (Han et al 1994b). ApoE immunoreactive cortical neurons can be detected with certain apoE antibodies at both the light and electron microscopic levels. Neurons containing immunoreactive apoE are commonly observed in aged rat and prosimians (Otolemur) but not in mice or baboons. ApoE immunoreactivity defined specific subsets of neurons, with punctate staining around soma, diffuse cytoplasmic staining, and granular staining. Ultrastructural studies of human and prosimian brain confirmed the cytoplasmic localization of apoE immunoreactivity (Han et al 1994b), as have recent studies on human inclusion body myositis (Askanas et al 1994b) (Figure 2).

The mechanisms of apoE entry into the cytoplasmic compartment of nerve cells are not yet understood. An analogous situation was recently reported in skeletal muscle from patients with sporadic and inherited forms of inclusion body myositis (Askanas et al 1994b). This disease demonstrates remarkable similarities to the neuropathology of AD. Muscle tissue from affected individuals contains intracellular amyloid fibrils made of Aβ and paired helical-like filaments (tubulofilaments) made of hyperphosphorylated tau (Askanas et al 1993, 1994a). Moreover, these muscle cells also contain cytoplasmic apoE, even though no evidence suggests that muscle cells produce apoE (Askanas et al 1994b). Interestingly, apoE decorates the paired helical-like filaments (Askanas et al 1994b). Taken together, these various lines of evidence indicate that in both AD and inclusion body myositis, apoE gains access to the cytoplasmic compartment of nerve cells and muscle cells. The sorting mechanisms responsible for the intracellular trafficking of apoE may also be responsible for the sorting of the amyloid precursor protein, with specific mutations associated with some cases of early-onset AD.

In vitro studies in tissue culture systems and in vivo studies in transgenic animal models are being used to characterize isoform-specific, time-dependent reactions favoring the formation of paired helical filaments. An isoform-specific role for apoE in intracellular metabolism related to paired helical filament formation has been proposed and is discussed below that takes into account prior observations of widespread intracellular localization of apoE in endosomes, peroxisomes, and cytoplasm in rat hepatocytes (Hamilton et al 1990) and current observations in human neurons (Han et al 1994a,b; Strittmatter et al 1993) and muscle (Askanas et al 1994b). Understanding the intraneuronal

metabolism of apoE may lead to the design and testing of rational neuropro-
tective treatments for late-onset AD.

## Irreversible, Isoform-Specific Binding of ApoE3 with the Microtubule Binding Proteins Tau and MAP2

Dementia in AD is generally accepted to be better correlated with NFT pathology than with the extent of Aβ deposition (Arriagada et al 1992). Neurofibrillary lesions contain paired helical filaments whose principal constituent is
hyper-phosphorylated tau, a MAP[see Goedert et al (1991) for review]. The
apoE-containing neurons may represent a stage in the life of the neuron, with
fully formed NFTs representing one of the end points shortly before death.
Because of the genetic relevance of *APOE4* and the presence of immunoreactive apoE in neurons containing NFTs, isoform-specific interactions of apoE
with tau are being studied.

In vitro, human apoE3 binds tau, forming a molecular complex that resists
dissociation by boiling in 2% sodium dodecyl sulfate (SDS) (Strittmatter et al
1994a,b). The apoE3-tau complex has an apparent molecular weight of approximately 105 kDa (tau-40 isoform, 68 kDa; apoE, 34 kDa). Irreversible
binding of tau and apoE3 is maximal within 30 min at 37°C and occurs between
pH 4.6–7.6. Binding of tau to apoE3 is detected down to $3 \times 10^{-8}$ M apoE.
ApoE4 does not irreversibly bind tau under identical conditions. The SDS-stable apoE3-tau complex is dissociated by boiling in the reducing agent β-mercaptoethanol. However, apoE3-tau was probably not complexed through disulfide bond formation because tau binds both the monomer and the homodimer
of apoE3. There is only insignificant SDS-stable binding of tau by apoE4 under
a variety of conditions, including increased duration of incubation, increased
concentration of apoE4, or pH values 4.6–7.6. Tau binds the 22-kDa amino-
terminal fragment of apoE3, which contains the LDL receptor-binding domain.
The 22-kDa amino-terminal fragment of apoE4 does not bind tau.

Paired helical filament tau is phosphorylated at a number of serine- and
threonine-proline sites. At least some of these sites are phosphorylated by
incubating recombinant tau with a crude rat brain extract and ATP. Recombinant tau-40 phosphorylated in this manner does not bind either apoE3 or apoE4,
even with a prolonged 12-h incubation. In AD, hyperphosphorylated tau is
believed to self-assemble into the paired helical filament by formation of
antiparallel dimers of the microtubule-binding repeat region. In vitro, these
microtubule-binding repeats of tau self-assemble into paired helical-like filaments (Crowther et al 1994). ApoE3 irreversibly binds these microtubule-binding repeat regions of tau, whereas apoE4 does not bind with this avidity
(Strittmatter et al 1994a,b).

MAP2c and tau are both members of a group of MAPs that bind microtubules and promote their assembly, and stabilize their polymerized structure (Goedert et al 1991). In vitro, apoE3 also irreversibly binds MAP2c, similar to its isoform-specific interaction with tau (Huang et al 1994). ApoE4 does not bind MAP2c under identical conditions. Binding of MAP2c by apoE3, forming a complex stable in SDS, is detectable within 1 h. However, no such binding of MAP2c by apoE4 is observed even after 4 h incubation. The MAP2c-apoE3 complex is detectable at apoE3 concentrations down to $3 \times 10^{-9}$ M and MAP 2c concentrations down to $3 \times 10^{-9}$ M. Tau and MAP2c contain a highly conserved microtubule-binding repeat region (Lewis et al 1988). This region consists of three or four conserved amino acid sequences (depending on the isoform), each 31 or 32 amino acids. MAP2c contains three copies of these microtubule-binding repeats, homologous to the microtubule-binding repeats of tau.

Isoform-specific interactions of apoE with tau and MAP2c may regulate intraneuronal metabolism in AD and alter the rate of formation of paired helical filaments and NFTs. Such isoform-specific interactions of apoE with tau and MAP2c may alter their metabolism and their function in microtubule assembly and stabilization and be central to disease mechanism.

## Irreversible, Isoform-Specific Binding of ApoE with Aβ Peptide

The senile, or neuritic, plaque is the other defining lesion in AD and contains Aβ peptide. ApoE accumulates extracellularly in the senile plaque and congophilic angiopathy of AD (Namba et al 1991, Strittmatter et al 1993a). In vitro, apoE avidly binds to synthetic Aβ peptide, the primary constituent of the senile plaque (Strittmatter et al 1993b). ApoE3 and apoE4 irreversibly bind Aβ peptide, resisting dissociation by SDS or guanidine hydrochloride (Strittmatter et al 1993b). ApoE4 binds Aβ peptide more rapidly and with a different pH dependence than does apoE3. The apoE3-Aβ peptide complex is first detectable after 2 h of incubation and increases over the next 24 h. In contrast the apoE4-Aβ complex is detected after 5 min of incubation.

Only a small percentage (less than 10%) of the total amount of apoE in the in vitro incubation binds Aβ peptide after 24 h, despite a large molar excess of Aβ peptide. The incomplete formation of the apoE-Aβ peptide complex could be due either to the slow association of these molecules or to modification of protein prior to binding. Addition of the reducing agents dithiothreitol or β-mercaptoethanol, either before or after incubation of apoE and Aβ peptide, prevents this SDS-stable binding, suggesting that oxidation may be required (Strittmatter et al 1993b). Oxygen-saturated buffer increases whereas a nitrogen-saturated buffer decreases, the rate of this SDS-stable binding. Binding of Aβ by apoE therefore appears to require the oxidation of apoE and can be prevented or reversed by reduction with dithiothreitol or β-mercaptoethanol.

The more rapid binding of Aβ by apoE4 than by apoE3 may be due to an increased rate of oxidation of apoE4 or other factors leading to differences in the oxidation of these isoforms.

The domain of apoE that binds Aβ peptide was determined by examining various recombinantly expressed apoE fragments (Strittmatter et al 1993b). Aβ4 peptide does not bind to the 22-kDa apoE3 fragment containing the amino terminus 199 amino acids. Binding of Aβ to apoE3$_{(1-244)}$ is very low or minimal. In contrast, apoE3$_{(1-266)}$ forms an SDS-stable Aβ peptide complex, which is further increased with apoE$_{(1-272)}$. Therefore, Aβ binding to apoE appears to require the domain of apoE between amino acids 244 to 272, within the region of apoE previously demonstrated to mediate binding to lipoprotein particles.

After days of in vitro incubation, apoE and Aβ peptide form unique fibrils, with apoE4 forming more abundant fibrils than apoE3 (Sanan et al 1994). Aβ(1–28) or Aβ(1–40) incubated with apoE forms simple 10-nm fibrils that are distinctly different than the twisted ribbons formed by Aβ alone. The presence of apoE on or in these simple fibrils was demonstrated using several monoclonal antibodies, each specific for a different region of apoE. Immuno-gold labeling showed that apoE associates with these fibrils along their full length, suggesting that apoE may be adsorbed to the outside of the simple fibrils or intercalated between the Aβ peptide monomers, perhaps forming a hybrid fibril.

In vitro differences in the binding of these apoE isoforms with Aβ peptide are paralleled by differences in vivo. In a study of the comparative neuropathology of brain tissue from AD patients homozygous for apoE4 or apoE3, congophilic staining of amyloid in senile plaques was greatly increased in apoE4 patients (Schmechel et al 1993). In addition, Aβ immunoreactivity in plaques was also increased in apoE4 homozygotes. Although patients homozygous for *APOE4* have denser and larger amyloid plaques, their duration of disease is not different than patients homozygous for *APOE3*. Therefore, deposition of Aβ relates to the *APOE4* allele as a phenotypic variable that appears independent of the course of the disease. This observation is not only for late-onset AD, but also for siblings carrying the APP717val-ile mutation, where the course and onset of the disease is also independent of amyloid deposition.

## Summary

The *APOE4* allele is a risk factor or susceptibility gene in late-onset familial and sporadic AD. Therefore, the mechanism of disease expression must involve metabolic effects that are isoform specific. Identifying isoform-specific inter-actions of apoE becomes critical in the mechanism of AD pathogenesis. Detailed characterization of the binding of the apoE isoforms with proteins and peptides relevant to the pathology of the disease may be critical in understanding disease pathogenesis. These critical isoform-specific interactions of apoE may involve interactions with proteins and peptides in the defining neuropathological lesions of the disease, the NFT and senile plaque. Other possible critical isoform-specific interactions include the mechanism of internalization, intracellular trafficking, and subsequent metabolism. In addition, differential posttranslational modifications of apoE isoforms may determine differences in metabolism contributing to the pathogenesis of the disease. Posttranslational modifications of apoE, such as oxidation, may confer several isoform-specific, biochemically distinct properties. Since Aβ peptide binds apoE in the lipoprotein-binding domain of the protein and not in the receptor-binding domain, apoE could target bound Aβ4 peptide to neurons via cell surface receptors. Internalization of the apoE-Aβ peptide complex into the cell, by the same route as the apoE-containing lipoproteins, would result in incorporation into primary lysosomes and pH-dependent dissociation. The demonstration of apoE in the cytoplasm of neurons, with isoform-specific interactions of apoE with the microtubule-binding proteins tau and MAP2 demonstrated in vitro, suggests additional testable hypotheses of disease pathogenesis. Isoform-specific differences in binding of apoE illustrate only part of the differential repertoire that could lead to disease pathogenesis. The mechanisms that modulate apoE binding to tau, MAP2, Aβ peptide, and the cell surface receptors may be important in determining the intracellular metabolism of these molecules or their deposition in the extracellular space. Studying these processes in vitro may provide important insights into disease mechanisms.

## Literature Cited

Amouyel P, Brousseau T, Fruchart JC, Dallongeville J. 1993. Apolipoprotein E-epsilon 4 allele and Alzheimer's disease [letter]. *Lancet* 342:1309

Anwar N, Lovestone S, Cheetham ME, Levy R, Powell JF. 1993. Apolipoprotein E-epsilon 4 allele and Alzheimer's disease [letter]. *Lancet* 342:1308–9

Armstrong TP, Hansen LA, Salmon DP, Masliah E, Pay M, et al. 1991. Rapidly progressive dementia in a patient with the Lewy body variant of Alzheimer's disease. *Neurology* 41:1178–80

Arriagada PV, Growdon JH, Hedley WE, Hyman BT. 1992. Neurofibrillary tangles but not senile plaques parallel duration and severity of Alzheimer's disease. *Neurology* 42:631–39

Askanas V, Engel WK, Alvarez RB. 1993. Enhanced detection of congo-red-positive amyloid deposits in muscle fibers of inclusion body myositis and brain of Alzheimer's disease using fluorescence technique. *Neurology* 43:1265–67

Askanas V, Engel WK, Bilak M, Alvarez RB, Selkoe DJ. 1994a. Twisted tubulofilaments of inclusion body myositis muscle resemble paired helical filaments of Alzheimer brain and contain hyperphosphorylated tau. *Am. J. Pathol.* 144:177–87

Askanas V, Mirabella M, Engel WK, Alvarez RB, Weisgraber KH. 1994b. Apolipoprotein E immunoreactive deposits in inclusion-body muscle diseases [letter]. *Lancet* 343: 364–65

Benjamin R, Leake A, Edwardson JA, McKeith IG, Ince PG. 1994. Apolipoprotein E genes in Lewy body and Parkinson's disease [letter]. *Lancet* 343:1565

Ben Shlomo Y, Lewis G, McKeigue PM. 1993. Apolipoprotein E-epsilon 4 allele and Alzheimer's disease [letter]. *Lancet* 342:1310

Biernat J, Mandelkow EM, Schroter C, Lichtenberg KB, Steiner B. 1992. The switch of tau protein to an Alzheimer-like state includes the phosphorylation of two serine-proline motifs upstream of the microtubule binding region. *EMBO J.* 11:1593–97

Bird TD, Lampe TH, Nemens EJ, Miner GW, Sumi SM, et al. 1988. Familial Alzheimer's disease in American descendants of the Volga Germans: probable genetic founder effect. *Ann. Neurol.* 23:25–31

Bird TD, Sumi SM, Nemens EJ, Nochlin D, Schellenberg G, et al. 1989. Phenotypic heterogeneity in familial Alzheimer's disease: a study of 24 kindreds. *Ann. Neurol.* 25:12–25

Borgaonkar DS, Schmidt LC, Martin SE, Kanzer MD, Edelsohn L, et al. 1993. Linkage of late-onset Alzheimer's disease with apolipoprotein E type 4 on chromosome 19 [letter]. *Lancet* 342:625

Bradley WA, Gilliam EB, Gotto AJ, Gianturco SH. 1982. Apolipoprotein-E degradation in human very low density lipoproteins by plasma protease(s): chemical and biological consequences. *Biochem. Biophys. Res. Commun.* 109:1360–67

Chartier HM, Crawford F, Houlden H, Warren A. 1991. Early-onset Alzheimer's disease caused by mutations at codon 717 of the beta-amyloid precursor protein gene. *Nature* 353:844–46

Chartier HM, Parfitt M, Legrain S, Perez TJ, Brousseau T, et al. 1994. Apolipoprotein E, epsilon 4 allele as a major risk factor for sporadic early- and late-onset forms of Alzheimer's disease: analysis of the 19q13.2 chromosomal region. *Hum. Mol. Genet.* 3: 569–74

Corder EH, Saunders AM, Strittmatter WJ, Schmechel DE, Gaskell PC, et al. 1993. Gene dose of apolipoprotein E type 4 allele and the risk of Alzheimer's disease in late onset families [see comments]. *Science* 261: 921–23

Corder EH, Saunders AM, Risch NJ, Strittmatter WJ, Schmechel DE, et al. 1994. Protective effect of apolipoprotein E type 2 allele for late onset Alzheimer disease. *Nat. Genet.* 7:180–83

Crews DE, Kamboh MI, Mancilha CJ, Kottke B. 1993. Population genetics of apolipoprotein A-4, E, and H polymorphisms in Yanomami Indians of northwestern Brazil: associations with lipids, lipoproteins, and carbohydrate metabolism. *Hum. Biol.* 65: 211–24

Crowther RA, Olesen OF, Smith MJ, Jakes R, Goedert M. 1994. Assembly of Alzheimer-like filaments from full-length tau protein. *FEBS Lett.* 337:135–38

Czech C, Monning V, Tienari PJ, Hartmann T, Masters C, et al. 1993. Apolipoprotein E-epsilon 4 allele and Alzheimer's disease [letter]. *Lancet* 342:1309–10

Galasko D, Saitoh T, Xia Y, Katzman R, Leon J, et al. 1994. The apolipoprotein E E4 allele is over-represented in the Lewy body variant of Alzheimer's disease. *Neurobiol. Aging* 15: S154

Ghiso J, Matsubara E, Koudinov A, Choi MN, Tomita M, et al. 1993. The cerebrospinal-fluid soluble form of Alzheimer's amyloid beta is complexed to SP-40,40 (apolipoprotein J), an inhibitor of the complement membrane-attack complex. *Biochem. J.* 293: 27–30

Goate A, Chartier HM, Mullan M, Brown J, et al. 1991. Segregation of a missense mutation in the amyloid precursor protein gene with familial Alzheimer's disease [see comments]. *Nature* 349:704–6

Goedert M, Sisodia SS, Price DL. 1991. Neurofibrillary tangles and beta-amyloid deposits in Alzheimer's disease [Review]. *Curr. Opin. Neurobiol.* 1:441–47

Goedert M, Spillantini MG, Cairns NJ, Crowther RA. 1992. Tau proteins of Alzheimer paired helical filaments: abnormal phosphorylation of all six brain isoforms. *Neuron* 8:159–68

Gregg RE, Ghiselli G, Brewer HJ. 1983. Apolipoprotein E Bethesda: a new variant of apolipoprotein E associated with type III hyperlipoproteinemia. *J. Clin. Endocrinol. Metab.* 57:969–74

Group ADC. 1993. Apolipoprotein E genotype and Alzheimer's disease [letter]. *Lancet* 342: 737–38

Hamilton RL, Wong JS, Guo LS, Krisans S, Havel RJ. 1990. Apolipoprotein E localization in rat hepatocytes by immunogold labeling of cryothin sections. *J. Lipid Res.* 31: 1589–603

Han SH, Einstein G, Weisgraber KH, Strittmatter WJ, Saunders AM, et al. 1994a. Apolipoprotein E is localized to the cytoplasm of human cortical neurons: a light and electron microscopic study. *J. Neuropathol. Exp. Neurol.* 53:535–44

Han SH, Hulette C, Saunders AM, Einstein G, Pericak-Vance M, et al. 1994b. Apolipoprotein E is present in hippocampal neurons without neurofibrillary tangles in Alzheimer's disease and in age-matched controls. *Exp. Neurol.* 128:13–26

Hansen L, Salmon D, Galasko D, Masliah E, Katzman R, et al. 1990. The Lewy body variant of Alzheimer's disease: a clinical and pathologic entity [see comments]. *Neurology* 40:1–8

Hansen LA, Galasko D, Samuel S, Xia Y, Chen X, et al. 1995. Apolipoprotein E e4 is associated with increased neurofibrillary pathology in the Lewy body variant of Alzheimer disease. *Ann. Neurol.* In press

Hansen LA, Masliah E, Galasko D, Terry RD. 1993. Plaque-only Alzheimer disease is usually the Lewy body variant, and vice versa. *J. Neuropathol. Exp. Neurol.* 52:648–54

Hardy J, Duff K. 1993. Heterogeneity in Alzheimer's disease [Review]. *Ann. Med.* 25:437–40

Hendriks L, van Duijn C, Cras P, Cruts M, Van HW, et al. 1992. Presenile dementia and cerebral hemorrhage linked to a mutation at codon 692 of the beta-amyloid precursor protein gene. *Nat. Genet.* 1:218–21

Houlden H, Crook R, Duff K, Collinge J, Roques P, et al. 1993. Confirmation that the apolipoprotein E4 allele is associated with late onset, familial Alzheimer's disease. *Neurodegeneration* 2:283–86

Huang DY, Goedert M, Jakes R, Weisgraber K, Garner C, et al. 1994. Isoform-specific interactions of apolipoprotein E with the microtubule associated protein MAP2c: implications for Alzheimer's disease. *Neurosci. Lett.* 182:55–58

Khachaturian ZS. 1985. Diagnosis of Alzheimer's disease. *Arch. Neurol.* 42:1097–105

Lander ES, Schork NJ. 1994. Genetic dissection of complex traits [Review]. *Science* 265:2037–48

Lannfelt L, Johnston J, Bogdanovich N, Cowburn R. 1994a. Amyloid precursor protein gene mutation at codon 670/671 in familial Alzheimer's disease in Sweden. *Biochem. Soc. Trans.* 22:176–79

Lannfelt L, Lilius L, Nastase M, Viitanen M, Fratiglioni L, et al. 1994b. Lack of association between apolipoprotein E allele epsilon 4 and sporadic Alzheimer's disease. *Neurosci. Lett.* 169:175–78

Lewis SA, Wang DH, Cowan NJ. 1988. Microtubule-associated protein MAP2 shares a microtubule binding motif with tau protein. *Science* 242:936–39

Liddell M, Williams J, Bayer A, Kaiser F, Owen M. 1994. Confirmation of association between the e4 allele of apolipoprotein E and Alzheimer's disease. *J. Med. Genet.* 31:197–200

Lucotte G, David F, Visvikis S, Leininger MB, Siest G, et al. 1993. Apolipoprotein E-epsilon 4 allele and Alzheimer's disease [letter]. *Lancet* 342:1309

Masliah E, Galasko D, Wiley CA, Hansen LA. 1990. Lobar atrophy with dense-core (brain stem type) Lewy bodies in a patient with dementia. *Acta Neuropathol.* 80:453–58

Masliah E, Mallory M, DeTeresa R, Alford M, Hansen L. 1993. Differing patterns of aberrant neuronal sprouting in Alzheimer's disease with and without Lewy bodies. *Brain Res.* 617:258–66

Mayeux R, Stern Y, Ottman R, Tatemichi TK, Tang MX, et al. 1993. The apolipoprotein epsilon 4 allele in patients with Alzheimer's disease. *Ann. Neurol.* 34:752–54

McKhann G, Drachman D, Folstein M, Katzman R, et al. 1984. Clinical diagnosis of Alzheimer's disease: report of the NINCDS-ADRDA Work Group under the auspices of Department of Health and Human Services Task Force on Alzheimer's Disease. *Neurology* 34:939–44

Murrell J, Farlow M, Ghetti B, Benson MD. 1991. A mutation in the amyloid precursor protein associated with hereditary Alzheimer's disease. *Science* 254:97–99

Namba Y, Tomonaga M, Kawasaki H, Otomo E, Ikeda K. 1991. Apolipoprotein E immunoreactivity in cerebral amyloid deposits and neurofibrillary tangles in Alzheimer's disease and kuru plaque amyloid in Creutzfeldt-Jakob disease. *Brain Res.* 541:163–66

Noguchi S, Murakami K, Yamada N. 1993. Apolipoprotein E genotype and Alzheimer's disease [letter][see comments]. *Lancet* 342:737

Okuizumi K, Onodera O, Tanaka H, Kobayashi H, et al. 1994. ApoE-epsilon 4 and early-onset Alzheimer's [letter]. *Nat. Genet.* 7:10–11

Payami H, Kaye J, Heston LL, Bird TD, Schellen GD. 1993. Apolipoprotein E genotype and Alzheimer's disease [letter]. *Lancet* 342:738

Pericak-Vance MA, Bebout JL, Gaskell PJ, Yamaoka LH, Hung WY, et al. 1991. Linkage studies in familial Alzheimer disease: evidence for chromosome 19 linkage. *Am. J. Hum. Genet.* 48:1034–50

Pericak-Vance MA, Yamaoka LH, Haynes CS, Speer MC, Haines JL, et al. 1988. Genetic linkage studies in Alzheimer's disease families. *Exp. Neurol.* 102:271–79

Pickering-Brown SM, Mann DM, Bourke JP, Roberts DA, Balderson D, et al. 1994. Apolipoprotein E4 and Alzheimer's disease pathology in Lewy body disease and in other beta-amyloid-forming diseases [letter]. *Lancet* 343:1155

Poirier J, Davignon J, Bouthillier D, Kogan S, Bertrand P, et al. 1993. Apolipoprotein E polymorphism and Alzheimer's disease [see comments]. *Lancet* 342:697–99

Rebeck GW, Reiter JS, Strickland DK, Hyman BT. 1993. Apolipoprotein E in sporadic Alzheimer's disease: allelic variation and receptor interactions. *Neuron* 11:575–80

Roses AD. 1994. Apolipoprotein E affects the rate of Alzheimer disease expression: beta-amyloid burden is a secondary consequence dependent on APOE genotype and duration of disease [see comments] [Review]. *J. Neuropathol. Exp. Neurol.* 53:429–37

Roses AD, Strittmatter WJ, Pericak VM, Corder EH, Saunders AM, et al. 1994. Clinical application of apolipoprotein E genotyping to Alzheimer's disease [letter]. *Lancet* 343: 1564–65

Sakoda S, Kuriyama M, Osame M, Takahashi K, Yamano T, et al. 1994. Apolipoprotein E allele ε4 and Alzheimer's disease. *Neurology* 44:2420

Sanan DA, Weisgraber KH, Russell SJ, Mahley RW, Hunag D, et al. 1994. Apolipoprotein E associates with beta amyloid peptide of Alzheimer's disease to form novel monofibrils. Isoform apoE4 associates more efficiently than apoE3. *J. Clin. Invest.* 94: 860–69

Saunders AM, Schmader K, Breitner JC, Benson MD, Brown WT, et al. 1993a. Apolipoprotein E epsilon 4 allele distributions in late-onset Alzheimer's disease and in other amyloid-forming diseases [see comments]. *Lancet* 342:710–11

Saunders AM, Strittmatter WJ, Schmechel D, St. George-Hyslop PH, Pericak-Vance, M, et al. 1993b. Association of apolipoprotein E allele epsilon 4 with late-onset familial and sporadic Alzheimer's disease. *Neurology* 43: 1467–72

Schellenberg GD, Anderson L, O'Dahl S, Wijsman EM, et al. 1991. APP717, APP693, and PRIP gene mutations are rare in Alzheimer disease [see comments]. *Am. J. Hum. Genet.* 49:511–17

Schellenberg GD, Bird TD, Wijsman EM, Orr HT, Anderson L, et al. 1992. Genetic linkage evidence for a familial Alzheimer's disease locus on chromosome 14. *Science* 258:668–71

Schellenberg GD, Deeb SS, Boehnke M, Bryant EM, Martin GM, et al. 1987. Association of an apolipoprotein CII allele with familial dementia of the Alzheimer type. *J. Neurogenet.* 4:97–108

Schmechel DE, Saunders AM, Strittmatter WJ, Crain BJ, Hulette CM, et al. 1993. Increased amyloid beta-peptide deposition in cerebral cortex as a consequence of apolipoprotein E genotype in late-onset Alzheimer disease. *Proc. Natl. Acad. Sci. USA* 90:9649–53

Selkoe DJ. 1994. Alzheimer's disease: a central role for amyloid [see comments] [Review]. *J. Neuropathol. Exp. Neurol.* 53:438–47

St. George-Hyslop PH, Haines JL, Farrer LA, Polinsky R, Van Broeckhoven C, et al. 1990. Genetic linkage studies suggest that Alzheimer's disease is not a single homogeneous disorder. FAD Collaborative Study Group [see comments]. *Nature* 347:194–97

St. George-Hyslop P, Haines J, Rogaev E, Martilla M, et al. 1992. Genetic evidence for a novel familial Alzheimer's disease locus on chromosome 14. *Nat. Genet.* 2:330–34

St. George-Hyslop P, McLachlan DC, Tsuda T, Rogaev E, Karlinsky H, et al. 1994. Alzheimer's disease and possible gene interaction [letter]. *Science* 263:537; Erratum: 1994. *Science* 263(5149):904

Strittmatter WJ, Saunders AM, Goedert M, Weisgraber KH, Dong LM, et al. 1994a. Isoform-specific interactions of apolipoprotein E with microtubule-associated protein tau: implications for Alzheimer disease. *Proc. Natl. Acad. Sci. USA* 91:11183–86

Strittmatter WJ, Saunders AM, Schmechel D, Pericak-Vance M, Enghild J, et al. 1993a. Apolipoprotein E: high-avidity binding to beta-amyloid and increased frequency of type 4 allele in late-onset familial Alzheimer disease. *Proc. Natl. Acad. Sci. USA* 90:1977–81

Strittmatter WJ, Weisgraber KH, Goedert M, Saunders AM, Huang D, et al. 1994b. Hypothesis: microtubule instability and paired helical filament formation in the Alzheimer disease brain are related to apolipoprotein E genotype [Review]. *Exp. Neurol.* 125:163–71

Strittmatter WJ, Weisgraber KH, Huang DY, Dong LM, Salveson GS, et al. 1993b. Binding of human apolipoprotein E to synthetic amyloid beta peptide: isoform-specific effects and implications for late-onset Alzheimer disease. *Proc. Natl. Acad. Sci. USA* 90: 8098–102

Tsuda T, Lopez R, Rogaeva EA, Freedman M, Rogaev E, et al. 1994. Are the associations between Alzheimer's disease and polymorphisms in the apolipoprotein E and the apolipoprotein CII genes due to linkage disequilibrium? *Ann. Neurol.* 36:97–100

Ueki A, Kawano M, Namba Y, Kawakami M, Ikeda K. 1993. A high frequency of apolipoprotein E4 isoprotein in Japanese patients with late-onset nonfamilial Alzheimer's disease. *Neurosci. Lett.* 163:166–68

Van Broeckhoven C, Backhovens H, Cruts M, Martin JJ, Crook R, et al. 1994. APOE genotype does not modulate age of onset in families with chromosome 14–encoded Alzheimer's disease. *Neurosci. Lett.* 169:179–80

van Duijn CM, de Knijff P, Cruts M, Wehnert A, Havekes LM, et al. 1994. Apolipoprotein E4 allele in a population-based study of early-onset Alzheimer's disease. *Nat. Genet.* 7:74–78

Weeks DE, Lange K. 1988. The affected-pedigree-member method of linkage analysis. *Am. J. Hum. Genet.* 42:315–26

Weisgraber KH. 1990. Apolipoprotein E distribution among human plasma lipoproteins: role of the cysteine-arginine interchange at residue 112. *J. Lipid Res.* 31:1503–11

Weisgraber KH. 1994. Apolipoprotein E: structure-function relationships [Review]. *Adv. Protein Chem.* 45:249–302

Weisgraber KH, Innerarity TL, Mahley RW. 1982. Abnormal lipoprotein receptor-binding activity of the human E apoprotein due to cysteine-arginine interchange at a single site. *J. Biol. Chem.* 257:2518–21

Weisgraber KH, Shinto LH. 1991. Identification of the disulfide-linked homodimer of apolipoprotein E3 in plasma. Impact on receptor binding activity. *J. Biol. Chem.* 266:12029–34

Wisniewski T, Frangione B. 1992. Apolipoprotein E: a pathological chaperone protein in patients with cerebral and systemic amyloid. *Neurosci. Lett.* 135:235–38

Wisniewski T, Haltia M, Ghiso J, Frangione B. 1991. Lewy bodies are immunoreactive with antibodies raised to gelsolin related amyloid-Finnish type. *Am. J. Pathol.* 138:1077–83

Yoshizawa T, Yamakawa KK, Komatsuzaki Y, Arinami T, Oguni E, et al. 1994. Dose-dependent association of apolipoprotein E allele epsilon 4 with late-onset, sporadic Alzheimer's disease. *Ann. Neurol.* 36:656–59

Yu CE, Payami H, Olson JM, Boehnke M, Wijsman EM, et al. 1994. The apolipoprotein E/CI/CII gene cluster and late-onset Alzheimer disease. *Am. J. Hum. Genet.* 54:631–42

*Annu. Rev. Neurosci. 1996. 19:79–107*
*Copyright © 1996 by Annual Reviews Inc. All rights reserved*

# TRINUCLEOTIDE REPEATS IN NEUROGENETIC DISORDERS

## H. L. Paulson and K. H. Fischbeck

Department of Neurology, University of Pennsylvania Medical School, 3400 Spruce Street, Philadelphia, Pennsylvania 19104

KEY WORDS:   neurodegeneration, genetic instability, polyglutamine repeats, Huntington's disease, expanded repeats

### ABSTRACT

Trinucleotide repeat expansion is increasingly recognized as a cause of neurogenetic diseases. To date, seven diseases have been identified as expanded repeat disorders: the fragile X syndrome of mental retardation (both *FRAXA* and *FRAXE* loci), myotonic dystrophy, X-linked spinal and bulbar muscular atrophy, Huntington's disease, spinocerebellar ataxia type 1, dentatorubral-pallidoluysian atrophy, and Machado-Joseph disease. All are neurologic disorders, affecting one or more regions of the neuraxis. Moreover, five of the seven (the last five above) are progressive neurodegenerative disorders whose strikingly similar mutations suggest a common mechanism of neuronal degeneration. In this article we discuss specific characteristics of each trinucleotide repeat disease, review their shared clinical and genetic features, and address possible molecular mechanisms underlying the neuropathology in each disease. Particular attention is paid to the neurodegenerative diseases, all of which are caused by CAG repeats encoding polyglutamine tracts in the disease gene protein.

## INTRODUCTION

Simple sequence repeats occur commonly in the human genome. Also called microsatellites or tandem repeats, these normally polymorphic sequences have proved useful as markers in linkage studies. In the past five years, it has become clear that they also underlie an entirely new class of human mutations. The expansion of one group of simple repeats, trinucleotide or triplet repeats, is now known to cause seven inherited disorders, a number that is likely to continue growing (Table 1). To date, the list includes the fragile X syndrome of mental retardation (both the *FRAXA* and *FRAXE* loci), myotonic dystrophy (DM), X-linked spinal and bulbar muscular atrophy (SBMA), Huntington's disease (HD), spinocerebellar ataxia type 1 (SCA1), dentatorubral-pallidoluysian atrophy (DRPLA), and Machado-Joseph disease (MJD).

0147-006X/96/0301-0079$08.00

**Table 1**  Trinucleotide repeat diseases

| Disease | Repeat | Repeat length | Gene product |
|---|---|---|---|
| Spinal and bulbar muscular atrophy | CAG | Normal: 11 to 34<br>Disease: 40 to 62 | Androgen receptor (transcription factor) |
| Fragile X syndrome (FRA-XA locus) | CGG | Normal: 6 to 52<br>Premutation: ~60 to 200<br>Full mutation: ~200 to >2000 | FMR-1 (RNA binding protein) |
| FRAXE mental retardation | GCC | Normal: 6 to 25<br>Disease: >200 | ? |
| Myotonic dystrophy | CTG | Normal: 5 to 37<br>Protomutation: ~50 to 180<br>Full mutation: ~200 to >2000 | myotonin protein kinase (serine-threonine kinase) |
| Huntington's disease | CAG | Normal: 11 to 34<br>Disease: 36 to 121 | huntingtin |
| Spinocerebellar ataxia type 1 | CAG | Normal: 6 to 39<br>Disease: 41 to 81 | ataxin-1 |
| Dentatorubral-pallidoluysian atrophy | CAG | Normal: 7 to 25<br>Disease: 49 to 88 | atrophin |
| Machado-Joseph disease | CAG | Normal: 13 to 36<br>Disease: 68 to 79 | MJD1 gene product |

Expanded repeat mutations do not adhere strictly to rules of Mendelian inheritance. They are unstable mutations that change size in successive generations and thus are especially intriguing to geneticists. But expanded repeat diseases should interest neuroscientists and neurologists as well. All of these diseases are neurologic disorders, affecting one or more regions of the neuraxis. Moreover, five of the seven (the last five above) are progressive neurodegenerative disorders whose strikingly similar mutations suggest a common mechanism of neuronal degeneration. In this article we discuss specific characteristics of each trinucleotide repeat disease and then review their shared clinical and genetic features. Finally, we address the molecular mechanisms that may underlie the neuropathology in each disease. Particular attention is paid to the neurodegenerative diseases, all of which are caused by CAG repeats encoding polyglutamine tracts in the disease gene protein. [For other recent reviews, see Bates & Lehrach (1994), Kuhl et al (1993), LaSpada et al (1994), Mandel (1994), Ross et al (1993), and Richards & Sutherland (1994).]

## GENERAL FEATURES AND CLASSIFICATION

Several generalizations can be made regarding the shared features of trinucleotide repeat diseases. 1. Inheritance is autosomal dominant or X-linked. 2. Disease severity varies widely and correlates directly with increasing repeat length. 3. Expansions arise from CG-rich triplet repeats that are polymorphic

**Table 2**   Classification of the trinucleotide repeat diseases

|  | Type 1 | Type 2 |
|---|---|---|
| Representative disorders | SBMA, HD, SCA1 DRPLA, MJD | FRAXA, FRAXE and DM |
| Repeat | CAG | CGG or CTG |
| Degree of instability | Moderate; small expansions | Highly unstable; prone to very large expansions |
| Repeat translated into protein? | Yes | No |
| Pattern of pathology | Specific neurons affected | Multisystem disorders |
| Proposed mechanism | Toxic gain of function | Altered expression of gene product or RNA |

in the normal population. 4. Expanded trinucleotide repeats are unstable, changing in size when transmitted to successive generations. In contrast, normal-sized repeats are usually transmitted stably despite their polymorphic nature. 5. Although either further expansions or contractions may occur with transmission of expanded repeats, expansions are more common. 6. The diseases frequently show anticipation—the tendency for disease severity to increase in successive generations of a family. Anticipation is explained by the tendency for expanded repeats to further enlarge with transmission, coupled with the fact that longer repeats correlate with more severe disease. 7. Parent-of-origin effects are common. With several of the neurodegenerative diseases, for example, the most severely affected individuals usually inherit the disease from the father. In contrast, the most severe forms of myotonic dystrophy and Fragile X syndrome are, with rare exceptions, maternally transmitted.

Despite their shared features, trinucleotide repeat diseases clearly fall into two groups (Table 2). The first, which we have designated Type 1 (LaSpada et al 1994), is comprised of the five progressive neurodegenerative disorders— SBMA, HD, SCA1, DRPLA, and MJD. These disorders are essentially confined to the nervous system (SBMA's hormonal effects excluded) and are caused by small, constrained CAG repeat expansions that encode polyglutamine tracts in the disease gene protein. These mutations likely alter one or more properties of the disease protein, leading (as is discussed later) to a gain of function that is particularly deleterious to neurons.

The second group, which we have called Type 2, includes DM, FRAXA, and FRAXE. These are multisystem disorders characterized by much larger expansions of trinucleotide repeats that fall outside the protein coding region. To date, these have been non-CAG repeats: CTG, CGG, and GCC in DM, FRAXA, and FRAXE, respectively. (CTG designates the coding strand in DM; the noncoding strand is CAG.) The underlying mechanism of disease does not seem to be altered protein function but instead altered expression of the gene in at least one (FRAXA) and possibly all three diseases. These are not progressive neurodegenerative disorders; instead, CNS manifestations of Type 2 diseases may in part reflect abnormalities in brain and neuromuscular development. Another feature distinguishing this group from CAG repeat diseases is marked somatic mosaicism. Whereas repeat sizes vary widely in somatic tissues from DM and FRAXA patients (Anvret et al 1993), relatively little somatic mosaicism occurs in the CAG repeat diseases (MacDonald et al 1993, Telenius et al 1994).

## THE DISEASES

Here we review the clinical, neuropathologic, and genetic features of each disease. First, the five Type 1 diseases are presented in order of discovery, followed by the Type 2 diseases.

### X-Linked Spinal and Bulbar Muscular Atrophy

SBMA, or Kennedy's disease, is a rare progressive neuromuscular disorder characterized by proximal muscle weakness, atrophy, and fasciculations (Kennedy et al 1968). The X-linked pattern of inheritance, adult onset, and bulbar involvement distinguish it from autosomal forms of spinal muscular atrophy (Harding et al 1982). Because SBMA patients often have signs of androgen insensitivity (gynecomastia, reduced fertility, and testicular atrophy), the androgen receptor (AR) gene became a candidate gene for SBMA once it was mapped to the same region of the X chromosome (Fischbeck et al 1991). La Spada et al (1991) found that a trinucleotide repeat, $(CAG)_n$, was expanded in the first exon of the AR gene in all SBMA patients but never in controls. This CAG repeat normally encodes a polyglutamine tract of 11–34 amino acids in the AR protein. In SBMA patients, the polyglutamine tract is lengthened to 40–62 amino acids.

SBMA is characterized pathologically by degeneration of anterior horn cells, bulbar neurons, and dorsal root ganglion cells (Sobue et al 1989). The AR is expressed in these neurons as well as in several other regions of the brain. What role androgens play in the normal function and viability of motor neurons is presently unclear, although Yu (1989) showed that androgens can attenuate neuronal degeneration following axotomy.

### Huntington's Disease

Probably the most common Type 1 disease, HD is an autosomal dominant neurodegenerative disorder characterized by involuntary movements, cognitive impairment, and emotional disturbance (Harper et al 1991, Wexler et al 1991). Symptoms of HD usually first appear in the third to fifth decades, but may begin in childhood or after age 60. Anticipation is not uncommon, and juvenile-onset cases are usually paternally transmitted.

As with all CAG repeat disorders, adult-onset HD begins insidiously and progresses slowly. The first motor sign of HD is usually chorea—uncontrolled flickering movements that become widespread and disabling as the disease progresses. In advanced disease, rigidity, bradykinesia, and dystonia may surface. Juvenile-onset disease differs from adult-onset disease in that juvenile patients often present with a combination of parkinsonian signs, dystonia, chorea, and seizures.

Degenerative changes in HD are most marked in the striatum. Cell death follows a gradient in three axes, occurring first in the tail of the caudate nucleus, then proceeding mediolaterally through the caudate and dorsolaterally through the putamen (Vonsattel et al 1985). Early in disease, there is selective degeneration of the most prevalent striatal neurons, the GABAergic medium spiny neurons that give rise to the striatal output to the globus pallidus and substantia nigra. In advanced stages neuronal loss becomes widespread, involving the cortex and cerebellum.

In 1983, the HD gene was mapped to chromosome 4 (Gusella et al 1983). Ten years later, the defective gene was identified after an intensive collaborative effort (Huntington's Disease Collaborative Group 1993). The defect is an expanded trinucleotide CAG repeat within the coding region of a large novel gene whose gene product is called huntingtin. The CAG repeat encodes a polyglutamine tract 11 to 34 amino acids long in normal individuals and 36 to 121 in HD patients (Andrew et al 1993, Duyao et al 1993, Huntington's Disease Collaborative Group 1993, Novelletto et al 1994, Snell et al 1993; reviewed in Goldberg et al 1994). A smaller polymorphic CGG repeat lies immediately 3' to the CAG repeat and encodes a polyproline stretch in the protein. The HD gene is critical for embryonic development, as targeted disruption of the gene results in embryonic lethality (Nasir et al 1995).

The primary structure of huntingtin does not provide obvious clues to its function. However, much information already exists on possible biochemical defects in HD [reviewed in Beal (1992) and Wexler et al (1991)]. Positron emission tomography studies, for example, have demonstrated decreased glucose metabolism in the basal ganglia of HD patients and asymptomatic at-risk individuals, and proton NMR spectroscopy has shown increased lactate in the brains of symptomatic patients. These findings suggest that neuronal energy metabolism is impaired in HD, and this may play a role in the neurodegenerative process. Beal (1992) has proposed that impaired energy metabolism underlies the cell death in many neurodegenerative disorders. Altered bioenergetics may result in slow excitotoxic injury as neurons, depleted of energy and partially depolarized, become vulnerable to the excitatory amino acid glutamate. Support for this model in HD has come from animal studies in which injections of either $N$-methyl-$D$-aspartate receptor agonists (e.g. quinolinic acid) or the mitochondrial toxin 3-nitropropionic acid result in pathologic changes resembling those seen in HD (Beal et al 1993, Ferrante et al 1993).

## Spinocerebellar Ataxia

SCA1 is an autosomal dominant neurodegenerative disease characterized by ataxia, ophthalmoparesis, and weakness (Harding 1993, Schut 1954). It is one of several genetically distinct disorders comprising the autosomal dominant cerebellar ataxias (Harding 1993, Rosenberg 1995), a group that includes at least one other trinucleotide repeat disorder, MJD (or SCA3). Many of the dominant hereditary ataxias may prove to be trinucleotide repeat disorders.

Like other CAG repeat disorders, SCA1 usually begins in the third to fifth decades and progresses over 10 to 20 years, although it can appear in childhood or much later in life. Anticipation has been reported (Schut 1954); juvenile-onset cases most often occur with paternal transmission. The disease typically begins with cerebellar signs such as clumsiness and ataxia but progresses to include dysarthria, bulbar dysfunction, oculomotor disturbance, and pyramidal tract signs. In its late stages, the disease may include dystonia and choreoathetosis, but dementia and parkinsonian signs are uncommon (Ranum et al 1994). Neuropathologic findings include neuronal loss in the cerebellum (principally Purkinje cells) and brain stem and degeneration of spinocerebellar tracts. Little is known about biochemical abnormalities in this disease.

The discovery of the genetic defect in SCA1, mapped earlier to chromosome 6, was facilitated by specifically screening genomic DNA for trinucleotide repeats. This screening identified a trinucleotide CAG repeat that is expanded in all SCA1 patients (Orr et al 1993). The SCA1 repeat is in the protein coding region of a novel gene, ataxin-1, and encodes a polyglutamine tract 6 to 39 repeats long in normal individuals versus 41 to 81 repeats long in affected individuals (Matilla et al 1993, Orr et al 1993, Ranum et al 1994). Predicted to be 87 kDa, ataxin-1 has no significant homology to other known proteins (Banfi et al 1993). One unusual feature of the ataxin-1 gene is that its transcript contains very large 5′ and 3′ untranslated regions. The 5′ region contains seven small exons and is subject to alternative splicing. These features suggest that the transcriptional and translational regulation of ataxin-1 is highly complicated. Untranslated regions of the ataxin-1 transcript could have biologic activity of their own, as has been shown for several 3′ untranslated regions of mRNAs (Rastinejad & Blau 1993).

Whereas the expanded SCA1 repeat is an uninterrupted sequence of CAG repeats, the normal SCA1 allele is usually interrupted midway by one to three

CAT trinucleotides (Chung et al 1993). These interruptions are thought to contribute to allele stability. Similar interruptions have been found in the FRAXA repeat (Eichler et al 1994, Fu et al 1991).

## Dentatorubral-Pallidoluysian Atrophy

The sixth trinucleotide repeat disease to be identified, DRPLA is a rare neuro-degenerative disorder characterized by ataxia, choreathetosis, dementia, myoclonus, and epilepsy (Naito and Oyanagi 1982, Smith et al 1958). Although DRPLA has clinically been divided into three subtypes—myoclonic-epilepsy, pseudo-Huntington's, and ataxic-choreoathetoid forms—the distinctions are not precise, and more than one form may be seen in a family. There is anticipation, with the severe early-onset cases most often paternally transmitted (Koide et al 1994, Komure et al 1995, Nagafuchi et al 1994b, Takahishi et al 1988). As with HD, DRPLA in juvenile-onset patients differs from the disease in adults: Younger patients are more likely to display progressive myoclonic-epilepsy and deteriorate more quickly.

The neuronal degeneration in DRPLA is widespread but particularly severe in the globus pallidus, dentatorubral system, and subthalamic nucleus (Luys body) (Neumann 1959, Takahishi et al 1988). Though DRPLA can resemble HD clinically, it usually is distinguished pathologically from HD by the presence of dentatorubral atrophy and the absence of significant striatal atrophy.

Because DRPLA shares clinical and genetic features with SCA1 and HD, two groups of investigators hypothesized that it, too, might be a trinucleotide repeat disorder, and analyzed a gene on chromosome 12 that had previously been shown to contain a polymorphic CAG repeat (Koide et al 1994, Nagafuchi et al 1994b). This CAG repeat proved to be expanded in all DRPLA patients, from its normal range of 7 to 25 copies to an expanded range of 49 to 88 copies (Koide et al 1994, Komure et al 1995, Nagafuchi et al 1994b). The repeat encodes a polyglutamine tract in the middle of a novel protein predicted to be 124 kDa (Nagafuchi et al 1994a). In addition to the polyglutamine tract, the predicted DRPLA gene product (also called atrophin-1) contains small homoproline and homoserine tracts upstream of the polyglutamine tract, and two stretches of arginine-glutamate dipeptides and one of alternating histidines downstream. Though the function of the DRPLA gene product is unknown, the dipeptide sequences are reminiscent of similar stretches found in a group of nuclear RNA-binding proteins and in components of the spliceosome (SR proteins).

DRPLA is most prevalent in Japan, where it has been extensively characterized. Since the discovery of the DRPLA repeat, Burke et al (1994) have determined that the Haw River Syndrome (HRS), a dominant neurodegenerative disease affecting a large African-American family in North Carolina, is also DRPLA. Though the genetic defect is identical, HRS differs clinically from other DRPLA in that it combines features of pseudo-Huntington's and myoclonic-epilepsy and is characterized by generalized seizures without myoclonus (the two are usually seen together in DRPLA). The pathology in HRS also involves more neuraxonal dystrophy, basal ganglia calcification, and demyelination than is typically seen in most DRPLA patients (Burke et al 1994). These differences illustrate that additional genetic factors may influence the phenotypic expression of this and, perhaps, other trinucleotide repeat diseases.

## Machado-Joseph Disease

One of the more common forms of spinocerebellar degeneration, MJD is often clinically indistinguishable from SCA1 (Harding 1993, Rosenberg 1992). This recently discovered expanded repeat disorder had already been assumed by many to be a CAG repeat disease because of its clinical and genetic characteristics (autosomal dominance, variable phenotype, progressive nature, and usual adult onset with anticipation). The genetic defect was identified by directly searching for candidate CAG repeat genes. Screening a brain cDNA library, Kawaguchi and colleagues (1994) isolated a cDNA that mapped to a region of chromosome 14 where MJD had already been localized. This cDNA, MJD1, contains a CAG repeat that is expanded in MJD patients and encodes a polyglutamine tract in the predicted gene product. In the limited number of normal and affected individuals screened so far, the CAG-polyglutamine repeat is 13 to 36 repeats in normal alleles and 68 to 79 repeats in disease alleles. A small hydrophilic protein of 359 amino acids, the predicted MJD1 gene product is unique among Type 1 disease genes in that its polyglutamine tract lies near the carboxyl terminus. Other than the polyglutamine tract, it contains no obvious homology to other known proteins.

## Fragile X Syndrome

Named after a folate-sensitive fragile site (the *FRAXA* locus) on the X chromosome, this X-linked disorder is the most common cause of inherited mental retardation. The mental retardation can vary from mild to severe and is associated with dysmorphic features such as enlarged ears, head, and testicles (Nussbaum & Ledbetter 1989). Although pathologic studies have demonstrated hypoplasia of the cerebellar vermis and white matter changes, no distinctive biochemical or cellular abnormality has been noted in the brain or elsewhere (Reiss et al 1991, Wisniewski et al 1991).

Until the discovery of the underlying mutation in fragile X, several aspects of its inheritance pattern were puzzling, including the fact that hemizygous males may be phenotypically normal (20%), yet females carrying the mutant allele are often affected (30%) (Sherman et al 1985). The risk of disease increases from one generation to the next, a phenomenon known as the Sherman paradox (Sherman et al 1985). This paradox is now explained by the tendency of the trinucleotide repeat to further expand with transmission (Fu et al 1991).

Cloning of the fragile site revealed that the primary genetic defect in fragile X syndrome is an expanded $(CGG)_n$ repeat (Kremer et al 1991, Oberle et al 1991, Verkerk et al 1991). The repeat is found in the 5' untranslated region of a gene called *FMR-1* (Figure 1) (Ashley et al 1993a). In the normal population, the repeat exists as a stable polymorphism of between 6 and 52 repeats, with a mode of 30. Repeats between 35 and 55 repeats fall into a gray zone in which alleles may be stable or unstable, depending on whether AGG triplets interrupt the CGG repeat (Eichler et al 1994). In *FRAXA* families, the repeat is expanded over a broad range, from premutations of ~60–200 repeats to full mutations of several thousand repeats or more. Expansions in the premutation range usually do not cause disease but are highly prone to expand to the full mutation, resulting in the full-blown manifestations of disease. Expansion is associated with increased methylation of both the repeat and an adjacent CpG island, leading to transcriptional silencing of the gene (Bell et al 1991, Hansen et al 1992, Pieretti et al 1991, Sutcliffe et al 1992). The CGG repeat expansion and subsequent methylation likely account for the chromosomal fragility at the *FRAXA* locus.

*FRAXE*    The fragile site associated with the CGG repeat in *FMR-1* is designated *FRAXA*. More recently Knight et al (1993) showed that a second folate-sensitive fragile site (*FRAXE*) 600 kb distal to FRAXA contained a $(GCC)_n$ repeat that is expanded in mildly retarded individuals. This second, less common site was demonstrated when cytogenetic screening uncovered a few affected families lacking the expanded *FRAXA* repeat and instead expressing the nearby *FRAXE* fragile site. The degree of mental retardation in these families seems milder than in *FRAXA* patients, although the sample size is small (Knight et al 1993). In the few *FRAXE*-positive families screened so far, the *FRAXE* GCC repeat is enlarged to greater than 200 copies in all mentally retarded males. In contrast, normal unrelated males have a trinucleotide repeat that is 6–25 copies in length. What gene the *FRAXE* trinucleotide resides in and how it causes disease are still unknown.

OTHER FRAGILE SITES    Recently, two more folate-sensitive fragile sites, *FRAXF* and *FRA16A,* were cloned. The *FRAXF* site, located on the X chromosome ~1200 kb distal to *FRAXA,* is a GCC repeat that is normally 6–29 repeats but expanded to 300–500 repeats in patients expressing the fragile site (Parrish et al 1994, Ritchie et al 1994). Whether the *FraXF* site is associated

# Type 1 Diseases

(CAG/polyglutamine disorders)

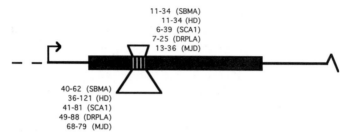

11-34 (SBMA)
11-34 (HD)
6-39 (SCA1)
7-25 (DRPLA)
13-36 (MJD)

40-62 (SBMA)
36-121 (HD)
41-81 (SCA1)
49-88 (DRPLA)
68-79 (MJD)

# Type 2 Diseases

Fragile X Syndrome

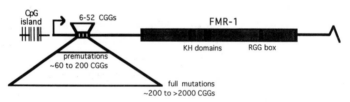

CpG island
6-52 CGGs
FMR-1
KH domains        RGG box
premutations ~60 to 200 CGGs
full mutations ~200 to >2000 CGGs

Myotonic Dystrophy

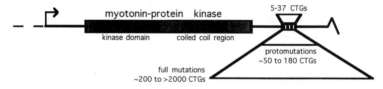

myotonin-protein kinase
5-37 CTGs
kinase domain        coiled coil region
protomutations ~50 to 180 CTGs
full mutations ~200 to >2000 CTGs

*Figure 1*  Location of expanded repeats in Type 1 and 2 trinucleotide repeat disorders. Shown are a schematic composite of the Type 1 disease genes and the *FMR-1* and myotonin protein kinase genes. Arrows depict transcription start sites, and shaded boxes depict protein coding regions. Striped boxes indicate the repeat; the ranges of normal and expanded repeats are indicated for each disease. (In FMR-1, the KH domains and RGG box are predicted to be RNA-binding motifs.)

with human disease is still unclear, although a few affected males have shown developmental delay.

*FRA16A* (chromosome 16) is the first autosomal fragile site to be cloned (Nancarrow et al 1994). As in *FRAXA*, *FRAXE*, and *FRAXF*, the molecular basis for the *FRA16A* site is a normally polymorphic $(CCG)_n$ repeat that is grossly expanded (1000–2000 repeats) in individuals expressing the fragile site. When expanded, the *FRA16A* locus and an adjacent CpG island become hypermethylated like the other fragile sites. Hypermethylation seems to be a consequence of expansion (Nancarrow et al 1994). *FRA16A* and other autosomal fragile sites are not associated with any known disease, perhaps because they have not been observed in homozygous form.

The striking similarities of the four cloned fragile sites suggest that all folate-sensitive fragile sites are due to expanded repeats. If resultant hypermethylation of the repeat is itself the cause of the observed chromosomal fragility at these sites, then the sequence of the repeat must always contain the CpG dinucleotide.

## Myotonic Dystrophy

An autosomal dominant disease, DM is characterized by progressive weakness and myotonia, often in association with cataracts, cardiac arrhythmias, endocrinopathy, and cognitive impairment (Harper 1989). The range in clinical severity is remarkably broad, and anticipation is common (Ashizawa et al 1992, Harley et al 1992, Harper et al 1989). Classic DM typically first appears in the second or third decades, with variable weakness and clinically evident myotonia (repetitive, rapid muscle firing after forceful contraction). Some patients, however, may be completely asymptomatic carriers or have only mild signs of disease, such as cataracts and electromyographically detectable myotonia. At the other extreme are patients, nearly always the offspring of affected women, who present with severe manifestations at birth (congenital DM). These patients are floppy as babies and usually mentally retarded. Pathologic studies of adults reveal distinct, but not specific, myopathic changes; in contrast, congenital cases show muscle fiber changes that suggest arrested development (Sarnat & Silbert 1976).

An international effort led to the identification of the defective gene on chromosome 19 (Aslandis et al 1992, Brook et al 1992, Buxton et al 1992, Fu et al 1992, Harley et al 1992, Mehadevan et al 1992). The defect is an expanded trinucleotide $(CTG)_n$ repeat in the 3' untranslated region of a gene called myotonin-protein kinase (Figure 1). Only 5 to 37 CTGs long in normal individuals, the CTG repeat is enlarged from ~50 to greater than several thousand repeats in affected individuals. As in fragile X, smaller protomutations (~50–180 repeats) cause mild or no disease but have a high probability of expanding to much larger repeats (~200 to >3000) in the next generation. The largest repeats (>1500) are associated with congenital DM. No mutations other than the expanded repeat have been identified in patients with DM, although a few cases without detectable CTG expansion have been reported (Thornton et al 1994).

## REPEAT INSTABILITY: CLINICAL ASPECTS

The discovery of expanded repeats has had an immediate clinical and scientific impact. The tests used for detecting expanded repeats are highly sensitive and specific [for example, see Kremer et al (1994)]. They permit certainty in diagnosis and eliminate the need for expensive ancillary tests (Redman et al 1993). Because the PCR-based repeat test is technically simpler and cheaper than the linkage analysis previously used for HD, requests for presymptomatic HD testing have recently increased. [Guidelines for HD testing were recently revised and may prove applicable to other trinucleotide repeat disorders (Broholm et al 1994, Hersch et al 1994).] In the near future, the ability to determine carrier status in presymptomatic individuals should facilitate trials aimed at disease prevention and may help researchers detect early biochemical changes in presymptomatic patients.

The discovery of expanded repeats has helped explain many of the unusual characteristics of these diseases (discussed individually below): variable phenotype, anticipation, parent-of-origin effects, and sporadic disease. In addition, repeat testing has helped determine the full range of clinical manifestations in each trinucleotide disorder. For example, amid the confusing nomenclature and overlapping syndromes of the hereditary ataxias, testing for the SCA1 repeat has better defined the clinical manifestations of SCA1 (Ranum et al 1994).

### Repeat Size and Clinical Correlation

For each trinucleotide repeat disease, there is a correlation between increasing repeat size and disease severity. This is best illustrated for Type 1 disorders, where larger repeats generally lead to earlier onset illness. In SCA1 (Figure 2) 66% of the variability in age of onset is accounted for by repeat length (Orr et al 1993, Ranum et al 1994). In other CAG repeat disorders it is slightly lower, for example, ~50% in HD (Andrew et al 1993, Doyu et al 1992, Duyao et al 1993, Igarishi et al 1992, Koide et al 1994, LaSpada et al 1992, Nagafuchi et al 1994, Novelletto et al 1994). As this correlation implies, juvenile-onset patients usually carry the largest repeats. Longer repeats are also associated with other aspects of disease, including earlier age of death in HD (Andrew et al 1993), more rapidly progressive disease in SCA1 (Orr et al 1993, Ranum et al 1994), and expression of the progressive myoclonus epilepsy subtype in DRPLA (Koide et al 1994, Nagafuchi et al 1994b).

It is important to note that repeat length accounts for only part of the variability in disease onset. Disease severity must partly be determined by other factors, including genetic background (Burke et al 1994, Ranum et al 1994). For this reason, clinicians should not use repeat length to predict disease

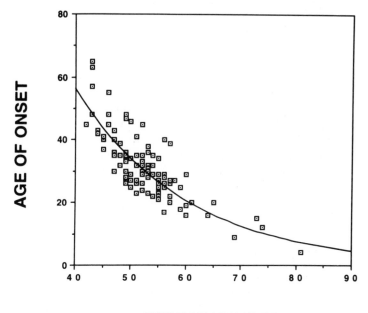

*Figure 2*  Inverse correlation between SCA1 repeat length and age of disease onset. In SCA1, 66% of the variability in age of onset is accounted for by repeat length. Other Type 1 diseases show a similar correlation between repeat length and age of onset. [Reprinted from Ranum et al (1994), with Dr. Ranum's permission.]

severity. At best, repeat length is only a partial predictor of age of onset, and for the most frequently observed lengths in HD (40–50 repeats), it is a poor predictor.

Type 2 disorders also show a correlation between repeat length and disease severity. In DM, for example, asymptomatic carriers typically have 60 to 120 repeats; classic adult-onset patients 200 to 400 repeats; childhood-onset patients 500 to 1500 repeats; and congenital DM 750 to >2000 repeats (Harley et al 1992, Redman et al 1993, Tsilfidis et al 1992). And in fragile X, the most profoundly retarded patients nearly always demonstrate dramatic repeat expansions, whereas smaller expansions cause milder disease (Fu et al 1991, Pieretti et al 1991). However, the correlation seen in these disorders is far from perfect. The lack of a precise correlation may be due in part to the extensive and variable somatic mosaicism seen in DM and fragile X. Repeat lengths are usually measured in peripheral blood, which may not accurately reflect repeat length in the most affected somatic tissues, such as muscle (Anvret et al 1993).

## Sporadic Disease

Before direct DNA testing, patients were occasionally seen with the clinical features of HD but without a family history of disease. Testing has now confirmed that these sporadic cases, up to 3% of all HD, are due to trinucleotide expansion of the HD gene (Goldberg et al 1993, Myers et al 1993). In other Type 1 disorders the frequency of sporadic cases is unknown but is probably no greater. Sporadic HD cases arise from a small pool of intermediate size alleles (30–38 trinucleotides) that are larger than those in the general population but smaller than those in most patients. These meiotically unstable, intermediate repeats are analogous to asymptomatic premutations of FRAXA and DM. Not large enough to cause disease in the carrier, they still carry the risk of expanding into the disease range when transmitted to the next generation. Most if not all sporadic HD is paternally transmitted.

## Parent-of-Origin Effect

For several Type 1 diseases (HD, DRPLA, and SCA1), severe juvenile-onset cases are usually paternally transmitted. This is due to greater repeat instability with paternal transmission (male bias) (Andrew et al 1993, Chung et al 1993, Koide et al 1994, Komure et al 1995, LaSpada et al 1992, Nagafuchi et al 1994, Novelletto et al 1994). With paternal transmission, intergenerational changes are larger (4 to 40 trinucleotides, versus typically just a few with maternal transmission), and expansions outnumber contractions (the reverse may be true for maternal transmission). Hence affected males are more likely to transmit a greatly enlarged CAG repeat of the size that typically causes juvenile-onset disease. Analysis of repeat size in sperm from HD patients suggests that this male bias is due to repeat instability during spermatogenesis (Duyao et al 1993, MacDonald et al 1993, Telenius et al 1994). Germ line instability in males also probably explains why sporadic cases of HD are paternally transmitted.

Congenital DM is nearly always maternally transmitted despite a male bias in the generation of new DM alleles (Ashizawa et al 1992, Bergoffen et al 1994, Wieringa 1994). This seems to occur, at least in part, because of constraints on the size of the repeat transmitted by males. An affected male with a very large CTG repeat—the size that might cause congenital DM—transmits a much smaller allele, usually in the premutation range. This may be due to selection against larger alleles during spermatogenesis. Since the dramatic expansions of CTG repeats appear to occur postzygotically in early embryogenesis (Jansen et al 1994, Wieringa 1994), it may instead be the case that the smaller repeats in sperm derive from embryonic tissues that did not undergo expansion. The end result is that affected females alone are able to transmit a large repeat that with further expansion (meiotically or postzygotically) leads to congenital DM.

Similar mechanisms may help explain the fact that only maternally transmitted *FRAXA* repeats progress to full mutations. As in DM, affected FRAXA males with a full mutation contain only premutation alleles in their sperm (Reyniers et al 1993), and the dramatic *FRAXA* expansions appear to occur postzygotically early in embryogenesis (Nelson & Warren 1993, Wohrle et al 1993). Postzygotic expansion may occur exclusively on maternal *FRAXA* alleles, suggesting that a form of imprinting also contributes to the parental bias in transmission (Sutherland & Richards 1993).

## REPEAT INSTABILITY: GENETIC ASPECTS

The major determinant of repeat instability appears to be repeat length itself. Normal-sized repeats have a low mutation rate that shows some length dependency [for example, a three- to fourfold increase in the mutation rate for the AR CAG repeat when going from 20–22 repeats to 28–31 repeats (Zhang et al 1994)]. However, only when repeats reach a critical size threshold, as they have in disease families, do they become highly unstable. Evidence suggests that this threshold is a stretch of roughly 35 uninterrupted trinucleotides. At the *FRAXA* locus, for example, marked instability begins when uninterrupted CGG repeat length reaches 34–38 repeats (Eichler et al 1994). This size coincides well with the upper limit of normal repeats for other trinucleotide repeat diseases (34, 34, 39, 25, 36, and 37 for SBMA, HD, SCA1, DRPLA, MJD, and DM, respectively). Moreover HD repeats of 30–38 constitute a pool of intermediate alleles that are predisposed to expand into the disease range, further suggesting that the critical threshold is in this range.

Once beyond this threshold length, trinucleotide repeats are highly mutable. For example, *FRAXA* alleles of 60–69 repeats have a 17% chance, and repeats greater than 90 a nearly 100% chance, of expanding to the full mutation when maternally transmitted (Fu et al 1991). In Type 1 disorders, where expansions seem constrained (perhaps because the repeat is translated into protein), a more modest degree of instability is observed during transmission (Andrew et al 1993, Chung et al 1993, Duyao et al 1993, Huntington's Disease Collaborative Group 1993, Koide et al 1994, LaSpada et al 1992, Nagafuchi et al 1994b, Novelletto et al 1993, Orr et al 1993, Snell et al 1993). In Type 2 diseases, a second size threshold exists (~60–70 uninterrupted repeats), beyond which repeats are prone to undergo hyperexpansion to the full mutation of several thousand repeats (Eichler et al 1994).

The existence of a similar threshold size for all trinucleotide repeat diseases suggests a common mechanism to expansion. Slipped-strand mispairing during replication has been proposed (Eichler et al 1994, Richards & Sutherland 1994) as a mechanism of mutation that could account for several features of trinucleotide expansion: Expansions and contractions occur in a polarized manner and by multiples of three base pairs, and interruptions within the repeat seem to confer stability. Okazaki fragment slippage during replication would provide an explanation for both small changes in repeat length (as seen in Type 1 repeats and Type 2 premutations) and the enormous hyperexpansions seen in Type 2 disorders. During replication of a threshold length repeat (~35 pure repeats), slippage could occur from only one end of the Okazaki fragment (the other end being anchored outside the repeat), resulting in small to moderate expansion. During replication of longer repeats (~70 repeats), the likelihood of Okazaki fragment synthesis beginning and ending within a repeat would increase. Thus slippage could occur from both ends of the unanchored Okazaki fragment, resulting (after several rounds of replication) in hyperexpansion. DNA mismatch repair mechanisms likely play a role in correcting errors caused by slippage, since defects in mismatch repair lead to increased simple sequence repeat instability in both yeast and humans [for review, see Modrich (1994)].

Although slipped-strand mispairing offers a testable model of DNA expansion, it leaves many questions unanswered. In Type 1 disorders, for example, why is marked repeat instability confined to the male germ line? Are the dramatic expansions of Type 2 repeats truly limited to an early point in embryogenesis, and if so, why? How do we reconcile the fact that for expanded repeats, further expansions outnumber contractions, while for normal repeats, contractions are more common (Zhang et al 1994)? And why, in mice containing a human AR transgene (Bingham et al 1995), is the expanded CAG repeat stably transmitted? Taken together, these questions make it clear that factors other than repeat length must play a significant role in repeat instability. [For more thorough discussions, see Eichler et al (1994), Nelson & Warren (1993), Richards & Sutherland (1994).]

## MOLECULAR MECHANISMS OF TRINUCLEOTIDE REPEAT DISEASES

### Fragile X

The mechanism of disease is more clearly understood for fragile X than for other trinucleotide repeat disorders. The fragile X phenotype is due to decreased expression of the FMR-1 protein (Pieretti et al 1991). As the CGG repeat number increases in the 5' region of the *FMR-1* gene, mRNA and protein levels decrease, and the fragile X syndrome worsens (Sutcliffe et al 1992). The degree of methylation correlates with disease severity, and hypermethylation may itself cause transcriptional silencing of the *FMR-1* gene (Bell et al 1991, McConkie-Rosell et al 1993, Nelson and Warren 1993, Pieretti et al

1991). A further argument that loss of *FMR-1* expression underlies disease is that small chromosomal deletions encompassing the *FMR-1* gene also lead to a fragile X phenotype (Gedeon et al 1992, Hirst et al 1995, Wohrle et al 1992). Transcriptional silencing seems to be the principle cause of decreased *FMR-1* expression, but translational inhibition may also play a role with the largest repeats (Feng et al 1995).

What is the role of FMR-1, the expression of which seems critical to normal brain function and development? FMR-1 is a member of a family of RNA-binding proteins (Ashley et al 1993b, Siomi et al 1993, Verheij et al 1993). It can bind its own and other human fetal brain mRNAs (Ashley et al 1993b), and alternatively spliced forms exist with distinct RNA-binding properties (Hoogeven et al 1994). The RNA-binding property of FMR-1 appears critical, as the fragile X phenotype can be generated by a point mutation that alters a highly conserved amino acid that is critical for RNA binding in related proteins (DeBoulle et al 1993). FMR-1 is widely expressed in fetal and adult tissues, most abundantly in neurons (Devys et al 1993). In the developing brain, FMR-1 is found initially in proliferating and migrating cells and later in differentiating neurons, with the highest levels in nucleus basalis magnocellularis and hippocampus (Abitbol et al 1993).

RNA-binding proteins are involved in the processing, transport, and translational control of mRNAs [see Simpson & Emeson (1996) in this volume]. Alternative splicing of transcripts helps to establish the intricate spatially and temporally restricted patterns of gene expression found in the brain. In addition, the targeting, localization, and translational control of mRNAs in subcellular domains of neurons (dendrites, for example) adds a further level of complexity to the regulation of neuronal gene expression. Which of these functions FMR-1 serves is presently unknown. Studies of *FMR-1* knockout mice, which display several phenotypic features of fragile X (Dutch-Belgian Consortium 1994), should help determine whether there is a specific class of RNAs that have processing, localization, or translation controlled by the FMR-1 protein.

The mechanism of disease in *FRAXE* may be similar to that in *FRAXA*. *FRAXE* disease severity also correlates with increasing repeat size and hypermethylation of a nearby CpG dinucleotide island (Knight et al 1993). It has been proposed that hypermethylation reduces expression of an as yet unidentified gene.

## Myotonic Dystrophy

The DM gene product was given the name myotonin-protein kinase because sequence analysis showed it to have homology to members of the serine-threonine protein kinase family (Pizzuti et al 1993, Wieringa 1994). Experiments with recombinant protein have confirmed that it has kinase activity, phosphorylating itself and other substrates at threonine and serine residues, although initial studies suggest that it differs from related serine-threonine protein kinases (Dunne et al 1994). Its amino acid sequence predicts both an α-helical coiled coil region and a hydrophobic domain near the carboxyl terminus. In muscle, the protein is predominantly localized to the neuromuscular junction (van der Ven et al 1993), although it is expressed elsewhere in muscle and in brain and heart (Jansen et al 1993). Several alternatively spliced RNAs exist that differ principally in the 5′ and 3′ regions of the transcript, leaving the kinase domain intact (Fu et al 1993). Although differential RNA processing occurs with this gene, what role it plays in the tissue-specific expression and function of myotonin-protein kinase is unknown (Pizzuti et al 1993).

Considerable focus has been directed at understanding the pathophysiology of muscle dysfunction in DM. Myotonia occurs in a number of rare inherited and acquired diseases. The underlying mechanism varies depending on the disorder, for example, chloride channel mutations in myotonia congenita and sodium channel mutations in hyperkalemic periodic paralysis (reviewed in Ptacek et al 1993). The primary electrophysiologic defect in DM is still unknown, although abnormal regulation of sodium channels has been proposed. The DM gene's kinase activity raises the possibility that in DM, myotonia may be due to altered phosphorylation of protein(s) involved in the muscle action potential (the sodium channel, for example) or excitation-contraction coupling. The normal endogenous substrates of myotonin protein kinase, however, have not yet been identified.

The muscle defects in DM cannot be due to changes in the protein itself, since the trinucleotide repeat lies in the 3′ untranslated region of the gene. Are they instead due to changes in the level of gene expression? Efforts to answer this question have been inconclusive. Studies performed with different techniques have shown either decreased or increased mRNA levels and reduced or essentially unchanged myotonin protein levels (Fu et al 1993, Novelli et al 1993, Sabouri et al 1993). These discrepancies need to be resolved before it can be determined whether altered expression plays a role in the pathogenesis of DM. The CTG expansion may also affect expression of other nearby genes, particularly in the full mutation range.

The biochemical defect in DM may occur at the RNA level instead. mRNA in the 3′ untranslated region is known to play a role in mRNA stability, cell proliferation, and differentiation (Rastinejad & Blau 1993). A greatly expanded GC-rich (CTG) repeat in the 3′ mRNA could perturb an important regulatory role for this part of the message, either by distorting its normal secondary structure or by titrating out necessary RNA-binding proteins (J Eberwine, personal communication). Such an effect could be dominant, thus accounting for DM's pattern of inheritance. An expanded repeat within the transcript could also lead to aberrant RNA processing of the DM transcript, for which there is some experimental evidence (Krahe et al 1994).

## CAG-Polyglutamine Diseases

In all five CAG repeat diseases, expanded CAG-polyglutamine tracts of about the same size cause neurodegeneration (Figure 3). Several findings support the view that expansion of CAG-polyglutamine tracts leads to a gain of function that may be toxic to neurons. First, the dominant inheritance seen in these disorders is more consistent with a gain, not loss, of function. HD in particular is a pure dominant disorder, since patients homozygous for the disease allele develop HD that is clinically indistinguishable from HD caused by a single disease allele (Wexler et al 1991). [SBMA follows an X-linked recessive pattern of inheritance, but in this case heterozygous females could be protected by lyonization (if the toxic effect is cell autonomous) or by low androgen levels (if the effect is ligand-dependent).] Second, in all these disorders the disease gene is expressed in the disease state, arguing against a simple decrease

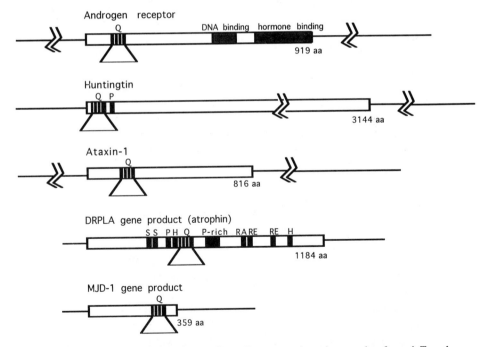

*Figure 3*  The Type 1 disease gene products. Shown are schematic transcripts for each Type 1 disease gene. Large open boxes indicate the protein coding domain, and smaller striped boxes the polyglutamine tracts. Shaded boxes represent regions of known function in the AR, and filled boxes represent homopolymeric or dipeptide domains (specified by the single letter amino acid code).

in gene expression as the cause of disease (Banfi et al 1994, Li et al 1993, Nagafuchi et al 1994a, Servadio et al 1995, Sharp et al 1995, Strong et al 1993, Trottier et al 1995, Warner et al 1992, Yazawa et al 1995). (MJD expression in the disease state has not yet been reported.) The fact that individuals missing one copy of the normal HD gene (because of chromosomal deletions or translocations) do not develop HD further suggests that disease is not caused by decreased gene expression. Third, the loss of function for one CAG repeat gene, the AR, leads to an entirely different disease, androgen insensitivity syndrome (testicular feminization), which does not have the neurologic manifestations of SBMA. Fourth, the recent development of a transgenic mouse model for SCA1 provides direct evidence supporting a gain of function: An expanded repeat SCA1 transgene causes cerebellar degeneration, whereas a normal repeat transgene does not (Burright et al 1995).

Any proposal to explain neurodegeneration caused by CAG-polyglutamine repeats must take into account several facts. 1. Cell death appears to be neuron specific despite the fact that the repeat-containing genes are widely expressed in nonneuronal cells (Banfi et al 1994, Li et al 1993, Nagafuchi et al 1994a, Strong et al 1993). 2. Distinct subsets of neurons die in each disorder, even though the disease gene product is expressed throughout the brain. (The androgen receptor may be an exception, since its neuronal expression is restricted to discrete regions of the brain and spinal cord.) 3. Degeneration occurs slowly, with little evidence of neuropathologic or biochemical abnormality before the clinical onset of disease. 4. The presumed neurotoxicity increases with repeat length, since longer repeats lead to earlier disease that in some cases, may progress more rapidly. 5. The diseases are caused by dissimilar genes that lack homology to one another outside of the CAG-polyglutamine repeats. There is no reason, based on sequence comparison and immunolocalization studies, to believe the encoded proteins serve similar cellular functions. 6. The gene products are all thought or known to be intracellular proteins, suggesting that the presumed toxic effect occurs intracellularly. Thus in summary, expansions in a group of dissimilar, ubiquitously expressed CAG repeat genes cause selective neurodegeneration that is dependent on the length of the CAG-polyglutamine repeat. Given the presumed common mechanism of neurodegeneration and the ubiquitous expression of CAG repeat genes, cell-specific factors must account for the death of certain neuronal populations in each disease.

The putative gain of function most likely occurs at the protein level, since the CAG repeat is translated into polyglutamine in the mature protein (this has been verified experimentally for DRPLA, HD, SBMA, and SCA1) (Servadio et al 1995, Sharp et al 1995, Trottier et al 1995, Yazawa et al 1995). However, a gain of function could also occur at the RNA level. For example, expanded CG-rich triplet repeats may form stable stem loop structures that could alter properties of the mRNA (J Eberwine, personal communication). Although the

SCA1 transgenic model supports a gain of function mechanism, whether this occurs at the protein or RNA level is unknown, and both theories warrant experimental testing.

What is the normal function of polyglutamine tracts? Increasing evidence indicates that polyglutamine domains are functionally important in regulatory proteins and transcription factors (Gerber et al 1994). Across species as evolutionarily distant as yeast, *Drosophila,* and humans, glutamine-rich domains and homopolymeric polyglutamine tracts have been found in common or ubiquitous transcription factors (e.g. TATA-binding protein, Sp1, AP1, glucocorticoid receptor), homeotic genes (e.g. *Antennapedia*), neurogenic genes (e.g. *Notch* and *mastermind*), and other developmentally important genes (e.g. embryonic polarity dorsal protein, *couch potato*) (Creusot et al 1988, Duboule et al 1987, Kao et al 1990, Mitchell & Tjian 1989, Wharton et al 1985). Mitchell & Tjian (1989) have proposed that polyglutamine tracts mediate transactivation by engaging in protein-protein interactions and thereby enabling formation of transcription activation complexes. Evidence for this hypothesis has come from deletion experiments. The yeast regulatory protein MCM1 contains a glutamine-rich region that is necessary for it to interact with the coregulatory protein alpha 1 and activate transcription (Bruhn et al 1992). When the glutamine-rich domains are deleted from Sp1, this transcription factor is also unable to form polymeric complexes and activate transcription of target genes (Pascal & Tjian 1991). More recently, studies of polyglutamine-containing fusion proteins directly demonstrated that polyglutamine stretches can activate transcription, optimally with lengths of 10–30 amino acids (Gerber et al 1994). It thus appears that polyglutamine tracts can serve as transactivation domains that probably function by modulating interactions with both homologous and heterologous transcription factors.

In fact, the only Type 1 disease gene product with a known function is a transcription factor, the AR. A member of the steroid and thyroid hormone receptor superfamily, the AR enters the nucleus upon binding androgen and activates transcription of androgen-responsive genes (Zhou et al 1994). Its polyglutamine tract is located in a large amino-terminal regulatory domain far removed from the DNA-binding and hormone-binding domains (Figure 3). The polyglutamine tract is not strictly necessary for androgen-induced gene activation. In nonneuronal cells, transcriptional activation by AR is increased when the polyglutamine tract is eliminated (Chamberlain et al 1994), and decreased when the polyglutamine tract is expanded into the disease range (Mhatre et al 1993). Whether this will also hold true in neurons or has any relationship to a toxic gain of function in neurons is unknown.

There is no direct experimental evidence to indicate that other Type 1 disease genes encode transcription factors. None of them contain definite DNA-binding motifs, although sequence analysis suggests atrophin may be an RNA-binding protein (Nagafuchi et al 1994a) and ataxin-1 localizes to the nucleus in most neurons (Servadio et al 1995). Although huntingtin contains a sequence resembling a leucine zipper (a motif present in one group of transcription factors), recent studies indicate that in neurons huntingtin localizes to nerve terminals, not the nucleus (DiFiglia et al 1995, Sharp et al 1995, Trottier et al 1995). Regardless of the function of other Type 1 disease genes, polyglutamine stretches within these genes likely serve to modulate interactions with other cellular proteins. Perutz et al (1994) have proposed that polyglutamine tracts can self-associate in the form of a polar zipper, and recently provided direct evidence supporting this model: Incorporating a small polyglutamine tract into a normally monomeric protein caused dimers and trimers to form (Stott et al 1995).

How might expansion of a polyglutamine stretch cause neurodegeneration? There are at least three plausible mechanisms, each involving changes in protein-protein interactions. The first could be called the aberrant transregulation model. In this model, expanding the polyglutamine tract in a regulatory protein would alter, perhaps subtly, the strength of its interactions with other coregulatory proteins, thereby disrupting what is normally a highly ordered and regulated molecular cascade. For example, the AR with an expanded polyglutamine tract might display altered binding to transcription associated factors, causing aberrant transcriptional activation of one or more genes in neurons. This in turn could lead to the neuron's receiving inappropriate signals, which in some cases may result in neuronal apoptosis (DiBenedetto & Pittman 1995).

The second model could be called the protein aggregation model. In this model, expanding the polyglutamine tract would cause changes in protein-protein interactions that are not subtle and not necessarily restricted to normal cointeracting proteins. Expanded polyglutamine tracts may tend to form an extended polar zipper involving tight noncovalent interactions between proteins (Perutz et al 1994). These could be nonproductive interactions that perturb the function of any involved proteins and interfere with degradation of the aggregated proteins. For example, the expanded AR could aggregate with itself or with other proteins. This not only would deplete a potentially androgen-dependent cell of functional AR but also might lead to the accumulation of insoluble proteinaceous material that is deleterious to neurons. An attractive feature of this model is that it resembles current theories to explain two other neurodegenerative disorders, Creutzfeld Jacob disease (insoluble, protease-resistant aggregates of the prion protein) and Alzheimer's disease (aggregates of β-amyloid and microtubule-associated proteins).

The third model, proposed by Green (1993), is the transglutaminase cross-linking model. As polyglutamine tracts expand beyond the size normally found in proteins (less than 30 amino acids), they may become unusually avid substrates for transglutaminase activity, resulting in the cross-linking of proteins within neurons. Either the cross-linked proteins themselves or the insoluble isodipeptide that is produced during subsequent proteolysis could be toxic to cells. Long-lived postmitotic cells, like neurons, would be most susceptible to the accumulation of cross-linked material. For example, the expanded AR might become cross-linked nonspecifically to itself and other proteins, effectively titrating out functional AR, perturbing nuclear and cellular processes, and leading to an accumulation of cross-linked protein or isodipeptide. Because the conformation and nuclear transport of the AR protein is controlled by androgen binding, the extent of cross-linking might be androgen dependent.

All three models can explain the repeat length dependence of the disease process. One can further speculate how each of the three could be specifically neurotoxic: In the first model one or more regulatory cascades that are neuron specific would be perturbed, and in the second and third models the long-lived postmitotic nature of neurons would make them uniquely susceptible to the accumulation of aggregated or cross-linked protein. The first model, however, would seem best at explaining the specific neuropathology seen in each CAG repeat disease; for each disease, the affected regulatory cascade(s) would be more important to cell survival in precisely those neurons that preferentially die. It is more difficult to explain how an accumulation of aggregated or cross-linked material—which should occur in most or all neurons, given the widespread CNS expression of CAG repeat genes—could lead in each disease to the death of different subsets of neurons. In order for the second and third models to explain the specific neuropathology of each disease, further disease-specific and cell-specific factors need to be invoked that result in a differential accumulation of (or sensitivity to) aggregated and cross-linked material or breakdown products, in distinct regions of the brain. No one has yet found evidence for aggregated, precipitated, or cross-linked protein in any of these diseases.

Intensive research is now underway in many laboratories to test aspects of these models. A full understanding of the molecular mechanisms at play will only come from research across a broad front, including transgenic and knock-out animal models, in vitro expression systems, yeast and bacterial systems designed to identify interacting proteins, and rigorous biochemical and immunocytochemical analyses of the normal and expanded proteins. Recent successes in creating an SCA1 transgenic model (Burright et al 1995) and in achieving targeted disruption of the HD gene [resulting in embryonic lethality (Nasir et al 1995)] are promising steps in this direction.

ACKNOWLEDGMENTS

We thank Brian Brooks, Peter Bingham, Peter Crino, and Jim Eberwine for critically reading the manuscript. H Paulson is supported by a Howard Hughes

Postdoctoral Research Fellowship for Physicians and a Neuropharmacology Research Fellowship from the American Academy of Neurology. K Fischbeck is supported by grants from the National Institutes of Health, the Muscular Dystrophy Association, the March of Dimes, and the Amyotrophic Lateral Sclerosis Association.

## Literature Cited

Abitbol M, Menini C, Delezoide A-L, Rhyner T, Vekemans M, et al. 1993. Nucleus basalis magnocellularis and hippocampus are the major sites of FMR-1 expression in the human fetal brain. *Nat. Genet.* 4:147–53

Andrew SE, Goldberg YP, Kremer B, Telenius H, Theilman J, et al. 1993. The relationship between trinucleotide (CAG) repeat length and clinical features of Huntington's disease. *Nat. Genet.* 4:398–403

Anvret M, Ahlberg G, Grandell U, Hedberg B, Johnson K, et al. 1993. Larger expansions of the CTG repeat in muscle compared to lymphocytes from patients with myotonic dystrophy. *Hum. Mol. Genet.* 2:1397–400

Ashizawa T, Dubel JR, Dunne PW, Dunne CJ, Fu H-Y, et al. 1992. Anticipation in myotonic dystrophy. II. Complex relationships between clinical findings and structure of the GCT repeat. *Neurology* 42:1877–83

Ashley CT, Sutcliffe JS, Kunst CB, Leiner HA, Eichler EE, et al. 1993a. Human and murine FMR-1: alternative splicing and translational initiation downstream of the CGG-repeat. *Nat. Genet.* 4:244–51

Ashley CT, Wilkinson KD, Reines D, Warren ST. 1993b. FMR-1 protein: conserved RNP family domains and selective RNA binding. *Science* 262:563–66

Aslanidis C, Jansen G, Amemiya C, Shutler G, Mahadevan M, et al. 1992. Cloning of the essential myotonic dystrophy region and mapping of the putative defect. *Nature* 355:548–51

Banfi S, Servadio A, Chung M, Kwiatkowski TJ, McCall AE, et al. 1994. Identification and characterization of the gene causing type 1 spinocerebellar ataxia. *Nat. Genet.* 7:513–20

Bates G, Lehrach H. 1994. Trinucleotide repeat expansions and human genetic disease. *BioEssays* 16:277–84

Beal MF. 1992. Does impairment of energy metabolism result in excitotoxic neuronal death in neurodegenerative diseases? *Neurology* 31:119–30

Beal MF, Brouillet E, Jenkins BG, Ferrante RJ, Kowall NW, et al. 1993. Neurochemical and histologic characterization of striatal excitotoxic lesions produced by the mitochondrial toxin 3-nitropropionic acid. *J. Neurosci.* 13:4181–92

Bell MV, Hirst MC, Nakahori Y, MacKinnon RN, Roche A, et al. 1991. Physical mapping across the fragile X: hypermethylation and clinical expression of the fragile X syndrome. *Cell* 64:861–66

Bergoffen J, Kant J, Sladky J, McDonald-McGinn M, Zackai E, Fischbeck KH. 1994. Paternal transmission of congenital myotonic dystrophy. *J. Med. Genet.* 31:518–20

Bingham PM, Scott MO, Wang S, McPhaul MJ, Wilson EM, et al. 1995. Stability of an expanded trinucleotide repeat in the androgen receptor gene in transgenic mice. *Nat. Genet.* 9:191–96

Broholm J, Cassiman J-J, Craufurd D, Falek A, Farmer-Little C, et al. 1994. Guidelines for the molecular genetics predictive test in Huntington's disease. *Neurology* 44:1533–36

Brook JD, McCurrach ME, Harley HG, Buckler AJ, Church D, et al. 1992. Molecular basis of myotonic dystrophy: Expansion of a trinucleotide (CTG) repeat at the 3' end of a transcript encoding a protein kinase family member. *Cell* 68:799–808

Bruhn L, Hwang-Shum JJ, Sprague GF. 1992. The N-terminal 96 residues of MCM1, a regulator of cell type-specific genes in *Saccharomyces cerevisiae*, are sufficient for DNA binding, transcription activation, and interaction with alpha 1. *Mol. Cell. Biol.* 12:3563–72

Burke JR, Wingfield MS, Lewis KE, Roses AD, Lee JE, et al. 1994. The Haw River syndrome: dentatorubral-pallidoluysian atrophy (DRPLA) in an African-American family. *Nat. Genet.* 7:521–24

Burright E, Clark HB, Servadio A, Matilla T, Feddersen RM, et al. 1995. SCA1 transgenic mice: a model for neurodegeneration caused by trinucleotide expansion. Submitted

Buxton J, Shelbourne P, Davies J, Jones C, Van Tongeren T, et al. 1992. Detection of an unstable fragment of DNA specific to individuals with myotonic dystrophy. *Nature* 355: 547–48

Chamberlain NL, Driver ED, Miesfeld RL. 1994. The length and location of CAG trinucleotide repeats in the androgen receptor N-terminal domain affect transactivation function. *Nucl. Acids Res.* 22:3181–86

Chung M, Ranum LPW, Duvick LA, Servadio A, Zoghbi HY, et al. 1993. Evidence for a mechanism predisposing to intergenerational CAG repeat instability in spinocerebellar ataxia type 1. *Nat. Genet.* 5:254–58

Creusot F, Verdiere J, Gaisne M, Slonimski PP. 1988. CYP1 (HAP1) regulator of oxygen-dependent gene expression in yeast. *J. Mol. Biol.* 204:263–76

De Boulle K, Verkerk AJMH, Reyniers E, Vits L, Hendricks J, et al. 1993. A point mutation in the *FMR-1* gene associated with fragile X mental retardation. *Nat. Genet.* 3:31–35

Devys D, Lutz Y, Rouyer N, Bellocq JP, Mandel J-L. 1993. The FMR-1 protein is cytoplasmic, most abundant in neurons, and appears normal in carriers of the fragile X premutation. *Nat. Genet.* 4:335–40

DiBenedetto AJ, Pittman RN. 1995. Death in the balance. *Perspect. Dev. Neurobiol.* In press

DiFiglia M, Sapp E, Chase K, Schwarz C, Meloni A, et al. 1995. Huntingtin is a cytoplasmic protein associated with vesicles in human and rat brain neurons. *Neuron* 14: 1075–81

Doyu M, Sobue G, Mukai E, Kachi T, Yasuda T, et al. 1992. Severity of X-linked recessive bulbospinal neuronopathy correlates with size of the tandem CAG repeat in androgen receptor gene. *Ann. Neurol.* 32:707–10

Duboule D, Haenlin M, Galliot B, Mohier E. 1987. DNA sequences homologous to the *Drosophila opa* repeat are present in murine mRNAs that are differentially expressed in fetuses and adult tissues. *Mol. Cell. Biol.* 7: 2003–6

Dunne PW, Walch ET, Epstein HF. 1994. Phosphorylation reactions of recombinant human myotonic dystrophy protein kinase and their inhibition. *Biochemistry* 33:10809–14

Dutch-Belgian Fragile X Consortium. 1994. Fmr1 knockout mice: a model to study fragile X mental retardation. *Cell* 78:23–33

Duyao MP, Ambrose C, Myers R, Novelletto A, Persichetti F, et al. 1993. Trinucleotide repeat instability and age of onset in Huntington's disease. *Nat. Genet.* 4:387–92

Eichler EE, Holden JJ, Popovich BW, Reiss AL, Snow K, et al. 1994. Length of uninterrupted CGG repeats determines instability in the FMR1 gene. *Nat. Genet.* 8:88–94

Feng Y, Zhiang F, Lokey LK, Chastain JL, Lakkis L, et al. 1995. Translational suppression by trinucleotide repeat expansion at *FMR1. Science* 268:731–34

Ferrante RJ, Kowall NW, Cipolloni PB, Storey E, Beal MF, et al. 1993. Excitotoxic lesions in primates as a model for Huntington's disease: histopathologic and neurochemical characterization. *Exp. Neurol.* 119:46–71

Fischbeck KH, Souders D, La Spada AR. 1991. A candidate gene for X-linked spinal muscular atrophy. *Adv. Neurol.* 56:209–13

Fu YH, Friedman DL, Richards S, Pearlman JA, Gibbs RA, et al. 1993. Decreased expression of myotonin-protein kinase messenger RNA and protein in adult form of myotonic dystrophy. *Science* 260:235–38

Fu YH, Kuhl DPA, Pizzuti A, Pieretti M, Sutcliffe JS, et al. 1991. Variation of the CGG repeat at the fragile X site results in genetic instability: resolution of the Sherman paradox. *Cell* 67:1047–58

Fu YH, Pizzuti A, Fenwick RG, King J, Rajnarayan S, et al. 1992. An unstable triplet repeat in a gene related to myotonic muscular dystrophy. *Science* 255:1256–58

Gedeon AK, Baker E, Robinson H, Partington MW, Gross B, et al. 1992. Fragile X syndrome without CCG amplification has an *FMR-1* deletion. *Nat. Genet.* 1:341–44

Gerber H-P, Seipel K, Georgiev O, Hofferer M, Hug M, et al. 1994. Transcriptional activation modulated by homopolymeric glutamine and proline stretches. *Science* 263: 808–11

Goldberg YP, Kremer B, Andrew SE, Theilmann J, Graham RK, et al. 1993. Molecular analysis of new mutations for Huntington's disease: intermediate alleles and sex of origin effects. *Nat. Genet.* 5:174–79

Goldberg YP, Telenius H, Hayden MR. 1994. The molecular genetics of Huntington's disease. *Curr. Opin. Neurol.* 7:325–32

Green H. 1993. Human genetic diseases due to codon reiteration: relationship to an evolutionary mechanism. *Cell* 74:955–56

Gusella JF, Wexler NS, Conneally PM, Naylor SL, Anderson MA, et al. 1983. A polymorphic marker genetically linked to Huntington's disease. *Nature* 306:234–38

Hansen RS, Gartler SM, Scott CR, Chen SH, Laird CD. 1992. Methylation analysis of CGG sites in the CpG island of the FMR1 gene. *Hum. Mol. Genet.* 1:571–78

Harding AE. 1993. Clinical features and classification of inherited ataxias. *Adv. Neurol.* 61:1–14

Harding AE, Thomas PK, Baraitser M, Bradbury PG, Morgan-Hughes JA, et al. 1982.

X-linked recessive bulbospinal neuropathy: a report of ten cases. *J. Neurol. Neurosurg. Psychiatry* 45:1012–19

Harley HG, Brook JD, Rundle SA, Crow S, Reardon W, et al. 1992. Expansion of an unstable DNA region and phenotypic variation in myotonic dystrophy. *Nature* 355:545–46

Harper PS. 1989. *Myotonic Dystrophy.* London/Philadelphia: Saunders. 2nd ed.

Harper PS, Morris MJ, Quarell O, eds. 1991. *Huntington's Disease.* Philadelphia: Saunders

Hersch S, Jones R, Koroshetz W, Quaid K. 1994. The neurogenetics genie: testing for the Huntington's disease mutation. *Neurology* 44:1369–73

Hirst M, Grewal P, Flannery A, Slatter R, Maher E, et al. 1995. Two new cases of FMR1 deletion associated with mental impairment. *Am. J. Hum. Genet.* 56:67–74

Hoogeveen AT, Verhey C, Bakker CE, de Graaf E, Willemsen R, et al. 1994. A tissue dependent processing and function of the FMR-1 gene product. *Am. J. Hum. Genet.* 55:A18 (Abstr.)

Huntington's Disease Collaborative Research Group. 1993. A novel gene containing a trinucleotide repeat that is expanded and unstable on Huntington's disease chromosomes. *Cell* 72:971–83

Igarashi S, Tanno Y, Onodera O, Yamazaki M, Sato S, et al. 1992. Strong correlation between the number of CAG repeats in androgen receptor genes and the clinical onset of features of spinal and bulbar muscular atrophy. *Neurology* 42:2300–2

Jansen G, Willems P, Coerwinkel M, Nillesen W, Smeets H, et al. 1994. Gonosomal mosaicism in myotonic dystrophy patients: involvement of mitotic events in $(CTG)_n$ repeat variation and selection against expansion in sperm. *Am. J. Hum. Genet.* 54:575–85

Kao CC, Lieberman PM, Schmidt MC, Zhou Q, Pei R, et al. 1990. Cloning of a transcriptionally active human TATA binding factor. *Science* 248:1646–50

Kawaguchi Y, Okamoto T, Taniwaki M, Aizawa M, Inoue M, et al. 1994. CAG expansions in a novel gene for Machado-Joseph disease at chromosome 14q32.1. *Nat. Genet.* 8:221–24

Kennedy WR, Alter M, Sung JH. 1968. Progressive proximal spinal and bulbar muscular atrophy of late onset: a sex-linked recessive trait. *Neurology* 18:671–80

Knight SJL, Flannery AV, Hirst MC, Campbell L, Christodoulou Z, et al. 1993. Trinucleotide repeat expansion and hypermethylation of a CpG island in FRAXE mental retardation. *Cell* 74:127–34

Koide R, Ikeuchi I, Onodera O, Tanaka H, Igarishi S, et al. 1994. Unstable expansion of CAG repeat in hereditary dentatorubral-pallidoluysian atrophy (DRPLA). *Nat. Genet.* 6:9–13

Komure O, Sano A, Nishino N, Yamaguchi N, Ueno S, et al. 1995. DNA analysis in hereditary dentatorubral-pallidoluysian atrophy: correlation between CAG repeat length and phenotypic variation and the molecular basis of anticipation. *Neurology* 45:143–49

Krahe R, Ashizawa T, Narayana L, Fununage VL, Carango P, et al. 1994. Effect of the myotonic dystrophy expanded $(CTG)_n$ repeat on the transcripts of DMPK alleles. *Am. J. Hum. Genet.* 55:A227 (Abstr.)

Kremer B, Goldberg P, Andrew SE, Theilman J, Telenius H, et al. 1994. A worldwide study of the Huntington's disease mutation: the sensitivity and specificity of measuring CAG repeats. *New Engl. J. Med.* 330:1401–6

Kremer EJ, Pritchard M, Lynch M, Yu S, Holman K, et al. 1991. Mapping of DNA instability at the fragile X to a trinucleotide repeat sequence $p(CGG)_n$. *Science* 252:1711–14

Kuhl D, Pizzuti A, Caskey CT. 1993. Triplet repeat mutations: an important etiology of genetic disease. *Curr. Opin. Neurol.* 13:55–85

LaSpada AR, Paulson HL, Fischbeck KH. 1994. Trinucleotide repeat expansion in neurological disease. *Ann. Neurol.* 36:814–22

La Spada AR, Roling D, Harding AE, Warner CL, Spiegel R, et al. 1992. Meiotic stability and genotype-phenotype correlation of the trinucleotide repeat in X-linked spinal and bulbar muscular atrophy. *Nat. Genet.* 2:301–4

La Spada AR, Wilson EM, Lubahn DB, Harding AE, Fischbeck KH. 1991. Androgen receptor gene mutations in X-linked spinal and bulbar muscular atrophy. *Nature* 352:77–79

Li SH, Schilling G, Young WS, Li X-J, Margolis RL, et al. 1993. Huntington's disease gene (IT15) is expressed widely in human and rat tissues. *Neuron* 11:985–93

MacDonald ME, Barnes G, Srinidi J, Duyao MP, Ambrose CM, et al. 1993. Gametic but not somatic instability of CAG repeat length in Huntington's disease. *J. Med. Genet.* 30:982–86

Mahadevan M, Tsilfidis C, Sabourin L, Shutler G, Amemiya C, et al. 1992. Myotonic dystrophy mutation: an unstable CTG repeat in the 3′ untranslated region of the gene. *Science* 255:1253–55

Mandel J-L. 1994. Trinucleotide diseases on the rise. *Nat. Genet.* 7:453–55

Matilla T, Volpini V, Genis D, Rosell J, Corral J, et al. 1993. Presymptomatic analysis of spinocerebellar ataxia type 1 (SCA1) via the expansion of the SCA1 CAG-repeat in a large pedigree displaying anticipation and parental male bias. *Hum. Mol. Genet.* 2:2123–28

McConkie-Rosell A, Lachiewicz AM, Spiridigliozzi GA, Tarleton J, Schoenwald S, et al. 1993. Evidence that methylation of the FMR-1 locus is responsible for variable phenotypic expression of the fragile X syndrome. *Am. J. Hum. Genet.* 53:800–9

Mhatre AN, Trifiro MA, Kaufman M, Kazemi-Esfarjani P, Figlewicz D, et al. 1993. Reduced transcriptional regulatory competence of the androgen receptor in X-linked spinal and bulbar muscular atrophy. *Nat. Genet.* 5:184–88

Mitchell PJ, Tjian R. 1989. Transcriptional regulation in mammalian cells by sequence-specific DNA binding proteins. *Science* 245:371–78

Modrich P. 1994. Mismatch repair, genetic stability, and cancer. *Science* 266:1959–60

Myers RH, MacDonald ME, Koroshetz WJ, Duyao MP, Ambrose CM, et al. 1993. De novo expansion of a $(CAG)_n$ repeat in sporadic Huntington's disease. *Nat. Genet.* 5:168–73

Nagafuchi S, Yanagisawa H, Ohsaki E, Shirayama T, Tadokoro K, et al. 1994a. Structure and expression of the gene responsible for the triplet repeat disorder, dentatorubral and pallidoluysian atrophy (DRPLA). *Nat. Genet.* 8:177–82

Nagafuchi S, Yanagisawa H, Sato K, Shirayama T, Ohsaki E, et al. 1994b. Dentatorubral and pallidoluysian atrophy expansion of an unstable CAG trinucleotide on chromosome 12p. *Nat. Genet.* 6:14–18

Naito H, Oyanagi S. 1982. Familial myoclonus epilepsy and choreoathetosis: hereditary dentatorubral-pallidoluysian atrophy. *Neurology* 32:798–807

Nancarrow JK, Kremer E, Holman K, Eyre H, Doggett NA, et al. 1994. Implications of FRA16A structure for the mechanism of chromosomal fragile site genesis. *Science* 264:1938–41

Nasir J, Floresco SB, O'Kusky JR, Dewart VM, Richman J, et al. 1995. Targeted disruption of the Huntington's disease gene results in embryonic lethality and behavioral and morphological changes in heterozygotes. *Cell* 81:811–23

Nelson DL, Warren ST. 1993. Trinucleotide repeat instability: when and where? *Nat. Genet.* 4:107–8

Neumann MN. 1959. Combined degeneration of globus pallidus and dentate nucleus and their projections. *Neurology* 9:430–38

Novelletto A, Persichetti F, Sabbadini G, Mandich P, Bellone E, et al. 1994. Analysis of the trinucleotide repeat expansion in Italian families affected with Huntington's disease. *Hum. Mol. Genet.* 3:93–98

Novelli G, Gennarelli M, Zelano G, Pizzati A, Fattorini C, et al. 1993. Failure in detecting mRNA transcripts from the mutated allele in myotonic dystrophy muscle. *Biochem. Mol. Biol. Int.* 29:291–97

Nussbaum RL, Ledbetter DH. 1989. The fragile X syndrome. In *Metabolic Basis of Inherited Diseases,* ed. CR Scriver, AL Beaudet, WS Sly, D Valle, pp. 327–42. New York: McGraw-Hill

Oberle I, Rousseau F, Heitz D, Kietz C, Devys D, et al. 1991. Instability of a 550-base pair DNA segment and abnormal methylation in fragile X syndrome. *Science* 252:1087–102

Orr HT, Chung M, Banfi S, Kwiatkowski TJ, Servadio A, et al. 1993. Expansion of an unstable trinucleotide CAG repeat in spinocerebellar ataxia type 1. *Nat. Genet.* 4:221–26

Parrish JE, Oostra BA, Verkerk AJ, Richards CS, Reynolds J, et al. 1994. Isolation of a GCC repeat showing expansion in FRAXF, a fragile site distal to FRAXA and FRAXE. *Nat. Genet.* 8:229–35

Pascal E, Tjian R. 1991. Different activation domains of Sp1 govern formation of multimers and mediate transcriptional synergism. *Genes Dev.* 5:1646–56

Perutz MF, Johnson T, Suzuki M, Finch JT. 1994. Glutamine repeats as polar zippers: their possible role in inherited neurodegenerative diseases. *Proc. Natl. Acad. Sci. USA* 91:5355–58

Pieretti M, Zhang F, Fu YH, Warren ST, Oostra BA, et al. 1991. Absence of expression of the FMR-1 gene in fragile X syndrome. *Cell* 66:817–22

Pizzuti A, Friedman DL, Caskey CT. 1993. The myotonic dystrophy gene. *Arch. Neurol.* 50:1173–79

Ptacek LJ, Johnson KJ, Griggs RC. 1993. Genetics and physiology of the myotonic muscle disorders. *New Engl. J. Med.* 328:482–89

Ranum LP, Chung M, Banfi S, Bryer A, Schut LJ, et al. 1994. Molecular and clinical correlations in spinocerebellar ataxia type 1: evidence for familial effects on the age of onset. *Am. J. Hum. Genet.* 55:244–52

Rastinejad F, Blau HM. 1993. Genetic complementation reveals a novel regulatory role for 3' untranslated regions in growth and differentiation. *Cell* 72:903–17

Redman JB, Fenwick RG, Fu Y-H, Pizzuti A, Caskey CT. 1993. Relationship between parental trinucleotide GCT repeat length and severity of myotonic dystrophy in offspring. *J. Am. Med. Assoc.* 269:1960–65

Reiss AL, Aylward E, Freund LS, Joshi PK, Bryan RN. 1991. Neuroanatomy of fragile X syndrome: the posterior fossa. *Ann. Neurol.* 29:26–32

Reyniers E, Vits L, De Boulle K, Van Roy B, Van Velzen D, et al. 1993. The full mutation in the FMR-1 gene of male fragile X patients is absent in their sperm. *Nat. Genet.* 4:143–46

Richards RI, Sutherland GR. 1994. Simple DNA is not repeated simply. *Nat. Genet.* 6: 114–16

Ritchie RJ, Knight SJL, Hirst MC, Grewal PK, Bobrow M, et al. 1994. The cloning of FRAXF: trinucleotide repeat expansion and methylation at a third fragile site in distal Xqter. *Hum. Mol. Genet.* 3:2115–21

Rosenberg RN. 1992. Machado-Joseph disease: an autosomal dominant motor system degeneration. *Mov. Disord.* 7:193–203

Rosenberg RN. 1995. Autosomal dominant cerebellar phenotypes: the genotype has settled the issue. *Neurology* 45:1–5

Ross CA, McInnis MG, Margolis RL, Li S-H. 1993. Genes with triplet repeats: candidate mediators of neuropsychiatric disorders. *Trends Neurosci.* 16:254–60

Sabouri LA, Mahadevan MS, Narang M, Lee DSE, Surh LC, et al. 1993. Effect of the myotonic dystrophy (DM) mutation on mRNA levels of the DM gene. *Nat. Genet.* 4:233–38

Sarnat HB, Silbert SW. 1976. Maturational arrest of fetal muscle in neonatal myotonic dystrophy. *Arch. Neurol.* 33:466–74

Schut JW. 1954. Hereditary ataxia: clinical study through six generations. *Arch. Neurol. Psychiatry* 63:535–67

Servadio A, Koshy B, Armstrong D, Antalffy B, Orr HT, Soghbi HY. 1995. Expression analysis of the ataxin-1 protein in tissues from normal and spinocerebellar ataxia type 1 individuals. *Nat. Genet.* 10:94–98

Sharp AH, Loev SJ, Schilling G, Li S-H, Li X-J, et al. 1995. Widespread expression of Huntington's disease gene (IT15) protein product. *Neuron* 14:1–20

Sherman SL, Jacobs PA, Morton NE, Froster-Iskenius U, Howard-Peebler PN, et al. 1985. Further segregation analysis of the fragile X syndrome with special reference to transmitting males. *Hum. Genet.* 69:289–99

Siomi H, Siomi MC, Nussbaum RL, Dreyfuss G. 1993. The protein product of the fragile X gene, FMR-1, has characteristics of an RNA binding protein. *Cell* 74:291–98

Simpson L, Emeson RB. 1996. RNA editing. *Annu. Rev. Neurosci.* 19:In press

Smith JK, Gonda VE, Malamud N. 1958. Unusual form of cerebellar ataxia: combined dentatorubral-pallidoluysian atrophy. *Neurology* 32:798–807

Snell RG, MacMillan JC, Cheadle JP, Fenton I, Lazarou PL, et al. 1993. Relationship between trinucleotide repeat expansion and phenotypic variation in Huntington's disease. *Nat. Genet.* 4:393–97

Sobue G, Hashizume Y, Mukai E, Hirayama M, Mitsuma T, et al. 1989. X-linked recessive bulbospinal neuronopathy: a clinico-pathological study. *Brain* 112:209–32

Stott K, Blackburn JM, Butler PJG, Perutz M.

1995. Incorporation of glutamine repeats makes protein oligomerize: implications for neurodegenerative diseases. *Proc. Natl. Acad. Sci. USA.* In press

Strong TV, Tagle DA, Valdes JM, Elmer LW, Boehm K, et al. 1993. Widespread expression of the human and rat Huntington's disease gene in brain and nonneural tissues. *Nat. Genet.* 5:259–65

Sutcliffe JS, Nelson DL, Zhang F, Pieretti M, Caskey CT, et al. 1992. DNA methylation represses FMR-1 transcription in fragile X syndrome. *Hum. Mol. Genet.* 1:397–400

Sutherland GR, Richards RI. 1993. Dynamic mutations on the move. *J. Med. Genet.* 30: 978–81

Takahashi H, Okama E, Naito H, Takeda S, Nakashima S, et al. 1988. Hereditary dentatorubral-pallidoluysian atrophy: clinical and pathologic variants in a family. *Neurology* 38:1065–70

Telenius H, Kremer B, Goldberg YP, Theilman J, Andrew SE, et al. 1994. Somatic and gonadal mosaicism of the Huntington disease gene CAG repeat in brain and sperm. *Nat. Genet.* 6:409–14

Thornton CA, Griggs RC, Moxley RT III. 1994. Myotonic dystrophy with no trinucleotide repeat expansion. *Ann. Neurol.* 35: 269–71

Trottier Y, Devys D, Imbert G, Saudou F, An I, et al. 1995. Cellular localization of the Huntington's disease protein and discrimination of the normal and mutated form. *Nat. Genet.* 10:104–10

Tsilfidis C, MacKenzie AE, Mettler G, Barcelo J, Korneluk RG. 1992. Correlation between CTG trinucleotide repeat length and frequency of severe congenital myotonic dystrophy. *Nat. Genet.* 1:192–95

van der Ven PF, Jansen G, van Kuppevelt TH, Perryman MB, Lupa M, et al. 1993. Myotonic dystrophy kinase is a component of neuromuscular junctions. *Hum. Mol. Genet.* 2:1889–94

Verheij C, Bakker CE, de Graff E, Keulemana J, Willemsen R, et al. 1993. Characterization and localization of the FMR-1 gene product associated with fragile-X syndrome. *Nature* 363:722–24

Verkerk AJ, Pieretti M, Sutcliffe JS, Fu YH, Kuhl DP, et al. 1991. Identification of a gene (FMR-1) containing a CGG repeat coincident with a breakpoint cluster region exhibiting length variation in fragile X syndrome. *Cell* 65:905–14

Vonsattel JP, Myers RH, Stevens TJ, Ferrante RJ, Bird ED, et al. 1985. Neuropathological classification of Huntington's disease. *J. Neuropath. Exp. Neurol.* 44:559–77

Warner CL, Griffen JE, Wilson JD, Jacobs LD, Murray KR, et al. 1992. X-linked spinomuscular atrophy; a kindred with associated abnormal androgen binding. *Neurology* 42: 2181–84

Wexler NS, Rose EA, Housman DE. 1991. Molecular approaches to hereditary diseases of the nervous system: Huntington's disease as a paradigm. *Annu. Rev. Neurosci.* 14:503–29

Wharton KA, Yedvobnick B, Finnerty VG, Artavanis-Tsakonas S. 1985. *opa:* a novel family of transcribed repeats shared by the Notch locus and other developmentally regulated loci in *D. melanogaster. Cell* 40:55–62

Wieringa B. 1994. Myotonic dystrophy reviewed: back to the future? *Hum. Mol. Genet.* 3:1–7

Wisniewski KE, Segan SM, Miezejeski CM, Serson EA, Rudelli RD. 1991. The fragile X syndrome: neurological, electrophysiological, and neuropathological abnormalities. *Am. J. Med. Genet.* 38:476–80

Wohrle D, Hennig I, Vogel W, Steinbach P. 1993. Mitotic stability of fragile X mutations in differentiated cells indicates early postconceptional trinucleotide repeat expansion. *Nat. Genet.* 4:140–42

Wohrle D, Kotzot D, Hirst MC, Manca A, Korn B, et al. 1992. A microdeletion of less than 250 kb, including the proximal part of the *FMR-1* gene and the fragile X site, in a male with the clinical phenotype of fragile X syndrome. *Am. J. Hum. Genet.* 51:299–306

Yazawa I, Nukina N, Hashida H, Goto J, Yamada M, Kanazawa I, et al. 1995. Abnormal gene product identified in hereditary dentatorubral-pallidoluysian atrophy (DRPLA) brain. *Nat. Genet.* 10:99–103

Yu W-HA. 1989. Administration of testosterone attenuates neuronal loss following axotomy in the brain-stem motor nuclei of female rats. *J. Neurosci.* 9:3908–14

Zhang L, Leeflang EP, Yu JY, Arnheim N. 1994. Studying human mutations by sperm typing: instability of CAG trinucleotide repeats in the human androgen receptor gene. *Nat. Genet.* 7:531–35

Zhou Z-X, Wong C-I, Sar M, Wilson EM. 1994. The androgen receptor: an overview. *Recent Prog. Hormone Res.* 49:249–74

*Annu. Rev. Neurosci. 1996. 19:109–39*

# INFEROTEMPORAL CORTEX AND OBJECT VISION

*Keiji Tanaka*

The Institute of Physical and Chemical Research (RIKEN), 2-1 Hirosawa, Wako-shi, Saitama, 351-01, Japan

KEY WORDS:    macaque monkey, extrastriate visual cortex, object vision, optical imaging, population coding

---

### ABSTRACT

Cells in area TE of the inferotemporal cortex of the monkey brain selectively respond to various moderately complex object features, and those that cluster in a columnar region that runs perpendicular to the cortical surface respond to similar features. Although cells within a column respond to similar features, their selectivity is not necessarily identical. The data of optical imaging in TE have suggested that the borders between neighboring columns are not discrete; a continuous mapping of complex feature space within a larger region contains several partially overlapped columns. This continuous mapping may be used for various computations, such as production of the image of the object at different viewing angles, illumination conditions, and articulation poses.

---

## Introduction

Recognizing objects by their visual images is a key function of the primate brain. This recognition is not a template matching between the input image and stored images but a flexible process in which considerable change in images—due to different illumination, viewing angle, and articulation of the object—can be tolerated. In addition, our visual system can deal with images of novel objects, based on previous visual experience of similar objects. Generalization may be an intrinsic property of the primate visual system. In this article, I discuss the neural organization essential for these flexible aspects of visual object recognition in the anterior part of the inferotemporal cortex.

The inferotemporal cortex (IT) of the monkey brain has been divided into subregions in several different manners. Our own division into posterior IT and anterior IT, based on the size of the receptive fields and the properties of responses (Tanaka et al 1991, Kobatake & Tanaka 1994), roughly corresponds to the previous cytoarchitectural division into TEO and TE (Iwai & Mishkin 1967; von Bonin & Bailey 1947, 1950): Posterior IT corresponds to TEO, and

109

anterior IT to TE. I use TEO and TE in this article because they are more popular.

TE receives visual information from the primary visual cortex (V1) through a serial pathway, which is called the ventral visual pathway (V1-V2-V4-TEO-TE). Although there are also jumping projections, such as that from V2 to TEO (Nakamura et al 1993) and that from V4 to the posterior part of TE (Saleem et al 1992), the step-by-step projections are more numerous. The IT projects to various brain sites outside the visual cortex, including the perirhinal cortex (areas 35 and 36), the prefrontal cortex, the amygdala, and the striatum of the basal ganglia. The projections to these targets are more numerous from TE, especially from the anterior part of TE, than from the areas at earlier stages (Iwai & Yukie 1987, Ungerleider et al 1989, Saleem et al 1993a, Cheng et al 1993, Suzuki & Amaral 1994). Therefore, there is a sequential cortical pathway from V1 to TE, and outputs from the pathway originate mainly in TE.

Monkeys that have had their TE bilaterally ablated showed severe but selective deficits in learning tasks that required the visual recognition of objects (Gross 1973, Dean 1976). These behavioral results, together with the above-described important anatomical position of TE, suggest that TE is the site of neural organization essential for the flexible properties of visual object recognition.

In this review, our own data are emphasized, and the citation of other references is selective. This selection is not based only on the value of the studies but also on their relevance to the subject. The readers should read other reviews to get an overview of studies in the IT, e.g. Rolls (1991), Miyashita (1993), Gross (1994), and Desimone et al (1994). In particular, mechanisms of short-term memory of object images are not discussed in this article. I first summarize the data from unit-recording experiments to show that cells in TE respond to moderately complex object features and that those that cluster in a columnar region respond to similar features. I then consider the process by which the selectivity is formed in the afferent pathways to TE. I introduce the data of optical imaging of TE in order to discuss the function of the TE columns. Finally, I consider how the concept of the object emerges in the brain. The selections of our recordings that are introduced in this article were all conducted in anesthetized preparation, and they were from the lateral part of TE, lateral to the anterior middle temporal sulcus (AMTS). This part is often referred to as TEd (dorsal part of TE).

## Stimulus Selectivity of Cells in TE

One obstacle in the study of neuronal mechanisms of object vision has been the difficulty in determining the stimulus selectivity of individual cells. There is a great variety of object features in the natural world, and we do not know how the brain scales down the dimension of this variety.

Single-unit recordings from TE were initiated by Gross and his colleagues (Gross et al 1969, 1972). They found that cells in TE had large receptive fields, most of which included the fovea, and that some cells responded specifically to a brush-like shape with many protrusions or to the silhouette of a hand. They extended the study of the stimulus selectivity by using two different methods: a constructive method and a reductive one. In the constructive method, they used Fourier descriptors that were defined by the number (frequency) and amplitude of periodic protrusions from a circle. Any contour shape can be reconstructed by linearly combining elementary Fourier descriptors of single frequency and amplitude. Some cells responded specifically to Fourier descriptors of a particular range of frequencies, with a considerable invariance for the overall size of the stimulus (Schwartz et al 1983). This method was not very promising, however, because the same group of authors found that the response of a TE cell to a composite contour was far from the linear combination of its responses to the elementary component contours (Albright & Gross 1990). Fourier descriptors are not the basis functions that the IT uses for the representation of object images.

The other direction that Gross's group pursued was reductive. They first presented many object stimuli for individual cells in order to find effective stimuli. Next, the images of the effective stimuli were simulated by paper cutouts to determine which features were critical for the activation (Desimone et al 1984).

We expanded this latter method and have developed a systematic reduction method with the aid of a specially designed image-processing computer system (Tanaka et al 1991; Fujita et al 1992; Kobatake & Tanaka 1994; Ito et al 1994, 1995). After spike activities from a single cell were isolated, many three-dimensional (3D) animal and plant imitations were presented to find the effective stimuli. Different aspects of the objects were presented with different orientations. Next, the images of the effective stimuli were recorded with a video camera and presented on a TV monitor by the computer to determine the most effective stimulus. Finally, to determine which feature or combination of features contained in the image was essential for the maximal activation, the image of the most effective object stimulus was simplified step-by-step while the activity of the cell was monitored. The minimal combination of features that evoked the maximal activation was determined as the critical feature for the cell. Figure 1 exemplifies the process for a cell for which the effective stimulus was reduced from the view of a water bottle to a combination of a vertical ellipse and a downward projection from the ellipse.

After the reduction was completed, the image was modified so that the selectivity could be further examined. Figure 2 exemplifies this latter process

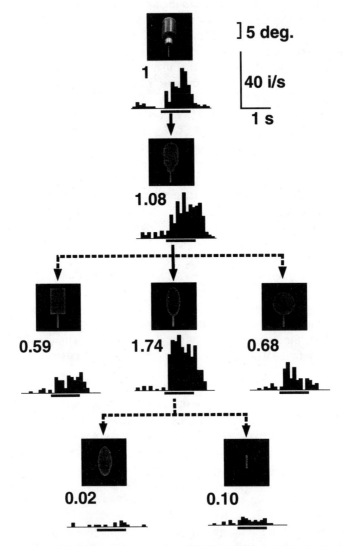

*Figure 1*  An example of the reduction process to determine the feature critical for the activation of cells in the ventral visual pathway. The responses were averaged over ten repetitions of the stimuli. The underlines indicate the period of stimulus presentation, and the numbers above histograms indicate the magnitude of the responses normalized by the response to the image of a water bottle.

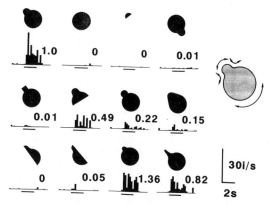

*Figure 2*  An example of further study of the selectivity after the reduction process was completed. This cell was recorded from TE.

for a cell, which is one of the cases in which the domain of selectivity was most clearly determined. The cell responded maximally to a pear model within the routine set of object stimuli, and the critical feature was determined as a rounded protrusion in the 10 o'clock direction from a rounded body with a concave smooth neck. The body or the head by itself did not evoke any responses (the first line in Figure 2). The head had to be rounded because the response disappeared when the rounded head was replaced by a square, and the body had to be rounded because the response decreased by 51% when the body was cut in half (the second line, left). The neck had to be smooth and concave because the response decreased by 78 or 85% when the neck had sharp corners or was straight (the second line, right). The critical feature was neither the right upper contour alone nor the left lower contour alone, because either half of the stimulus did not evoke responses (the third line, left). The width and length of the projection were not very critical (the third line, right).

By determining the critical features for hundreds of cells in TEd, we concluded that most cells in TEd required moderately complex features for their activation, e.g. the 16 examples in Figure 3. The critical features were more complex than orientation, size, color, and simple texture, which are known to be extracted and represented by cells in V1. Some of the features were moderately complex shapes, while others were combinations of such shapes with color or texture. Responses were selective for the contrast polarity of the shapes: The contrast reversal of the critical feature reduced the response by >50% in 60% of tested cells, and replacement of the solid critical features by

*Figure 3*   Sixteen examples of the critical features of cells in TE. They are moderately complex.

line drawings of the contour reduced the response by >50% in 70% of tested cells (Ito et al 1994).

## Invariance or Selectivity for Position, Orientation, and Size

Gross et al (1972), in their pioneering experiments, found that cells in TEd had large receptive fields across which a stimulus kept evoking responses. By using a set of shape stimuli composed of the critical feature independently determined for individual cells by the reduction method and several other shape stimuli made by modifying the critical feature, we demonstrated that the selectivity for shape was mostly preserved over the large receptive fields (Ito et al 1995). In contrast, responses of TEd cells to their critical features were more selective for the orientation in the frontoparallel plane (Tanaka et al 1991) and for the size of the stimuli (Ito et al 1995).

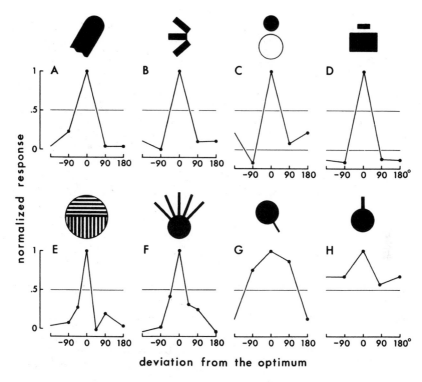

*Figure 4*  Tuning of responses of eight TE cells for the orientation of stimulus.

Figure 4 shows the data of eight cells for the tuning of responses for the orientation in the frontoparallel plane. The eight cells were selected so that they represent the general properties of cells in TEd. Rotation of the critical feature by 90° decreased the response by >50% for most cells (Figure 4A–F). The tuning of the remaining cells was broader: The response was reduced by >50% by a rotation of 180° (Figure 4G), or the cell showed <50% change (Figure 4H).

The effects of changes in stimulus size varied more among cells than did position or orientation. Twenty-one percent of TEd cells tested responded to size ranges of more than four octaves of the critical features, whereas 43% responded to size ranges of less than two octaves. Tuning curves of four TEd cells, two from each group, are shown in Figure 5. The tuned cells may be in the course of constructing the size-invariant responses, or alternatively, there are both size-dependent and -independent processing of images in TEd.

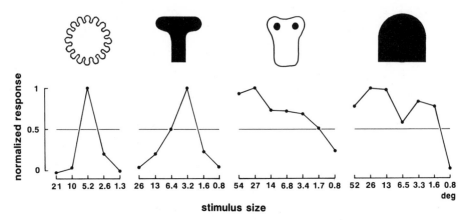

*Figure 5*  Tuning of responses of four TE cells for the size of stimulus.

## Columnar Organization in TE

How are cells with various critical features distributed in TE? Is there a columnar organization like that found in V1? By simultaneously recording, with a single electrode, from two or more TEd cells, we have found that cells located at nearby positions in the cortex have a similar stimulus selectivity (Fujita et al 1992). The critical feature of one isolated cell was determined by using the procedure described above, while responses of another isolated cell, or nonisolated multiunits, were simultaneously recorded. In most cases, the second cell responded to the optimal and suboptimal stimuli of the first cell. The selectivity of the two cells varied slightly, however, in that the maximal response was evoked by slightly different stimuli, or the mode of the decrease in response was different when the stimulus was changed from the optimal stimulus. Figure 6 shows an example of the latter case.

To determine the spatial extent of the clustering of cells with similar selectivity, we examined responses of cells recorded successively along long penetrations made perpendicular or oblique to the cortical surface (Fujita et al 1992). The critical feature of a cell located at the middle of the penetration was determined first. A set of stimuli, including the optimal feature for the first cell, its rotated versions, and ineffective control stimuli, were made, and cells recorded at different positions along the penetration were tested only with the fixed set of stimuli. As in the example shown in Figure 7, cells recorded along the perpendicular penetrations commonly responded to the critical feature of the first cell or some related stimuli. The span of the commonly responsive cells covered nearly the entire thickness from layer 2 to 6. The

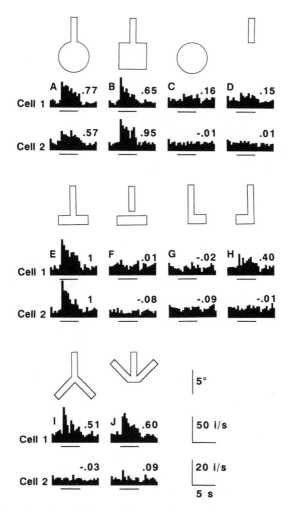

*Figure 6*  An example of simultaneous recording from two nearby neurons in TE.

situation was different in the penetrations made oblique to the cortical surface. The cells that were commonly responsive to the critical feature of the first cell or its related versions were limited to a short span around the first cell. The horizontal extent of the span averaged 400 μm. The cells outside the span either were not responsive to any of the stimuli included in the set or responded to some stimuli that were not effective for the first cell and were included in the set as ineffective control stimuli.

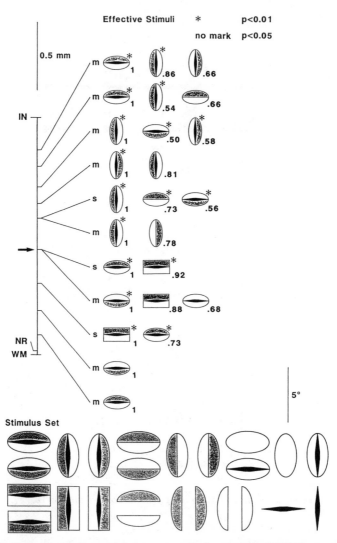

*Figure 7* Responses of cells recorded along a perpendicular penetration in TE. The responsiveness of the cells was tested with the set of stimuli shown at the bottom, which were made in reference to the critical feature of the first cell indicated by the arrow. Effective stimuli are listed separately for individual recording sites, in the order of effectiveness. "m" indicates recording from multiunits, and "s" from single units.

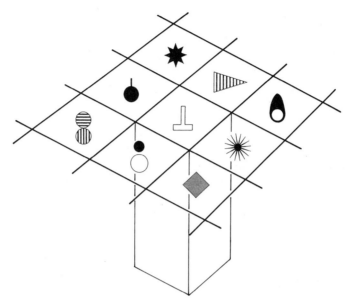

*Figure 8*   Schematic drawing of columnar organization in TE.

The TEd region is thus composed of columnar modules in which cells respond to similar features (Figure 8). Cells within the same column respond to similar features, but cells in different columns respond to different features. The width of a columnar module across the cortical surface may be slightly greater than 400 μm. The span of a column along an oblique penetration should be smaller than the real size of the column if the penetration crosses its periphery. The number of modules, which was estimated by a division of the whole surface area of TEd into 500 × 500 μm squares, was 1300.

## Organization of Afferents to TE

The selective responses to complex features, which were first found in TE cells, have been traced to earlier stages in the afferent pathway to TE. We have found that cells requiring such complex features for the maximal activation were already present in TEO and V4 (Kobatake & Tanaka 1994), although their proportion was small. Gallant et al (1993) also found that there were cells that responded preferentially to concentric or hyperbolic stripes rather than to straight stripes. The optimal features that we found in these areas were much more divergent, though they included concentric stripes.

To compare such cells in TEO and V4 with cells in TE, we compared responses of individual cells to a fixed set of stimuli of simple features to their responses to individually determined critical features (Kobatake & Tanaka 1994). The set stimuli were composed of 16 bars of 4 different orientations with a 45° interval and 2 different sizes (0.5° by 10° and 0.5° by 2°) and 16 colored squares of 4 different colors and 2 different sizes (0.5° by 0.5° and 2.5° by 2.5°). Stimuli both darker than and lighter than the background were included. This set evoked some good, though not maximal, responses in cells in V2 and V4 that showed selectivity only in the domain of orientation or color, and size. Either most cells in TE did not respond to the simple stimuli included in the set, or the responses were negligible compared with their responses to the individually determined complex critical features (Figure 9, *left*), whereas cells in TEO and V4 showed divergent properties in responses to the simple stimuli. Some TEO and V4 cells showed no or negligible responses to any of the simple stimuli as cells in TE; some others showed moderately strong responses to some of the simple stimuli in addition to the maximum response to the complex critical features (Figure 9, *right*); and the remaining cells even maximally responded to some of the simple stimuli. Figure 10 shows the proportion of three groups of cells that were classified by setting arbitrary borders at 0.25 and 0.75 in the magnitude of the maximum response to the simple stimuli normalized by the overall maximum response of the cell.

TEO and V4 were thus characterized by the mixture of cells with various levels of selectivity. We may take this mixture of various cells as evidence that selectivity is constructed through local networks in these regions. If we randomly sample cells from a local network in which the selective responses to complex features are constructed by integrating simple features, the sample should include cells with various levels of selectivity. Cells located close to the input end should be maximally activated by simple features, cells close to the output end should respond only to the complex features, and cells at intermediate stages should show some intermediate properties. The areas that satisfy this condition were TEO and V4. It may be proposed that the selectivity to features of medium complexity is mainly constructed in local networks in TEO and V4.

Although we did not find clear evidence of selectivity to complex features in V2, slightly stronger responses to a complex pattern than the best simple stimulus (bar or grating) may not be unusual among cells in V1 (Lehky et al 1992). Selectivity to moderately complex features may gradually develop throughout the lower stages and become apparent in V4 and TEO.

The anatomical organization of the forward projection from TEO to TE is consistent with the idea that the selectivity to moderately complex features is already developed in the circuit up to TEO. We injected an anterograde tracer, PHA-L, into a single small region (the horizontal width of the injection sites

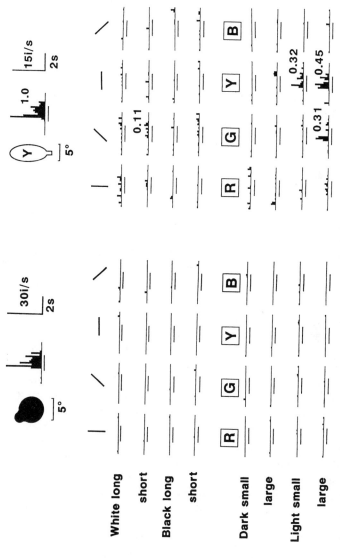

*Figure 9* Responses of a TE cell (*left*) and TEO cell (*right*) to a set of simple stimuli. Their responses to individually determined critical features are shown at the top.

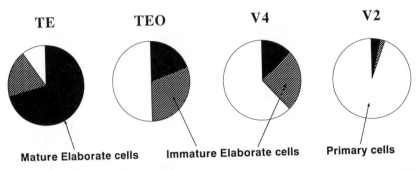

*Figure 10* Proportion of cells with different levels of selectivity to complex features. The maximum response to the simple stimuli was <0.25 of the response to the complex critical feature for mature elaborate cells, between 0.25 and 0.75 for immature elaborate cells, and >0.75 for primary cells.

was 330 to 600 µm) of the part of TEO representing the central visual field, and observed labeled axon terminals in TE (Saleem et al 1993b). Labeled terminals were nearly limited to three to five focal regions in TE (Figure 11). In each of the projection foci, the labeled terminals were not limited to the middle layers, but were distributed to form a columnar region encompassing from layer 1 to 6. The horizontal width of the columnar foci was 200 to 380 µm, which was slightly smaller than the physiologically determined width of columns in TEd. As noted above, the receptive fields of cells in TE are large and usually include the fovea; no retinotopical organization has been found in TE. Thus, the specificity of the TEO to TE connections should be defined in the feature space but not in the retinotopical space. Outputs from a single site of TEO may carry information about a particular complex feature, and they are sent, therefore, only to a limited number of foci in TE.

Although the overall distribution of labeled terminals was elongated to form a columnar region, this does not necessarily mean that individual axons make arbors in columnar regions. Indeed, single axons reconstructed from serial sections were heterogenous in shape. Some axons terminated exclusively in layer 4 and the bottom of layer 3. Some other axons terminated only in layers 1 and 2. Single axons also terminated in elongated regions, including both the middle and superficial layers. A single site in TEO may send many different kinds of information about a complex feature to a column in TE, and the different kinds of information interact with each other through the local network within the TE column.

Thus, there are two things first achieved in TE: One is the columnar organization, namely, arrangement of cells with overlapping and slightly different selectivity in local columnar regions; and the other is the invariance of responses for the stimulus position. The receptive fields of cells in TE are large

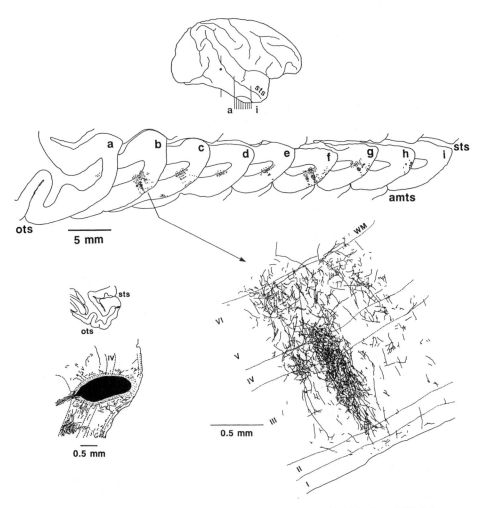

*Figure 11*   Distribution of labeled axon terminals in TE after a single focal injection of PHA-L into TEO.

and include the fovea, and the selectivity of responses is essentially constant throughout the large receptive fields. A significant part of cells in TEO and V4 respond to moderately complex stimuli, as in TE. However, the receptive fields of the cells in TEO and V4 are still much smaller than those of cells in TE and are retinotopically organized (Boussaoud et al 1991, Kobatake & Tanaka 1994). This means that there are two steps in the formation of cells responding to integrated features with invariance to changes in stimulus position. First, the selectivity is constructed for stimuli at a particular retinal position in TEO and V4; then, the invariance is achieved in TE by obtaining inputs of the same selectivity but with the receptive fields at different retinal positions.

One problem with this two-step structure is that individual cells or columns in TE each require a set of input cells with receptive fields distributed over the large receptive fields of the TE cells. Because the central visual field is magnified in TEO (Boussaoud et al 1991, Kobatake & Tanaka 1994), cells in the peripheral TEO may not be sufficiently numerous to extract the great number of integrated features. I would raise the possibility that the inputs from the peripheral TEO convey information on primitive features and that the selectivity in the periphery is constructed at synapses of TE cells. The selectivity can be generalized in TE cells from the central to the peripheral visual field by modifying the synapses of the peripheral inputs according to the generalized Hebbian rule (Földiák 1991) whenever the object containing the critical feature moves from the center to the periphery in the visual field. The selective inputs from the central TEO are used as seeds. This hypothesis can be tested by examining properties of responses of cells in the part of TEO representing the peripheral visual field.

## Optical Imaging of the Columnar Organization

To further study the spatial properties of columnar organization in TE, we have introduced the technique of optical imaging (Wang et al 1994). The intrinsic signals, which are thought to originate in the increase of deoxidized hemoglobin in capillaries around elevated neuronal activities (Frostig et al 1990), were measured. The cortical surface was exposed and illuminated with red light tuned to 605 nm with 10-nm bandwidth. Activated neuronal tissue takes oxygen from hemoglobin, so the density of deoxidized hemoglobin in nearby capillaries increases. Because deoxidized hemoglobin absorbs much more light than oxidized hemoglobin at that wavelength, the region of cortex with elevated neuronal activities becomes darker in the reflected image.

To find visual stimuli that would be effective for the part of TE later exposed for optical imaging and to establish the relation between the optical changes and elevated neuronal activities in TE, we combined electrophysiological recordings from single cells with the optical imaging. The single-cell recordings were conducted in separate sessions prior to the optical imaging session. The critical features were determined for 15 to 25 cells recorded in 6 to 8 penetrations at different sites. The optical imaging was performed with the critical features and some other control stimuli. Five to 25 different stimuli were used and each of them was presented 24 or 40 times each for 4 s. The raw images for individual stimuli were divided by the image obtained while

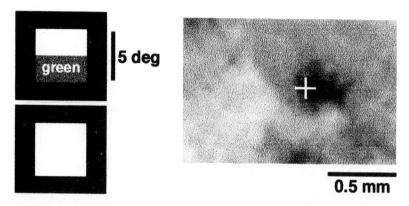

*Figure 12*  Optical imaging of a column responding to a combination of a green rectangle and white one. The image was obtained by dividing the image obtained while the monkey saw the stimulus by the image during stimulation with a white square. The cross mark indicates the site of the electrode penetration at which the stimulus was determined as the critical feature of one cell.

the monkey saw a blank screen without stimuli to remove the basic level unrelated with the visual stimuli, and the images for the critical features were divided by those for their null component stimuli that were not effective for the activation of cells to remove the activation due to the component features.

Each of the critical features determined in the preceding unit-recording sessions activated one to seven dark spots within the imaged region of TEd (3.3 mm by 6.1 mm). The locations of the spots were different for different features, and one of the spots covered the site of the electrode penetration from which the critical feature was determined (Figure 12). The average diameter of individual spots was 490 μm, which roughly coincided with the width of columns in TE inferred by the unit-recording experiments (Fujita et al 1991). The features should have activated a large proportion of cells within the region to evoke the observable metabolic change. Thus, clustering of cells that responded to a moderately complex feature was confirmed.

The set of visual stimuli used in one block of optical imaging included three critical features, which were determined for three different cells recorded in the same penetration. Two of them were combinations of two regions with different luminosity of the same or similar color, and the third one included a gradation of luminosity (Figure 13). The three stimuli all evoked dark spots around the site of the electrode penetration. All of the spots covered the site of the penetration, but they extended to different directions from the site (Figure 13). Each of the spots was about 500 μm in size, and the size of the overall region was 1100 μm. The three stimuli are similar to each other in that they commonly contain a change in luminosity.

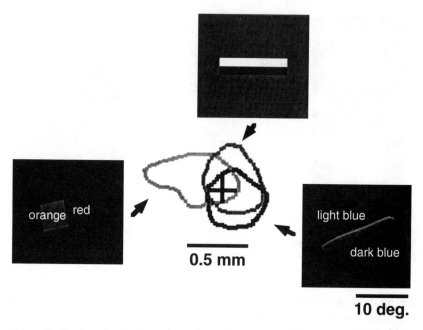

*Figure 13* Overlap of activation spots evoked by three critical features determined for three different cells recorded in the same penetration. Only the outlines of the spots at 1/*e* drop from the peak are shown. The cross mark indicates the site of the electrode penetration.

Partially overlapped activation by similar stimuli was also observed with a series of faces of different viewing angles. All of the five cells recorded in an electrode penetration selectively responded to the sight of a face. The image of a face could be simplified to a combination of the eyes and nose for one of them, but we failed to find simpler stimuli for the remaining four cells. Three of the five maximally responded to front faces, and the other two maximally responded to profiles. In the optical imaging experiment, all of the face stimuli evoked activation spots around the penetration (we failed to recover the exact location of the electrode penetration in this case), and the center position of the spots systematically moved in one direction as the face turned from the left profile to the right profile through the front and 45° faces. Individual spots were 300 to 400 μm in size and the overall region covered by them was 800 μm in the long axis along which the center of spots moved.

These facts may suggest that several columns that represent different but related features overlap with one another and as a whole compose a larger-scale unit in TEd. The face case further suggests that the arrangement of features within the larger-scale unit has some rule, that is, some space of complex

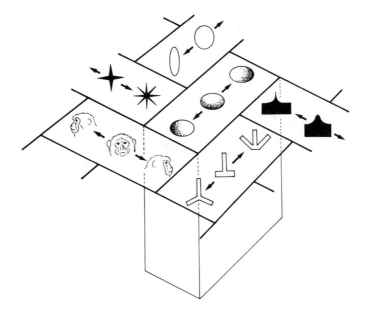

*Figure 14*   Revised scheme of columnar organization in TE.

features is continuously mapped (Figure 14). Whether the mapping is continuous throughout a large part of TEd or discontinuous between the units that are probably around 1 mm in size is still unknown. Considering that the dimension of the feature space that TEd should represent is so high, the latter is more likely.

## Changeability of the Selectivity in the Adult

The selectivity to complex critical features can change as a result of changes in the visual environment in the adult. We have trained two adult monkeys to discriminate 28 moderately complex shapes with a stand-alone apparatus—including a display, a computer, and a touch screen—by using a task similar to the delayed-matching-to-sample (Kobatake et al 1992, 1993). One stimulus, randomly selected from the 28 stimuli, appeared on the display (sample stimulus) and disappeared by the monkey's touching it. After a 16-s delay with a blank screen, 5 shapes, including the sample, appeared on the display. The monkey had to select the sample and touch it to get a drop of juice. After a year of training, the monkeys were prepared for repeated recordings. The recordings were performed from cells in TEd under anesthesia. We determined for individual cells the best stimulus from the set of animal and plant models that we had previously prepared to investigate the critical features in naive monkeys. The response to the best object stimulus was then compared with responses of the same cell to the shape stimuli used in the training. We did not perform the reduction process in this experiment for the sake of time.

In TE of the trained monkeys, about 25% of cells gave a maximum response to some of the stimuli used in the training. Conversely, 5% of TE cells in untrained animals responded maximally to these stimuli. These results indicate that the number of cells that responded to training stimuli increased owing to the 1-year-long discrimination training. However, the spatial organization of the modified cells in the cortex has yet to be studied. We do not know whether new columns were formed for the discrimination of training stimuli or whether cells distributed over many columns present since before the training were tuned to the training stimuli. Whether or not similar changes happened in TEO and V4, and the time course of the change also are still unknown.

Sakai & Miyashita (1991) have shown effects of discriminatory training in the adult on the stimulus selectivity of TE cells, although indirectly. The task paradigm they used in the training was associative pair matching. The stimuli were composites of Fourier descriptors. They arbitrarily made 12 pairs of stimuli, and the monkeys were asked to select the member of the pair in response to the other member of the pair. One of the stimuli appeared as the sample, and after a delay period, two stimuli, including the stimulus paired to the sample, appeared. The monkey had to touch the paired stimulus to get a reward. After training for about a month, through which the association was learned, recording of TE cells was started by using the same task paradigm. Some cells responded to the two stimuli composing the pairs, and the pairing was shown to be significantly more frequent than that expected by chance. Considering that the pairs of stimuli were made arbitrarily from infinite possibilities, the dual responsiveness of the cells to the paired stimuli should have been formed through the adult learning.

There are two studies that show that responses of cells in TE and surrounding regions change within the course of recording from the same single cell. Miller and colleagues (Miller & Desimone 1991, Li et al 1993) found that as the newly introduced stimulus became familiar, responses of cells at the border region between TE and the perirhinal cortex to the stimuli gradually decreased. This effect is discriminated from the habituation of responses to successive presentation of the same stimulus, because the decrease occurred even after several intervening presentations of different stimuli. However, because the changes that Miller and colleagues observed were opposite to those we observed after the long training, the two phenomena are not likely to be related. Rolls et al (1989) found that responses of cells in TE and the ventral bank of the superior temporal sulcus to faces changed rather rapidly after an introduction of a new set of faces. The changes of the responses to faces included both an increase and a decrease of the responses. Because the immediate changes that Rolls et al observed were relative changes among responses to similar stimuli (faces), they may be different from the changes of stimulus selectivity after long-lasting training, which should be changes among more different stimuli.

## Functions of the TE Columns

The columnar organization suggests that an object feature is not represented by activity of a single cell but by the activity of many cells within a single columnar module. Representation by multiple cells in a columnar module, in which the selectivity varied from cell to cell while effective stimuli largely overlapped, can satisfy two apparently conflicting requirements in visual recognition: robustness to subtle changes in input images and preciseness of representation. Whereas the image of an object projected to the retina changes owing to changes in illumination, viewing angle, and articulation of the object, the global organization of outputs from TE should be little changes. The clustering of cells with overlapping and slightly different selectivity works as a buffer to absorb the changes.

The representation by multiple cells with overlapping selectivity can be more precise than a mere summation of representation by individual cells. A similar argument has been made for hyperacuity (Erickson 1968, Snippe & Koenderink 1992). The position of the receptive fields changes gradually in the retina, with a large overlap among nearby cells. By taking the difference between the activity of nearby cells, an acuity much smaller than the size of the receptive fields is produced. A similar mechanism to that in retinal space may work in feature space with largely overlapping and gradually changing selectivity, as suggested by Edelman (1995). A subtle change in a particular feature, which does not markedly change the activity of individual cells, can be coded by the differences in the activity of cells with overlapping and slightly different selectivity.

The function of the columnar organization in TEd may go beyond the discrimination of input images. The optical imaging experiments suggested that there is a continuous mapping of features within cortical units about 1 mm in size across the cortical surface. There may be a twofold functional significance to this continuous mapping. First, an evenly distributed variety of cell properties is made along the feature axis. The continuous mapping may be a tool to make the full divergence without omission (Malach 1994, Purves et al 1992). Second, computations are conducted involving the varied features based on the local neuronal connections between the cells representing the varied features. The computations may be to transfer the image of an object for 3D rotations or production of the image at different illumination conditions or at different articulation poses.

A series of studies recently performed in slices of the rat motor cortex suggest that there are two kinds of connections between pyramidal neurons through their axon collaterals (Thomson & Deuchars 1994). Pyramidal cells located within a narrow columnar region 50 to 100 μm in width are tightly connected by synapses on the basal dendrites or the proximal part of the apical dendrites. They tend to fire together by sharp-rising big excitatory postsynaptic potentials (EPSPs) exerted through these connections. Another anatomical structure gives similar response properties to pyramidal cells within a narrow column. The shafts of their apical dendrites get close to compose a bundle, on which input axons may make synapses without discrimination (Peters & Yilmaz 1993). This narrow column corresponds to the "minicolumn" of Mountcastle (1978). In contrast, pyramidal cells with a longer horizontal distance are connected by synapses at the distal part of the apical dendrites. The EPSPs are small and slowly rising, although they are long-lasting: They may contain NMDA-type glutamate receptors.

Taken together, we may draw a schema of area TE that cells within the minicolumn compose a unit by receiving common inputs and exciting each other and that nearby minicolumns exert long-lasting but weak excitatory inputs to each other. After a minicolumn is activated by the retinal visual input, subthreshold activation propagates from it to nearby minicolumns, forming a pattern of activation with a focus. The focus of activation may move from one minicolumn to another, through interaction with distant activation foci in TE or interaction with the other brain sites (as will be described in the section on object recognition by activities distributed over the brain). This mechanism may be used for various kinds of computation that the visual system has to conduct to realize the flexibility of visual recognition, such as transfer of the image of an object for 3D rotations, production of the image at different illumination conditions, and in the case of faces, production of the image with different expressions. Thus, the columnar organization of TE may provide an overlapping and continuous representation of object features, upon which various kinds of calculations can be performed.

## Binding Activities in Distant Columns

Because object features to which individual TE cells respond are only moderately complex and because cells within a single column respond to similar features, the calculation performed within a column can provide only information on partial (but not necessarily local) features of object images. To represent the whole image of an object, calculation in several different columns must be combined. This evokes the problem of *binding*, that is, how to discriminate different sets of activity when there are more than two objects in the nearby retinal positions. The receptive fields of TE cells are too large to discriminate different objects according to their retinal positions. I examine in a later section whether there are single cells anywhere in the brain that represent the concept of objects through their activity alone. The problem of binding exists regardless of the presence or absence of such concept units in brain sites beyond TE. The concept units, if present, have to discriminate different sets of TE activity originating in different objects.

One possible mechanism to solve the problem is the synchronization of firings (Engel et al 1992, Singer 1993). If firing of cells that originates from the image of the same object is synchronized and if firing that originates from different objects is desynchronized, the different sets of firings will be discriminated from each other. Firing synchronized with oscillations has been found between cells in the cat visual cortex, and some context dependency of the synchronization has also been reported. Although oscillating firing has not been found in TE (Young et al 1992, Tovee & Rolls 1992), nonperiodic synchronization may be present in TE.

Another possible mechanism of binding in TE is selection by attention (Crick 1984). We can pay attention to only one, or a few at most, object at a time. If the representation of features of an attended object is enhanced and that of other objects is suppressed, the binding problem will disappear. This mechanism is likely to be working, because strong effects of attention have been found on responses of TE cells (Richmond & Sato 1987, Moran & Desimone 1985, Spitzer et al 1988, Chelazi et al 1993).

A third possibility is that the set of distant activity in TE originating from a single object is combined by making loops of activity with activity in retinotopically organized former stages in the ventral pathway. TE projects back to TEO, V4, V2, and even V1 (Rockland et al 1994, Rockland & Van Hoesen 1994). There are also step-by-step feedback projections. Different sets of activity originating from different objects are discriminated by the position of combined activity on the retinotopical maps. Kawato et al (1991, 1993) have indicated the importance of feedback projections from a similar viewpoint.

## Responses to Complex Object Features in Other Brain Sites

Is there any brain site that contains cells selectively responding to more integrated features than the critical features determined in TEd? One group of candidates are the brain sites to which TEd projects: the polysensory area in the anterior part of the superior temporal sulcus (STPa), the prefrontal cortex, the perirhinal cortex, the amygdala, and the striatum of the basal ganglia. Visually responsive cells have been reported in all of these sites.

Cells that selectively responded to the sight of a face were found in STPa in the early 1980s (Bruce et al 1981; Yamane et al 1988; Perrett et al 1982,

1987; Young & Yamane 1992; Rolls 1992), and such cells, which were so-called face neurons, have been extensively studied. There are reports that such cells are also present in TE itself (Baylis et al 1987, Tanaka et al 1991), area TG (Nakamura et al 1994), and the amygdala (Leonard et al 1985, Nakamura et al 1992). The meaning of selective varied among these studies: In some of the studies only several non–face stimuli were presented, and most of the studies did not test partial features of the image of a face. However, a few of them used a scrambled face, which was made with scrambled patches of the picture of a face, and showed that the scrambled face was not effective (Bruce et al 1981). A few studies found that there were cells that were not activated by a face without the eye or by the eye only (Perrett et al 1982, Rolls et al 1985). We found systematically arranged columns in TEd that respond to different views of faces. These data suggest that there are cells that require all the essential features that compose the image of a face. The image of a face is more complex than the other features represented by cells in TEd.

The presence of face neurons cannot be generalized to the representation of other objects. Faces of monkeys are special for monkeys, and also those of human beings for laboratory monkeys, in that faces are important media for social communication between individuals. Discrimination of faces from other objects is only a preliminary stage to represent information of expressions or features of individual faces. This view is supported by the fact that there are cells that respond to the view of body movements or hand actions in STPa (Perrett et al 1985, 1989, 1992; Oram & Perrett 1994). Body movements and hand actions often express important information about the relation between individuals in the scene or between the individual in the scene and the observer. These groups of cells in STPa may be specially prepared for social communication.

Responses specific to the color or shape of visual stimuli have been found in the principal region and inferior convexity of the prefrontal cortex (Fuster et al 1982, Watanabe 1986, Wilson et al 1993). Watanabe (1986) found specific responses by using green and red squares, or a disk and perpendicular grating. Wilson et al (1993) examined cells in the inferior convexity with a larger variety of stimuli, including faces. Some of the cells that Wilson et al studied specifically responded to a face, although the feature critical for the activation was not identified in this study. Interestingly, responses of many of the cells that Watanabe studied were not determined by the attributes of the stimuli but by the temporal meaning of the stimulus attributes in the behavioral frameworks. Watanabe demonstrated this by changing the behavioral meaning of the same stimuli by the cue signal presented prior to the stimuli. The main computation conducted in the prefrontal cortex on the images of visual stimuli may be to relate the inputs to the temporal behavioral frameworks.

Visual responses with various level of stimulus selectivity have been found in the amygdala (Ono et al 1983, Nishijo et al 1988, Nakamura et al 1992). Many cells responded selectively to a category of objects that might cause a certain kind of emotion. This view is consistent with a general idea that the amygdala is essential for emotion (Turner et al 1980). A main function of the amygdala, from the viewpoint of visual object recognition, seems to be the association of sensory inputs to different kinds of emotion. A question is whether the association takes place after inputs representing a particular object are denoted by activity of a single group of cells. Nishijo et al (1988) and Nakamura et al (1992) reported that some cells in the amygdala responded selectively to particular objects. However, because the variety of reference stimuli used in these studies was not very extensive, the tuning property of the cells may not be sharp enough, and a kind of population coding such as that in TEd is directly associated with different kinds of emotion.

Miller & Desimone (1991) and Nakamura et al (1994) have conducted recordings from the perirhinal cortex. Miller & Desimone used six arbitrarily chosen pictures of natural objects and found that most of them activated the cells, although their effectiveness was different. These facts may suggest that selectivity of cells in the perirhinal cortex is more gradual than that of cells in TEd. Nakamura et al examined cells in TE as well as the perirhinal cortex and the TG part of it (polar cortex) with a larger variety of stimuli and concluded that cells in the perirhinal cortex and TG were as selective as those in TE.

The data of our anatomical studies, in which an anterograde tracer, PHA-L, was injected into a single site in TEd and the ventral part of TE (TEv), indicated that the projection from TE to the perirhinal cortex diverges: A single site in TE projects to a large part of area 36 of the perirhinal cortex (Saleem et al 1993a, 1994). Because this divergent projection was found regardless of the injection site in TE, a single site in the perirhinal cortex should receive convergent inputs from multiple sites in TE. This anatomical structure provides the opportunity for interaction between different features represented by distant columns in TE. Information processing in the perirhinal cortex may be more associative than discriminative.

The recent finding by Miyashita et al (1994) is consistent with this idea. They trained monkeys for the association of two different pictures (see section on changeability of the selectivity in the adult). The commissural connections of the monkey had been destroyed. After the association of several pairs was learned, the perirhinal cortex and the entorhinal cortex that connects the perirhinal cortex to the hippocampus were destroyed by injecting ibotenic acid. There were TE cells that responded to two paired stimuli before the lesion, but such cells disappeared after the lesion. This disappearance indicates that the dual responsiveness to two paired stimuli was underlined by a network that included the perirhinal cortex. The associative aspect of information processing in the perirhinal cortex may also underlie its importance for performing the delayed-matching-to-sample (DMS) with visual images of objects (Murray & Mishkin 1986, Zola-Morgan et al 1989, Meunier et al 1993, Eacott et al 1994).

Finally, we need to discuss TEv. The part of TE medial to the anterior middle temporal sulcus (AMTS) has been referred to as TEv. Martin-Elkins & Horel (1992) and Yukie et al (1992) found that afferent pathways to TEd and TEv are separate: TEd receives inputs via the lateral part of the posterior IT, whereas TEv received inputs from regions at the ventral surface. Iwai & Yukie (1987) found that TEd and TEv have different projection patterns to the amygdala. We have also recently found that TEd and TEv have different projection patterns to the perirhinal cortex, prefrontal cortex, and striatum, as well as to the amygdala (Cheng et al 1993, Saleem et al 1994). TEd and TEv thus seem to represent inferotemporal stages of parallel pathways. Because projections to the perirhinal cortex, which is thought to be important for visual DMS, are more numerous from TEv than from TEd (Saleem et al 1994), it is hypothesized that TEv is more involved in visual memory than TEd. The stimulus selectivity of cells in TEv may be different from that of cells in TEd, although there have been no studies that explicitly conducted the comparison.

In summary, there have been no firm findings of cells that responded only to features more integrated than the critical features found for TEd cells, except for the cells in STPa responding to faces. The cells in STPa are thought to be specially prepared to convey information for social communication among monkeys. There is some evidence suggesting that cells in the prefrontal cortex are organized to relate visual stimuli to temporal behavioral frames and that those in the amygdala are organized to relate visual stimuli to specific emotion of the monkey.

## Object Recognition by Activities Distributed over the Brain

Sakata and his colleagues have recently found shape-selective cell activities in the parietal areas in the intraparietal sulcus (Taira et al 1990, Sakata & Kusunoki 1992). Many cells in the lateral bank of the sulcus selectively responded to visual images of switches that the monkey was trained to manipulate. The monkey reached the switches and pulled them. The switches varied in shape, so the monkey shaped its hand differently before it reached for different switches. Because the discharges started when the monkey saw the switches and because the discharges decreased when the task was performed in a dark room, the discharges should be caused, at least partially, by the visual inputs.

Sakata and colleagues further found that some cells in the more posterior part of the sulcus responded to more primitive features of stimuli, including the 3D orientation of a pole and the 3D tilt of a plane. They presented the stimuli on a computer graphic system with binocular disparity (Kusunoki et al 1993; Tanaka et al 1992, 1994). The responses were reduced when the binocular disparity was removed, which indicates that the binocular disparity gave an essential cue for the responses.

Taken together, there seems to be a flow of information of 3D shapes of objects that the monkey may manipulate. These findings coincide with the proposition by Goodale and colleagues (1991, 1992), based on human clinical data, that the dorsal pathway leading to the parietal cortex is responsible for visuo-motor control, but not for spatial vision (Mishkin et al 1983). To manipulate an object, the object's 3D structure should be perceived. There may be an independent representation of the shape of objects in the dorsal pathway that is independent of the representation of objects in the ventral pathway; and probably only course shape is represented. This dorsal representation of objects may influence the representation of objects in the ventral pathway through the indirect connection via the parahippocampal structures (Van Hoesen 1982, Suzuki & Amaral 1995) or via regions in the superior temporal sulcus (Seltzer & Pandya 1978, 1984, 1989, 1994).

The accumulated findings favor the idea that no cognition units represent the concept of objects; instead the concept of objects is found only in the activities distributed over various regions in the brain. When the visual image of an object is given, it is processed in the ventral visual pathway, and the representation containing similarity to other objects and association with images of the same object under different conditions is reconstructed there. This representation in the ventral visual pathway utilizes the population coding in two levels. First, the image of the object is represented by a combination of multiple partial (but not necessarily local) features designated by different columns in TE. Because the presence of the partial features is represented in an analog manner, the combinatorial representation can be understood as a combination of similarity to different prototypes of object images (Edelman 1995). The second level of population coding is in the representation of partial features. The features are represented by multiple cells within a TE column that has overlapping selectivity. This level of population coding has been extensively discussed in the section on functions of the TE columns. Triggered by inputs from the IT, emotional information of the object is read out in the amygdala; association with other objects is read out through the perirhinal cortex; and behavioral significance emerges in the prefrontal cortex. The visual image of the object is also processed in the dorsal pathway, and information necessary for the monkey to manipulate the object is read out in the parietal cortex. All this recovered information of the object, distributed over the brain, may represent the concept of the object.

## Conclusions

Cells in area TE of the IT selectively respond to various moderately complex object features, and those that respond to similar features cluster in a columnar region elongated perpendicular to the cortical surface. Although cells within a column respond to similar features, their selectivity is not necessarily identical. The data of optical imaging in TE have suggested that the borders between neighboring columns are not discrete; there is a continuous mapping of complex feature space within a larger region containing several partially overlapped columns. This continuous mapping may be used for various computations, such as production of the image of the object at different viewing angles, illumination conditions, and articulation poses.

## Literature Cited

Albright TD, Gross CG. 1990. Do inferior temporal cortex neurons encode shape by acting as Fourier Descriptor filters? *Proc. Int. Conf. Fuzzy Logic & Neural Networks, Izuka, Japan,* pp. 375–78

Baylis GC, Rolls ET, Leonard CM. 1987. Functional subdivisions of the temporal lobe neocortex. *J. Neurosci.* 7:330–42

Boussaoud D, Desimone R, Ungerleider LG. 1991. Visual topography of area TEO in the macaque. *J. Comp. Neurol.* 306:554–75

Bruce C, Desimone R, Gross CG. 1981. Visual properties of neurons in a polysensory area in superior temporal sulcus of the macaque. *J. Neurophysiol.* 46:369–84

Chelazzi L, Miller EK, Duncan J, Desimone R. 1993. A neural basis for visual search in inferior temporal cortex. *Nature* 363:345–47

Cheng K, Saleem KS, Tanaka K. 1993. PHA-L study of the subcortical projections of the macaque inferotemporal cortex. *Soc. Neurosci. Abstr.* 19:971

Crick F. 1984. Function of the thalamic reticular complex: the searchlight hypothesis. *Proc. Natl. Acad. Sci. USA* 81:4586–90

Dean P. 1976. Effects of inferotemporal lesions on the behavior of monkeys. *Psychol. Bull.* 83:41–71

Desimone R, Albright TD, Gross CG, Bruce C. 1984. Stimulus-selective properties of inferior temporal neurons in the macaque. *J. Neurosci.* 4:2051–62

Desimone R, Miller EK, Chelazzi L. 1994. The interaction of neural systems for attention and memory. In *Large Scale Theories of the Brain,* ed. C Koch, pp. 75–91. Cambridge/London: MIT Press. 343 pp.

Eacott MJ, Gaffan D, Murray EA. 1994. Preserved recognition memory for small sets, and impaired stimulus identification for large

sets, following rhinal cortex ablations in monkeys. *Eur. J. Neurosci.* 6:1466–78

Edelman S. 1995. Representation, similarity and the chorus of prototypes. In *Minds and Machines.* 5:45–68

Engel AK, Konig P, Kreiter AK, Schillen TB, Singer W. 1992. Temporal coding in the visual cortex: new vistas on integration in the nervous system. *Trends Neurosci.* 15:218–26

Erickson RP. 1968. Stimulus coding in topographic and nontopographic afferent modalities: on the significance of the activity of individual neurons. *Psychol. Rev.* 75:447–65

Földiák P. 1991. Learning invariance from transformaion sequences. *Neural Comput.* 3:194–200

Frostig RD, Lieke DE, Ts'o DY, Grinvald A. 1990. Cortical functional architecture and local coupling between neuronal activity and the microstimulation revealed by in vivo high-resolution optical imaging of intrinsic signals. *Proc. Natl. Acad. Sci. USA* 87:6082–86

Fujita I, Tanaka K, Ito M, Cheng K. 1992. Columns for visual features of objects in monkey inferotemporal cortex. *Nature* 360:343–46

Fuster JM, Bauer RH, Jervey JP. 1982. Cellular discharge in the dorsolateral prefrontal cortex of the monkey in cognitive tasks. *Exp. Neurol.* 77:679–94

Gallant JL, Braun J, Van Essen DC. 1993. Selectivity for polar, hyperbolic, and Cartesian gratings in Macaque visual cortex. *Science* 259:100–3

Goodale MA, Milner AD. 1992. Separate visual pathways for perception and action. *Trends Neurosci.* 15:20–25

Goodale MA, Milner AD, Jakobson LS, Carey DP. 1991. A neurological dissociation between perceiving objects and grasping them. *Nature* 349:154–56

Gross CG. 1973. Visual functions of inferotemporal cortex. In *Handbook of Sensory Physiology*, ed. R Jung, 7(Part 3B): 451–82. Berlin: Springer-Verlag

Gross CG. 1994. How inferior temporal cortex became a visual area. *Cereb. Cortex* 5:455–69

Gross CG, Bender DB, Rocha-Miranda CE. 1969. Visual receptive fields of neurons in inferotemporal cortex of the monkey. *Science* 166:1303–6

Gross CG, Rocha-Miranda CE, Bender DB. 1972. Visual properties of neurons in inferotemporal cortex of the macaque. *J. Neurophysiol.* 35:96–111

Ito M, Fujita I, Tamura H, Tanaka K. 1994. Processing of contrast polarity of visual images in inferotemporal cortex of the macaque monkey. *Cereb. Cortex* 5:499–508

Ito M, Tamura H, Fujita I, Tanaka K. 1995. Size and position invariance of neuronal responses in monkey inferotemporal cortex. *J. Neurophysiol.* 73:218–26

Iwai E, Mishkin M. 1969. Further evidence on the locus of the visual area in the temporal lobe of the monkey. *Exp. Neurol.* 25:585–94

Iwai E, Yukie M. 1987. Amygdalofugal and amygdalopetal connections with modality-specific visual cortical areas in macaques (*Macaca fuscata, M. mulatta*, and *M. fascicularis*). *J. Comp. Neurol.* 261:362–87

Kawato M, Hayakawa H, Inui T. 1993. A forward-inverse optics model of reciprocal connections between visual cortical areas. *Network* 4:415–22

Kawato M, Inui T, Hongo S, Hayakawa H. 1991. Computational theory and neural network models of interaction between visual cortical areas. *ATR Technical Report TR-A-0105.* Kyoto, ATR

Kobatake E, Tanaka K. 1994. Neuronal selectivities to complex object features in the ventral visual pathway of the macaque cerebral cortex. *J. Neurophysiol.* 71:856–67

Kobatake E, Tanaka K, Tamori Y. 1992. Long-term learning changes the stimulus selectivity of cells in the inferotemporal cortex of adult monkeys. *Neurosci. Res.* S17:S237 (Suppl.)

Kobatake E, Tanaka K, Wang G, Tamori Y. 1993. Effects of adult learning on the stimulus selectivity of cells in the inferotemporal cortex. *Soc. Neurosci. Abstr.* 19:975

Kusunoki M, Tanaka Y, Ohtsuka H, Ishiyama K, Sakata H. 1993. Selectivity of the parietal visual neurons in the axis orientation of objects in space. *Soc. Neurosci. Abstr.* 19:770

Lehky SR, Sejnowski TJ, Desimone R. 1992. Predicting responses of nonlinear neurons in monkey striate cortex to complex patterns. *J. Neurosci.* 12:3568–81

Leonard CM, Rolls ET, Wilson FAW, Baylis GC. 1985. Neurons in the amygdala of the monkey with responses selective for faces. *Behav. Brain Res.* 15:159–76

Li L, Miller EK, Desimone R. 1993. The representation of stimulus familiarity in anterior inferior temporal cortex. *J. Neurophysiol.* 69:1918–29

Malach R. 1994. Cortical columns as devices for maximizing neuronal diversity. *Trends Neurosci.* 17:101–4

Martin-Elkins CL, Horel JA. 1992. Cortical afferents to behaviorally defined regions of the inferior temporal and parahippocampal gyri as demonstrated by WGA-HRP. *J. Comp. Neurol.* 321:177–92

Meunier M, Bachevalier J, Mishkin M, Murray EA. 1993. Effects on visual recognition of combined and separate ablations of the entorhinal and perirhinal cortex in rhesus monkeys. *J. Neurosci.* 13:5418–32

Miller EK, Li L, Desimone R. 1991. A neural mechanism for working and recognition memory in inferior temporal cortex. *Science* 254:1377–79

Mishkin M, Ungerleider LG, Macko KA. 1983. Object vision and spatial vision: two cortical pathways. *Trends Neurosci.* 6:414–17

Miyashita Y. 1993. Inferior temporal cortex: where visual perception meets memory. *Annu. Rev. Neurosci.* 16:245–63

Miyashita Y, Higuchi S, Zhou X, Okuno H, Hasegawa I. 1994. Disruption of backward connections from the rhinal cortex impairs visual associative coding of neurons in inferotemporal cortex of monkeys. *Soc. Neurosci. Abstr.* 20:428

Moran J, Desimone R. 1985. Selective attention gates visual processing in the extrastriate cortex. *Science* 229:782–84

Mountcasstle VB. 1978. An organizing principle for cerebral function: the unit module and the distributed system. In *The Mindful Brain,* ed. VB Mountcastle, GM Edelman, pp. 7–50. Cambridge, MA: MIT Press

Murray EA, Mishkin M. 1986. Visual recognition in monkeys following rhinal cortical ablations combined with either amygdalectomy or hippocampectomy. *J. Neurosci.* 6:1991–2003

Nakamura H, Gattass R, Desimone R, Ungerleider LG. 1993. The modular organization of projections from areas V1 and V2 to areas V4 and TEO in macaques. *J. Neurosci.* 13:3681–91

Nakamura K, Matsumoto K, Mikami A, Kubota K. 1994. Visual response properties of single neurons in the temporal pole of behaving monkeys. *J. Neurophysiol.* 71:1206–21

Nakamura K, Mikami A, Kubota K. 1992. Activity of single neurons in the monkey amygdala during performance of a visual discrimination task. *J. Neurophysiol.* 67:1447–63

Nishijo H, Ono T, Nishino H. 1988. Single neuron responses in amygdala of alert monkey during complex sensory stimulation with affective significance. *J. Neurosci.* 8:3570–83

Ono T, Fukuda M, Nishino H, Sasaki K, Muramoto K. 1983. Anygdaloid neuronal responses to complex visual stimuli in an operant feeding situation in the monkey. *Brain Res. Bull.* 11:515–18

Oram MW, Perrett DI. 1994. Responses of anterior superior temporal polysensory (STPa) neurons to "biological motion" stimuli. *J. Cogn. Neurosci.* 6:99–116

Perrett DI, Harries MH, Bevan R, Thomas S, Benson PJ, et al. 1989. Frameworks of analysis for the neural representation of animate objects and actions. *J. Exp. Biol.* 146:87–113

Perrett DI, Hietanen JK, Oram MW, Benson PJ. 1992. Organization and functions of cells responsive to faces in the temporal cortex. *Philos. Trans. R. Soc. London Ser. B* 335:23–30

Perrett DI, Mistlin AJ, Chitty AJ. 1987. Visual neurones responsive to faces. *Trends Neurosci.* 10:358–64

Perrett DI, Rolls ET, Caan W. 1982. Visual neurones responsive to faces in the monkey temporal cortex. *Exp. Brain Res.* 47:329–42

Perrett DI, Smith PAJ, Mistlin AJ, Chitty AJ, Head AS, et al. 1985. Visual analysis of body movements by neurones in the temporal cortex of the macaque monkey: a preliminary report. *Behav. Brain Res.* 16:153–70

Peters A, Yilmaz E. 1993. Neuronal organization in area 17 of cat visual cortex. *Cereb. Cortex* 3:49–68

Purves D, Riddle DR, LaMantia A-S. 1992. Iterated patterns of brain circuitry (or how the cortex gets its spots). *Trends Neurosci.* 15:362–68

Richmond BJ, Sato T. 1987. Enhancement of inferior temporal neurons during visual discrimination. *J. Neurophysiol.* 58:1292–306

Rockland KS, Saleem KS, Tanaka K. 1994. Divergent feedback connections from areas V4 and TEO in the macaque. *Vis. Neurosci.* 11:579–600

Rockland KS, Van Hoesen GW. 1994. Direct temporal-occipital feedback connections to striate cortex (V1) in the macaque monkey. *Cereb. Cortex* 4:300–13

Rolls ET. 1991. Neural organiation of higher visual functions. *Curr. Opin. Neurobiol.* 1:274–78

Rolls ET. 1992. Neurophysiological mechanisms underlying face processing within and beyond the temporal cortical visual areas.

*Philos. Trans. R. Soc. London Ser. B* 335:11–21

Rolls ET, Baylis GC, Hasselmo ME, Nalwa V. 1989. The effect of learning on the face-selective responses of neurons in the cortex in the superior temporal sulcus of the monkey. *Exp. Brain Res.* 76:153–64

Rolls ET, Baylis GC, Leonard CM. 1985. Role of low and high spatial frequencies in the face-selective responses of neurons in the cortex in the superior temporal sulcus in the monkey. *Vis. Res.* 25:1021–35

Sakai K, Miyashita Y. 1991. Neural organization for the long-term memory of paired associates. *Nature* 354:152–55

Sakata H, Kusunoki M. 1992. Organiation of space perception: neural representation of three-dimensional space in the posterior parietal cortex. *Curr. Opin. Neurobiol.* 2:170–74

Saleem KS, Cheng K, Suzuki W, Tanaka K. 1994. Differential projection from ventral and dorsal parts of the anterior TE to perirhinal cortex in the macaque monkey: PHAL study. *Neurosci. Res.* S19:S201 (Suppl.)

Saleem KS, Cheng K, Tanaka K. 1993a. Organization of projection from the anterior TEA to the perirhinal (areas 35/36) and frontal cortices in the macaque inferotemporal cortex. *Soc. Neurosci. Abstr.* 19:971

Saleem KS, Tanaka K, Rockland KS. 1992. PHA-L study of connections from TEO and V4 to TE in the monkey visual cortex. *Soc. Neurosci. Abstr.* 18:294

Saleem KS, Tanaka K, Rockland KS. 1993b. Specific and columnar projection from area TEO to TE in the macaque inferotemporal cortex. *Cereb. Cortex* 3:454–64

Schwartz EI, Desimone R, Albright TD, Gross CG. 1983. Shape recognition and inferior temporal neurons. *Proc. Natl. Acad. Sci. USA* 80:5776–78

Seltzer B, Pandya DN. 1978. Afferent cortical connections and architectonics of the superior temporal sulcus and surrounding cortex in the rhesus monkey. *Brain Res.* 149:1–24

Seltzer B, Pandya DN. 1984. Further observations on parieto-temporal connections in the rhesus monkey. *Exp. Brain Res.* 55:301–12

Seltzer B, Pandya DN. 1989. Intrinsic connections and architectonics of the superior temporal sulcus in the rhesus monkey. *J. Comp. Neurol.* 290:451–71

Seltzer B, Pandya DN. 1994. Parietal, temporal, and occipital projections to cortex of the superior temporal sulcus in the rhesus monkey: a retrograde tracer study. *J. Comp. Neurol.* 343:445–63

Singer W. 1993. Synchronization of cortical activity and its putative role in information processing and learning. *Annu. Rev. Physiol.* 55:349–74

Snippe HP, Koenderink JJ. 1992. Discrimination thresholds for channel-coded systems. *Biol. Cybern.* 66:543–51

Spitzer H, Desimone R, Moran J. 1988. Increased attention enhances both behavioral and neuronal performance. *Science* 240:338–40

Suzuki WA, Amaral DG. 1994. Perirhinal and parahippocampal cortices of the macaque monkey: cortical afferents. *J. Comp. Neurol.* 350:497–533

Taira M, Mine S, Georgopoulos AP, Murata A, Sakata H. 1990. Parietal cortex neurons of the monkey related to the visual guidance of hand movement. *Exp. Brain Res.* 83:29–36

Tanaka K, Saito H, Fukada Y, Moriya M. 1991. Coding visual images of objects in the inferotemporal cortex of the macaque monkey. *J. Neurophysiol.* 66:170–89

Tanaka Y, Kusunoki M, Ohtsuka H, Takiura K, Sakata H. 1992. Analysis of three dimensional directional selectivity of the monkey parietal depth-movement-sensitive neurons using a stereoscopic computer display system. *Neurosci. Res.* S17:S238 (Suppl.)

Tanaka Y, Murata A, Taira M, Shikata E, Sakata H. 1994. Responses of the parietal visual neurons to stereoscopic stimuli on the computer graphic display in alert monkeys. *Neurosci. Res.* S19:S200 (Suppl.)

Thomson AM, Deuchars J. 1994. Temporal and spatial properties of local circuits in neocortex. *Trends Neurosci.* 17:119–26

Tovee MJ, Rolls ET. 1992. Oscillatory activity is not evident in the primate temporal visual cortex with static stimuli. *NeuroReport* 3:369–72

Turner BH, Mishkin M, Knapp M. 1980. Organization of the amygdalopetal projections from modality-specific cortical association areas in the monkey. *J. Comp. Neurol.* 191:515–43

Ungerleider LG, Gaffan D, Pelak VS. 1989. Projections from inferior temporal cortex to prefrontal cortex via the uncinate fascicle in rhesus monkeys. *Exp. Brain Res.* 76:473–84

Van Hoesen GW. 1982. The parahippocampal gyrus: new observations regarding its cortical connections in the monkey. *Trends Neurosci.* 5:345–50

von Bonin G, Bailey P. 1947. *The Neocortex of Macaca Mulatta.* Urbana, IL: Univ. Ill. Press

von Bonin G, Bailey P. 1950. *The Isocortex of the Chimpanzee.* Urbana, IL: Univ. Ill. Press

Wang G, Tanaka K, Tanifuji M. 1994. Optical imaging of functional organization in macaque inferotemporal cortex. *Soc. Neurosci. Abstr.* 20:316

Watanabe M. 1986. Prefrontal unit activity during delayed conditional Go/No-Go discrimination in the monkey. I. relation to the stimulus. *Brain Res.* 382:1–14

Wilson FAW, Scalaidhe SPO, Goldman-Rakic PS. 1993. Dissociation of object and spatial processing domains in primate prefrontal cortex. *Science* 260:1955–58

Yamane S, Kaji S, Kawano K. 1988. What facial features activate face neurons in the inferotemporal cortex of the monkey? *Exp. Brain Res.* 73:209–14

Young MP, Tanaka K, Yamane S. 1992. On oscillating neuronal responses in the visual cortex of the monkey. *J. Neurophysiol.* 67:1464–74

Young MP, Yamane S. 1992. Sparse population coding of faces in the inferotemporal cortex. *Science* 29:1327–31

Yukie M, Hikosaka K, Iwai E. 1992. Organization of cortical visual projections to the dorsal and ventral parts of area TE of the inferotemporal cortex in macaques. *Soc. Neurosci. Abstr.* 18:294

Zola-Morgan S, Squire LR, Amaral DG, Suzuki WA. 1989. Lesions of perirhinal and parahippocampal cortex that spare the amygdala and hippocampal formation produce severe memory impairment. *J. Neurosci.* 9:4355–70

*Annu. Rev. Neurosci. 1996. 19:141–64*

# SODIUM CHANNEL DEFECTS IN MYOTONIA AND PERIODIC PARALYSIS

*Stephen C. Cannon*

Department of Neurobiology, Harvard Medical School, and Neurology Department, Massachusetts General Hospital, Boston, Massachusetts 02114

KEY WORDS:   muscle, ion channel, hyperkalemia, paramyotonia

### ABSTRACT

Myotonias and periodic paralyses constitute a diverse group of inherited disorders of muscle in which the primary defect is an alteration in the electrical excitability of the muscle fiber. The ion channel defects underlying these excitability derangements have recently been elucidated at the molecular and functional levels. This review focuses on sodium channel mutations that disrupt inactivation and thereby cause both the enhanced excitability of myotonia (muscle stiffness due to repetitive discharges) and the inexcitability resulting from depolarization during attacks of paralysis.

## INTRODUCTION

The myotonias and periodic paralyses constitute a diverse group of heritable disorders of muscle excitability. Their dramatic presentation—muscle stiffness or recurrent attacks of paralysis, respectively—has fascinated clinicians and physiologists alike for the past century. Derangements in the electrical excitability of the sarcolemma were identified over 50 years ago for some members of this group. In the past five years, three different ion channels and one putative kinase have been identified as the sites of the causative molecular defect. This review focuses on a subset of the myotonias and periodic paralyses that are caused by mutations in the $\alpha$ subunit of a skeletal muscle sodium channel.

## SPECTRUM OF MYOTONIA AND PERIODIC PARALYSIS

### Abnormalities in Skeletal Muscle Excitability

Myotonia and paralysis lie at opposite ends of the spectrum of muscle excitability. Myotonia is a pathological state of hyperexcitability that is manifested

141

as delayed relaxation of muscle force after voluntary contraction (action myotonia). An affected patient usually complains of painless stiffness. A light mechanical tap elicits a persistent dimpling of the muscle belly that takes several seconds or minutes to resolve (percussion myotonia). Electromyography reveals a characteristic "dive-bomber" pattern of repetitive discharges that wax and wane in amplitude and frequency (reviewed in Lipicky 1979, Rüdel & Lehmann-Horn 1985). These myotonic bursts may last for several seconds to a minute or more. The repetitive firing persists in the presence of nondepolarizing neuromuscular blockers, such as curare, which proves that the excitation is myogenic in origin (Brown & Harvey 1939). The temporal correlation between the duration of myotonic runs and the slow decay in muscle force has been interpreted as evidence that the abnormal relaxation is caused by myotonic discharges. The severity of myotonia varies considerably not only from patient to patient, but also from moment to moment for an affected individual (Riggs & Griggs 1979, Barchi 1992). Sudden forceful effort after rest elicits maximal symptoms, and with repeated trials of the same movement, the myotonia diminishes (warm-up phenomenon).

In episodes of periodic paralysis or weakness, affected muscle is hypo- or even inexcitable. Between attacks muscle strength is normal, as are muscle bulk, tone, and tendon reflexes. Paralytic spells are episodic, rather than having any regular periodicity, and usually last for minutes to hours (Riggs & Griggs 1979). Even with severe limb and truncal weakness that renders someone bedridden, respiratory compromise is extremely rare. Consciousness is not impaired; seizures or cardiac dysrhythmia do not occur; and all sensory modalities remain unaffected. Thus the abnormality in electrical excitability is limited to skeletal muscle. During an attack, affected muscles are depolarized at $-60$ to $-40$ mV from a normal resting potential of about $-95$ mV (Creutzfeldt et al 1963, Lehmann-Horn et al 1983). Interictally, the resting potential may be slightly less than normal (about 5 to 10 mV). Depolarized fibers are inexcitable and will not generate an action potential nor twitch in response to direct galvanic stimulation. The major defect appears to be limited to the electrical excitability of the sarcolemma. There is no evidence for abnormalities of the contractile proteins, and calcium-induced twitches in skinned fibers are indistinguishable from those in normal fibers (Ruff 1991).

## Clinical Disorders

The periodic paralyses and myotonias have been delineated into separate clinical entities (Lipicky 1979, Rüdel & Lehmann-Horn 1993). This review is focused on the subset caused by mutations of the sodium channel α subunit, and only a brief summary is made of related disorders that have been the subject of recent reviews (Rüdel & Lehmann-Horn 1985, Ptacek et al 1993,

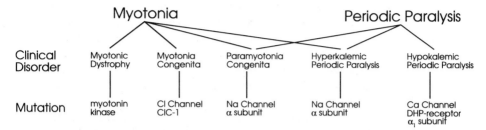

*Figure 1* Spectrum of altered excitability among the myotonias and periodic paralyses. The heritable myotonias and periodic paralyses are arranged along a continuum from the hyperexcitability of myotonia (*left*) to the inexcitability of paralysis (*right*). The disorders in the middle have features of both myotonia and paralysis. The mutant gene product implicated as the causative lesion is listed for each disorder.

Hoffman et al 1995). As with the excitability abnormalities, the clinical disorders can be viewed as a continuum along a spectrum from myotonic to paralytic disorders (Figure 1). Each member of this group is inherited in a Mendelian fashion, and all but one follow an autosomal-dominant pattern.

At one extreme is myotonic dystrophy, which is unique among this group in that it is the only disease associated with dystrophy (progressive loss of muscle) and affects multiple organ systems. The progressive weakness is usually much more severe than the myotonia, which may be subclinical. The genetic defect in this autosomal-dominant disorder is an expansion of a trinucleotide repeat (CTG) on 19q13.2–13.3 in the 3′ untranslated region of a gene coding for myotonin kinase, that by sequence consensus appears to be a cAMP-dependent serine-threonine kinase (Brook et al 1992; one of six groups that simultaneously located the defect). The tantalizing hypothesis is that aberrant phosphorylation of ion channel(s) produces an abnormality of membrane excitability. One study on biopsy material has suggested an alteration in Na channel inactivation (Franke et al 1990), but this needs confirmation and further characterization.

Myotonia congenita is the prototypic myotonic disorder. Autosomal-dominant (Thomsen's disease) and the more severe autosomal-recessive (Becker's recessive generalized myotonia) forms are characterized by prominent electromyographic and symptomatic myotonia. Weakness does not occur, and the muscle may become hypertrophied in the recessive type due to the "exercise" accomplished by severe myotonia. A homologue of the dominant form occurs in the goat, and a classic series of experiments by Bryant and colleagues (Lipicky & Bryant 1966, Adrian & Bryant 1974) proved that the membrane defect was a reduction in the chloride conductance. The genetic defect was first identified in a murine homologue of the recessive form, the arrested development of righting response, or ADR, mouse. A transposon is inserted at an exon-intron boundary of the major skeletal muscle chloride channel, ClC-1 (Steinmeyer et al 1991). The insertion completely disrupts the mRNA 3′ to the ninth transmembrane-spanning segment and presumably codes for a nonexpressing truncated channel. Subsequently, point mutations and deletions in ClC-1 have been identified in humans with either the recessive (Kock et al 1992, Heine et al 1994) or the dominant form (George et al 1993, Steinmeyer et al 1994). The mechanism for the different patterns of expression associated with a particular mutation remains unknown, but coexpression of mutant and wild-type mRNAs causes a stoichiometric reduction in chloride current that is consistent with a homomultimeric structure for the channel (Steinmeyer et al 1994). In recordings from myoballs cultured from patients with Becker's recessive myotonia congenita, Fahlke et al (1993) identified a 50% reduction in the conductance of an ~30-pS chloride channel. The interpretation is confounded by the fact that when heterologously expressed in embryonic kidney cells, wild-type ClC-1 forms a 1-pS channel (Pusch et al 1994). Thus how a reduction in one of several types of chloride conductances in human myotubes relates to a mutation in the gene for ClC-1 is uncertain.

Two variants of periodic paralysis were distinguished clinically on the basis of the serum $K^+$ at the time of an attack: hypokalemic and hyperkalemic periodic paralysis, HypoPP and HyperPP, respectively (Buruma & Schipperheyn 1979). Both forms are inherited autosomal dominantly, but for HypoPP, penetrance is reduced in females. HypoPP lies completely within the paralytic end of the spectrum. Affected individuals become severely weak from maneuvers that lower serum $[K^+]$, such as carbohydrate ingestion or diuretic use. The muscles are flaccid, depolarized, and inexcitable (Rüdel et al 1984). Strength returns within hours, and $K^+$ ingestion hastens recovery. Myotonia does not occur in association with HypoPP. By using random markers (Fontaine et al 1994) and positional cloning (Jurkatt-Rott et al 1994, Ptacek et al 1994), researchers recently established that HypoPP is caused by point mutations in the gene encoding the DHP-receptor $\alpha_1$ subunit, an L-type skeletal muscle Ca channel localized to the T tubule and an effector of excitation-contraction coupling (Tanabe et al 1988). Interestingly, all of the mutations occur in arginine residues within the voltage-sensing S4 segments. The functional consequences of these defects, how they lead to membrane inexcitability, and why attacks are triggered by hypokalemia remain to be elucidated.

The clinical presentation is more variable with HyperPP than for HypoPP. In HyperPP, serum $[K^+]$ measured during a spontaneous attack is usually high but may be normal, and a more consistent finding is weakness produced by provocative testing with $K^+$ loading. Attacks may occur daily or only sporadically over a lifetime and are triggered by rest after exercise, fasting, or ingestion of $K^+$. In many families, affected individuals also have myotonia, clinical as well as electromyographic, especially with the onset or resolution of the weak-

ness. Some families with otherwise characteristic HyperPP do not have myotonia. Thus HyperPP occupies the middle portion of the spectrum of excitability, with mixed features of both myotonia and paralysis. After the fourth or
fifth decade, some patients develop a slowly progressive chronic proximal limb
weakness. Histologically, fibers are remarkably normal in appearance but may
contain perinuclear vacuoles and increased variability of fiber size (Bradley
et al 1990). HyperPP also occurs in American quarter horses (Cox 1985). The
excessive spontaneous muscle activity produces increased muscle definition
and hypertrophy, which are desirable features in show horses. Selective breeding practices led to such a rapid dissemination of HyperPP within the quarter
horse gene pool that up to 1 in 50 animals may be affected (Naylor et al 1992).

A closely-related disease, paramyotonia congenita (PMC), was initially
regarded as a distinct disorder characterized by paradoxical myotonia that
worsened with repeated movement (paramyotonia) and by muscle stiffness
that was aggravated by cold. In some families weakness occurred with extreme
cold, while in others $K^+$ administration worsened the myotonia. DeSilva et al
(1990) reported individuals that had features of HyperPP or PMC, but not both,
within the same family. These similarities led many to hypothesize that HyperPP and PMC were caused by a common genetic defect. Finally, other
variants with some features of both HyperPP and PMC occur consistently in
affected members of the same family. These disorders (myotonia fluctuans,
potassium-aggravated myotonia, normokalemic periodic paralysis, acetazo-
lamide-responsive myotonia congenita) are now considered by most neurologists to be part of the HyperPP-PMC complex (Rüdel & Lehmann-Horn 1993,
Ricker et al 1994).

The first major insights into the cause of HyperPP and PMC were made in
the 1980s by Lehmann-Horn, Rüdel, Ricker, and colleagues (Lehmann-Horn
et al 1981, Lehmann-Horn et al 1983, Lehmann-Horn et al 1987). Using the
three-electrode voltage clamp on whole muscle fibers obtained from intercostal
biopsies, they showed that a noninactivating tetrodotoxin (TTX)-sensitive current produces the aberrant depolarization in HyperPP and PMC. These results
directed everyone's attention toward the voltage-gated sodium channel as the
source of the aberrant sarcolemmal excitability.

## SODIUM CHANNEL MUTATIONS ASSOCIATED WITH HyperPP, PMC, AND RELATED DISORDERS

The association between a sodium channel gene and heritable muscle disorders
was initially established by genetic linkage. Rather than using random allelic
markers, a candidate gene approach was possible because of the implication
of a sodium channel defect from Lehmann-Horn's work (1983). Sodium channels in skeletal muscle are heterodimers composed of a large (260 kDa)

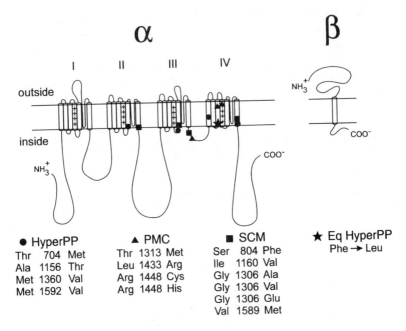

*Figure 2* Schematic diagram of the skeletal muscle sodium channel and the location of disease-associated mutations. The polypeptide backbones of the α and β subunits are depicted as a transmembrane-folding diagram. Every mutation is in the α subunit, and there are no known disorders caused by β subunit mutations. See text and recent reviews for references (Rüdel & Lehmann-Horn 1993, Barchi 1995). Abbreviations: HyperPP, hyperkalemic periodic paralysis; PMC, paramyotonia congenita; SCM, sodium channel myotonia; and Eq HyperPP, equine hyperkalemic periodic paralysis.

pore-forming α subunit and a smaller (38 kDa) noncovalently associated β subunit (Kraner et al 1985, Barchi 1995). The α subunit has four homologous domains, I–IV, each of which is thought to contain six transmembrane segments, S1–S6 (Noda et al 1984). Natural and site-directed mutageneses in Na, Ca, and K channels have provided a conceptual framework of structure-function relations for voltage-gated channels (for review, see Catterall 1992). The residues between S5 and S6 from each domain contribute to formation of the ion-conducting pore. Positively charged residues, Arg and Lys, in the S4 segments form the voltage sensor. Unique to Na channels, a 50-residue cytoplasmic loop between domains III and IV is critical for rapid inactivation and may act as a hinged lid that latches via hydrophobic interactions with the pore (Patton et al 1992, West et al 1992). Trimmer et al (1989) cloned an α subunit isoform selectively expressed in adult innervated skeletal muscle (SkM1 or μ1) from the rat. By using this rat sequence, a portion of the human α subunit cDNA was amplified and used as a probe in a restriction fragment polymorphism analysis. A specific allele of the α subunit and the occurrence of disease cosegregated in four generations of a family with HyperPP plus myotonia far greater than expected by chance, LOD = 4.00 at a recombination fraction of 0 (Fontaine et al 1990). Subsequently, linkage has also been established between the gene for human SkM1 (SCN4A) at 17q23.1–25.3 and HyperPP without myotonia, PMC, or any other variant in the HyperPP-PMC group (Ebers et al 1991; Ptacek et al 1991b,c). More than 50 unrelated families with these disorders have shown genetic linkage to SCN4A, with no recombinations. These results imply that SCN4A or some nearby gene at 17q23 is the causative genetic defect in these disorders.

At least 16 different point mutations in the sodium channel α subunit gene have been identified in HyperPP, PMC, and their variants (for recent reviews, see Rüdel & Lehmann-Horn 1993, Barchi 1995). In each case, the mutation is a transversion of a single nucleotide. The mutations are all within coding regions of the gene and cause substitutions at highly conserved residues (Figure 2). The molecular genetic basis for assigning these changes to mutations rather than to benign polymorphisms is fourfold: (a) All affected members of a family carry a mutant allele; (b) none of the unaffected members has the mutation; (c) screens of large samples of human genomic DNA from the general population do not contain these changes; and (d) affected residues are conserved in Na channels from different tissues and species.

Rüdel & Lehmann-Horn (1993) have found support for a genotype-phenotype correlation within these disorders. However, the clinical assessment for many pedigrees is incomplete (for example, provocative $K^+$-challenge is not always done), and ambiguities in the clinical phenotype have caused the same kindred to be assigned conflicting diagnoses. The Met1592Val substitution in S6 of domain IV (IVS6) and the Thr704Met in IIS5 account for the majority of cases of HyperPP (Ptacek et al 1991a, Rojas et al 1991). Affected individuals have K-aggravated weakness and no cold sensitivity. Other mutations listed as HyperPP associated in Figure 2 are either very rare or do not have all the clinical features of classical HyperPP. The most consistently occurring mutations associated with PMC are Thr1313Met in the III-IV cytoplasmic linker (McClatchey et al 1992) and Arg1448Cys or Arg1448His in IVS4 (Ptacek et al 1992). These individuals have cold-induced myotonia. Syndromes with mixed features of HyperPP and PMC have been lumped together as sodium channel myotonias (SCM). A cluster of three SCM mutations occurs in a single Gly in the III-IV linker: Gly1306Ala, Gly1306Val, and Gly1306Glu (McClatchey et al 1992, Lerche et al 1993). These mutations cause myotonia without weakness, and the intensity of symptoms correlates with the severity of the residue substitution Glu > Val > Ala.

In many instances, the same mutation has been found in unrelated families.

Haplotype analysis of dinucleotide repeats has confirmed the lack of a common founder and suggests that the same mutation has arisen multiple times (Wang et al 1993). Conversely, in equine HyperPP all affected animals can be traced to a single sire. Substitution of a Leu for a conserved Phe at the cytoplasmic end of IVS3 accounts for all equine cases tested to date (Rudolph et al 1992).

The molecular genetic data provide very compelling evidence that the HyperPP-PMC disorders are caused by mutations in the $\alpha$ subunit of the skeletal muscle sodium channel. Except for one family with a HyperPP-like phenotype (Wang et al 1993), all studies for which informative testing was possible have implicated a sodium channel defect. Further proof that these amino acid changes are causative mutations, especially for some of the more benign-appearing substitutions, has been provided by measuring sodium currents in affected myotubes or heterologous expression systems.

# ABNORMALITIES IN SODIUM CHANNEL FUNCTION

## Rapid Inactivation Is Impaired

The primary defect in Na channel function caused by the HyperPP-PMC–associated mutations is an impairment of rapid inactivation. Several distinct types of inactivation defect have been observed for specific mutations.

HyperPP MUTATIONS CAUSE A SMALL PERSISTENT Na CURRENT    The steady-state open probability for Na channels in depolarized muscle is normally very small, on the order of 0.001 (Patlak & Ortiz 1986). In response to depolarization, channels open after a brief delay and close within a millisecond to an inactive state from which further openings are exceedingly rare (Aldrich & Stevens 1987). Repolarization is necessary for channels to recover from the inactive to closed state(s).

In cell-attached patch recordings from cultured myotubes containing the Met1592Val HyperPP mutation (Cannon et al 1991), depolarization occasionally elicited repetitive openings and closings of Na channels (Figure 3). This impairment of inactivation produced a steady-state open probability of ~0.025 in depolarized patches (ensembled average, Figure 3). The abnormal persistent Na current is about 8% of the peak and is detectable in whole-cell recordings of heterologously expressed mutants (Cannon & Strittmatter 1993). Sequential test depolarizations of patches containing only one Na channel showed that mutants inactivate normally most of the time. The persistent current is derived from intermittent failure of inactivation that occurs in clusters of consecutive trials (Figure 3). Thus the Met1592Val mutation causes a slow transition, or shift, in gating between normal and noninactivating modes. The mutation does

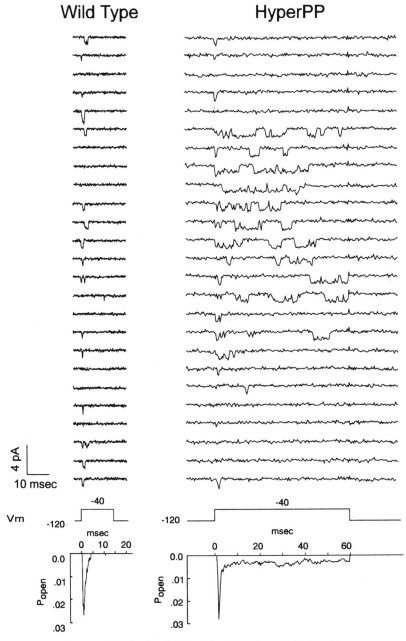

*Figure 3* Impairment of inactivation in HyperPP sodium channels. Unitary Na currents were elicited with depolarizing pulses in cell-attached patches on normal and HyperPP (Met1592Val) myotubes. The latency to opening and the current amplitude were unchanged, but inactivation was disrupted, as evidenced by reopenings (downward current deflection) and prolonged open times. The clustering of noninactivating behavior in consecutive trials suggests a modal switch in gating. Ensemble averages (*bottom*) show the increased steady-state open probability caused by disruption of inactivation. (Adapted from Cannon et al 1991.)

not give rise to a new gating mode. Bursts of reopenings do occur in wild-type channels, although about 20-fold less frequently, and clustering of 2 or 3 noninactivating trials occurs with regularity only in mutants. The open time distribution for channels switching between modes contains two exponential components, with the second being about 4 times slower than the normal 0.25 ms component at −20 mV (Cannon & Strittmatter 1993). This implies that the noninactivating mode is associated with a kinetically distinct open state. In contrast to the persistent Na current, the inactivating component displays normal kinetics. In whole-cell recordings of Met1592Val mutants expressed in HEK cells, neither the voltage dependence of current decay ($\tau_h$) nor the voltage dependence of steady-state inactivation ($h_\infty$) is altered.

Other HyperPP-associated mutants also have an abnormal persistent current. The Phe to Leu substitution in IVS3 of equine HyperPP causes an eightfold increase in steady-state open probability, and bursts of reopenings occur in clusters of consecutive trials (Cannon et al 1995). A persistent Na current was observed for the other common HyperPP mutation in humans, Thr704Met, but a modal switch in gating was not present. Rather, a sprinkling of late openings occurred in most depolarizations, and the open time distribution was fit by a single slow exponential (Cannon & Strittmatter 1993). Whole-cell currents for heterologously expressed Thr704Met channels also had a ~10 mV hyperpolarizing shift in the voltage dependence of activation, which may contribute to a persistent "window" current at modest depolarization (Cummins et al 1993). The primary defect for the Val1589Met mutation, which causes $K^+$-aggravated myotonia, is also an increased persistent Na current (Mitrovic et al 1994), similar to Met1592Val.

The Thr704Met mutant is the only tested mutation from Figure 2, including the two PMC-associated Arg mutations in IVS4 (Chahine et al 1994), known to affect activation (Cummins et al 1993, Yang et al 1994). Data from both studies show an unequivocal 10 mV hyperpolarizing shift in the I–V or G(V) curves and shorter times to normalized peak current, but interpretation of these whole-cell measurements is problematic. Due to coupling between the kinetics of opening and inactivation, a change in the transition rate from the open to the inactive states alone is sufficient to cause these phenomena (Gonoi & Hille 1987). Even first latency data from single-channel records do not provide an unbiased measurement of activation when the inactive state is nonabsorbing. C → I → O transitions can be misinterpreted as evidence for slowed activation.

MYOTONIC MUTATIONS ALTER RATE OF ENTRY AND VOLTAGE DEPENDENCE OF INACTIVATION    For most of the myotonia-associated mutations, the predominant defect is an alteration in the rate and voltage dependence of inactivation. Chahine et al (1994) showed that two heterologously expressed PMC mutations, Arg1448His and Arg1448Cys, caused a slowing and a reduced voltage dependence of inactivation. The difference was most pronounced in the −20 to +20 mV range, where the time constant of $I_{Na}$ decay ($\tau_h$) fell from 2.5 to 0.3 ms in wild-type channels but remained relatively voltage-independent at 2.5 ms for either mutant. Although cooling worsens myotonia in PMC, no significant difference in the temperature sensitivity of $\tau_h$ occurred for either mutation at Arg1448 (Chahine et al 1994). The voltage dependence of steady-state inactivation—$h_\infty(V)$—was also diminished with a twofold reduction in slope and a 5- to 10-mV shift in the hyperpolarizing direction. Steady-state Na currents measured in whole-cell recordings were not distinguishably larger than those observed for wild-type channels. Recovery from inactivation was accelerated twofold at −70 mV for either mutant. Although the positively-charged Arg and Lys residues in S4 segments most likely form part of the voltage-sensing mechanism (Stühmer et al 1989), the highly voltage-dependent process of activation was not altered by either mutation. These same investigators (Yang et al 1994) reported similar inactivation defects: slowed macroscopic rate, reduced voltage dependence, and enhanced rate of recovery for several other myotonia-associated mutants—Thr1313Met, Ala1156Thr, and Leu1433Arg.

Some myotonia-associated mutants produce both an increased persistent Na current and a slowed inactivation. Three different mutations at a single Gly in the III-IV linker are associated with myotonic disorders (Figure 2). Cell-attached patch recordings from acutely dissociated biopsy specimens showed an increased steady-state Na current (Mitrovic et al 1994). The amplitude of the persistent current varied from no change (Gly → Ala) to 4.5-fold (Gly → Val) and 9-fold (Gly → Glu) larger than normal. Similar to the mutations at Arg 1448, $\tau_h$ was slower and had a reduced voltage dependence.

The consequences of mutations in the III-IV linker have also been studied in heterologous expression systems (Hayward et al 1995). Representative currents in Figure 4 illustrate slowed rates of inactivation that occur in these myotonic disorders. Of the mutations at Gly1306, the Gly → Glu substitution causes the most severe disruption of inactivation. The decay in macroscopic current is slowed twofold (Figure 4) and steady-state inactivation is shifted to the right (depolarizing) by 10 mV. Myotonia is worsened by cold for the Thr1313Met mutation but not for the substitutions at Gly1306. Interestingly, although $\tau_h$ had comparable temperature sensitivities for wild-type, Gly1306Glu, and Thr1313Met constructs ($Q_{10}$ = 3.4, 2.7, 3.5), the amplitude of the steady-state Na current increased tremendously with cooling for Thr1313Met but not for Gly1306Glu or wild type (LJ Hayward, RH Brown & SC Cannon, unpublished results). This difference may contribute to the temperature sensitivity of PMC in the Thr1313Met mutation.

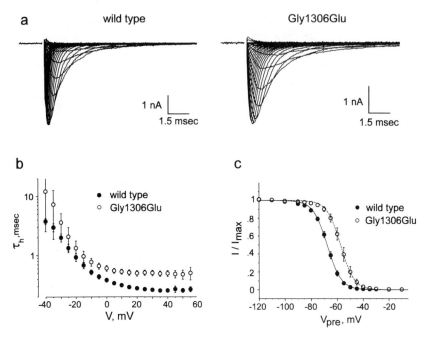

*Figure 4* Disruption of macroscopic inactivation by the Gly → Glu mutation in the III-IV linker. Whole-cell Na currents were recorded from HEK cells cotransfected with the β subunit and either wild-type or Gly1306Glu α subunit cDNA. (*a*) Currents were elicited by a series of test depolarizations from −80 to +80 mV in increments of 5 mV from a holding potential of −120 mV. (*b*) Rate of macroscopic current decay, $\tau_h$, is twofold slower in mutants. (*c*) Steady-state inactivation by a 300-ms prepulse, $h_\infty$, is shifted by 10 mV in the depolarizing direction. (Adapted from Hayward et al 1995.)

## Is The Inactivation Defect K Dependent?

A controversial issue has been whether extracellular [K+] affects the gating of Na channel mutants that occur in K-sensitive clinical phenotypes. Such an effect could contribute to the episodic nature of the K-induced weakness in HyperPP. We have observed a [$K_o^+$]-dependent increase in the amplitude of the aberrant persistent Na current for two HyperPP-associated mutations: Met1592Val (Cannon et al 1991) and the Phe-to-Leu substitution in IVS3 of equine HyperPP (Cannon et al 1995). The steady-state open probability of mutant channels was increased about threefold for currents recorded with 10-mM K+ in the recording pipette (extracellular face of the channel) compared to ones with 0 mM.

The $K_o$-dependent impairment of inactivation has been observed in muscle preparations only. When the same mutation, Met1592Val, was introduced into rat SkM1 cDNA and heterologously expressed in HEK cells, an aberrant persistent Na current was observed in whole-cell recordings. However, the amplitude of the persistent current was abnormally large even in 1-mM $[K_o^+]$ and did not increase for cells superfused with higher $[K_o^+]$ (Cannon & Strittmatter 1993). Thus a specific inactivation defect is reproduced in the heterologous system, but there is no detectable $K_o$-effect. The reason for the discrepancy remains unknown. Several other mutations associated with K-aggravated clinical phenotypes have been tested in heterologous expression systems, and none have had K-dependent inactivation defects (Cummins et al 1993, Chahine et al 1994, Mitrovic et al 1994). Attempts to reconstitute a $K_o$-effect in Met1592Val expressed in HEK cells by cotransfection with a β subunit (LJ Hayward & SC Cannon, unpublished data) or by using the human SkM1 cDNA (SC Cannon & AL George, unpublished data) were unsuccessful.

The notion that an extracellular cation could alter ion channel gating kinetics is not without precedent. Submillimolar amounts of $Zn^{2+}$ slow the kinetics of activation for several types of K channel (Spires & Begenisich 1994). The mechanism cannot be explained by screening of surface charge (Armstrong & Lopez-Barneo 1987), and an allosteric gating site has been proposed (Neyton & Pelleschi 1991).

Whatever the mechanism, extracellular influences can alter Na channel inactivation, a process usually attributed to cytoplasmic domains. Polypeptide toxins from sea anemone or α scorpion toxins impede inactivation when applied extracellularly (for review, see Catterall 1992). Mutations near the extracellular mouth of the pore can slow the decay of macroscopic Na currents (Tomaselli et al 1995). Finally, mutations in the extracellular amino terminus of the β subunit disrupt the β-induced shift to the fast inactivating mode for Na channels expressed in Xenopus oocytes (Chen & Cannon 1994). Thus changes in extracellular $K^+$ could conceivably alter the gating kinetics of mutant Na channels.

## Structure-Function Correlations for the HyperPP-PMC Mutations

At first glance, the distribution of disease-associated mutations is spread widely throughout the primary structure of the α subunit (Figure 2) so that no consistent functional defect can be predicted. When viewed from the other perspective, however, coherent motifs emerge from the knowledge that all of these mutations primarily affect inactivation.

Four of the 15 mutations lie in the III-IV linker, a cytoplasmic loop proposed to act as an inactivation gate (Stühmer et al 1989), or hinged lid (Patton et al 1992, West et al 1992), that occludes the open pore. Mutations here are therefore expected to alter inactivation. Residue 1306 is the first of a pair of glycines that may confer a critical flexibility to the hinge. Mutations with successively larger side groups at this residue caused progressively severe impairment of inactivation (Mitrovic et al 1994, Hayward et al 1995). The Thr mutation at 1313 is adjacent to a triplet, Ile-Phe-Met, that may form a latch to keep the inactivation gate shut by hydrophobic interactions with the pore (West et al 1992, Eaholtz et al 1994). One possible mechanism for the particularly large persistent Na current in the Thr1313Met mutant is an increased off rate caused by destabilization of the hydrophobic interaction. All of the mutations within the III-IV loop are associated with predominantly myotonic, rather than paralytic, syndromes.

Six mutations are located at the cytoplasmic end of S5 or S6 segments (Figure 2). By analogy to K channels, these regions are postulated to lie near the inner vestibule of the pore and may contribute to the receptor site for an inactivation gate (Isacoff et al 1991). Consequently, an inactivation defect from these mutations is also consistent with current models. This group of mutations accounts for the vast majority of families with HyperPP (Thr704Met and Met1592Val). As with HyperPP, the myotonic disorders caused by mutations at the cytoplasmic end of S5 and S6 (Ser804Phe, Ile1160Val, and Val1589Met) are also aggravated by raised extracellular $K^+$. This phenomenon could simply be caused by a $K_o$-induced (Nernstian) depolarization of the resting potential, which would then open Na channels and reveal an inactivation defect. Alternatively, these mutations at the inner mouth of the pore may destabilize the inactive state and thereby unmask a $K_o^+$ effect on inactivation.

Three mutations are in the outermost Arg of IVS4 or an adjacent region of S3. These mutations of the putative voltage sensor are expected to affect the voltage dependence of activation. However, only small changes occurred in the first latency to opening. Despite normal activation, inactivation was impaired, as demonstrated by slowed Na current decay and reduced apparent voltage dependence of $\tau_h$ (Chahine et al 1994). Based on single-channel data, these authors concluded that closing from the open to the inactive states was impaired, which reduced the normal coupling between activation and inactivation. These mutations all cause cold-aggravated phenotypes.

Several additional features of the distribution of mutations within the primary structure of the $\alpha$ subunit deserve comment. No mutations are in the pore-forming S5-S6 loops, and only one report (Lehmann-Horn et al 1991) of alteration in unitary conductance has been published. Except for possibly the Thr704Met mutation, none of the 11 mutants tested to date have significantly affected activation. Finally, there is a paucity of mutations in the amino end of the $\alpha$ subunit. This may be a feature of the structure-function constraints of the channel, or it could represent an ascertainment bias, because molecular genetic screening has been more efficient at the 5' end of the gene where the exons are larger.

# FUNCTIONAL CONSEQUENCES OF AN INACTIVATION DEFECT

The lack of a convenient animal model and the short-lived viability of biopsied muscle fibers in these very rare diseases have created major impediments to studying the functional consequences of the Na channel defects in the HyperPP-PMC complex. To date, a transgenic model is not available. If Na channel mutations are to be accepted as the cause of weakness or myotonia, then it is essential to establish that the inactivation defects identified by recording ionic currents are sufficient to cause the clinical phenotype. We have used two approaches to explore the consequences of impaired Na channel inactivation in skeletal muscle: a toxin-based in vitro model (Cannon & Corey 1993) and a computer simulation (Cannon et al 1993).

Within the clinical spectrum of HyperPP-PMC, the predominant symptom in some families is myotonia, while in other families the phenotypic expression of disease is paralysis. Are these distinct clinical phenotypes caused by specific types of Na channel inactivation defect? As detailed above, specific mutations are consistently associated with a particular clinical phenotype (Rüdel & Lehmann-Horn 1993). In addition, the functional consequences of each $\alpha$ subunit mutation can be grouped into distinct classes of inactivation defect. Thus it seems plausible that some types of inactivation defect may cause myotonia, while other types produce paralysis. The ability to simulate specific types of inactivation defect in the theoretical model has provided an opportunity to test this hypothesis (Cannon 1994).

## A Small Persistent Sodium Current Is Sufficient to Cause Myotonia or Paralysis

In HyperPP the predominant functional defect is a small persistent Na current (Figure 3). The steady-state open probability is about 0.025, which is small but 25 times larger than normal. A comparable persistent Na current was induced in rat fast-twitch muscle by application of anemone toxin (ATXII) in vitro (Cannon & Corey 1993). Single-channel recordings established that micromolar amounts of ATXII increased the steady-state open probability of Na channels in rat skeletal muscle from 0.0011 to 0.018 (Figure 5a). By comparison, this is in the low range for the persistent currents observed in HyperPP myotubes and corresponds to a $I_{ss}/I_{peak}$ measured in whole-cell experiments of about 0.05. When whole-muscle preparations were bathed in micromolar ATXII, the relaxation in twitch tension was slowed by an order of magnitude (Figure 5b). Injection of a long-duration current pulse to a single fiber normally elicits one or two action potentials. In ATXII-exposed muscle, a single pulse elicited a train of repetitive discharges that waxed and waned in frequency (Figure 5c). Thus a small persistent Na current with an open probability on the order of 0.01–0.02 is sufficient to produce mechanical and electrical myotonia. In other studies with more potent highly-purified ATXII, the toxin caused paralysis in mammalian (Alsen et al 1981) and amphibian skeletal muscle (Khan et al 1986).

The mechanism by which a small persistent Na current leads to myotonia or paralysis has been further defined by computer simulation for a model muscle cell (Cannon et al 1993). The model consisted of two electrically-coupled membrane compartments to simulate the sarcolemma and T tubules. Each compartment contained voltage-gated conductances that were approximated by the Hodgkin-Huxley formalism. Finally, changes in [K+] for the T-tubule space were estimated from a balance between current flow across the T-tubule membrane and passive diffusion into the bulk solution. In agreement with the animal model, the theoretical model generated myotonic discharges when even a small proportion of the Na conductance failed to inactivate (Figure 6). Loss of inactivation in a small fraction of Na channels, 0.008 or greater, was sufficient to generate myotonic runs. With slightly more noninactivating Na channels (0.025 or greater), the train of discharges decayed in amplitude as the resting potential shifted to a maintained depolarization at about −40 mV. From the depolarized state, an action potential cannot be elicited. The overwhelming majority of Na channels are inactivated, thereby causing flaccid paralysis in an inexcitable fiber.

Loss of inactivation in a small proportion of Na channels is more likely to cause paralysis than a primarily myotonic disorder. The computer simulations show that sustained trains of myotonic discharges occur for only a narrow range of persistent current (Cannon et al 1993). Muscle behavior is normal if the fraction of noninactivating channels, $f$, is less than 0.008. When $f$ is between 0.008 and 0.02, myotonia occurs. In the range 0.03 to 0.04, the initial response is myotonia, but progressive K+ accumulation in the T tubule leads to paralysis (Figure 6). More severe disruption of inactivation, $f > 0.04$, will cause paralysis without myotonia. These simulation results are consistent with the genotype-phenotype correlations and functional defects identified in specific mutants. The two most common causes of HyperPP—Met1592Val and Thr704Met—

----------------------------------------------------------------→

*Figure 5* A small proportion of noninactivating Na channels is sufficient to produce myotonia. The inactivation defect in HyperPP was mimicked by application of 10 μM of anemone toxin (ATXII) to rat muscle in vitro. (*a*) Unitary Na currents were recorded from cell-attached patches. ATXII partially disrupts inactivation and causes a steady-state open probability of 0.02 (averaged traces), comparable to the defect in HyperPP. (*b*) Isometric twitch responses were elicited by a 5 ms shock that was 1.5 times threshold intensity. Relaxation in force is markedly prolonged in toxin-exposed muscle, similar to kinograms from patients with myotonia. (*c*) Microelectrode injection of current elicits one or two spikes in normal muscle. A small defect in Na channel inactivation causes a run of myotonic discharges that persists beyond the stimulus. (Adapted from Cannon & Corey 1993.)

cause abnormally large persistent Na currents (in voltage-clamp, ~5% of the initial peak) (Cannon & Strittmatter 1993). Conversely, the Val1589Met mutation causes a smaller noninactivating Na current (3% of peak) and leads to myotonia without significant weakness (Mitrovic et al 1994).

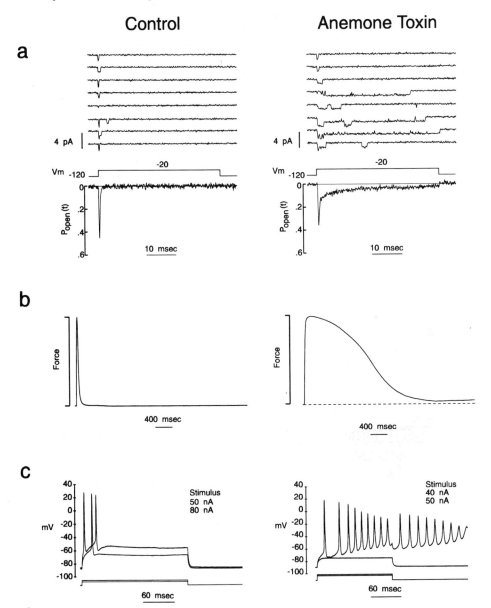

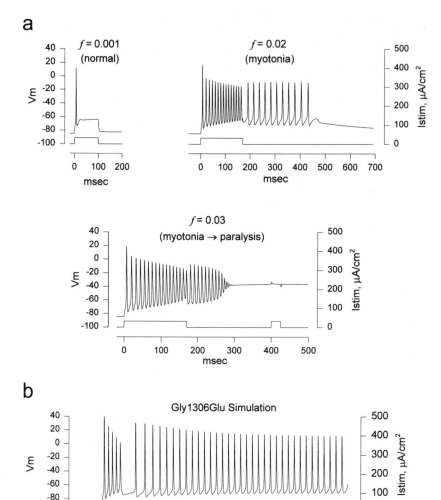

*Figure 6*  Simulated responses in a model muscle cell with different types of inactivation defect for the Na current. (*a*) To simulate HyperPP (Figure 3), a fraction of the Na current, *f*, activated normally but failed to inactivate. Small persistent Na currents (*f* ~ 0.02) cause brief runs of myotonic discharges. With any larger disruption of inactivation, the discharges decay to a depolarized state from which the system is refractory from firing action potentials (*f* = 0.03). (*b*) Prolonged trains of myotonic discharges occur when Na current inactivation is slowed, and its steady-state voltage dependence is shifted in the depolarizing direction. Model parameters were set to simulate the defects measured for the Gly1306Glu mutant in Figure 4. (See Cannon et al 1993 for details of the model.)

This model explains the dominant pattern of expression in HyperPP. In molecular genetic terms, the mutation actually causes a gain-of-function, which is manifest as a partial loss of inactivation. If it is small, the aberrant persistent Na current is sufficient to produce the enhanced excitability of myotonia. With a slightly larger fraction of noninactivating channels, the persistent inward current depolarizes the membrane and thereby inactivates Na channels from the normal allele (as well as most of the mutants). Thus, as a consequence of its effect on the resting potential, the mutant Na channel behaves as a dominant negative to cause paralysis.

## An Intact T-Tubule System Is Required for the Generation of Myotonia

Repetitive discharges, continuing beyond the stimulus period, are the hallmark of myotonia. In studies on myotonic goats, Adrian & Bryant (1974) concluded that myotonic discharges are driven by cumulative after-depolarizations. For both normal and myotonic fibers, a series of discharges leaves the membrane more depolarized than the original resting potential (about 1 mV per impulse). This after-depolarization decays with a half-time of about 0.5 s and is thought to arise from activity-driven accumulation of $K^+$ in the transverse tubule, an extracellular space. For normal muscle the membrane simply repolarizes, but the reduced Cl conductance in the myotonic goat lowers the threshold for spike generation such that an after-depolarization elicits myotonic discharges. In support of this hypothesis, detubulation with osmotic shock abolished both after-depolarization and after-discharges. The same mechanism drives the myotonic discharges in muscle fibers with impaired Na channel inactivation. When ATXII-exposed fibers are detubulated, repetitive discharges are elicited while the stimulus is applied, but a progressive depolarization, after-depolarization, and after-discharges do not occur (Cannon & Corey 1993). Thus the integrity of the T tubule may be a requirement in common with all causes of myotonia. Either Cl channel or Na channel mutations can enhance the excitability of the membrane, but a depolarizing influence from $K^+$-laden T tubules is needed to sustain the run of myotonic discharges.

## Slowed Inactivation with an Altered Voltage Dependence May Cause a Predilection for Myotonia

A distinct form of inactivation defect has been observed in Na channel mutants associated with primarily myotonic disorders. As detailed above, the defects in common with these mutations include a decrease in the rate of macroscopic inactivation, altered voltage dependence, and faster recovery (Chahine et al 1994, Yang et al 1994). Within this group, only the Thr1313Met mutant has a large noninactivating component. For all the myotonia-associated mutants, however, either a depolarizing shift or a reduced voltage dependence of steady-state inactivation, $h_\infty(V)$, produces a persistent "window current." The window occurs at mild depolarizations, in the $-60$ to $-50$ mV range, sufficient to activate Na channels but where inactivation is incomplete. Our computer simulation was used to investigate the consequences of these alterations in Na channel inactivation (Cannon 1994). Figure 6b shows the responses obtained when Na channel gating parameters were adjusted to simulate the behavior of the Gly1306Glu mutant shown in Figure 4. Long runs of myotonic discharges were triggered by suprathreshold current stimuli. Damped oscillations to a stable depolarized membrane potential did not occur. Thus there was no paralysis.

Variations in the model parameters to simulate the range of Na channel inactivation defects observed for the myotonia-associated mutants always produced repetitive discharges without paralysis. The window current rapidly diminishes for depolarizations positive to $-50$ mV. Therefore, unlike HyperPP, there is no persistent current to give rise to a depolarized resting potential at $-45$ to $-40$ mV. The alteration in $h_\infty$ that produces the window current also promotes oscillations (repetitive discharges) in the $-70$ to $-50$ mV range, because fewer Na channels are inactivated than normal. The slowed rate of macroscopic inactivation, $\tau_h$, causes equilibrium potentials in the $-55$ to $-40$ mV range to be unstable. In other words, slowed inactivation creates a dynamic instability that causes oscillations and prevents plateau depolarizations. Interestingly, the accelerated rate of recovery from inactivation observed in experimental mutants did not appear to play a critical role in the generation of myotonic behavior. Simulations with normal recovery rates at voltages less than $-60$ mV but with altered $\tau_h$ and $h_\infty$ produced myotonia, whereas accelerated recovery at hyperpolarized potentials in combination with normal $\tau_h$ and $h_\infty$ did not.

## PROSPECTUS

There remains little doubt that HyperPP, PMC, and related disorders are caused by mutations in the $\alpha$ subunit of the skeletal muscle Na channel. In comparison to work on other familial diseases, progress has been unusually rapid over the past five years due to the convergence of information form multiple disciplines. The search for the causative mutation (Fontaine et al 1990) was tremendously accelerated because a candidate gene was implicated by the physiologic identification of an abnormal Na current (Lehmann-Horn et al 1981, Lehmann-Horn et al 1983) and by the previous cloning of a rat skeletal muscle $\alpha$ subunit (Trimmer et al 1989). The second major advantage was that 40 years of research had established the normal function of Na channels so that the rather subtle defects caused by the mutations in HyperPP-PMC could be identified.

Finally, the ability to reconstitute the clinical phenotype in animal (Cannon & Corey 1993) and computer (Cannon et al 1993, Cannon 1994) models provides powerful confirmation that the identified defects are likely to be causative.

Many details remain to be filled in, especially in the realm of understanding the rich spectrum of clinical phenomena. What causes the cold sensitivity in PMC? What is the mechanism of the warm-up and paramyotonia phenomena? If the K aggravation of weakness or myotonia is produced by (Nernstian) membrane depolarization, why isn't every variant sensitive to K? Does this imply that a subset of mutations causes K-dependent inactivation defects? What features of exercise are protective against HyperPP, despite elevated $K^+$? The answers to these question will probably await the development of transgenic animals for the manipulation and study of whole muscle or whole animal behavior.

Despite these gaps in knowledge, our understanding of the pathophysiology of HyperPP-PMC stands out as one of the best examples for which the process can be traced over several magnitudes of scale: from point mutation to altered protein function to cellular and clinical phenotype.

ACKNOWLEDGMENTS

Work from the author's laboratory was supported by the Howard Hughes Medical Institute, the Muscular Dystrophy Association, the Alfred P. Sloan Foundation, and the NIH (AR42703). David P. Corey was an instrumental collaborator during the early stages of these investigations and later generously shared laboratory facilities.

## Literature Cited

Adrian RH, Bryant SH. 1974. On the repetitive discharge in myotonic muscle fibers. *J. Physiol.* 235:103–31

Aldrich RW, Stevens CF. 1987. Voltage-dependent gating of single sodium channels from mammalian neuroblastoma cells. *J. Neurosci.* 7:418–31

Alsen C, Harris JB, Tesseraux I. 1981. Mechanical and electrophysiological effects of sea anemone (*Anemonia sulcata*) toxins on rat innervated and denervated skeletal muscle. *Br. J. Pharmacol.* 74:61–71

Armstrong C, Lopez-Barneo J. 1987. External calcium ions are required for potassium channel gating in squid neurons. *Science* 236:712–14

Barchi RL. 1992. The nondystrophic myotonic syndromes. In *Handbook of Clinical Neurology*, ed. LP Rowland, S DiMauro, 62:261–86. New York: Elsevier Sci.

Barchi RL. 1995. Molecular pathology of the skeletal muscle sodium channel. *Annu. Rev. Physiol.* 57:355–85

Bradley WG, Taylor R, Rice DR, Hausmanowa-Petruewicz I, Adelman LS, et al. 1990. Progressive myopathy in hyperkalemic periodic paralysis. *Arch. Neurol.* 47:1013–17

Brook JD, McCurrach ME, Harley HG, Buckler AJ, Church D, et al. 1992. Molecular basis of myotonic dystrophy: expansion of a trinucleotide (CTG) repeat located at the 3' end of a transcript encoding a protein kinase family member. *Cell* 68:799–808

Brown GL, Harvey AM. 1939. Congenital myotonia in the goat. *Brain* 62:341–63

Buruma OJS, Schipperheyn J. 1979. Periodic paralysis. In *Diseases of Muscle, Part 2. Handbook of Clinical Neurology,* ed. PJ Vinken, GW Bruyn, SP Ringel, 41:147–74. New York: Elsevier

Cannon SC. 1994. A predilection for myotonia or paralysis based on different defects in Na channel inactivation. *J. Gen. Physiol.* 104:20 (Abstr.)

Cannon SC, Brown RH, Corey DP. 1991. A sodium channel defect in hyperkalemic periodic paralysis: potassium-induced failure of inactivation. *Neuron* 6:619–26

Cannon SC, Brown RH, Corey DP. 1993. Theoretical reconstruction of myotonia and paralysis caused by incomplete inactivation of sodium channels. *Biophys. J.* 65:270–88

Cannon SC, Corey DP. 1993. Loss of sodium channel inactivation by Anemone toxin (ATX II) mimics the myotonic state in hyperkalemic periodic paralysis. *J. Physiol.* 466:501–20

Cannon SC, Hayward LJ, Beech J, Brown RH. 1995. Sodium channel inactivation is impaired in equine hyperkalemic periodic paralysis. *J. Neurophysiol.* 73:1892–99

Cannon SC, Strittmatter SM. 1993. Functional expression of sodium channel mutations identified in families with periodic paralysis. *Neuron* 10:317–26

Catterall WA. 1992. Cellular and molecular biology of voltage-gated sodium channels. *Physiol. Rev.* 72:S15–S48

Chahine M, George AL, Zhou M, Ji S, Sun W, et al. 1994. Sodium channel mutations in paramyotonia congenita uncouple inactivation from activation. *Neuron* 12:281–94

Chen CC, Cannon SC. 1994. β subunit modulation of Na channel inactivation does not occur from the cytoplasmic side. *Biophysical J.* 66:A243

Cox JH. 1985. An episodic weakness in four horses associated with intermittent serum hyperkalemia and the similarity of the disease to hyperkalemic periodic paralysis in man. *Proc. Am. Assoc. Equine Pract.* 21:383–91

Creutzfeldt OD, Abbott PC, Fowler WM, Pearson CM. 1963. Muscle membrane potentials in episodica adynamia. *Electroenceph. Clin. Neurophysiol.* 15:508–15

Cummins TR, Zhou J, Sigworth FJ, Ukomadu U, Stephan M, et al. 1993. Functional consequences of a sodium channel mutation causing hyperkalemic periodic paralysis. *Neuron* 10:667–78

DeSilva SM, Kuncl RW, Griffin JW, Cornblath DR, Chavoustie S. 1990. Paramyotonia congenita or hyperkalemic periodic paralysis?

Clinical and electrophysiological features of each entity in one family. *Muscle Nerve* 13:21–26

Eaholtz G, Scheuer T, Catterall WA. 1994. Restoration of inactivation and block of open sodium channels by an inactivation gate peptide. *Neuron* 12:1041–48

Ebers GC, George AL, Barchi RL, Ting-Passador SS, Kallen RG, et al. 1991. Paramyotonia congenita and hyperkalemic periodic paralysis are linked to the adult muscle sodium channel gene. *Ann. Neurol.* 30:810–16

Fahlke C, Zachar E, Rüdel R. 1993. Chloride channels with reduced single-channel conductance in recessive myotonia congenita. *Neuron* 10:225–32

Fontaine B, Khurana TS, Hoffman EP, Bruns GAP, Haines JL, et al. 1990. Hyperkalemic periodic paralysis and the adult muscle sodium channel α-subunit gene. *Science* 250:1000–2

Fontaine B, Vale-Santos J, Jurkatt-Rott K, Reboul F, Plassart E, et al. 1994. Mapping of the hypokalemic periodic paralysis (HypoPP) locus to chromosome 1q31–32 in three European families. *Nat. Genet.* 6:267–72

Franke C, Hatt H, Iaizzo PA, Lehmann-Horn F. 1990. Characteristics of $Na^+$ channels and $Cl^-$ conductance in resealed muscle fibre segments from patients with myotonic dystrophy. *J. Physiol.* 425:391–405

George AL, Crackower MA, Abdalla JA, Hudson AJ, Ebers GC. 1993. Molecular basis of Thomsen's disease (autosomal dominant myotonia congenita). *Nat. Genet.* 3:305–10

Gonoi T, Hille B. 1987. Inactivation modifiers discriminate among gating models. *J. Gen. Physiol.* 89:253–74

Hayward LJ, Brown RH, Cannon SC. 1995. Inactivation defects in sodium channel III-IV linker mutations associated with myotonia. *Biophys. J.* 68:A154

Heine R, George AL, Pika U, Deymeer F, Rüdel R, Lehmann-Horn F. 1994. Proof of a non-functional muscle chloride channel in recessive myotonia congenita (Becker) by detection of a 4 base pair deletion. *Hum. Mol. Genet.* 3:1123–28

Hoffman EP, Lehmann-Horn F, Rüdel R. 1995. Overexcited or inactive: ion channels in muscle disease. *Cell* 80:681–86

Isacoff EY, Jan YN, Jan LY. 1991. Putative receptor for the cytoplasmic inactivation gate in the Shaker $K^+$ channel. *Nature* 353:86–90

Jurkatt-Rott K, Lehmann-Horn F, Albaz A, Heine R, Gregg RG, et al. 1994. A calcium channel mutation causing hypokalemic periodic paralysis. *Hum. Mol. Genet.* 3:1415–19

Khan AR, Lemeignan M, Molgo J. 1986. Effects of toxin II from the sea anemone *Anemonia sulcata* on contractile and electrical responses of frog skeletal muscle fibres. *Toxicon* 24:373–84

Koch MC, Steinmeyer K, Lorenz C, Ricker K, Wolf F, et al. 1992. The skeletal muscle chloride channel in dominant and recessive human myotonia. *Science* 257:797–800

Kraner SD, Tanaka JC, Barchi RL. 1985. Purification and functional reconstitution of the voltage-sensitive sodium channel from rabbit T-tubular membranes. *J. Biol. Chem.* 25: 6341–47

Lehmann-Horn F, Iaizzo PA, Hatt H, Franke C. 1991. Altered gating and conductance of Na$^+$ channels in hyperkalemic periodic paralysis. *Pflügers Arch.* 418:297–99

Lehmann-Horn F, Kuther G, Ricker K, Grafe P, Ballanyi K, Rüdel R. 1987. Adynamia episodica hereditaria with myotonia: a noninactivating sodium current and the effect of extracellular pH. *Muscle Nerve* 10:363–74

Lehmann-Horn F, Rüdel R, Dengler R, Lorkovic H, Haass A, Ricker K. 1981. Membrane defects in paramyotonia congenita with and without myotonia in a warm environment. *Muscle Nerve* 4:396–406

Lehmann-Horn F, Rüdel R, Ricker K, Lorkovic H, Dengler R, Hopf HC. 1983. Two cases of adynamia episodica hereditaria: in vitro investigation of muscle membrane and contractile parameters. *Muscle Nerve* 6:113–21

Lerche H, Heine R, Pika U, George AL, Mitrovic N, et al. 1993. Human sodium channel myotonia: slowed channel inactivation due to substitutions for a glycine within the III-IV linker. *J. Physiol.* 470:13–22

Lipicky RJ. 1979. Myotonic syndromes other than myotonic dystrophy. In *Handbook of Clinical Neurology*, ed. PJ Vinken, GW Bruyn, 40:533–71. New York: Elsevier

Lipicky RJ, Bryant SH. 1966. Sodium, potassium, and chloride fluxes in intercostal muscle from normal goats and goats with hereditary myotonia congenita. *J. Gen. Physiol.* 50:89–111

McClatchey AI, Van den Bergh P, Pericak-Vance MA, Raskind W, Verellen C, et al. 1992. Temperature-sensitive mutations in the III-IV cytoplasmic loop region of the skeletal muscle sodium channel gene in paramyotonia congenita. *Cell* 68:769–74

Mitrovic N, George AL, Heine R, Wagner S, Pika U, et al. 1994. K$^+$-aggravated myotonia: destabilization of the inactivated state of the human muscle Na$^+$ channel by the V1589M mutation. *J. Physiol.* 478:395–402

Naylor JM, Robinson JA, Bertone J. 1992. Familial incidence of hyperkalemic periodic paralysis in quarter horses. *J. Am. Vet. Med. Assoc.* 200:340–43

Neyton J, Pelleschi M. 1991. Multi-ion occupancy alters gating in high conductance Ca$^{2+}$-activated K$^+$ channels. *J. Gen. Physiol.* 97:641–65

Noda M, Shimizu S, Tanabe T, Takai T, Kayano T, et al. 1984. Primary structure of *Electrophorus electricus* sodium channel deduced from cDNA sequence. *Nature* 312: 121–27

Patlak J, Ortiz M. 1986. Two modes of gating during late Na$^+$ channel currents in frog sartorius muscle. *J. Gen. Physiol.* 87:305–26

Patton DE, West JW, Catterall WA, Goldin AL. 1992. Amino acid residues required for fast Na$^+$-channel inactivation: charge neutralizations and deletions in the III-IV linker. *Proc. Natl. Acad. Sci. USA* 89:10905–9

Ptacek LJ, George AL, Barchi RL, Griggs RC, Riggs JE, et al. 1992. Mutations in an S4 segment of the adult skeletal muscle sodium channel cause paramyotonia congenita. *Neuron* 8:891–97

Ptacek LJ, George AL, Griggs RC, Tawil R, Kallen RG, et al. 1991a. Identification of a mutation in the gene causing hyperkalemic periodic paralysis. *Cell* 67:1021–27

Ptacek LJ, Johnson KJ, Griggs RC. 1993. Genetics and physiology of the myotonic muscle disorders. *New Engl. J. Med.* 328:482–89

Ptacek LJ, Tawil R, Griggs RG, Engel AG, Layzer RB, et al. 1994. Dihydropyridine receptor mutations cause hypokalemic periodic paralysis. *Cell* 77:863–68

Ptacek LJ, Trimmer JS, Agnew WS, Roberts JW, Petajan JH, Leppert M. 1991b. Paramyotonia congenita and hyperkalemic periodic paralysis map to the same sodium-channel gene locus. *Am. J. Hum. Genet.* 49: 851–54

Ptacek LJ, Tyler F, Trimmer JS, Agnew WS, Leppert M. 1991c. Analysis in a large hyperkalemic periodic paralysis pedigree supports tight linkage to a sodium channel locus. *Am. J. Hum. Genet.* 49:378–82

Pusch M, Steinmeyer K, Jentsch TJ. 1994. Low single channel conductance of the major skeletal muscle chloride channel, ClC-1. *Biophys. J.* 66:149–52

Ricker K, Moxley RT, Heine R, Lehmann-Horn F. 1994. Myotonia fluctuans a third type of muscle sodium channel disease. *Arch. Neurol.* 51:1095–102

Riggs JE, Griggs RC. 1979. Diagnosis and treatment of the periodic paralyses. *Clin. Neuropharmacol.* 4:123–38

Rojas CV, Wang J, Hoffman EP, Powell BR, Brown RH. 1991. A methionine to valine mutation in the skeletal muscle sodium channel alpha-subunit in human hyperkalemic periodic paralysis. *Nature* 54:387–89

Rüdel R, Lehmann-Horn F. 1985. Membrane changes in cells from myotonia patients. *Physiol. Rev.* 65:310–56

Rüdel R, Lehmann-Horn F. 1993. Genotype-phenotype correlations in human skeletal muscle sodium channel diseases. *Arch. Neurol.* 50:1241–48

Rüdel R, Lehmann-Horn F, Ricker K, Küther G. 1984. Hypokalemic periodic paralysis: in vitro investigation of muscle fiber membrane parameters. *Muscle Nerve* 7:110–20

Rudolph JA, Spier SJ, Byrns G, Rojas CV, Dernoco D, Hoffman EP. 1992. Periodic paralysis in quarter horses: a sodium channel mutation disseminated by selective breeding. *Nat. Genet.* 2:144–47

Ruff RL. 1991. Calcium-tension relationships of muscle fibers from patients with periodic paralysis. *Muscle Nerve* 14:838–44

Spires S, Begenisich T. 1994. Modulation of potassium channel gating by external divalent cations. *J. Gen. Physiol.* 104:675–92

Steinmeyer K, Klocke R, Ortland C, Gronemeier M, Jockusch H, et al. 1991. Inactivation of muscle chloride channel by transposon insertion in myotonic mice. *Nature* 354:304–8

Steinmeyer K, Lorenz C, Pusch M, Koch MC, Jentsch T. 1994. Multimeric structure of ClC-1 chloride channel revealed by mutations in dominant myotonia congenita (Thomsen). *EMBO J.* 13:737–43

Stühmer W, Conti F, Suzuki H, Wang X, Noda M, et al. 1989. Structural parts involved in activation and inactivation of the sodium channel. *Nature* 339:597–603

Tanabe T, Beam KG, Powell JA, Numa S. 1988. Restoration of excitation-contraction coupling and slow calcium current in dysgenic muscle by dihydropyridine receptor complementary DNA. *Nature* 336:134–39

Tomaselli GF, Chiamvimonvat N, Nuss HB, Balser JR, Perez-Garcia MT, et al. 1995. A mutation in the pore of the sodium channel alters gating. *Biophys. J.* 68:1814–27

Trimmer JS, Cooperman SS, Tomiko SA, Zhou J, Crean SM, et al. 1989. Primary structure and functional expression of a mammalian skeletal muscle sodium channel. *Neuron* 3:33–49

Wang J, Zhou J, Todorovic SM, Feero WG, Barany F, et al. 1993. Molecular genetic and genetic correlations in sodium channelopathies: lack of founder effect and evidence for a second gene. *Am. J. Hum. Genet.* 52:1074–84

West JW, Patton DE, Scheuer T, Wang Y, Goldin AL, Catterall WA. 1992. A cluster of hydrophobic amino acid residues required for fast Na-channel inactivation. *Proc. Natl. Acad. Sci. USA* 89:10910–14

Yang N, Ji S, Zhou M, Ptacek L, Barchi RL, et al. 1994. Sodium channel mutations in paramyotonia congenita exhibit similar biophysical phenotypes in vitro. *Proc. Natl. Acad. Sci. USA* 91:12785–89

Annu. Rev. Neurosci. 1996. 19:165–86
Copyright © 1996 by Annual Reviews Inc. All rights reserved

# ACTIVE PROPERTIES OF NEURONAL DENDRITES

*Daniel Johnston, Jeffrey C. Magee, Costa M. Colbert, and Brian R. Christie*

Division of Neuroscience, Baylor College of Medicine, 1 Baylor Plaza, Houston, Texas 77030

KEY WORDS:     patch clamp, hippocampus, fluorescence imaging, calcium channels, sodium channels

## ABSTRACT

Dendrites of neurons in the central nervous system are the principal sites for excitatory synaptic input. Although little is known about their function, two disparate perspectives have arisen to describe the activity patterns inherent to these diverse tree-like structures. Dendrites are thus considered either passive or active in their role in integrating synaptic inputs. This review follows the history of dendritic research from before the turn of the century to the present, with a primary focus on the hippocampus. A number of recent techniques, including high-speed fluorescence imaging and dendritic patch clamping, have provided new information and perspectives about the active properties of dendrites. The results support previous notions about the dendritic propagation of action potentials and also indicate which types of voltage-gated sodium and calcium channels are expressed and functionally active in dendrites. Possible roles for the active properties of dendrites in synaptic plasticity and integration are also discussed.

## INTRODUCTION

The shapes and sizes of dendrites of different neurons are very diverse (see Ramón-Moliner 1968). Some dendrites are unipolar, while others are multipolar; some have many orders of branching, while others have only one or two orders of branching; and some branch primarily in two dimensions, while others have complex three-dimensional structures. Out of the $10^{12}$ neurons in the human nervous system, there are ~10,000 distinct morphological classes. The size and complexity of dendritic arbors increase during development (Rihn & Claiborne 1990), and, in particular, when animals are reared in complex sensory environments (Greenough 1975). These data suggest that dendritic size and branching patterns are important features of normal development and function. This conclusion is strengthened by the fact that dendritic structure is

165

altered in specific ways in patients with certain neurological and psychiatric disorders (Abebe et al 1991; Mehraein et al 1975; Purpura 1974, 1975; Scheibel 1970; Scheibel & Conrad 1993; Scheibel & Scheibel 1973).

The majority of synapses, both excitatory and inhibitory, terminate on dendrites, so it has long been assumed that dendrites somehow integrate (i.e. coordinate and blend into a unified whole) these myriad inputs to produce an output of the neuron. The properties of dendrites that provide this integrative function and the nature of the integration itself, however, are poorly understood.

Sometimes areas of science are cyclic. That is, formerly hot issues can lie dormant for awhile only to be rediscovered later when newer techniques become available. This certainly appears to be the case for the study of neuronal dendrites. Because many of the questions being addressed today have been asked for many years, we think it is appropriate to begin this review with a brief summary of the history of research on neuronal dendrites.

## HISTORY OF UNDERSTANDING DENDRITES

The first detailed description of dendrites, or protoplasmic prolongations as they were originally called, was provided by Camillo Golgi in his landmark paper of 1873 (for a more thorough discussion, see Shepherd 1991). The function of these processes was not clear to Golgi; thus he suggested that they played a nutritive role for the neuron. Such notions were debated until just before the turn of the century when Ramón y Cajal, arguably the founder of modern neuroscience, took the Golgi stain and put the neuron doctrine and the role of dendrites as the site of synaptic connections between neurons on a solid scientific basis. Also, in 1889 Wilhelm His introduced the term dendrite to replace the rather awkward term protoplasmic prolongations (Table 1).

**Table 1** Summary of important milestones in the study of dendrites

| Dates | Techniques | Findings |
|-------|------------|----------|
| 1870 | Golgi stain | Dendritic morphology |
| 1930s | Extracellular fields | Slow cortical waves |
| 1950s | Intracellular recordings | EPSPs, passive dendrites |
| 1959 | Cable theory | Short electrotonic length |
| 1960s | Quantitative microanatomy | Dendritic dimensions |
| 1970s | Intradendritic recordings | Dendritic spikes |
| 1980s | Detailed modeling | Active and passive |
| 1980s | Molecular biology | Molecular diversity of channels |
| 1990s | Fluorescence imaging | $[Ca^{2+}]_i$ entry in dendrites |
| 1990s | Dendritic patching | Channel properties |

One of the concepts put forth by Cajal was that of "dynamic polarization," that is, a unidirectional flow of information from the synapse to the dendrite, forward to the soma, and then out the axon. An underlying hypothesis in this and other work was that for integration to take place, there must be a graded sum of excitatory and inhibitory influences; otherwise the neuron and its dendrites would function merely as relays with no modification or addition of information along the way (see Llinás 1975). Most of the studies of dendrites were anatomic in nature until the 1930s when axonologists began to apply knowledge gained from peripheral nerves to the problem of electrical signals in the cortex. Stimulation of the optic nerve or the surface of the cortex produced a negative potential at the surface that was too slow to result from the axon spikes that had been recorded from isolated peripheral nerves (Adrian 1936, Bartley & Bishop 1933). Bishop and colleagues (Bishop & Clare 1952, O'Leary & Bishop 1943), using large recording electrodes in the dendritic layers of visual cortex, proposed that these negative surface waves were non-conducted, stationary potentials, which Eccles (1951) suggested were synaptic potentials in the dendrites. These and other results led Bishop to conclude that "the chief and most characteristic functions of nervous and other excitable tissues are performed by means of graded responses" (Bishop 1956). This viewpoint of graded and electrotonically propagating responses in dendrites was in complete agreement with the dynamic polarization postulate of Cajal and the idea of neuronal integration of Sherrington (1906). For the most part, this viewpoint was widely accepted (but see Chang 1951).

With the advent of intracellular recordings with microelectrodes, excitatory and inhibitory postsynaptic potentials (EPSPs and IPSPs) were recorded from many types of neurons. Properties of synapses, such as equilibrium potentials and reversal potentials, along with properties of the postsynaptic cell, such as membrane time and space constants, were determined. This new information provided a better foundation for understanding neuronal integration via spatial and temporal summation of passive synaptic potentials in dendrites (see Grundfest 1957). In fact, Bishop (1956) and Grundfest (1958) suggested that dendrites were composed of more primitive membrane than axons and thus lacked mechanisms for excitability. Hence, the electrotonic or passive properties of dendrites became of great interest as a means of understanding dendritic integration of synaptic signals. The prevailing view during this time—which was supported by work on motorneurons (see Eccles 1964) and sensory receptors (Eyzaguirre & Kuffler 1955, Katz 1950)—was that the action potential was initiated in the axon hillock region as a result of the graded or algebraic summation of EPSPs and IPSPs occurring in various parts of the neuron. Because of the relatively short space constant that had been calculated for motorneurons (cf Coombs et al 1955), several researchers predicted that EPSPs from synapses on the distal portions of dendrites would be ineffective in firing a neuron because of severe electrotonic decrement of their amplitudes (Eccles 1964, Lorente de Nó 1938, 1947; Lorente de Nó & Condouris 1959). These researchers proposed that only synapses on the proximal dendrites were capable of influencing cell firing.

In 1957 Rall published the first of a series of papers that would prove to revolutionize the field of dendrite physiology (reviewed in Rall 1977, Segev et al 1995). Coming from a physics background, Rall utilized the mathematics of cable theory (first developed for transatlantic telegraph cables) to provide a theoretical framework for the study of neurons and their dendrites. Contrary to Eccles and Lorente de Nó, he concluded that both the time and space constants of motorneurons were greater than originally estimated and that synapses on distal dendrites could indeed contribute to somatic depolarization. There were many other unique insights into dendritic function that were derived from Rall's neuron models, and the mathematical modeling of neurons and dendrites continues to have a major impact on neuroscience today.

As part of the detailed modeling of neurons, the necessity arose for quantitative microanatomy, that is, accurate measurements of dendritic diameters and branch lengths. This was initially provided for cortical neurons by Lux & Pollen (1966) and for motorneurons by Nelson & Lux (1970) and Burke & ten Bruggencate (1971). Much subsequent work has also been devoted to this issue, using a variety of neurons (e.g. see Desmond & Levy 1984, Larkman 1991, Major et al 1994, Shelton 1985, Wilson 1992). Recently, the availability of computer programs that handle many of the mathematical details of modeling neurons has allowed many investigators to contribute insights into the function of dendrites (Hines 1993, Wilson et al 1989).

Although the prevailing view in the 1950s and early 1960s (reviewed in Purpura 1967) was that dendrites were entirely passive and merely summed excitatory and inhibitory influences, evidence had been accumulating, beginning with the work of Chang (1951), that action potentials could propagate in dendrites. Fatt (1957a,b) demonstrated that antidromic action potentials invaded the soma and propagated into the dendrites of motorneurons. Transection of axons (chromatolysed motorneurons) also led to the appearance of dendritic action potentials (Eccles et al 1958). Using laminar analyses of field potentials in rabbit hippocampus, several investigators concluded that action potentials could be initiated in the dendrites by synaptic stimulation and that these action potentials would then propagate in both directions at velocities of around 0.3 m per s (Andersen 1960, Cragg & Hamlyn 1955, Fujita & Sakata 1962). Spencer & Kandel (1961) reported so-called fast prepotentials from intracellular recordings in the somata of hippocampal pyramidal neurons and proposed that they were action potentials initiated in dendrites. Direct intracellular recordings from dendrites of cerebellar Purkinje cells (Llinás & Nicholson 1971), cortical neurons (Houchin 1973), and hippocampal neurons (Wong et al 1979) provided the necessary evidence that action potentials could indeed be initiated and propagated in dendrites. With this new information, dendrites could no longer be considered completely passive.

Outside of the knowledge that some dendrites can generate spikes, very little is known about the active, or voltage-dependent, properties of dendrites. Until recently, dendrites were relatively inaccessible to study. Action potentials could be recorded from dendrites, but more specific questions remained unanswered. For example, what types of ion channels are present in dendrites, and where are they located? How do active properties alter the concept of synaptic integration? Where are action potentials first initiated under different conditions?

For the remainder of this review we describe recent findings concerning the active properties of dendrites. Most of the recent work in this area involves the techniques of high-speed fluorescence imaging and dendritic patch clamping. We begin with a discussion of hippocampal pyramidal neurons and follow with a brief review of the distinct properties of cerebellar Purkinje cells and neocortical pyramidal neurons.

## HIPPOCAMPAL PYRAMIDAL NEURON DENDRITES

### Passive Properties

With the advent of the in vitro slice preparation and intracellular recordings, a number of studies explored the passive electrotonic properties of hippocampal neurons (CA1 and CA3 pyramidal neurons and dentate granule cells) (Brown et al 1981, Durand et al 1983, Johnston 1981). Although the neurons and dendrites are not expected to be entirely passive, knowledge about the electrotonic structure of a neuron is a necessary first step in understanding the spread of electrical signals. The overall conclusions from these studies were that hippocampal neurons have long membrane time constants ($\tau_m$), large input resistances, and high specific resistivities, and the electrotonic length ($L$) of the dendrites is significantly less than one. Moreover, the "electrical compactness" of the neurons was introduced (Brown et al 1981, Johnston & Brown 1983); an electrotonically compact neuron is one in which the attenuation of DC or slowly changing potentials is minimal. The use of patch-clamp techniques (Spruston & Johnston 1992) and accurate morphometric reconstructions (e.g. Major et al 1994) added refinements to these general overall conclusions.

What is the significance for synaptic integration of short $L$ and long $\tau_m$? For example, will synaptic potentials generated in distal dendrites be attenuated at all upon reaching the soma? Spruston et al (1993) addressed this question and demonstrated that distal EPSPs will indeed diminish in amplitude by passive propagation in dendrites. Dendritic EPSPs may attenuate fivefold or more upon reaching the soma in pyramidal neurons. The explanation for this apparent paradox, that is, the fact that a neuron can be electrotonically compact and yet still produce large attenuations of potentials in its dendrites, is that a neuron may be compact for DC potentials but electrically "loose" for AC or rapidly changing signals such as those that occur during EPSPs (Spruston et al 1994). These findings are important not only for understanding passive spread of electrical signals in dendrites, but also suggest a possible role for voltage-gated channels in dendrites (see below).

## Active Properties

ACTION POTENTIALS IN HIPPOCAMPAL DENDRITES    As discussed above, there is considerable evidence for the initiation and propagation of action potentials in hippocampal pyramidal neuron dendrites. The evidence is based on field potentials (Andersen 1960, Andersen et al 1966, Cragg & Hamlyn 1955, Fujita & Sakata 1962), current source-density (CSD) analysis of field potentials (Herreras 1990), and intracellular recordings of intact (Wong et al 1979) and isolated dendrites (Bernardo et al 1982). Antidromic action potentials propagate well into the dendrites (Miyakawa & Kato 1986, Richardson et al 1987, Turner et al 1991), and several reports suggest that synaptic input to the dendrites (specifically Schaffer collateral input to CA1) can initiate action potentials that propagate in both directions from the site of input (Andersen 1960, Herreras 1990). This issue of the site of initiation of action potentials from synaptic stimulation has been readdressed in the past few years with somewhat differing results. Andersen et al (1987) looked at action potential thresholds from distal and proximal inputs and found a lower threshold for distal inputs. Colling & Wheal (1994) injected QX314, a $Na^+$ channel blocker, intrasomatically and found that action potentials triggered by proximal inputs were blocked first before those triggered by distal inputs. Both results suggest that action potentials can be initiated in dendrites (see also Regehr et al 1993, Regehr & Armstrong 1994).

Several other groups have reached different conclusions, however. Richardson et al (1987) did a careful CSD analysis of antidromic vs orthodromic evoked action potentials and found that in both cases the action potential was initiated first in the soma-axon hillock region and then propagated into the dendrites. Miyakawa & Kato (1986) made similar conclusions. Taube & Schwartzkroin (1988) also concluded that the action potential triggered by orthodromic stimulation was initiated in the soma region, although they also identified a long-lasting inward current in the proximal dendritic region. Turner et al (1989), using local application of tetrodotoxin (TTX), concluded that there were two sites of action potential initiation in response to Schaffer collateral stimulation: the soma-axon hillock region and, at higher stimulus intensities, the proximal apical dendrites. They also observed the low-threshold, slow inward current in the proximal dendrites first reported by Taube & Schwartzkroin (1988) and further demonstrated its sensitivity to TTX. Spruston et al (1995b) have recently used dual-patch recordings, one on the soma and one on the dendrites, to suggest that the lowest threshold for spike initiation is in the soma-axon hillock region.

This brief review of action potentials in hippocampal dendrites (mostly CA1) clearly demonstrates that dendrites can sustain the active propagation of action potentials. (This fact becomes even clearer in the following sections.) Nevertheless, the site of initiation of the action potential from synaptic activity in the apical dendrites is still a matter of some controversy. Sufficient data seem to support the conclusion that action potentials can, under certain conditions, be initiated first in the dendrites. Much (but not all) of the work supporting this conclusion, however, was done in vivo in anesthetized animals and the effects of anesthesia, which usually involve enhanced inhibition, on action potential initiation have not been assessed (but see Turner et al 1989). One must also keep in mind that the conditions of slice preparations (and thus the properties of neurons) can vary among laboratories (e.g. age of animals, temperature of slices, submerged vs interface slices, ionic composition of media, and surgical procedures are all potential variables). In vitro conditions are also different from those in vivo in terms of health of the cells, amount of spontaneous activity, and the effects on cells of cutting their axons and dendrites during the slicing procedures. Any or all of these factors may explain some of the differing results found in the literature. The question of the initial site for initiating orthodromic action potentials in the intact animal under physiological conditions therefore remains open.

VOLTAGE-GATED ION CHANNELS IN HIPPOCAMPAL DENDRITES    Although action potentials can initiate and propagate in dendrites of hippocampal pyramidal neurons, very little information was available until recently concerning the ion channels that sustain them. Furthermore, there was little direct evidence concerning whether subthreshold synaptic potentials activate voltage-gated ion channels in dendrites and thus influence synaptic integration.

*Dendritic Na+ channels and Ca2+ influx*    Recent advances in two well-established recording techniques (fluorescence imaging and patch clamping in brain slices) have permitted a more direct and quantitative study of dendritic properties. For imaging, the development of highly efficient fluorescent probes (Tsien 1989) and high-speed imaging techniques (Lasser-Ross et al 1991) have made it possible to image changes in Na+ or Ca2+ concentration throughout the dendritic tree every 10 ms or less (Lasser-Ross et al 1991, Lev-Ram et al 1992). Several groups have used this technique to investigate the distribution of changes in $[Ca^{2+}]_i$ during action potentials and synaptic potentials (Jaffe et al 1992; Miyakawa et al 1992; Regehr et al 1989; Regehr & Tank 1990, 1992). A consistent finding is that during trains of action potentials there is a characteristic M-shaped profile of $[Ca^{2+}]_i$ across the neuron in which the rise in $[Ca^{2+}]_i$ is small in the soma, highest in the proximal apical and basal dendrites, and small again in the distal apical dendrites. If the action potentials are made very broad by the addition of the $K^+$-channel blocker tetraethyl ammonium (TEA), then the distribution of the rise in $[Ca^{2+}]_i$ is more uniform, and large increases can be seen at the very tips of the dendrites (Jaffe et al 1992). This result strongly suggests that voltage-gated $Ca^{2+}$ channels are present throughout the dendritic tree.

The distribution of $Ca^{2+}$ influx in response to a train of normal-shaped action potentials (i.e. the M pattern) does not appear to be due to the distribution of $Ca^{2+}$ channels but instead to the propagation into the dendrites of trains of action potentials. Imaging with a $Na^+$-sensitive dye yielded a similar M pattern, suggesting that the influx of $Ca^{2+}$ into the dendrites is driven by $Na^+$-dependent action potentials (Jaffe et al 1992; 1994b). Recent results by Spruston et al (1995b) found that during a train of action potentials, a frequency-dependent decrease in amplitude occurs, which is particularly prominent at branch points. These authors hypothesize that $Na^+$-channel inactivation or $K^+$-channel activation during the action potential train allows only the first few spikes in a train to actively propagate the entire extent of the dendrites (i.e. >300 µm from the soma) (see also Jaffe et al 1992). Although $Ca^{2+}$-dependent action potentials in hippocampal neurons and dendrites have been demonstrated by Schwartzkroin & Slawsky (1977) and Bernardo et al (1982), the imaging results place a greater emphasis on $Na^+$ channels in dendrites and on the resulting propagation of $Na^+$-dependent action potentials for driving $Ca^{2+}$ in- flux through voltage-gated channels.

A recent advance in the application of patch-clamping techniques to brain slices has been particularly useful for directly identifying the types and relative densities of $Na^+$ and $Ca^{2+}$ channels in dendritic membrane. The use of infrared illumination, high-resolution video detectors, and differential interference contrast optics has allowed the visualization of living dendrites in brain slices of unstained tissue (Stuart et al 1993). With the ability to visualize these structures, whole-cell, cell-attached, and excised-patch techniques can be applied to study the electrical properties of hippocampal and neocortical dendrites directly (Magee & Johnston 1995a,b; Spruston et al 1995a,b; Stuart & Sakmann 1994). When cell-attached recordings are made, $Na^+$ channels are found in every patch made from neuronal somata and dendrites (up to 350 µm from the soma) (see Magee & Johnston 1995b). Such recordings are limited to branch diameters of >1 µm, so smaller diameter branches may have fewer $Na^+$ channels. Nevertheless, the properties of these channels appear to be quite uniform across the soma-dendritic axis. They have conductances of 15 pS and rapid activation kinetics and are opened by depolarizations of ~15 mV from rest. An interesting feature of these channels is that inactivation has both a fast and a slow phase, with late openings of long duration recorded up to 100 ms after the onset of the depolarization. The most parsimonious conclusion from these studies is that there is a single type of $Na^+$ channel present throughout the neuron, and it has two modes of inactivation. It is possible, however, that there are two channels with different inactivation properties, but they would have to be together in constant proportion at all sites on the neuron.

The density of $Na^+$ channels (or at least the minimum number of $Na^+$ channels per patch) is remarkably constant, even up to 350 $\mu$m from the soma. A slight decrease in density at distances greater than 200 $\mu$m from the soma, however, may be present. Channel density in the dendrites increases with development, in particular, among animals from 2 to 8 weeks of age. The presence of $Na^+$ channels in distal dendrites helps explain the back-propagation of action potentials (at least for single action potentials) well into the dendritic tree, which has been observed with field, intracellular microelectrode, and whole-cell patch recordings (see above).

As mentioned above, single action potentials propagate in dendrites more effectively than trains of action potentials. Single action potentials are, however, significantly filtered as they propagate throughout the dendrites (Spruston et al 1995b, Stuart & Sakmann 1994, Turner et al 1991). These phenomena become particularly intriguing considering that $Na^+$-channel density is fairly uniform across the somato-dendritic axis. If dendritic $Na^+$ channels allow for the regenerative propagation of action potentials well into the dendritic tree, how are spike amplitude and duration and the number of spikes fired during a train altered in the dendrites? Although this is obviously still an open question, dendritic $Na^+$-channel density is probably low enough that dendritic action potentials are only weakly regenerative. As a result, the safety factor for action potential propagation in the dendrites is rather low, and action potentials propagating throughout the dendritic arbor are still susceptible to the effects of cable filtering.

*Dendritic $Ca^{2+}$-channel types*    Whole-cell recordings from isolated somata and cell-attached patch recordings from cell bodies have demonstrated multiple types of voltage-gated $Ca^{2+}$ channels in hippocampal neurons (Eliot & Johnston 1994, Fisher et al 1990, Kay & Wong 1987, Mintz et al 1992, Mogul et al 1993). Much of the evidence for multiple channel types is pharmacological. There are several toxins isolated from spiders and snails that are specific blockers for different types of $Ca^{2+}$ channels. For example, $\omega$-conotoxin-GVIA ($\omega$-CgTX-GVIA) blocks the N-type, $\omega$-agatoxin-IVA ($\omega$-Aga-IVA) blocks the P-type, and $\omega$-conotoxin-MVIIC ($\omega$-CgTX-MVIIC) blocks the N-, P-, and

Q-type channels (Bean 1989, Dunlap et al 1995, Eliot & Johnston 1994, Tsien et al 1988). Dihydropyridines such as nimodipine block L-type channels. Also, divalent cations such as $Ni^{2+}$ and $Cd^{2+}$ have some differential selectivity for blocking low-voltage activated (LVA) channels (e.g. T-type) and certain high-voltage activated (HVA) channels. Given that action potentials propagating into the dendrites trigger $Ca^{2+}$ entry through voltage-gated channels, these drugs were used to help identify the type(s) of channels responsible for the spike-triggered $Ca^{2+}$ signals measured by fluorescence imaging.

When trains of action potentials are triggered by current injection into the soma, they propagate into the first 200 μm of the apical dendrites (Jaffe et al 1992). Christie et al (1995) measured the resulting rise in $[Ca^{2+}]_i$ from such trains before and after applying one of the $Ca^{2+}$ channel blockers. The results suggest that L-, N-, P-, Q-, R-, and T-type channels mainly contribute to spike-triggered $Ca^{2+}$ entry into the soma and first 50 μm of the apical dendrites. On the other hand, spike-triggered $Ca^{2+}$ entry into dendritic sites more distal than about 100 μm from the soma appears to be due primarily to LVA T-type channels and HVA channels sensitive to $Ni^{2+}$ [e.g. the R-type (Randall et al 1993)]. These results were the first to suggest a heterogeneous distribution of different types of voltage-gated $Ca^{2+}$ channels within the soma and dendrites of hippocampal neurons by using physiological measurements and were in partial agreement with immunocytochemical studies (Westenbroek et al 1990). Although these results address the question of which $Ca^{2+}$ channel types are opened by back-propagating action potentials in the dendrites, they are less useful for determining the density and distribution of channels because the amplitude and width of the action potential is not constant throughout the dendrites.

Recordings of single $Ca^{2+}$ channels in dendrite-attached patches support the general conclusions from the fluorescence-imaging experiments. That is, there appears to be a differential distribution of $Ca^{2+}$ channel types in the soma and dendrites. In the soma, recordings of L-, N-, P-, Q-, R-, and T-type $Ca^{2+}$ currents have been made, and single channels corresponding to most of these whole-cell currents have been identified (Eliot & Johnston 1994, Fisher et al 1990, Magee & Johnston 1995b, Mintz et al 1992, Mogul et al 1993). Fewer types of $Ca^{2+}$ channels appear to be present in the dendrites than in the soma. The predominant channel types appear to be a LVA channel of ~9 pS conductance and a HVA channel of ~17 pS (both measured in 110 mM $Ba^{2+}$). These channels are most prominent at dendritic sites greater than 100 μm from the soma. A channel corresponding to the L-type $Ca^{2+}$ channel is observed at fairly high density in the first 50 μm from the soma, but at much lower density in more distal dendritic patches. An N-type channel is also occasionally observed in dendrite-attached patches. An interesting finding about $Ca^{2+}$ channel density is that it is also fairly constant from the soma out to at least 350 μm along the major dendritic branches. The types of channels contributing to the constant density, however, vary a great deal between soma and dendrites.

The LVA and HVA channels encountered in the dendrites are sensitive to low concentrations of $Ni^{2+}$ and amiloride, but insensitive to dihydropyridines and $\omega$-CgTX-MVIIC, suggesting that they are T- and R-type channels, respectively. The classification of single $Ca^{2+}$ channels into particular types, however, is fraught with difficulties, especially since many of the different types of $Ca^{2+}$ channels have similar electrophysiological properties. These conclusions should therefore be considered tentative, even though they are consistent with currently available information (see Figure 1).

*Synaptic activation of dendritic voltage-gated channels*    When trains of action potentials are elicited by synaptic potentials, the distribution of $[Ca^{2+}]_i$ across the dendrites is similar to that which occurs from trains of antidromic action potentials (Miyakawa et al 1992). The amount of $Ca^{2+}$ entry and the distribution of $[Ca^{2+}]_i$ from orthodromic trains is relatively unaffected by addition of the NMDA receptor antagonist APV. This result suggests that most of the $Ca^{2+}$ entry from orthodomic action potentials is through voltage-gated $Ca^{2+}$ channels (Miyakawa et al 1992). Subthreshold synaptic stimulation also produces $Ca^{2+}$ entry through voltage-gated $Ca^{2+}$ channels, but the rise in $[Ca^{2+}]_i$ is much smaller and more highly localized to the site of synaptic input (Jaffe et al 1994a,b; Magee & Johnston 1995a; Magee et al 1995; Miyakawa et al 1992). Although $Ca^{2+}$ entry also occurs through NMDA receptors (Connor et al 1994, MacDermott et al 1986, Malinow et al 1994), such $Ca^{2+}$ entry is sometimes difficult to distinguish from that mediated by voltage-gated $Ca^{2+}$ channels. Antagonists that block NMDA receptors will also reduce the local synaptic depolarization, which in turn activates voltage-gated $Ca^{2+}$ channels (Miyakawa et al 1992). A reasonable conclusion of the fluorescence-imaging experiments is that subthreshold synaptic inputs will trigger local $Ca^{2+}$ entry through NMDA receptors and voltage-gated $Ca^{2+}$ channels. If the synaptic potentials are of sufficient amplitude to trigger action potentials, then a more widespread distribution of $Ca^{2+}$ entry into the neuron will result.

Subthreshold EPSPs can directly activate $Na^+$ and LVA $Ca^{2+}$ channels found in dendrites (Magee & Johnston 1995a). When a dendrite-attached patch recording is made during an EPSP, $Na^+$ channels and the LVA $Ca^{2+}$ channel open during the rising phase, peak, and falling phase of the EPSP. Thus, $Na^+$ and $Ca^{2+}$ channels can contribute to the amplitude and kinetics of EPSPs at least locally in the dendrites if not more globally throughout the neuron. Moreover, the activation of the LVA $Ca^{2+}$ channel is increased by prior hyperpolarization such as might occur with IPSPs or after-hyperpolarizations. These findings have important functional implications for concepts of synaptic integration (see next section). The HVA $Ca^{2+}$ channels, on the other hand, only

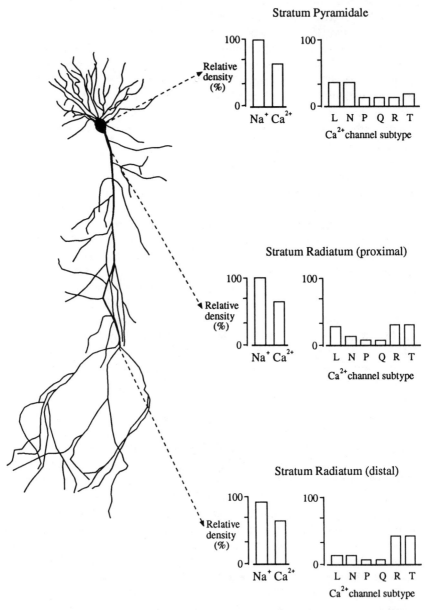

*Figure 1*  Hypothesized distribution of voltage-gated Na⁺ and Ca²⁺ channels in hippocampal, CA1 pyramidal neurons. The bar graphs, which are based on fluorescence imaging and dendritic patch clamping, represent the approximate relative density of the channels in the soma and proximal and distal apical dendrites.

appear to open during back-propagating action potentials or during other conditions where the membrane potential is depolarized from rest by more than about 20 mV. The dendrites therefore appear to contain at least two functionally distinct types of $Ca^{2+}$ channels—one that opens with events near the resting potential and one that opens only with larger depolarizations.

## DENDRITES OF OTHER NEURONS

Although less is known about the dendrites of most other types of neurons, there are nevertheless some interesting differences and similarities between the dendrites of these neurons and those of hippocampal neurons. The first intradendritic recordings of action potentials were made in cerebellar Purkinje cells by Llinás & Nicholson (1971) (see also Llinás 1988). These dendritic spikes were subsequently shown to be due to voltage-gated $Ca^{2+}$ channels in dendrites (specifically the P-type) (see Llinás et al 1992). Intracellular recordings and fluorescence-imaging experiments indicated that $Na^+$-dependent action potentials initiated in the soma-axon hillock region do not propagate very far into the dendrites (Lasser-Ross & Ross 1992, Llinás & Sugimori 1980). Using whole-cell clamping of the soma and synaptic stimulation in the dendrites, Regher et al (1992) measured $Na^+$ action currents and suggested that $Na^+$-dependent action potentials may, under certain conditions, be activated in dendrites. Stuart & Hausser (1994) reevaluated the question of $Na^+$ spikes in Purkinje cell dendrites using dual patch recordings and concluded that $Na^+$ channel density declines in the dendrites with distance from the soma and that $Na^+$ spikes spread passively from the soma into the dendrites. The lack of $Na^+$-dependent action potentials in Purkinje cell dendrites is thus distinctly different from hippocampal pyramidal neurons.

The interest in active properties of neocortical dendrites began with the suggestion of Chang (1951) (see above) that cortical field potentials were due in part to active spike propagation in dendrites. Houchin (1973) published a short report in which intracellular recordings of action potentials were made from dendrites of two cortical neurons. More extensive dendritic recordings using microelectrodes have been made in vivo by Pockberger (1991) and in vitro by Amitai et al (1993). Whole-cell patch clamping of dendrites has recently been performed by Kim & Conners (1993) and Stuart & Sakmann (1994). Fluorescence imaging of neocortical cells was done by Yuste et al (1994) and Markram & Sakmann (1994, 1995), the latter in conjunction with dendritic patch clamping.

The general conclusion from these studies is that neocortical dendrites are capable of sustaining action potentials, both $Na^+$ and $Ca^{2+}$ dependent. The work of Kim & Conners (1993) and Markram & Sakmann (1994, 1995) suggest multiple types of $Ca^{2+}$ channels in dendrites. The $Ca^{2+}$ channels appear to be distributed throughout the dendritic tree (Yuste et al 1994), although the density of the different types is unknown. Na$^+$ channels are also widely distributed in cortical dendrites (Stuart & Sakmann 1994), but the normal site of initiation of the Na$^+$ spike is somewhat controversial (see Regehr et al 1993, Regehr & Armstrong 1994). Stuart & Sakmann (1994) show rather convincingly that the Na$^+$ spike is first initiated in the soma-axon hillock region and then propagates into the dendrites. The dendritic properties of neocortical pyramidal neurons, in general, appear to be similar to those for hippocampal neurons.

# FUNCTIONAL PROPERTIES OF ACTIVE DENDRITES

Most of this section on the functional properties of active dendrites will consist of ideas, speculations, and extrapolations of existing data, because, unfortunately, there is very little direct information on how the active properties of dendrites participate in the normal functioning of a neuron. The two general areas for discussion involve synaptic integration and plasticity and modulation.

## Synaptic Integration

The most prominent idea for the function of voltage-gated (inward current) channels in dendrites relates to the boosting of distal synaptic events. Although Rall showed that based on purely passive properties, distal EPSPs do not decrease to zero in the soma, as had first been suggested, they will nevertheless decrease in amplitude significantly. Voltage-gated Na$^+$ and Ca$^{2+}$ channels opened by EPSPs (see above) would obviously counter these effects of passive decay by boosting or amplifying the magnitude of EPSPs en route to the soma.

This argument, however, is somewhat somatocentric, i.e. the emphasis is on how dendrites might amplify events so that they are bigger in the soma. An alternative viewpoint is that these channels are more important for dendritic interactions in the immediate vicinity of the synaptic inputs. For example, voltage-gated channels will alter the local input resistance and time constant, which in turn would influence both spatial and temporal summation of EPSPs and IPSPs. The interactions among synapses might also be highly nonlinear (Shepherd et al 1985). Multiple EPSPs occurring on the same branch and within a narrow time window might activate voltage-gated channels and produce a much bigger response than would occur if they were on separate branches or occurred outside this time window. This possibility has been formalized, which led to the hypothesis that dendrites with active properties are coincidence detectors (Jaslove 1992, Softky 1994; but see also Shadlen & Newsome 1994). Both the duration of the window for coincidence detection and the degree of nonlinear interaction will depend on the active properties of the dendrites. If action potentials are initiated in dendrites, then the coincidence detection can take place on individual dendritic branches to produce an output of the neuron.

In addition to providing an output, the action potentials elicited by synaptic input, whether initiated at the local synaptic sites or at the axon hillock and back-propagated into the dendrites, would essentially reset the dendritic membrane potential. This would have the effect of eliminating synaptic summation occurring within a time window of the action potential and its associated after potentials. The action potential might also provide a feedback or an associative signal to other synapses active just before or just after the action potential (see also next section). For example, NMDA receptors bound by glutamate released from active synapses could be opened rapidly by a back-propagating action potential that removes the voltage-dependent $Mg^{2+}$ block of the receptors. The off rate of glutamate from NMDA receptors is slow, so an action potential occurring within several 100 ms (perhaps up to even a second or more) (see Lester et al 1990, Spruston et al 1995a) of an active synapse could produce $Ca^{2+}$ influx through NMDA receptors.

Most of our discussion has revolved around $Ca^{2+}$ and $Na^+$ channels. This is because other dendritic channels such as $K^+$ channels have not been studied to any great extent. $K^+$ channels may influence the falling phase of EPSPs (Cassell & McLachlan 1986, Nicoll et al 1993). This would obviously also affect spatial and temporal summation. Much depends on the type of $K^+$ channel present in dendrites, however. A $Ca^{2+}$-dependent $K^+$ channel would only be important as a consequence of $Ca^{2+}$ influx (Lancaster & Zucker 1994). Transient $K^+$ currents would tend to reduce rapidly changing depolarizations (low-pass filter), while slowly activating $K^+$ currents would act as a high-pass filter for depolarizing events. $K^+$ channels may also help linearize functions relating frequency of synaptic input to current in the soma (Bernander et al 1994). Inward rectifiers help establish the resting potential, so different densities of $K^+$ selective or nonselective rectifiers from the soma could lead to different resting potentials in the dendrites (Holt & Eatock 1995).

## Plasticity and Modulation

The activation of voltage-gated channels by either action potentials or subthreshold synaptic events will alter the concentration of $Ca^{2+}$ in the dendrites. $Ca^{2+}$ is a ubiquitous second messenger that has been implicated in postsynaptically-induced forms of synaptic plasticity such as short-term potentiation (STP) (Malenka 1991), long-term potentiation (LTP) (both NMDA receptor dependent and NMDA receptor independent), and long-term depression (LTD) (Bliss & Collingridge 1993, Christie et al 1994, Johnston et al 1992). Different levels of $[Ca^{2+}]_i$ may produce different types of plasticity (Artola & Singer 1993, Lisman 1989). For example, small increases in $[Ca^{2+}]_i$ may elicit LTD;

larger increases, STP; still larger increases, NMDA LTP; and finally the largest increases, non–NMDA LTP (Teyler et al 1994). The type of plasticity, however, may also be dependent on where the changes in $[Ca^{2+}]_i$ occur, the degree to which they are compartmentalized, and whether or not they are in conjunction with presynaptic activity (i.e. Hebbian vs anti-Hebbian) (Holmes & Levy 1990, Lisman 1989). In any event, the voltage-gated $Na^+$ and $Ca^{2+}$ channels in dendrites discussed above could play important roles in these processes.

The changes in $[Ca^{2+}]_i$ produced by synaptic input should be separated into two categories: subthreshold and superthreshold. The subthreshold entry of $Ca^{2+}$ will occur through low-threshold $Ca^{2+}$ channels. The degree of activation of these channels will be determined by the amplitude of the EPSP, the amount of simultaneous activation of $Na^+$ channels, and the prior membrane potential of the local region of the dendrite (this will determine the amount of steady-state inactivation of the channels prior to the EPSP). The entry of $Ca^{2+}$ under such conditions would be highly localized (Eilers & Augustine 1995; Jaffe et al 1994a,b; Magee & Johnston 1995a; Magee et al 1995). In contrast, the entry of $Ca^{2+}$ in response to a back-propagating action potential will take place throughout the dendritic tree and will provide a global signal to the neuron involving the activation of both low- and high-threshold $Ca^{2+}$ channels (Christie et al 1995).

Because $[Ca^{2+}]_i$ is an important second messenger, it could have myriad effects on the cell. These include direct activation of $K^+$ and $Cl^-$ channels (Lancaster & Adams 1986, Owen et al 1984); release of internal $Ca^{2+}$ stores (Llano et al 1994, Sharp et al 1993); modulation of voltage-gated $Ca^{2+}$, $Na^+$, and $K^+$ channels (Chetkovich et al 1991, Kirkwood et al 1991, Li et al 1993, Marrion et al 1991); and modulation of ligand-gated channels such as those for AMPA, NMDA, and GABA (Lonart et al 1992; Medina et al 1994; Rosenmund et al 1994, 1995; Stelzer 1992). Modulation of ion channels in dendrites by second-messenger systems that are activated by neurotransmitters such as norepinephrine, serotonin, acetylcholine, and various neuroactive peptides is also likely to occur (Dunlap et al 1987, Fisher & Johnston 1990, Levitan 1994, Tsien et al 1988). The long-term changes in dendritic, voltage-gated channels may play a role in some of the synaptic plasticities mentioned above or in the integrative properties of the dendrites. For example, the local and global changes in $[Ca^{2+}]_i$ may primarily be for establishing, and dynamically changing, the strengths of individual synapses. On the other hand, the $Na^+$ channels may be more important as a means of normalizing the amplitudes of EPSPs across the spatial extent of the dendrites. In fact, neural network models suggest that the information storage capacity of a neuron would benefit somewhat if synapses of similar strength had comparable effects on the firing of the neurons regardless of where the synapses were located on the dendritic tree (E Cook & D Johnston, in preparation).

# SUMMARY AND CONCLUSIONS

Dendrites of many neurons in the central nervous system clearly have at least some active properties. It would not be too surprising if there were a great deal of variability among types of neurons. For example, dendrites of motorneurons are less excitable than pyramidal neurons. Also, Purkinje cell dendrites may not sustain $Na^+$-dependent action potentials as has been demonstrated for neocortical and hippocampal pyramidal neurons. Very little is known about the properties of dendrites of other primary neurons or interneurons.

Hippocampal neurons propagate $Na^+$-dependent action potentials from the soma-axon hillock to distal regions of at least the main dendritic branches. These action potentials may also, under certain conditions, be initiated near synaptic inputs and propagate in both directions. The action potentials also open voltage-gated $Ca^{2+}$ channels, thereby leading to a widespread influx of $Ca^{2+}$. The type and distribution of $Na^+$ and $Ca^{2+}$ channels appears to determine the nature of this propagation.

The dendrites of pyramidal neurons predominantly contain one type of $Na^+$ channel and multiple types of $Ca^{2+}$ channels. The $Ca^{2+}$ channels fall mainly into two classes, a LVA channel that can be activated (along with the $Na^+$ channel) by subthreshold EPSPs and a HVA channel that is opened by back-propagating action potentials. Very little is known about the types of $K^+$ channels in dendrites.

The voltage-gated $Na^+$ and $Ca^{2+}$ channels in dendrites could significantly affect synaptic integration both locally and throughout the neuron. The back-propagating action potentials may also provide a feedback or associative signal to synapses in the dendrites that an output of the neuron has occurred. This might take place through increases in $Ca^{2+}$ concentration, lead to modulation of ion channels, and form the basis of various types of synaptic plasticity.

Individual neurons are complex units of integration that dynamically change with their environment. As discussed in this review, dendrites are the principle structures responsible for synaptic integration and for the changes in synaptic strengths that take place as a function of the activity of the neuron. It is our belief that before one can hope to understand the operation of systems of neurons fully, one must first be able to describe the function of the basic unit of the nervous system, that is, the single neuron and its associated dendritic tree.

ACKNOWLEDGEMENTS

Portions of the work performed in the authors' laboratories were supported by USPHS grants MH44754, MH48432, and NS11535.

## Literature Cited

Abebe T, Austin R, Forsythe S, Scheibel A. 1991. Hippocampal pyramidal cell disarray in schizophrenia as a bilateral phenomenon. *Arch. Gen. Psychiatry* 48:413–17

Adrian E. 1936. The spread of activity in the cerebral cortex. *J. Physiol.* 88:127–61

Amitai Y, Friedman A, Connors BW, Gutnick M. 1993. Regenerative activity in apical dendrites of pyramidal cells in neocortex. *Cereb. Cortex* 3:26–38

Andersen P. 1960. Interhippocampal impulses. II. Apical dendritic activation of CA1 neurons. *Acta Physiol. Scand.* 48:178–208

Andersen P, Holmqvist B, Voorhoeve P. 1966. Excitatory synapses on hippocampal apical dendrites activated by entorhinal stimulation. *Acta Physiol. Scand.* 66:461–72

Andersen P, Storm J, Wheal H. 1987. Thresholds of action potentials evoked by synapses on the dendrites of pyramidal cells in the rat hippocampus in vitro. *J. Physiol.* 383:509–26

Artola A, Singer W. 1993. Long-term depression of excitatory synaptic transmission and its relationship to long-term potentiation. *Trends Neurosci.* 16:480–87

Bartley SH, Bishop G. 1933. Factors determining the form of the electrical response from the optic cortex of the rabbit. *Am. J. Physiol.* 103:173–84

Bean B. 1989. Classes of calcium channels in vertebrate cells. *Annu. Rev. Physiol.* 51:367–84

Bernander O, Koch C, Douglas R. 1994. Amplification and linearization of distal synaptic input to cortical pyramidal cells. *J. Neurophysiol.* 72:2743–53

Bernardo LS, Masukawa LM, Prince D. 1982. Electrophysiology of isolated hippocampal pyramidal dendrites *J. Neurosci.* 2:1614–22

Bishop G. 1956. Natural history of the nerve impulse. *Physiol. Rev.* 36:376–99

Bishop GH, Clare M. 1952. Sites of origin of electric potentials in striate cortex. *J. Neurophysiol.* 15:201–20

Bliss TP, Collingridge G. 1993. A synaptic model of memory: long-term potentiation in the hippocampus. *Nature* 361:31–39

Brown TH, Fricke RA, Perkel D. 1981. Passive electrical constants in three classes of hippocampal neurons. *J. Neurophysiol.* 46:812–27

Burke RE, ten Bruggencate G. 1971. Electronic

characteristics of alpha motoneurones of varying size. *J. Physiol.* 212:1–20

Cassell JF, McLachlan E. 1986. The effect of a transient outward current (IA) on synaptic potentials in sympathetic ganglion cells of the guinea-pig. *J. Physiol.* 374:273–88

Chang H-T. 1951. Dendritic potential of cortical neurons produced by direct electrical stimulation of the cerebral cortex. *J. Neurophysiol.* 14:1–21

Chetkovich D, Gray R, Johnston D, Sweatt J. 1991. N-methyl-D-aspartate receptor activation increases cAMP levels and voltage-gated $Ca^{2+}$ channel activity in area CA1 of hippocampus. *Proc. Natl. Acad. Sci. USA* 88:6467–71

Christie BR, Eliot LS, Ito K, Miyakawa H, Johnston D. 1995. Different $Ca^{2+}$ channels in soma and dendrites of hippocampal pyramidal neurons mediate spike-induced $Ca^{2+}$ influx. *J. Neurophysiol.* 73:2553–57

Christie BR, Kerr DS, Abraham W. 1994. Flip side of synaptic plasticity: long-term depression mechanisms in the hippocampus. *Hippocampus* 4:127–35

Colling SB, Wheal H. 1994. Fast sodium action potentials are generated in the distal apical dendrites of rat hippocampal CA1 pyramidal cells. *Neurosci. Lett.* 172:73–76

Connor JA, Miller LP, Petrozzino J, Muller W. 1994. Calcium signaling in dendritic spines of hippocampal neurons. *J. Neurobiol.* 25:234–42

Coombs JS, Eccles JC, Fatt P. 1955. The electrical properties of the motoneurone membrane. *J. Physiol.* 130:291–325

Cragg BG, Hamlyn L. 1955. Action potentials of the pyramidal neurones in the hippocampus of the rabbit. *J. Physiol.* 129:608–27

Desmond NL, Levy W. 1984. Dendritic caliber and the 3/2 power relationship of dentate granule cells. *J. Comp. Neurol.* 227:589–96

Dunlap K, Holz GG, Rane S. 1987. G proteins as regulators of ion channel function. *Trends Neurosci.* 10:241–44

Dunlap K, Luebke JI, Turner T. 1995. Exocytotic $Ca^{2+}$ channels in mammalian central neurons. *Trends Neurosci.* 18:89–98

Durand D, Carlen PL, Gurevich N, Ho A, Kunov H. 1983. Electrotonic parameters of rat dentate granule cells measured using short current pulses and HRP staining. *J. Neurophysiol.* 50:1080–97

Eccles J. 1951. Interpretation of action potentials evoked in the cerebral cortex. *Electroencephalogr. Clin. Neurophysiol.* 3:449–64

Eccles J. 1964. *The Physiology of Synapses.* Berlin: Springer-Verlag

Eccles JC, Libet B, Young R. 1958. The behaviour of chromatolysed motoneurones studied by intracellular recording. *J. Physiol.* 143:11–40

Eilers J, Augustine G. 1995. Subthreshold synaptic $Ca^{2+}$ signalling in fine dendrites and spines of cerebellar Purkinje neurons. *Nature* 373:155–58

Eliot LS, Johnston D. 1994. Multiple components of calcium current in acutely dissociated dentate gyrus granule neurons. *J. Neurophysiol.* 72:762–77

Eyzaguirre C, Kuffler S. 1955. Processes of excitation in the dendrites and in the soma of single isolated sensory nerve cells of the lobster and crayfish. *J. Gen. Physiol.* 39:87–119

Fatt P. 1957a. Electric potentials occurring around a neurone during its antidromic activation. *J. Neurophysiol.* 20:27–60

Fatt P. 1957b. Sequence of events in synaptic activation of a motoneurone. *J. Neurophysiol.* 20:61–80

Fisher RE, Gray R, Johnston D. 1990. Properties and distribution of single voltage-gated calcium channels in adult hippocampal neurons. *J. Neurophysiol.* 64:91–104

Fisher R, Johnston D. 1990. Differential modulation of single voltage-gated calcium channels by cholinergic and adrenergic agonists in adult hippocampal neurons. *J. Neurophysiol.* 64:1291–302

Fujita Y, Sakata H. 1962. Electrophysiological properties of CA1 and CA2 apical dendrites of rabbit hippocampus. *J. Neurophysiol.* 25:209–22

Greenough W. 1975. Experimental modification of the developing brain. *Am. Sci.* 63:37–46

Grundfest H. 1957. Electrical inexcitability of synapses and some consequences in the central nervous system. *Physiol. Rev.* 37:337–61

Grundfest H. 1958. Electrophysiology and pharmacology of dendrites. *Electroencephalogr. Clin. Neurophysiol. Suppl.* 10:22–41

Herreras O. 1990. Propagating dendritic action potential mediates synaptic transmission in CA1 pyramidal cells in situ. *J. Neurophysiol.* 64:1429–41

Hines M. 1993. NEURON: A program for simulation of nerve equations. In *Neural Systems: Analysis and Modeling,* ed. F Eeckman, pp. 127–36. Norwell, MA: Kluwer

Holmes WR, Levy W. 1990. Insights into associative long-term potentiation from computational models of NMDA receptor-mediated calcium influx and intracellular

calcium concentration changes. *J. Neurophysiol.* 63:1148–68

Holt JR, Eatock RJ. 1995. Inwardly rectifying currents of saccular hair cells from the leopard frog. *J. Neurophysiol.* 73:1484–502

Houchin J. 1973. Procion Yellow electrodes for intracellular recording and staining of neurones in the somatosensory cortex of the rat. *J. Physiol.* 232:67–69

Jaffe DB, Fisher SA, Brown T. 1994a. Confocal laser scanning microscopy reveals voltage-gated calcium signals within hippocampal dendritic spines. *J. Neurobiol.* 25:220–33

Jaffe DB, Johnston D, Lasser-Ross N, Lisman JE, Miyakawa H, Ross W. 1992. The spread of $Na^+$ spikes determines the pattern of dendritic $Ca^{2+}$ entry into hippocampal neurons. *Nature* 357:244–46

Jaffe DB, Ross WN, Lisman JE, Lasser-Ross N, Miyakawa H, Johnston D. 1994b. A model for dendritic $Ca^{2+}$ accumulation in hippocampal pyramidal neurons based on fluorescence imaging measurements. *J. Neurophysiol.* 71:1065–77

Jaslove S. 1992. The integrative properties of spiny distal dendrites. *Neuroscience* 47:495–519

Johnston D. 1981. Passive cable properties of hippocampal CA3 pyramidal neurons. *Cell. Molec. Neurobiol.* 1:41–55

Johnston D, Brown T. 1983. Interpretation of voltage-clamp measurements in hippocampal neurons. *J. Neurophysiol.* 50:464–86

Johnston D, Williams S, Jaffe D, Gray R. 1992. NMDA-receptor-independent long-term potentiation. *Annu. Rev. Physiol.* 54:489–505

Katz B. 1950. Depolarization of sensory terminals and the initiation of impulses in the muscle spindle. *J. Physiol.* 111:261–82

Kay AR, Wong RS. 1987. Calcium current activation kinetics in isolated pyramidal neurones of the CA1 region of the mature guinea pig hippocampus. *J. Physiol.* 392:603–16

Kim HG, Connors B. 1993. Apical dendrites of the neocortex: correlation between sodium- and calcium-dependent spiking and pyramidal cell morphology. *J. Neurosci.* 13:5301–11

Kirkwood A, Simmons MA, Mather RJ, Lisman J. 1991. Muscarinic suppression of the M-current is mediated by a rise in internal $Ca^{2+}$ concentration. *Neuron* 6:1009–14

Lancaster B, Adams P. 1986. Calcium-dependent current generating the afterhyperpolarization of hippocampal neurons. *J. Neurophysiol.* 55:1268–82

Lancaster B, Zucker R. 1994. Photolytic manipulation of $Ca^{2+}$ and the time course of slow, $Ca^{2+}$-activated $K^+$ current in rat hippocampal neurones. *J. Physiol.* 475:229–39

Larkman A. 1991. Dendritic morphology of pyramidal neurones of the visual cortex of the rat: I. Branching patterns. *J. Comp. Neurol.* 306:307–19

Lasser-Ross N, Miyakawa H, Lev-Ram V, Young SR, Ross W. 1991. High time resolution fluorescence imaging with a CCD camera. *J. Neurosci. Methods* 36:253–61

Lasser-Ross N, Ross W. 1992. Imaging voltage and synaptically activated sodium transients in cerebellar Purkinje cells. *Proc. R. Soc. London Ser. B* 247:35–39

Lester RJ, Clements JD, Westbrook GL, Jahr C. 1990. Channel kinetics determine the time course of NMDA receptor-mediated synaptic currents. *Nature* 346:565–67

Levitan I. 1994. Modulation of ion channels by protein phosphorylation and dephosphorylation. *Annu. Rev. Physiol.* 56:193–212

Lev-Ram V, Miyakawa H, Lasser-Ross N, Ross W. 1992. Calcium transients in cerebellar Purkinje neurons evoked by intracellular stimulation. *J. Neurophysiol.* 68(4):1167–77

Li M, West JW, Numann R, Murphy BJ, Scheuer T, Catterall W. 1993. Convergent regulation of sodium channels by protein kinase C and cAMP-dependent protein kinase. *Science* 261:1439–42

Lisman J. 1989. A mechanism for the Hebb and the anti-Hebb processes underlying learning and memory. *Proc. Natl. Acad. Sci. USA* 86:9574–78

Llano I, DiPolo R, Marty A. 1994. Calcium-induced calcium release in cerebellar Purkinje cells. *Neuron* 12:663–73

Llinás R. 1975. Electroresponsive properties of dendrites in central neurons. In *Advances in Neurology*, ed. GW Kreutzberg, pp. 1–13. New York: Raven

Llinás R. 1988. The intrinsic eletrophysiological properties of mammalian neurons: insights into central nervous system function. *Science* 242:1654–64

Llinás R, Nicholson C. 1971. Electrophysiological properties of dendrites and somata in alligator Purkinje cells. *J. Neurophysiol.* 34:532–51

Llinás R, Sugimori M. 1980. Electrophysiological properties of in vitro Purkinje cell somata in mammalian cerebellar slices. *J. Physiol.* 305:171–95

Llinás R, Sugimori M, Hillman DE, Cherksey B. 1992. Distribution and functional significance of the P-type, voltage-dependent $Ca^{2+}$ channels in the mammalian central nervous system. *Trends Neurosci.* 15:351–55

Lonart G, Alagarsamy S, Radhika R, Wang J, Johnson K. 1992. Inhibition of the phospholipase C-linked metabotropic glutamate receeptor by 2-amino-3-phosphonopropionate is dependent on extracellular calcium. *J. Neurochem.* 59:772–75

Lorente de Nó R. 1938. Synaptic stimulation of motoneurons as a local process. *J. Neurophysiol.* 1:194–206

Lorente de Nó R. 1947. A study of nerve physiology. *Stud. Rockefeller Inst. Med. Res.* 32: 384–483

Lorente de Nó R, Condouris G. 1959. Decremental conduction in peripheral nerve. Integration of stimuli in the neuron. *Proc. Natl. Acad. Sci. USA* 45:592–617

Lux HD, Pollen D. 1966. Electrical constants of neurons in the motor cortex of the cat. *J. Neurophysiol.* 29:207–20

MacDermott AB, Mayer ML, Westbrook GL, Smith SJ, Barker J. 1986. NMDA-receptor activation increases cytoplasmic calcium concentrations in cultured spinal cord neurones. *Nature* 321:519–22

Magee JC, Christofi G, Miyakawa H, Christie B, Lasser-Ross N, Johnston D. 1995. Subthreshold synaptic activation of voltage-gated $Ca^{2+}$ channels mediates a localized $Ca^{2+}$ influx into dendrites of hippocampal pyramidal neurons. *J. Neurophysiol.* 74:1335–42

Magee JC, Johnston D. 1995a. Synaptic activation of voltage-gated channels in the dendrites of hippocampal pyramidal neurons. *Science* 268:301–4

Magee JC, Johnston D. 1995b. Characterization of single voltage-gated $Na^+$ and $Ca^{2+}$ channels in apical dendrites of rat CA1 pyramidal neurons. *J. Physiol.* 481(1):67–90

Major G, Larkman AU, Jonas P, Sakmann B, Jack JB. 1994. Detailed passive cable models of whole-cell recorded CA3 pyramidal neurons in rat hippocampal slices. *J. Neurosci.* 14:4613–38

Malenka R. 1991. The role of postsynaptic calcium in the induction of long-term potentiation. *Mol. Neurobiol.* 5:289–95

Malinow R, Otmakhov N, Blum KI, Lisman J. 1994. Visualizing hippocampal synaptic function by optical detection of $Ca^{2+}$ entry through the N-methyl-D-aspartate channel. *Proc. Natl. Acad. Sci. USA* 91:8170–74

Markram H, Helm J, Sakmann B. 1995. Dendritic calcium transients evoked by single back-propagating action potentials in rat neocortical pyramidal neurons. *J. Physiol.* 485:1–20

Markram H, Sakmann B. 1994. Calcium transients in dendrites of neocortical neurons evoked by single subthreshold excitatory postsynaptic potentials via low-voltage-activated calcium channels. *Proc. Natl. Acad. Sci. USA* 91:5207–11

Marrion NV, Zucker RS, Marsh SJ, Adams P. 1991. Modulation of M-current by intracellular $Ca^{2+}$. *Neuron* 6:533–45

Medina I, Filippova N, Barbin G, Ben-Ari Y, Bregestovski P. 1994. Kainate-induced inactivation of NMDA currents via an elevation of intracellular $Ca^{2+}$ in hippocampal neurons. *J. Neurophysiol.* 72:456–65

Mehraein P, Yamada M, Tarnowska-Dziduszko E. 1975. Quantitative study on dendrites and dendritic spines in Alzheimer's disease and senile dementia. In *Advances in Neurology*, ed. GW Kreutzberg, pp. 453–58. New York: Raven

Mintz IM, Adams ME, Bean B. 1992. P-type calcium channels in rat central and peripheral neurons. *Neuron* 9:85–95

Miyakawa H, Kato H. 1986. Active properties of dendritic membrane examined by current source density analysis in hippocampal CA1 pyramidal neurons. *Brain Res.* 399: 303–9

Miyakawa H, Ross WN, Jaffe D, Callaway JC, Lasser-Ross N, et al. 1992. Synaptically activated increases in $Ca^{2+}$ concentration in hippocampal CA1 pyramidal cells are primarily due to voltage-gated $Ca^{2+}$ channels. *Neuron* 9:1163–73

Mogul DJ, Adams ME, Fox A. 1993. Differential activation of adenosine receptors decreases N-type but potentiates P-type $Ca^{2+}$ current in hippocampal CA3 neurons. *Neuron* 10:327–34

Nelson PG, Lux H. 1970. Some electrical measurements of motoneuron parameters. *Biophys. J.* 10:55–73

Nicoll A, Larkman A, Blakemore C. 1993. Modulation of EPSP shape and efficacy by intrinsic membrane conductances in rat neocortical pyramidal neurons in vitro. *J. Physiol.* 468:693–710

O'Leary JL, Bishop G. 1943. Analysis of potential sources in the optic lobe of duck and goose. *J. Cell. Comp. Physiol.* 22:73–87

Owen DG, Segal M, Barker J. 1984. A Ca-dependent Cl⁻ conductance in cultured mouse spinal neurons. *Nature* 311:567–70

Pockberger H. 1991. Electrophysiological and morphological properties of rat motor cortex neurons in vivo. *Brain Res.* 539:181–90

Purpura D. 1967. Comparative physiology of dendrites. In *The Neurosciences*, ed. GC Quarton, T Melnechuk, FO Schmitt, pp. 372–875. New York: Rockefeller Univ. Press

Purpura D. 1974. Dendritic spine "dysgenesis" and mental retardation. *Science* 186:1126–28

Purpura D. 1975. Dendritic differentiation in human cerebral cortex: normal and aberrant development patterns. In *Advances in Neurology*, ed. GW Kreutzberg, pp. 91–116. New York: Raven

Rall W. 1977. Core conductor theory and cable properties of neurons. In *Handbook of Physiology. The Nervous System*, ed. ER Kandel, pp. 39–97. Bethesda: Am. Physiol. Soc.

Ramon-Moliner E. 1968. The morphology of dendrites. In *The Structure And Function of Nervous Tissue*, ed. GH Bourne, pp. 205–67. New York: Academic

Randall AD, Wendland B, Schweizer F, Miljanich G, Adams ME, Tsien R. 1993. Five pharmacologically distinct high volt-

age-activated $Ca^{2+}$ channels in cerebellar granule cells. *Soc. Neurosci. Abstr.* 19:1478

Regehr W, Kehoe J, Ascher P, Armstrong C. 1993. Synaptically triggered action potentials in dendrites. *Neuron* 11:145–51

Regehr WG, Armstrong C. 1994. Where does it all begin? *Curr. Biol.* 4:436–39

Regehr WG, Connor JA, Tank D. 1989. Optical imaging of calcium accumulation in hippocampal pyramidal cells during synaptic activation. *Nature* 341:533–36

Regehr WG, Konnerth A, Armstrong C. 1992. Sodium action potentials in the dendrites of cerebellar Purkinje cells. *Proc. Natl. Acad. Sci. USA* 89:5492–96

Regehr WG, Tank D. 1990. Postsynaptic NMDA receptor-mediated calcium accumulation in hippocampal CA1 pyramidal cell dendrites. *Nature* 345:807–10

Regehr WG, Tank D. 1992. Calcium concentration dynamics produced by synaptic activation of CA1 hippocampal pyramidal cells. *J. Neurosci.* 12(11):4202–23

Richardson TL, Turner RW, Miller J. 1987. Action-potential discharge in hippocampal CA1 pyramidal neurons: current source-density analysis. *J. Neurophysiol.* 58:981–96

Rihn LL, Claiborne B. 1990. Dendritic growth and regression in rat dentate granule cells during late postnatal development. *Dev. Brain Res.* 54:115–24

Rosenmund C, Carr DW, Bergeson SE, Nilaver G, Scott JD, Westbrook G. 1994. Anchoring of protein kinase A is required for modulation of AMPA/kainate receptors on hippocampal neurons. *Nature* 368:853–56

Rosenmund C, Feltz A, Westbrook G. 1995. Calcium-dependent inactivation of synaptic NMDA receptors. *J. Neurophysiol.* 73:427–30

Scheibel A. 1970. Development of axonal and dendritic neuropil as a function of evolving behavior. In *The Neurosciences: Fourth Study Program*, ed. G Adelman, FO Schmitt, pp. 381–98. Cambridge: MIT Press

Scheibel AB, Conrad A. 1993. Hippocampal dysgenesis in mutant mouse and schizophrenic man: Is there a relationship? *Schizophr. Bull.* 19:21–33

Scheibel ME, Scheibel S. 1973. Hippocampal pathology in temporal lobe epilepsy. A Golgi survey. In *Epilepsy: Its Phenomena in Man*, ed. MAB Brazier, pp. 315–37. New York: Academic

Schwartzkroin PA, Slawsky M. 1977. Probable calcium spikes in hippocampal neurons. *Brain Res.* 135:157–61

Segev I, Rinzel J, Shepherd G. 1995. *The Theoretical Foundation of Dendritic Function*, Cambridge: MIT Press

Shadlen MN, Newsome WT. 1995. Noise, neural codes and cortical organization. *Curr. Op. Neurobiol.* 4:569–79

Sharp AH, McPherson PS, Dawson TM, Aoki C, Campbell KP, Snyder S. 1993. Differential immunohistochemical localization of inositol 1,4,5-trisphosphate- and ryanodine-sensitive $Ca^{2+}$ release channels in rat brain. *J. Neurosci.* 13:3051–63

Shelton D. 1985. Membrane resistivity estimated for the Purkinje neuron by means of a passive computer model. *Neuroscience* 14: 111–31

Shepherd G. 1991. In *Foundations of the Neuron Doctrine*, New York: Oxford Univ. Press

Shepherd GM, Brayton RK, Miller JP, Segev I, Rinzel J, Rall W. 1985. Signal enhancement in distal cortical dendrites by means of interactions between active dendritic spines. *Proc. Natl. Acad. Sci. USA* 82:2192–95

Sherrington C. 1906. *The Integrative Action of the Nervous System.* New Haven: Yale Univ. Press

Softky W. 1994. Sub-millisecond coincidence detection in active dendritic trees. *Neuroscience* 58:13–41

Spencer WA, Kandel E. 1961. Electrophysiology of hippocampal neurons: IV. Fast prepotentials. *J. Neurophysiol.* 24:272–85

Spruston N, Jaffe DB, Johnston D. 1994. Dendritic attenuation of synaptic potentials and currents: the role of passive membrane properties. *Trends Neurosci.* 17:161–66

Spruston N, Jaffe DB, Williams SH, Johnston D. 1993. Voltage- and space-clamp errors associated with the measurement of electronically remote synaptic events. *J. Neurophysiol.* 70:781–802

Spruston N, Johnston D. 1992. Perforated patch-clamp analysis of the passive membrane properties of three classes of hippocampal neurons. *J. Neurophysiol.* 67: 508–29

Spruston N, Jonas P, Sakmann B. 1995a. Dendritic glutamate receptor channels in rat hippocampal CA3 and CA1 pyramidal neurons. *J. Physiol.* 482:325–52

Spruston N, Schiller Y, Stuart G, Sakmann B. 1995b. Activity-dependent action potential invasion and calcium influx into hippocampal CA1 dendrites. *Science* 268:297–300

Stelzer A. 1992. Intracellular regulation of $GABA_A$-receptor function. In *Ion Channels*, ed. T Narahashi, pp. 83–136. New York: Plenum

Stuart G, Hausser M. 1994. Initiation and spread of sodium action potentials in cerebellar Purkinje cells. *Neuron* 13:703–12

Stuart GJ, Dodt HU, Sakmann B. 1993. Patch-clamp recordings from the soma and dendrites of neurones in brain slices using infrared video microscopy. *Pflügers Arch.* 423:511–18

Stuart GJ, Sakmann B. 1994. Active propagation of somatic action potentials into neocortical pyramidal cell dendrites. *Nature* 367: 69–72

Taube JS, Schwartzkroin P. 1988. Mechanisms of long-term potentiation: a current-source density analysis. *J. Neurosci.* 8:1645–55

Teyler TJ, Cavus I, Coussens C, DiScenna P, Grover L, et al. 1994. Multideterminant role of calcium in hippocampal synaptic plasticity. *Hippocampus* 4:623–34

Tsien R. 1989. Fluorescent probes of cell signaling. *Annu. Rev. Neurosci.* 12:227–54

Tsien RW, Lipscombe D, Madison DV, Bley KR, Fox A. 1988. Multiple types of neuronal calcium channels and their selective modulation. *Trends Neurosci.* 11:431–38

Turner RW, Meyers DR, Barker J. 1989. Localization of tetrodotoxin-sensitive field potentials of CA1 pyramidal cells in the rat hippocampus. *J. Neurophysiol.* 62:1375–87

Turner RW, Meyers DR, Richardson TL, Barker J. 1991. The site for initiation of action potential discharge over the somatodendritic axis of rat hippocampal CA1 pyramidal neurons. *J. Neurosci.* 11:2270–80

Westenbroek RE, Ahlijanian MK, Catterall W. 1990. Clustering of L-type $Ca^{2+}$ channels at the base of major dendrites in hippocampal pyramidal neurons. *Nature* 347:281–84

Wilson C. 1992. Dendritic morphology, inward rectification, and the functional properties of neostriatal neurons. In *Single Neuron Computation*, ed. T McKenna, J Davis, SF Zornetzer, pp. 141–71. Boston: Academic

Wilson MA, Bhalla US, Uhley JD, Bower J. 1989. GENESIS: a system for simulating neural networks. In *Advances in Neural Information Processing Systems 1*, ed. DS Touretzky, pp. 485–92. San Mateo, CA: Morgan Kaufmann

Wong RS, Prince DA, Basbaum A. 1979. Intradendritic recordings from hippocampal neurons. *Proc. Natl. Acad. Sci. USA* 76:986–90

Yuste R, Gutnick MJ, Saar D, Delaney KR, Tank D. 1994. $Ca^{2+}$ accumulations in dendrites of neocortical pyramidal neurons: an apical band and evidence of two functional compartments. *Neuron* 13:23–43

Annu. Rev. Neurosci. 1996. 19:187–217

# NEURONAL INTERMEDIATE FILAMENTS

*Michael K. Lee*[1#] *and Don W. Cleveland*[1,2*]

Departments of [1]Biological Chemistry and [2]Neuroscience, The Johns Hopkins University, School of Medicine, 725 N. Wolfe Street, Baltimore, Maryland 21205

KEY WORDS:    neurofilaments, radial growth, axonal transport, motor neuron disease, axonal cytoskeleton

---

## ABSTRACT

Neurofilaments (NFs) are the most abundant structural components in large-diameter myelinated axons. Assembled as obligate heteropolymers requiring NF-L and substoichiometric amounts of NF-M and/or NF-H, NF investment into axons is essential for establishment of axonal caliber, itself a key determinant of conduction velocity. Use of transgenic mice to increase axonal accumulation of NFs or to express mutant NFs subunits has proven that aberrant organization or assembly of NFs is sufficient to cause disease arising from selective dysfunction and degeneration of motor neurons. Because aberrant accumulation of NFs is a common pathology in a series of motor neuron diseases—including amyotrophic lateral sclerosis—NF misaccumulation, and the resultant disruption in axonal transport, is probably a key intermediate in the pathogenesis of these diseases.

---

## INTRODUCTION

Nineteenth century anatomists (Valentin 1836, Purkinje 1838, Schulze 1871, Cajal 1899) were the first to describe fibrous networks (then named neurofibrils) within neurons. Subsequent development of silver staining methods in the late nineteenth (Apathy 1897) and early twentieth (Bielschowsky 1902, Cajal 1903) centuries allowed visualization of neurofibrils with great clarity. Using electron microscopy, Schmitt (1968a, b) found that neurofibrils were comprised of ~10-nm diameter filaments, which were called neurofilaments

---

[#]Present address: Neuropathology Laboratory and Department of Pathology, Johns Hopkins University School of Medicine, 720 Rutland Avenue, Baltimore, Maryland 21205

[*]Present address: Ludwig Institute for Cancer Research and Departments of Medicine and Neuroscience, University of California at San Diego, 9500 Gilman Drive, La Jolla, California 92093

187

0147-006X/96/0301-0187$08.00

(NFs). NFs are now known to be members of the intermediate filament (IF) family, whose name derives from the characteristic diameter of 8–10 nm, which is intermediate between actin filaments (6 nm) and microtubules (24 nm). As major components in the cytoplasm of most eukaryotic cells, roles for IFs in cell architecture, stability, and differentiation have long been postulated and have now been demonstrated for keratins (see Fuchs & Weber 1994 for review) and NFs (see below).

We review here efforts proving that neuronal IFs (nIFs) are heteropolymers composed of multiple subunits with distinct biochemical properties and expression patterns. We also review efforts to dissect assembly properties and filament organization and the influence of glycosylation and phosphorylation on these properties. Finally, we review how the use of classical genetics and transgenic mice has led to proof that NFs are essential for establishing the diameter of myelinated axons and that defects in NF assembly and organization, often seen in human neurodegenerative disease, can directly lead to motor neuron dysfunction and death. For a more detailed review of IF structure and properties, we refer readers elsewhere (Fuchs & Weber 1994).

## DIVERSITY OF nIF PROTEINS

### Mammalian nIF Proteins

The current family of mammalian nIFs for which complete sequence information is available consists of six members: the three NF subunits [NF-L (66 kDa) (Geisler et al 1985, Lewis & Cowan 1986, Julien et al 1987, Chin & Liem 1989), NF-M (95–100 kDa) (Levy et al 1987, Napolitano et al 1987, Myers et al 1987), and NF-H (110–115 kDa) (Lees et al 1988, Julien et al 1988, Dautigny et al 1988, Chin & Liem 1990)], α-internexin (~60 kDa) (Kaplan et al 1990, Chien & Liem 1994), peripherin (~50 kDa) (Leonard et al 1988, Thompson & Ziff 1989), and nestin (~200 kDa) (Lendhal et al 1990, Dahlstrand et al 1992). As displayed in Figure 1, all six share a characteristic ~310–amino acid α-helical domain containing a hydrophobic heptad repeat essential for assembly. Flanking this central rod are globular head and tail segments, which are markedly divergent in length and sequence among the various subunits. Intron-exon position and sequence similarities show all to be in the type-IV IF class [see Fuchs & Weber (1994) for definition of the five classes], except peripherin, which along with muscle-specific desmin and the widely expressed vimentin comprises the self-assembly–competent type-III IF class.

Among the nIFs, the NF triplet proteins consist of NF-L, NF-M, and NF-H. Largely due to their abundance in mammalian neurons, NFs were the first type of nIFs to be identified (Hoffman & Lasek 1975). Like other IFs, NFs are long,

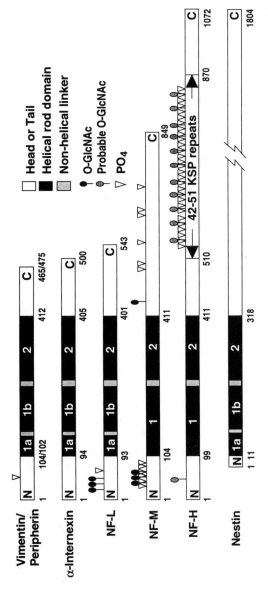

*Figure 1*  Schematic of mammalian nIF subunits. For all nIF subunits, variable head and tail domains (*open boxes*) flank the conserved α-helical rod domain (*black boxes*). The rod domains are subdivided by a short nonhelical linker (*gray boxes*) into coils 1 and 2. Except for NF-M and NF-H, coil 1 is further divided into coils 1a and 1b by another nonhelical linker. Known *O*-linked glycosylation sites are indicated by the black balloons, and the probable glycosylation sites are indicated by the gray balloons (Dong et al 1993). Known phosphorylation sites in NF subunits are indicated by the inverted triangles (see text for references). The amino acid numbers indicate the approximate delineations of the head, rod, and tail domains derived from sequences of hamster vimentin (Quax et al 1983); rat peripherin (Thompson & Ziff 1989); rat α-internexin (Kaplan et al 1990); murine NF-L (Lewis & Cowan 1986); murine NF-M (Levy et al 1987); murine NF-H (Julien et al 1988); and rat nestin (Lendhal et al 1990).

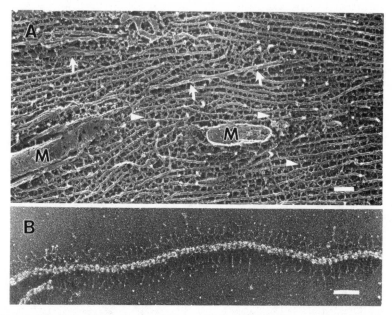

*Figure 2*  NFs in a myelinated axon and in vitro. (*A*) Quick freeze, deep etch view of axoplasm from a large myelinated axon from *Xenopus laevis*. Arrowheads point to cross bridges between ~10-nm diameter NF. Arrows point to the only two microtubules in the field. (M, mitochondria; Bar, 100 μm.) [Reproduced with permission from Hirokawa (1982).] (*B*) Native NF from bovine spinal cord viewed by glycerol spraying and rotary shadowing. (Bar, 100 μm.) [Reproduced with permission from Troncoso et al (1990).]

unbranched filaments (Figure 2*B*) that in large myelinated axons accumulate as the most abundant structures (see Figure 2*A*), outnumbering microtubules 5- to 10-fold. Unlike other IFs, NFs have characteristic "side arms" (Hirokawa et al 1984) extending from the filament backbone (Figure 2*B*) that appear to form bridges between filaments (see Figure 2*A*). Although the head domains of NF subunits do not share obvious amino acid sequence homologies, they are rich in serines and threonines. Phosphorylation and *O*-glycosylation (summarized in Figure 1) of these residues are believed to be important for in vivo regulation of NF assembly. Although the rod domain of NF-L is similar to that of type-III IFs (Geisler et al 1985), with an interruption of the heptad repeat in coil 1, this domain in NF-M (Myers et al 1987, Levy et al 1987) and NF-H (Lees et al 1988, Julien et al 1988) is a continuous coil. The most distinctive features of the NF subunits are the carboxyl-terminal tail domains. For NF-L, this region is highly acidic, with many glutamic acid residues comprising a segment sometimes referred to as the E segment (Shaw 1991).

NF-M has a longer tail domain, and this too contains E segments as well as segments rich in glutamic acid and lysines. The NF-H tail is distinct from other IFs, owing to the presence of between 42 and 51 repeats of lysine-serine-proline (KSP). Although the number and the distribution of repeats differ among vertebrates, the serines in the KSP domains are heavily phosphorylated in axons (Julien & Mushynski 1982, 1983; Carden et al 1985; Lee et al 1988; Elhanany et al 1994). NF-M also contains a few KSP motifs. Their number and position are not conserved (Shaw 1991), but they are phosphorylated in axons (Xu et al 1992). For a more detailed discussion of sequence motifs in NF subunits, we refer readers elsewhere (Shaw 1991).

At least two other IF proteins, α-internexin (Pachter & Liem 1985, Fliegner et al 1990, Kaplan et al 1990) and peripherin (Portier et al 1983b, Escurat et al 1988, Leonard et al 1988, Parysek & Goldman 1988, Gorham et al 1990), are also expressed in subsets of terminally differentiated neurons. First purified from rat optic nerve and spinal cord as an NF-associated, IF-like protein (Pachter & Liem 1985), α-internexin has been shown to self-assemble into 10-nm filaments in vitro (Kaplan et al 1990), and cloning of the gene shows it to be a type-IV IF subunit (Fliegner et al 1990, Chien & Liem 1994). The tail domain of α-internexin is enriched in glutamate and lysines, and both head and tail segments have short stretches highly homologous to NF-M; but like NF-L, the rod domain contains an interruption within coil 1. This yields a subunit with a hybrid nature, a feature that may contribute to special assembly properties of α-internexin (see below).

Peripherin was initially identified as a 56- to 57-kDa protein that was present in neuroblastoma and rat pheochromocytoma PC12 cells and that exhibited the characteristics of an IF protein (Portier et al 1983a, b; Parysek & Goldman 1987, 1988; Leonard et al 1988). Its sequence (and intron-exon placement) firmly established it as a type-III IF subunit (Leonard et al 1988, Thompson & Ziff 1989).

The remaining nIF subunit, nestin, is a transient component of neurons that is expressed only in multipotent neuroectodermal cells prior to terminal differentiation (Lendhal et al 1990, Zimmerman et al 1994). Rodent nestin is the largest known vertebrate IF subunit, with almost no head domain (11 amino acids) and a long (~150-kDa) tail domain (Lendhal et al 1990), which contains up to 35 repeats of an acidic 11–amino acid motif (Lendhal et al 1990, Dahlstrand et al 1992).

## nIFs in Lower Vertebrates and Invertebrates

Examination of nIFs in lower vertebrates and invertebrates has revealed the presence of nIF proteins similar to mammalian nIF subunits. With a few exceptions, most lower vertebrate and invertebrate neurons contain a low-molecular-weight and one or more high-molecular-weight NF-like proteins (Lasek et al 1985). However, arthropods seem to lack nIFs (Lasek et al 1985). Genes encoding proteins homologous to NF-M, peripherin, and α-internexin have been identified in *Xenopus laevis* (Sharpe 1988, Sharpe et al 1989, Charnas et al 1992) and goldfish (Glasgow et al 1994a,b). Novel nIFs that do not have obvious vertebrate homologues have also been found, including the goldfish protein plastin, which is a type-III nIF induced during optic nerve regeneration (Glasgow et al 1992), and the 1744–amino acid nIF-like protein tanabin, which is localized in a subset of axons and growth cones of *X. laevis* (Hemmati-Brivanlou et al 1992). Immunological studies indicate the presence of NF-L–like nIF in both amphibia (Quitschke & Schechter 1986, Szaro & Gainer 1988) and fish (Quitschke et al 1985). In primitive chordates such as *Ciona* (Mencarelli et al 1987) and *Lamprey* (Pleasure et al 1989), only a single high-molecular-weight nIF has been found, and this subunit has characteristics of all three mammalian NF subunits, including a long tail domain reminiscent of NF-H and NF-M.

Squid nIFs consist of subunits with apparent molecular weights of 60, 70, 220, and >500 kDa (Roslansky et al 1980, Szaro et al 1991, Way et al 1992). These subunits arise from alternative splicing of a single gene (Szaro et al 1991, Way et al 1992). Unlike the vertebrate NFs, the rod domain of these squid nIFs resembles nuclear lamins an additional 42 amino acids (Szaro et al 1991, Way et al 1992). Although intron placement suggests an origin from the type-III IF family, the tail of the largest subunit contains a repeated KSP motif similar to that in mammalian NF-M and NF-H (Way et al 1992). Finally, the mollusk *Aplysia* apparently has only low-molecular-weight nIFs, which is an exception to the general rule that invertebrate neurons contain a nIF subunit with a tail domain containing KSP phosphorylation repeats (Lasek et al 1985, Mencarelli et al 1987).

## EXPRESSION OF VERTEBRATE nIF PROTEINS FOLLOWS TERMINAL DIFFERENTIATION AND INCREASES AS NEURONS MATURE

Expression of different types of nIFs in mature neurons is neither mutually exclusive nor entirely overlapping. The earliest IFs associated with neuronal development are nestin (Lendhal et al 1990) and vimentin (Tapscott et al 1981, Cochard & Paulin 1984), both of which are expressed at high levels in neuroectoderm cells that have the potential to develop into both neurons and glia. Concomitant with the postmitotic differentiation of neurons, nestin (Lendhal et al 1990) and vimentin (Tapscott et al 1981, Cochard & Paulin 1984) expression is silenced, and other nIFs begin to accumulate. Nestin is also transiently expressed in embryonic and neonatal muscle (Lendhal et al 1990, Zimmerman et al 1994).

α-Internexin and NFs are expressed in many central and peripheral nervous system neurons (Kaplan et al 1990, Fliegner et al 1994). Although NF-L and NF-M are associated with terminal mitotic division of neuroblasts in specific neuronal populations, such as spinal motor neurons (Tapscott et al 1981, Cochard & Paulin 1984), accumulation of both seems to lag slightly behind the initial appearance of α-internexin in most other neurons (Kaplan et al 1990, Fliegner et al 1994). When neurons are actively migrating and elaborating neurites, NF-H expression is generally below detectable levels (Carden et al 1987). In mature central nervous system neurons, the distribution of α-internexin is similar to that of NF-L (Kaplan et al 1990). However, neurons with smaller axons express α-internexin abundantly, whereas the larger neurons primarily have NFs (Kaplan et al 1990, Fliegner et al 1994). In rat cerebellum, α-internexin accumulates in parallel fibers, but NFs are undetectable (Kaplan et al 1990). The generality of this finding, however, will require further study because α-internexin has not been detected in human cerebellar parallel fibers (Yachnis et al 1993).

Peripherin is limited to the neurons of neural crest origin or to those neurons that will extend their axons outside of the central neural axis (Parysek & Goldman 1988, Escurat et al 1990, Gorham et al 1990, Troy et al 1990a). In peripheral sensory and autonomic ganglia, the largest neurons express NFs, while the smallest neurons express peripherin; somewhat medium-sized neurons express both (Escurat et al 1990, Gorham et al 1990, Troy et al 1990a, Goldstein et al 1991). In general, peripherin appears later than NF-L (Escurat et al 1990, Gorham et al 1990).

NF accumulation, although detectable during neurite elongation, becomes robust only after synapse formation and concomitant with myelination (Carden et al 1987, Schlaepfer & Bruce 1990). In large motor and sensory neurons, the increase in NF expression is accompanied by marked decreases in α-internexin and peripherin (Escurat et al 1990, Gorham et al 1990, Fliegner et al 1994). However, high levels of either persist in the smaller neurons that accumulate few NFs. The developmental pattern of nIF expression is recapitulated during axonal regrowth following injury (Dahl et al 1988, Hoffman & Cleveland 1988, Oblinger et al 1989, Troy et al 1990b). Although NFs are widely held to be neuron specific, NF-M and NF-L may be expressed in immature Schwann cells (Kelly et al 1992, Roberson et al 1992), and NF-H in T lymphocytes (Murphy et al 1993) and embryonic heart muscles [which also express peripherin (Belecky-Adams et al 1993)]. The significance of this nonneuronal expression is unclear.

## REGULATORY ELEMENTS FOR nIF EXPRESSION

Search for the basis of the nearly neuron-specific expression of nIF has focused on identifying regulatory elements in the 5'-promoter regions of nIF genes. This has lead to only limited success. In the case of NF subunits (Zopf et al

1990, Shneidman et al 1992, Yazdanbakhsh et al 1993) and α-internexin (Ching & Liem 1991), the efforts have been hindered by the fact that when 5′-promoter regions have been used to drive the expression of a reporter protein [e.g. bacterial chloramphenicol acetyltransferase (CAT)], little or no difference in expression of the reporter has been seen between neuronal and nonneuronal cells in culture. Although multiple potential regulatory elements have been identified by sequence and in vitro binding assays for promoter regions from *NF-L* (Shneidman et al 1992, Yazdanbakhsh et al 1993, Pospelov et al 1994), *NF-M* (Zopf et al 1990, Elder et al 1992c, Shneidman et al 1992), and *NF-H* (Elder et al 1992a,b; Shneidman et al 1992), no clear cell- or tissue-specific signals have been identified. Rather, the promoter regions of NF genes contain a proximal basal promoter, a weak proximal neuronal enhancer, and a strong distal negative element (Shneidman et al 1992, Yazdanbakhsh et al 1993)— each of which may be necessary for expression of NF genes in brain but not in other tissues (Shneidman et al 1992). For peripherin, DNA transfection into PC12 cells has identified two NGF-sensitive positive regulatory elements and a constitutive negative element in the proximal promoter (Thompson et al 1992), while additional transfections into neuroblastoma lines have identified a negative regulatory element between −98 and −46 that may be important for expression exclusive to neurons (Desmarais et al 1992).

The ultimate utility of these in vitro efforts has been questioned by in vivo approaches that have demonstrated that key regulator elements for both the level and neuronal specificity of expression lie downstream of the transcription initiation sites of nIF genes. In transgenic mice, neither the 5-kb rat (Yazdan-bakhsh et al 1993) nor the 2.3-kb human (Beaudet et al 1992) *NF-L* promoter yields neuron-specific expression of a CAT reporter gene. Further, the level of expression achieved using just the *NF-L* promoter is very low compared to the level of endogenous NF-L expression (Beaudet et al 1992, Yazdanbakhsh et al 1993). However, relatively high levels of specific neuronal expression of human NF-L can be achieved in transgenic mice by using constructs that include only 300 bp of the promoter sequence (minimal promoter), as long as it is linked to the coding sequences and introns (Beaudet et al 1992, 1993). Expression exclusively in neurons requires the combined presence of the segment between 300 and 190 bases 5′ to the transcription initiation site and the intragenic transcribed sequences: Excluding either component results in ectopic expression (Beaudet et al 1992, Yazdanbakhsh et al 1993). Whether such an *NF-L*–promoted transgene can faithfully replicate the developmental stage-specific expression of endogenous *NF-L* is unknown. Similar require-ments also exist for the regulation of peripherin (Belecky-Adams et al 1993) and nestin (Zimmerman et al 1994), where the presence of both 5′-promoter and intragenic sequences are necessary to replicate the expression pattern of the endogenous genes.

Although the *NF-L* promoter alone cannot yield the quantitative level of endogenous NF-L expression in transgenic mice, it can produce significant neuronal expression. When linked to a cDNA encoding the amyloid Aβ peptide that accumulates in plaques of Alzheimer's patients, the 1.8-kb murine *NF-L* promoter produced a level of Aβ mRNA in brain neurons that was equal to the mRNA encoding the endogenous amyloid precursor protein, and this led to neurodegeneration (LaFerla et al 1995).

An important quantitative contribution to NF expression level also arises posttranscriptionally. NGF induces NF-subunit expression in PC12 cells, and subsequent inhibition of transcription has shown that the increase in NF expression is due in part to NGF-dependent increase in the half-lives of the NF-encoding mRNAs (Lindenbaum et al 1988, Ikenaka et al 1990). Similar stabilization of NF transcripts was seen during postnatal maturation of cultured dorsal root ganglia neurons (Schwartz et al 1994). Julien and colleagues have also shown, using mice expressing a human *NF-L* transgene, that a 3- to 5-fold increase in NF-L mRNA corresponds to only a 10 to 50% increase in NF-L protein in the central nervous system (Beaudet et al 1993). Another example is a mutant strain of quail, in which a point mutation that creates a premature translation terminator in the *NF-L* gene (truncating 80% of the carboxyl-terminal polypeptide sequence) results in significant reduction in both NF-L and NF-H mRNAs, albeit the former is reduced to a greater extent. Furthermore, although the NF-M mRNA level is nearly normal, accumulation of the NF-M polypeptide is minimal (Ohara et al 1993).

## ASSEMBLY OF nIF AND MODULATION BY POSTTRANSLATIONAL MODIFICATIONS

### NFs Are Obligate Heteropolymers

Because of their abundance in most mature mammalian neurons and their extraordinary stability at neutral pH and physiological ionic strength, NFs were easily purified by differential sedimentation (Schlaepfer & Freeman 1978; Liem et al 1978, 1979) (see Figure 2*B*). This yielded relative molar amounts for NF-L, NF-M, and NF-H of approximately 5:2:1. Although both NF-M and NF-H contain the central α-helical rod domains typical of all IFs, NF-M, NF-H, and combinations of NF-M and NF-H fail to reassemble without NF-L (Geisler & Weber 1981, Liem & Hutchison 1982, Gardner et al 1984, Balin & Lee 1991, Balin et al 1991), while NF-L alone produces short, smooth-walled filaments (Geisler & Weber 1981; Liem & Hutchison 1982; Gardner et al 1984; Hisanaga & Hirokawa 1988, 1990b; Hisanaga et al 1990b; Heins et al 1993) without the protrusions characteristic of in vivo NFs. Because only antibodies directed against the tail domains of NF-M and NF-H decorate the side arms both in vivo and in vitro (Willard & Simon 1981, Hirokawa et al 1984, Mulligan et al 1991) and because reassembly in the presence of NF-H or NF-M restores side arms with dimensions almost identical to those present in isolated NFs or NFs in situ (Hisanaga & Hirokawa 1988), the side arms are the tails of NF-M and NF-H. This led to the view that the NF backbone was composed of NF-L, with NF-M and NF-H associated onto the NF-L filaments. However, immunoelectron microscopy with antibodies to the head, rod, or tail domains of individual NF subunits indicated that all three NF subunits are incorporated integrally into filaments (Balin & Lee 1991, Balin et al 1991). Further, analysis of early assembly and disassembly intermediates, represented by tetramers stable in 2-M urea, showed that NF-L can associate with a stochiometric amount of NF-M or NF-H, supporting the idea that all three NF subunits are integral components of NF (Cohlberg et al 1995). Assembly properties of bacterially derived NF-L subunits showed that the head is required to promote lateral association of protofilaments, while the tail limits lateral interaction to yield a filament of ~10 nm (Heins et al 1993). An unsettling aspect of all of these efforts is that reassembly in vitro requires a nonphysiologically low pH.

Assembly properties have now been reexamined in an in vivo context by using DNA transfection and transgenic mice. As expected, all three NF subunits can individually coassemble with vimentin into cytoplasmic filament networks (Chin & Liem 1989, 1990; Gill et al 1990; Wong & Cleveland 1990; Chin et al 1991; Muma & Cork 1993; Sacher et al 1994). Competence for coassembly with vimentin requires both the head and rod domains, but most of the tail domains are dispensable (Gill et al 1990, Wong & Cleveland 1990, Chin et al 1991). However, the surprise was that in a cell line without an endogenous cytoplasmic IF network [SW 13⁻ (Sarria et al 1990)] none of the NF subunits can assemble de novo (Lee et al 1993, Ching & Liem 1993). Instead, assembly requires NF-L and a substoichiometric amount of either NF-M or NF-H (Lee et al 1993). Assembly synergy between NF-L and NF-M requires interactions provided by both the head and rod (Lee et al 1993, Ching & Liem 1993), since full length NF-L or NF-M can restore assembly competence to tailless, but not headless, NF-M or NF-L, respectively. Further, because hybrid subunits composed of NF-L head and rod and NF-H tail assemble with NF-M (Lee et al 1993), steric interaction between the unusually long tails of NF-H and NF-M are not the source of their self-assembly incompetence in vitro.

The requirement for multiple NF subunits in de novo cytoplasmic IF assembly has also been demonstrated by using baculovirus to force expression of mammalian NF proteins in insect cells (Nakagawa et al 1995). Although high levels of NF-L can lead to homopolymeric assembly, much of the NF-L was found in nonfilamentous aggregates. In contrast, coexpression of NF-M with

NF-L eliminated nonfilamentous aggregates and produced long bundles of filaments with interfilament spacing of ~32 nm. A series of carboxyl-terminal truncations proved that filament-filament distances can be mediated by the NF-M tail.

Transgenic mice have been used to establish that NF are obligate heteropolymers in a true in vivo context by using transgenes, whose promoter elements direct expression of NF-L or NF-M in oligodendrocytes (Lee et al 1993). These cells, which myelinate central nervous system axons, do not normally express cytoplasmic IFs. Extended arrays of filaments were not observable in oligodendrocytes expressing only NF-L or only NF-M; however, in animals carrying both *NF-L* and *NF-M* transgenes, closely spaced bundles of 10-nm filaments were readily apparent (Lee et al 1993).

These results collectively indicate that the NFs in vivo are obligate heteropolymers requiring NF-L and either NF-M or NF-H. They also unequivocally establish that both NF-M and NF-H directly participate in the assembly and organization of NFs. The obligate heteropolymeric nature of NFs resembles the properties of keratin filaments, but with an important distinction: Keratin filaments are stoichiometric heterodimers requiring one Type-I and one Type-II keratin, whereas NF heteropolymers can apparently accommodate a wide range of NF-subunit ratios. This point is potentially significant because the subunit ratios do change markedly during neuronal development (e.g. Nixon & Shea 1992; and references therein). Whether the requirement for heteropolymerization acts at the dimer, tetramer, or higher-order oligomer level has not yet been determined, although cross-linking experiments (Carden & Eagles 1986, Cohlberg et al 1995) and antibody labeling of filaments assembled in vitro (Balin et al 1991) are most easily explained by NF-L–NF-M or NF-L–NF-H heterodimers.

## α-*Internexin Self-Assembles*

Unlike the other type-IV IF subunits, α-internexin can assemble into an extended filament network in the absence of other IF proteins and efficiently coassemble with other type-IV and type-III subunits (Ching & Liem 1993). Although the biochemical bases for the obligate heteropolymeric assembly of NFs and the homopolymeric assembly of the highly related α-internexin have not been proven, sequence comparison reveals that the rod domain of α-internexin is organized into 1a, 1b, and 2 segments (Fliegner et al 1990) (Figure 1) similar to those in NF-L, whereas the rod 1 domains of NF-M and NF-H do not have obvious subdivisions. On the other hand, the head and tail domains of α-internexin contain sequences homologous to NF-M (Fliegner et al 1990). Thus, the apparent hybrid nature of α-internexin may underlie its competence for homopolymer assembly in vivo.

## Properties of Peripherin and Nestin

The in vivo and in vitro self-assembly properties of nestin are unknown, except that in cultured cells expressing both nestin and vimentin, nestin is able to coassemble with vimentin into IF networks in cells (Lendhal et al 1990). Assembly properties of peripherin have not been shown, although from its extensive homology to other type-III IFs, it would be predicted to self-assemble in vivo and in vitro. Additionally, the proximal half of the tail domain of peripherin, like the other type-III subunits, can directly interact in vitro with lamin B (Djabali et al 1991). Indirect evidence suggests this may also be true in vivo (Djabali et al 1991), although how this could happen if lamins are really on the inside of the nuclear envelope while peripherin is on the outside remains unanswered.

## Modulating nIF Assembly Through Phosphorylation or Glycosylation of Head Domains

Like other IF subunits (see Fuchs & Weber 1994), deletions of NF-L (Gill et al 1990, Chin et al 1991, Ching & Liem 1993, Lee et al 1993) and NF-M (Wong & Cleveland 1990, Chin et al 1991, Ching & Liem 1993, Lee et al 1993) have shown that the head and rod, but not the tail, are essential for assembly. Further, the head domains of vimentin (Ando et al 1989, Evans 1989, Inagaki et al 1989, Chou et al 1990), NFs (Hisanaga et al 1990a, 1994; Sihag & Nixon 1990, 1991), and lamins (Peter et al 1990) are known to be substrates for kinases in vivo, and phosphorylation of this region either inhibits assembly (Hisanaga et al 1990a, 1994; Sihag & Nixon 1990, 1991) or causes disassembly of filaments (Ando et al 1989, Evans 1989, Inagaki et al 1989, Chou et al 1990, Hisanaga et al 1990a), both in vitro and in vivo. For the NF subunits, major in vivo phosphorylation sites in the amino-terminal head domain have been identified within [Ser55 on NF-L and within a phosphopeptide starting at Ser44 on NF-M (Sihag & Nixon 1990, 1991)] or adjacent [a phosphopeptide starting at Ser25 on NF-M (Sihag & Nixon 1990)] to the head sequences essential for in vivo NF assembly. The phosphate on Ser55 of NF-L displays rapid turnover immediately following NF-L synthesis in neurons (Sihag & Nixon 1991), raising the interesting possibility that phosphorylation may block premature assembly prior to transport into neurites. Moreover, the amino-terminal head regions of both NF-L and NF-M isolated from rat spinal cord are posttranslationally modified by addition of $O$-linked $N$-acetylglucosamine moieties (Dong et al 1993). Three of four identified sites lie in the amino-terminal head region (see Figure 1), and all three are located within (Ser27 on NF-L and Thr48 on NF-M) or near (Thr21 on NF-L) the domain essential for in vivo NF assembly. Similar modifications are present in the

NF-H head and tail, although the precise site(s) have not yet been identified (Dong et al 1993).

## AXONAL TRANSPORT OF nIFs

nIFs are moved into and through axons via slow axonal transport, the rate for which varies between 0.25 and 3 mm/day (Hoffman & Lasek 1975, Black & Lasek 1980, Hoffman et al 1983, Willard & Simon 1983, Nixon & Logvinenko 1986). The velocity is dependent on a number of factors, including the group of neurons, the age of the animal, and the location along the nerve. For example, motor neurons transport NFs at a faster rate than retinal ganglion cells (Lasek et al 1993). In the same neuronal group, NFs move along more rapidly in immature or regenerating axons than in mature axons (Hoffman & Lasek 1980, Hoffman et al 1983, Willard & Simon 1983, McQuarrie et al 1989) and in proximal axons than in distal axons (Watson et al 1989b).

The initial analyses, measuring only the peak of the transported component, concluded that all three NF subunits were transported together in a single wave (Hoffman & Lasek 1975, Black & Lasek 1980). This led to the proposal (Hoffman & Lasek 1975, Black & Lasek 1980) that NFs moved as polymers containing all three subunits. More recent reexamination of this has revealed that there are two pools of NFs in axons, one moving at the traditionally measured slow axonal transport rate and the other essentially stationary (Nixon & Logvinenko 1986, Lasek et al 1992). There is disagreement regarding the fraction of filaments in the slower pool, estimated by one group to be as much a third of the total newly synthesized pool (Nixon & Logvinenko 1986) and by another to be a negligible proportion (Lasek et al 1992). Whichever estimate is closer to the actual amount, even with a seemingly small deposition of newly synthesized NFs to the relatively stationary pool, the steady-state stationary pool can reach a significant proportion of the total axonal NFs. For example, if deposition of NFs from the pool moving at the traditional slow rate is a mere 3% along the full length of the axon and if the difference in transport rates is 25-fold (0.75 mm/day vs 0.03 mm/day), then at steady state the essentially stationary pool could reach more than 40% of the total NFs present in the axon.

There is also disagreement over whether the moving form of NFs is filaments (as envisioned by the initial Lasek models) or subunits or oligomers. In the classical axonal transport paradigm following injection into cell bodies, the wave of transported NF subunits are in the insoluble, cytoskeletal compartment (Lasek 1986, Lasek et al 1992), a finding consistent with transport as filaments. On the other hand, after injection and equilibration of fluorescently tagged NF-L in cultured dorsal root ganglion neurons, fluorescence recovery following zonal photobleaching occurs without translocation of the bleached zone

(Okabe et al 1993). Hence, at least in this example, most of the filaments must be relatively stationary. This experiment was interpreted to support oligomers as the transported form, since the bulk polymer did not move and free subunits should diffuse much too rapidly to provide the coherent wave seen in axonal transport experiments. However, as the moving phase was not detected, NF polymers could also be the moving components, with their movement obscured by the more abundant stationary pool.

The mechanism by which NFs, and microtubules for that matter, are transported remains unsettled. Lasek and colleagues have observed that over a short distance some NF subunits move along axons at a rate as fast as 72 to 144 mm/day (Lasek et al 1993). From this, they suggested that NFs may be transported in discrete steps, alternating between short, fast-moving strides and relatively long pauses. Taking advantage of delayed degeneration upon transection of sciatic nerves in mice with the *ola* mutation, Griffin and colleagues observed that NFs as well as other cytoskeletal proteins redistribute towards both distal and proximal transected ends (Watson et al 1993). While this raises the possibility of retrograde slow transport in normal axons (Watson et al 1993), it may also represent a simple loss of axonal polarity resulting from, for example, changes in microtubule orientation in these transected axons. In any event, these results provide support for the idea that the motor(s) responsible for the slow axonal transport may be similar to motors responsible for the fast axonal transport components (i.e. the kinesin family and cytoplasmic dynein). As in the fast transport model, the overall rate of transport for any specific component(s) depends on the association and dissociation rate of the transported component with the motor, which may intrinsically move at a much higher speed than the effective transport rate. Such a model is also consistent with a stationary pool of NFs, with the actual rate and efficiency of transport determined by exchange of subunits between stationary and moving phases.

## DYNAMICS OF nIFs

Despite the traditional view of IFs as static, nIFs can be highly dynamic structures. The first evidence for this arose from a demonstration that filaments assembled from NF-L tagged for fluorescence energy transfer rapidly exchange subunits between filaments in vitro (Angelides et al 1989). In vivo support for both the dynamic nature of NFs and the presence of stationary NFs in neurons emerged from microinjecting a fluorescently tagged NF-L subunit into cultured dorsal root ganglia neurons. After allowing equilibration with neurite NFs and then photobleaching a segment, recovery of fluorescence was found to occur with a half-time of 40 min (Okabe et al 1993). This rate is remarkably rapid when viewed in the context of an expected static array and the very small soluble pool of subunits available for exchange. Further, following microinjection of NF-L tagged with biotin, immunoelectron microscopy proved that the injected subunits incorporate along the full length of existing NFs. This is consistent with a mechanism in which the transported subunits constantly exchange with the existing stationary NF network. The rate and the extent of the exchange could be regulated by phosphorylation and dephosphorylation on the head residues, as discussed above. Microinjection of tagged NF-H led to similar findings (Takeda et al 1994), with the important exception that the exchange of photobleached NF-H subunits occurred faster than with NF-L (recovery half-times in this study of ~19 vs ~35 min, respectively). This was interpreted to represent preferential association of NF-H at the periphery of the filament (Takeda et al 1994). However, the exchangeable components, likely to be the tetramers seen as early assembly intermediates (Cohlberg et al 1995), may have differential dynamic properties depending on whether they are NF-L homotetramers or NF-L–NF-M or NF-L–NF-H heterotetramers. This latter possibility is particularly attractive because subunit exchange occurs all along the filament (Okabe et al 1993, Takeda et al 1994) and because all three NF subunits directly participate in assembly in vivo (Ching & Liem 1993, Lee et al 1993, Nakagawa et al 1995) and appear to be integral components of the filament (Balin & Lee 1991, Balin et al 1991).

Both the level and phosphorylation state of NF-H have been proposed to regulate the rate of NF transport. The slowing of axonal transport during development correlates with an increase in the expression of NF subunits, particularly NF-H (Willard & Simon 1983). Phosphorylation of the NF-H tail domain occurs following entry of NFs into axons and increases during transport (Glicksman et al 1987, Nixon et al 1987, Oblinger 1987). The extent of phosphorylation of the KSP repeat domain also correlates with the slowing of the transport rate (Watson et al 1989a, 1991; Nixon et al 1994a,b) and with transition of transported NFs into the nearly stationary pool of axonal NFs (Nixon & Logvinenko 1986, Lewis & Nixon 1988). Further, a two- to threefold increase in total neuronal NF-H level in transgenic mice carrying a human NF-H transgene leads to accumulations of NFs in the cell body (Cote et al 1993), mediated by slowing of NF transport into axons (Collard et al 1995).

Two mechanisms through which NF-H phosphorylation may slow NF transport have been proposed. In the first, NFs in axons are strongly associated with each other (Gilbert 1975), and the NF-H tail domains form cross bridges between NFs (Willard & Simon 1983, Hirokawa et al 1984, Hisanaga & Hirokawa 1988). Despite some appealing features, NF-H–containing native or reassembled filaments do not appear to cross-link in vitro (Hisanaga & Hirokawa 1989), and analysis of filament-filament spacing does not reveal a fixed interfilament distance (Price et al 1988, Hsieh et al 1994b) that would be predicted if there were a significant in vivo association. An alternative idea is that phosphorylation of the NF-H tail favors dissociation of NFs from the axonal transport machinery. This idea is supported by the demonstration that dephosphorylated NF-H associates with microtubules and tubulin, but phosphorylation of the tail abolishes this affinity (Hisanaga & Hirokawa 1990a; Miyasaka et al 1993).

As to what happens to NFs after arriving at the nerve terminus, it is believed that they are degraded upon arrival at the axon terminus. Indeed, NFs can be degraded by a variety of proteases, particularly those activated by calcium (Banik et al 1983, Kamakura et al 1983, Schlaepfer et al 1985, Gallant et al 1986, Nixon et al 1986, Banay-Schwartz et al 1987, Pant 1988, Johnson et al 1991, Greenwood et al 1993). Direct support for distal proteolysis comes from injection of the protease inhibitor leupeptin into the optic tectum of goldfish (Roots 1983): This leads to accumulation of NFs at the nerve terminals. Whether active degradation of NFs occurs only at the terminus is unclear, but phosphorylation (Goldstein et al 1987, Pant 1988) and lower free-calcium levels (Glass et al 1994) seem to protect from degradation in intact axons. An additional consideration is that under normal conditions axonal NFs must be very long-lived proteins: At 1–2 mm/day rate of slow transport (Black & Lasek 1980, Hoffman et al 1983, Willard & Simon 1983, Nixon & Logvinenko 1986), a 1–2 year transit time is required for the meter-long axons of the human sciatic nerve.

## ROLE OF NF IN RADIAL GROWTH OF AXONS AND MODULATION OF NF FUNCTION BY PHOSPHORYLATION

Although NFs are minor components during initial neurite elongation, after successful synapse formation NF expression is markedly elevated, myelination begins, and the fully elongated axons increase in diameter by up to 10-fold (volume increases by up to 100-fold!). In this second growth phase, referred to as radial growth, NFs become the most abundant cytoskeletal element, often exceeding the number of microtubules by up to an order of magnitude (e.g. see Figure 2A). Earlier correlative evidence had shown a linear relationship between NF number and cross-sectional area throughout normal radial growth (Friede & Samorajski 1970; Weiss & Mayr 1971; Hoffman et al 1984, 1985a) and during regrowth following axonal injury (Hoffman et al 1985b, 1987), observations that strongly suggested that NFs are intrinsic determinants of radial growth. This increase in axonal diameter (caliber) is important for normal nerve function because caliber is a principal determinant of the speed at which nerve impulses are propagated along the axon.

The importance of NFs in specifying normal axonal caliber has recently been proven unequivocally by analysis of a recessive mutation (quv) in a Japanese quail that lacks NFs as the result of a premature translation terminator in the *NF-L* gene. The absence of NFs leads to severe inhibition of normal radial growth in these animals (Yamasaki et al 1991, 1992; Ohara et al 1993), with a consequent reduction in axonal conduction velocity (Sakaguchi et al 1993). In addition to proving the importance of NFs in normal radial growth, the survival of this mutant quail demonstrates that NFs are not essential for viability, but are required for proper neural functioning as indicated by generalized ataxia and quivering exhibited by animals lacking NFs (Yamasaki et al 1991). Further, in mice, total inhibition of NF transport into axons by expression of a NF-H–$\beta$-galactosidase fusion protein results in severe inhibition of radial growth of large myelinated axons, although here no overt phenotype is apparent, despite complete absence of axonal NF (Eyer & Peterson 1994).

The mechanism through which NFs mediate increases in axonal size remains unsettled. The linear correlation between NF number and axonal cross-sectional area initially suggested that the axon expanded or contracted so as to maintain a constant density of NFs (Friede & Samorajski 1970; Hoffman et al 1985b, 1987). However, analysis of transgenic mice that express elevated levels of wild-type NF subunits has disproven this simple view. Increasing NF-L levels leads to increases in the number and density of NFs and an inhibition of radial growth in motor and sensory neurons of the sciatic nerve (Monteiro et al 1990). Twofold increases in NF-M levels lead to a corresponding reduction in NF-H and strongly inhibited radial growth, despite an unchanged level of either NF-L or filament number (Wong et al 1995). The diminution in NF-H is most easily explained by a competition between NF-M and NF-H for coassembly (or cotransport) with a limiting amount of NF-L. Despite reduction in NF-H, nearest-neighbor spacing between filaments was unaffected. A plausible explanation is that interactions between adjacent filaments are mediated by the NF-M tail, whereas the NF-H tail mediates radial growth through longer-range interactions.

A slightly differing result emerged from raising NF-M levels by using a human *NF-M* transgene. This transgene was expressed primarily in the central nervous system, particularly the neocortex where human NF-M accumulated to a level comparable to that of endogenous mouse NF-M (Tu et al 1995). This resulted in a doubling of NF-L and a twofold reduction in fully phosphorylated NF-H, possibly because competition between NF-H and NF-M reduced the assembled proportion of NF-H. Unlike the situation in the sciatic nerve, the packing density of NFs was increased, supporting a role for phosphorylated NF-H in mediating filament spacing. Expression of NF-M deleted in 12 of 13 KSP repeats in the tail did not affect NF-H levels or its phosphorylation state, but did yield an increase in packing density, suggesting that sequences in the NF-H tail are essential for mediating radial growth.

Phosphorylation of NF-H and NF-M appears to play a critical role in organizing NFs and in their ability to mediate control of caliber. Direct support for this has come from the myelin-deficient trembler mouse: The NFs are less phosphorylated and are more closely spaced in axons (de Waegh et al 1992). Cole et al (1994) reported similar findings for transgenic mice in which hypomyelination was achieved by selectively killing Schwann cells. Further, similar correlations among myelination, phosphorylation, and caliber are found in the initial segment of axons [from dorsal root ganglion (Hsieh et al 1994a) and retinal ganglion neurons (Nixon et al 1994b)] and in nodes of Ranvier (Hsieh et al 1994a). Myelination does not occur in these segments, and NFs are both less phosphorylated and more closely packed than in adjacent myelinated segments of the same axons (Hsieh et al 1994a, Nixon et al 1994b). These data also suggest that the phosphorylation process is ultimately linked to, and may be regulated by, myelination.

## NF KINASES

The likelihood that phosphorylation of the NF-H tail domain plays a central role in modulating neurofilament transport, dynamics, and organization within the axon has prompted an extensive search for the kinase(s) responsible. Several kinases are reportedly associated with NFs, including A kinase (Hisanaga et al 1994), cyclic AMP–dependent kinase (Dosemeci and Pant 1992), and casein kinase I (Link et al 1993), while another, NFAK115, was identified by its affinity for the NF-H tail (Xiao & Montiero 1994). PK40 (Roder & Ingram 1991), casein kinase I–like kinase (Link et al 1993), and NFAK115 (Xiao & Monteiro 1994) have been shown to phosphorylate the NF-H tail, and they may represent distinct classes of KSP-domain–specific kinases. Some of the phosphorylation sites by A kinase are on the heads of NF-L and NF-M, and this can cause disassembly of reassembled NFs (Hisanaga et al 1994).

In addition, following the initial observation that a cdc2-like, cyclin-dependent kinase (CDK) can phosphorylate a subset of KSP repeats in the tail of NF-H (Hisanaga et al 1991), several kinases with a CDK-like activity and/or sequence homologies have been identified (Hellmich et al 1992; Lew et al 1992, 1993; Miyasaka et al 1993; Shetty et al 1993). Multiple CDK-like kinases are expressed in nervous tissues (Myerson et al 1992), and some are either similar or identical to CDK5 (Myerson et al 1992). In situ hybridization showed that one of the CDK-like kinases (distinct from CDK5) is expressed predominantly in neurons (Hellmich et al 1992) and that another CDK-like kinase purified from nervous tissue can phosphorylate peptides containing a cdc2 consensus phosphorylation site [X(S/T)PXK] (Lew et al 1993, Miyasaka et al 1993, Shetty et al 1993). In one case (Miyasaka et al 1993, Shetty et al 1993), the purified CDK was shown to specifically phosphorylate the KSPXK motifs present in the carboxyl-terminal tail of both NF-H and NF-M. Although the activity of neural CDK-like kinases, including CDK5, are not regulated by cyclin, three activators of CDK5 [p35 (Tsai et al 1994), p25 (Lew et al 1994), and p67 (Shetty et al 1995)] have been identified and cloned. Although CDK5 is expressed in nonneural tissues (Myerson et al 1992, Lew et al 1994), both p35 and p25 are expressed predominantly in nervous tissues (Lew et al 1994, Tsai et al 1994). Immunocytochemical analysis has shown that p67 is expressed exclusively in neurons (Shetty et al 1995).

Despite their promise, the CDK-like kinases only phosphorylate a small subset of available KSP sites (approximately 10%) (Miyasaka et al 1993, Shetty et al 1993), and no kinase has yet been proven to play a significant role in in vivo phosphorylation of NF-H. Indeed, an appealing view is that neurons may express multiple CDKs (Myerson et al 1992), as well as multiple regulators of the kinases.

## NF ABNORMALITIES IN THE PATHOGENESIS OF MOTOR NEURON DISEASES

In addition to a function supporting the growth and maintenance of axonal caliber in normal neurons, NFs have long been suspected to play a role in the pathogenesis of several types of neurodegenerative diseases, including the motor neuron diseases amyotrophic lateral sclerosis (ALS), infantile spinal muscular atrophy (SMA), and hereditary sensory motor neuropathy. The common clinical symptom of these diseases is progressive failure of motor neurons, which in turn causes atrophy of the skeletal muscles innervated by the dying neurons, ultimately culminating in paralysis and death. Although various degrees of motor neuron loss are seen as the major pathological hallmark, for none of these diseases is the pathogenic progression understood. High-resolution analysis derived from the combination of silver staining (Hirano et al 1967, Carpenter 1968, Delisle & Carpenter 1984), immunocytochemistry (Vitto & Nixon 1986, Schmidt et al 1987, Sobue et al 1990), and electron microscopic examination (Takahashi et al 1972; Hirano et al 1984a,b; Sasaki & Maruyama 1992) has clearly shown that abnormal assembly and accumulation of NFs in motor neuron cell bodies and axons are common, conspicuous findings in most cases of human motor neuron disease [reviewed by Hirano (1991)]; for an example from a sporadic ALS patient, see Figure 3. Further, abnormal accumulations of NFs are common in most spontaneously occurring motor neuron diseases in animals (Cork et al 1988) and hereditary motor neurons disease in dogs (Cork et al 1990, Muma & Cork 1993).

A central unresolved question has been whether aberrant accumulations of NFs are merely by-products of the pathogenic process or active participants in motor neuron dysfunction. Partial resolution of this issue came from two studies in which forced overexpression of NF subunits in transgenic mouse

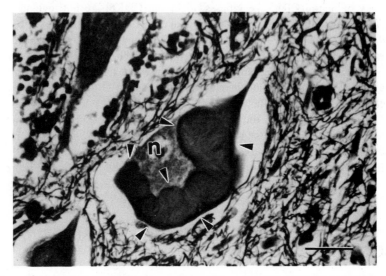

*Figure 3* Accumulation of NFs in motor neurons from a sporadic ALS patient whose disease progressed rapidly. Most of the perikaryal volume in an anterior horn motor neuron is occupied by abnormal aggregates of NF (denoted by the arrowheads). (n, the nucleus.) The section has been stained with silver. (Bar, 20 μm.) [Reproduced with permission from Hirano (1988).]

models was shown to cause selective motor neuron dysfunction. Elevating expression of wild-type mouse NF-L to about approximately four times the normal level produced large accumulations of NFs in the motor neuron cell bodies and proximal axons, motor neuron dysfunction, and neurogenic atrophy of skeletal muscle that usually leads to death by three to four weeks of age (Xu et al 1993). A few animals survived and eventually recovered concomitant with silencing of the transgene expression in motor neurons. Similarly, transgenic mice expressing human NF-H mRNA at three to four times the level of endogenous mouse NF-H mRNA in spinal cord display perikaryal accumulations of NFs in motor and sensory neurons, proximal axonal swellings, distal axonal atrophy (shrunken axons with reduced NF content), and neurogenic atrophy of skeletal muscles arising from motor neuron dysfunction (Cote et al 1993). The distal axonal atrophy arises from slowing of axonal transport (Collard et al 1995), but no motor neuron death occurs, even in aged animals. Onset of overt pathology is slower than in the NF-L transgenic mice. In light of the >150 amino acid differences between murine (Julien et al 1988) and human (Lees et al 1988) NF-H, an unresolved ambiguity is whether the human NF-H functions in mice as a wild-type or a mutant subunit. In a third set of transgenic animals, severe NF accumulation in cell bodies of motor neurons (caused by the expression of NF-H linked to almost the entirety of β-galactosidase) is accompanied by nearly 25% loss of ventral root motor axons by one year of age (Eyer & Peterson 1994).

Even more compelling evidence that aberrant NF can cause not only dysfunction, but selective death, of motor neurons has come from a final set of transgenic animals expressing a helix disrupting proline to glutamic acid mutation in the highly conserved sequence at the end of the rod domain of NF-L (Lee et al 1994). Despite the ability of this mutant to disrupt assembly of extended filament networks in transfected cells, NF assembly is not disrupted in the animals. Rather, mutant expression at the level appropriate for a dominantly inherited disease leads to massive, selective degeneration and death of spinal motor neurons accompanied by abnormal masses of NF in perikarya and proximal axons. This in turn results in severe neurogenic atrophy of skeletal muscles. As is also true in human ALS (Kawamura et al 1981), the largest (NF-rich) sensory axons are also selectively lost in the NF-L mutant–expressing mice.

Collectively, these transgenic animal models (Cote et al 1993, Xu et al 1993, Lee et al 1994) have established that alterations in NF are sufficient to produce the pathological changes encountered in human ALS. These data, along with the prominent NF misaccumulation in human disease (Figure 3), further promote the suggestion that abnormal NF accumulation is a central pathological intermediary leading to subsequent axonal swelling and degeneration. Strong support for this view has also emerged from examination of familial ALS caused by mutation in Zn,Cu superoxide dismutase 1 (SOD1). Point mutations in this ubiquitously expressed enzyme have been proven (Deng et al 1993, Rosen et al 1993) to cause ~15% of familial ALS (~1.5% of all incidences of ALS). One of these mutations [with valine substituted for alanine at position 4 (T Siddique, personal communication)] leads to prominent neurofilamentous accumulation in motor neurons, as reported in a detailed examination of one large familial ALS kindred (Hirano et al 1967, 1984b). Pathological examination of a patient with a different SOD1 mutation (threonine substituted for isoleucine at position 113) again reveals aberrant neurofilamentous aggregates as prominent pathological features (MB Clark & G Rouleau, personal communication). It has been proposed that SOD1 mutation leads to oxidative damage that in turn triggers motor neuron degeneration (Beckman et al 1993, Deng et al 1993, Rosen et al 1993). Since NFs are among the most abundant axonal components and are very long lived, we note that NF subunits are likely candidates to accumulate such damage, ultimately leading to aberrant filament assembly and/or organization that then retards or blocks axonal transport.

How NF misaccumulation may occur in human disease is unproven. Overproduction cannot be excluded, although this was not observed in hereditary spinal muscular atrophy in dogs (Muma & Cork 1993). A defect in NF degradation is unlikely, as this would be expected to result in distal NF accumulation, a feature that has not been reported. The most plausible cause for filaments to accumulate in proximal axons and cell bodies is a defect(s) in the slow axonal transport machinery or directly in an NF subunit that affects filament structure and/or organization. Consistent with the latter of these, two mutations in the NF-H gene (leading to deletions of either 1 or 34 amino acids in the KSP tail domain of NF-H) have been found in a small percentage (5 of 356) of sporadic ALS patients, but not in over 300 normal individuals (Figlewicz et al 1994). Such data, along with the transgenic mice models, suggest that mutations in NF subunits may be primary causes of or contributory factors in ALS. If this is true, they would be expected to appear in familial ALS patients, but this has not yet been reported. Neither is it known whether any instances of familial ALS are linked to chromosome 8 [containing *NF-L* and *NF-M* genes (Hurst et al 1987)] or chromosome 22 [containing the *NF-H* gene (Lieberburg et al 1989)]. The idea that NFs play a key pathogenic role also offers an attractive explanation for the late onset (age 40–50 years) of ALS; for example, in mice NFs naturally increase fourfold during aging (J Marszalek, Z-S Xu & DW Cleveland, unpublished data).

## SUMMARY AND CONCLUSIONS

Significant progress has been made in elucidating the factors that regulate nIF assembly and dynamics, particularly for NFs. NFs are now known to be obligate heteropolymers requiring NF-L and either NF-M or NF-H. Photobleaching efforts have clearly demonstrated that filaments within neurites are dynamic, not static as had been inferred from their resistance to dilution-induced disassembly. Posttranslational modifications of phosphorylation and glycosylation probably affect assembly properties, but precisely how still must be determined. What is now certain, however, is that in the normal context, NFs are an intrinsic determinant of radial growth of axons. Misaccumulation of NFs is not only a frequent hallmark of motor neuron disease, but construction and analysis of a series of transgenic mice have proven that defects in NFs can be primary causes of motor neuron disease. Because two mutations in NF-H have been found in patients with ALS, the most common human motor neuron disease, alterations in NFs may represent primary or contributory causes for a proportion of ALS.

ACKNOWLEDGMENT

The authors gratefully acknowledge the support of NIH grant NS27036.

## Literature Cited

Ando S, Tanabe K, Gonda Y, Sato C, Inagaki M. 1989. Domain- and sequence-specific phosphorylation of vimentin induces disassembly of the filament structure. *Biochemistry* 28:2974–79

Angelides KJ, Smith KE, Takeda M. 1989. Assembly and exchange of intermediate filament proteins of neurons: neurofilaments are dynamic structures. *J. Cell Biol.* 108:1495–506

Apathy S. 1897. Nach welcher Richtung hin soll die Nervenlehre reformiert werden. *Biol. Zentralal.* 9:527–648

Balin BJ, Clark EA, Trojanowski JQ, Lee VM. 1991. Neurofilament reassembly in vitro: biochemical, morphological and immunoelectron microscopic studies employing monoclonal antibodies to defined epitopes. *Brain Res.* 556:181–95

Balin BJ, Lee VM. 1991. Individual neurofilament subunits reassembled in vitro exhibit unique biochemical, morphological and immunological properties. *Brain Res.* 556:196–208

Banay-Schwartz M, Dahl D, Hui KS, Lajtha A. 1987. The breakdown of the individual neurofilament proteins by cathepsin D. *Neurochem. Res.* 12:361–67

Banik NL, Hogan EL, Jenkins MG, McDonald JK, McAlhaney WW, Sostek MB. 1983. Purification of a calcium-activated neutral protease from bovine brain. *Neurochem. Res.* 8:1389–405

Beaudet L, Charron G, Houle D, Tretjakoff I, Peterson A, Julien JP. 1992. Intragenic regulatory elements contribute to transcriptional control of the neurofilament light gene. *Gene* 116:205–14

Beaudet L, Cote F, Houle D, Julien JP. 1993. Different posttranscriptional controls for the human neurofilament light and heavy genes in transgenic mice. *Mol. Brain Res.* 18:23–31

Beckman JS, Carson M, Smith CD, Koppenol WH. 1993. ALS, SOD and peroxynitrite. *Nature* 364:584

Belecky-Adams T, Wight DC, Kopchick JJ, Parysek LM. 1993. Intragenic sequences are required for cell type–specific and injury-induced expression of the rat peripherin gene. *J. Neurosci.* 13:5056–65

Bielschowsky M. 1902. Die Silberimpragnation der Axencylinder. *Neurol. Zentralbl.* 21:570–84

Black MM, Lasek RJ. 1980. Slow components of axonal transport: two cytoskeletal networks. *J. Cell Biol.* 86:616–23

Cajal SR. 1899. Textura del sistema nervioso del hombre y de los vertebrados. Madrid: Nicholas Moya

Cajal SR. 1903. Embryogenesis of neurofibrils. *Trab. Lab Investigaciones Biol. Univ. Madrid* 2:219–25

Carden MJ, Eagles PA. 1986. Chemical cross-linking analyses of ox neurofilaments. *Biochem. J.* 234:587–91

Carden MJ, Schlaepfer WW, Lee VM. 1985. The structure, biochemical properties, and immunogenicity of neurofilament peripheral regions are determined by phosphorylation state. *J. Biol. Chem.* 260:9805–17

Carden MJ, Trojanowski JQ, Schlaepfer WW, Lee VM. 1987. Two-stage expression of neurofilament polypeptides during rat neurogenesis with early establishment of adult phosphorylation patterns. *J. Neurosci.* 7:3489–504

Carpenter S. 1968. Proximal axonal enlargement in motor neuron disease. *Neurology* 18:841–51

Charnas LR, Szaro BG, Gainer H. 1992. Identification and developmental expression of a novel low molecular weight neuronal intermediate filament protein expressed in *Xenopus laevis*. *J. Neurosci.* 12:3010–24

Chien CL, Liem RKH. 1994. Intermediate filament protein alpha-internexin. *Gene* 149:289–92

Chin SS, Macioce P, Liem RK. 1991. Effects of truncated neurofilament proteins on the endogenous intermediate filaments in transfected fibroblasts. *J. Cell Sci.* 99:335–50

Chin SS, Liem RK. 1989. Expression of rat neurofilament proteins NF-L and NF-M in transfected non-neuronal cells. *Eur. J. Cell Biol.* 50:475–90

Chin SS, Liem RK. 1990. Transfected rat high-molecular-weight neurofilament (NF-H) coassembles with vimentin in a predominantly nonphosphorylated form. *J. Neurosci.* 10:3714–26

Ching GY, Liem RK. 1991. Structure of the gene for the neuronal intermediate filament protein alpha-internexin and functional analysis of its promoter. *J. Biol. Chem.* 266:19459–68

Ching GY, Liem RK. 1993. Assembly of type IV neuronal intermediate filaments in non-neuronal cells in the absence of preexisting cytoplasmic intermediate filaments. *J. Cell Biol.* 122:1323–35

Chou CH, Rosevear E, Goldman RD. 1990. Phosphorylation and disassembly of intermediate filaments in mitotic cells. *Proc. Natl. Acad. Sci. USA* 86:1885–89

Cochard P, Paulin D. 1984. Initial expression of neurofilaments and vimentin in the central and peripheral nervous system of the mouse embryo in vivo. *J. Neurosci.* 4:2080–94

Collard J-F, Cote F, Julien J-P. 1995. Defective axonal transport in a transgenic mouse model of amyotrophic lateral sclerosis. *Nature* 375: 61–64

Cohlberg JA, Hajarian H, Tran T, Alipourjeddi P, Noveen A. 1995. Neurofilament protein heterotetramers as assembly intermediates. *J. Biol. Chem.* 270:9334–39

Cole JS, Messing A, Trojanowski JQ, Lee VMY. 1994. Modulation of axon diameter and neurofilaments by hypomyelinating Schwann cells in transgenic mice. *J. Neurosci.* 14:6956–66

Cork LC, Price DL, Griffin JW, Sack GH Jr. 1990. Hereditary canine spinal muscular atrophy: canine motor neuron disease. *Can. J. Vet. Res.* 54:77–82

Cork LC, Troncoso JC, Klavano GG, Johnson ES, Sternberger LA, et al. 1988. Neurofilamentous abnormalities in motor neurons in spontaneously occurring animal disorders. *J. Neuropathol. Exp. Neurol.* 47:420–31

Cote F, Collard JF, Julien JP. 1993. Progressive neuronopathy in transgenic mice expressing the human neurofilament heavy gene: a mouse model of amyotrophic lateral sclerosis. *Cell* 73:35–46

Dahl D, Labkovsky B, Bignami A. 1988. Neurofilament phosphorylation in axons and perikarya: immunofluorescence study of the rat spinal cord and dorsal root ganglia with monoclonal antibodies. *J. Comp. Neurol.* 271:445–50

Dahlstrand J, Zimmerman LB, McKay RD, Lendahl U. 1992. Characterization of the human nestin gene reveals a close evolutionary relationship to neurofilaments. *J. Cell Sci.* 103:589–97

Dautigny A, Pham-Dinh D, Roussel C, Felix JM, Nussbaum JL, Jolles P. 1988. The large neurofilament subunit (NF-H) of the rat: cDNA cloning and in situ detection. *Biochem. Biophys. Res. Commun.* 154:1099–106

de Waegh SM, Lee VM, Brady ST. 1992. Local modulation of neurofilament phosphorylation, axonal caliber, and slow axonal transport by myelinating Schwann cells. *Cell* 68:451–63

Delisle MB, Carpenter S. 1984. Neurofibrillary axonal swellings and amyotropic lateral sclerosis. *J. Neurol. Sci.* 63:241–50

Deng H-X, Hentati A, Tainer JA, Iqbal Z, Cayabyab A, et al. 1993. Amyotrophic lateral sclerosis and structural defects in Cu,Zn superoxide dismutase. *Science* 261: 1047–51

Desmarais D, Filion M, Lapointe L, Royal A. 1992. Cell-specific transcription of the peripherin gene in neuronal cell lines involves a *cis*-acting element surrounding the TATA box. *EMBO J.* 11:2971–80

Djabali K, Portier MM, Gros F, Blobel G, Georgatos SD. 1991. Network antibodies identify nuclear lamin B as a physiological

attachment site for peripherin intermediate filaments. *Cell* 64:109–21

Dong DL, Xu ZS, Chevrier MR, Cotter RJ, Cleveland DW, Hart GW. 1993. Glycosylation of mammalian neurofilaments. Localization of multiple O-linked N-acetylglucosamine moieties on neurofilament polypeptides L and M. *J. Biol. Chem.* 268: 16679–87

Dosemeci A, Pant HC. 1992. Association of cAMP-dependent protein kinase with neurofilaments. *Biochem. J.* 282:477–81

Elder GA, Liang Z, Lee N, Friedrich VL Jr, Lazzarini RA. 1992a. Novel DNA binding proteins participate in the regulation of human neurofilament H gene expression. *Mol. Brain Res.* 15:85–98

Elder GA, Liang Z, Li C, Lazzarini RA. 1992b. Targeting of Sp1 to a non-Sp1 site in the human neurofilament (H) promoter via an intermediary DNA-binding protein. *Nucleic Acids Res.* 20:6281–85

Elder GA, Liang Z, Snyder SE, Lazzarini RA. 1992c. Multiple nuclear factors interact with the promoter of the human neurofilament M gene. *Mol. Brain Res.* 15:99–107

Elhanany E, Jaffe H, Link WT, Sheeley DM, Gainer H, Pant HC. 1994. Identification of endogenously phosphorylated KSP sites in the high-molecular-weight rat neurofilament protein. *J. Neurochem.* 63:2324–35

Escurat M, Djabali K, Gumpel M, Gros F, Portier MM. 1990. Differential expression of two neuronal intermediate-filament proteins, peripherin and the low-molecular-mass neurofilament protein (NF-L), during the development of the rat. *J. Neurosci.* 10:764–84

Escurat M, Gumpel M, Lachapelle F, Gros F, Portier MM. 1988. Comparative expression of 2 intermediate filament proteins, peripherin and the 68 kDa neurofilament protein, during embryonal development of the rat. *C. R. Acad. Sci. Ser. C* 306:447–56

Evans RM. 1989. Phosphorylation of vimentin in mitotically selected cells. In vitro cAMP-independent kinase and calcium stimulated phosphatase activity. *J. Cell Biol.* 108:67–78

Eyer J, Peterson A. 1994. Neurofilament-deficient axons and perikaryal aggregates in viable transgenic mice expressing a neurofilament-beta-galactosidase fusion protein. *Neuron* 12:389–405

Figlewicz DA, Krizus A, Martinoli MG, Meininger V, Dib M, et al. 1994. Variants of the heavy neurofilament subunit are associated with the development of amyotrophic lateral sclerosis. *Human Mol. Genet.* 3: 1757–61

Fliegner KH, Ching GY, Liem RK. 1990. The predicted amino acid sequence of alpha-internexin is that of a novel neuronal intermediate filament protein. *EMBO J.* 9:749–55

Fliegner KH, Kaplan MP, Wood TL, Pintar JE,

Liem RK. 1994. Expression of the gene for the neuronal intermediate filament protein alpha-internexin coincides with the onset of neuronal differentiation in the developing rat nervous system. *J. Comp. Neurol.* 342:161–73

Friede RL, Samorajski T. 1970. Axon caliber related to neurofilaments and microtubules in sciatic nerve fibers of rats and mice. *Anat. Rec.* 167:379–87

Fuchs E, Weber K. 1994. Intermediate filaments: structure, dynamics, function and disease. *Annu. Rev. Biochem.* 63:345–82

Gallant PE, Pant HC, Pruss RM, Gainer H. 1986. Calcium-activated proteolysis of neurofilament proteins in the squid giant neuron. *J. Neurochem.* 46:1573–81

Gardner EE, Dahl D, Bignami A. 1984. Formation of 10-nanometer filaments from the 150K-dalton neurofilament protein in vitro. *J. Neurosci. Res.* 11:145–55

Geisler N, Plessmann U, Weber K. 1985. The complete amino acid sequence of the major mammalian neurofilament protein (NF-L). *FEBS Lett.* 182:475–78

Geisler N, Weber K. 1981. Self-assembly in vitro of the 68,000 molecular weight component of the mammalian neurofilament triplet proteins into intermediate-sized filaments. *J. Mol. Biol.* 151:565–71

Gilbert DS. 1975. Axoplasmic architecture and physical properties as seen in the Myxicola giant axon. *J. Physiol.* 253:257–301

Gill SR, Wong PC, Monteiro MJ, Cleveland DW. 1990. Assembly properties of dominant and recessive mutations in the small mouse neurofilament (NF-L) subunit. *J. Cell Biol.* 111:2005–19

Glasgow E, Druger RK, Fuchs C, Lane WS, Schechter N. 1994a. Molecular cloning of gefiltin (ON1): serial expression of two new neurofilament mRNAs during optic nerve regeneration. *EMBO J.* 13:297–305

Glasgow E, Druger RK, Levine EM, Fuchs C, Schechter N. 1992. Plasticin, a novel type III neurofilament protein from goldfish retina: increased expression during optic nerve regeneration. *Neuron* 9:373–81

Glasgow E, Hall CM, Schechter N. 1994b. Organization, sequence, and expression of a gene encoding goldfish neurofilament medium protein. *J. Neurochem.* 63:52–61

Glass JD, Schryer BL, Griffin JW. 1994. Calcium-mediated degeneration of the axonal cytoskeleton in the *ola* mouse. *J. Neurochem.* 62:2472–75

Glicksman MA, Soppet D, Willard MB. 1987. Posttranslational modification of neurofilament polypeptides in rabbit retina. *J. Neurobiol.* 18:167–96

Goldstein ME, House SB, Gainer H. 1991. NF-L and peripherin immunoreactivities define distinct classes of rat sensory ganglion cells. *J. Neurosci. Res.* 30:92–104

Goldstein ME, Sternberger NH, Sternberger LA. 1987. Phosphorylation protects neurofilaments against proteolysis. *J. Neuroimmunol.* 14:149–60

Gorham JD, Baker H, Kegler D, Ziff EB. 1990. The expression of the neuronal intermediate filament protein peripherin in the rat embryo. *Dev. Brain Res.* 57:235–48

Greenwood JA, Troncoso JC, Costello AC, Johnson GV. 1993. Phosphorylation modulates calpain-mediated proteolysis and calmodulin binding of the 200-kDa and 160-kDa neurofilament proteins. *J. Neurochem.* 61:191–99

Heins S, Wong PC, Muller S, Goldie K, Cleveland DW, Aebi U. 1993. The rod domain of NF-L determines neurofilament architecture, whereas the end domains specify filament assembly and network formation. *J. Cell Biol.* 123:1517–33

Hellmich MR, Pant HC, Wada E, Battey JF. 1992. Neuronal cdc2-like kinase: a cdc2-related protein kinase with predominantly neuronal expression. *Proc. Natl. Acad. Sci. USA* 89:10867–71

Hemmati-Brivanlou A, Mann RW, Harland RM. 1992. A protein expressed in the growth cones of embryonic vertebrate neurons defines a new class of intermediate filament protein. *Neuron* 9:417–28

Hirano A. 1988. *Color Atlas of Pathology of the Nervous System.* New York: Igaku-shoin

Hirano A. 1991. Cytopathology of amyotrophic lateral sclerosis. In *Advances in Neurology.* Vol. 56. *Amyotrophic Lateral Sclerosis and Other Motor Neuron Diseases,* ed. LP Rowland, pp. 91–101. New York: Raven

Hirano A, Donnenfeld H, Sasaki S, Nakano I. 1984a. Fine structural observations of neurofilamentous changes in amyotrophic lateral sclerosis. *J. Neuropathol. Exp. Neurol.* 43: 461–70

Hirano A, Kurland LT, Sayre GP. 1967. Familial amyotoropic lateral sclerosis, a subgroup characterized by posterior and spinocellebellar tract involvement and hyaline inclusions in the anterior horn cells. *Arch. Neurol.* 16: 232–43

Hirano A, Nakano I, Kurland LT, Mulder DW, Holley PW, Saccomanno G. 1984b. Fine structural study of neurofibrillary changes in a family with amyotrophic lateral sclerosis. *J. Neuropathol. Exp. Neurol.* 43:471–80

Hirokawa N. 1982. Cross-linker system between neurofilaments, microtubules, and membranous organelles in frog axons revealed by the quick-freeze, deep-etching method. *J. Cell Biol.* 94:129–42

Hirokawa N, Glicksman MA, Willard MB. 1984. Organization of mammalian neurofilament polypeptides within the neuronal cytoskeleton. *J. Cell Biol.* 98:1523–36

Hisanaga S, Gonda Y, Inagaki M, Ikai A, Hirokawa N. 1990a. Effects of phosphorylation of the neurofilament L protein on filamentous structures. *Cell Regul.* 1:237–48

Hisanaga S, Hirokawa N. 1988. Structure of the peripheral domains of neurofilaments revealed by low angle rotary shadowing. *J. Mol. Biol.* 202:297–305

Hisanaga S, Hirokawa N. 1989. The effects of dephosphorylation on the structure of the projections of neurofilament. *J. Neurosci.* 9: 959–66

Hisanaga S, Hirokawa N. 1990a. Dephosphorylation-induced interactions of neurofilaments with microtubules. *J. Biol. Chem.* 265: 21852–58

Hisanaga S, Hirokawa N. 1990b. Molecular architecture of the neurofilament. II. Reassembly process of neurofilament L protein in vitro. *J. Mol. Biol.* 211:871–82

Hisanaga S, Ikai A, Hirokawa N. 1990b. Molecular architecture of the neurofilament. I. Subunit arrangement of neurofilament L protein in the intermediate-sized filament. *J. Mol. Biol.* 211:857–69

Hisanaga S, Kusubata M, Okumura E, Kishimoto T. 1991. Phosphorylation of neurofilament H subunit at the tail domain by CDC2 kinase dissociates the association to microtubules. *J. Biol. Chem.* 266:21798–803

Hisanaga S, Matsuoka Y, Nishizawa K, Saito T, Inagaki M, et al. 1994. Phosphorylation of native and reassembled neurofilaments composed of NF-L, NF-M, NF-H by the catalytic subunit of cAMP-dependent protein kinase. *J. Neurochem.* 63:387–89

Hoffman PN, Cleveland DW. 1988. Neurofilament and tubulin expression recapitulates the developmental program during axonal regeneration: induction of a specific beta-tubulin isotype. *Proc. Natl. Acad. Sci. USA* 85:4530–33

Hoffman PN, Cleveland DW, Griffin JW, Landes PW, Cowan NJ, Price DL. 1987. Neurofilament gene expression: a major determinant of axonal caliber. *Proc. Natl. Acad. Sci. USA* 84:3472–76

Hoffman PN, Griffin JW, Gold BG, Price DL. 1985a. Slowing of neurofilament transport and the radial growth of developing nerve fibers. *J. Neurosci.* 5:2920–29

Hoffman PN, Griffin JW, Price DL. 1984. Control of axonal caliber by neurofilament transport. *J. Cell Biol.* 99:705–14

Hoffman PN, Lasek RJ. 1975. The slow component of axonal transport. Identification of major structural polypeptides of the axon and their generality among mammalian neurons. *J. Cell Biol.* 66:351–66

Hoffman PN, Lasek RJ. 1980. Axonal transport of the cytoskeleton in regenerating motor neurons: constancy and change. *Brain Res.* 202:317–33

Hoffman PN, Lasek RJ, Griffin JW, Price DL.

1983. Slowing of the axonal transport of neurofilament proteins during development. *J. Neurosci.* 3:1694–700

Hoffman PN, Thompson GW, Griffin JW, Price DL. 1985b. Changes in neurofilament transport coincide temporally with alterations in the caliber of axons in regenerating motor fibers. *J. Cell Biol.* 101:1332–40

Hsieh S-T, Crawford TO, Griffin JW. 1994b. Neurofilament distribution and organization in the myelinated axons of the peripheral nervous system. *Brain Res.* 642:316–26

Hsieh S-T, Kidd GJ, Crawford TO, Xu Z-S, Trapp BD, et al. 1994a. Regional modulation of neurofilament organization by myelination in normal axons. *J. Neurosci.* 14:6392–401

Hurst J, Flavell D, Julien JP, Meijer D, Mushynski W, Grosveld F. 1987. The human neurofilament gene (NEFL) is located on the short arm of chromosome 8. *Cytogenet. Cell Genet.* 45:30–32

Ikenaka K, Nakahira K, Takayama C, Wada K, Hatanaka H, Mikoshiba K. 1990. Nerve growth factor rapidly induces expression of the 68-kDa neurofilament gene by posttranscriptional modification in PC12h-R cells. *J. Biol. Chem.* 265:19782–85

Inagaki M, Nishi Y, Nishizawa K, Matsuyama M, Sato C. 1989. Site-specific phosphorylation induces disassembly of vimentin filaments in vitro. *Nature* 328:649–52

Johnson GV, Greenwood JA, Costello AC, Troncoso JC. 1991. The regulatory role of calmodulin in the proteolysis of individual neurofilament proteins by calpain. *Neurochem. Res.* 16:869–73

Julien JP, Cote F, Beaudet L, Sidky M, Flavell D, et al. 1988. Sequence and structure of the mouse gene coding for the largest neurofilament subunit. *Gene* 68:307–14

Julien JP, Grosveld F, Yazdanbaksh K, Flavell D, Meijer D, Mushynski W. 1987. The structure of a human neurofilament gene (NF-L): a unique exon-intron organization in the intermediate filament gene family. *Biochim. Biophys. Acta* 909:10–20

Julien JP, Mushynski WE. 1982. Multiple phosphorylation sites in mammalian neurofilament polypeptides. *J. Biol. Chem.* 257: 10467–70

Julien JP, Mushynski WE. 1983. The distribution of phosphorylation sites among identified proteolytic fragments of mammalian neurofilaments. *J. Biol. Chem.* 258:4019–25

Kamakura K, Ishiura S, Sugita H, Toyokura Y. 1983. Identification of $Ca^{2+}$-activated neutral protease in the peripheral nerve and its effects on neurofilament degeneration. *J. Neurochem.* 40:908–13

Kaplan MP, Chin SS, Fliegner KH, Liem RK. 1990. Alpha-internexin, a novel neuronal intermediate filament protein, precedes the low molecular weight neurofilament protein (NF-L) in the developing rat brain. *J. Neurosci.* 10:2735–48

Kawamura Y, Dyck PJ, Shimono M, Okazaki H, Tateishi J, Doi H. 1981. Morphometric comparison of the vulnerability of peripheral motor and sensory neurons in amyotrophic lateral sclerosis. *J. Neuropathol. Exp. Neurol.* 40:667–75

Kelly BM, Gillespie CS, Sherman DL, Brophy PJ. 1992. Schwann cells of the myelin-forming phenotype express neurofilament protein NF-M. *J. Cell Biol.* 118:397–410

LaFerla FM, Tinkle BT, Bieberich CJ, Haudenschild CC, Jay G. 1995. The Alzheimer's Aβ peptide induced neurodegeneration and apoptotic cell death in transgenic mice. *Nat. Genet.* 9:21–30

Lasek RJ. 1986. Polymer sliding in axons. *J. Cell Sci. Suppl.* 5:161–79

Lasek RJ, Paggi P, Katz MJ. 1992. Slow axonal transport mechanisms move neurofilaments relentlessly in mouse optic axons. *J. Cell Biol.* 117:607–16

Lasek RJ, Paggi P, Katz MJ. 1993. The maximum rate of neurofilament transport in axons: a view of molecular transport mechanisms continuously engaged. *Brain Res.* 616:58–64

Lasek RJ, Phillips L, Katz MJ, Autilio-Gambetti L. 1985. Function and evolution of neurofilament proteins. *Ann. NY Acad. Sci.* 455:462–78

Lee MK, Marszalek JR, Cleveland DW. 1994. A mutant neurofilament subunit causes massive, selective motor neuron death: implications for the pathogenesis of human motor neuron disease. *Neuron* 13:975–88

Lee MK, Xu Z, Wong PC, Cleveland DW. 1993. Neurofilaments are obligate heteropolymers in vivo. *J. Cell Biol.* 122:1337–50

Lee VM, Otvos L Jr, Carden MJ, Dietzschold B, Lazzarini RA. 1988. Identification of the major multiphosphorylation site in mammalian neurofilaments. *Proc. Natl. Acad. Sci. USA* 85:1998–2002

Lees JF, Shneidman PS, Skuntz SF, Carden MJ, Lazzarini RA. 1988. The structure and organization of the human heavy neurofilament subunit (NF-H) and the gene encoding it. *EMBO J.* 7:1947–55

Lendhal U, Zimmerman L, McKay R. 1990. CNS stem cells express a new class of intermediate protein. *Cell* 60:585–95

Leonard D, Gorham J, Cole P, Green L, Ziff E. 1988. A nerve growth factor-regulated messenger RNA encodes a new intermediate filament protein. *J. Cell Biol.* 106:181–93

Levy E, Liem RK, D'Eustachio P, Cowan NJ. 1987. Structure and evolutionary origin of the gene encoding mouse NF-M, the middle-molecular-mass neurofilament protein. *Eur. J. Biochem.* 166:71–77

Lew J, Beaudette K, Litwin CM, Wang JH. 1993. Purification and characterization of a novel proline-directed kinase from bovine brain. *J. Biol. Chem.* 267:13383–90

Lew J, Huang QQ, Qi Z, Winkfein RJ, Aebersold R, et al. 1994. A brain-specific activator of cyclin-dependent kinase 5. *Nature* 371:423–26

Lew J, Winkfein RJ, Paudel HK, Wang JH. 1992. Brain proline-directed protein kinase is a neurofilament kinase which displays high sequence homology to p34cdc2. *J. Biol. Chem.* 267:25922–26

Lewis SA, Cowan NJ. 1986. Anomolous placement of introns in a member of the intermediate filament multigene family: an evolutionary condundrum. *Mol. Cell. Biol.* 6:1529–34

Lewis SE, Nixon RA. 1988. Multiple phosphorylated variants of the high molecular mass subunit of neurofilaments in axons of retinal cell neurons: characterization and evidence for their differential association with stationary and moving neurofilaments. *J. Cell Biol.* 107:2689–701

Lieberburg I, Spinner N, Snyder S, Anderson J, Goldgaber D, et al. 1989. Cloning of a cDNA encoding the rat high molecular weight neurofilament peptide (NF-H): developmental and tissue expression in the rat, and mapping of its human homologue to chromosomes 1 and 22. *Proc. Natl. Acad. Sci. USA* 86:2463–67

Liem RK, Hutchison SB. 1982. Purification of individual components of the neurofilament triplet: filament assembly from the 70,000-dalton subunit. *Biochemistry* 21:3221–26

Liem RK, Selkoe DJ, Yen SH, Salomon G, Shelanski ML. 1979. New insights on the composition of neurofilaments. *Res. Publ. Assoc. Res. Nerv. Ment. Dis.* 57:145–52

Liem RKH, Yen SH, Salomon GD, Shelanski ML. 1978. Intermediate filaments in nervous tissues. *J. Cell Biol.* 79:637–45

Lindenbaum MH, Carbonetto S, Grosveld F, Flavell D, Mushynski WE. 1988. Transcriptional and post-transcriptional effects of nerve growth factor on expression of the three neurofilament subunits in PC-12 cells. *J. Biol. Chem.* 263:5662–67

Link WT, Dosemeci A, Floyd CC, Pant HC. 1993. Bovine neurofilament-enriched preparations contain kinase activity similar to casein kinase I—neurofilament phosphorylation by casein kinase I (CKI). *Neurosci. Lett.* 151:89–93

McQuarrie IG, Brady ST, Lasek RJ. 1989. Retardation in the slow axonal transport of cytoskeletal elements during maturation and aging. *Neurobiol. Aging* 10:359–65

Mencarelli C, Bugnoli M, Contorni M, Moscatelli A, Ruggiero P, Pallini V. 1987. Phosphorylated epitopes of neurofilaments have been conserved during chordate evolution. *Biochem. Biophys. Res. Commun.* 149:807–14

Miyasaka H, Okabe S, Ishiguro K, Uchida T, Hirokawa N. 1993. Interaction of the tail domain of high molecular weight subunits of neurofilaments with the COOH-terminal region of tubulin and its regulation by tau protein kinase II. *J. Biol. Chem.* 268:22695–702

Monteiro MJ, Hoffman PN, Gearhart JD, Cleveland DW. 1990. Expression of NF-L in both neuronal and nonneuronal cells of transgenic mice: increased neurofilament density in axons without affecting caliber. *J. Cell Biol.* 111:1543–57

Mulligan L, Balin BJ, Lee VM, Ip W. 1991. Antibody labeling of bovine neurofilaments: implications on the structure of neurofilament sidearms. *J. Struct. Biol.* 106:145–60

Muma NA, Cork LC. 1993. Alterations in neurofilament mRNA in hereditary canine spinal muscular atrophy. *Lab. Invest.* 69: 436–42

Murphy A, Breen KC, Long A, Feighery C, Casey EB, Kelleher D. 1993. Neurofilament expression in human T lymphocytes. *Immunology* 79:167–70

Myers MW, Lazzarini RA, Lee VM, Schlaepfer WW, Nelson DL. 1987. The human mid-size neurofilament subunit: a repeated protein sequence and the relationship of its gene to the intermediate filament gene family. *EMBO J.* 6:1617–26

Myerson M, Enders GH, Wu CL, Su LK, Gorka C, et al. 1992. A family of human cdc2-related protein kinases. *EMBO J.* 11: 2909–17

Nakagawa T, Chen J, Zhang Z, Kanai Y, Hirokawa N. 1995. Two distinct functions of the carboxyl terminal tail domain of NF-M upon neurofilament assembly: crossbridge formation and longitudinal elongation of filaments. *J. Cell Biol.* 129:411–29

Napolitano EW, Chin SS, Colman DR, Liem RK. 1987. Complete amino acid sequence and in vitro expression of rat NF-M, the middle molecular weight neurofilament protein. *J. Neurosci.* 7:2590–99

Nixon RA, Lewis SE, Marotta CA. 1987. Post-translational modification of neurofilament proteins by phosphate during axoplasmic transport in retinal ganglion cell neurons. *J. Neurosci.* 7:1145–58

Nixon RA, Lewis SE, Mercken M, Sihag RK. 1994a. Orthophosphate and methionine label separate pools of neurofilaments with markedly different axonal transport kinetics in mouse retinal ganglion cells in vivo. *Neurochem. Res.* 19:1445–53

Nixon RA, Logvinenko KB. 1986. Multiple fates of newly synthesized neurofilament proteins: evidence for a stationary neurofilament network distributed nonuniformly along axons of retinal ganglion cell neurons. *J. Cell Biol.* 102:647–59

Nixon RA, Paskevich PA, Sihag RK, Thayer CY. 1994b. Phosphorylation on carboxyl terminus domains of neurofilament proteins in retinal ganglion cell neurons in vivo: influences on regional neurofilament accumulation, interneurofilament spacing, and axon caliber. *J. Cell Biol.* 126:1031–46

Nixon RA, Quackenbush R, Vitto A. 1986. Multiple calcium-activated neutral proteinases (CANP) in mouse retinal ganglion cell neurons: specificities for endogenous neuronal substrates and comparison to purified brain CANP. *J. Neurosci.* 6:1252–63

Nixon RA, Shea TB. 1992. Dynamics of neuronal intermediate filaments: a developmental perspective. *Cell Motil. Cytoskeleton* 22:81–91

Oblinger MM. 1987. Characterization of post-translational processing of the mammalian high-molecular-weight neurofilament protein in vivo. *J. Neurosci.* 7:2510–21

Oblinger MM, Wong J, Parysek LM. 1989. Axotomy-induced changes in the expression of a type III neuronal intermediate filament gene. *J. Neurosci.* 9:3766–75

Ohara O, Gahara Y, Miyake T, Teraoka H, Kitamura T. 1993. Neurofilament deficiency in quail caused by nonsense mutation in neurofilament-L gene. *J. Cell Biol.* 121:387–95

Okabe S, Miyasaka H, Hirokawa N. 1993. Dynamics of the neuronal intermediate filaments. *J. Cell Biol.* 121:375–86

Pachter JS, Liem RK. 1985. alpha-Internexin, a 66-kD intermediate filament-binding protein from mammalian central nervous tissues. *J. Cell Biol.* 101:1316–22

Pant HC. 1988. Dephosphorylation of neurofilament proteins enhances their susceptibility to degradation by calpain. *Biochem. J.* 256:665–68

Parysek LM, Goldman RD. 1987. Characterization of intermediate filaments in PC12 cells. *J. Neurosci.* 7:781–91

Parysek LM, Goldman RD. 1988. Distribution of a novel 57 kDa intermediate filament (IF) protein in the nervous system. *J. Neurosci.* 8:555–63

Peter M, Nakagawa J, Doree M, Ley JC, Goldman RD. 1990. In vitro disassembly of nuclear lamina and M-phase-specific phosphorylation of lamins by cdc2 kinase. *Cell* 61:591–602

Pleasure SJ, Selzer ME, Lee VM. 1989. Lamprey neurofilaments combine in one subunit the features of each mammalian NF triplet protein but are highly phosphorylated only in large axons. *J. Neurosci.* 9:698–709

Portier MM, Brachet P, Croizat B, Gros F. 1983a. Regulation of peripherin in mouse neuroblastoma and rat PC12 pheochromocytoma cell lines. *Dev. Neurosci.* 6:215–26

Portier MM, de Nechaud B, Gros F. 1983b. Peripherin, a new member of the intermediate filament protein family. *Dev. Neurosci.* 6:335–44

Pospelov VA, Pospelova TV, Julien JP. 1994. AP-1 and Krox-24 transcription factors activate the neurofilament light gene promoter in P19 embryonal carcinoma cells. *Cell Growth Differ.* 5:187–96

Price RL, Paggi P, Lasek RJ, Katz MJ. 1988. Neurofilaments are spaced randomly in the radial dimension of axons. *J. Neurocytol.* 17: 55–62

Purkinje JE. 1838. Untersuchungen aus der Nerven und Himanatomie. *Ber. Versamml. deut. Naturforsch. u. Arzte (Prague)* p. 177

Quax W, Egberts W-V, Hendriks W, Quax-Jeuken Y, Bloemendal H. 1983. The structure of vimentin gene. *Cell* 35:215–23

Quitschke W, Jones PS, Schechter N. 1985. A survey of intermediate filament proteins in optic nerve and spinal cord: evidence for differential expression. *J. Neurochem.* 44: 1465–76

Quitschke W, Schechter N. 1986. 62K proteins constitute the major neurofilament proteins in the frog optic nerve. *J. Neurochem.* 46: 986–89

Roberson MD, Toews AD, Goodrum JF, Morell P. 1992. Neurofilament and tubulin mRNA expression in Schwann cells. *J. Neurosci. Res.* 33:156–62

Roder HM, Ingram VM. 1991. Two novel kinases phosphorylate tau and the KSP site of heavy neurofilament subunits in high stoichiometric ratios. *J. Neurosci.* 11:3325–43

Roots BI. 1983. Neurofilament accumulation induced in synapses by leupeptin. *Science* 221:971–72

Rosen DR, Siddique T, Patterson D, Figlewicz DA, Sapp P, et al. 1993. Mutations in Cu/Zn superoxide dismutase gene are associated with familial amyotropic lateral sclerosis. *Nature* 362:59–62

Roslansky PF, Cornell-Bell A, Rice RV, Adelman WJ Jr. 1980. Polypeptide composition of squid neurofilaments. *Proc. Natl. Acad. Sci. USA* 77:404–8

Sacher MG, Athlan ES, Mushynski WE. 1994. Increased phosphorylation of the amino-terminal domain of the low molecular weight neurofilament subunit in okadaic acid-treated neurons. *J. Biol. Chem.* 269:18480–84

Sakaguchi T, Okada M, Kitamura T, Kawasaki K. 1993. Reduced diameter and conduction velocity of myelinated fibers in the sciatic nerve of a neurofilament-deficient mutant quail. *Neurosci. Lett.* 153:65–68

Sarria A, Nordeen S, Evans R. 1990. Regulated expression of vimentin cDNA in the presence and absence of preexisiting vimentin filament network. *J. Cell Biol.* 111:553–65

Sasaki S, Maruyama S. 1992. Increase in diameter of the axonal initial segment is an early change in amyotrophic lateral sclerosis. *J. Neurol. Sci.* 110:114–20

Schlaepfer WW, Bruce J. 1990. Simultaneous up-regulation of neurofilament proteins during the postnatal development of the rat nervous system. *J. Neurosci. Res.* 25:39–49

Schlaepfer WW, Freeman LA. 1978. Neurofilament proteins of rat peripheral nerve and spinal cord. *J. Cell Biol.* 78:653–62

Schlaepfer WW, Lee C, Lee VM, Zimmerman UJ. 1985. An immunoblot study of neurofilament degradation in situ and during calcium-activated proteolysis. *J. Neurochem.* 44: 502–9

Schmidt ML, Carden MJ, Lee VM, Trojanowski JQ. 1987. Phosphate dependent and independent neurofilament epitopes in the axonal swellings of patients with motor neuron disease and controls. *Lab. Invest.* 56: 282–94

Schmitt FO. 1968a. Fibrous proteins–neuronal organelles. *Proc. Natl. Acad. Sci. USA* 60: 1092–101

Schmitt FO. 1968b. II. The molecular biology of neuronal fibrous proteins. *Neurosci. Res. Prog. Bull.* 6:119–44

Schulze M. 1871. Allegeines uber die Structurelemente des Nerven systems. In *Handbuch der Lehre von den Gewegen*, ed. S Stricker, pp. 108–36. Leipzig

Schwartz ML, Shneidman PS, Bruce J, Schlaepfer WW. 1994. Stabilization of neurofilament transcripts during postnatal development. *Mol. Brain Res.* 27:215–20

Sharpe CR. 1988. Developmental expression of a neurofilament-M and two vimentin-like genes in *Xenopus laevis*. *Development* 103: 269–77

Sharpe CR, Pluck A, Gurdon JB. 1989. XIF3, a *Xenopus* peripherin gene, requires an inductive signal for enhanced expression in anterior neural tissue. *Development* 107: 701–14

Shaw G. 1991. Neurofilament proteins. In *The Neuronal Cytoskeleton*, ed. R Burgoyne, pp. 185–214. New York: Wiley-Liss

Shetty KT, Kaech S, Link WT, Jaffe H, Flores CM, et al. 1995. Molecular characterization of a neuronal-specific protein that stimulates the activity of cdk5. *J. Neurochem.* 64:1988–95

Shetty KT, Link WT, Pant HC. 1993. cdc2-like kinase from rat spinal cord specifically phosphorylates KSPXK motifs in neurofilament proteins: isolation and characterization. *Proc. Natl. Acad. Sci. USA* 90:6844–48

Shneidman PS, Bruce J, Schwartz ML, Schlaepfer WW. 1992. Negative regulatory regions are present upstream in the three mouse neurofilament genes. *Mol. Brain Res.* 13:127–38

Sihag RK, Nixon RA. 1990. Phosphorylation of the amino-terminal head domain of the middle molecular mass 145-kDa subunit of neurofilaments. Evidence for regulation by second messenger-dependent protein kinases. *J. Biol. Chem.* 265:4166–71

Sihag RK, Nixon RA. 1991. Identification of Ser-55 as a major protein kinase A phosphorylation site on the 70-kDa subunit of neurofilaments. Early turnover during axonal transport. *J. Biol. Chem.* 266:18861–67

Sobue G, Hashizume Y, Yasuda T, Mukai E, Kumagai T, et al. 1990. Phosphorylated high molecular weight neurofilament protein in lower motor neurons in amyotrophic lateral sclerosis and other neurodegenerative diseases involving ventral horn cells. *Acta Neuropathol.* 79:402–8

Szaro BG, Gainer H. 1988. Identities, antigenic determinants, and topographic distributions of neurofilament proteins in the nervous systems of adult frogs and tadpoles of *Xenopus laevis. J. Comp. Neurol.* 273: 344–58

Szaro BG, Pant HC, Way J, Battey J. 1991. Squid low molecular weight neurofilament proteins are a novel class of neurofilament protein. A nuclear lamin-like core and multiple distinct proteins formed by alternative RNA processing. *J. Biol. Chem.* 266:15035–41

Takahashi K, Nakamura H, Okada E. 1972. Hereditary amyotrophic lateral sclerosis, histochemical and electron microscopic study of hyaline inclusions in motor neurons. *Arch. Neurol.* 27:292–99

Takeda S, Okabe S, Funakoshi T, Hirokawa N. 1994. Differential dynamics of neurofilament-H protein and neurofilament-L protein in neurons. *J. Cell Biol.* 127:173–85

Tapscott SJ, Bennett GS, Holtzer H. 1981. Neuronal precursor cells in the chick neural tube express neurofilament proteins. *Nature* 292:836–38

Thompson MA, Lee E, Lawe D, Gizang-Ginsberg E, Ziff EB. 1992. Nerve growth factor-induced derepression of peripherin gene expression is associated with alterations in proteins binding to a negative regulatory element. *Mol. Cell Biol.* 12:2501–13

Thompson MA, Ziff EB. 1989. Structure of the gene encoding peripherin, an NGF-regulated neuronal-specific type III intermediate filament protein. *Neuron* 2:1043–53

Troncoso JC, March JL, Haner M, Aebi U. 1990. Effect of aluminum and other multivalent cations on neurofilaments in vitro: an electron microscopic study. *J. Struct. Biol.* 103:2–12

Troy CM, Brown K, Greene LA, Shelanski

ML. 1990a. Ontogeny of the neuronal intermediate filament protein, peripherin, in the mouse embryo. *Neuroscience* 36:217–37

Troy CM, Muma NA, Greene LA, Price DL, Shelanski ML. 1990b. Regulation of peripherin and neurofilament expression in regenerating rat motor neurons. *Brain Res.* 529:232–38

Tsai LH, Delalle I, Caviness VS, Chae T, Harlow E. 1994. p35 is a neural-specific regulatory subunit of cyclin-dependent kinase 5. *Nature* 371:419–23

Tu PH, Elder G, Lazzarini RA, Nelson D, Trojanowski JA, Lee VMY. 1995. Overexpression of the human NF-M subunit in trans- genic mice modifies the level of endogenous NF-L and the phosphorylation state of NF-H subunits. *J. Cell Biol.* 129:1629–40

Valentin G. 1836. Uber den Verlauf und die letzten Enden der Nerven. *Nova Acta physico-medica Academiae Caesareae Leopoldino-Carolinae* 18:51–71

Vitto A, Nixon RA. 1986. Calcium-activated neutral proteinase of human brain: subunit structure and enzymatic properties of multiple molecular forms. *J. Neurochem.* 47: 1039–51

Watson DF, Fittro KP, Hoffman PN, Griffin JW. 1991. Phosphorylation-related immunoreactivity and the rate of transport of neurofilaments in chronic 2,5-hexanedione intoxication. *Brain Res.* 539:103–9

Watson DF, Glass JD, Griffin JW. 1993. Redistribution of cytoskeletal proteins in mammalian axons disconnected from their cell bodies. *J. Neurosci.* 13:4354–60

Watson DF, Griffin JW, Fittro KP, Hoffman PN. 1989a. Phosphorylation-dependent immunoreactivity of neurofilaments increases during axonal maturation and beta,beta'-iminodipropionitrile intoxication. *J. Neurochem.* 53:1818–29

Watson DF, Hoffman PN, Fittro KP, Griffin JW. 1989b. Neurofilament and tubulin transport slows along the course of mature motor axons. *Brain Res.* 477:225–32

Way J, Hellmich MR, Jaffe H, Szaro B, Pant HC, et al. 1992. A high-molecular-weight squid neurofilament protein contains a lamin-like rod domain and a tail domain with Lys-Ser-Pro repeats. *Proc. Natl. Acad. Sci. USA* 89:6963–67

Weiss PA, Mayr R. 1971. Organelles in neuroplasmic ("axonal") flow: neurofilaments. *Proc. Natl. Acad. Sci. USA* 68:846–50

Willard M, Simon C. 1981. Antibody decoration of neurofilaments. *J. Cell Biol.* 89:198–205

Willard M, Simon C. 1983. Modulations of neurofilament axonal transport during the development of rabbit retinal ganglion cells. *Cell* 35:551–59

Wong PC, Cleveland DW. 1990. Characterization of dominant and recessive assembly-defective mutations in mouse neurofilament NF-M. *J. Cell Biol.* 111:1987–2003

Wong PC, Marszelak J, Crawford TO, Xu Z-S, Hsieh S-T, Griffin JW, Cleveland DW. 1995. Increasing NF-M expression reduces axonal NF-H, inhibits radial growth and results in neurofilamentous accumulation in motor neurons. *J. Cell Biol.* 130:1413–22

Xiao J, Monteiro MJ. 1994. Identification and characterization of a novel (115 kDa) neurofilament-associated kinase. *J. Neurosci.* 14: 1820–33

Xu Z, Cork LC, Griffin JW, Cleveland DW. 1993. Involvement of neurofilaments in motor neuron disease. *J. Cell Sci. Suppl.* 17: 101–8

Xu ZS, Liu WS, Willard MB. 1992. Identification of six phosphorylation sites in the COOH-terminal tail region of the rat neurofilament protein M. *J Biol. Chem.* 267:4467–71

Yachnis AT, Rorke LB, Lee VM, Trojanowski JQ. 1993. Expression of neuronal and glial polypeptides during histogenesis of the human cerebellar cortex including observations on the dentate nucleus. *J. Comp. Neurol.* 334: 356–69

Yamasaki H, Bennett GS, Itakura C, Mizutani M. 1992. Defective expression of neurofilament protein subunits in hereditary hypotrophic axonopathy of quail. *Lab. Invest.* 66: 734–43

Yamasaki H, Itakura C, Mizutani M. 1991. Hereditary hypotrophic axonopathy with neurofilament deficiency in a mutant strain of the Japanese quail. *Acta Neuropathol.* 82:427–34

Yazdanbakhsh K, Fraser P, Kioussis D, Vidal M, Grosveld F, Lindenbaum M. 1993. Functional analysis of the human neurofilament light chain gene promoter. *Nucleic Acids Res.* 21:455–61

Zimmerman L, Parr B, Lendahl U, Cunningham M, McKay R, et al. 1994. Independent regulatory elements in the nestin gene direct transgene expression to neural stem cells or muscle precursors. *Neuron* 12:11–24

Zopf D, Dineva B, Betz H, Gundelfinger ED. 1990. Isolation of the chicken middle-molecular weight neurofilament (NF-M) gene and characterization of its promoter. *Nucleic Acids Res.* 18:521–29

Annu. Rev. Neurosci. 1996. 19:219–33

# NEUROTRANSMITTER RELEASE

## Gary Matthews

Department of Neurobiology and Behavior, State University of New York, Stony Brook, New York 11794-5230

KEY WORDS:   synaptic transmission, exocytosis, synaptic vesicles, endocytosis, calcium channels

### ABSTRACT

Synaptic vesicle exocytosis is rapid and highly localized, which are features that arise from the organization of the presynaptic active zone, where vesicle fusion occurs. Colocalization of calcium channels with the proteins making up the vesicle docking machinery at the active zone, combined with the low affinity and high cooperativity of the calcium sensor for vesicle fusion, allows vesicles to fuse with short delay after a presynaptic action potential. Evidence suggests that the calcium concentration driving synaptic vesicle fusion corresponds to the high level (50–100 $\mu M$) achieved only within the microdomain of elevated calcium near the inner mouth of open calcium channels. Retrieval of synaptic vesicle membrane by endocytosis is also regulated by internal calcium but at much lower concentrations. Endocytosis occurs rapidly after exocytosis if internal calcium is near the basal level but is inhibited by elevated internal calcium (0.5–1 $\mu M$).

## INTRODUCTION

Neurotransmitter is released from synaptic terminals via exocytosis as synaptic vesicles fuse with the presynaptic plasma membrane and liberate their contents into the synaptic cleft. Exocytosis at synapses differs from other types of exocytosis primarily in the speed and the precise spatial localization of the secretion event. To maintain high temporal and spatial fidelity of information transfer, release of neurotransmitter at a synapse must be tied to depolarization on a submillisecond time scale and must be precisely targeted to the appropriate receptor sites of the postsynaptic neuron. This review briefly summarizes recent progress in understanding the molecular and physiological processes that allow for rapid and focal release of neurotransmitter during fast synaptic transmission.

On the molecular front, proteins associated with synaptic vesicles and with

219

0147-006X/96/0301-0219$08.00

the plasma membrane and cytoplasm of synaptic terminals have been identified and their interactions have been characterized. A picture is emerging of the molecular machine by which vesicles are docked at release sites and membrane fusion is triggered in response to calcium influx. Synaptic vesicle exocytosis likely shares molecular features with other, more general types of membrane trafficking (Bennett & Scheller 1993, Ferro-Novick & Jahn 1994). On the physiological front, direct information about the membrane fusion process and its regulation in single synaptic terminals has become feasible. Because several recent reviews have focused on synaptic proteins (Jahn & Südhof 1993, Bark & Wilson 1994, Bajjalieh & Scheller 1995) and proteins of secretory pathways (Rothblatt et al 1994), this review emphasizes physiological findings. Even then, space limitations preclude a comprehensive review, and I apologize for omissions. For instance, readers are referred to Van der Kloot & Molgó (1994) for a detailed review of quantal release at the vertebrate neuromuscular junction. Also, I focus on fast release of neurotransmitter from synaptic terminals and exclude coverage of neuroendocrine cells (e.g. Thomas et al 1990, Chow et al 1992) and neurosecretory nerve terminals (e.g. Fidler Lim et al 1990, Lindau et al 1992).

## MICRODOMAINS OF HIGH INTERNAL CALCIUM NEAR ACTIVE ZONES

Many properties of fast neurotransmitter release can be understood in terms of the underlying arrangement of the release machinery into a discrete locus for vesicle docking and fusion, the active zone. At the neuromuscular junction, the principal sites of vesicle fusion correspond to specialized presynaptic active zones that parallel the junctional folds bearing the acetylcholine receptors of the postsynaptic muscle cell (Heuser et al 1974, Ceccarelli et al 1979, Heuser & Reese 1981). Similar structural specializations are also found in CNS synapses (Akert et al 1972, Pfenninger & Rovainen 1974, Venzin et al 1977). A key feature of the active zone is that voltage-activated calcium channels, through which $Ca^{2+}$ enters the terminal to trigger transmitter release, are also localized to the active zones (Robitaille et al 1990, Cohen et al 1991). When $Ca^{2+}$ enters through discrete channels, a nonuniform distribution of $Ca^{2+}$ under the membrane is expected (e.g. Chad & Eckert 1984, Fogelson & Zucker 1985), with the concentration being much higher near channels. Because $Ca^{2+}$ enters through the open channel pore faster than it can diffuse away into the cytoplasm, a microdomain of elevated $[Ca^{2+}]_i$ is created near the channel. Hence, the colocalization of calcium channels and vesicle fusion sites suggests that the relevant level of $[Ca^{2+}]_i$ for triggering exocytosis is much higher than the average concentration achieved throughout the terminal during activation of calcium current (Smith & Augustine 1988). If the separation between the inner mouth of the channel and the calcium sensor for vesicle fusion is <50 nm, diffusion models predict that [$Ca^{2+}$] at the sensor would rise within microseconds to levels approaching 100 μM or higher. Similar concentrations can also be achieved at somewhat greater distances if calcium channels form clusters that allow summation of the calcium contributions from individual channels.

Because microdomains of high $Ca^{2+}$ are so tiny, conventional calcium imaging techniques are unable to detect them. However, Llinàs et al (1992) were able to visualize microdomains directly in the squid giant synapse by injecting into the presynaptic terminal a calcium-dependent photoprotein, $n$-aequorin-J, which has reduced sensitivity to $Ca^{2+}$. During presynaptic action potentials, punctate flashes of light were observed, localized to regions that probably correspond to presynaptic active zones. Because $n$-aequorin-J produces light only at high concentrations of $Ca^{2+}$, the light flashes correspond to regions where action potentials elevated [$Ca^{2+}$]$_i$ to levels estimated by Llinàs et al (1992) to exceed 100 μM.

## CALCIUM DEPENDENCE OF EXOCYTOSIS IN SYNAPTIC TERMINALS

Several lines of evidence indicate that the calcium concentration necessary to drive exocytosis in synaptic terminals corresponds to the high levels expected near the inner mouth of open calcium channels. Some of this evidence derives from capacitance measurements, which give a direct view of exocytosis in single synaptic terminals. Before discussing these experiments, a brief description of the basis of capacitance measurements is useful.

### Membrane Capacitance as an Index of Exocytosis and Endocytosis

When secretory granules fuse with the plasma membrane during exocytosis, the surface area of the secreting cell increases. Conversely, when membrane is retrieved by endocytosis, surface area decreases. These changes in membrane area can be followed in single cells by using electrical measurement of the cell capacitance, which under favorable geometrical conditions is directly proportional to surface area. In single cells or single terminals, the capacitance can be measured under whole-cell voltage clamp by analyzing the current responses to sinusoidal or pulsatile voltage stimuli. For a technical discussion of the procedure and its limitations, see Lindau & Neher (1988). Capacitance measurements have proved valuable in elucidating mechanisms of exocytosis in various types of secretory cell, including adrenal chromaffin cells (Neher & Marty 1982), mast cells (Fernandez et al 1984, Almers & Neher 1987), and pituitary cells (Thomas et al 1990, Fidler Lim et al 1990, Lindau et al 1992).

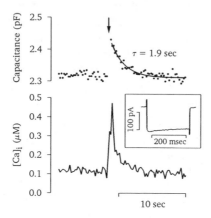

*Figure 1* Capacitance change elicited by calcium current in an isolated synaptic terminal of a retinal bipolar neuron. The terminal was approximately 10 μm in diameter. The upper trace shows the membrane capacitance, and the lower trace shows the simultaneously determined intraterminal calcium concentration, monitored with Fura-2 fluorescence. At the arrow, a 250-ms depolarization from the holding potential of −60 to 0 mV was given, eliciting the calcium current shown on expanded time scale in the inset. Depolarization elicited a jump in capacitance, followed by a return to baseline along an exponential time course (solid line in upper trace) with a time constant of 1.9 s. [Figure modified from von Gersdorff & Matthews (1994b).]

In neurons, capacitance measurements have been applied in giant synaptic terminals of retinal bipolar neurons (von Gersdorff & Matthews 1994a,b), in retinal photoreceptors (Rieke & Schwartz 1994), and in saccular hair cells (Parsons et al 1994).

Figure 1 shows the change in membrane capacitance observed in the synaptic terminal of a retinal bipolar neuron in response to brief depolarization. These terminals have a single type of slowly-inactivating, dihydropyridine-sensitive calcium channel (Heidelberger & Matthews 1992) (see inset of Figure 1), which has been shown to control glutamate release from the terminal (Tachibana et al 1993). Activation of calcium current elicited a rapid increase in capacitance, followed by return to baseline. The increase in capacitance is taken to represent calcium-dependent fusion of synaptic vesicles, and the return to baseline is taken to represent endocytosis, as membrane is retrieved in preparation for vesicle recycling. Endocytosis is discussed later. The rate of exocytosis during activation of calcium current and the size of the readily releasable pool of vesicles were determined by measuring the capacitance jump in response to depolarizations of varying duration (Figure 2). The capacitance response increased exponentially with pulse duration up to a limit of approximately 150 fF, corresponding to about 2000 synaptic vesicles with a diameter of 50 nm, or ≈30 vesicles at each of the ≈60 ribbon-type synaptic active zones

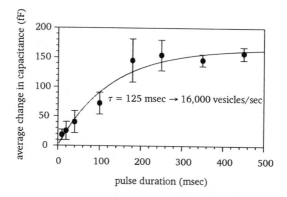

*Figure 2* Dependence of the size of the capacitance response on the duration of depolarization in synaptic terminals of retinal bipolar neurons. Data points show averages of 5 to 27 terminals (±s.e.m.). The change in capacitance was measured just after the depolarization. The solid line shows an exponential rise, fitted by a least-squares criterion. The exponential time constant corresponds to the release of about 16,000 vesicles/s at the initial part of the curve, assuming an initial vesicle pool of about 2000 vesicles. [Figure modified from von Gersdorff & Matthews (1994a).]

of the bipolar-cell terminal. This pool may represent the vesicles docked at ribbons, ready for rapid release upon activation of calcium current. Note that the estimated size of the vesicle pool depends strongly on vesicle diameter and would increase to about 4000 vesicles if the actual diameter is 35 nm rather than 50 nm. The exponential time constant obtained from the data in Figure 2 was 128 ms. For a pool size of 2000 vesicles, this yields an initial fusion rate of about 16,000 vesicles/s during activation of calcium current. A similar rate of vesicle fusion was estimated in saccular hair cells (Parsons et al 1994), but in hair cells the capacitance response continued to increase with pulse duration up to 2 s, whereas in bipolar cells the response saturated within a few hundred milliseconds. This suggests that the size of the releasable pool is larger in hair cells.

## Evidence That Synaptic Vesicle Fusion Requires High Internal Calcium

Exogenous $Ca^{2+}$ buffers added to the presynaptic terminal reduce the postsynaptic response at the squid giant synapse (Adler et al 1991) and inhibit depolarization-induced capacitance responses in bipolar-cell synaptic terminals (von Gersdorff & Matthews 1994). In both cases, however, high concentrations of fast buffers (BAPTA or BAPTA derivatives) were required to affect transmitter release. The slower $Ca^{2+}$ chelator, EGTA, had little effect in either synaptic terminal. Modeling of $Ca^{2+}$ diffusion and binding to buffers shows that EGTA is ineffective at buffering $Ca^{2+}$ in the near vicinity of channel pores and that high concentrations of BAPTA are required to buffer $[Ca^{2+}]_i$ within microdomains (Stern 1992). Thus, the effects of EGTA and BAPTA are consistent with the calcium sensor for secretion being located within the microdomain of high $Ca^{2+}$ near open calcium channels.

Exocytosis can also be triggered when $Ca^{2+}$ is elevated with calcium ionophores rather than when voltage-dependent calcium channels are activated. Verhage et al (1991) showed that in synaptosomes, the calcium ionophore ionomycin effectively released neuropeptides and catecholamines but was much less effective at releasing amino acid transmitters. Calcium influx via voltage-dependent calcium channels released all transmitter types equally well. This suggests that the amino acid transmitters, which are focally released by fusion of small clear-core synaptic vesicles at active zones, require a higher level of $[Ca^{2+}]_i$ for secretion than do peptides and catecholamines, which are released by fusion of large dense-core vesicles at sites distant from the active zones. Ionomycin was able to raise the average $Ca^{2+}$ in the synaptosomes into the range sufficient to activate the high-affinity calcium sensors for the dense-core vesicles but could not achieve the high microdomain levels required for fusion of small, clear-core vesicles. These differences in calcium requirement may arise from different distances between calcium channels and calcium sensors, which may in turn reflect the functional demands for rapid and focal release of fast neurotransmitter, on the one hand, and more diffuse release of slower neuromodulatory substances, on the other.

von Gersdorff & Matthews (1994a) were able to elicit capacitance increases in bipolar-cell synaptic terminals by rapid application of high concentrations of ionomycin, but only if $[Ca^{2+}]_i$ was elevated to 20 μM or greater. The maximum rate of change of the capacitance produced by ionomycin corresponds to fusion of 480 vesicles/s, compared with about 16,000 vesicles/s when exocytosis is activated by calcium current. Thus, as with synaptosomes (Verhage et al 1991), ionomycin is much less effective than calcium channels in activating exocytosis. The higher rate of exocytosis with calcium current suggests that $[Ca^{2+}]_i$ achieved at release sites by activation of calcium channels in bipolar-cell synaptic terminals is greater than the 20–50 μM level attained with ionomycin.

In whole-cell patch clamp, internal $Ca^{2+}$ can also be elevated by dialysis with patch-pipette solutions containing buffered levels of elevated $[Ca^{2+}]$ (Augustine & Neher 1992). In bipolar-cell synaptic terminals, calcium current typically drives average cytoplasmic $[Ca^{2+}]_i$ to 0.5–1 μM. However, dialysis of terminals with solutions containing 1.4 μM $[Ca^{2+}]$ (ten times the resting level) had no effect on membrane capacitance. This shows that the bulk level of $[Ca^{2+}]_i$ produced by calcium current is insufficient to stimulate exocytosis (von Gersdorff & Matthews 1994a). Pipette solutions containing 20 μM $[Ca^{2+}]$

were also ineffective, but dialysis with 52 μM $[Ca^{2+}]$ did increase capacitance (von Gersdorff & Matthews 1994a). This again indicates that $[Ca^{2+}]_i$ in the range of 50 μM is required to trigger synaptic vesicle exocytosis in bipolar-cell synaptic terminals.

## Calcium Dependence of the Rate of Exocytosis

Dialysis of buffered $[Ca^{2+}]$ via a patch pipette increases $[Ca^{2+}]_i$ slowly and does not allow conclusions about the rate of exocytosis after a rapid jump in $[Ca^{2+}]_i$. However, flash photolysis can produce rapid increases in $[Ca^{2+}]_i$ by liberating $Ca^{2+}$ from photolyzable chelators, such as DM-nitrophen (Kaplan & Ellis Davis 1988). This strategy was used by Heidelberger et al (1994) to determine the kinetics of the capacitance increase in bipolar-cell synaptic terminals after a sudden increase in $[Ca^{2+}]_i$. Figure 3 shows capacitance changes in a synaptic terminal following UV flashes of two different intensities. With a dimmer flash (Figure 3A), $[Ca^{2+}]_i$ increased to a peak of 10–20 μM (measured with the low-affinity calcium indicator Furaptra), but there was no change in capacitance. Subsequently, a brighter flash elevated $[Ca^{2+}]_i$ to >100 μM and evoked a rapid increase in capacitance (Figure 3B). In general, the threshold $[Ca^{2+}]_i$ for eliciting a detectable capacitance increase was 10–20 μM, and the rate of rise of capacitance increased steeply with $[Ca^{2+}]_i$ above this level, reaching the maximal rate constant of about 3000/s at $[Ca^{2+}]_i$ of 100–200 μM. This maximal rate corresponds to >$10^6$ vesicles/s. Thus, if $[Ca^{2+}]_i$ at the active zone rises to >100 μM after onset of presynaptic depolarization, there is a high probability that vesicles docked at the active zone will fuse with the plasma membrane in less than a millisecond [mean latency 0.35 ms, as estimated by Heidelberger et al (1994)]. The observed fast rate of exocytosis and the low affinity of the process for $Ca^{2+}$ are well-suited for the rapid release of neurotransmitter from synaptic terminals.

# SYNAPTIC PROTEINS

## The SNARE Hypothesis of Vesicle Docking and Fusion

To achieve vesicle fusion with very short latency after $Ca^{2+}$ influx during presynaptic depolarization, synaptic vesicles must be docked near the calcium channels at the active zone and primed for fusion. Also, precise docking is necessary to ensure that fusion will occur at the correct location, just opposite the neurotransmitter receptors of the postsynaptic neuron. In this light, synaptic vesicle docking and fusion can be viewed as a special case of more general cellular mechanisms of membrane trafficking (for reviews, see Bennett & Scheller 1993, Jahn & Südhof 1993, Bark & Wilson 1994, Ferro-Novick & Jahn 1994). Selective targeting of a particular vesicle to a particular membrane

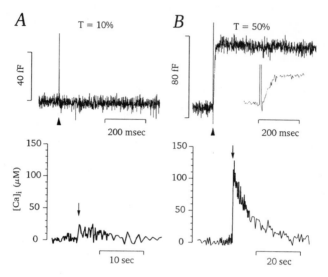

*Figure 3* Capacitance changes elicited by release of caged calcium in a single synaptic terminal of a retinal bipolar neuron. The upper traces show the membrane capacitance on a fast time scale, while the lower traces show $[Ca^{2+}]_i$ on a slower time scale (note different time scale calibrations for lower and upper traces in both $A$ and $B$). The timing of the brief UV uncaging flash is indicated by arrowheads in the upper traces and by arrows in the lower traces. Terminals were loaded with 10 mM calcium-saturated DM-nitrophen (caged calcium) and with the low-affinity, fluorescent calcium indicator, Furaptra, to monitor the resulting change in $[Ca^{2+}]_i$ after the flash. (*A*) A 10% transmission quartz neutral density filter was placed in the photolysis light path to produce a dim uncaging flash. No change in capacitance was observed. (*B*) A 50% transmission filter replaced the T = 10% filter in the photolysis light path. A larger change in $[Ca^{2+}]_i$ was elicited, and capacitance rapidly increased after the flash. The inset in the upper trace shows the rise of capacitance on a faster time scale (the rise was exponential with a time constant of approximately 5 ms). [Data from Heidelberger et al (1994).]

destination is thought to be mediated by specific combinations of vesicle and target membrane proteins (called SNAP receptors, or SNAREs) that interact to dock vesicles at the appropriate fusion site (Söllner et al 1993). For synaptic vesicles, synaptobrevin (also called VAMP, for vesicle-associated membrane protein) is proposed to play the role of the vesicle SNARE, while in the plasma membrane of the terminal the corresponding target SNARE is syntaxin, which is likely complexed with 25-kDa synaptosomal-associated protein (SNAP-25). The combination of synaptobrevin with syntaxin and SNAP-25 forms a backbone for the binding of *N*-ethylmaleimide–sensitive factor (NSF) and soluble NSF-attachment proteins (SNAPs; unfortunately the same acronym as SNAP-25), which are cytoplasmic proteins required for vesicle fusion during membrane trafficking in the Golgi apparatus.

Syntaxin, SNAP-25, synaptobrevin, NSF, and SNAPs combine to form a complex that presumably bridges the gap between the synaptic vesicle and the plasma membrane, tethering the vesicle at the active zone. In nonregulated membrane fusion (e.g. in the Golgi apparatus), aggregation of the homologous proteins both docks vesicles and initiates fusion. In synaptic transmission, however, fusion is prevented until $Ca^{2+}$ influx occurs. This suggests the presence of a calcium-dependent regulator, and one candidate for this regulator is the vesicle membrane protein synaptotagmin. The cytoplasmic portion of synaptotagmin includes two regions with high homology to the calcium- and phospholipid-binding domains (C2 domains) of protein kinase C (Perin et al 1990). Also, synaptotagmin binds to syntaxin, so synaptotagmin is a part of the putative docking-fusion complex. Taken together, this suggests that binding of $Ca^{2+}$ to the C2 domains may alter the interaction of synaptotagmin with membrane lipids and/or syntaxin so that calcium-bound synaptotagmin either initiates fusion or allows other components of the complex to initiate fusion of the docked synaptic vesicle. A role for synaptotagmin in exocytosis is also indicated by its interaction with neurexins (Petrenko 1993), a family of plasma-membrane proteins that are receptors for α-latrotoxin. α-Latrotoxin is a component of black widow spider venom, which dramatically increases the rate of spontaneous transmitter release (see Van der Kloot & Molgó 1994), possibly via neurexin's interaction with synaptotagmin.

Readers desiring more detailed information about the biochemical and molecular properties of the proteins involved in membrane trafficking and vesicle fusion (including many not mentioned here) are referred to Rothblatt et al (1994) and to the literature cited in the reviews by Bark & Wilson (1994), Ferro-Novick & Jahn (1994), and Bajjalieh & Scheller (1995). Also, a pictorial overview of presynaptic proteins is provided by Augustine (1994).

## Calcium Channels Are Part of the Vesicle-Docking Complex

A key functional feature of the active zone is the colocalization of voltage-activated calcium channels with the vesicle-docking and release machinery. Thus, it is particularly interesting that N-type calcium channels bind to syntaxin (Bennett et al 1992), placing the calcium channel in molecular proximity to the proposed vesicle-docking complex at the active zone. In addition, synaptotagmin binds to syntaxin and copurifies with the syntaxin-calcium channel complex (Lévêque et al 1994), so the putative calcium sensor is also strategically located with respect to the source of $Ca^{2+}$ influx. This molecular arrangement, with the calcium sensor within the macromolecular complex that includes the calcium channel, may provide the structural basis for the high levels of $[Ca^{2+}]_i$ required to drive exocytosis in synaptic terminals (Heidelberger et al 1994, von Gersdorff & Matthews 1994a).

## Physiological Effects of Synaptic Proteins

One indication that the synaptic proteins described above play functional roles in neurotransmitter release is the targeting of these proteins by the clostridial neurotoxins—tetanus toxin and botulinum toxins. These bacterial toxins, which cause paralysis by blocking neurotransmission, are proteases that selectively cleave components of the putative vesicle fusion complex (for review, see Jahn & Niemann 1994). For example, tetanus toxin and botulinum toxins B, D, F, and G cleave synaptobrevin, whereas botulinum toxins A and E selectively attack SNAP-25, and botulinum toxin C cleaves syntaxin. Toxin treatment interferes with the formation of complexes among synaptobrevin, SNAP-25, and syntaxin in vitro (Hayashi et al 1994), and this inhibition of complex formation may account for the blockade of transmitter release by the toxins, in accordance with the SNARE hypothesis. However, electron microscopic observation of the squid giant synapse after blockade of transmission with tetanus toxin showed that the number of vesicles docked at active zones was greater in toxin-treated terminals (Hunt et al 1994), not smaller as might be expected under the simplest form of the SNARE hypothesis for the targeting of synaptic vesicles to active zones. Failure to observe a decrease in docking after cleavage of synaptobrevin suggests either that synaptobrevin is not involved in vesicle docking or that there are redundant docking mechanisms in addition to synaptobrevin.

Another approach to test the roles of the presynaptic proteins in transmitter release is to interfere with individual proteins, either by mutations of the corresponding genes or by microinjection of competitive peptide fragments or antibodies. For synaptotagmin, all three variations have been employed (reviewed by DeBello et al 1993; also see Geppert et al 1994). The accumulated evidence indicates that synaptotagmin is indeed involved in neurotransmitter release, because injected antibodies and peptide fragments from the C2 regions of synaptotagmin reduce transmitter release and because animals with defects in synaptotagmin gene expression show impairments of synaptic transmission. Anatomical evidence indicates that the interference with transmission is at a stage after synaptic-vesicle docking, but at this writing the exact point(s) in the release process that require synaptotagmin are uncertain.

In addition to the tetanus-toxin experiments mentioned above, Hunt et al (1994) also found that the cytoplasmic portion of synaptobrevin inhibited transmitter release when injected into squid presynaptic terminals. This result was interpreted as being due to competition between the exogenous protein fragment and endogenous synaptobrevin for a protein-protein binding site required for vesicle fusion. A role for SNAPs is also suggested by experiments in which exogenous SNAPs were found to potentiate transmission at the squid giant synapse, while injection of SNAP peptide fragments inhibited transmission (DeBello et al 1995).

## MEMBRANE RETRIEVAL

One of the striking features of capacitance responses to $Ca^{2+}$ influx in synaptic terminals is the rapid return of the capacitance to baseline following cessation of stimulation (see Figure 1). This retrieval of membrane presumably reflects the first step in the vesicle recycling process (Heuser & Reese 1973). Experiments using electron microscopy or the fluorescent membrane probe FM 1-43 (Betz & Bewick 1992, Ryan et al 1993) give estimates for membrane retrieval times in the tens of seconds. This is much longer than the time constant of 1–2 s for return of membrane capacitance to rest after brief depolarization (<250 ms) in synaptic terminals of retinal bipolar neurons (Figure 1) (von Gersdorff & Matthews 1994a). However, strong stimulation is typically used in anatomical experiments to elicit large release and recycling, which is more readily detectable. Von Gersdorff & Matthews (1994a) showed that after prolonged stimulation comparable to that used to study recycling in anatomical experiments, the time constant of retrieval slowed approximately tenfold, placing it in good agreement with the rates estimated from activity-dependent staining. Thus, two components of membrane retrieval follow bouts of exocytosis: a rapid retrieval seen after brief stimuli and a slower retrieval observed following repetitive stimuli sufficient to exhaust the pool of releasable vesicles in the terminal.

Membrane retrieval via pinching off of coated vesicles is an important component of vesicle recycling at synaptic terminals (Heuser & Reese 1973, 1981). The vesicle coat consists of a lattice of clathrin light and heavy chains, which polymerize at the site of endocytosis to form the coated pits from which the coated vesicles derive (reviewed in Brodsky et al 1991). Because the speed of appearance of clathrin-coated pits in anatomical experiments corresponds reasonably to the slower component of endocytosis (time constant 20–30 s) observed in capacitance measurements, this slower component may represent clathrin-mediated endocytosis. The faster component (time constant 1–2 s) may be too rapid to be mediated via assembly of clathrin lattices and budding of coated pits. This raises the possibility of a different molecular mechanism for rapid retrieval, such as reversibility of the exocytosis machinery.

One possible explanation for the slower endocytosis after strong stimulation is that endocytosis is inhibited by elevated $[Ca^{2+}]_i$. After a brief depolarization, $[Ca^{2+}]_i$ typically reaches a peak of <400 nM in bipolar-cell synaptic terminals (e.g. Figure 1), while prolonged or repetitive stimuli raise $[Ca^{2+}]_i$ to 1000 nM or more (von Gersdorff & Matthews 1994a). When background $[Ca^{2+}]_i$ was elevated into the 400–1000 nM range by dialysis with high-$Ca^{2+}$ pipette solu-

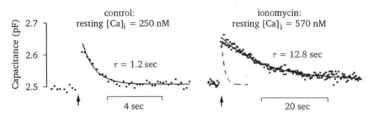

*Figure 4* Elevated $[Ca^{2+}]_i$ slows endocytosis in synaptic terminals of retinal bipolar neurons. Capacitance increases were elicited by single 250-ms depolarizations, given at the arrows, and $[Ca^{2+}]_i$ was monitored with Fura-2. The left trace shows the capacitance response at 250-nM $[Ca^{2+}]_i$, with exponential recovery with a time constant of 1.2 s. After external application of ionomycin elevated $[Ca^{2+}]_i$ to 570 nM (*right trace*), the recovery of capacitance after the response slowed to a time constant of 12.8 s. Note the different time scales for the two traces. The dashed trace shows the exponential recovery of the control trace replotted on the time scale of the right-hand trace. [Figure modified from von Gersdorff & Matthews (1994b).]

tions or by ionomycin, endocytosis slowed dramatically (von Gersdorff & Matthews 1994b). An example is shown in Figure 4, which shows capacitance responses to single 250-ms depolarizations at normal resting $[Ca^{2+}]_i$ and after $[Ca^{2+}]_i$ was increased with ionomycin. Dialysis experiments showed that the slowing of retrieval rate was graded with $[Ca^{2+}]_i$, with half-inhibition at about 500 nM. At $[Ca^{2+}]_i > 1 \mu M$, endocytosis was completely inhibited, and capacitance responses took on a staircase appearance. High $Ca^{2+}$ eliminated recovery after both brief depolarizations and after prolonged stimulation. This calcium-dependent modulation of recycling may be important in the regulation of vesicle availability during repetitive stimulation in synaptic terminals. Although this effect of $Ca^{2+}$ is in the right direction to contribute to the slowing of exocytosis after strong stimulation, the return of $[Ca^{2+}]_i$ to the resting level after prolonged stimulation was too rapid to account entirely for the observed slowing of endocytosis (von Gersdorff & Matthews 1994b). Other factors evidently also contribute to the tenfold slowing of retrieval after repetitive stimulation, such as a transition from clathrin-independent to clathrin-dependent endocytosis.

In addition to clathrin, another protein known to be important in endocytosis is the GTP-binding protein dynamin (also called dephosphin) (reviewed in Robinson et al 1994). Several lines of evidence indicate that GTPase activity of dynamin is required for endocytosis. The *Drosophila* mutant *shibire,* with mutations in the gene coding for dynamin, exhibits paralysis because of failure of endocytosis at synaptic terminals (Kosaka & Ikeda 1983), and *shibire* mutations have been linked to defects in the GTPase domain of dynamin (van der Bliek & Meyerowitz 1991). Also, mutations in the GTP-binding domain of dynamin block endocytosis via coated pits in mammalian cells (van der

Bliek et al 1993). In addition, inhibition of GTPase activity by GTP-γS blocks the pinching off of clathrin-coated invaginations in synaptic terminals and leads to the formation of long invaginations of the plasma membrane, ringed by dynamin (Takei et al 1995). These findings led Kelly (1995) to suggest that dynamin forms a molecular noose at the neck of invaginating coated pits and that GTP hydrolysis is required to constrict the noose and promote membrane fission. The GTPase activity of dynamin, and thus presumably its ability to promote endocytosis, is increased by phosphorylation, and dynamin is dephosphorylated in response to $Ca^{2+}$ influx during depolarization (reviewed in Robinson et al 1994). Calcium-dependent dephosphorylation of dynamin, leading to reduced GTPase activity and slower endocytosis, is thus an attractive mechanism to explain the calcium-dependent slowing of endocytosis in synaptic terminals (see Figure 4) (von Gersdorff & Matthews 1994b).

## SUMMARY

Rapid fusion of synaptic vesicles with the plasma membrane of the synaptic terminal during synaptic transmission is triggered by high concentrations of internal $Ca^{2+}$ achieved near calcium channels at the active zone. The calcium sensor that drives synaptic-vesicle exocytosis is characterized by a fast on-rate and high cooperativity for $Ca^{2+}$ binding, but low affinity. The combination of these kinetic properties plus the high-$[Ca^{2+}]_i$ microdomain at the active zone accounts for the rapid release of neurotransmitter upon depolarization. The placement of the synaptic vesicle at the appropriate docking site at the active zone is probably mediated via a complex of synaptic-vesicle, cytoplasmic, and plasma-membrane proteins homologous to proteins involved in general membrane trafficking. Similarly, mechanisms involved in the retrieval of the membrane components of fused vesicles also appear to share some features with other types of endocytosis. The exact roles of the various synaptic proteins in neurotransmitter release and in vesicle recycling have not yet been definitively established, but the continuing interplay among molecular, cell biological, and electrophysiological approaches makes it likely that a detailed mechanism for synaptic-vesicle exocytosis and endocytosis will soon emerge.

ACKNOWLEDGMENTS

Supported by NIH grants EY03821 and EY08673 and by a Research Award from the Alexander von Humboldt Foundation. I thank Drs. Ruth Heidelberger, Henrique von Gersdorff, and Erwin Neher for helpful discussions.

## Literature Cited

Adler EM, Augustine GJ, Duffy SN, Charlton MP. 1991. Alien intracellular calcium chelators attenuate neurotransmitter release at the squid giant synapse. *J. Neurosci.* 11:1496–507

Akert K, Pfenninger D, Sandri C, Moor H. 1972. Freeze-etching and cytochemistry of vesicles and membrane complexes in synapses of the central nervous system. In *Structure and Function of Synapses,* ed. GD Pappas, DP Purpura, pp. 67–86. New York: Raven

Almers W, Neher E. 1987. Gradual and stepwise changes in the membrane capacitance of rat peritoneal mast cells. *J. Physiol.* 386: 205–17

Augustine G. 1994. Proteins of the presynaptic terminal. [Poster]. *Trends Neurosci.* Vol. 17

Augustine GJ, Neher E. 1992. Calcium requirements for secretion in bovine chromaffin cells. *J. Physiol.* 450:247–71

Bajjalieh SM, Scheller RH. 1995. The biochemistry of neurotransmitter secretion. *J. Biol. Chem.* 270:1971–74

Bark IC, Wilson MC. 1994. Regulated vesicular fusion in neurons: snapping together the details. *Proc. Natl. Acad. Sci. USA* 91:4621–24

Bennett MK, Calakos N, Scheller RH. 1992. Syntaxin: a synaptic protein implicated in docking of synaptic vesicles at presynaptic active zones. *Science* 257:255–59

Bennett MK, Scheller RH. 1993. The molecular machinery for secretion is conserved from yeast to neurons. *Proc. Natl. Acad. Sci. USA* 90:2559–63

Betz WJ, Bewick GS. 1992. Optical analysis of synaptic vesicle recycling at the frog neuromuscular junction. *Science* 255:200–3

Brodsky FM, Hill BL, Acton SL, Näthke I, Wong DH, et al. 1991. Clathrin light chains: arrays of protein motifs that regulate coated-vesicle dynamics. *Trends Biochem. Sci.* 16: 208–13

Ceccarelli B, Grohouaz F, Hurlbut WP. 1979. Freeze fracture studies of frog neuromuscular junctions during intense release of neurotransmitter. II. Effects of electrical stimulation and high potassium. *J. Cell Biol.* 81: 178–92

Chad JE, Eckert R. 1984. Calcium domains associated with individual channels can account for anomalous voltage relations of Ca-dependent responses. *Biophys. J.* 45:993–99

Chow RH, von Rüden L, Neher E. 1992. Delay in vesicle fusion revealed by electrochemical monitoring of single secretory events in adrenal chromaffin cells. *Nature* 356:60–63

Cohen MW, Jones OT, Angelides KJ. 1991. Distribution of $Ca^{2+}$ channels on frog motor nerve terminals revealed by fluorescent $\omega$-conotoxin. *J. Neurosci.* 11:1032–39

DeBello WM, Betz H, Augustine GJ. 1993. Synaptotagmin and neurotransmitter release. *Cell* 74:947–50

DeBello WM, O'Connor V, Dresbach T, Whiteheart SW, Wang SSH, et al. 1995. SNAP-mediated protein-protein interactions essential for neurotransmitter release. *Nature* 373:626–30

Fernandez JM, Neher E, Gomperts BD. 1984. Capacitance measurements reveal stepwise fusion events in degranulating mast cells. *Nature* 312:453–55

Ferro-Novick S, Jahn R. 1994. Vesicle fusion from yeast to man. *Nature* 370:191–93

Fidler Lim N, Nowycky MC, Bookman RJ. 1990. Direct measurement of exocytosis and calcium currents in single vertebrate nerve terminals. *Nature* 344:449–51

Fogelson AL, Zucker RS. 1985. Presynaptic calcium diffusion from various arrays of single channels. *Biophys. J.* 48:1003–17

Geppert M, Goda Y, Hammer RE, Li C, Rosahl TW, et al. 1994. Synaptotagmin I: a major $Ca^{2+}$ sensor for transmitter release at a central synapse. *Cell* 79:717–27

Hayashi T, McMahon H, Yamasaki S, Binz T, Hata Y, et al. 1994. Synaptic vesicle membrane fusion complex: action of clostridial neurotoxins on assembly. *EMBO J.* 13: 5051–61

Heidelberger R, Heinemann C, Neher E, Matthews G. 1994. Calcium dependence of the rate of exocytosis in a synaptic terminal. *Nature* 371:513–15

Heidelberger R, Matthews G. 1992. Calcium influx and calcium current in single synaptic terminals of goldfish retinal bipolar neurons. *J. Physiol.* 447:235–56

Heuser JE, Reese TS. 1973. Evidence for recycling of synaptic vesicle membrane during transmitter release at the frog neuromuscular junction. *J. Cell Biol.* 57:315–44

Heuser JE, Reese TS. 1981. Structural changes after transmitter release at the frog neuromuscular junction. *J. Cell Biol.* 88: 564–80

Heuser JE, Reese TS, Landis DMD. 1974. Functional changes in frog neuromuscular junctions studied with freeze-fracture. *J. Neurocytol.* 3:108–31

Hunt JM, Bommert K, Charlton MP, Kistner A, Habermann E, et al. 1994. A post-docking role for synaptobrevin in synaptic vesicle fusion. *Neuron* 12:1269–79

Jahn R, Niemann H. 1994. Molecular mechanisms of clostridial neurotoxins. *Ann. NY Acad. Sci.* 733:245–55

Jahn R, Südhof TC. 1993. Synaptic vesicle traffic: rush hour in the nerve terminal. *J. Neurochem.* 61:12–21

Kaplan JH, Ellis Davis GCR. 1988. Photolabile chelators for the rapid photorelease of divalent cations. *Proc. Natl. Acad. Sci. USA* 85:6571–75

Kelly RB. 1995. Ringing necks with dynamin. *Nature* 374:116–17

Kosaka T, Ikeda K. 1983. Possible temperature-dependent blockage of synaptic vesicle recycling induced by a single gene mutation in *Drosophila*. *J. Neurobiol.* 14:207–25

Lévêque C, El Far O, Martin-Moutot N, Sato K, Kato R, et al. 1994. Purification of the N-type calcium channel associated with syntaxin and synaptotagmin. A complex implicated in synaptic vesicle exocytosis. *J. Biol. Chem.* 269:6306–12

Lindau M, Neher E. 1988. Patch-clamp techniques for time-resolved capacitance measurements in single cells. *Pflügers Arch.* 411:137–46

Lindau M, Stuenkel EL, Nordmann JJ. 1992. Depolarization, intracellular calcium and exocytosis in single invertebrate nerve endings. *Biophys. J.* 61:19–30

Llinàs R, Sugimori M, Silver RB. 1992. Microdomains of high calcium concentration in a presynaptic terminal. *Science* 256:677–79

Neher E, Marty A. 1982. Discrete changes of cell membrane capacitance observed under conditions of enhanced secretion in bovine adrenal chromaffin cells. *Proc. Natl. Acad. Sci. USA* 79:6712–16

Parsons TD, Lenzi D, Almers W, Roberts WM. 1994. Calcium-triggered exocytosis and endocytosis in an isolated presynaptic cell: capacitance measurements in saccular hair cells. *Neuron* 13:875–83

Perin MS, Fried VA, Mignery GA, Jahn R, Südhof TC. 1990. Phospholipid binding by a synaptic vesicle protein homologous to the regulatory region of protein kinase C. *Nature* 345:260–63

Petrenko AG. 1993. α-Latrotoxin receptor. Implications in nerve terminal function. *FEBS Lett.* 325:81–85

Pfenninger KH, Rovainen CM. 1974. Stimulation and calcium dependence of vesicle attachment sites in the presynaptic membranes. A freeze-cleave study of the lamprey spinal cord. *Brain Res.* 72:1–23

Rieke F, Schwartz EA. 1994. A cGMP-gated current can control exocytosis at cone synapses. *Neuron* 13:863–73

Robinson PJ, Liu J-P, Powell KA, Fykse EM, Südhof TC. 1994. Phosphorylation of dynamin I and synaptic-vesicle recycling. *Trends Neurosci.* 17:348–53

Robitaille R, Adler EM, Charlton MP. 1990. Strategic location of calcium channels at transmitter release sites of frog neuromuscular synapses. *Neuron* 5:773–79

Rothblatt J, Novick P, Stevens T, eds. 1994. *Guidebook to the Secretory Pathway.* Oxford: Oxford Univ. Press

Ryan TA, Reuter H, Wendland B, Schweizer FE, Tsien RW, Smith SJ. 1993. The kinetics of synaptic vesicle recycling measured at single presynaptic boutons. *Neuron* 11:713–24

Smith SJ, Augustine GJ. 1988. Calcium ions, active zones and synaptic transmitter release. *Trends Neurosci.* 11:458–64

Söllner T, Bennett MK, Whiteheart SW, Scheller RH, Rothman JE. 1993. A protein assembly-disassembly pathway in vitro that may correspond to sequential steps of synaptic vesicle docking, activation, and fusion. *Cell* 75:409–18

Stern MD. 1992. Buffering of calcium in the vicinity of a channel pore. *Cell Calcium* 13:183–92

Tachibana M, Okada T, Arimura T, Kobayashi K, Piccolino M. 1993. Dihydropyridine-sensitive calcium current mediates neurotransmitter release from bipolar cells of the goldfish retina. *J. Neurosci.* 13:2898–909

Takei K, McPherson PS, Schmid SL, De Camilli P. 1995. Tubular membrane invaginations coated by dynamin rings are induced by GTP-γS in nerve terminals. *Nature* 374:186–90

Thomas P, Suprenant A, Almers W. 1990. Cytosolic $Ca^{2+}$, exocytosis, and endocytosis in single melanotrophs of the rat pituitary. *Neuron* 5:723–33

van der Bliek AM, Meyerowitz EM. 1991. Dynamin-like protein encoded by the *Drosophila shibire* gene associated with vesicular traffic. *Nature* 351:411–14

van der Bliek AM, Redelmeier TE, Damke H, Tisdale EJ, Meyerowitz EM, Schmid SL. 1993. Mutations in human dynamin block an intermediate stage in coated vesicle formation. *J. Cell Biol.* 122:553–63

Van der Kloot W, Molgó J. 1994. Quantal acetylcholine release at the vertebrate neuromuscular junction. *Physiol. Rev.* 74:899–991

Venzin M, Sandri C, Akert K, Wyss YR. 1977. Membrane associated particles of the presynaptic active zone in rat spinal cord. A morphometric analysis. *Brain Res.* 130:393–404

Verhage M, McMahon HT, Ghijsen WEJM, Boomsma F, Scholten G, et al. 1991. Differential release of amino acids, neuropeptides, and catecholamines from isolated nerve terminals. *Neuron* 6:517–24

von Gersdorff H, Matthews G. 1994a. Dynamics of synaptic vesicle fusion and membrane retrieval in synaptic terminals. *Nature* 367:735–39

von Gersdorff H, Matthews G. 1994b. Inhibition of endocytosis by elevated internal calcium in a synaptic terminal. *Nature* 370:652–55

*Annu. Rev. Neurosci. 1996. 19:235-63*

# STRUCTURE AND FUNCTION OF CYCLIC NUCLEOTIDE–GATED CHANNELS

## William N. Zagotta

Department of Physiology and Biophysics, Howard Hughes Medical Institute, University of Washington School of Medicine, Box 357370, Seattle, Washington 98195-7370

## Steven A. Siegelbaum

Department of Pharmacology, Center for Neurobiology and Behavior, Howard Hughes Medical Institute, Columbia University, 722 W. 168th Street, New York, New York 10032

KEY WORDS:    photoreceptor, olfaction, potassium channel, sodium channel, calcium channel

### ABSTRACT

Cyclic nucleotide–gated (CNG) channels play important roles in both visual (Yau & Baylor 1989) and olfactory (Zufall et al 1994) signal transduction. The cloning of the gene coding for a rod photoreceptor channel (Kaupp et al 1989) and the subsequent cloning of related genes from olfactory neurons (Dhallan et al 1990, Ludwig et al 1990, Goulding et al 1992) has sparked much progress over the past several years in elucidating the structural bases for the function of the CNG channels. One of the surprising features to emerge from these cloning studies was that the CNG channels are structurally homologous to the voltage-gated channels (Jan & Jan 1990) despite the fact that the CNG channels are gated by the binding of a ligand—cAMP or cGMP—and not by voltage. In this review we focus on recent studies of the relationship between the structure and function of the CNG channels. Given the homology between CNG channels and voltage-gated channels, such studies are likely to provide important general information about the structure and function of a wide variety of channel types.

Ion channels that are directly activated by cyclic nucleotides were first discovered in the plasma membrane of retinal photoreceptors (Fesenko et al 1985). The closing of a CNG channel in the outer segment of photoreceptors repre-

235

0147-006X/96/0301-0235$08.00

sents a fundamental step in the transduction of light energy into an electrical signal [see reviews by Stryer (1986) and Yau & Baylor (1989)]. Photoactivated rhodopsin activates a phosphodiesterase, which catalyzes the hydrolysis of guanosine 3′,5′-cyclic monophosphate (cGMP). The fall in the cytosolic concentration of cGMP then closes a cGMP-activated channel in the membrane of the outer segment (Fesenko et al 1985). The closing of this cation-selective channel causes a hyperpolarization of the photoreceptor, leading to a decrease in transmitter release. The transduction of odorant signals by the olfactory epithelium also depends on a CNG channel (Nakamura & Gold 1987). In this case the odorant receptors activate adenylyl cyclase, which generates adenosine 3′,5′-cyclic monophosphate (cAMP) and opens a cAMP-activated channel [see reviews by Lancet (1986) and Zufall et al (1994)]. This leads to a depolarization of the olfactory neuron and the generation of an action potential.

The CNG channels exhibit varying selectivities for cGMP and cAMP, reflecting their physiological role. By cooperatively binding multiple cyclic nucleotide molecules, these channels become exquisitely sensitive to changes in the levels of cytosolic cyclic nucleotide (Haynes et al 1986, Zimmerman & Baylor 1986), allowing, for example, the rod CNG channel to faithfully detect and signal the drop in cGMP concentration resulting from the absorption of a single photon (Baylor et al 1979). Whereas CNG channels were originally thought to be static sensors of cyclic nucleotide concentration, they have recently been found to open at a concentration of nucleotide that can be tuned by physiological stimuli. The apparent ligand affinity of the olfactory and rod channels can be modulated by $Ca^{2+}$ (Kramer & Siegelbaum 1992, Lynch & Lindemann 1994), $Ca^{2+}$-calmodulin (Chen et al 1994, Chen & Yau 1994, Gordon et al 1995b, Hsu & Molday 1993, Hsu & Molday 1994, Liu et al 1994), phosphorylation (Gordon et al 1992), diacylglycerol analogues (Gordon et al 1995a), and transition metal divalent cations (Gordon & Zagotta 1995a,b; Ildefonse & Bennett 1991; Karpen et al 1993).

The CNG channels are permeable to both monovalent and divalent cations. The $Ca^{2+}$ permeability of these channels has several important physiological consequences for sensory signal transduction. In photoreceptors, the light-induced closure of the CNG channels leads to a decrease in internal $Ca^{2+}$, which underlies adaptation to levels of illumination (Yau & Baylor 1989, Matthews et al 1988, Nakatani & Yau 1988). In the dark, $Ca^{2+}$ influx through the CNG channels can directly contribute to transmitter release (Rieke & Schwartz 1994). In olfactory neurons, odorant-induced activation of the CNG channels leads to an increase in $Ca^{2+}$ influx, which contributes to adaptation (Zufall et al 1994). In addition, the $Ca^{2+}$ influx can activate a depolarizing $Ca^{2+}$-activated $Cl^-$ channel, which helps amplify the response to odorants (Kleene 1993, Kurahashi & Kaneko 1993, Kurahashi & Yau 1993, Lowe & Gold 1993). Because the permeation of $Ca^{2+}$ and $Mg^{2+}$ is so slow, these divalent cations effectively block the flow of monovalent cations, making the single-channel conductance quite small under physiological conditions [see below and reviews by Yau & Baylor (1989) and Zufall et al (1994)].

CNG channels have also been demonstrated in a variety of other nonsensory tissues, such as the heart, testis, and kidney (Ahmad et al 1990; Biel et al 1993, 1994; DiFrancesco & Tortora 1991; Distler et al 1994; Marunaka et al 1991; Weyand et al 1994). $Ca^{2+}$ influx through the CNG channels may play a role in sperm chemotaxis (Biel et al 1994, Weyand et al 1994).

## MOLECULAR CLONING OF CNG CHANNELS

Kaupp and coworkers first isolated a cDNA clone for a subunit of the bovine rod cGMP-activated channel that coded for a protein that contained 690 amino acids (Kaupp et al 1989). Based on sequence homology to the rod CNG channel, clones for other CNG channels have been isolated from a variety of different tissues and species (Baumann et al 1994; Biel et al 1993, 1994; Bönigk et al 1993; Bradley et al 1994; Chen et al 1993; Dhallan et al 1990, 1992; Goulding et al 1992; Liman & Buck 1994; Ludwig et al 1990; Weyand et al 1994). These channel clones appear to come from a set of distinct but structurally related genes with tissue- and species-specific expression (Distler et al 1994). The rod CNG channel appears to be primarily expressed in the rod photoreceptors. A distinct CNG channel gene was first cloned from olfactory epithelium (Dhallan et al 1990, Goulding et al 1992, Ludwig et al 1990) and subsequently from rabbit aorta (Biel et al 1993). A third gene for CNG channels has been separately cloned from the cone photoreceptors (Bönigk et al 1993), the testis (Weyand et al 1994), and the kidney (Biel et al 1994). These cDNA clones code for a subunit of CNG channels, referred to as the α-subunit (or subunit 1) that can, by itself, produce functional channels when exogenously expressed in either *Xenopus* oocytes or a human embryonic kidney cell line (HEK293). In addition, clones for a separate β-subunit, or subunit 2, have been isolated for both the rod (Chen et al 1993) and olfactory (Bradley et al 1994, Liman & Buck 1994) CNG channels. These β-subunits do not express by themselves, but when coexpressed with their corresponding α-subunit, they produce channels with altered permeation, pharmacology, and/ or cyclic nucleotide selectivity (see below).

The primary structure of the CNG channel polypeptide is diagrammed in Figure 1A and compared to that of the voltage-dependent potassium channel from the *Shaker* locus in *Drosophila* (Tempel et al 1987). Like the other ion channels that have been cloned, the rod cGMP-activated channel contains a number of hydrophobic domains, shown as filled boxes, thought to represent transmembrane segments. In addition, the cGMP-activated channel contains some sequence similarity to the voltage-gated channels, particularly in two

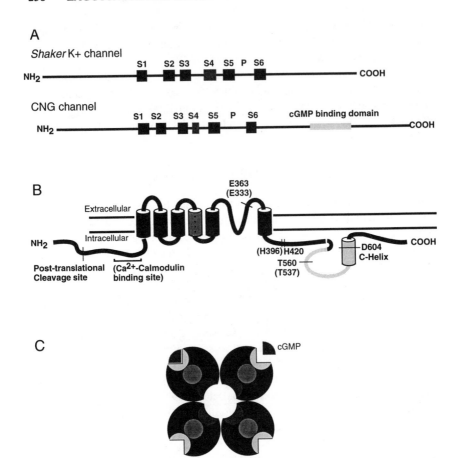

*Figure 1*  Summary of the structure of CNG channels. (*A*) Primary structure of the CNG channels compared to that of the *Shaker* potassium channel. The black boxes indicate hydrophobic segments thought to represent transmembrane domains. The region between S5 and S6 in each channel (P region) is thought to line the ion conducting pore. The cyclic nucleotide binding domain is shown by a shaded box in the carboxyl-terminal region of the CNG channel. (*B*) Putative transmembrane topology of the CNG channels. The positions of several important functional regions in the bovine rod channel are labeled. The amino acid positions in the olfactory channel are shown in parentheses. (*C*) Putative tetrameric arrangement of the subunits around a centrally located pore.

regions: the S4 segment (Jan & Jan 1990), thought to represent the voltage sensor for the voltage-dependent transitions in these channels, and the P region (Guy et al 1991, Goulding et al 1992), thought to comprise part of the channel pore. On the basis of this sequence comparison, the cGMP-activated channel appears more closely related to the voltage-gated family of channels than to the ligand-gated channels. The sequence similarity suggests that the transmembrane topology for the CNG channels is similar to that proposed for the voltage-dependent channels (Figure 1B). Consistent with this interpretation, immunocytochemistry studies have shown that the amino terminal domain is located on the cytoplasmic side of the membrane (Molday et al 1991), and the beginning of the P region is located on the extracellular side of the membrane (Wohlfart et al 1992). Although the subunit composition of the functional channels is unknown, based on their homology to the tetrameric voltage-dependent potassium channels (Mackinnon 1991) and a Hill coefficient occasionally greater than 3, the CNG channels may exist as tetramers (Figure 1C). When expressed exogenously from only a single subunit the channel is homomultimeric. However, the native channel is likely to be a heteromultimer containing some combination of α and β subunits (see below).

The cDNA of the bovine rod cGMP-activated channel predicts a sequence in the amino-terminal domain that is very highly charged, owing to the presence of many lysine and glutamic acid residues. Molday et al (1991) have suggested that the channel is posttranslationally modified by specific cleavage and removal of the first 92 amino acids in the amino terminal domain. This processing of the channel polypeptide would account for the difference in molecular mass between the protein predicted from the cDNA clone (80 kDa) and the protein isolated from bovine retina (63 kDa). Consistent with this hypothesis, immunofluorescent and immunogold labeling have confirmed that the 63-kDa form of the channel is present in rod outer segments (Molday et al 1991).

While the channels discussed above are quite similar both structurally and functionally, other, more distantly related channels, whose gating is altered by the direct binding of cyclic nucleotides, also have been identified. A family of delayed rectifying potassium channels related to *Drosophila eag* (Ludwig et al 1994, Warmke et al 1991, Warmke & Ganetzky 1994) exhibits voltage-dependent gating that is modulated by the direct binding of cAMP (Bruggemann et al 1993). Furthermore, the voltage dependence of an inwardly rectifying potassium channel from *Arabidopsis* (KAT1) is modulated by cGMP (Hoshi 1995). The sequences of the *eag* and KAT1 channels contain a region that exhibits sequence similarity to other cyclic nucleotide-binding proteins (Anderson et al 1992, Guy et al 1991), supporting a direct action of cyclic nucleotides on these channels. In addition, DiFrancesco & Tortora (1991) have suggested that cAMP has a direct action on an inwardly rectifying nonselective cation channel in the sino-atrial node of the heart that is responsible for pacemaker activity. These more distantly related CNG channels suggest an evolutionary connection between the voltage-dependent potassium channels and the CNG channels.

# MOLECULAR STUDIES ON THE CHANNEL PORE

How do ions permeate the CNG channels and how do the channels select among different ions? This question is of physiological importance because permeation has two important functions. First, it generates membrane depolarization, which is critical for electrical signaling. Second, as discussed above, the channel provides a pathway for $Ca^{2+}$ entry, which can then affect a number of intracellular targets.

## Single Channel Properties

The first recordings of cyclic nucleotide–induced currents in normal extracellular solution failed to resolve single-channel events (Fesenko et al 1985, Nakamura & Gold 1987). Noise measurements suggested that the unitary conductance was <0.1 pS, raising the question as to whether they were indeed channels (Fesenko et al 1985). Subsequent studies demonstrated that the low conductance was due to a rapid and profound block of the channels by divalent cations (Haynes et al 1986, Zimmerman and Baylor 1986). Upon removal of external $Mg^{2+}$ and $Ca^{2+}$, clear single-channel currents could be detected, although even then the channel openings tended to be very brief and flickery. Estimates for single-channel conductance in divalent cation free solution range from around 20 to 60 pS (Yau & Baylor 1989, Zufall et al 1994, Sesti et al 1994). Yau & Baylor (1989) have suggested that the reduction in single-channel conductance due to block by external divalent cations may have an important physiological role in improving the signal-to-noise ratio of sensory transduction. This is because the current generated by many small-conductance channels will be relatively less noisy than the current generated by a few large-conductance channels.

## Similarity of Pore Region Structure of CNG and Voltage-Gated $K^+$ Channels

Our understanding of ion permeation through the CNG channels has been greatly aided by structural insights provided by the molecular cloning of these channels. One of the most surprising results to emerge from inspection of the deduced amino acid sequence of CNG channels was the presence of a region that was homologous to the P region (or H5 or SS1-SS2) of voltage-gated $K^+$ channels (Figure 2), which links the extracellular ends of the S5 and S6 transmembrane regions (Guy et al 1991, Goulding et al 1992).

In $K^+$ channels, the P region has been shown to be the major determinant of ion selectivity (see Miller 1991 for review). To explain how this relatively short stretch of 20 amino acids can control ion permeation (Hartmann et al 1991, Yool & Schwartz 1991) and affect both extracellular and intracellular

A

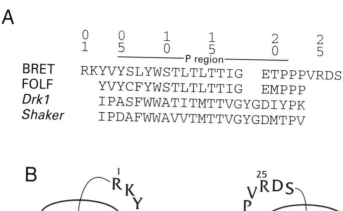

|  | 0 1 | 0 5 | 1 0 | 1 5 | 2 0 | 2 5 |
|---|---|---|---|---|---|---|

------P region------

BRET  RKYVYSLYWSTLTLTTIG  ETPPPVRDS
FOLF  YVYCFYWSTLTLTTIG  EMPPP
Drk1  IPASFWWATITMTTVGYGDIYPK
Shaker  IPDAFWWAVVTMTTVGYGDMTPV

B

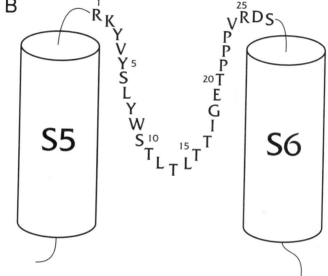

*Figure 2* Structure of the pore region of CNG and related channels. (*A*) Amino acid sequence alignment of the bovine retinal CNG channel (BRET) amino acid residues R345–S371 (numbered 1 to 27 for convenience), the catfish olfactory CNG channel (FOLF) residues Y317–P337, and the residues of the P region of two voltage-gated potassium channels, *drk1* (Frech et al 1989) and *Shaker* (Tempel et al 1987). (*B*) The transmembrane topology of the BRET P-domain residues according to the β-hairpin model of the potassium channel P region.

tetraethylammonium binding (Yellen et al 1991), it was initially proposed that the P region dips into and out of the membrane, crossing it twice as a β hairpin. Subsequent results have shown that the neighboring S5 and S6 segments can also affect ion permeation and, in particular, participate in the binding of intracellular channel blockers (Choi et al 1993, Kirsch et al 1993, Slesinger et al 1993, Lopez et al 1994). This has led to the view that the S5 and S6 segments form the internal vestibule of the channel, whereas the P region may form the narrowest region of the channel, which serves as the ion selectivity filter.

The similarity of the amino acid sequence between $K^+$ channels and CNG channels in the P region is quite surprising because the CNG channels are nonselective cation channels. Although they are good at discriminating cations from anions, they show little or no discrimination between $Na^+$ and $K^+$ [for photoreceptor channels, see Yau & Baylor (1989), Kaupp et al (1989), Furman & Tanaka (1990), Menini (1990), Picones & Korenbrot (1992), Sesti et al (1994, 1995); for olfactory channels, Kolesnikov et al (1990), Frings et al (1992), Goulding et al (1993)]. Close inspection of the amino acid sequences of the P regions from $K^+$ channels and CNG channels revealed a deletion of two amino acids, tyrosine and glycine, in the CNG channels (Figure 2). In an elegant study, Heginbotham et al (1992) deleted these two residues in Shaker $K^+$ channels. Remarkably, the mutant channels now displayed many of the properties associated with the pore of the CNG channels. These deletion mutants lost their selectivity for $K^+$ and became sensitive to block by external $Ca^{2+}$ and $Mg^{2+}$ with high affinity.

Goulding et al (1993) provided direct evidence that the P region forms the CNG channel pore exchanging the P regions between a bovine retinal and catfish olfactory channel. Compared with the olfactory channel, the retinal channel has a smaller single-channel conductance (20 vs 55 pS), a higher degree of selectivity among monovalent cations, and a smaller apparent pore diameter (5.8 vs 6.3 Å; determined from organic cation permeability) (see also Picco & Menini 1993). A chimeric retinal channel containing the olfactory P region displayed the permeation properties appropriate to the olfactory channel. The difference in single-channel conductance could be accounted for quantitatively on the basis of the difference in pore diameter. The fact that the P region determined apparent pore diameter supports the view that it forms the narrowest part of the pore, where the ion selectivity filter resides.

A further connection between the pores of voltage-gated $K^+$ and CNG channels comes from a study of the effects of a peptide derived from the amino terminus of Shaker $K^+$ channels, which mediates rapid inactivation in these channels (Zagotta et al 1990). Isacoff et al (1991) proposed that the S4-S5 loop serves as the binding site for this inactivation peptide in Shaker $K^+$ channels. Although the CNG channels do not show any sequence similarity with the $K^+$ channels in the S4-S5 loop, they too are blocked by the inactivation peptide with high affinity (Kramer et al 1994). Using chimeric retinal and olfactory channels, which differ in their affinity for the peptide 100-fold, Kramer et al (1994) showed that the P region was likely to serve as the peptide-binding site in the CNG channels. Subsequent studies have found that the P region may also participate in inactivation peptide binding to $K^+$ channels (Yool & Schwartz 1995).

How can we explain the differences in permeability properties between $K^+$ channels and CNG channels based on the deletion of a tyrosine and glycine residue? Given the larger diameter of the CNG channel pore compared to that of the $K^+$ channel pore, it is tempting to speculate that this deletion is responsible for the dilation of the pore. Unfortunately, attempts to convert a CNG channel to a $K^+$-selective channel with a YG insertion have met with little success (EH Goulding, GR Tibbs, & SA Siegelbaum, unpublished data).

## CNG Channels Share Many Permeation Properties with Voltage-Gated Calcium Channels

Although the pore of the CNG channels resembles voltage-gated $K^+$ channels most closely in terms of its amino acid sequence, the CNG channel permeation properties are more like those of voltage-gated L-type $Ca^{2+}$ channels. Similar to $Ca^{2+}$ channels, the CNG channels are permeable to both $Ca^{2+}$ and monovalent cations (Yau & Baylor 1989). The permeability of calcium through the photoreceptor and olfactory neuron channels is physiologically significant. Approximately 15% of the current entering rod photoreceptors in the dark is carried by calcium (Yau & Nakatani 1985). Recent estimates suggest that the ratio of calcium to $Na^+$ permeability is around three for the rod photoreceptor channel and around seven for the cone photoreceptor channel, contributing to the more rapid light adaptation seen in cones (Picones & Korenbrot 1995a). Frog olfactory neuron channels are also reported to have a higher permeability to $Ca^{2+}$ than to $Na^+$ (Kolesnikov et al 1990).

In addition to permeating the CNG channels, $Ca^{2+}$ also profoundly blocks the flow of current carried by monovalent cations through these channels [for photoreceptors, see Yau & Baylor (1989), Colamartino et al (1991), Zimmerman & Baylor (1992), Tanaka & Furman (1993), Picones & Korenbrot (1992); for olfactory channels, see Zufall et al (1994)], similar to the behavior of the voltage-gated $Ca^{2+}$ channels (Almers & McCleskey 1984, Hess & Tsien 1984). In $Ca^{2+}$ channels, this behavior is thought to be due to multi-ion occupancy in the pore. Although many of the permeability properties of CNG channels can be explained by a single ion binding site within the pore that binds divalent cations with a higher affinity than monovalent cations (Zimmerman & Baylor 1992), other evidence favors multi-ion occupancy (Furman & Tanaka 1990, Sesti et al 1995). In particular, Sesti et al (1995) report an anomalous mole fraction effect for the rod channel in which the conductance of the channel to a mixture of $Li^+$ and $Cs^+$ was greater than the conductance seen with either $Cs^+$ or $Li^+$ alone. Such effects can best be explained by multi-ion channels.

The CNG channels also resemble voltage-gated $Ca^{2+}$ channels in that they are blocked by organic $Ca^{2+}$ antagonists. The olfactory channels are blocked by nifedipine (Zufall & Firestein 1993), l-*cis*-diltiazem, and D600 (Frings et al 1992), although at 100-fold higher concentrations than required for blocking $Ca^{2+}$ channels. Photoreceptor channels are inhibited by l-*cis*-diltiazem with a relatively high affinity (Stern et al 1986). Finally, CNG channels resemble $Ca^{2+}$ channels (Pietrobon et al 1988, Prod'hom et al 1987) in that they display a distinct subconductance state that is mediated by a rapid partial blocking effect of external protons, readily seen in the cloned catfish olfactory channel (Goulding et al 1992) and native salamander olfactory channel (Zufall & Firestein 1993). This block, like that in $Ca^{2+}$ channels, is relieved by raising the external pH and by membrane depolarization. Although the proton-induced subconductance state is less evident in photoreceptor channels, it is likely to be present but obscured by more rapid kinetics. This rapid, short-lived subconductance state is independent of cGMP concentration; thus it is distinct from a longer-lived subconductance state observed at low concentrations of cGMP in photoreceptor channels that is thought to be due to the opening of partially liganded channels (Ildefonse & Bennett 1991, Taylor & Baylor 1995; also see below). Both photoreceptor (Tanaka 1993) and olfactory (Frings et al 1992) channels are also blocked by internal protons. The internal proton–binding site appears also to serve as the binding site for internal divalent cations (Tanaka 1993).

## *Role of a Conserved Glutamate in Controlling Ion Permeation*

Where are the divalent cation–blocking sites in the pore? Where are the multiple sites that bind permeant ions, allowing the channel to be multiply occupied? Where is the proton-binding site? Remarkably, the answer to many of these seemingly distinct questions appears to involve a single residue, the acidic glutamate in the P region (E19, according to the numbering scheme of Figure 2; rod channel: E363; catfish olfactory channel: E333). Root & MacKinnon (1993) and Eismann et al (1994) found that mutation of this residue to neutral residues nearly abolished block by external $Mg^{2+}$. Both groups report relatively little effect on block by internal $Mg^{2+}$, suggesting the presence of a distinct divalent cation–binding site near the internal mouth of the channel. The mutations also altered the open channel current-voltage relation, causing a pronounced outward rectification, consistent with an electrostatic role of E19 in increasing the local external cation concentration. Sesti et al (1995) showed that such mutations converted the behavior of the channel from that of a multi-ion pore to that of a single-ion pore because the anomalous mole fraction effect was no longer observed.

Root & MacKinnon (1994) demonstrated that this same glutamate residue (E19) was also responsible for external proton block. In a careful analysis, they were able to distinguish three open conductance states in the catfish CNG channel, a fully open state and two subconductance states, corresponding to channels with zero, one or two protons bound. Because the CNG channels are likely to be tetramers, the channels should contain four equivalent glutamates. To explain the presence of only two apparent proton-binding sites, Root & MacKinnon proposed that each binding site consisted of two glutamates forming a carboxyl-carboxylate pair that share a single proton. The influence of neighboring negative charges explains why the pKa of the site (7.6) is much higher than that of a free glutamate carboxyl side chain (4.3). Given the proximity of the two binding sites, one might have expected that the binding of the first proton would inhibit the binding of the second due to electrostatic effects. However, the binding events were independent, suggesting that a carboxyl-carboxylate was never free but was occupied by a monovalent alkali cation when deprotonated. The aligned glutamate residues in calcium channels have been shown to be important for calcium selectivity (Heinemann et al 1992, Yang et al 1993) and may function as the site of multi-ion occupancy.

## Effect of Subunit Composition on Channel Properties

The subunit composition of the CNG channels has a pronounced effect on their single-channel current characteristics (Table 1). The native rod channels display very rapid and flickery channel openings, even in the absence of divalent cations, which is indicative of very brief open and closed times. In contrast, the cloned α subunit of the rod CNG channels displays much longer openings when expressed in *Xenopus* oocytes. Although the β subunit is not expressed when injected alone into *Xenopus* oocytes, when it is coexpressed with the α subunit, the channel openings now become flickery, like the native rod channel (Chen et al 1993). Coexpression with the β subunit also appears to reduce sensitivity to block by external divalent cations to a level similar to that of the native channel. The β subunit contains a glycine instead of a glutamate at the position corresponding to residue E19, which is consistent with a role for this residue in divalent cation block. Finally, the β subunit also increases the sensitivity of the expressed channel to l-*cis*-diltiazem to a level similar to that seen in native channels.

The rat olfactory β subunit does not yield functional channels when expressed on its own. However, coexpression of the β subunit with the α subunit yields channels whose properties are distinct from the channels expressed by the α subunit alone but resemble more closely the native rat olfactory channel (see Table 1) (Bradley et al 1994, Liman & Buck 1994). Compared to channels composed of only α subunits, channels containing both α and β subunits exhibit decreased divalent cation block, increased flickering of channel openings, and decreased apparent single-channel conductance (Table 1). The β subunit contains an aspartate residue in place of the glutamate residue responsible for divalent cation block. Because the negatively charged aspartate also

**Table 1**  Comparison of properties of cloned and native CNG channels

| Species | Composition | $\gamma$ (pS) | $K_{1/2}(\mu M)$ cAMP | cGMP | $n$ (Hill Coefficient) cAMP | cGMP | References |
|---|---|---|---|---|---|---|---|
| Olfactory channels | | | | | | | |
| Catfish | Native | 44 | 3.4 | 2.5 | 1.4 | 1.3 | Bruch & Teeter 1990, Goulding et al 1992 |
| | $\alpha$ subunit | 55 | 64 | 75 | 1.5 | 1.2 | Goulding et al 1994 |
| Rat | Native | 12–15 | 4 | 1.8 | 1.8 | 1.3 | Frings et al 1992 |
| | $\alpha$ subunit | 35 | 55 | 2 | 2.6 | 2.5 | Dhallan et al 1990 |
| | $\alpha + \beta$ subunits | 19 | 14 | 4 | 1.5 | 1.4 | Liman & Buck 1994, Bradley et al 1994 |
| Rod channels | | | | | | | |
| Bovine | Reconstituted | 26 | — | 31 | — | 2.3 | Hanke et al 1988 |
| | Native | 6 | — | 165 | — | 2 | Quandt et al 1991 |
| | $\alpha$ subunit | 20 | 2000 | 60 | 1.2 | 2 | Kaupp et al 1989, Goulding et al 1994 |
| Human | $\alpha$ subunit | 25–30 | — | 60 | — | 2 | Chen et al 1993 |
| | $\alpha + \beta$ subunits | Flick-ery | — | 60 | — | 2 | Chen et al 1993 |

permits divalent cation block (Root & MacKinnon 1993), other residues may be important for the differences in single-channel properties.

## CYCLIC NUCLEOTIDE–BINDING SITE

Based on experiments demonstrating a direct activation of the CNG channels by cyclic nucleotides applied to the inside of the membrane in inside-out patches (Fesenko et al 1985, Nakamura & Gold 1987), retinal and olfactory channels were thought to contain a cytoplasmic-binding site for cyclic nucleotides. Such a binding site was subsequently identified near the carboxy terminus in the predicted amino acid sequence of the cloned rod (Kaupp et al 1989) and olfactory channels (Dhallan et al 1990, Ludwig et al 1990, Goulding et al 1992). As shown in Figure 3, the carboxy terminus of these channels contains a highly conserved stretch of approximately 120 amino acids that is homologous to the cyclic nucleotide–binding domains of other proteins, including the cAMP- and cGMP-dependent protein kinases and the catabolite-activating protein (CAP), a bacterial transcription factor (Shabb & Corbin 1992). The structure of the cAMP-binding site of CAP has been determined (Weber & Steitz 1987). It consists of a short amino-terminal $\alpha$ helix (A helix) preceding an eight-stranded antiparallel β roll, followed by two α helixes (B and C). cAMP binds within a pocket formed by the β roll and the C helix (Figure 3). Although the overall sequence identity among the cyclic nucleotide–binding proteins over this region is fairly low (around 20%), certain key residues that make important contacts with the bound cAMP or occur at turns between adjacent β strands in CAP are conserved (Figure 3). Kumar & Weber (1992) have constructed a structural model for the retinal channel–binding site based on this sequence homology (Figure 4A).

## Cyclic Nucleotide Selectivity

Although the retinal and olfactory CNG channels exhibit a high degree of sequence similarity (over 80% amino acid identity) in the putative binding region, the native channels exhibit different cyclic nucleotide selectivities. For the retinal channel, cGMP is a much more potent and effective agonist than cAMP (Fesenko et al 1985). For the native olfactory channel, cAMP and cGMP have very similar effects (Nakamura & Gold 1987, Zufall et al 1994). The retinal channel shows a 20- to 50-fold higher apparent affinity for cGMP than for cAMP, as determined by $K_{1/2}$ values from dose-response curves (concentration required for half of maximal activation) (Table 1). In contrast, the $K_{1/2}$ values of native olfactory channels for cAMP and cGMP are nearly identical (Zufall et al 1994) (Table 1). For the retinal channel, cAMP acts as a partial agonist, activating only a fraction of the current that is activated by cGMP, even at maximal concentrations. For the olfactory channels, both cGMP and cAMP are full agonists, producing similar maximal responses. Presumably these differences between retinal and olfactory channels reflect the primary physiological role of cGMP in phototransduction and cAMP in olfactory signaling. Why the olfactory channel has retained a high sensitivity to cGMP is unknown.

The cyclic nucleotide responses of the cloned and expressed α subunit of the olfactory CNG channel differ in certain respects from the responses of the native channels (Table 1). Both the rat (Dhallan et al 1990) and bovine olfactory (Altenhofen et al 1991) α subunits, expressed in cell lines or *Xenopus* oocytes, show identical maximal responses to cAMP and cGMP, similar to the native channels. However, the cloned channels exhibit a $K_{1/2}$ value for cAMP that is 25- to 40-fold higher than that for cGMP (Table 1). In contrast, the cloned catfish α subunit, when expressed in *Xenopus* oocytes, responds to cAMP and cGMP with nearly identical $K_{1/2}$ values and maximal responses, which is similar to the behavior of the native catfish olfactory channel (Goulding et al 1992). However, the absolute $K_{1/2}$ values for the cloned catfish channel are 15- to 20-fold higher than those of the native channel. Part of these discrepancies may be due to the fact that the native channels are heteromul-

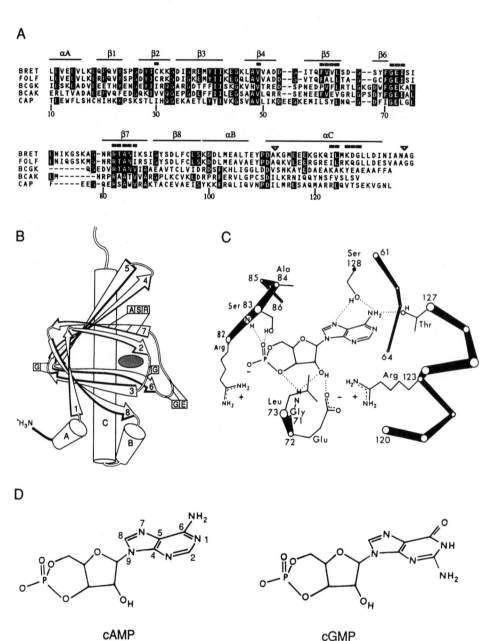

timers. Thus, coexpression of the rat olfactory $\alpha$ subunit with the rat olfactory $\beta$ subunit shifts the $K_{1/2}$ values for cAMP to lower values (Table 1), although the $K_{1/2}$ values for cAMP are still threefold higher than those in the native channels (Bradley et al 1994, Liman & Buck 1994). The remaining difference may reflect posttranslational modification due to phosphorylation (Gordon et al 1992) or interaction with $Ca^{2+}$-calmodulin (Hsu & Molday 1993, Chen & Yau 1994).

## Molecular Basis for Cyclic Nucleotide Selectivity

How do the structures of the retinal and olfactory channel–binding sites account for the different selectivity of these channels for activation by cAMP and cGMP? cAMP and cGMP only differ in structure at two positions on their purine rings (Figure 3). Several groups have tried to infer the properties of the cyclic nucleotide–binding site by studying the effectiveness of various cyclic nucleotide analogues. The only analogues with a higher apparent affinity for the channel than those of the native compounds involve substitutions at the 8′ position on the purine ring, with bulky hydrophobic substituents producing the highest apparent affinity (Tanaka et al 1989, Brown et al 1993, Scott & Tanaka 1995). This suggests the presence of a hydrophobic pocket within the binding site. Recently, Brown et al (1995) have confirmed the presence of such a pocket. Using an 8′-substituted photoaffinity analogue of cGMP, they labeled the rod channel at valine 524, valine 525, and alanine 526 on the $\beta 4$ strand of the $\beta$ roll. In CAP, a valine homologous to V524 is a site of contact with cAMP. Tanaka and colleagues have shown that interactions with the C2 and C6 positions on the purine ring are critical because substitutions at these positions greatly diminish activity (Tanaka et al 1989, Scott & Tanaka 1995).

---

*Figure 3*  Structure of the cyclic nucleotide–binding domain of CAP and cyclic nucleotides. (*A*) Amino acid sequence alignment of the bovine retinal CNG channel (BRET) amino acid residues L485–G615 (Kaupp et al 1989), the catfish olfactory CNG channel (FOLF) amino acid residues L455–G585 (Goulding et al 1992), the bovine cGMP-dependent kinase domain RII amino acid residues I228–A352 (Takio et al 1984), and the bovine cAMP-dependent kinase domain RII amino acid residues E261–V379 (Titani et al 1984) with the cyclic nucleotide–binding domain of the catabolite activating protein (CAP) (Weber & Steitz 1987). Amino acid residue numbering of CAP is shown beneath the sequence alignment. Lines above the sequence alignment indicate amino acid residues that form $\alpha$ (A–C) helices or $\beta$ (1–8) strands in the CAP crystal structure. Squares indicate amino acid residues that are close to cAMP bound to CAP. Triangles indicate limits of region in which amino acids residues were substituted in the C-helix chimeras of Goulding et al (1994). (*B*) Schematic drawing showing structure of one subunit of a CAP dimer (modified from Shabb & Corbin 1992). The numbered arrows repesent $\beta$ strands, and the lettered cylinders represent $\alpha$ helices. The binding site for cAMP in the $\beta$ roll is indicated by a shaded circle. (*C*) Schematic representation of cAMP (in the anticonformation) in CAP–binding site of one subunit. Hydrogen bonds are indicated as dotted lines, and charged groups that form ionic interactions are shown. Ser-128 from the adjacent subunit in the CAP dimer is shown as a black dot. (*D*) Schematic representation of the molecular structure of cAMP and cGMP.

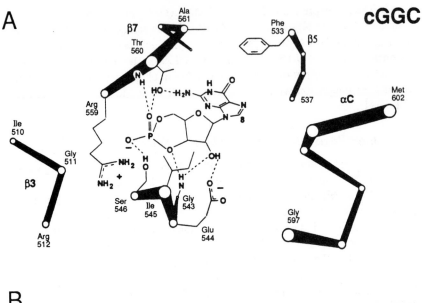

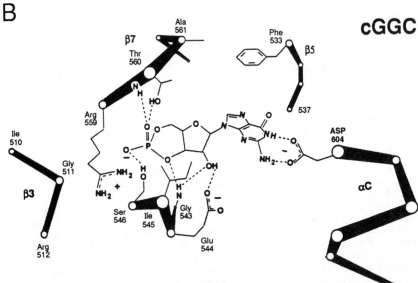

*Figure 4* Two models for the binding of cGMP to the retinal channel. (*A*) cGMP binding in the syn conformation. An important hydrogen bond is made between Thr560 and the 2-amino group of cGMP. (From Kumar & Weber 1992). (*B*) cGMP binding in the anticonformation. In the open state, Asp604 on the C helix forms a pair of hydrogen bonds with the N1 and N2 groups of cGMP.

The cyclic phosphate is also important because Sp-cAMPS is an agonist of the olfactory channel, whereas Rp-cAMPS is an antagonist (Kramer & Tibbs 1995).

Studies on protein kinases pointed to the importance of a particular contact site in the binding region, located on the β7 strand, whose sequence includes the residues RTA in the cGMP-dependent protein kinases, also present in the CNG channels, and RAA in the cAMP-dependent protein kinases (Figure 4A). In CAP, the positively charged arginine (R) makes an ionic contact with a negatively charged exocyclic oxygen. Molecular modeling studies on cGMP-dependent protein kinase suggested that the hydroxyl group of the threonine (T) residue formed an important hydrogen bond with the amino group attached to C2 on the guanine ring of cGMP (Weber et al 1989). cAMP does not contain this amino group, so the hydrogen bond cannot form. Mutation of the first alanine (A) to a threonine in the cAMP-dependent kinase greatly increased the binding of cGMP with little effect on the binding of cAMP (Shabb et al 1990). Kaupp and colleagues have provided evidence that a similar interaction occurs in the CNG channels (Altenhofen et al 1991). Thus, mutation of the homologous threonine (T560) to alanine, decreases the affinity of the channel for cGMP, although it also decreases its affinity for cAMP. However, this residue cannot account for differences in cyclic nucleotide selectivity between the olfactory and retinal channels, since all CNG channels cloned so far contain a threonine at this position. Even so, some channels (e.g. the catfish olfactory channel) show the same apparent affinity for cAMP and cGMP. Moreover, cGMP can only make this hydrogen bond with the threonine residue when it binds in the syn conformation (in which the guanine group rotates to lie above the ribose). In CAP, however, cyclic nucleotides are known to bind in the anticonformation (in which the guanine rotates away from the ribose).

A different mechanism for cyclic nucleotide selectivity, involving the long C helix, has been suggested by two sets of recent experiments (Figure 4B) (Goulding et al 1994, Varnum et al 1995). In CAP, the purine ring of cAMP points towards the C helix and makes important hydrogen bonds with it. Because the only differences between cAMP and cGMP lie in their purine ring substitutions, the C helix was reasoned to be a likely site for selectivity. The importance of this region was confirmed in experiments using chimeric channels in which the C helixes of the bovine retinal and catfish olfactory channels were exchanged (Goulding et al 1994). Replacing the C helix of the retinal channel with that of the olfactory channel resulted in a nonselective chimera that was activated equally well by cAMP and cGMP. Conversely, replacing the C helix of the olfactory channel with that of the retinal channel generated a chimera that was selectively activated by cGMP (the $K_{1/2}$ for cGMP was 100-fold lower than that for cAMP).

Which residues in the C helix are responsible for ligand selectivity? Varnum et al (1995) have shown that aspartate residue 604 (D604) in the bovine retinal channel C helix appears to play a critical role in the selective activation by cGMP. In CAP, the homologous residue is a threonine (T127) that makes an important hydrogen bond with the N6 amino group of cAMP (Figure 3B). In the wild-type retinal channel, the rank order of efficacy for activation by cyclic nucleotides, i.e. the maximal current elicited at saturating concentrations, is cGMP > cIMP (cyclic inosine monophosphate) > cAMP. Upon mutation of D604 to a neutral polar residue—such as glutamine, which is present in the catfish olfactory channel—the order changes to cGMP > cAMP > cIMP. Upon substitution of a nonpolar residue—such as methionine, which is present in the β subunit of the rat olfactory channel—the order becomes inverted: cAMP > cIMP > cGMP. Varnum et al proposed that the negatively charged carboxylic acid side chain forms a pair of hydrogen bonds with the N1 and N2 hydrogen atoms on the purine ring (Figure 4B). This type of interaction has been shown to occur in high-affinity GTP-binding proteins such as the α subunit of transducin (Noel et al 1993), EF-Tu (Jurnak 1985), and H-Ras (Pai et al 1989). An unfavorable interaction is proposed to occur between Asp604 and a pair of unshared electrons at the N1 position of cAMP. These substitutions at Asp604 affect the maximal response (efficacy) more than the affinity, suggesting that this residue may not play a role in the initial contact with cGMP but may instead contribute to the allosteric stabilization of the nucleotide in the open state of the channel (see below).

## MOLECULAR MECHANISM FOR ACTIVATION BY CYCLIC NUCLEOTIDES

Although they are homologous to the voltage-gated channels, both retinal and olfactory CNG channels are activated by the binding of cyclic nucleotides and not by voltage. However, in the presence of cyclic nucleotides, channel opening does exhibit a weak dependence on membrane voltage. Several groups have reported that depolarization increases channel opening and causes a small decrease in the $K_{1/2}$ for cyclic nucleotides (see Figure 5) (Yau & Baylor 1989, Lynch & Lindemann 1994).

### Kinetic Models of Gating

The gating kinetics of CNG channels and the basis for their voltage dependence have been studied by several investigators. Karpen et al (1988) showed that in the native channel from rod photoreceptors, with fast jumps in cGMP concentration, the speed of the activation process increases with the final cGMP concentration and is close to the diffusion-controlled limit. At high cGMP concentrations, the rate of opening reached a maximal value, owing to

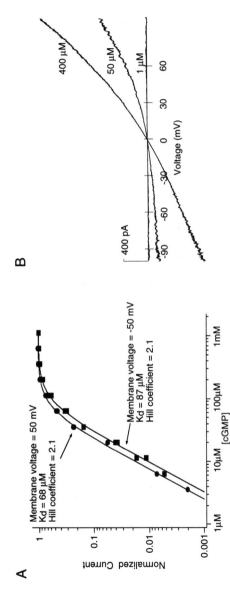

*Figure 5*  Dependence of the activation of the rod CNG channel on the concentration of cGMP. (*A*) Dose-response relations for the rod CNG channel at –50 mV (*squares*) and +50 mV (*circles*). Each set of data has been fit with the Hill equation with the indicated parameters. (*B*) Current through rod CNG channels during voltage ramps from –100 to +100 mV at the indicated cGMP concentrations. The current in the absence of cGMP has been subtracted from these traces.

gating transitions after ligand binding. At a fixed concentration of cGMP, the opening of the channels could also be increased by depolarization, although the maximum voltage sensitivity (roughly an e-fold increase in conductance for a 120-mV depolarization) is much less than the sensitivity of voltage-gated channels. A similarly low voltage sensitivity has been reported for olfactory CNG channels (Goulding et al 1992). Karpen et al described their results by using a model in which three diffusion-controlled ligand-binding steps are followed by a rapid, weakly voltage-dependent closed-open transition of the fully liganded channel (in which depolarization favors the open state), as shown in the scheme below.

$$C_1 \underset{}{\overset{cGMP}{\rightleftharpoons}} C_2 \underset{}{\overset{cGMP}{\rightleftharpoons}} C_3 \underset{}{\overset{cGMP}{\rightleftharpoons}} C_4 \underset{}{\overset{\Delta V}{\rightleftharpoons}} O$$

The model accounts for the steady-state dose-response relations at different voltages (Figure 5A). In particular, it correctly predicts a voltage-dependent apparent affinity and a Hill coefficient greater than one due to the requirement that three molecules of cGMP must bind before the channel can open. In addition, the model predicts an apparent cooperativity in the binding of cGMP due largely to a final closed-open transition that strongly favors the open state. This mechanism also provides a simple explanation for the change in the voltage-dependence of channel opening seen with different cGMP concentrations (Figure 5B). At saturating concentrations of cGMP, the steady-state current-voltage relation is nearly linear because the channels are open with a very high probability even at negative voltage. Thus, any depolarization-induced shift in the C4-to-O reaction towards the open state can produce only a very small increase in the open probability. However, at low concentrations of cGMP, when the open probability is low, voltage-dependent changes in the C4-to-O equilibrium can produce much larger changes in the open probability. This makes the steady-state current-voltage relation rectify much more strongly at low cGMP concentrations. A similar effect is expected for mutations that destabilize the C4-to-O transition.

This mechanism is supported by studies of the behavior of single channels (Matthews & Watanabe 1987, 1988; Haynes & Yau 1990; Taylor & Baylor 1995). At low or moderate concentrations of cGMP, the single-channel openings occur in short bursts separated by relatively long closed durations. The open and closed durations within a burst have been shown to be independent of cGMP concentration, consistent with a final closed-open transition. Furthermore the duration of the closed interval between bursts decreases with increasing cGMP concentration, which is as expected if these closures represent partially liganded channels. However, the model does not predict the occurrence of multiple single-channel conductance levels that are generally observed in single-channel recordings from rod cGMP-activated channels (Haynes et al 1986, Haynes & Yau 1990, Ildefonse & Bennett 1991, Taylor & Baylor 1995, Zimmerman & Baylor 1986). Recently Taylor & Baylor (1995) have shown that a major subconductance level is prominent at low cGMP concentrations but not at high cGMP concentrations, suggesting that the subconductance state may represent a partially liganded form of the channel. Ildefonse & Bennett have reported the occurrence of as many as four different conductance levels in rod channels reconstituted into planer lipid bilayers, possibly corresponding to channels containing one to four ligands (Ildefonse & Bennett 1991).

Single olfactory CNG channel kinetics exhibit broad similarities, but also important differences, with retinal channel kinetics. The native salamander channel shows both brief openings and prolonged bursts, with a mean open time of ~1.5 ms (Zufall et al 1991). The cloned catfish channel shows a burst duration of around 5 ms in response to a low concentration of cAMP (10 µM) that increases up to 10 ms at high concentrations of cAMP (100 µM) (Goulding et al 1992). In contrast to the retinal channel, the olfactory channel subconductance state does not depend on cyclic nucleotide concentration (Goulding et al 1992). One striking feature of the olfactory channel is that it exhibits very prolonged bursts of openings that outlast brief periods of cAMP application by several hundred milliseconds, which may underlie prolonged responses to brief applications of odorants (Zufall et al 1993).

Recent evidence suggests that the CNG channels can open spontaneously in the absence of cyclic nucleotides (Picones & Korenbrot 1995b, Tibbs et al 1995). This suggests that CNG channel activation may best be described by an allosteric model proposed by Stryer (1987).

$$C_1 \overset{cGMP}{\rightleftharpoons} C_2 \overset{cGMP}{\rightleftharpoons} C_3 \overset{cGMP}{\rightleftharpoons} C_4 \overset{cGMP}{\rightleftharpoons} C_5$$

$$\updownarrow \qquad \updownarrow \qquad \updownarrow \qquad \updownarrow \qquad \updownarrow$$

$$O_1 \overset{cGMP}{\rightleftharpoons} O_2 \overset{cGMP}{\rightleftharpoons} O_3 \overset{cGMP}{\rightleftharpoons} O_4 \overset{cGMP}{\rightleftharpoons} O_5$$

For this scheme, the channel is considered to exist in two quaternary structures, closed (C) and open (O). The affinity of the open channel for cGMP is much greater than the affinity of the closed channel for cGMP. The concerted conformational change between these structures is promoted by the binding of cGMP to each subunit. The greater the number of cGMP molecules that bind, the more stable the concerted conformational change. Within each structure, the binding of cGMP to each subunit is considered to be independent of the number of other bound subunits. This model is derived from a more general model for allosteric transitions in proteins proposed by Monod, Wyman, and

Changeux (Monod et al 1965) for the binding of oxygen to hemoglobin. It is intriguing because it provides a simple molecular explanation for the observed Hill coefficient for cyclic nucleotides, binding cooperativity, and the multiple open states. In addition, the model contains four cyclic nucleotide–binding sites, consistent with a tetrameric structure of the channel. Researchers have recently shown this model to be consistent with the steady-state behavior of rod and olfactory channels (Goulding et al 1994, Ildefonse & Bennett 1991).

## Domains of the Channel Involved in the Allosteric Conformational Change

The above kinetic schemes have proved useful in determining the molecular events that underlie the gating transitions and modulation of gating. As indicated earlier, the X-ray crystal structure of CAP provides a good candidate structure for cyclic nucleotide binding in the CNG channels. However, as in any allosteric enzyme, the CNG channel–binding site must exist in at least two different conformations: one that binds cyclic nucleotide relatively weakly when the channel is closed and one that binds more tightly when the channel is open. Varnum et al (1995) suggested a molecular mechanism for the conformational change in the binding site during the opening allosteric transition in rod CNG channels. In this mechanism, the cyclic nucleotide binds to the closed channel primarily through interactions between the β-roll structure and the ribose and exo-cyclic phosphate of the cyclic nucleotide. The opening allosteric conformational change is then thought to involve a movement of the β roll relative to the Chelix, thereby permitting the formation of a pair of hydrogen bonds between the carboxylate of D604 and the N1 and N2 of the guanine ring (Figure 4B). This mechanism is supported by evidence that the C-helix region confers cyclic nucleotide specificity in bovine rod and catfish olfactory CNG channels (Goulding et al 1994) and that mutations of D604 in the C helix produce a dramatic effect on the stability of the allosteric transition in a cyclic nucleotide–dependent way (Varnum et al 1995). Weber & Steitz (1987) proposed a similar mechanism for the conformational change in CAP that permits binding to DNA.

The mechanism for how the conformational change in the cyclic nucleotide–binding site is coupled to channel opening is unknown. However several regions of the channel have been suggested to be involved. Alterations in the amino-terminal region are particularly influential on the allosteric transition (Chen & Yau 1994, Goulding et al 1994, Gordon & Zagotta 1995b). The rod and olfactory CNG channels differ markedly in the free energy of their allosteric opening transition; opening in the olfactory channels is much more energetically favorable than opening in the rod channels. Chimeric channels in which the amino-terminal domain of the olfactory channel was replaced by a rod amino-terminal domain exhibit much less favorable opening; conversely, rod channels with an olfactory amino-terminal domain open much more favorably (Goulding et al 1994, Gordon & Zagotta 1995b). Substitution into the rod channel of a region from the olfactory channel beginning midway along the amino-terminal region and extending through the S2-S3 linker (N-S2 domain) caused the opening transition of the chimeric rod channel to become even more favorable, similar to the opening transition of the olfactory channel (Goulding et al 1994). The amino-terminal domain and S1 segments have been implicated in the process of subunit assembly in voltage-gated $K^+$ channels (Li et al 1992, Shen et al 1993, Babila et al 1994). These results led Goulding et al (1994) to suggest that channel opening may involve a change in subunit-subunit orientation, mediated by a change in the strength of intersubunit bonds between N-S2 domains. Such a change in quaternary structure is similar to the allosteric conformational change in hemoglobin (Perutz 1990) and the gating of gap junction channels (Unwin & Ennis 1984).

## MOLECULAR MECHANISM OF CHANNEL MODULATION

Recent studies have shown that the amino-terminal region of the CNG channels is also important for the modulation of olfactory CNG channels by $Ca^{2+}$-calmodulin (Chen & Yau 1994). $Ca^{2+}$-calmodulin causes a decrease in the apparent cyclic nucleotide affinity of both rod (Hsu & Molday 1993, Chen et al 1994, Hsu & Molday 1994, Gordon et al 1995b) and olfactory (Chen et al 1994, Liu et al 1994) channels. This modulation occurs by the direct binding of $Ca^{2+}$-calmodulin to the channels and does not involve the action of a kinase. Chen et al (1994) have shown that $Ca^{2+}$-calmodulin binds to a specific domain on the amino-terminal region of the olfactory channel (see Figure 1). Furthermore, they have shown that the decrease in apparent cyclic nucleotide affinity arises from a decrease in the stability of the allosteric opening transition, consistent with the effect of mutations in the amino-terminal region discussed above. The modulation of the rod channel also involves $Ca^{2+}$-calmodulin binding, but here the binding is to the $\beta$ subunit and not to the $\alpha$ subunit (Chen et al 1994).

The gating properties of the CNG channels can also be modulated by various transition metal divalent cations including $Ni^{2+}$, $Zn^{2+}$, $Cd^{2+}$, $Co^{2+}$, and $Mn^{2+}$ (Ildefonse & Bennett 1991; Karpen et al 1993; Gordon & Zagotta 1995a,b). For example, a low concentration of intracellular $Ni^{2+}$ (<1 $\mu$M) potentiates the response of rod CNG channels. This potentiation is manifested as a dramatic increase in the apparent affinity for cGMP and cAMP and an almost 50-fold increase in the maximal current observed with cAMP. These effects can be explained by a mechanism in which $Ni^{2+}$ binds to the channel primarily when it is open, thereby promoting the allosteric conformational change induced by cyclic nucleotide binding (Gordon & Zagotta 1995a). In the olfactory CNG channels, $Ni^{2+}$ produces the opposite effect, an inhibition, that can be explained by a mechanism in which $Ni^{2+}$ primarily binds to the channel when it is closed (Gordon & Zagotta 1995b). Using chimeras between the rod and olfactory channels, Gordon & Zagotta (1995b) localized the binding site for the potentiating effect of $Ni^{2+}$ on rod channels to a single histidine residue (H420) that is located just past the S6 transmembrane domain and should lie near the intracellular mouth of the rod channel (Figure 1). They localized the binding site for the inhibitory effect of $Ni^{2+}$ on the olfactory channel to a single histidine residue (H396) at a position just three amino acids downstream from the homologous potentiation site in the rod channel. Finding two residues located only three amino acids apart in the primary sequence, with opposite state dependence to their binding, suggested that this region of the channel, which links the transmembrane domains to the binding domain of the carboxy terminus, undergoes a movement during the opening transition (Gordon & Zagotta 1995a,b).

## CONCLUSION

Several recent studies have defined the role of different domains involved in gating and ion permeation in CNG channels. Several lines of evidence suggest that channel opening involves an allosteric change in the cyclic nucleotide–binding domain, which is coupled to a conformational change that opens the channel. Regions in the amino terminus and transmembrane domain of the channel determine the free energy change associated with gating, whereas regions in the carboxy terminus are associated with ligand binding. These domains are linked by a third region that follows the S6 domain, which also changes its structure during gating. In the binding domain, the C helix has been implicated in both cyclic nucleotide recognition and channel activation. The P region connecting the S5 and S6 segments forms the pore and ionic selectivity filter of the CNG channels, similar to its role in the voltage-gated channels.

The similarity in structure between the voltage-gated and cyclic nucleotide–gated channels raises some interesting questions. Does an allosteric conformational change also open the voltage-gated channels? Why are the CNG channels so insensitive to voltage? How can similar P regions give rise to highly K-selective or -nonselective pores? Finally, what is the precise mechanism coupling a stimulus, whether it be voltage or ligand binding, to gating?

ACKNOWLEDGMENTS

We thank Eric O'Dell for helping to prepare the manuscript and Gareth Tibbs for his helpful comments. We also thank many authors for sending us reprints and preprints of unpublished work. WNZ and SAS are supported by the Howard Hughes Medical Institute.

Any *Annual Review* chapter, as well as any article cited in an *Annual Review* chapter, may be purchased from the Annual Reviews Preprints and Reprints service. 1-800-347-8007; 415-259-5017; email: arpr@class.org

## Literature Cited

Ahmad I, Redmond LJ, Barnstable CJ. 1990. Developmental and tissue-specific expression of the rod photoreceptor cGMP-gated ion channel gene. *Biochem. Biophys. Res. Commun.* 173:463–70

Almers W, McCleskey EW. 1984. Nonselective conductance in calcium channels of frog muscle: calcium selectivity in a single-file pore. *J. Physiol.* 353:585–608

Altenhofen W, Ludwig J, Eismann E, Kraus W, Bönigk W, Kaupp UB. 1991. Control of ligand specificity in cyclic nucleotide-gated channels from rod photoreceptors and olfactory epithelium. *Proc. Natl. Acad. Sci. USA* 88:9868–72

Anderson JA, Huprikar SS, Kochian LV, Lucas WJ, Gaber RF. 1992. Functional expression of a probable *Arabidopsis thaliana* potassium channel in *Saccharomyces cerevisiae*. *Proc. Natl. Acad. Sci. USA* 89:3736–40

Babila T, Moscucci A, Wang H, Weaver FE, Koren G. 1994. Assembly of mammalian voltage-gated potassium channels: evidence for an important role of the first transmembrane segment. *Neuron* 12:615–26

Baumann A, Frings S, Godde M, Seifert R, Kaupp UB. 1994. Primary structure and functional expression of a *Drosophila* cyclic nucleotide-gated channel present in eyes and antennae. *EMBO J.* 13:5040–50

Baylor DA, Lamb TD, Yau K-W. 1979. Responses of retinal rods to single photons. *J. Physiol.* 288:613

Biel M, Altenhofen W, Hullin R, Ludwig J, Freichel M, et al. 1993. Primary structure and functional expression of a cyclic nucleotide-gated channel from rabbit aorta. *FEBS Lett.* 329:134–38

Biel M, Zong X, Distler M, Bosse E, Klugbauer N, et al. 1994. Another member of the cyclic nucleotide-gated channel family, expressed in testis, kidney, and heart. *Proc. Natl. Acad. Sci. USA* 91:3505–9

Bönigk W, Altenhofen W, Muller F, Dose A, Illing M, et al. 1993. Rod and cone photoreceptor cells express distinct genes for cGMP-gated channels. *Neuron* 10:865–77

Bradley J, Li J, Davidson N, Lester HA, Zinn K. 1994. Heteromeric olfactory cyclic nucleotide-gated channels: a subunit that confers increased sensitivity to cAMP. *Proc. Natl. Acad. Sci. USA* 91:8890–94

Brown RL, Bert RJ, Evans FE, Karpen JW. 1993. Activation of retinal rod cGMP-gated channels: What makes for an effective 8-substituted derivative of cGMP? *Biochemistry* 32:10089–95

Brown LR, Gramling R, Bert RJ, Karpen JW. 1995. Cyclic GMP contact point within the 63-kDa subunit and a 240-kDa subunit protein of retinal rod cGMP-activated channels. *Biochemistry.* In press

Bruggemann A, Pardo LA, Stuhmer W, Pongs O. 1993. Ether-à-go-go encodes a voltage-gated channel permeable to $K^+$ and $Ca^{2+}$ and modulated by cAMP. *Nature* 365:445–48

Chen TY, Illing M, Molday LL, Hsu YT, Yau KW, Molday RS. 1994. Subunit 2 (or beta) of retinal rod cGMP-gated cation channel is a component of the 240-kDa channel-associated protein and mediates $Ca^{2+}$-calmodulin modulation. *Proc. Natl. Acad. Sci. USA* 91:11757–61

Chen TY, Peng YW, Dhallan RS, Ahamed B, Reed RR, Yau KW. 1993. A new subunit of the cyclic nucleotide-gated cation channel in retinal rods. *Nature* 362:764–67

Chen TY, Yau KW. 1994. Direct modulation by $Ca^{2+}$-calmodulin of cyclic nucleotide-activated channel of rat olfactory receptor neurons. *Nature* 368:545–48

Choi KL, Mossman C, Aube J, Yellen G. 1993. The internal quaternary ammonium receptor site of *Shaker* potassium channels. *Neuron* 10:533–41

Colamartino G, Menini A, Torre V. 1991. Blockage and permeation of divalent cations through the cyclic GMP-activated channel from tiger salamander retinal rods. *J. Physiol.* 440:189–206

Cowan SW, Schirmer T, Rummel G, Steiert M, Ghosh R, et al. 1992. Crystal structures explain functional properties of two *E. coli* porins. *Nature* 358:727–33

Dhallan RS, Macke JP, Eddy RL, Shows TB, Reed RR, et al. 1992. Human rod photoreceptor cGMP-gated channel: amino acid sequence, gene structure, and functional expression. *J. Neurosci.* 12:3248–56

Dhallan RS, Yau KW, Schrader KA, Reed RR. 1990. Primary structure and functional expression of a cyclic nucleotide-activated channel from olfactory neurons. *Nature* 347:184–87

DiFrancesco D, Tortora P. 1991. Direct activation of cardiac pacemaker channels by intracellular cyclic AMP. *Nature* 351:145–47

Distler M, Biel M, Flockerzi V, Hofmann F. 1994. Expression of cyclic nucleotide-gated cation channels in non-sensory tissues and cells. *Neuropharmacology* 33:1275–82

Eismann E, Müller F, Heinemann SH, Kaupp UB. 1994. A single negative charge within the pore region of a cGMP-gated channel controls rectification, $Ca^{2+}$ blockage, and ionic selectivity. *Proc. Natl. Acad. Sci. USA* 91:1109–13

Fesenko EE, Kolesnikov SS, Lyubarsky AL. 1985. Induction by cyclic GMP of cationic conductance in plasma membrane of retinal rod outer segment. *Nature* 313:310–13

Frech G, VanDongen A, Shuster G, Brown AM, Joho RH. 1989. *Nature* 340:642–45

Frech GC, VanDongen AM, Shuster G, Brown AM, Joho RH. 1989. A novel potassium channel with delayed rectifier properties isolated from rat brain by expression cloning. *Nature* 340:642–45

Frings S, Lynch JW, Lindemann B. 1992. Properties of cyclic nucleotide-gated channels mediating olfactory transduction: activation, selectivity and blockage. *J. Gen. Physiol.* 100:45–67

Furman RE, Tanaka JC. 1990. Monovalent selectivity of the cyclic guanosine monophosphate-activated ion channel. *J. Gen. Physiol.* 96:57–82

Gordon SE, Brautigan DL, Zimmerman AL. 1992. Protein phosphatases modulate the apparent agonist affinity of the light-regulated ion channel in retinal rods. *Neuron* 9:739–48

Gordon SE, Downing-Park J, Tam B, Zimmerman AL. 1995a. Diacylglycerol analogs inhibit the rod cGMP-gated channel by a phosphorylation-independent mechanism. *Biophys. J.* 69:409–17

Gordon SE, Downing-Park J, Zimmerman AL. 1995b. Modulation of the cGMP-gated channel in frog rods by calmodulin and an endogenous inhibitory factor. *J. Physiol.* In press

Gordon SE, Zagotta WN. 1995a. A histidine residue associated with the gate of the cyclic nucleotide-activated channels in rod photoreceptors. *Neuron* 14:177–83

Gordon SE, Zagotta WN. 1995b. Localization of regions affecting an allosteric transition in cyclic nucleotide-activated channels. *Neuron* 14:857–64

Goulding EH, Ngai J, Kramer RH, Colicos S,

Axel R, et al. 1992. Molecular cloning and single-channel properties of the cyclic nucleotide-gated channel from catfish olfactory neurons. *Neuron* 8:45–58

Goulding EH, Tibbs GR, Liu D, Siegelbaum SA. 1993. Role of H5 domain in determining pore diameter and ion permeation through cyclic nucleotide-gated channels. *Nature* 364:61–64

Goulding EH, Tibbs GR, Siegelbaum SA. 1994. Molecular mechanism of cyclic-nucleotide-gated channel activation. *Nature* 372:369–74

Guy HR, Durell SR, Warmke J, Drysdale R, Ganetzky B. 1991. Similarities in amino acid sequences of *Drosophila eag* and cyclic nucleotide-gated channels. *Science* 254:730

Hanke W, Cook NJ, Kaupp UB, Kaupp UB. 1988. cGMP-dependent channel protein from photoreceptor membranes: single channel activity of the purified and reconstituted protein. *Proc. Natl. Acad. Sci. USA* 85:94–98

Hartmann HA, Kirsch GE, Drewe JA, Taglialatela M, Joho RH, Brown AM. 1991. Exchange of conduction pathways between two related $K^+$ channels. *Science* 251:942–44

Haynes LW, Kay AR, Yau KW. 1986. Single cyclic GMP-activated channel activity in excised patches of rod outer segment membrane. *Nature* 321:66–70

Haynes LW, Yau K-W. 1990. Single-channel measurement from the cyclic GMP-activated conductance of catfish retinal cones. *J. Physiol.* 429:451–81

Heginbotham L, Abramson T, MacKinnon R. 1992. A functional connection between the pores of distantly related ion channels as revealed by mutant $K^+$ channels. *Science* 258:1152–55

Heinemann SH, Terlau H, Stühmer W, Imoto K, Numa S. 1992. Calcium channel characteristics conferred on the sodium channel by single mutations. *Nature* 356:441–43

Hess P, Tsien RW. 1984. Mechanism of ion permeation through calcium channels. *Nature* 309:453–56

Hoshi T. 1995. Regulation of voltage-dependence of the KAT1 channel by intracellular factors. *J. Gen. Physiol.* 105:309–28

Hsu YT, Molday RS. 1993. Modulation of the cGMP-gated channel of rod photoreceptor cells by calmodulin. *Nature* 361:76–79

Hsu YT, Molday RS. 1994. Interaction of calmodulin with the cyclic GMP-gated channel of rod photoreceptor cells. Modulation of activity, affinity purification, and localization. *J. Biol. Chem.* 269:29765–70

Ildefonse M, Bennett N. 1991. Single-channel study of the cGMP-dependent conductance of retinal rods from incorporation of native vesicles into planar lipid bilayers. *J. Membr. Biol.* 123:133–47

Isacoff EY, Jan Y-N, Jan LY. 1991. Putative

receptor for the cytoplasmic gate in the *Shaker* K+ channel. *Nature* 353:86–90

Jan LY, Jan YN. 1990. A superfamily of ion channels. *Nature* 345:672

Jurnak F. 1985. Structure of the GDP domain of EF-Ti and location of the amino acids homologous to *ras* oncogene proteins. *Science* 230:32–36

Karpen JW, Brown RL, Stryer L, Baylor DA. 1993. Interactions between divalent cations and the gating machinery of cyclic GMP-activated channels in salamander retinal rods. *J. Gen. Physiol.* 101:1–25

Karpen JW, Zimmerman AL, Stryer L, Baylor DA. 1988. Gating kinetics of the cyclic-GMP-activated channel of retinal rods: flash photolysis and voltage-jump studies. *Proc. Natl. Acad. Sci. USA* 85:1287–91

Kaupp UB, Niidome T, Tanabe T, Terada S, Bönigk W, et al. 1989. Primary structure and functional expression from complementary DNA of the rod photoreceptor cyclic GMP-gated channel. *Nature* 342:762–66

Kirsch GE, Shieh CC, Drewe JA, Vener DF, Brown AM. 1993. Segmental exchanges define 4-aminopyridine binding and the inner mouth of K+ pores. *Neuron* 11:503–12

Kleene SJ. 1993. Origin of the chloride current in olfactory transduction. *Neuron* 11:123–32

Kolesnikov SS, Zhainazarov AB, Kosolapov AV. 1990 Cyclic nucleotide-activated channels in the frog olfactory receptor plasma membrane. *FEBS Lett.* 266:96–98

Kramer RH, Goulding E, Siegelbaum SA. 1994. Potassium channel inactivation peptide blocks cyclic nucleotide-gated channels by binding to the conserved pore domain. *Neuron* 12:655–62

Kramer RH, Siegelbaum SA. 1992. Intracellular Ca2+ regulates the sensitivity of cyclic nucleotide-gated channels in olfactory receptor neurons. *Neuron* 9:897–906

Kramer RH, Tibbs GR. 1995. Molecular determinants of agonist and antagonist action on cyclic nucleotide-gated ion channels. *Biophys. J.* 68:A253 (Abstr.)

Kumar VD, Weber IT. 1992. Molecular model of the cyclic GMP-binding domain of the cyclic GMP-gated ion channel. *Biochemistry* 31:4643–49

Kurahashi T, Kaneko A. 1993. Gating properties of the cAMP-gated channel in toad olfactory receptor cells. *J. Physiol.* 466: 287–302

Kurahashi T, Yau K-W. 1993. Co-existence of cationic and chloride components in odorant-induced current of vertebrate olfactory receptor cells. *Nature* 363:71–74

Lancet D. 1986. Vertebrate olfactory reception. *Annu. Rev. Neurosci.* 9:329–55

Li M, Jan YN, Jan LY. 1992. Specification of subunit assembly by the hydrophilic amino-terminal domain of the *Shaker* potassium channel. *Science* 257:1225–30

Liman ER, Buck LB. 1994. A second subunit of the olfactory cyclic nucleotide-gated channel confers high sensitivity to cAMP. *Neuron* 13:611–21

Liu M, Chen TY, Ahamed B, Li J, Yau KW. 1994. Calcium-calmodulin modulation of the olfactory cyclic nucleotide-gated cation channel. *Science* 266:1348–54

Lopez GA, Jan YN, Jan LY. 1994. Evidence that the S6 segment of the *Shaker* voltage-gated K+ channel comprises part of the pore. *Nature* 367:179–82

Lowe G, Gold GH. 1993. Nonlinear amplification by calcium-dependent chloride channels in olfactory receptor cells. *Nature* 366:283–86

Ludwig J, Margalit T, Eismann E, Lancet D, Kaupp UB. 1990. Primary structure of cAMP-gated channel from bovine olfactory epithelium. *FEBS Lett.* 270:24–29

Ludwig J, Terlau H, Wunder F, Bruggemann A, Pardo LA, et al. 1994. Functional expression of a rat homologue of the voltage gated ether à go-go potassium channel reveals differences in selectivity and activation kinetics between the *Drosophila* channel and its mammalian counterpart. *EMBO J.* 13:4451–58

Lynch JW, Lindemann B. 1994. Cyclic nucleotide-gated channels of rat olfactory receptor cells: divalent cations control the sensitivity to cAMP. *J. Gen. Physiol.* 103: 87–106

MacKinnon R. 1991. Determination of the subunit stoichiometry of a voltage-activated potassium channel. *Nature* 350:232–35

Marunaka Y, Ohara A, Matsumoto P, Eaton DC. 1991. Cyclic GMP-activated channel activity in renal epithelial cells (A6). *Biochim. Biophys. Acta* 1070:152–56

Matthews G, Watanabe S. 1987. Properties of ion channels closed by light and opened by guanosine 3′,5′-cyclic monophosphate in toad retinal rods. *J. Physiol.* 389:691–715

Matthews G, Watanabe S. 1988. Activation of single ion channels from toad retinal rod inner segments by cyclic GMP: concentration dependence. *J. Physiol.* 403:389–405

Matthews HR, Murphy RL, Fain GL, Lamb TD. 1988. Photoreceptor light adaptation is mediated by cytoplasmic calcium concentration. *Nature* 334:67–69

Menini A. 1990. Currents carried by monovalent cations through cyclic GMP-activated channels in excised patches from salamander rods. *J. Physiol.* 424:167–85

Miller C. 1991. 1990: annus mirabilis of potassium channels. *Science* 252:1092–96

Molday RS, Molday LL, Dose A, Clark LI, Illing M, et al. 1991. The cGMP-gated channel of the rod photoreceptor cell: characterization and orientation of the amino terminus. *J. Biol. Chem.* 266:21917–22

Monod J, Wyman J, Changeux JP. 1965. On the nature of allosteric transitions: a plausible model. *J. Mol. Biol.* 12:88–118

Nakamura T, Gold GH. 1987. A cyclic nucleotide-gated conductance in olfactory receptor cilia. *Nature* 325:442–44

Nakatani K, Yau KW. 1988. Calcium and light adaptation in retinal rods and cones. *Nature* 334:69–71

Noel JP, Hamm H, Sigler P. 1993. The 2.2 Å structure of transducin-α complexed with GTP-γ-S. *Nature* 366:654–63

Pai EF, Kabsch W, Krengel U, Holmes K, Jophn J, Wittinghofer A. 1989. Structure of the guanine-nucleotide-binding domain of the Ha-*ras* oncogene product p21 in the triphosphate conformation. *Nature* 341:209–14

Perutz MF. 1990. Haemoglobin. Molecular inventiveness. *Nature* 348:583–84

Picco C, Menini A. 1993. The permeability of the cGMP-activated channel to organic cations in retinal rods of the tiger salamander. *J. Physiol.* 460:741–58

Picones A, Korenbrot JI. 1992. Permeation and interaction of monovalent cations with the cGMP-gated channel of cone photoreceptors. *J. Gen. Physiol.* 100:647–73

Picones A, Korenbrot JI. 1995a. Permeability and interaction of Ca++ with cGMP-gated ion channels differ in retinal rod and cone photoreceptors. *Biophys. J.* 69:120–27

Picones A, Korenbrot JI. 1995b. Spontaneous, ligand-independent activity of the cGMP-gated ion channels in cone photoreceptors of fish. *J. Physiol.* 485:699–714

Pietrobon D, Prod B, Hess P. 1988. Conformational changes associated with ion permeation in L-type calcium channels. *Nature* 333:373–76

Prod'hom B, Pietrobon D, Hess P. 1987. Direct measurement of proton transfer rates to a group controlling the dihydropyridine-sensitive Ca²⁺ channel. *Nature* 329:243–46

Rieke F, Schwartz EA. 1994. A cGMP-gated current can control exocytosis at cone synapses. *Neuron* 13:863–73

Root MJ, MacKinnon R. 1993. Identification of an external divalent cation-binding site in the pore of a cGMP-activated channel. *Neuron* 11:459–66

Root MJ, MacKinnon R. 1994. Two identical noninteracting sites in an ion channel revealed by proton transfer. *Science* 265:1852–56

Scott SP, Tanaka JC. 1995. Molecular interactions of 3′,5′-cyclic purine analogues with the binding site of retinal rod ion channels. *Biochemistry* 34:2338–47

Sesti F, Eismann E, Kaupp UB, Nizzari M, Torre V. 1995. The multi-ion nature of the cGMP-gated channel from vertebrate rods. *J. Physiol.* In press

Sesti F, Straforini M, Lamb TD, Torre V. 1994. Gating, selectivity and blockage of single channels activated by cyclic GMP in retinal rods of the tiger salamander. *J. Physiol.* 474:203–22

Shabb JB, Corbin JD. 1992. Cyclic nucleotide-binding domains in proteins having diverse functions. *J. Biol. Chem.* 267:5723–26

Shabb JB, Ng L, Corbin JD. 1990. One amino acid change produces a high affinity cGMP-binding site in cAMP-dependent protein kinase. *J. Biol. Chem.* 265:16031–34

Shen NV, Chen X, Boyer MM, Pfaffinger PJ. 1993. Deletion analysis of K⁺ channel assembly. *Neuron* 11:67–76

Slesinger PA, Jan YN, Jan LY. 1993. The S4-S5 loop contributes to the ion-selective pore of potassium channels. *Neuron* 11:739–49

Stern JH, Kaupp UB, MacLeish PR. 1986. Control of the light-regulated current in rod photoreceptors by cyclic GMP, calcium and *l-cis*-diltiazem. *Proc. Natl. Acad. Sci. USA* 83:1163–67

Stryer L. 1986. Cyclic GMP cascade of vision. *Annu. Rev. Neurosci.* 9:87–119

Stryer L. 1987. Visual transduction: design and recurring motifs. *Chem. Scr.* 27B:161–71

Takio K, Wade RD, Smith SB, Krebs EG, Walsh KA, Titani K. 1984. Guanosine cyclic 3′,5′-phosphate dependent protein kinase, a chimeric protein homologous with two separate protein families. *Biochemistry* 23:4207–18

Tanaka JC. 1993. The effects of protons on 3′,5′-cGMP-activated currents in photoreceptor patches. *Biophys. J.* 65:2517–23

Tanaka JC, Eccleston JF, Furman RE. 1989. Photoreceptor channel activation by nucleotide derivatives. *Biochemistry* 28:2776–84

Tanaka JC, Furman RE. 1993. Divalent effects on cGMP-activated currents in excised patches from amphibian photoreceptors. *J. Mem. Biol.* 131:245–56

Taylor WR, Baylor DA. 1995. Conductance and kinetics of single cGMP-activated channels in salamander rod outer segments. *J. Physiol.* 483:567–82

Tempel B, Papazian D, Schwarz T, Jan YN, Jan LY. 1987. Sequence of a probable potassium channel component encoded at *Shaker* locus of *Drosophila*. *Science* 237:770–75

Tibbs GR, Goulding EH, Siegelbaum SA. 1995. Spontaneous opening of cyclic nucleotide-gated channels supports an allosteric model of activation. *Biophys. J.* 68:A253 (Abstr.)

Titani K, Sasagawa T, Ericsson LH, Kumar S, Smith SB, et al. 1984. Amino acid sequence of the regulatory subunit of bovine type I adenosine cyclic 3′,5′-phosphate dependent protein kinase. *Biochemistry* 23:4193–99

Unwin PN, Ennis PD. 1984. Two configurations of a channel-forming membrane protein. *Nature* 307:609–13

Varnum MD, Black KD, Zagotta WN. 1995. Molecular mechanism for ligand discrimination of cyclic nucleotide-gated channels. *Neuron* 15:In press

Warmke J, Drysdale R, Ganetzky B. 1991. A distinct potassium channel polypeptide encoded by the *Drosophila eag* locus. *Science* 252:1560–62

Warmke JW, Ganetzky B. 1994. A family of potassium channel genes related to *eag* in *Drosophila* and mammals. *Proc. Natl. Acad. Sci. USA* 91:3438–42

Weber I, Shabb J, Corbin J. 1989. Predicted structures of cGMP-binding domains of the cGMP-dependent protein kinase: a key alanine/threonine difference in evolutionary divergence of cAMP and cGMP binding sites. *Biochemistry* 28:6122–27

Weber IT, Steitz TA. 1987. Structure of a complex of catabolite gene activator protein and cyclic AMP refined at 2.5 Å resolution. *J. Mol. Biol.* 198:311–26

Weyand I, Godde M, Frings S, Weiner J, Muller F, et al. 1994. Cloning and functional expression of a cyclic-nucleotide-gated channel from mammalian sperm. *Nature* 368:859–63

Wohlfart P, Haase W, Molday RS, Cook NJ. 1992. Antibodies against synthetic peptides used to determine the topology and site of glycosylation of the cGMP-gated channel from bovine rod photoreceptors. *J. Biol. Chem.* 267:644–48

Yang J, Ellinor PT, Sather WA, Zhang J-F, Tsien RW. 1993. Molecular determinants of $Ca^{2+}$ selectivity and ion permeation in L-type $Ca^{2+}$ channels. *Nature* 366:158–61

Yau K-W, Baylor DA. 1989. Cyclic GMP-activated conductance of retinal photoreceptor cells. *Annu. Rev. Neurosci.* 12:289–327

Yau K-W, Nakatani K. 1985. Light-induced reduction of cytoplasmic free calcium in retinal rod outer segment. *Nature* 313:579–82

Yellen G, Jurman ME, Abramson T, MacKinnon R. 1991. Mutations affecting internal TEA blockage identify the probable pore-forming region of a K channel. *Science* 251:939–44

Yool AJ, Schwarz TL. 1991. Alteration of ionic selectivity of a $K^+$ channel by mutation of the H5 region. *Nature* 349:700–4

Yool AJ, Schwarz TL. 1995. Interactions of the H5 pore region and hydroxylamine with N-type inactivation in the *Shaker* $K^+$ channel. *Biophys. J.* 68:448–58

Zagotta WN, Hoshi Y, Aldrich RW. 1990. Restoration of inactivation in mutants of *Shaker* potassium channels by a peptide derived from ShB. *Science* 250:568–71

Zimmerman AL, Baylor DA. 1986. Cyclic GMP-sensitive conductance of retinal rods consists of aqueous pores. *Nature* 321:70–72

Zimmerman AL, Baylor DA. 1992. Cation interactions within the cyclic GMP-activated channel of retinal rods form the tiger salamander. *J. Physiol.* 449:759–83

Zufall F, Firestein S. 1993. Divalent cations block the cyclic nucleotide-gated channel of olfactory receptor neurons. *J. Neurophys.* 69:1758–68

Zufall F, Firestein S, Shepherd GM. 1991. Analysis of single cyclic nucleotide-gated channels in olfactory receptor cells. *J. Neurosci.* 11:3573–80

Zufall F, Firestein S, Shepherd GM. 1994. Cyclic nucleotide-gated ion channels and sensory transduction in olfactory receptor neurons. *Annu. Rev. Biophys. Biomol. Struct.* 23:577–607

Zufall F, Hatt H, Firestein S. 1993. Rapid application and removal of second messengers to cyclic nucleotide-gated channels from olfactory epithelium. *Proc. Natl. Acad. Sci. USA* 90:9335–39

Zufall F, Shepherd GM, Firestein S. 1991. Inhibition of the olfactory cyclic nucleotide gated ion channel by intracellular calcium. *Proc. R. Soc. London Ser. B* 246:225–30

*Annu. Rev. Neurosci. 1996. 19:265–87*

# GENE TRANSFER TO NEURONS USING HERPES SIMPLEX VIRUS–BASED VECTORS

*D. J. Fink[1,2,3], N. A. DeLuca[2], W. F. Goins[2], and J. C. Glorioso[2]*

Departments of [1]Neurology and [2]Molecular Genetics & Biochemistry, University of Pittsburgh, Pittsburgh, Pennsylvania 15261, and [3]VA Medical Center, Pittsburgh, Pennsylvania 15240

KEY WORDS:   gene transfer, gene therapy, neurons

## ABSTRACT

One important outgrowth of molecular medicine is the development of technologies for the transfer of therapeutic genes to cells in culture and tissues in vivo, which promises to revolutionize both experimental biomedical science and the clinical practice of medicine. Fundamental obstacles must still be overcome to create safe and efficient gene delivery vectors specifically designed for individual tissue types, and special strategies will be required for direct in vivo gene transfer to neurons because these cells are postmitotic and cannot be removed for transduction.

Herpes simplex virus type 1 (HSV-1), a neurotropic virus that naturally establishes a latent state in neurons, has many unique features that make it suitable as a gene transfer vector for the nervous system. In this review we describe the molecular biology of HSV-1, strategies for reducing potential pathogenesis of the recombinant vector, and methods for expressing transgenes from the vector genome. Gene transfer experiments using recombinant HSV-1–based vectors and defective HSV-1 vectors (amplicons) for gene transfer are also described and evaluated in terms of efficiency and safety.

## INTRODUCTION

Advances in recombinant DNA technology have led directly to major advances in scientific understanding of pathologic processes and in our ability to intervene in preventing the progression of disease. Recombinant proteins created by genetic engineering can be used to study the effect of specific polypeptides on cell development or physiology, and transgenic animal models can be bred

265

to express (or to block expression of) gene products in vivo in order to study the role of those products in whole-animal biology. Emerging technologies now allow gene transfer to mature cells in vitro and in vivo, but application of this technology to neuroscience has been limited, principally because the methods typically employed allow the transfer of genes only into dividing cells and because the most practical approaches to gene delivery to the brain, for example, must involve direct gene transfer to this organ in vivo.

The most common methods of gene transfer employ retroviral vectors. In its natural life cycle, the RNA-containing retrovirus enters the cell, and the cDNA copy of the viral genome produced by its reverse transcriptase is incorporated into the cellular genome during mitosis. Defective retroviral vectors, created by deleting essential viral structural genes, are propagated in producer cell lines engineered to provide those essential gene products from the cellular genome. A foreign gene, along with appropriate promoter and enhancer elements and a selectable marker, can be inserted into the vector genome—which when packaged in the retroviral coat, remains capable of entering the nucleus and recombining into the chromosome of dividing cells in culture (Miller 1992). These incorporated genetic elements lacking structural genes are transcribed from the cellular genome to produce the foreign gene mRNA and protein, without the production of the viral structural genes or infectious viral particles. Although essential normal cell gene expression theoretically may be disrupted by insertional mutagenesis or, conversely, potentially harmful genes may be activated by the juxtaposition of the retroviral promoter elements to adjacent cellular genes, these are rare events. Transfer of gene products into dividing cells in culture by using retroviruses has proven a powerful tool (Anderson 1992, Miller 1992), which in practice is limited only by the ability to purify appropriately high titers of the vector and the availability of transducible cells for transplant. The inability of these vectors to infect nondividing cells, however, makes them inappropriate for postmitotic primary neurons in culture, or for direct gene transfer into neuronal tissue in vivo, and has led to the development of alternate vectors for such studies.

Adenovirus (AV) is a 38-kb DNA virus, which in natural infection enters cells independent of their stage in the cell cycle. Gene expression proceeds in a cascade fashion, with activation of early genes leading to expression of late genes and the production of infectious viral particles. Deletion of essential early genetic elements (typically E1a-E1b) leads to defective viruses that can be propagated in producer cell lines that provide the E1a-E1b transcription factor gene products in *trans*. Foreign genes inserted along with appropriate promoter and enhancer elements into the defective adenoviral genome can be expressed in a wide variety of cell types, including pulmonary epithelium, muscle, and cells of neuronal origin (Acsadi et al 1994, Akli et al 1993, Davidson et al 1993, Le Gal La Salle et al 1993, Quantin et al 1992, Ragot et al 1993, Vincent et al 1993, Yang et al 1994). AV can be propagated to high titers [$10^{10}$–$10^{12}$ plaque-forming units/ml (pfu/ml)] with a substantial amount of foreign DNA, up to 10 kb, inserted. The major limitation in the development of AV vectors is that despite deletion of the E1a-E1b genes, there is "leaky" expression of late viral genes, leading to cytopathic effects when infected at high multiplicity of infection (MOI) and, in vivo, the development of immune responses to those viral gene products (Yang et al 1994). The further development of these vectors will require the removal of additional essential genes and the creation of complementing cell lines in which the complementing viral genes are integrated into the cellular genome under control of inducible promoters due to toxicity of the viral gene products. The induction of the complementing gene on infection must be sufficient to propagate the defective vector to high titer. If problems related to vector backbone gene expression can be overcome, adenoviral vectors should prove to be useful for gene transfer to glial cells of the nervous system since the virus infects neurons poorly.

Adeno-associated viruses (AAVs) are naturally defective small DNA viruses that are known to be nonpathogenic. In natural infections AAV can enter cell nuclei and integrate in a site-specific manner into chromosome 19, although AAV vectors containing foreign gene sequences have not been found to exhibit this property. These viruses are difficult to grow to high titer using current technologies, but are capable of gene delivery and expression in nondividing cells, including neurons (Kaplitt et al 1994b). The long-term persistence and expression from these vectors has not yet been fully explored, but they may prove to be powerful tools in the expression of foreign genes in the brain. Their main limitations include their small packaging potential for foreign DNA and the lack of packaging cell lines necessitating the use of helper viruses for preparation of AAV stocks, which also contain the helper particles, and the difficulty associated with production of high titer stocks, since these viruses must be concentrated for use as a vector.

Our laboratories have focused on the development of herpes simplex virus type 1 (HSV-1)–based vectors for gene transfer into nondividing cells of neuronal origin, both in vitro and in vivo. HSV-1 is a double-stranded DNA virus that has a wide cellular host range but during natural infection is neurotropic. Natural viral infection begins in epithelial cells of the skin or mucous membrane, after which the virus is carried by retrograde axonal transport from the site of original inoculation to the neuronal perikaryon, where it may either begin a cycle of lytic replication or enter a latent state in which the viral genome persists without the expression of any viral proteins, possibly for the life of the host. Latently infected neurons function normally and are not rejected by the host immune response. Although latent genomes are capable of reactivating—under the influence of stresses such as hyperthermia, axotomy, or steroids—to reenter the lytic cycle and produce recurrent infection at the site of initial inoculation, understanding of the molecular biology of the HSV-1 has allowed the construction of vectors that are incapable of reactivation. In addition, vectors can be engineered that are incapable of initial replication in the epithelium or in neuronal cells but are nonetheless capable of establishing the latent state.

This review summarizes the salient features of the molecular biology of HSV-1, progress that has been made in curbing the potential toxicity of HSV-1 vectors, and observations on the use of HSV-1 vectors for transgene expression in neuronal cells in vitro and in vivo.

## MOLECULAR BIOLOGY OF HSV-1

Physically, HSV-1 is an enveloped particle containing an icosahedral-shaped capsid surrounded by a layer of proteins referred to as tegument (Figure 1A). The envelope contains approximately 12 glycoproteins [glycoproteins B (gB) through M (gM)], and initial attachment of the enveloped particle to the cell surface occurs through nonspecific charge interactions between at least three of these glycoproteins (gB, gC, and gD) and heparan sulfate moieties on the plasma membrane. This nonspecific step is followed by specific recognition of an unidentified cellular receptor, thought to be mediated by gD, and fusion of the virus envelope with the cell membrane requiring the essential glycoproteins gB, gD, and gH (for review, see Roizman & Sears 1990, Spear 1993). Virus capsids are internalized and transported to the nucleus along microtubules by retrograde axonal transport, a process that is complex and not completely understood. Following reactivation, capsids are transported in an anterograde direction along the same cytoskeletal framework (Stevens 1989). Following direct intracranial inoculation in experimental models, specific viral strains display a predilection for anterograde, retrograde, or bidirectional transport along axonal pathways within the brain (Bak et al 1977, McFarland et al 1986). Upon reaching the nucleus, viral DNA is released through a capsid penton into the nucleus via a nuclear pore to begin the eclipse phase—during which no viral particles can be detected by electron microscopy—of the replication cycle.

The HSV-1 genome is a linear double-stranded DNA molecule composed of two unique segments [unique long ($U_L$) and unique short ($U_S$)] each flanked by a pair of inverted repeat ($I_R$) segments (Figure 1B). More than 75 distinct genes are encoded in 152 kb of viral DNA, and they are expressed in a well-ordered temporal cascade of immediate early (IE or α), followed by early (E or β), and subsequently late (L or γ) gene products (Figure 2). The initial expression of IE genes does not require de novo viral protein synthesis but is enhanced by at least one tegument protein (referred to as VP16, Vmw65, or αTIF), which enters the nucleus along with the viral genome and, in collabo-

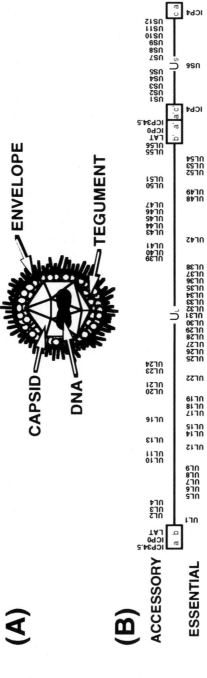

*Figure 1*   (*A*) Schematic illustration of the HSV-1 virion, showing the capsid, tegument, and glycoprotein-containing lipid envelope. (*B*) Schematic representation of the HSV-1 genome, showing the unique long ($U_L$) and unique short ($U_S$) segments, each bounded by inverted repeat ($I_R$) elements. The location of the essential genes, which are required for viral replication in vitro, and the nonessential, or accessory genes, which may be deleted without affecting replication in vitro, are indicated.

**(A)**

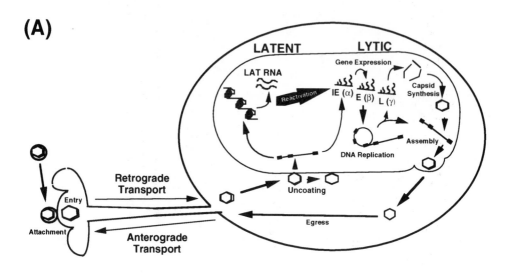

**(B)**

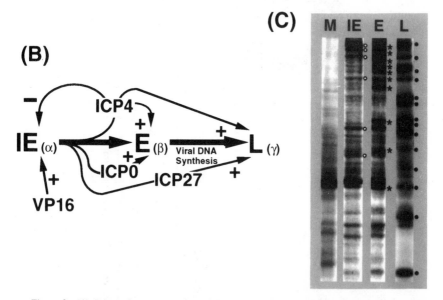

*Figure 2* (A) Schematic representation of the HSV-1 life cycle displaying the lytic and latent phases. (B) The HSV-1 lytic gene regulatory cascade. Expression of essential IE (α) genes is required for subsequent expression of E (β) and L (γ) class genes, but the establishment of latency does not require viral replication or IE class gene expression. (C) HSV-1 gene expression within the host cell during lytic infection. HSV-1–encoded polypeptides specific for different stages in the regulatory cascade are denoted.

ration with a cellular factor (octamer binding protein one or Oct 1), binds to a consensus TAATGARAT sequence in IE gene promoter enhancers (Gaffney et al 1985; Mackem & Roizman 1982a,b). Another tegument protein [virion host shut-off (vhs)] assists viral replication by degrading cellular mRNA and interfering with host cell protein synthesis (Oroskar & Read 1989).

There are five IE genes, a subset of which are responsible for directing a well-ordered temporal cascade of coordinated viral gene expression that culminates in viral replication (Honess & Roizman 1974). Two of these IE gene products, infected cell polypeptide 4 (ICP4) and ICP27, control the expression of E and L genes through both transcriptional and posttranscriptional mechanisms and are essential for viral replication (DeLuca & Schaffer 1985, Dixon & Schaffer 1980, Preston 1979a, 1979b, Sacks et al 1985, Watson & Clements 1980). ICP0 is also an activator of E and L genes, although it is not a DNA-binding protein, and has been demonstrated to be capable of transactivating a variety of other non-HSV promoters (Everett 1984, Gelman & Silverstein 1985, Quinlan & Knipe 1985, Sacks & Schaffer 1987, Stow & Stow 1986).

E gene products are primarily responsible for viral DNA synthesis and include the enzymes necessary to produce the appropriate nucleotide pools in nondividing cells [e.g. thymidine kinase (TK), ribonucleotide reductase (RR)] and the enzymes and DNA-binding proteins that are essential for DNA replication (e.g. DNA polymerase and origin-binding products). Viral DNA synthesis proceeds via a rolling circle mechanism that produces head-to-tail concatemers of the HSV-1 genome, during which process the $U_L$ and $U_S$ segments may independently invert their orientation by homologous recombination between $I_R$ flanking elements (Jacob et al 1979).

Expression of L gene products, which include many of the structural proteins of the capsid, tegument, and envelope, is triggered by the IE gene products ICP4 and ICP27 only after viral DNA synthesis has occurred. Viral DNA is cleaved into genome-length units and packaged into the capsid through the recognition of packaging sequences located in the $I_R$s and referred to as the "a" sequence. Viral tegument proteins are then added through a process that is as yet poorly understood, and the envelope is acquired as the viral particle (capsid and tegument) buds through a modified patch of nuclear membrane containing the viral glycoproteins. The lytic cycle is rapid, resulting in cell death in as little as 10 h in vitro.

Experimentally, viral replication in vitro is usually carried out in African Green Monkey kidney (Vero) cells, and the viral genes have been categorized according to whether they are required or not for the production of infectious viral particles in these highly permissive cell culture conditions (see Figure 1B). Nonessential or accessory genes contribute to the ability of the virus to effectively replicate and spread within the host in vivo. Two of the IE genes

(*ICP4* and *ICP27*) are essential, as are at least 20 structural genes (principally of the L class) and the 7 viral gene products involved in viral DNA replication. Recombinant vectors lacking essential genes can be propagated only in cells that provide the essential gene product from the cellular genome in *trans,* while viruses deleted in nonessential genes are easily propagated in unmodified Vero cells. Deletion of essential IE genes creates vectors that can be propagated in complementing cell lines (see below) yet are not capable of expressing E or L class genes in noncomplementing cells such as neurons. Inspection of the physical arrangement of essential and accessory genes in the genome, shown in Figure 1*B,* reveals that from the *UL53* gene to the right-hand terminal repeat sequence there are only three essential genes (*ICP4, ICP27,* and gD or *US6*), so it should be possible to accommodate up to 40–50 kb of foreign sequences in a modified HSV-1 genome after large-scale deletion of nonessential genes and complementation of these three essential genes. Few of the genes in the HSV-1 genome are spliced, so limited deletion of specific genetic elements can be accomplished precisely using the natural virus recombination machinery.

## HSV-1 LATENCY

In an alternate pathway to lytic infection, the virus may establish a life-long latent state in neurons (Figure 2). The persistence of HSV-1 genomes in the absence of infection was originally suspected because of the propensity of epithelial HSV-1 infections to recur in the same dermatomal distribution as the initial infection and because reactivated virus can be recovered from latently infected peripheral ganglia after explantation into tissue culture and cocultivation with Vero cells. No viral particles nor any detectable viral proteins are found in latently infected ganglia in vivo (Stevens 1989), although a single region of the HSV-1 genome does remain transcriptionally active, producing a family of latency-associated transcripts (LATs) (Croen et al 1987, Deatly et al 1988, Rock et al 1987, Spivack & Fraser 1987, Stevens et al 1987). The major LAT species is a 2-kb non-polyadenylated RNA that remains intranuclear and is transcribed from the repeat regions flanking the $U_L$ component of the genome (Figure 1*B*), with its last 723 bases overlapping and complementary to the 3' end of the *ICP0* gene, which is encoded off of the opposite strand. The 2-kb LAT appears to be a highly stable intron spliced from a large poly(A)$^+$ mRNA that initiates 28 bp downstream of a TATA element and extends 8.3 kb downstream to the nearest polyadenylation site (Devi-Rao et al 1991, Dobson et al 1989, Farrell et al 1991), although the putative 8.3-kb LAT transcript has not been identified in latently infected neurons and the anticipated splice donor-acceptor sites have yet to be confirmed. Two other LAT species of 1.45 and 1.5 kb can be detected during latency (Spivack & Fraser 1987, 1988; Spivack et al 1991; Wagner et al 1988) and may be the result of splicing events preceding the removal of the 2-kb LAT by splicing. During latency, the HSV-1 genome persists as a circular or concatemeric element (no termini are detectable) that is intranuclear but not integrated into the cellular genome, and appears to be bound by nucleosomes (Deshmane & Fraser 1989; Efstathiou et al 1986; Mellerick & Fraser 1987; Rock & Fraser 1983, 1985).

The biologic significance of LAT expression is unclear, though recently some consensus has begun to emerge. Studies using deletion and insertion mutants in the LAT region have demonstrated clearly that detectable amounts of LATs are unnecessary for the establishment or maintenance of latency (Fareed & Spivack 1994, Hill et al 1990, Ho & Mocarski 1989, Javier et al 1988, Leib et al 1989, Steiner et al 1989), although in specific animal model systems and with individual strains of the virus, LAT deletion mutants demonstrate delayed kinetics of reactivation (Hill et al 1990, Leib et al 1989, Sawtell & Thompson 1992, Steiner et al 1989, Trousdale et al 1991). These studies have not allowed unambiguous determination of whether establishment of latency by LAT deletion mutants may be quantitatively impaired, and the mechanism of the effect on reactivation kinetics remains unknown. The significance of LAT expression for HSV-1 vector development is twofold: First, LAT expression is one marker of the persistence of latent genomes that could serve as platforms for foreign gene expression within neurons, and second, the production of LAT RNA from the latent genomes demonstrates that RNA can be produced from those genomes and suggests that the same promoter elements might be exploited for the expression of foreign genes from the latent viral genome. However, the transcriptional control of LAT gene expression is complex.

The nearest TATA box and basal transcriptional regulatory sequences lie in a promoter element (LAP1) approximately 700–1300 bp upstream of the 5′ end of the 2-kb LAT (Dobson et al 1989), and as yet undefined upstream sequences that increase promoter activity in some cell lines of neuronal origin (Batchelor & O'Hare 1990, 1992; Zwaagstra et al 1990). A second promoter, designated LAP2 (Goins et al 1994), lies between LAP1 and the 5′ end of the 2-kb LAT and has sequence homology with many eukaryotic housekeeping gene promoters. In vivo, a viral mutant (KOS/29) deleted in LAP1 establishes latency in trigeminal ganglia (TG) without producing LATs detectable by Northern blot analysis (Dobson et al 1989, Nicosia et al 1993), although LATs are detectable by reverse transcriptase PCR (RT-PCR) (X Chen & JC Glorioso, manuscript in preparation). Viral mutants in which the *lacZ* reporter gene is juxtaposed to the LAP2 promoter express β-galactosidase after the establishment of latency in the TG (Goins et al 1994), whereas LAP1 without LAP2 sequences is unable to drive reporter gene expression in the TG during latency (Lokensgard et al 1994, Margolis et al 1993). Thus, both LAP1 and LAP2 sequences may be required to drive full LAT expression, or to drive foreign gene expression from the context of the latent viral genome in vivo. A more detailed consideration of gene expression from the context of a latent HSV-1 vector genome follows.

Viral replication is not required for the establishment of latency. $ICP4^-$ mutants persist in TG after peripheral inoculation and in the brain after direct intracranial inoculation; genomes are detectable by PCR, and LAT expression is detectable by RT-PCR and in situ hybridization (Dobson et al 1990, Katz et al 1990; DJ Fink & JC Glorioso, unpublished data).

# EXPRESSION OF GENE PRODUCTS FROM HSV-1 VECTORS IN VITRO

Eukaryotic genes can be inserted into the HSV-1 genome and expressed from that context in cells in culture. Shih et al (1984) first demonstrated that the hepatitis B virus surface antigen gene driven by HSV-1 IE or E promoters expressed the gene product in vitro, with kinetics characteristic of viral IE and E genes. Other researchers later demonstrated that globin genes could be expressed by their own promoters in a similar situation (Panning & Smiley 1989, Smiley et al 1987) and that hypoxanthine guanine phophoribosyltrans- ferase (HPRT) could be driven by the viral thymidine kinase (TK) promoter (Palella et al 1988). A wide variety of promoters have now been demonstrated to drive production of foreign gene products transiently, including the HSV promoters for TK, US3, and gC; the IE promoter of the human cytomegalovirus (HCMV IEp); mammalian promoters such as that of nerve-specific enolase (NSE); neurofilament-L (NF-L); and various pol III promoters or hybrid pro- moter constructs (discussed below).

While HSV-1 naturally enters the latent state in neurons during infections in vivo, in tissue culture the situation is more complex. Wilcox, Johnson, and colleagues (Wilcox et al 1990, 1992; Wilcox & Johnson 1987, 1988) showed that rat sympathetic neurons in culture exposed to HSV-1 at high MOI (5 pfu per neuron) in the presence of the antiviral drug acyclovir and nerve growth factor (NGF) contain persistent viral genomes in virtually all the neurons with no detectable production of viral antigens or infectious particles. This state, which can be maintained up to five weeks, depends on the presence of NGF in the culture medium, because removal of NGF or treatment with 6-hydroxy- dopamine and colchicine reactivates the pseudo-latent genomes. However, attempts to achieve long-term expression of a reporter gene (lacZ) in rat CNS neurons in culture by using the HCMV-IEp to drive gene expression produced only transient expression, most likely due to the loss of infected cells (Creedon et al 1993). Therefore expression of transgenes by using HSV-1 vectors in neuronal cell culture would be greatly facilitated by removal of viral genes, which would force the virus into a latent state without the use of inhibitors and in the absence of viral toxicity.

The principal strategy for eliminating cytotoxicity of HSV-1–based vectors is to engineer the vector genome to block viral gene expression at an early stage and thus abort the cascade of HSV-1 lytic gene expression. DeLuca et al (1985) created a virus (d120) deleted in the essential IE gene *ICP4*. This mutant can be propagated only in a cell line (E5 cells) engineered to provide the *ICP4* gene product in *trans*, and although the *ICP4⁻* virus particles enter neuronal or other cells normally, they do not produce E or L gene products or infectious viral progeny in those cells (Figure 3*A*). Nonetheless infection of neuronal and glial cells in vitro with an *ICP4⁻* mutant caused alteration in cell morphology and fragmentation of chromatin (Johnson et al 1992, 1994), suggesting that the other IE gene products, which do not require *ICP4* to transactivate their expression, may have toxic effects. That the viral particles themselves are not toxic to cells has been demonstrated by the observation that infection with UV-irradiated particles (thus rendered incapable of viral gene expression) does not cause pathologic changes (Leiden et al 1980). The potential culprits contributing to toxicity are the IE gene products; ICP27, a regulatory protein that affects HSV-1 gene expression at the transcriptional and posttranscriptional level (McCarthy et al 1989, Rice & Knipe 1990, Smith et al 1992); ICP22, a protein that affects the phosphorylation of the carboxy-terminal domain of cellular RNA polymerase II (Rice et al 1994); ICP47, which interferes with the processing of major histocompatibility complex antigen 1 (MHC 1) molecules (York et al 1994); and ICP0, a promiscuous transactivator of gene expression (Everett 1984, 1987; Gelman & Silverstein 1985). In addition, virion associated proteins, e.g. VP16 (discussed above) and UL41 (vhs), which nonspecifically degrade mRNA molecules (Oroskar & Read 1989), may adversely affect host cell metabolism. The *ICP6* gene, which codes for the large subunit of ribonucleotide reductase and additionally possesses protein kinase activity (Chung et al 1989), is also expressed soon after viral infection at both IE and maximally at E times in an ICP0-dependent manner and may contribute to viral toxicity.

Systematic deletion of individual gene products has allowed the determination of cytotoxicity of each of these components. Starting with an *ICP4*-deleted background, further deletion of *ICP22* and *ICP47* or the removal of *UL41* does not further reduce cytotoxicity in vitro (NA DeLuca, personal communication; Johnson et al 1992, 1994). However, a virus deleted in *ICP4* and *ICP27* is less toxic than viruses deleted in *ICP4* or *ICP27* alone, and the further elimination of *ICP6* and *UL41* increased the survival of infected cells 10- to 30-fold over *ICP4* deletion mutant alone (NA DeLuca, personal communication). The residual toxicity is probably due to *ICP0*, and efforts are underway to generate a vector that is additionally deleted for this function.

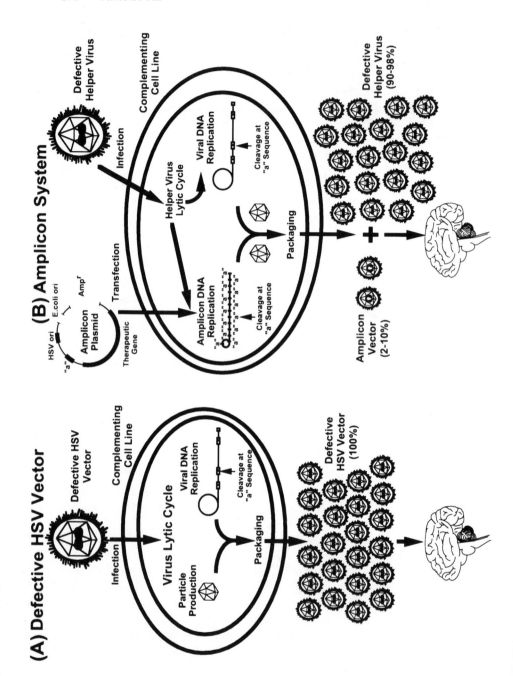

(A) Defective HSV Vector

(B) Amplicon System

# AMPLICONS

An alternative to the use of modified whole-virus vectors is to employ the HSV-1 capsid and envelope to deliver plasmids containing the foreign gene, thereby exploiting the natural ability of the virion to deliver DNA to the nucleus of neuronal cells. Often referred to as defective HSV-1 vectors, amplicons are plasmids engineered to contain an HSV-1 origin of replication and an HSV-1 packaging site ("a" sequence) along with a bacterial origin of replication (Spaete & Frenkel 1982). The hybrid plasmid is propagated in bacteria, and then cotransfected with a defective HSV-1 helper recombinant, creating a population of HSV-1 particles that contain either the defective HSV-1 genome or concatemers of the plasmid packaged into the HSV-1 capsid (Figure 3B). Efficient expression of a reporter gene in primary cultures of neurons from the PNS and the CNS (Geller & Breakefield 1988, Geller & Freese 1990) has been demonstrated with amplicons. Whereas the initial studies employed temperature-sensitive HSV-1 mutants for the packaging HSV-1 helper virus, subsequent studies have typically employed $ICP4^-$ mutants that are incapable of replication at any temperature.

Subsequent experiments have demonstrated the transfer and expression of the gene for a constitutively active adenylate cyclase into PC12 cells with increases in cAMP, PKA activity, and protein phosphorylation (Geller et al 1993); the low-affinity human NGF receptor into primary neurons in culture (Battleman & Geller 1993); and the transfer of NGF receptor into cell lines and cultured neurons with demonstrated alterations in the response to exogenous NGF (Geschwind et al 1994). In addition, expression of β-galactosidase in cells of the hippocampus in organotypic slice cultures was achieved using amplicons (Casaccia-Bonnefil et al 1993).

Several technical issues remain to be resolved with the amplicon system. After a single passage, the ratio of amplicon to helper virus is generally low, so repeated passage is required until an optimal ratio of amplicon-to-helper virus is achieved. With each passage there is a chance of recombination between amplicon, helper viral genomes, and the complementing gene in the cell, resulting in the potential production of replication competent virus. Indeed a background of replication competent virus in preparations that have undergone repeated passage has been estimated at a frequency of $10^{-5}$ (During et al 1994, Geller et al 1990), which accounts for the death of some animals (During

←

*Figure 3*    (A) Production of defective full-length HSV-1–based vectors is carried out in cell lines that are engineered to provide the deleted essential genes in *trans*. These vectors are incapable of replicating in neurons because of the missing essential genes. (B) Amplicons are propagated in bacteria (using the bacterial origin of replication) and then transfected into a complementing cell line that is infected with defective helper HSV-1, thus producing particles consisting either of amplicon concatemers (about 150 kb in length) or defective HSV-1.

et al 1994) and may be related to the persistent expression of the transgene in surviving animals. The problem of recombination after repeated passage poses severe limitations for the use of these types of vectors in human applications. The results reviewed above suggest that amplicons may be useful for some experimental gene transfer applications in animals, particularly if only short-term expression is required. If, however, efficient methods can be developed for amplicon packaging free of helper virus and measures to prevent recombination can be devised, then the approach of using amplicons could become quite attractive.

## EXPRESSION OF TRANSGENES FROM HSV-1 VECTORS IN VIVO

Because HSV-1 naturally establishes a latent state in vivo, expression of foreign genes from the viral genome in the animal might be more successful than in vitro studies predict. Palella et al (1989) demonstrated that the HPRT gene under the control of the HSV TK promoter expressed HPRT mRNA in the brains of mice after intracranial inoculation. However, the virus employed, attenuated only by the deletion of the HSV-1 TK function, replicated in the brain leading to the death of the animals. Subsequently, reporter genes were inserted into a variety of viral mutants of reduced pathogenicity, including deletions in accessory functions such as the US3 protein kinase gene (Fink et al 1992), ribonucleotide reductase (Ramakrishnan et al 1994), various glycoproteins (Chrisp et al 1989, Meignier et al 1988, Sunstrum et al 1988), and ICP34.5 (MacLean et al 1991, Whitley et al 1993). Animals uniformly survive injection with these attenuated vectors, but reporter gene expression detected by X-gal staining and immunocytochemistry is generally restricted to the first few days after inoculation, whether *lacZ* expression is driven by viral lytic cycle gene promoters, other viral promoters such as retroviral LTRs or the HCMV IEp, pol III promoters (CA Meaney & JC Glorioso, unpublished data), housekeeping gene promoters known to be active in brain such as HPRT (KA Lee & JC Glorioso, unpublished data), or mammalian promoters such as NSE (Andersen et al 1992) or NF (MA Bender & JC Glorioso, unpublished data). Similar results have also been seen with $ICP4^-$ mutants, which are entirely incapable of viral replication in vivo.

There are two possible explanations for this loss of gene expression in the CNS: Either the viral genomes that serve as the platform for gene expression are being lost from the brain, or the genomes persist but the promoter-enhancer elements are being silenced. In order to evaluate the first issue, we have used competitive quantitative PCR (CQPCR) to determine the number of viral genomes in the hippocampus of rats injected with a ribonucleotide reductase–deficient ($RR^-$) mutant vector (Ramakrishnan et al 1994). Although the number of genomes present decreased from two to seven days after inoculation (most likely representing the loss of viral genomes that were not taken up into neurons), it remained constant one to eight weeks after inoculation. At 8 weeks, a single 10-micron slice of injected hippocampus contained $1 \times 10^5$ genomes, with approximately 15 copies of LAT per genome, as determined by CQ RT-PCR. This suggests that sufficient genomes persist after inoculation into brain to serve as an adequate platform for transgene expression. In recent studies we have used the alternate latency promoter LAP2, consisting of the sequences between LAP1 and the 5′ end of the stable LAT, to drive *lacZ* expression from an *ICP4⁻* vector backbone with the LAP2-*lacZ* cassette placed either in the LAT or *gC* loci. Although β-galactosidase could not be detected by X-gal staining or immunocytochemistry, *lacZ* mRNA was detectable by RT-PCR at two and four weeks after stereotaxic inoculation into the hippocampus (DJ Fink & JC Glorioso, unpublished data). This suggests that the LAP2 promoter may serve to drive low-level gene expression in the CNS during viral latency.

In contrast to the situation in the CNS, long-term expression of reporter genes in the PNS has been achievable. *lacZ* expression has been demonstrated in dorsal root ganglion (DRG) or TG neurons after introducing the Moloney murine leukemia retroviral (MoMuLV) LTR into the *ICP4* locus (Dobson et al 1990), the MoMuLV LTR with the LAP1 upstream sequence in the *gC* locus (Lokensgard et al 1994), or the neuronal-specific sodium channel promoter (D Leib, personal communication) or the natural viral latency promoter system (Goins et al 1994).

We used in situ PCR of infected TG sections following corneal scarification to define the cells within the ganglion that harbor latent viral genomes. Compared to in situ hybridization detection of LAT RNA, substantially more neurons harbored latent virus than expressed detectable LATs (Ramakrishnan et al 1994). Several potential strategies are under investigation to exploit those genomes for the expression of the transgene (Figure 4). One such strategy is the use of border sequences, which cause an association of the vector genome with the nuclear matrix (*mar* elements), an event that promotes transcription. One of these *mar* element–containing vector recombinants that has the HIV LTR promoter *lacZ* reporter gene construct flanked by these elements also showed *lacZ* RNA expression by RT-PCR at four weeks after inoculation (M Levine, personal communication).

Initial studies with intracranial inoculation of amplicons carrying the *lacZ* reporter gene demonstrated only transient expression of β-galactosidase; a few positive cells persist up to one month (During et al 1994, Freese et al 1990), although with the enkephalin promoter driving *lacZ* expression in an amplicon, Kaplitt et al (1994a) reported long-term expression of β-galactosidase. The restricted cellular distribution of long-term *lacZ* expression suggested a cell-

# (A) Transient Expression

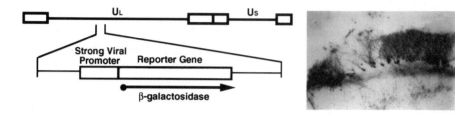

# (B) Long-term Expression

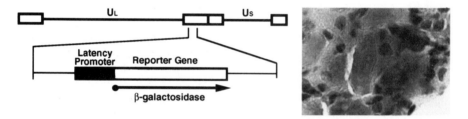

# (C) Autoregulatable Expression

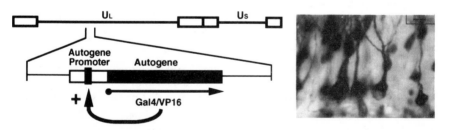

*Figure 4* Strategies for driving transgene expression from the context of the latent HSV genome. (*A*) HSV-1 vector–mediated transient expression of β-galactosidase in brain. (*B*) HSV-1 vector–mediated long-term β-galactosidase expression in the PNS. (*C*) Autoregulatable expression of the Gal4-VP16 transcriptional activator in the CNS.

specific promoter effect, although the distribution of amplicons or replicating recombinant viruses in that study was not determined. Several studies have demonstrated the expression of bioactive peptides in vivo. Federoff et al (1992) used an amplicon to transfer the NGF gene to cervical sympathetic ganglia, demonstrating production of NGF mRNA by Northern blot analysis and an apparent biologic effect on the response to axotomy in those cells. More recently, amplicon-mediated transfer of the tyrosine hydroxylase (TH) gene into the striatum of 6-hydroxydopamine–lesioned rats was shown to reduce apomorphine-induced rotational behavior up to one year after inoculation (During et al 1994).

## POTENTIAL TARGET DISEASES

What diseases might be treated by HSV-1–mediated gene transfer, and what strategies might be employed? One attractive class of targets are the neurodegenerative diseases typified by Parkinson's disease, Huntington's disease, and Alzheimer's disease—complex conditions whose pathophysiology, despite major advances in recent years, remains poorly understood. Three different approaches are being developed. The first, exemplified by experimental treatment in animal models of Parkinson's disease, uses gene transfer to deliver macromolecules that may provide symptomatic benefit. Transplantation of myoblasts transduced with the TH gene (Jiao et al 1993) or direct transfer of the TH gene with amplicons (During et al 1994), presumably resulting in excess local production of dopamine and mimicking the clinical approach of replacement therapy with the dopamine precursor L-DOPA, results in amelioration of apomorphine-induced rotational behavior in 6-hydroxydopamine–lesioned rats. We have also shown that an HSV-1 vector carrying the TH gene effectively transforms B103 cells in vitro to produce dopamine (MA Bender & JC Glorioso, unpublished data). Although the clinical treatment of Parkinson's disease, which requires exquisite regulation of dopamine level may require more sophisticated means of controlling TH activity or release from the TH transduced cells, the paradigm that exploits gene transfer to achieve focal production of a therapeutic macromolecule may also be applied to other diseases.

The second paradigm is typified by Huntington's disease, a dominantly inherited degeneration of cells in the caudate and putamen. The disease results from a dominant pathologic effect of expanded trinucleotide repeats in the *huntington* gene on the short arm of chromosome 4. Although the gene has been cloned, the identity and function of the gene product and the molecular mechanisms by which it causes neuronal degeneration are unknown. Nonetheless, strategies based on antisense activities might be developed to block the effect of the toxic gene product and thus prevent the disease phenotype from developing in presymptomatic patients identified by genetic testing.

The third paradigm is typified by Alzheimer's disease. Most cases are apparently sporadic, the inherited cases genetically are diverse, and the degeneration affects several classes of neurons with distinct phenotypes. Thus neither a single gene product nor a single neurotransmitter would be therapeutic. However, were a single trophic factor (NGF, for instance) effective in preventing degeneration of neurons in the disease or an enzyme identified that blocked the final common pathway leading to the accumulation of amyloid to prove to be effective in preventing the pathologic changes, either of these genes could be delivered to at-risk brain regions by using gene transfer.

A second class of neurologic diseases are those that might be treated by transient transgene expression, exemplified by brain tumors and multiple sclerosis. Animal studies have shown that the HSV-1 TK gene can be used to activate a prodrug (gancyclovir) that kills both tumor cells containing the TK gene and surrounding tumor cells as a result of a bystander effect by using either retrovirus or an HSV-1 vector to deliver the TK gene (Martuza et al 1991, Takamiya et al 1992). Human trials using a TK-containing retrovirus producer cell line have already been initiated, but the early results are not as impressive as the animal studies (K Culver, personal communication). HSV-1 vectors could be used to deliver cytokine genes in addition to the TK gene, in order to enhance the immune response against the tumor. In either case, gene transfer would be used to achieve focal high-level expression of the transgenes, but gene expression should be temporally restricted. In a similar fashion, one might envision treating acute attacks of multiple sclerosis with immunomodulatory peptides generated from a gene transfer vector. The advantages of gene transfer would be the focal transient production of high levels of the cytokine in a situation where it is unlikely that long-term expression of the immunomodulatory agents would be advantageous.

## SUMMARY

The era of molecular biology has opened new avenues for both experimental manipulation and treatment of disease. Until recently, experimental manipulation of postmitotic primary neuronal cells in culture was limited by the vectors developed. Using currently available HSV-1 vectors and HSV-1–based amplicons, expressing foreign genes transiently in neurons in order to study the effects of those gene products on cell function and physiology should be possible. Controlling for potential direct effects of the vector itself will be critical, but the overall prospects appear promising.

Gene transfer directly into the nervous system by using these types of vectors has begun on an experimental level and is likely to be extended to human trials before the end of the decade. Use of HSV-1 vectors for gene therapy applications requiring transient gene expression, such as treatment of tumors, is immanent. For long-term application, two major issues remain to be resolved. The first issue concerns appropriate engineering of the vector genome to create an entirely apathogenic vector. Progress in this direction has been achieved by selectively deleting essential and nonessential genes from the HSV-1 genome. The second major issue is the appropriate choice of promoter-enhancer elements and their placement in the vector backbone to achieve long-term expression

Solutions to both problems may result from the creation of deletion mutants that fail to express viral products and from studies detailing the mechanisms of latency gene expression. Insight into this latter activity should provide strategies to improve the level of transgene expression during latency.

---

Any *Annual Review* chapter, as well as any article cited in an *Annual Review* chapter, may be purchased from the Annual Reviews Preprints and Reprints service.
1-800-347-8007; 415-259-5017; email: arpr@class.org

---

## Literature Cited

Acsadi G, Jani A, Massie B, Simoneau M, Holland P, et al. 1994. A differential efficiency of adenovirus-mediated in vivo gene transfer into skeletal muscle cells of different maturity. *Hum. Mol. Genet.* 3:579–84

Akli S, Cailland C, Vigne E, Stratford-Perricaudet LD, Poenaru L, et al. 1993. Transfer of a foreign gene into the brain using adenovirus vectors. *Nat. Genet.* 3: 224–28

Andersen JK, Garber DA, Meaney CA, Breakfield XO. 1992. Gene transfer into mammalian central nervous system using herpes virus vectors: extended expression of bacterial *lacZ* in neurons using the neuron-specific enolase promoter. *Hum. Gene Ther.* 3:487–99

Anderson WF. 1992. Human gene therapy. *Science* 256:803–13

Bak IJ, Markham CH, Cook ML. 1977. Intraaxonal transport of herpes simplex virus in the rat central nervous system. *Brain Res.* 136:415–29

Batchelor AH, O'Hare PO. 1990. Regulation of cell-type-specific activity of a promoter located upstream of the latency-associated transcript of herpes simplex virus type 1. *J. Virol.* 64:3269–79

Batchelor AH, O'Hare PO. 1992. Localization of *cis*-acting sequence requirements in the promoter of the latency-associated transcript of herpes simplex virus type 1 required for

cell-type-specific activity. *J. Virol.* 66:3573–82

Battleman DS, Geller AI. 1993. HSV-1 vector-mediated gene transfer of the human nerve growth factor receptor p75hNGFR defines high-affinity NGF binding. *J. Neurosci.* 13: 941–51

Casaccia-Bonnefil P, Benedikz E, Shen H, Stelzer A, Edelstein D, et al. 1993. Localized gene transfer into organotypic hippocampal slice cultures and acute hippocampal slices. *J. Neurosci. Methods* 50:341–51

Chrisp CE, Sunstrum JC, Averill DR, Levine M, Glorioso JC. 1989. Characterization of encephalitis in adult mice induced by intracerebral inoculation of herpes simplex virus type 1 (KOS) and comparison with mutants showing decreased virulence. *Lab. Invest.* 60:822–30

Chung TP, Wymer JP, Smith CC, Kulka M, Aurelian L. 1989. Protein kinase activity associated with the large subunit of herpes simplex virus type 2 ribonucleotide reductase (ICP6). *J. Virol.* 63:3389–98

Creedon DJ, Greenland LS, Johnson EM Jr. 1993. Characterization of a defective herpes simplex virus vector for ectopic gene expression in cultured neurons. *Soc. Neurosci.* 19(1): 886 (Abstr.)

Croen KD, Ostrove JM, Dragovic LJ, Smialek JE, Straus SE. 1987. Latent herpes simplex virus in human trigeminal ganglia. Detection of an immediate early gene "anti-sense" transcript by in situ hybridization. *New Engl. J. Med.* 317:1427–32

Davidson BL, Allen EE, Kozarsky KF, Wilson JM, Roessler BJ. 1993. A model system for in vivo gene transfer into the central nervous system using an adenoviral vector. *Nat. Genet.* 3:219–23

Deatly AM, Spivack JG, Lavi E, O'Boyle D, Fraser NW. 1988. Latent herpes simplex virus type 1 transcripts in peripheral and central nervous systems tissues of mice map to similar regions of the viral genome. *J. Virol.* 62:749–56

DeLuca NA, McCarthy AM, Schaffer PA. 1985. Isolation and characterization of deletion mutants of herpes simplex virus type 1 in the gene encoding immediate-early regulatory protein ICP4. *J. Virol.* 56:558–70

DeLuca NA, Schaffer PA. 1985. Activation of immediate-early, early, and late promoters by temperature-sensitive and wild-type forms of herpes simplex virus type 1 protein ICP4. *Mol. Cell Biol.* 5:558–70

Deshmane SL, Fraser NW. 1989. During latency, herpes simplex virus type 1 DNA is associated with nucleosomes in a chromation structure. *J. Virol.* 63:943–47

Devi-Rao GB, Goddart SA, Hecht LM, Rochford R, Rice MK, Wagner EK. 1991. Relationship between polyadenylated and nonpolyadenylated, HSV type 1 latency-associated transcripts. *J. Gen. Virol.* 65: 2179–90

Dixon RAF, Schaffer PA. 1980. Fine-structure mapping and functional analysis of temperature-sensitive mutants in the gene encoding the herpes simplex virus type 1 immediate early protein VP175. *J. Virol.* 36:189–203

Dobson AT, Margolis TP, Sederati F, Stevens JG, Feldman TL. 1990. A latent, nonpathogenic HSV-1-derived vector stably expresses beta-galactosidase in mouse neurons. *Neuron* 5:353–60

Dobson AT, Sederati F, Devi-Rao G, Flanagan WM, Farrell MJ, et al. 1989. Identification of the latency-associated transcript promoter by expression of rabbit β-globin mRNA in mouse sensory nerve ganglia latently infected with a recombinant herpes simplex virus. *J. Virol.* 63:3844–51

During MJ, Naegele JR, O'Malley KL, Geller AI. 1994. Long-term behavioral recovery in Parkinsonian rats by an HSV vector expressing tyrosine hydroxylase. *Science* 266:1399–403

Efstathiou S, Minson AC, Field HJ, Anderson JR, Wildy P. 1986. Detection of herpes simplex virus-specific DNA sequences in latently infected mice and in human. *J. Virol.* 57:446–55

Everett RD. 1984. Transactivation of transcription by herpes virus products: requirements

for two HSV-1 immediate-early polypetides for maximum activity. *EMBO J.* 3:3135–41

Everett RD. 1987. The regulation of transcription of viral and cellular genes by herpesvirus immediate-early gene products. *Anticancer Res.* 7:580–604

Fareed MU, Spivack JG. 1994. Two open reading frames (ORF1 and ORF2) within the 2.0-kilobase latency-associated transcript of herpes simplex virus type 1 are not essential for reactivation from latency. *J. Virol.* 68: 8071–81

Farrell MJ, Dobson AT, Feldman LT. 1991. Herpes simplex virus latency-associated transcript is a stable intron. *Proc. Natl. Acad. Sci. USA* 88:790–94

Federoff HJ, Geschwind MD, Geller AI, Kessler JA. 1992. Expression of nerve growth factor in vivo from a defective herpes simplex virus 1 vector prevents effects of axotomy on sympathetic ganglia. *Proc. Natl. Acad. Sci. USA* 89:1636–40

Fink DJ, Sternberg LR, Weber PC, Mata M, Goins WF, Glorioso JC. 1992. In vivo expression of β-galactosidase in hippocampal neurons by HSV-mediated gene transfer. *Hum. Gene Ther.* 3:11–19

Freese A, Geller AI, Neve R. 1990. HSV-1 vector mediated neuronal gene delivery. Strategies for molecular neuroscience and neurology. *Biochem. Pharm.* 40:2189–99

Gaffney DF, McLauchlin J, Whitton JL, Clements JB. 1985. A modular system for the assay of transcription regulatory signals: the sequence TAATGARAT is required for herpes simplex virus immediate early gene activation. *Nucleic Acids Res.* 13: 7847–63

Geller AI, Breakefield XO. 1988. A defective HSV-1 vector expresses *Escherichia coli* β-galactosidase in cultured peripheral neurons. *Science* 241:1667–69

Geller AI, During MJ, Haycock JW, Freese A, Neve R. 1993. Long-term increases in neurotransmitter release from neuronal cells expressing a constitutively active adenylate cyclase from a herpes simplex virus type 1 vector. *Proc. Natl. Acad. Sci. USA* 90:7603–7

Geller AI, Freese A. 1990. Infection of cultured central nervous system neurons with a defective herpes simplex virus 1 vector results in stable expression of *Escherichia coli* β-galactosidase. *Proc. Natl. Acad. Sci. USA* 87: 1149–53

Geller AI, Keyomarsi K, Bryan J, Pardee AB. 1990. An efficient deletion mutant packaging system for defective herpes simplex virus vectors: potential applications to human gene therapy and neuronal physiology. *Proc. Natl. Acad. Sci. USA* 87:8950–54

Gelman IH, Silverstein S. 1985. Identification of immediate-early genes from herpes simplex virus that transactivate the virus thymidine kinase gene. *Proc. Natl. Acad. Sci. USA* 82:5265–69

Geschwind MD, Kessler JA, Geller AI, Federoff HJ. 1994. Transfer of the nerve growth factor gene into cell lines and cultured neurons using a defective herpes simplex virus vector. *Brain Res.* 24:327–35

Goins WF, Sternberg LR, Croen KD, Krause PR, Hendricks RL, et al. 1994. A novel latency-active promoter is contained within the herpes simplex virus type 1 U$_L$ flanking repeats. *J. Virol.* 68:2239–52

Hill JM, Sedarati F, Javier RT, Wagner EK, Stevens JG. 1990. Herpes simplex virus latent phase transcription facilitates in vivo reactivation. *Virology* 174:117–25

Ho DY, Mocarski ES. 1989. Herpes simplex virus latent RNA (LAT) is not required for latent infection in the mouse. *Proc. Natl. Acad. Sci. USA* 87:7596–600

Honess RW, Roizman B. 1974. Regulation of herpes simplex virus macromolecular synthesis. I. Cascade regulation of the synthesis of three groups of viral proteins. *J. Virol.* 14:8–19

Jacob RJ, Morse LS, Roizman B. 1979. Anatomy of herpes simplex virus DNA. XII. Accumulation of head-to-tail concatemers in nuclei of infected cells and their role in the generation of the four isomeric arrangements of viral DNA. *J. Virol.* 29:448–57

Javier RT, Stevens JG, Dissette VB, Wagner EK. 1988. A herpes simplex virus transcript abundant in latently infected neurons is dispensable for establishment of the latent state. *Virology* 166:254–57

Jiao S, Gurevich V, Wolff JA. 1993. Long-term correction of rat model of Parkinson's disease by gene therapy. *Nature* 362:450–53

Johnson PA, Miyanohara A, Levine F, Cahill T, Friedmann T. 1992. Cytotoxicity of a replication-defective mutant herpes simplex virus type 1. *J. Virol.* 66:2952–65

Johnson PA, Wang MJ, Friedmann T. 1994. Improved cell survival by the reduction of immediate-early gene expression in replication-defective mutants of herpes simplex virus type 1 but not by mutation of the virion host shutoff function. *J. Virol.* 68:6347–62

Kaplitt MG, Kwong AD, Kleopoulos SP, Mobbs CV, Rabkin SD, Pfaff DW. 1994a. Preproenkephalin promoter yields region-specific and long-term expression in adult brain after direct in vivo gene transfer via a defective herpes simplex viral vector. *Proc. Natl. Acad. Sci. USA* 91:8979–83

Kaplitt MG, Leone P, Samulski RJ, Xiao X, Pfaff DW, et al. 1994b. Long-term gene expression and phenotype correction using adeno-associated virus vectors in the mammalian brain. *Nat. Genet.* 8:148–54

Katz JP, Bodin ET, Coen DM. 1990. Quantita-tive polymerase chain reaction analysis of herpes simplex virus DNA in ganglia of mice infected with replication-incompetent mutants. *J. Virol.* 64:4288–95

Le Gal La Salle G, Robert JJ, Berrard S, Ridoux V, Stratford-Perricaudet LD, et al. 1993. An adenovirus vector for gene transfer into neurons and glia in the brain. *Science* 259:988–90

Leib DA, Coen MD, Bogard CL, Hicks KA, Yager DR, Knipe DM. 1989. Immediate-early regulatory gene mutants define different stages in the establishment and reactivation of herpes simplex virus latency. *J. Virol.* 63:759–68

Leiden JM, Frenkel N, Rapp F. 1980. Identification of the herpes simplex virus DNA sequences present in six herpes simples virus thymidine kinase-transformed mouse cell lines. *J. Virol.* 33:272–85

Lokensgard JR, Bloom DC, Dobson AT, Feldman TL. 1994. Long-term promoter activity during herpes simplex virus latency. *J. Virol.* 68:7148–58

Mackem S, Roizman B. 1982a. Differentiation between alpha promoter and regulatory regions of herpes simplex virus type 1: the functional domains and sequence of a movable alpha regulator. *Proc. Natl. Acad. Sci. USA* 79:4917–21

Mackem S, Roizman B. 1982b. Structural features of the herpes simplex virus alpha gene 4, 0, and 27 promoter-regulatory sequences which confer alpha regulation on chimeric thymidine kinase. *J. Virol.* 44:939–49

MacLean AR, ul-Fareed M, Robertson L, Harland J, Brown SM. 1991. Herpes simplex virus type 1 deletion variants 1714 and 1716 pinpoint neurovirulenece-related sequences in Glasgow strain 17+ between immediate early gene 1 and the "a" sequence. *J. Gen. Virol.* 72:631–39

Margolis TP, Bloom DC, Dobson AT, Feldman LT, Stevens JG. 1993. Decreased reporter gene expression during latent infection with herpes simplex virus latency-associated transcript promoter constructs. *Virology* 197:585–92

Martuza RL, Malick A, Markert JM, Ruffner KL, Coen DM. 1991. Experimental therapy of human glioma by means of a genetically engineered virus mutant. *Science* 252:854–86

McCarthy AM, McMahan L, Schaffer PA. 1989. Herpes simplex virus type 1 ICP27 deletion mutants exhibit altered patterns of transcription and are DNA deficient. *J. Virol.* 63:18–27

McFarland DJ, Sikora E, Hotchkin J. 1986. The production of focal herpes encephalitis in mice by stereotaxic inoculation of virus. Anatomical and behavioral effects. *J. Neurol. Sci.* 72:307–18

Meignier B, Longnecker R, Mavromara-Nazos P, Sears AE, Roizman B. 1988. Virulence of and establishment of latency by genetically engineered deletion mutants of herpes simplex virus type 1. *Virology* 162:251–54

Mellerick DM, Fraser NW. 1987. Physical state of the latent herpes simplex virus genome in a mouse model system: evidence suggesting an episomal state. *Virology* 158:265–75

Miller AD. 1992. Retroviral vectors. *Curr. Top. Microbiol. Immunol.* 158:1–24

Nicosia M, Deshmane SL, Zabolotny JM, Valyi-Nagy T, Fraser NW. 1993. Herpes simplex virus type 1 latency-associated transcript (LAT) promoter deletion mutants can express a 2-kilobase transcript mapping to the LAT region. *J. Virol.* 67:7276–83

Oroskar AA, Read GS. 1989. Control of mRNA stability by the virion host shutoff function of herpes simplex virus. *J. Virol.* 63:1897–906

Palella TD, Hidaka Y, Silverman LJ, Levine M, Glorioso JC, Kelley WM. 1989. Expression of human HPRT mRNA in brains of mice infected with a recombinant herpes simplex virus type 1 vector. *Gene* 80:137–44

Palella TD, Silverman LJ, Schroll CT, Homa FL, Levine M, Kelley WM. 1988. Herpes simplex virus-mediated human hypoxanthine-guanine phosphoribosyl-transferase gene transfer into neuronal cells. *Mol. Cell Biol.* 8:457–60

Panning B, Smiley JR. 1989. Regulation of cellular genes transduced by herpes simplex virus. *J. Virol.* 63:1929–37

Preston CM. 1979a. Control of herpes simplex virus type 1 mRNA synthesis in cells infected with wild-type virus or the temperature-sensitive mutant tsK. *J. Virol.* 29:275–84

Preston C. 1979b. Abnormal properties of an immediate early polypeptide in cells infected with the herpes simplex virus type 1 mutant tsK. *J. Virol.* 32:357–69

Quantin B, Perricaudet LD, Tajbakhsh S, Mandel J-L. 1992. Adenovirus as an expression vector in muscle cells in vivo. *Proc. Natl. Acad. Sci. USA* 89:2581–84

Quinlan MP, Knipe DM. 1985. Stimulation of expression of a herpes simplex virus DNA-binding protein by two viral functions. *Mol. Cell. Biol.* 5:957–63

Ragot T, Vincent N, Chafey P, Gilgenkrantz H, Couton D, et al. 1993. Efficient adenovirus-mediated transfer of a human minidystrophin gene to skeletal muscle of mdx mice. *Nature* 321:647–50

Ramakrishnan R, Fink DJ, Jiang G, Desai P, Glorioso JC, Levine M. 1994. Competitive quantitative polymerase chain reaction (PCR) analysis of herpes simplex virus type 1 DNA and LAT RNA in latently infected cells of brain. *J. Virol.* 68:1864–70

Rice SA, Knipe DM. 1990. Genetic evidence for two different transactivation functions of the herpes simplex virus and protein ICP27. *J. Virol.* 64:1704–15

Rice SA, Long MC, Lam V, Spencer CA. 1994. RNA polymerase II is aberrantly phosphorylated and localized to viral replication compartments following herpes simplex virus infection. *J. Virol.* 68:988–1001

Rock DL, Fraser NW. 1983. Detection of HSV-1 genome in the central nervous system of latently infected mice. *Nature* 302:523–25

Rock DL, Fraser NW. 1985. Latent herpes simplex virus type 1 DNA contains two copies of the virion DNA joint region. *J. Virol.* 55:849–52

Rock DL, Nesburn AB, Ghiasi H, Ong J, Lewis TL, et al. 1987. Detection of latency-related viral RNAs in trigeminal ganglia of rabbits infect with herpes simplex virus type 1. *J. Virol.* 61:3820–26

Roizman B, Sears AE. 1990. Herpes simplex viruses and their replication. In *Virology*, ed. B Fields, pp. 1795–841. New York: Raven

Sacks WR, Greene CC, Aschman DP, Schaffer PA. 1985. Herpes simplex virus type 1 ICP27 is essential regulatory protein. *J. Virol.* 55:796–805

Sacks WR, Schaffer PA. 1987. Deletion mutants in the gene encoding the herpes simplex virus type 1 immediate-early protein ICP0 exhibit impaired growth in cell culture. *J. Virol.* 61:829–39

Sawtell NM, Thompson RL. 1992. Herpes simplex virus type 1 latency-associated transcription unit promotes anatomical site-dependent establishment and reactivation from latency. *J. Virol.* 66:2157–69

Shih M-F, Arsenakis M, Tiollais P, Roizman B. 1984. Expression of hepatitis B virus S gene by herpes simplex virus type 1 vectors carrying alpha- and beta-regulated gene chimeras. *Proc. Natl. Acad. Sci. USA* 81:5867–70

Smiley JR, Smibert C, Everett RD. 1987. Expression of a cellular gene cloned in herpes simplex virus: rabbit beta-globin is regulated as an early viral gene in infected fibroblasts. *J. Virol.* 61:2368–77

Smith IL, Hardwicke MA, Sandri-Goldin RM. 1992. Evidence that herpes simplex virus immediate early protein ICP27 acts post-transcriptionally during infection to regulate gene expression. *Virology* 168:74–86

Spaete RR, Frenkel N. 1982. The herpes simplex virus amplicon: a new eucaryotic defective-virus cloning-amplifying vector. *Cell* 30:295–304

Spear PG. 1993. Membrane fusion induced by herpes simplex virus. In *Viral Fusion Mechanisms*, ed. J Bentz, pp. 201–32. Boca Raton: CRC

Spivack JG, Fraser NW. 1987. Detection of

herpes simplex virus type 1 transcripts during latent infection in mice. *J. Virol.* 61: 3841–47

Spivack JG, Fraser NW. 1988. Expression of herpes simplex virus type 1 latency-associated transcripts in trigeminal ganglia of mice during acute infection and reactivation of latent infection. *J. Virol.* 62:1479–85

Spivack JG, Woods GM, Fraser NW. 1991. Identification of a novel latency-specific splice donor signal within HSV type 1 2.0-kilobase latency-associated transcript (LAT): translation inhibition of LAT open reading frames by the intron within the 2.0-kilobase LAT. *J. Virol.* 65:6800–10

Steiner I, Spivack JG, Lirette RP, Brown SM, MacLean AR, Subak-Sharpe JH. 1989. Herpes simplex virus type 1 latency-associated transcripts are evidently not essential for latent infection. *EMBO J.* 8:505–11

Stevens JG. 1989. Human herpesviruses: a consideration of the latent state. *Microbiol. Rev.* 53:318–32

Stevens JG, Wagner EK, Devi-Rao GB, Cook ML, Feldman LT. 1987. RNA complementary to a herpesviruses α gene mRNA is prominent in latently infected neurons. *Science* 255:1056–59

Stow ND, Stow EC. 1986. Isolation and characterization of a herpes simplex virus type 1 mutant containing a deletion within the gene encoding the immediate early polypeptide Vmw 110. *J. Gen. Virol.* 67: 2571–85

Sunstrum JC, Chrisp CE, Levine M, Glorioso JC. 1988. Pathogenicity of glycoprotein C negative mutants in the mouse central nervous system. *Virus Res.* 11:17–32

Takamiya Y, Short MP, Ezzeddine ZD, Moolton FL, Breakefield XO, Martuza RL. 1992. Gene therapy of malignant brain tumors: a rat glioma line bearing the herpes simplex virus type 1 thymidine kinase gene and wild type retrovirus kills other tumor cells. *J. Neurosci. Res.* 33:493–503

Trousdale MD, Steiner I, Spivack JG, Deshmane SM, Brown S, MacLean AR. 1991. In vivo and in vitro reactivation impairment of a herpes simplex virus type 1 latency-associated transcript variant in a rabbit eye model. *J. Virol.* 65:6989–93

Vincent N, Ragot T, Gilgenkrantz H, Couton D, Chafey P, et al. 1993. Long-term correction of mouse dystrophic degeneration by adenovirus-mediated transfer of a mini-dystrophin gene. *Nat. Genet.* 5:130–34

Wagner EK, Flanagan WM, Devi-Rao GB, Zhang YF, Hill JM, et al. 1988. The herpes simplex virus latency-associated transcript is spliced during the latent phase of infection. *J. Virol.* 62:4577–85

Watson RJ, Clements JB. 1980. A herpes simplex virus type 1 function continuously required for early and late virus RNA synthesis. *Nature* 285:329–30

Whitley RJ, Kern ER, Chatterjee S, Chou J, Roizman B. 1993. Replication establishment of latency, and induced reactivation of herpes simplex virus γ₁ 34.5 deletion mutants in rodent models. *J. Clin. Invest.* 91:2837–43

Wilcox CL, Crnic LS, Pizer LI. 1992. Replication, latent infection, and reactivation in neuronal culture with a herpes simplex virus thymidine kinase-negative mutant. *Virology* 187:348–52

Wilcox CL, Johnson EM Jr. 1987. Nerve growth factor deprivation results in the reactivation of latent herpes simplex virus in vitro. *J. Virol.* 61:2311–15

Wilcox CL, Johnson EM Jr. 1988. Characterization of nerve growth factor-dependent herpes simplex virus latency in neurons in vitro. *J. Virol.* 62:393–99

Wilcox CL, Smith RL, Freed CR, Johnson EM Jr. 1990. Nerve growth factor-dependence of herpes simplex virus latency in peripheral sympathetic and sensory neurons in vitro. *J. Neurosci.* 10:1268–75

Yang Y, Nunes FA, Berencis K, Gonczol E, Engelhardt JF, Wilson JM. 1994. Inactivation of E2a in recombinant adenoviruses improves the prospect for gene therapy in cystic fibrosis. *Nat. Genet.* 7:362–69

York IA, Roo C, Andrews DW, Riddell SR, Graham L, Johnson DC. 1994. A cytosolic herpes simplex virus protein inhibits antigen presentation to CD8+ T lymphocytes. *Cell* 77:525–35

Zwaagstra JC, Ghiasi H, Slanina SM, Nesburn AB, Wheatley SC, et al. 1990. Activity of herpes simplex virus type 1 latency-associated transcript (LAT) promoter in neuron-derived cells: evidence for neuron specificity and for a large LAT transcript. *J. Virol.* 64: 5019–28

Annu. Rev. Neuroscience. 1996. 19:289–317
Copyright © 1996 by Annual Reviews Inc. All rights reserved

# PHYSIOLOGY OF THE NEUROTROPHINS

*Gary R. Lewin and Yves-Alain Barde*

Max-Planck Institute for Psychiatry, Department of Neurobiochemistry, D-82152 Planegg-Martinsried, Germany

KEYWORDS:     nerve growth factor, brain-derived neurotrophic factor, neurotrophin-3, sensory neurons, sympathetic neurons

## ABSTRACT

The neurotrophins are a small group of dimeric proteins that profoundly affect the development of the nervous system of vertebrates. Recent studies have established clear correlations between the survival requirements for different neurotrophins of functionally distinct subsets of sensory neurons. The biological role of the neurotrophins is not limited to the prevention of programmed cell death of specific groups of neurons during development. Neurotrophin-3 in particular seems to act on neurons well before the period of target innervation and of normally occurring cell death. In animals lacking functional neurotrophin or receptor genes, neuronal numbers do not seem to be massively reduced in the CNS, unlike in the PNS. Finally, rapid actions of neurotrophins on synaptic efficacy, as well as the regulation of their mRNAs by electrical activity, suggest that neurotrophins might play important roles in regulating neuronal connectivity in the developing and in the adult central nervous system.

## INTRODUCTION

The term neurotrophin refers to a family of proteins that have common structural features. Although the first family member, nerve growth factor (NGF), was identified four decades ago, only in the past six years has evidence been provided for the existence of additional neurotrophins. Compared to other growth factor families, the neurotrophins are of special interest to the neurobiologist because they exert their biological actions primarily on cells of the nervous system. The early availability of NGF, present in miraculous amounts in the submandibular gland of the adult male mouse, allowed much progress to be made in the delineation of its physiology. Over the past four years, the availability of recombinant, non-NGF neurotrophins has helped to make substantial in vivo work possible. More recently, gene targeting experiments, in which the neurotrophins or their receptor genes have been deleted, and anti-

289

body deprivation experiments have considerably expanded our knowledge of the biological role of these proteins.

This review is not intended to be comprehensive. Several recent reviews are available to readers needing more information on structural aspects of neurotrophins and of their receptors, on signal transduction following receptor binding, and on mRNA distributions (Barbacid 1994, Ibáñez 1994, Kaplan & Stephens 1994, Bothwell 1995). Our focus is on the physiology of neurotrophins and on the specificity of their biological actions in several distinct systems. Primary sensory neurons receive special emphasis because all the neurotrophins exert biological effects on these neurons.

# NEUROTROPHINS AND THEIR RECEPTORS

## Neurotrophins: Nomenclature

NGF, brain-derived neurotrophic factor (BDNF), and neurotrophin-3 (NT-3) are encoded by three distinct genes that have been identified in all higher vertebrates examined, including teleost fishes. NT-4 was first discovered in *Xenopus laevis* (Hallböök et al 1991), and it supports the survival of chick neurons, whereas NT-5 (Berkemeier et al 1991), thought to be the mammalian equivalent of NT-4, does not (Davies et al 1993a). Whether or not NT-4 and NT-5 are the same protein in different species is presently unclear. Very little is known about the role of NT-4 in *X. laevis,* and no NT-4 sequence has been identified in the chick. Thus, the designation NT-4/5 seems appropriate because these two proteins share properties in several in vitro assay systems. NT-6 has been identified in the teleost fish *Xiphophorus* (Götz et al 1994). It shows interesting, novel biochemical characteristics (see below), but whether or not NT-6 exists in mammals remains unclear.

Sequence comparisons, as well as consideration of the evolution of the PNS, suggest that BDNF and NT-3 might have evolved before NGF (Barde 1994). In NT-3, the replacement of only seven amino acids at strategic locations leads to a neurotrophin that displays the full biological activity of NT-3, BDNF, and NGF (Urfer et al 1994), thereby supporting the view that NT-3 may have been the neurotrophin ancestor gene.

## Structure and Biochemistry

All neurotrophins have similar biochemical characteristics. They are secretory proteins that are synthesized as precursor proteins that may provide an appropriate environment for the formation of disulfide bridges in mature neurotrophins linking the six cysteine residues (1-4, 2-5, 3-6) in NGF and BDNF (Acklin et al 1993), which are found at identical positions in all neurotrophins. Although the precursor sequences of the neurotrophins diverge significantly more than the mature part, they do not seem to be specific and can be exchanged with no measurable consequences on the formation of homo- or heterodimers, as examined for BDNF and NT-3 (Jungbluth et al 1994). In NGF, only limited segments of the prosequence seem to be essential for biosynthesis (Suter et al 1991). The mature (or processed) part of all six neurotrophins shows a high degree of sequence similarity, and approximately 50% of the amino acids are common to all six neurotrophins. The crystal structure of the murine NGF dimer could be resolved at 2.3 Å (McDonald et al 1991). The NGF monomer consists of 3 antiparallel β strands forming a large flat surface that is covered in the dimer by the corresponding surface of the second monomer. Most of the hydrophobic residues conserved in all neurotrophins contribute to this hydrophobic surface, and the overall three-dimensional structure of all neurotrophins may be similar. The neurotrophin protomers are not linked with disulfide bridges. The prediction that neurotrophins can form heterodimers has been tested. For example, BDNF–NT-3 can form stable heterodimers following either denaturation-renaturation of homodimer mixtures or coexpression of both genes in one cell (Radziejewski & Robinson 1993, Arakawa et al 1994, Jungbluth et al 1994, Heymach & Shooter 1995, Robinson et al 1995). BDNF–NT-3 heterodimers display the combined biological activity of both BDNF and NT-3 homodimers, but with a substantially lower potency, even when tested on neurons expressing both NT-3 and BDNF receptors (Jungbluth et al 1994). Whether such heterodimers are functionally relevant in vivo remains to be examined.

NT-6 has a sequence insert that is rich in glycine and basic amino acids. This insert is three times the size of an insert located at an identical position in mammalian NT-4/5, and it seems to confer to NT-6 the property to bind heparin, a characteristic not seen with the other neurotrophins (Götz et al 1994).

## Receptors

THE LOW-AFFINITY NEUROTROPHIN RECEPTOR p75   NGF, BDNF, NT-3, and NT-4/5 bind to a molecule that was initially described as the NGF receptor (D Johnson et al 1986, Radeke et al 1987). A cluster of basic amino acids located at one end of the neurotrophin dimer seems to be largely responsible for binding to the negatively charged p75 (Ibáñez et al 1992, Ryden et al 1995). Detailed kinetic and equilibrium studies comparing NGF, BDNF, and NT-3 reveal that the binding characteristics of these neurotrophins to p75 expressed in cell lines differ, both in terms of rate constants and of co-operativity, while the dissociation constants are roughly similar, between $10^{-9}$ M (NGF) and $4 \times 10^{-10}$ M (for NT-3 dimers) (Rodríguez-Tébar et al 1990, 1992; see also Bothwell 1995). The significance of the binding to p75 is still difficult to appreciate, because up until recently, no biochemical changes could be demonstrated to occur following neurotrophin binding. However, Dobrowsky et al (1994) have recently shown that in a glioma cell line, NGF can activate sphingomyelin hydrolysis following binding to p75, resulting in decreased thymidine incorporation. Whether or not this signaling function is important in vivo remains to be determined. If it is important, alternative signaling pathways must exist, because the selective deletion of *p75* by homologous recombination does not lead to a dramatic phenotype in the wide variety of tissues expressing p75 (Lee et al 1992). The phenotype of *p75 –/–* mice is somewhat reminiscent of a partial NGF deprivation (Lee et al 1992, 1994; Davies et al 1993b). This phenotype could be accounted for by another function of p75, which is to accelerate the on rate of binding to the tyrosine kinase receptor trkA, thus leading to high-affinity NGF receptors (Mahadeo et al 1994). These effects become measurable when the levels of p75 exceed those of *trkA* by about 10:1, and they might explain the numerous recent observations that biological responses can be elicited at lower concentrations of NGF when p75 is present or can be recruited in responsive cells (Chao 1994).

THE NEUROTROPHIN TYROSINE KINASE RECEPTORS    In mammals, three different tyrosine kinases receptors have been identified—designated trkA, -B, and -C. Expression of these receptors in cell lines has led to the conclusion that these receptors primarily bind NGF, BDNF and NT-4/5, and NT-3, respectively (Barbacid 1994). All trk genes code for more than one transcript, leading to a variety of functionally different receptor proteins; the most dramatic examples are forms of trkB and trkC (not of trkA) that completely lack the tyrosine kinase domain (Barbacid 1994). Null mutations introduced in the tyrosine kinase domains of each *trk* receptor gene have provided conclusive evidence that these receptors are necessary components in mediating the effects of the neurotrophins (Snider 1994). Thus, the phenotypes of such animals reveal striking similarities with those in which the *NGF, BDNF,* and *NT-3* genes have been deleted. Therefore, these receptors largely determine the specificity of neurotrophin effects (Barbacid 1994, Snider 1994).

Several observations indicate that the biological effects mediated by NT-4/5 and BDNF, while largely overlapping, are not identical, which suggests that neurons can discriminate between these neurotrophins. One of the clearest examples of this is that unlike BDNF, mammalian NT-4/5 does not support the survival of chick neurons (Davies et al 1993a). Using cell lines expressing both trkB and p75 and a NT-4/5 mutant without the ability to bind to p75, it has been shown that signaling through trkB by this mutant is less efficient than with wild-type NT-4/5 (Ryden et al 1995). In contrast, a BDNF mutant that does not bind to p75 shows no difference in its ability to activate trkB when compared with wild-type BDNF (Ryden et al 1995). A related observation has been reported for trkA, which is more readily activated by NT-3 when binding to p75 is prevented or when p75 is absent (Benedetti et al 1993, Clary & Reichardt 1994). Taken together, these observations suggest that one role of p75 might be to assist the trk receptors in discriminating their ligands.

NEUROTROPHIN RECEPTORS ON NEURONS    Some important questions as to the exact nature of neurotrophin receptors on neurons remain open. In particular, studies performed with PNS neurons have consistently indicated that the binding characteristics of NGF, BDNF, or NT-3 do not fully correspond with those observed with trk receptors expressed in fibroblastic cell lines. For example, the binding affinity of BDNF is about 100 times lower for chick trkB expressed in a cell line than it is for BDNF high-affinity sites on chick sensory neurons (Dechant et al 1993a). At present, the molecular details accounting for these discrepancies are still unclear. Also, all neuronal receptors may not have the same characteristics. In particular, binding studies performed with BDNF and NT-3 and cerebellar granule cells indicate that these neuronal receptors have affinities close to those observed with trkB or trkC expressed in fibroblasts (Lindholm et al 1993). However, this is not a general property of CNS neurons, because BDNF- and NT-3–specific high-affinity sites, closely resembling those described on PNS neurons, have been identified on chick retinal neurons (Rodríguez-Tébar et al 1993). Among other discrepancies between trk receptors in cell lines and neurons has been the observation that the formation of ras-GTP in response to BDNF markedly decreases with time in cerebellar or sensory neurons following prolonged exposure of these neurons to BDNF. Such is not the case when similar experiments are performed using trkB expressed in a cell line, where no BDNF desensitization occurs (Carter et al 1995).

Although trk receptors seem to be involved in the vast majority of the biological responses to neurotrophins, one should not assume that all biological responses to a neurotrophin imply the presence of a trk molecule on the responsive cells. For example, the migration of Schwann cells on sciatic explants seems to be mediated by the activation of p75. No full-length trk receptors have been reported on such cells, and the effects elicited by NGF can be blocked by application of antibodies to p75 (Anton et al 1994).

Finally, neurotrophin receptors are widely expressed in the PNS and CNS, both during development and in the adult. Their activation can lead to a wide variety of responses (see below), and these differences probably result from the activation of distinct second-messenger pathways in different cellular environments. Work using cell lines has shown that the second-messenger cascades activated by neurotrophins are various and complex (Kaplan & Stephens 1994), and this complexity is likely to be much higher in neurons, depending on whether they are still dividing, just exiting mitosis, long postmitotic, or lesioned.

## PERIPHERAL NERVOUS SYSTEM

### *Sensory Neurons*

EARLY DEVELOPMENT    Although NGF and trkA are expressed comparatively late in development, much in line with the role of NGF as a target-derived survival factor (Williams et al 1995), studies on the localization of NT-3 and trkC have indicated that both are expressed substantially earlier. For example, NT-3 mRNA and immunoreactivity are found in the neural tube, and trkC mRNA is seen in the neural tube before its closure, as well as on migrating neural crest cells (Tessarollo et al 1993, Williams et al 1993, Elkabes et al 1994, Kahane & Kalcheim 1994, Williams et al 1995). Experiments in vitro indicate that NT-3 exerts a mitogenic effect on newly migrated neural crest cells (Kalcheim et al 1992, Pinco et al 1993) and increases the number of neurons that differentiate when early dorsal root ganglia (DRGs) are cultured (Wright et al 1992). In vivo, analysis of quail embryos deprived of NT-3 by using a blocking antibody indicated a 34% loss of neurons in the nodose ganglion and the DRG at embryonic day 5 (E5) and E6 respectively, which is before the period of target-selected cell death in these ganglia (Gaese et al 1994). In *NT-3 −/−* mice, loss of trkC-expressing neurons has been noted in the DRG as early as E11.5, and the large NT-3–responsive neurons, normally present in the mouse DRG at E12.5, are absent (Tessarollo et al 1994, Kucera et al 1995b).

If NT-3 acts as a mitogen in vivo, as the in vitro experiments seem to indicate, application of excess NT-3 during the proliferative period in sensory ganglia might be expected to increase neuronal numbers. But in fact, recent experiments in which NT-3 levels were already increased at E3 revealed that neuronal numbers were actually markedly decreased in these ganglia. The mechanism underlying this surprising reduction may result from NT-3 prematurely taking neurons out of the cell cycle (Ockel et al 1996). Recent observations with rat sympathetic neurons in vitro support the idea that NT-3 can act as a mitogenic stop signal for rat sympathetic neuroblasts (Verdi & Anderson 1994). Thus, in vivo, the relative levels of NT-3 may regulate the number of neurons generated by determining when their precursors stop dividing.

The decrease in cell numbers seen after antibody application might result from the need of NT-3 as a survival factor for those neurons already projecting an axon to the peripheral or the central target. This explanation is supported by the results of in vitro experiments indicating that when neurons are isolated during the period of early axonal outgrowth in vivo, a majority of sensory neurons of the trigeminal ganglion depend for survival in culture on NT-3 (or BDNF). When the targets have been reached in vivo, a majority of these neurons then become dependent on NGF for survival (Davies 1994). Also, the long intersegmental projections formed by the first DRG neurons in the spinal cord (as early as E4 in the chick) are absent in anti-NT-3 treated embryos (Eide et al 1994, Eide & Glover 1995), suggesting that neural tube–derived NT-3 might be required for the survival of neurons forming intersegmental projections. The results obtained with *NT-3* –/– animals also indicate that sensory neurons are lost before programmed cell death, as the dramatic neuronal losses in the DRGs and the trigeminal ganglia are far higher than those expected from the number of *trkC*–expressing neurons in the DRG during programmed cell death (Mu et al 1993, Ernfors et al 1994b, Fariñas et al 1994).

A series of other observations suggest further possible roles for neurotrophins in the early embryo. In particular, studies with cultured neural crest cells show that BDNF can bias the differentiation of neural crest cells toward a sensory neuron type lineage (Kalcheim & Gendreau 1988, Sieber-Blum 1991). Also, the death of young neurons in vivo when a membrane is inserted between the neural tube and the DRG anlage can be prevented when such membranes are coated with BDNF and laminin (Kalcheim et al 1987), suggesting an early role for BDNF. Early analyses of *BDNF* –/– and *trkB* –/– mice might help to understand the in vivo relevance of these observations. Indeed, the lack of antibodies blocking the biological activity of BDNF has precluded acute in vivo deprivation experiments such as those that have helped us to understand the role of NGF and NT-3 during early development.

PERIOD OF TARGET-DEPENDENT NEURONAL SURVIVAL    In the mature animal, sensory neurons are physiologically and anatomically heterogenous (see Figure 1). Several lines of evidence indicate that this heterogeneity starts to develop around the time of target innervation, or even before (Fitzgerald 1987, Smith & Frank 1988). The roles of the neurotrophins in the development of these sensory neuron subtypes from target innervation onward is examined below.

*NGF*    NGF is not required for the survival of the entire sensory neuron population during normal development, raising the question as to which neurons require this factor. Neurochemical, anatomical, and behavioral evidence have long suggested that sensory neurons concerned with nociceptive information are lost in the absence of NGF induced by antibody injections (Lewin & Mendell 1993). In addition, the results of the gene knockout experiments [of either *NGF* or *trkA* (Crowley et al 1994, Smeyne et al 1994)] are so far in very good agreement with the results of the NGF-antibody experiments, illustrating the specificity of the antibody used. In mammals, as many as 70–80% of sensory neurons in the DRG or the trigeminal ganglion are dependent on NGF during development (Johnson et al 1980, Pearson et al 1983), and the vast majority of them have axon terminals in lamina I and II of the spinal cord

| SOMA SIZE | LARGE/MEDIUM | | SMALL | |
|---|---|---|---|---|
| AXON SIZE | Large Myelinated | Small Thinly Myelinated | Small Thinly Myelinated | Small Unmyelinated |
| TARGETS AND MODALITY | SKIN | SKIN | | SKIN |
| | slowly adapting mechanoreceptor (NT-3, BDNF) | D-hair receptor (NT-3) | | polymodal nociceptor (NGF) |
| | rapidly adapting mechanoreceptor (?) | mechanonociceptors (NGF) | | mechanonociceptors (NGF?) |
| | MUSCLE | MUSCLE | | MUSCLE |
| | spindle afferents (NT-3) | Nociceptors (NGF?) | | Nociceptors (NGF?) |
| | Golgi tendon organ afferents (NT-3) | | | |

*Figure 1*  The anatomical and physiological heterogeneity of sensory neurons of the DRG in mature animals. On a simple anatomical level, these cells can be separated in rodents into large light and small dark neurons (*top*). However, on the basis of axonal caliber and myelination, the cells can be further subdivided into three categories: large myelinated (Aβ-fibers), small thinly myelinated (Aδ-fibers), and small unmyelinated (C-fibers). These categories do not, however, predict the tissue innervated by the fiber or its sensory modality. This is illustrated by the classification of sensory neurons into various physiological types (*bottom*). The stimuli that best excite the neuron (adequate stimulus) are shown in parentheses. Where appropriate, an indication is also given of the neurotrophic factor that has been shown to exert biological actions on these fibers during development or in maturity.

(Ruit et al 1992, Crowley et al 1994, Smeyne et al 1994). However, only about 40% of mature DRG neurons appear to express high-affinity binding sites for NGF and sufficient levels of trkA receptors to be detected by in situ hybridization and immunocytochemistry (Verge et al 1989, 1992; Averill et al 1994). In bird embryos, only around 30% of the DRG neurons appear to be absolutely dependent on NGF during embryogenesis (Rohrer et al 1988). As in mammals, the majority of the sensory neurons in birds are nociceptive neurons that only respond to high-threshold, potentially damaging stimuli (Koltzenburg et al 1994). Approximately 50% of these nociceptors are polymodal in nature, i.e. they respond to more than one type of noxious stimuli (e.g. mechanical stimuli and pain-generating chemicals), while the remaining neurons only respond to noxious mechanical stimuli (Lewin & Mendell 1993). After anti-NGF treatment of chick embryos, essentially all chemo-sensitive and polymodal nociceptors are missing, but mechanonociceptors remain (Lewin et al 1994b). These results, together with the fact that lamina I and II are not innervated after anti-NGF treatment (Eide et al 1994), suggest that only one functional subtype of nociceptor is absolutely dependent on peripherally-derived NGF in the chick embryo. The numerical discrepancy between birds and mammals (when the numbers of neurons missing after NGF-antibody deprivation are compared) seems to correlate with the fact that most, if not all, nociceptors in mammals are polymodal, in that they respond to chemicals as well as to other stimuli (Kessler et al 1992). These results indicate that in vivo, the number of NGF-dependent sensory neurons is higher in mammals than in birds.

*Other neurotrophins* A number of studies have indicated that sensory neurons not responding to NGF can be supported by other neurotrophins when taken out of the embryos during the period of normally occurring cell death. Thus for example, placode-derived neurons [which are not NGF responsive (Davies & Lindsay 1985)] can be supported by BDNF or NT-3. Even among neural crest–derived neurons, the BDNF-responsive population only partially overlaps the NGF-responsive one, and a pure proprioceptive neuronal population responds to BDNF and NT-3, but not to NGF (Hohn et al 1990). In the DRG, the survival of many retrogradely labeled, muscle-innervating sensory neurons is specifically supported by NT-3 (Hory-Lee et al 1993, LoPresti & Scott 1994), suggesting that muscle afferents require NT-3 during the period of naturally occurring cell death. Indeed, the analysis of *NT-3 −/−* mice indicates that muscle spindle afferents are missing, together with Golgi tendon organ afferents (Ernfors et al 1994b). Furthermore, *NT-3 +/−* mice show a 50% loss of proprioceptive end organs, suggesting that the limited availability of NT-3 regulates the extent of proprioceptor survival (Ernfors et al 1994b). However, the peripheral targets, i.e. the intrafusal fibers of muscle spindles and the Golgi tendon organs, are unlikely to be the source of NT-3 during programmed cell death, because the end organs seem to be induced by the sensory fibers about two days after the cell death period in chick (Toutant 1982, Milburn 1984). Indeed, recent more detailed analysis of the *NT-3 −/−* animals indicates that sensory axons never reach the muscle spindles (Kucera et al 1995b). In the chick, peripheral administration of antibodies to NT-3 leads to the loss of muscle spindle afferents, suggesting that NT-3 may act as a target-derived factor (Oakley et al 1995). It thus seems that NT-3 may have a role in promoting the differentiation of muscle afferents early on in development, whereas later, it may become a target-derived survival factor for these neurons.

At least in mammals, sensory neurons innervating the skin are much more numerous than muscle afferents, and in adult rats, about 15% of cutaneous afferents express *trkC* mRNA (McMahon et al 1994). This suggests that a numerically large population of these afferents could respond to NT-3 in development. Indeed in chick embryos, many such skin-innervating neurons depend on NT-3 in primary culture (Hory-Lee et al 1993, LoPresti & Scott 1994). Electrophysiological analysis of *NT-3 +/−* animals indicates that slowly adapting mechanoreceptors and D-hair afferents appear to depend on NT-3 (Lewin et al 1995).

BDNF has been known for some time to rescue many sensory neurons from naturally occurring cell death in vivo (Hofer & Barde 1988). These include substantial numbers of dorsal root ganglion neurons, as well as of placode-derived nodose ganglion neurons, which innervate a large range of tissues in the viscera. The physiological types of most BDNF-dependent neurons remain to be established, but recent data on rodents suggest that the tyrosine hydroxylase positive neurons of the petrosal ganglion that innervate the carotid body are BDNF and NT-4/5 dependent (Hertzberg et al 1994). In the DRG and the nodose ganglion of *BDNF −/−* mice, about 35% of the neurons are lost (Ernfors et al 1994a, Jones et al 1994). Since there is no obvious loss of the end organs of muscle proprioceptors, in the DRG, the missing neurons in the mutant are assumed to innervate cutaneous targets. Receptor localization studies in the DRG indicate that *trkB* is predominantly expressed on medium- to large-sized neurons in adult animals (Mu et al 1993, Wright & Snider 1995). Interestingly, one study indicates extensive colocalization of *trkB* with *trkA* in the DRG neurons innervating the viscera (McMahon et al 1994).

The primary sensory neurons of the vestibular and auditory systems depend entirely on BDNF and NT-3 for their survival. Thus, in animals lacking both the *BDNF* and *NT-3* genes, essentially all neurons are lost in the vestibular and spiral ganglia (Ernfors et al 1995). A detailed analysis of the single mutants also revealed that the vestibular neurons depend mostly on BDNF. Conversely, the majority of spiral neurons depend on NT-3, except for the small population (5%) of the type 2 spiral neurons that innervate the outer hair cells. This population is selectively lost in *BDNF −/−* animals (Ernfors et al 1995).

NEUROTROPHIN REGULATION OF SENSORY PHENOTYPES    Changes in the levels of neurotrophins leading to no changes in cell numbers but to changes in neuronal phenotype have also been reported during development (Ritter et al 1991, Lewin et al 1992a). Thus, reducing the availability of NGF in neonatal rats by antibody injections leads to the replacement of nociceptive neurons, identified electrophysiologically, by neurons responding to low-threshold stimulation of hairs in the skin, so-called D-hair afferents (see Figure 1). These myelinated afferents have the same axonal caliber as the nociceptive neurons, and a critical period for the effects of the antibody could be defined between postnatal day 5 and 10. These results can be accounted for by a switch in neuronal phenotype induced by NGF deprivation, because they occur in the absence of changes in neuronal numbers. In addition, the decrease in nociceptors is compensated for by an increase in the numbers of D-hair afferents (Lewin et al 1992a). Subsequent studies on unmyelinated afferents showed similar changes, indicating that mechanoheat nociceptors were dramatically reduced in number by this same NGF deprivation (Lewin & Mendell 1994).

These data raise the interesting possibility that the availability of NGF plays a critical role in determining the phenotypic fate of the NGF-responsive sensory neurons. Although virtually nothing is known about such roles for the other neurotrophins, recent experiments indicate that NT-3 could act similarly (Lewin et al 1994a). Thus in chick embryos, continuous application of NT-3 (secreted by implanted, engineered cells) forces most of the cutaneous sensory neurons to lose nociceptive receptive field properties to take on a low-threshold phenotype (Lewin et al 1994a). At the low NT-3 levels used, these changes occur without changes in neuronal numbers. This phenotypic conversion is accompanied by a loss of innervation of lamina II, which is exclusively concerned with nociceptive inputs (Eide et al 1994). Thus, NT-3 treatment induces a coordinated change in the anatomical and physiological phenotype of these neurones. These results suggest that the availability of NT-3 may influence how many sensory neurons take on a low-threshold (or touch receptor) phenotype, much like NGF determines the numbers of nociceptors (Lewin & Mendell 1993; see also Kucera et al 1995b). Whether or not BDNF and NT4/5 play similar roles for other sensory phenotypes is unknown.

NEUROTROPHIN REGULATION OF DIFFERENTIATED PHENOTYPES    The importance of NGF for mature sensory neurons has been extensively investigated in adult animals autoimmunized against their own NGF (EM Johnson et al 1986). This work indicates that mature sensory neurons do not appear to require NGF derived from the periphery for their continued survival in intact animals (Rich et al 1984). This observation is supported by the results of experiments with mature sensory neurons in culture, which do not require any exogenous neurotrophic factor for survival (Lindsay 1988, Acheson 1995). In addition, the levels of neuropeptides used by sensory neurons as neurotransmitters or neuromodulators, calcitonin gene–related peptide (CGRP) and substance P in particular, are dramatically regulated by the availability of NGF (Otten 1984, Lindsay & Harmar 1989). These peptides are contained in a subset of NGF-responsive DRG neurons that appear to have high-affinity NGF receptors and to express high levels of *p75* mRNA and of *trkA* mRNA (Verge et al 1989, 1992). However, it is still unclear whether this subset of small NGF-responsive neurons corresponds to a physiologically defined subset of nociceptors. The tight regulation of these neuropeptides by NGF has recently been shown to be an integral part of this neurotrophin's role in the processing of painful stimuli following tissue injury (see below) (Lewin & Mendell 1993, McMahon et al 1995). The action potential configuration of some nociceptive sensory neurons can also be regulated by NGF in vivo and in vitro (Ritter & Mendell 1992, Shen & Crain 1994). The relevance of this for sensory neuron function is as yet unclear.

Little is known about the physiological importance of the newer neurotrophins for the function of mature sensory neurons. Like NGF however, these neurotrophins continue to be expressed in the targets of sensory neurons into adulthood, and in some cases, their levels actually increase, suggesting that they might still have a role for the differentiated function of sensory neurons (Maisonpierre et al 1990). BDNF, NT-3, and NT-4/5 are retrogradely transported by adult sensory neurons (Distefano et al 1992; Zhou & Rush 1993, 1994). The possibility also remains that the non-NGF neurotrophins may be required in vivo for the continued survival of some mature sensory neurons (Acheson et al 1995).

In *BDNF* –/+ animals, specific functional impairments in sensory neurons innervating the skin have recently been observed. Massively elevated mechanical thresholds were observed when recording from slowly-adapting cutaneous afferents, probably innervating touch domes. This impairment of function appears to be present in the absence of changes in the number of these afferents, suggesting a specific regulatory role of BDNF on mechanical threshold (Carroll et al 1994). These observations strongly suggest that neurotrophins in adult animals can regulate specific aspects of sensory neuron function.

NERVE GROWTH FACTOR: A LINK BETWEEN INFLAMMATION AND PAIN    Enough evidence has been accumulated to implicate NGF as a critically important chemical that links tissue injury to changes in sensory neuron function that can lead to pain (Lewin 1995). The crucial observation here is that blockade of NGF activity with antibodies prevents both heat and mechanical hyperalgesia, which normally follows tissue inflammation (Lewin et al 1994b, Woolf et al 1994). The up-regulation of NGF levels after inflammation is thought to mediate an increase in neuropeptide synthesis, which then leads to the hyperalgesia (Donnerer et al 1992, Lewin & Mendell 1993, Lewin et al 1994b, Lewin 1995). It appears that the action of NGF here is to increase the synaptic efficacy of sensory neuron connections in the spinal cord, which in part mediates the hyperalgesia (Lewin et al 1992b, 1994b). This model is supported by the fact that exogenous NGF can cause hyperalgesia in animals and humans (Lewin et al 1993, Petty et al 1994). If the other neurotrophins are regulated by inflammatory mediators, physiologically relevant changes may occur in the function of their target sensory neurons as well. Recent results indicate that in visceral tissues, BDNF mRNA levels are dramatically increased by inflammatory injury (SB McMahon, unpublished observations). The functional importance of this observation for the regulation of visceral afferent function remains to be determined.

NEUROTROPHINS AND PERIPHERAL NERVE REGENERATION    As well as promoting the survival of sensory neurons in primary culture, all neurotrophins promote neurite extension, suggesting that they may be necessary for mature sensory neurons to regenerate their axons after nerve injury in vivo. Additional indirect support for this view came from studies indicating a dramatic up-regulation of *NGF* and *BDNF* (but not *NT-3*) mRNA levels in the distal stump of transected sciatic nerves in the adult rat (Heumann et al 1987, Meyer et al 1992, Funakoshi et al 1993). Finally, NGF has been shown to promote axonal regeneration when placed in chambers through which a nerve is allowed to regenerate (Hollowell et al 1990). This and other studies of this type clearly show the effects of NGF, but it appears unlikely that they are direct, because at least some of the axonal regeneration is by large myelinated axons, which do not normally respond to NGF (Lewin et al 1993) and might not express the appropriate receptors (McMahon et al 1994). Are any of the neurotrophins directly involved and required in peripheral nerve regeneration? Surprisingly, direct and convincing evidence is still lacking. On the contrary, antibody experiments indicate that NGF does not seem to be required for nerve regeneration but that it is required for collateral sprouting of nociceptive afferent terminals in their targets, which are known to be NGF sensitive (Diamond et al 1987; 1992a,b). As it is difficult to positively demonstrate an adequate penetration of sufficient amounts of NGF antibodies in the adult lesioned nerve, experiments with adult transgenic animals lacking the *NGF* or *BDNF* gene would be useful to settle this issue.

Irrespective of their participation in nerve regeneration, the neurotrophins are considered potentially useful agents to treat peripheral neuropathies. NGF, for example, has been proposed as a therapeutic agent for neuropathies commonly accompanying diabetes (Brewster et al 1994). The known hyperalgesic side effects of NGF may, however, pose some problems for the use of this factor (Lewin et al 1993, Petty et al 1994). Recently, NT-3 has been used to treat animal models of large fiber neuropathies induced by cisplatin or large doses of pyridoxin, and initial results indicate that it can substantially reverse the anatomical and electrophysiological deficits in these animals (Gao et al 1994, Helgren et al 1994).

## Sympathetic Neurons

Following the seminal observation in newborn rodents treated with NGF antibodies, NGF is known to be required for the development of the sympathetic paravertebral chain (Cohen 1960, Levi-Montalcini & Booker 1960). Thus, more than 90% of the sympathetic neurons disappear in the superior cervical ganglion of the mouse following antibody injection at birth, a finding confirmed in studies of animals lacking functional NGF or trkA genes (Crowley et al 1994, Smeyne et al 1994). But much like NGF-responsive sensory neurons (see above), many sympathetic neurons also first go through a phase of NT-3 dependence. For example, the survival of very young sympathetic neurons is supported in vitro by NT-3 (both in rodents and chick) and only later by NGF (Birren et al 1993, Dechant et al 1993b). Also, these cells have high-affinity receptors for NT-3, but not for NGF at early embryonic ages (Rodríguez-Tébar & Rohrer 1991, Dechant et al 1993b). Verdi & Anderson (1994) suggested that NT-3 actually induces the expression of trkA mRNA in rodents. Most importantly, in the NT-3 −/− animals, about 50% of the neurons are missing in the superior cervical ganglion (Ernfors et al 1994b, Fariñas et al 1994). Thus, these data show striking parallels with observations made with primary sensory neurons, and in the peripheral nervous system, all NGF-dependent neurons may first go through a phase during which they depend on NT-3 (Davies 1994).

In the adult animal, sympathetic neurons still require NGF for their continued survival, as shown by large losses in the superior cervical ganglion of animals immunized with NGF (Gorin & Johnson 1980). This continuous dependence on NGF for survival of the sympathetic neurons is unlike the situation with NGF-sensitive sensory neurons (Gorin & Johnson 1980).

# CENTRAL NERVOUS SYSTEM

## Motoneurons

Following axotomy in newborn rodents, many motoneurons are eliminated (Kuno 1990). However, a significant proportion can be rescued by the administration of BDNF (Sendtner et al 1992, Yan et al 1992, Koliatsos et al 1993). In vitro studies with purified rat motoneurons also indicate that at low concentrations, BDNF, NT-4/5, and NT-3 can prevent cell death (Henderson et al 1993). However, a molecule structurally unrelated to the neurotrophins and belonging to the TGF-β superfamily, the so-called glial–cell line–derived neurotrophic factor (GDNF), seems to support the survival of motoneurons at significantly lower concentrations than the neurotrophins are able to (Henderson et al 1994). Although there is little doubt that neurotrophins can prevent the death of motoneurons following injury, it is still unclear if BDNF or any of the other neurotrophins are of unique importance in preventing the death of motoneurons during development. No decrease in motoneuron numbers could be observed in *BDNF –/–* or in *BDNF –/–* and *NT-4/5 –/–* animals (Ernfors et al 1994a, Jones et al 1994, Conover et al 1995, Liu et al 1995). In contrast, the *trkB –/–* animals have been reported to show a substantial loss of motoneurons in the facial nucleus and the spinal cord (Klein et al 1993). These mice died within the first postnatal week. However, upon continuous breeding into a C56-Bl6 background, the *trkB –/–* animals survive for up to three weeks and do not display significant losses of motoneurons. Whether the differences in motoneuron survival in the *trkB –/–* animals are due to differences in the genetic background of these mice remains to be determined (I Silos-Santiago & M Barbacid, personal communication). With the *NT-3 –/–* and, especially, with the *trkC –/–* animals, the published data could be interpreted to indicate that the gamma motoneurons might be missing, perhaps as a consequence of the muscle spindle loss (Klein et al 1994, Kucera et al 1995a).

It is still unclear if BDNF (or other neurotrophins) can significantly prevent naturally occurring cell death. So far, one study shows that the administration of BDNF during normal development leads to a modest prevention of motoneuron death: About one third of the lumbo-sacral motoneurons that are normally eliminated between E6 and E10 can be rescued by the addition of 5 μg BDNF daily, whereas NGF is without effect (Oppenheim et al 1992). Quantitatively very similar effects were recently noted following the administration of GDNF, and it is unclear if the GDNF- and BDNF-responsive populations are the same (Oppenheim et al 1995). In any case, it appears the majority of motoneurons cannot be rescued during normal development by the addition of BDNF or indeed of any single other neurotrophic factor.

Finally, although the effects of BDNF are clear-cut following axotomy or deafferentation, they do not seem to be long lasting. Indeed, three weeks after section of the sciatic nerve and in spite of the repeated addition of BDNF, motoneuron numbers are not different from untreated animals (Eriksson et al 1994, Vejsada et al 1995).

## Retinal Ganglion Cells

Studies in rat, chick, and *X. laevis* have indicated that BDNF prevents the death of cultured retinal ganglion cells (JE Johnson et al 1986, Rodríguez-

Tébar et al 1989, Cohen-Cory & Fraser 1994). Also in vivo, intraoccular injections of BDNF and NT-4/5 rescue significant numbers of identified retinal ganglion cells that would have otherwise died following section of the adult rat optic nerve (Mansour-Robaey et al 1994). However, although the cell bodies of the ganglion cells were rescued, their axons fail to exit the eye cup. It is still unclear if programmed cell death can be significantly prevented during development by the addition of BDNF, or of any other neurotrophin. Preliminary experiments performed in chick embryos indicate that cell death in the retinal ganglion cell layer cannot be significantly prevented by BDNF (Cellerino et al 1995), and only transient effects of NT-4/5 have been noted in the rat (Cui & Harvey 1994). In *BDNF –/–* animals, no gross defects in the neural retina have been noted (Jones et al 1994), and the number of axons in the optic nerve appears to be normal 13 days postnatally (A Cellerino, unpublished observations). So far, no changes have been reported in the retina or in the optic nerve of *NT-3 –/–* mice. However, treatment of newborn rats with NT-3–blocking antibodies causes a substantial reduction in the number of oligodendrocytes in the optic nerve (Barres et al 1994). In vitro, NT-3 has been shown to collaborate with PDGF to promote the clonal expansion of oligodendrocyte precursors (Barres et al 1994).

## Isthmo-Optic Nucleus

One model CNS structure where programmed cell death during development has been studied extensively is the isthmo-optic nucleus (ION) of birds (Clarke 1992). In the chick, about 60% of the neurons normally die between E12 and E17, which is the period in which they innervate their targets (the amacrine cells) in the retina (Clarke 1992). Many of these neurons can be rescued by adding more targets (Boydston & Sohal 1979), which suggests that some target-derived influence might regulate their survival. This is in contrast to the situation with retinal ganglion cells, where extra targets made available by fetal monocular enucleation in cats do not lead to more retinal ganglion cell survival in the contralateral eye (Sretavan & Shatz 1986). If BDNF or (to a lesser extent) NT-3 are injected into the eye or given systemically, normally occurring cell death can be substantially prevented in the ION (Von Bartheld et al 1994). The neurons can also be shown to retrogradely transport both neurotrophins from the eye (Von Bartheld et al 1994), and *BDNF* mRNA has been detected in the developing chick retina (Herzog et al 1994). NGF is less efficiently transported by the neurons, but remarkably, it increases normally occurring cell death in the ION when administered into the eye. p75 seems to be involved in this phenomenon, because the effects can be blocked with antibodies to p75 (Von Bartheld et al 1994). A direct involvement of p75 in mediating cell death has also been proposed (Rabizadeh et al 1993, Barrett & Bartlett 1994).

## *Septo-Hippocampal System*

In the mammalian CNS, the most extensively studied example of neurotrophin influence is the rodent septo-hippocampal system. Following the accidental observation that NGF can be retrogradely transported by cholinergic neurons in the basal forebrain when injected into the hippocampus (Schwab et al 1979), lesion studies indicated that injections of NGF could prevent the death of axotomized septal neurons (Hefti 1986), as well as prevent their atrophy and loss of cholinergic marker enzymes (Peterson et al 1990; Naumann et al 1992, 1994). NGF is also able to reverse the atrophy of these neurons normally seen in a sizable proportion of aging rats and to improve their behavioral performance (Fischer et al 1987). These results point to a role of NGF as a maintenance factor for these neurons in the adult. Further support for this idea comes from *NGF –/–* animals that survive when transgenic animals expressing the *NGF* gene under the control of a keratin promotor (Phillips et al 1994). Phillips et al (1994) have noted a marked reduction in the terminal fields of the forebrain cholinergic neurons in the hippocampus of such animals.

The physiological role of NGF during the development of the basal forebrain cholinergic neurons is still far from clear. Preliminary analyses of *NGF –/–* animals revealed that most basal forebrain neurons seem to be present and to express a relatively normal cholinergic phenotype (Crowley et al 1994). Curiously, such is not the case with the *trkA –/–* animals, which show a marked reduction of acetylcholinesterase (Smeyne et al 1994). It thus seems that in rodents, the majority of the basal forebrain cholinergic neurons do not absolutely require NGF for survival during development, even though they clearly respond to this factor in vitro and in vivo after lesion (Hefti 1986, Hartikka & Hefti 1988). NGF in the adult may therefore only be required for maintenance of their cholinergic phenotype, as well as for the terminal field size. Thus for these neurons, as for the motoneurons and for the rodent retinal ganglion cells, neurotrophins are not absolutely required to regulate target-dependent cell death in the CNS. Also, compensatory mechanisms might be brought into play when one particular gene is deleted, and this possibility will not be easy to investigate in view of the variety of factors able to support (at least in culture) the survival of the same neuronal population. For example, BDNF, interleukin-6, and NGF have all been shown to support the survival of basal forebrain cholinergic neurons (Hama et al 1989, Alderson et al 1990, Hartikka & Hefti 1988).

## NEUROTROPHINS AND SYNAPTIC PLASTICITY

Synaptic plasticity as discussed here refers to the functional modification of connections within the developing or mature CNS, often accompanied by anatomical changes in the wiring of such connections. One of the first observations suggesting a role for neurotrophins in synaptic plasticity came from the PNS. Here NGF was shown to be capable of preventing the loss of functional synaptic connections made by preganglionic efferents onto sympathetic neurons after their axons were cut (Purves & Njå 1976, Purves 1988). The presence of the neurotrophins and their receptors in the developing and mature brain, together with the results of experiments indicating that neurotrophins levels may be regulated by activity (Gall & Isackson 1989, Zafra et al 1990), has led to investigations about their potential roles in plasticity in the visual system during development and in the hippocampal formation in more mature animals.

## Visual System

Ocular dominance in the visual cortex is established as a result of the activity-dependent sorting of thalamic afferents. Monocular deprivation has long been known to allow dominance of afferent input from the spared eye in the visual cortex. Maffei et al (1992) have shown this phenomenon to be influenced by intraventricular injections of NGF. Although most cortical neurons normally respond only to the nondeprived eye, they show a normal ocular dominance pattern after NGF administration (Maffei et al 1992). Follow-up experiments with NGF antibodies indicated that neurons in the visual cortex have a reduced acuity, i.e. larger receptive fields (Berardi et al 1994). Although ocular dominance plasticity itself is not influenced by NGF deprivation, the critical developmental period over which ocular dominance plasticity could be elicited was extended (Domenici et al 1995). One interpretation of these experiments is that in the cortex, NGF normally acts to stabilize synaptic contacts of thalamic afferents. In this scenario, NGF deprivation would leave the cortical network more prone to activity-dependent modifications, and conversely, exogenous NGF would prematurely stabilize contacts and prevent plasticity. The underlying cellular mechanisms of these changes are not well understood. In cats, although NGF administration does not affect the normal segregation of thalamic afferents into eye-specific patterns, BDNF or NT-4/5 injections completely prevent the segregation of thalamic afferents in the developing system (Cabelli et al 1995).

During development of the chick tectum, *BDNF* mRNA levels increase during the period of innervation by the retinal ganglion cells, and this increase is regulated by the activity of the retinal afferents at the time of the first contact between the axons of the retinal ganglion cells and the tectum (Herzog et al 1994). Castrén et al (1992) have shown that *BDNF* mRNA levels in the newborn and adult rat are regulated by activity arising from the retina. What the respective roles of NGF and BDNF exactly are in this classical model of developmental plasticity remains unclear, as is a possible role of other neurotrophins.

## Hippocampal Plasticity

BDNF, NGF, and NT-3, as well as *trkB* and *trkC,* are expressed at comparatively high levels in the adult hippocampus (Klein et al 1989, Hofer et al 1990, Phillips et al 1990, Wetmore et al 1990, Lamballe et al 1991, Maisonpierre et al 1990), and the regulation of their synthesis and their biological effects has been extensively studied in this structure.

FAST EFFECTS OF NEUROTROPHINS    Classically, NGF has been thought to affect synaptic connectivity by influencing the growth of neuronal processes. Thus, NGF has been shown to influence neuronal soma size, dendritic complexity (Purves 1988, Ruit et al 1990), and the number and size of synaptic contacts in the brain (Garofalo et al 1992). Other neurotrophins may have similar effects on different CNS populations, and long-term morphological changes are of obvious importance in functional connectivity (Lisman & Harris 1993, Pierce & Lewin 1994). However, recent experiments indicated additional mechanisms by which neurotrophic factors could influence synaptic function. Thus, observations with NT-3 and BDNF show that they could influence synaptic function over a short time scale of minutes or less. For example, within minutes bath application of NT-3 or BDNF in cultures containing amphibian spinal cord neurons and muscle cells induces an increase of the spontaneous and the impulse-invoked synaptic activity recorded in the muscle cells (Lohof et al 1993). These effects seem to be presynaptic in origin and to require the activation of a kinase. The relatively slow time course of the effects suggests either a very slow mobilization of additional synaptic vessicles or the formation of new functional contacts through growth cone expansion. Berninger et al (1993) subsequently observed that BDNF or NT-3 could rapidly increase intracellular calcium in cultured hippocampal neurons, an obvious possible mechanism for increasing transmitter release. Following local application of BDNF, such effects can be observed within seconds (B Berninger & H Thoenen, personal communication). Similar potentiation events to those observed by Lohof et al (1993) have also been observed in cultures of hippocampal neurons (Leßmann et al 1994) and in more intact preparations, such as slices of visual cortex and hippocampus (Carmignoto et al 1993, Kang & Schuman 1995, Knipper et al 1994). In addition, Kim et al (1994) reported that NT-3 increases the activity of hippocampal neurons in culture, possibly by decreasing GABAergic transmission.

Most recently, application of neurotrophins (NGF, BDNF, NT-3, or NT-4/5) into the intact brain of adult animals has been shown to produce remarkably powerful, acute, and coordinated effects on the activity of whole neuronal ensembles, manifested as seizure activity in some cases (Berzarghi et al 1995). These data indicate that neurotrophins can have acute and powerful effects, presumably on synaptic efficacy even in the intact the brain.

## Regulation and Release

Early experiments in the hippocampus showed that limbic seizures lead to increased *NGF* mRNA expression by hippocampal neurons (Gall & Isackson 1989). Subsequent work indicated that *BDNF* mRNA levels are also dramatically up-regulated by hippocampal seizure activity induced by chemical and electrical stimuli (Zafra et al 1990, Ernfors et al 1991, Isackson et al 1991). Interestingly, the time course, the distribution, and the extent of mRNA induction are quite different for *BDNF* and *NGF; NT-3* mRNA levels, if anything may decrease (Ernfors et al 1991, Isackson et al 1991, Rocamora et al 1992, Humpel et al 1993). The seizure models used in many of these studies can lead to considerable morphological and physiological changes in the hippocampus. Much less radical electrical stimuli that lead to long-term potentiation in the CA1 or CA3 regions of the hippocampus also increase neurotrophin expression (Patterson et al 1992, Castrén et al 1993). The increased levels of *BDNF* mRNA also lead to increased basal and evoked levels of the BDNF protein (Nawa et al 1995).

Remarkably little is yet known regarding the mechanisms of release of neurotrophins by neurons. Recent work has shown that NGF is secreted by hippocampal neurons in culture and in slices, and it appears to be released by a novel pathway that is activity dependent but, surprisingly, independent of extracellular calcium (Blöchl & Thoenen 1995). However, it requires extracellular sodium, and the release can be blocked by tetrodotoxin (Blöchl & Thoenen 1995). If the other neurotrophins are released in the brain by similar mechanisms, the extracellular availability of the neurotrophins may be, at least in part, determined by the activity of the neuronal ensembles.

It thus appears that neurotrophin levels can be increased in the brain by events that are known to lead to synaptic plasticity. The neurotrophins in turn can themselves have powerful effects on the strength of synaptic connections. Therefore, the existence of mechanisms efficiently removing the neurotrophins from the extracellular space appears to be required. Splice variants of *trkB* and *trkC* mRNA that entirely lack the tyrosine kinase domains have been detected predominantly on nonneuronal cells (Barbacid 1994). These variants code for truncated receptors, possibly allowing for the rapid and selective elimination of neurotrophins, as such receptors have been shown not only to bind the neurotrophins, but also to rapidly eliminate them by internalization (Biffo et al 1995). This role would be analogous to uptake systems described for classical neurotransmitters.

HIPPOCAMPAL LONG-TERM POTENTIATION    So far, most experiments suggesting a role of neurotrophins in hippocampal synaptic plasticity have been performed in vitro, and when performed in vivo are pharmacological in nature.

However, the consequences of neurotrophin removal also begin to be assessed with regard to long-term potentiation (LTP) in the hippocampus. In *BDNF* mouse mutants, a dramatic attenuation of LTP has been observed in the CA1 subfield; the incidence and the amplitude of the potentiation seen after stimulation of the so-called Schaffer collaterals are reduced by about 70% (Korte et al 1995). Surprisingly, about the same reduction is observed in both heterozygous and homozygous animals (Korte et al 1995). This latter finding suggests that the levels of BDNF in the CNS are normally present in limiting amounts in normal animals and that small reductions of BDNF levels can lead to impairments in synaptic function. However, the direct involvement of BDNF in the induction of LTP remains to be demonstrated.

## CONCLUSIONS

Although it is clear that the different neurotrophins are required for the survival of functionally defined and distinct subsets of sensory neurons, the non-NGF neurotrophins, and especially NT-3, seem to regulate neuronal numbers during development in ways that are different from, and in addition to, the classical target-derived model. Most, if not all, NGF-dependent neurons in the PNS seem to show a previous developmental dependence on NT-3. NT-3 and its functional receptor trkC are expressed very early during development, substantially before the period of peripheral target innervation, and NT-3 seems to regulate neuron numbers by influencing the generation of neurons from precursors.

With mouse mutants carrying deleted neurotrophin genes, the effects of neurotrophins on cell numbers can be readily demonstrated in the PNS, but not in the CNS, even though many CNS neurons have receptors for, and respond to, the neurotrophins. It is unclear at present whether this is due to compensatory effects by other gene products. Neuronal development is also influenced by neurotrophin availability in ways other than the regulation of cell numbers. In particular, the functional identity of neurons can be influenced by the availability of neurotrophins. NT-3 appears to bias sensory neurons towards a low-threshold mechanoreceptor phenotype, while NGF moves them toward a more nociceptive phenotype. The functional phenotype of CNS neurons may also be affected in similar ways.

Recent data obtained on the rapid actions of neurotrophins on synaptic efficacy and their powerful in vivo effects on the developing visual system indicate that the neurotrophins might play specific and important roles in regulating neuronal connectivity and synaptic strength in the brain. This possibility is especially attractive in view of the fact that the levels of neurotrophins are affected by neuronal activity. The role of neurotrophins in neuronal connectivity is likely to represent a major focus of research in the coming years.

ACKNOWLEDGMENTS

We thank our colleagues at the Max-Planck Institute for Psychiatry for valuable comments, in particular Alessandro Cellerino, Johnathan Cooper, and Hans Thoenen.

## Literature Cited

Acheson A, Conover JC, Fandl JP, DeChiara TM, Russell M, et al. 1995. A BDNF autocrine loop in adult sensory neurons prevents cell death. *Nature* 374:450–53

Acklin C, Stoney K, Rosenfeld RA, Miller JA, Rohde MF, et al. 1993. Recombinant human brain-derived neurotrophic factor (rHuBDNF). *Int. J. Peptide Protein Res.* 41:548–56

Alderson RF, Alterman AL, Barde Y-A, Lindsay RM. 1990. Brain-derived neurotrophic factor increases survival and differentiated functions of rat septal cholinergic neurons in culture. *Neuron* 5:297–306

Anton ES, Weskamp G, Reichardt LF, Matthew WD. 1994. Nerve growth factor and its low-affinity receptor promote Schwann cell migration. *Proc. Natl. Acad. Sci. USA* 91:2795–99

Arakawa T, Haniu M, Narhi LO, Miller JA, Talvenheimo J, et al. 1994. Formation of heterodimers from three neurotrophins, nerve growth factor, neurotrophin-3, and brain-derived neurotrophic factor. *J. Biol. Chem.* 269:27833–39

Averill S, McMahon SB, Clary DO, Reichardt LF, Priestley JVP. 1994. trkA immunohistochemistry in adult rat dorsal root ganglia. *Eur. J. Neurosci.* 7:1484–94

Barbacid M. 1994. The trk family of neurotrophin receptors. *J. Neurobiol.* 25:1386–403

Barde Y-A. 1994. Neurotrophic factors: an evolutionary perspective. *J. Neurobiol.* 25:1329–33

Barres BA, Raff MC, Gaese F, Bartke I, Dechant G, et al. 1994. A crucial role for neurotrophin-3 in oligodendrocyte development. *Nature* 367:371–75

Barrett GL, Bartlett PF. 1994. The p75 nerve growth factor receptor mediates survival or death depending on the stage of sensory neuron development. *Proc. Natl. Acad. Sci. USA* 91:6501–5

Benedetti M, Levi A, Chao MV. 1993. Differential expression of nerve growth factor receptors leads to altered binding affinity and neurotrophin responsiveness. *Proc. Natl. Acad. Sci. USA* 90:7859–63

Berardi N, Cellerino A, Domenici L, Fagiolini M, Pizzorusso T, et al. 1994. Monoclonal antibodies to nerve growth factor affect the postnatal development of the visual system. *Proc. Natl. Acad. Sci. USA* 91:684–88

Berkemeier LR, Winslow JW, Kaplan DR, Nikolics K, Goeddel DV, et al. 1991. Neurotrophin-5: a novel neurotrophic factor that activates trk and trkB. *Neuron* 7:857–66

Berninger B, Garcia DE, Inagaki N, Hahnel C, Lindholm D. 1993. BDNF and NT-3 induce intracellular $Ca^{2+}$ elevation in hippocampal neurones. *NeuroReport* 4:1303–6

Berzaghi MP, Gutierrez R, Heinemann U, Lindholm D, Thoenen H. 1995. Neurotrophins induce acute transmitter-mediated changes in brain electrical activity. *Abstr. Soc. Neurosci.* 21:226.3

Biffo S, Offenhäuser N, Carter BD, Barde Y-A. 1995. Selective binding and internalisation by truncated receptors restrict the availability of BDNF during development. *Development* 121:2461–70

Birren SJ, Lo L, Anderson DJ. 1993. Sympathetic neuroblasts undergo a developmental switch in trophic dependence. *Development* 119:597–610

Blöchl A, Thoenen H. 1995. Characterization of nerve growth factor (NGF) release from hippocampal neurons: evidence for a constitutive and an unconventional sodium-dependent regulated pathway. *Eur. J. Neurosci.* 7:1220–28

Bothwell M. 1995. Functional interactions of neurotrophins and neurotrophin receptors. *Annu. Rev. Neurosci.* 18:223–53

Boydston WR, Sohal GS. 1979. Grafting of additional periphery reduces embryonic loss of neurons. *Brain Res.* 178:403–10

Brewster WJ, Fernyhough P, Diemel LT, Mohiuddin L, Tomlinson DR. 1994. Diabetic neuropathy, nerve growth factor and other neurotrophic factors. *Trends Neurosci.* 17: 321–25

Cabelli RJ, Hohn A, Shatz CJ. 1995. Inhibition of ocular dominance column formation by infusion of NT-4/5 or BDNF. *Science* 267: 1662–66

Carmignoto G, Negro A, Vicini S. 1993. NGF and BDNF modulate excitatory synapses in rat visual cortical neurons. *Abstr. Soc. Neurosci.* 19:690.9

Carroll P, Lewin GR, Koltzenburg M, Toyka KV, Wolf E, et al. 1994. Analysis of mice carrying targeted mutations of the BDNF gene. *Abstr. Soc. Neurosci.* 20:451.5

Carter B, Zirrgiebel U, Barde Y-A. 1995. Differential regulation of p21$^{ras}$ activation in neurons by nerve growth factor and brain-derived growth factor. *J. Biol. Chem.* 270: 21751–57

Castrén E, Pitkanen M, Sirvio J, Parsadanian A, Lindholm D, et al. 1993. The induction of LTP increases BDNF and NGF mRNA but decreases NT-3 mRNA in the dentate gyrus. *NeuroReport* 4:895–98

Castrén E, Zafra F, Thoenen H, Lindholm D. 1992. Light regulates expression of brain-derived neurotrophic factor mRNA in rat visual cortex. *Proc. Natl. Acad. Sci. USA* 89:9444–48

Cellerino A, Strohmaier C, Barde Y-A. 1995. Brain-derived neurotrophic factor and the developing chick retina. In *Life and Death in the Nervous System*, ed. C Ibáñez, pp. 131–39. London: Elsevier

Chao MV. 1994. The p75 neurotrophin receptor. *J. Neurobiol.* 25:1373–85

Clarke PGH. 1992. Neuron death in the developing avian isthmo-optic nucleus, and its relation to the establishment of functional circuitry. *J. Neurobiol.* 23:1140–58

Clary DO, Reichardt LF. 1994. An alternatively spliced form of the nerve growth factor receptor TrkA confers an enhanced response to neurotrophin 3. *Proc. Natl. Acad. Sci. USA* 91:11133–37

Cohen S. 1960. Purification of a nerve-growth promoting protein from the mouse salivary gland and its neuro-cytotoxic antiserum. *Proc. Natl. Acad. Sci. USA* 46:302–11

Cohen-Cory S, Fraser SE. 1994. BDNF in the development of the visual system of *Xenopus. Neuron* 12:747–61

Conover JC, Erickson JT, Katz DM, Bianchi LM, Poueymirou WT, et al. 1995. Neuronal deficits, not involving motor neurons, in mice lacking BDNF and/or NT4. *Nature* 375:235–38

Crowley C, Spencer SD, Nishimura MC, Chen KS, Pitts-Meeks, et al. 1994. Mice lacking nerve growth factor display perinatal loss of sensory and sympathetic neurons yet develop basal forebrain cholinergic neurons. *Cell* 76: 1001–11

Cui Q, Harvey AR. 1994. NT-4/5 reduces naturally occurring retinal ganglion cell death in neonatal rats. *NeuroReport* 5:1882–84

Davies AM. 1994. The role of neurotrophins in the developing nervous system. *J. Neurobiol.* 25:1334–48

Davies AM, Horton A, Burton LE, Schmelzer C, Vandlen R, et al. 1993a. Neurotrophin-4/5 is a mammalian-specific survival factor for distinct populations of sensory neurons. *J. Neurosci.* 13:4961–67

Davies AM, Lee K-F, Jaenisch R. 1993b. p75-deficient trigeminal sensory neurons have an altered response to NGF but not to other neurotrophins. *Neuron* 11:565–74

Davies AM, Lindsay RM. 1985. The cranial sensory ganglia in culture: differences in the response of placode-derived and neural crest-derived neurons to nerve growth factor. *Dev. Biol.* 111:62–72

Davies AM, Thoenen H, Barde Y-A. 1986. The response of chick sensory neurons to brain-derived neurotrophic factor. *J. Neurosci.* 6: 1897–904

Dechant G, Biffo S, Okazawa H, Kolbeck R, Pottgiesser J, et al. 1993a. Expression and binding characteristics of the BDNF receptor chick trkB. *Development* 119:545–58

Dechant G, Rodríguez-Tébar A, Kolbeck R, Barde Y-A. 1993b. Specific high-affinity receptors for neurotrophin-3 on sympathetic neurons. *J. Neurosci.* 13:2610–16

Diamond J, Coughlin M, MacIntyre L, Holmer M, Visheau B, et al. 1987. Evidence that endogenous beta nerve growth factor is responsible for the collateral sprouting but not the regeneration of nociceptive axons in adult rats. *Proc. Natl. Acad. Sci. USA* 84: 6596–600

Diamond J, Foerster A, Holmes M, Coughlin M. 1992a. Sensory nerves in adult rats regenerate and restore sensory function to the skin independently of endogenous NGF. *J. Neurosci.* 12:1467–76

Diamond J, Holmes M, Coughlin M. 1992b. Endogenous NGF and nerve impulses regulate the collateral sprouting of sensory axons in the skin of the adult rat. *J. Neurosci.* 12: 1454–66

Distefano PS, Friedman B, Radziejewski C, et al. 1992. The neurotrophins BDNF, NT-3 and NGF display distinct patterns of retrograde axonal transport in peripheral and central neurons. *Neuron* 8:983–93

Dobrowsky RT, Werner MH, Castellino AM, Chao MV, Hannun YA. 1994. Activation of the sphingomyelin cycle through the low-affinity neurotrophin receptor. *Science* 265: 1596–99

Domenici L, Cellerino A, Berardi N, Cattaneo A, Maffei L. 1995. Antibodies to nerve growth factor (NGF) prolong the sensitive period for monocular deprivation in the rat. *NeuroReport* 5:2041–44

Donnerer J, Schuligoi R, Stein C. 1992. Increased content and transport of substance P and calcitonin gene-related peptide in sensory nerves innervating inflamed tissue: evidence for a regulatory function of nerve growth factor in vivo. *Neuroscience* 49:693–98

Eide A-L, Glover JC. 1995. The development of longitudinal projection patterns of lumbar primary sensory afferents in the chicken embryo. *J. Comp. Neurol.* 353:247–59

Eide A-L, Lewin GR, Barde Y-A. 1994. Influence of neurotrophins on the development of primary afferent projections in the chick spinal cord. *Abstr. Soc. Neurosci.* 20:452.9

Elkabes S, Dreyfus CF, Schaar DG, Black IB. 1994. Embryonic sensory development: local expression of neurotrophin-3 and target expression of nerve growth factor. *J. Comp. Neurol.* 341:204–13

Eriksson NP, Lindsay RM, Aldskogius H. 1994. BDNF and NT-3 rescue sensory but not motoneurones following axotomy in the neonate. *NeuroReport* 5:1445–48

Ernfors P, Bengzon J, Kokaia Z, Persson H, Lindvall O. 1991. Increased levels of messenger RNAs for neurotrophic factors in the brain during kindling epileptogenesis. *Neuron* 7:165–76

Ernfors P, Lee K-F, Jaenisch R. 1994a. Mice lacking brain-derived neurotrophic factor develop with sensory deficits. *Nature* 368:147–50

Ernfors P, Lee K-F, Kucera J, Jaenisch R. 1994b. Lack of neurotrophin-3 leads to deficiencies in the peripheral nervous system and loss of limb proprioceptive afferents. *Cell* 77:503–12

Ernfors P, Van de Water, T, Loring J, Jaenisch R. 1995. Complementary roles of BDNF and NT-3 in vestibular and auditory development. *Neuron* 14:1153–64

Fariñas I, Jones KR, Backus C, Wang X-Y, Reichardt LF. 1994. Severe sensory and sympathetic deficits in mice lacking neurotrophin-3. *Nature* 369:658–61

Fischer W, Wictorin K, Björklund A, Williams LR, Varon S, et al. 1987. Amelioration of cholinergic neuron atrophy and spatial memory impairment in aged rats by nerve growth factor. *Nature* 329:65–68

Fitzgerald M. 1987. Spontaneous and evoked activity of fetal primary afferents in vivo. *Nature* 326:603–5

Funakoshi H, Frisen J, Barbany G, Timmusk T, Zachrisson O, et al. 1993. Differential expression of mRNAs for neurotrophins and their receptors after axotomy of the sciatic nerve. *J. Cell Biol.* 123:455–65

Gaese F, Kolbeck R, Barde Y-A. 1994. Sensory ganglia require neurotrophin-3 early in development. *Development* 120:1613–19

Gall CM, Isackson PJ. 1989. Limbic seizures increase neuronal production of messenger RNA for nerve growth factor. *Science* 245:758–61

Gao W-Q, Dybdal N, Schmelzer C, Murnane A, Hefti F, et al. 1994. Neurotrophin-3 reverses cisplatin-induced peripheral neuropathy. *Abstr. Soc. Neurosci.* 20:193.8

Garofalo L, Ribeiro-da-Silva A, Cuello AC. 1992. Nerve growth factor-induced synaptogenesis and hypertrophy of cortical cholinergic terminals. *Proc. Natl. Acad. Sci. USA* 89:2639–43

Gorin PD, Johnson EM. 1980. Effects of long-term nerve growth factor deprivation on the nervous system of the adult rat: an autoimmune approach. *Brain Res.* 198:27–42

Götz R, Köster R, Winkler C, Raulf F, Lottspeich F, et al. 1994. Neurotrophin-6 is a new member of the nerve growth factor family. *Nature* 372:266–69

Hallböök F, Ibáñez CF, Persson H. 1991. Evolutionary studies of the nerve growth factor family reveal a novel member abundantly expressed in *Xenopus* ovary. *Neuron* 6:845–58

Hama T, Miyamoto M, Tsukui H, Nishio C, Hatanaka H. 1989. Interleukin-6 as a neurotrophic factor for promoting the survival of cultured basal forebrain cholinergic neurons from postnatal rats. *Neurosci. Lett.* 104:340–44

Hartikka J, Hefti F. 1988. Development of septal cholinergic neurons in culture: plating density and glial cells modulate effects of NGF on survival, fiber growth, and expression of transmitter-specific enzymes. *J. Neurosci.* 8:2967–85

Hefti F. 1986. Nerve growth factor promotes survival of septal cholinergic neurons after fimbrial transections. *J. Neurosci.* 6:2155–62

Helgren ME, Torrento K, Distefano PS, Curtis R, Lindsay RM, et al. 1994. NT-3 attenuates proprioceptive deficits in the adult rat with a large fiber neuropathy. *Abstr. Soc. Neurosci.* 20:455.20

Henderson CE, Camu W, Mettling C, Gouin A, Poulsen K, et al. 1993. Neurotrophins promote motor neuron survival and are present in embryonic limb bud. *Nature* 363:266–70

Henderson CE, Phillips HS, Pollock RA, Davies AM, Lemeulle C, et al. 1994. GDNF: a potent survival factor for motoneurons present in peripheral nerve and muscle. *Science* 266:1062–64

Hertzberg T, Fan G, Finley JCW, Erickson JT, Katz DM, et al. 1994. BDNF supports mammalian chemoafferent neurons in vitro and following peripheral target removal in vivo. *Dev. Biol.* 166:801–11

Herzog K-H, Bailey K, Barde Y-A. 1994. Expression of the BDNF gene in the developing visual system of the chick. *Development* 120: 1643–49

Heumann R, Korsching S, Bandtlow C, Thoenen H. 1987. Changes of nerve growth factor synthesis in non-neuronal cells in responses to sciatic nerve transection. *J. Cell Biol.* 104:1623–31

Heymach JV Jr, Shooter EM. 1995. The biosynthesis of neurotrophin heterodimers by transfected mammalian cells. *J. Biol. Chem.* 270:12297–304

Hofer M, Barde Y-A. 1988. Brain-derived neurotrophic factor prevents neuronal death in vivo. *Nature* 331:261–62

Hofer M, Pagliusi SR, Hohn A, Leibrock J, Barde Y-A. 1990. Regional distribution of brain-derived neurotrophic factor mRNA in the adult mouse brain. *EMBO J.* 9:2459–64

Hohn A, Leibrock J, Bailey K, Barde Y-A. 1990. Identification and characterization of a novel member of the nerve growth factor/brain-derived neurotrophic factor family. *Nature* 344:339–41

Hollowell JP, Villadiego A, Rich KM. 1990. Sciatic nerve regeneration across gaps within silicone chambers: long-term effects of NGF and consideration of axonal branching. *Exp. Neurol.* 110:45–51

Hory-Lee F, Russell M, Lindsay RM, Frank E. 1993. Neurotrophin 3 supports the survival of developing muscle sensory neurons in culture. *Proc. Natl. Acad. Sci. USA* 90:2613–17

Humpel C, Wetmore C, Olson L. 1993. Regulation of brain-derived neurotrophic factor messenger RNA and protein at the cellular level in pentylenetetrazol-induced epileptic seizures. *Neuroscience* 53:909–18

Ibáñez CF. 1994. Structure-function relationships in the neurotrophin family. *J. Neurobiol.* 25:1349–61

Ibáñez CF, Ebendal T, Barbany G, Murray-Rust J, Blundell TL, et al. 1992. Disruption of the low affinity receptor-binding site in NGF allows neuronal survival and differentiation by binding to trk gene product. *Cell* 69:329–41

Isackson PJ, Towner MD, Huntsman MM. 1991. Comparison of mammalian, chicken and *Xenopus* brain-derived neurotrophic factor coding sequences. *FEBS Lett.* 285:260–64

Johnson D, Lanahan A, Buck CR, Sehgal A, Morgan C, et al. 1986. Expression and structure of the human NGF receptor. *Cell* 47: 545–54

Johnson EM, Gorin PD, Brandeis LD, Pearson J. 1980. Dorsal root ganglion neurons are destroyed by exposure in utero to maternal antibodies to nerve growth factor. *Science* 210:916–18

Johnson EM, Rich KM, Yip HK. 1986. The role of NGF in sensory neurons in vivo. *Trends Neurosci.* 9:33–37

Johnson JE, Barde Y-A, Schwab M, Thoenen H. 1986. Brain-derived neurotrophic factor supports the survival of cultured rat retinal ganglion cells. *J. Neurosci.* 6:3031–38

Jones KR, Fariñas I, Backus C, Reichardt LF. 1994. Targeted disruption of the BDNF gene perturbs brain and sensory neuron development but not motor neuron development. *Cell* 76:989–99

Jungbluth S, Bailey K, Barde Y-A. 1994. Purification and characterisation of a brain-derived neurotrophic factor/neurotrophin-3 (BDNF/NT-3) heterodimer. *Eur. J. Biochem.* 221:677–85

Kahane N, Kalcheim C. 1994. Expression of trkC receptors mRNA during development of the avian nervous system. *J. Neurobiol.* 25:571–84

Kalcheim C, Barde Y-A, Thoenen H, Le-Douarin NM. 1987. In vivo effect of brain-derived neurotrophic factor on the survival of developing dorsal root ganglion cells. *EMBO J.* 6:2871–73

Kalcheim C, Carmeli C, Rosenthal A. 1992. Neurotrophin 3 is a mitogen for cultured neural crest cells. *Proc. Natl. Acad. Sci. USA* 89:1661–65

Kalcheim C, Gendreau M. 1988. Brain-derived neurotrophic factor stimulates survival and neuronal differentiation in cultured avian neural crest. *Dev. Brain Res.* 41:79–86

Kang HJ, Schuman EM. 1995. Long lasting neurotrophin-induced enhancement of synaptic transmission in the adult hippocampus. *Science* 267:1658–62

Kaplan DR, Stephens RM. 1994. Neurotrophin signal transduction by the trk receptor. *J. Neurobiol.* 25:1404–17

Kessler W, Kirchhoff C, Reeh PW, Handwerker HO. 1992. Excitation of cutaneous afferent nerve endings in vitro by a combination of inflammatory mediators and conditioning effect of substance P. *Exp. Brain Res.* 91:467–76

Kim HG, Wang T, Olafsson P, Lu B. 1994. Neurotrophin 3 potentiates neuronal activity and inhibits gamma-aminobutyratergic synaptic transmission in cortical neurons. *Proc. Natl. Acad. Sci. USA* 91:12341–45

Klein R, Parada LF, Coulier F, Barbacid M. 1989. trkB, a novel tyrosine protein kinase receptor expressed during mouse neural development. *EMBO J.* 8:3701–9

Klein R, Silos-Santiago I, Smeyne RJ, Lira SA, Brambilla R, et al. 1994. Disruption of the neurotrophin-3 receptor gene trkC eliminates Ia muscle afferents and results in abnormal movements. *Nature* 368:249–51

Klein R, Smeyne RJ, Wurst W, Long LK, Auerbach BA, et al. 1993. Targeted disruption of the trkB neurotrophin receptor gene results

314    LEWIN & BARDE

in nervous system lesions and neonatal death. *Cell* 75:113–22

Knipper M, Leung LS, Zhao D, Rylett RJ. 1994. Short-term modulation of glutamatergic synapses in adult rat hippocampus by NGF. *NeuroReport* 5:2433–36

Koliatsos VE, Clatterbuck RE, Winslow JW, Cayouette MU, Price DL, et al. 1993. Evidence that brain-derived neurotrophic factor is a trophic factor for motor neurons in vivo. *Neuron* 10:359–67

Koltzenburg M, Lewin GR, Barde Y-A. 1994. Single unit recordings of chick cutaneous sensory neurons in vitro. *Abstr. Soc. Neurosci.* 20:452.8

Korte M, Carroll P, Wolf E, Brem G, Thoenen H, et al. 1995. Hippocampal long-term potentiation is impaired in mice lacking brain-derived neurotrophic factor. *Proc. Natl. Acad. Sci. USA.* 192:8856–60

Kucera J, Ernfors P, Walro JM, Jaenisch R. 1995a. Reduction in the number of spinal motoneurons in NT-3 deficient mice. *Neuroscience.* In press.

Kucera J, Fan G, Jaenisch R, Linnarsson S, Ernfors P.1995b. Dependence of developing Ia afferents on neurotrophin-3. *J. Comp. Neurol.* In press.

Kuno M. 1990. Target dependence of motoneuronal survival: the current status. *Neurosci. Res.* 9:155–72

Lamballe F, Klein R, Barbacid M. 1991. trkC, a new member of the trk family of tyrosine protein kinases, is a receptor for neurotrophin-3. *Cell* 66:967–79

Lee K-F, Davies AM, Jaenisch R. 1994. p75-deficient embryonic dorsal root sensory and neonatal sympathetic neurons display a decreased sensitivity to NGF. *Development* 120:1027–33

Lee K-F, Li E, Huber LJ, Landis SC, Sharpe AH, et al. 1992. Targeted mutation of the p75 low affinity NGF receptor gene leads to deficits in the peripheral sensory nervous system. *Cell* 69:737–49

Levi-Montalcini R, Booker B. 1960. Destruction of the sympathetic ganglia in mammals by an antiserum to a nerve growth protein. *Proc. Natl. Acad. Sci. USA* 46:384–91

Lewin GR. 1995. Neurotrophic factors and pain. *Semin. Neurosci.* 7(4):227–32

Lewin GR, Koltzenburg M, Toyka KV, Barde Y-A. 1994a. Involvement of neurotrophins in the phenotypic specification of chick cutaneous afferents. *Abstr. Soc. Neurosci.* 20:452.7

Lewin GR, Mendell LM. 1993. Nerve growth factor and nociception. *Trends Neurosci.* 16:353–59

Lewin GR, Mendell LM. 1994. Regulation of cutaneous C-fiber heat nociceptors by nerve growth factor in the developing rat. *J. Neurophysiol.* 71:941–49

Lewin GR, Ritter AM, Mendell LM. 1992a. On the role of nerve growth factor in the development of myelinated nociceptors. *J. Neurosci.* 12:1896–905

Lewin GR, Ritter AM, Mendell LM. 1993. Nerve growth factor-induced hyperalgesia in the neonatal and adult rat. *J. Neurosci.* 13:2136–48

Lewin GR, Rueff A, Mendell LM. 1994b. Peripheral and central mechanisms of NGF-induced hyperalgesia. *Eur. J. Neurosci.* 6:1903–1912

Lewin GR, Winter J, McMahon SB. 1992b. Regulation of afferent connectivity in the adult spinal cord by nerve growth factor. *Eur. J. Neurosci.* 4:700–7

Leβmann V, Gottmann K, Heumann R. 1994. BDNF and NT-4/5 enhance glutamatergic synaptic transmission in cultured hippocampal neurones. *NeuroReport* 6:21–25

Lindholm D, Dechant G, Heisenberg C-P, Thoenen H. 1993. Brain-derived neurotrophic factor is a survival factor for cultured rat cerebellar granule neurons and protects them against glutamate-induced neurotoxicity. *Eur. J. Neurosci.* 5:1455–64

Lindsay RM. 1988. Nerve growth factors (NGF, BDNF) enhance axonal regeneration but are not required for survival of adult sensory neurons. *J. Neurosci.* 8:2394–405

Lindsay RM, Harmar AJ. 1989. Nerve growth factor regulates expression of neuropeptide genes in adult sensory neurons. *Nature* 337:362–64

Lindsay RM, Thoenen H, Barde Y-A. 1985. Placode and neural crest-derived sensory neurons are responsive at early developmental stages to brain-derived neurotrophic factor. *Dev. Biol.* 112:319–28

Lisman JE, Harris KM. 1993. Quantal analysis of synaptic anatomy-integrating two views of hippocampal plasticity. *Trends Neurosci.* 16:141–47

Liu X, Ernfors P, Wu H, Jaenisch R. 1995. Sensory but not motor neuron deficits in mice lacking NT-4 and BDNF. *Nature* 375:238–41

Lohof AM, Ip NY, Poo M. 1993. Potentiation of developing neuromuscular synapses by the neurotrophins NT-3 and BDNF. *Nature* 363:350–53

LoPresti P, Scott SA. 1994. Target specificity and size of avian sensory neurons supported in vitro by nerve growth factor, brain-derived neurotrophic factor, and neurotrophin-3. *J. Neurobiol.* 25:1613–24

Maffei L, Berardi N, Domenici L, Parisi V, Pizzorusso T, et al. 1992. Nerve growth factor (NGF) prevents the shift in ocular dominance distribution of visual cortical neurons in monocularly deprived rats. *J. Neurosci.* 12:4651–62

Mahadeo D, Kaplan L, Chao MV, Hempstead

BL. 1994. High affinity nerve growth factor binding displays a faster rate of association than p140trk binding. Implications for multi-subunit polypeptide receptors. *J. Biol. Chem.* 269:6884–91

Maisonpierre PC, Belluscio L, Friedman B, Alderson RF, Wiegand SJ, et al. 1990. NT-3, BDNF, and NGF in the developing rat nervous system: parallel as well as reciprocal patterns of expression. *Neuron* 5:501–9

Mansour-Robaey S, Clarke DB, Wang Y-C, Bray GM, Aguayo AT. 1994. Effects of ocular injury and administration of brain-derived neurotrophic factor on survival and regrowth of axotomized retinal ganglion cells. *Proc. Natl. Acad. Sci. USA* 91:1632–36

McDonald NQ, Lapatto R, Murray-Rust J, Gunning J, Wlodawer A, et al. 1991. New protein fold revealed by a 2.3-Å resolution crystal structure of nerve growth factor. *Nature* 354:411–14

McMahon SB, Armanini MP, Ling LH, Phillips HS. 1994. Expression and coexpression of Trk receptors in subpopulations of adult primary sensory neurons projecting to identified peripheral targets. *Neuron* 12:1161–71

McMahon SB, Bennett DHL, Priestley JVP, Shelton DL. 1995. The biological effects of endogenous NGF in adult sensory neurons revealed by Trk A IgG fusion molecules. *Nat. Med.* 1:776–80

Meyer M, Matsuoka I, Wetmore C, Olson L, Thoenen H. 1992. Enhanced synthesis of brain-derived neurotrophic factor (BDNF) in the lesioned peripheral nerve: different mechanisms are responsible for the regulation of BDNF and NGF mRNA. *J. Cell Biol.* 119:45–54

Milburn A. 1984. Stages in the development of cat muscle spindles. *J. Embryol. Exp. Morphol.* 82:177–216

Mu X, Silos-Santiago I, Carroll SL, Snider WD. 1993. Neurotrophin receptor genes are expressed in distinct patterns in developing dorsal root ganglia. *J. Neurosci.* 13:4029–41

Naumann T, Kermer P, Seydewitz V, Ortmann R, Damato F, et al. 1994. Is there a long-lasting effect of a short-term nerve growth factor application on axotomized rat septohippocampal neurons? *Neurosci. Lett.* 173:213–15

Naumann T, Peterson GM, Frotscher M. 1992. Fine structure of rat septohippocampal neurons. II. A time course analysis following axotomy. *J. Comp. Neurol.* 325:219–42

Nawa H, Carnahan J, Gall C. 1995. BDNF protein measured by a novel enzyme immunoassay in normal brain and after seizure: partial disagreement with mRNA levels. *Eur. J. Neurosci.* 7:1527–36

Oakley RA, Garner AS, Large TH, Frank E. 1995. Muscle sensory neurons require neurotrophin-3 from peripheral tissues during

the period of normal cell death. *Development* 121:1341–50

Ockel M, Lewin GR, BardeY-A. 1996. In vivo effects of neurotrophin-3 during sensory neurogenesis. *Development.* In press

Oppenheim RW, Houenou LJ, Johnson JE, Lin L-F, Li L, et al. 1995. Developing motor neurons rescued from programmed and axotomy-induced cell death by GDNF. *Nature* 373:344–46

Oppenheim RW, Qin-Wei Y, Prevette D, Yan Q. 1992. Brain-derived neurotrophic factor rescues developing avian motoneurons from cell death. *Nature* 360:755–57

Otten U. 1984. Nerve growth factor and the peptidergic sensory neurons. *Trends Pharmacol. Sci.* 5:307–10

Patterson SL, Grover LM, Schwartzkroin PA, Bothwell M. 1992. Neurotrophin expression in rat hippocampal slices: a stimulus paradigm inducing LTP in CA1 evokes increases in BDNF and NT-3 mRNAs. *Neuron* 9:1081–88

Pearson J, Johnson EM, Brandeis L. 1983. Effects of antibodies to nerve growth factor on intrauterine development of derivatives of cranial neural crest and placode in the guinea pig. *Dev. Biol.* 96:32–36

Peterson GM, Lanford GW, Powell EW. 1990. Fate of septohippocampal neurons following fimbria-fornix transection: a time course analysis. *Brain Res. Bull.* 25:129–37

Petty BG, Cornblath DR, Adornato BT, Chandhry V, Flexner C, et al. 1994. The effect of systemically administered recombinant human nerve growth factor in healthy human subjects. *Ann. Neurol.* 36:244–46

Phillips HS, Hains JM, Laramee GR, Rosenthal A, Winslow JW, et al. 1990. Widespread expression of BDNF but not NT3 by target areas of basal forebrain cholinergic neurons. *Science* 250:290–94

Phillips HS, Nishimura MC, Chen KS, Crowley C, Spencer S, et al. 1994. Rescue of NGF-knockout mice by peripheral expression of NGF. *Abstr. Soc. Neurosci.* 20:358.9

Pierce JP, Lewin GR. 1994. An ultrastructural size principle. *Neuroscience* 58:441–46

Pinco O, Carmeli C, Rosenthal A, Kalcheim C. 1993. Neurotrophin-3 affects proliferation and differentiation of distinct neural crest cells and is present in the early neural tube of avian embryos. *J. Neurobiol.* 24:1626–41

Purves D. 1988. *Body and Brain. A Trophic Theory of Neural Connections.* Cambridge, MA: Harvard Univ. Press

Purves D, Njå A. 1976. Effect of nerve growth factor on synaptic depression after axotomy. *Nature* 260:535–36

Rabizadeh S, Oh J, Zhong L, Yang J, Bitler CM, et al. 1993. Induction of apoptosis by the low-affinity NGF receptor. *Science* 261:345–48

Radeke MJ, Misko TP, Hsu C, Herzenberg LA, Shooter EM, et al. 1987. Gene transfer and molecular cloning of the rat nerve growth factor receptor: a new class of receptors. *Nature* 325:593–97

Radziejewski C, Robinson RC. 1993. Heterodimers of the neurotrophic factors: formation, isolation, and differential stability. *Biochemistry* 32:13350–56

Rich KM, Yip HK, Osborne PA, Schmidt RE, Johnson EM, et al. 1984. Role of nerve growth factor in the adult dorsal root ganglia neuron and its response to injury. *J. Comp. Neurol.* 230:110–18

Ritter AM, Lewin GR, Kremer NE, Mendell LM. 1991. Requirement for nerve growth factor in the development of myelinated nociceptors in vivo. *Nature* 350:500–2

Ritter AM, Mendell LM. 1992. Somal membrane properties of physiologically identified sensory neurons in the rat: effects of nerve growth factor. *J. Neurophysiol.* 68:2033–41

Robinson RC, Radziejewski C, Stuart DI, and Jones EY. 1995. Structure of the brain-derived neurotrophic factor/neurotrophin-3 heterodimer. *Biochemistry* 34:4139–46

Rocamora N, Palacios JM, Mengod G. 1992. Limbic seizures induce a differential regulation of the expression of nerve growth factor, brain-derived neurotrophic factor and neurotrophin-3, in the rat hippocampus. *Mol. Brain Res.* 13:27–33

Rodríguez-Tébar A, Dechant G, Barde Y-A. 1990. Binding of brain-derived neurotrophic factor to the nerve growth factor receptor. *Neuron* 4:487–92

Rodríguez-Tébar A, Dechant G, Götz R, Barde Y-A. 1992. Binding of neurotrophin-3 to its neuronal receptors and interactions with nerve growth factor and brain-derived neurotrophic factor. *EMBO J.* 11:917–22

Rodríguez-Tébar A, De la Rosa EJ, Arribas A. 1993. Neurotrophin-3 receptors in the developing chicken retina. *Eur. J. Biochem.* 211:789–94

Rodríguez-Tébar A, Jeffrey PL, Thoenen H, Barde Y-A. 1989. The survival of chick retinal ganglion cells in response to brain-derived neurotrophic factor depends on their embryonic age. *Dev. Biol.* 136:296–303

Rodríguez-Tébar A, Rohrer H. 1991. Retinoic acid induces NGF-dependent survival response and high-affinity NGF receptors in immature chick sympathetic neurons. *Development* 112:813–20

Rohrer H, Hofer M, Hellweg R, Korsching S, Stehle AD, et al. 1988. Antibodies against mouse nerve growth factor interfere in vivo with the development of avian sensory and sympathetic neurones. *Development* 103:545–52

Ruit KG, Elliott JL, Osborne PA, Yan Q, Snider WD, et al. 1992. Selective dependence of mammalian dorsal root ganglion neurons on nerve growth factor during embryonic development. *Neuron* 8:573–87

Ruit KG, Osborne PA, Schmidt RE, Yan Q, Snider WD, et al. 1990. Nerve growth factor regulates sympathetic ganglion cell morphology and survival in the adult mouse. *J. Neurosci.* 10:2412–19

Ryden M, Murray-Rust J, Glass D, Ilag LL, Trupp M et al. 1995. Functional analysis of mutant neurotrophins deficient in low-affinity binding reveals a role for p75$^{LNGFR}$ in NT-4 signalling. *EMBO J.* 14:1979–90

Schwab ME, Otten U, Agid Y, Thoenen H. 1979. Nerve growth factor (NGF) in the rat CNS: absence of specific retrograde axonal transport and tyrosine hydroxylase induction in locus coeruleus and substantia nigra. *Brain Res.* 168:473–83

Sendtner M, Holtmann B, Kolbeck R, Thoenen H, Barde Y-A. 1992. Brain-derived neurotrophic factor prevents the death of motoneurons in newborn rats after nerve section. *Nature* 360:757–59

Shen K-F, Crain SM. 1994. Nerve growth factor rapidly prolongs the action potential of mature sensory ganglion neurons in culture, and this effect requires activation of Gs-coupled excitatory kappa-opioid receptors on these cells. *J. Neurosci.* 14:5570–79

Sieber-Blum M. 1991. Role of the neurotrophic factors BDNF and NGF in the commitment of pluripotent neural crest cells. *Neuron* 6:949–55

Smeyne RJ, Klein R, Schnapp A, Long LK, Bryant S, et al. 1994. Severe sensory and sympathetic neuropathies in mice carrying a disrupted Trk/NGF receptor gene. *Nature* 368:246–49

Smith CL, Frank E. 1988. Peripheral specification of sensory connections in the spinal cord. *Brain Behav. Evol.* 31:227–42

Snider WD. 1994. Functions of the neurotrophins during nervous system development: what the knockouts are teaching us. *Cell* 77:627–38

Sretavan DW, Shatz CJ. 1986. Prenatal development of cat retinogeniculate axon arbors in the absence of binocular interactions. *J. Neurosci.* 6:990–1003

Suter U, Heymach JV Jr, Shooter EM. 1991. Two conserved domains in the NGF propeptide are necessary and sufficient for the biosynthesis of correctly processed and biologically active NGF. *EMBO J.* 10:2395–400

Tessarollo L, Tsoulfas P, Martin-Zanca D, Gilbert DJ, Jenkins NA, et al. 1993. trkC, a receptor for neurotrophin-3, is widely expressed in the developing nervous system and in non-neuronal tissues. *Development* 118:463–75

Tessarollo L, Vogel KS, Palko ME, Reid SW,

Parada LF. 1994. Targeted mutation in the neurotrophin-3 gene results in loss of muscle sensory neurons. *Proc. Natl. Acad. Sci. USA* 91:11844–48

Toutant M. 1982. Quantitative and histochemical aspects of the differentiation of muscle spindles in the anterior latissimus dorsi of the developing chick. *Anat. Embryol.* 163:475–85

Urfer R, Tsoulfas P, Soppet D, Escandón E, Parada LF, et al. 1994. The binding epitopes of neurotrophin-3 to its receptors trkC and gp75 and the design of a multifunctional human neurotrophin. *EMBO J.* 13:5896–909

Vejsada R, Sagot Y, Kato AC. 1995. Quantitative comparison of the transient rescue effects of neurotrophic factors on axotomized motoneurons in vivo. *Eur. J. Neurosci.* 7:108–15

Verdi JM, Anderson DJ. 1994. Neurotrophins regulate sequential changes in neurotrophin receptor expression by sympathetic neuroblasts. *Neuron* 13:1359–72

Verge VMK, Merlio J-P, Grondin J, Ernfors P, Persson H, et al. 1992. Colocalization of NGF binding sites, trk mRNA, and low-affinity NGF receptor mRNA in primary sensory neurons: responses to injury and infusion of NGF. *J. Neurosci.* 12:4011–22

Verge VMK, Richardson PM, Benoit R, Riopelle RJ. 1989. Histochemical characterization of sensory neurons with high-affinity receptors for nerve growth factor. *J. Neurocytol.* 18:583–91

Von Bartheld CS, Kinoshita Y, Prevette D, Yin Q-W, Oppenheim RW, et al. 1994. Positive and negative effects of neurotrophins on the isthmo-optic nucleus in chick embryos. *Neuron* 12:639–54

Wetmore C, Ernfors P, Persson H, Olson L. 1990. Localization of brain-derived neurotrophic factor mRNA to neurons in the brain

by in situ hybridization. *Exp. Neurol.* 109:141–52

Williams R, Bäckström A, Ebendal T, Hallböök F. 1993. Molecular cloning and cellular localization of trkC in the chicken embryo. *Dev. Brain Res.* 75:235–52

Williams R, Bäckström A, Kullander K. 1995. Developmentally regulated expression of mRNA for neurotrophin high-affinity (trk) receptors within chick trigeminal sensory neurons. *Eur. J. Neurosci.* 7:116–28

Woolf CJ, Safieh-Garabedian B, Ma Q-P, Crilly P, Winter J. 1994. Nerve growth factor contributes to the generation of inflammatory sensory hypersensitivity. *Neuroscience* 62:327–31

Wright DE, Snider WD. 1995. Neurotrophin receptor mRNA expression defines distinct populations of neurons in rat dorsal root ganglia. *J. Comp. Neurol.* 351:329–38

Wright EM, Vogel KS, Davies AM. 1992. Neurotrophic factors promote the maturation of developing sensory neurons before they become dependent on these factors for survival. *Neuron* 9:139–50

Yan Q, Elliott J, Snider WD. 1992. Brain-derived neurotrophic factor rescues spinal motor neurons from axotomy-induced cell death. *Nature* 360:753–55

Zafra F, Hengerer B, Leibrock J, Thoenen H, Lindholm D. 1990. Activity dependent regulation of BDNF and NGF mRNAs in the rat hippocampus is mediated by non-NMDA glutamate receptors. *EMBO J.* 9:3545–50

Zhou X-F, Rush RA. 1993. Localization of neurotrophin-3-like immunoreactivity in peripheral tissues of the rat. *Brain Res.* 621:189–99

Zhou X-F, Rush RA. 1994. Localization of neurotrophin-3-like immunoreactivity in the rat central nervous system. *Brain Res.* 643:162–72

*Annu. Rev. Neurosci. 1996. 19:319–40*

# ADDICTIVE DRUGS AND BRAIN STIMULATION REWARD

*R. A. Wise*

Center for Studies in Behavioral Neurobiology and Department of Psychology,
Concordia University, Montreal, Canada, H3G 1M8

KEY WORDS:   reward, addiction, brain stimulation, stimulants, opiates

---

### ABSTRACT

Direct electrical or chemical stimulation of specific brain regions can establish response habits similar to those established by natural rewards such as food or sexual contact. Cocaine, mu and delta opiates, nicotine, phencyclidine, and cannabis each have actions that summate with rewarding electrical stimulation of the medial forebrain bundle (MFB). The reward-potentiating effects of amphetamine and opiates are associated with central sites of action where these drugs also have their direct rewarding effects, suggesting common mechanisms for drug reward per se and for drug potentiation of brain stimulation reward. The central sites at which these and perhaps other drugs of abuse potentiate brain stimulation reward and are rewarding in their own right are consistent with the hypothesis that the laboratory reward of brain stimulation and the pharmacological rewards of addictive drugs are habit forming because they act in the brain circuits that subserve more natural and biologically significant rewards.

---

## INTRODUCTION

In 1953, Olds & Milner discovered that rats learned to return to the portions of their environment where they had been given direct electrical stimulation of the septal area of the brain. This demonstration of a learned place preference suggested to Olds & Milner that the stimulation was rewarding, and they subsequently confirmed that they could train rats to lever-press, by making short pulse trains of septal brain stimulation contingent upon this arbitrary response (Olds & Milner 1954). This finding established that the stimulation

319

could serve as an "operant reinforcer"[1] (Skinner 1938), and subsequent studies have shown that brain stimulation reward[2] establishes and maintains response habits in patterns very similar to those established and maintained by natural rewards such as food and water (Beninger et al 1977, Trowill et al 1969). Similarly, intravenous drug rewards establish and maintain response habits similar to those established and maintained by natural rewards (Johanson 1978, Spealman & Goldberg 1978, Thompson 1968, Woods & Schuster 1971; but see Wise 1987). This should not be surprising; the brain mechanisms that make animals susceptible to brain stimulation reward evolved long before the human inventions that made intracranial self-stimulation or drug addiction[3] possible. Indeed, endogenous opioid neurotransmitters have apparently played a role in the control of behavior by food throughout most of our evolutionary history (Josefsson & Johansson 1979, Kavaliers & Hirst 1987).

Olds & Milner (1954) suggested immediately that rewarding stimulation probably activated brain circuitry relevant to the pursuit of natural incentives such as food and sexual contact; Olds (1956) expanded this view in his influential paper "Reward Centers in the Brain." While Olds' (1956; Olds et al 1971) discussions of reward and drive centers in the brain suggested a specialized and dominant role for the hypothalamus in reward function, he soon came to appreciate that diffuse, multisynaptic systems (Olds & Olds 1965)

---

[1]Skinner (1937) asserted that there were two kinds of learning, each established by its own form of reinforcement. In addition to the reinforcement identified in 1903 by Pavlov (1927), who first used the term to refer to the "stamping in" (Thorndike's 1898 phrase) of stimulus-stimulus associations, Skinner [like Thorndike (1898, 1933) before him] argued for an independent form of reinforcement, which he labeled operant reinforcement and considered essential in the learning of response habits. Although it remains an open question which kind of reinforcement is critical to the learning of operant or instrumental habits (see Bindra 1974), operant reinforcement is usually implied when the unqualified word reinforcement is used in the motivation literature.

[2]Behavioral pharmacologists usually prefer Skinner's specialized term reinforcement to the more familiar term reward. However, rewarding brain stimulation, rewarding drug injections, and also natural rewards (such as a salted peanut) have important incentive motivational properties that go beyond the strict properties attributed to operant reinforcement (Bindra 1974). In addition, their ability to establish conditioned place preferences reflects Pavlovian and not operant reinforcement (Wise 1989). Thus the specialized term reinforcement should only be used where the incentive motivational, Pavlovian reinforcement, and operant reinforcement effects can be clearly dissociated; they cannot be dissociated in most lever-pressing experiments, particularly those involving brain stimulation or drug reward (Wise 1989).

[3]Although self-intoxication occurs in lower animals (Siegel 1989), true addiction in the natural state does not. The serious forms of drug addiction depend on human skills and inventions—e.g. the use of fire, pipes, and cigarette papers; the use of the hypodermic syringe and needle; agricultural skills for the harvesting and curing of tobacco; the ability to synthesize or purify drugs; the ability to concentrate, store, and transport alcoholic beverages. Simple ingestion of drugs from botanical sources does not produce the rapid onset of drug effect or the high brain levels of drug that are associated with the levels of compulsive drug self-administration that we traditionally identify as addiction.

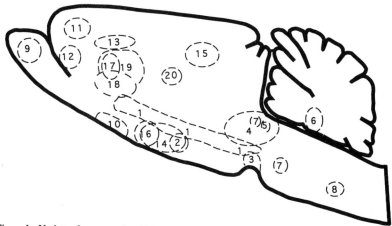

*Figure 1*   Variety of structures in which rewarding effects of brain stimulation have been tentatively or approximately localized in the rat brain. This list is not exhaustive. In only a few cases have the boundaries of reward sites been precisely localized in relation to the boundaries of the anatomical structures with which they have become associated: 1. medial forebrain bundle sites, including the anterior, posterior, and lateral hypothalamus; 2. ventromedial hypothalamus; 3. substantia nigra (zona compacta) and ventral tegmental area; 4. midline mesencephalon, including the regions of the dorsal and medial raphe nuclei; 5. region of locus coeruleus; 6. deep cerebellar nuclei and decussation of the brachium conjunctivum; 7. regions of the mesencephalic and motor nuclei of the trigeminal nerve; 8. nucleus of the solitary tract; 9. olfactory bulb; 10. olfactory tubercle; 11. medial frontal cortex; 12. sulcal frontal cortex; 13. anterior cingulate cortex; 14. entorhinal cortex; 15. hippocampus; 16. amygdala; 17. medial and lateral septal regions; 18. nucleus accumbens; 19. caudate nucleus; and 20. dorso-medial thalamus.

and subsystems (Olds 1958, Olds et al 1971) dominated both natural reward function and the phenomenon of brain stimulation reward. Robert Heath, who studied the rewarding and pleasure-producing effects of brain stimulation in human patients (Heath 1963, 1972), also posited a distributed, multisynaptic reward circuit (Heath 1975). Differences in opinion remain as to whether the various known reward sites (Figure 1) represent the way stations of multiple reward circuits organized in parallel (e.g. Phillips 1984) or a single system (or small number of systems) linking multiple reward sites in series (Wise & Bozarth 1984). However, there is increasing agreement that the substrates of brain stimulation reward play important roles in the habit-forming properties of natural rewards (Conover & Shizgal 1994, Wise 1982) and a variety of drugs of abuse (Fibiger 1978, Gardner 1992, Kornetsky & Esposito 1979, Stein 1968, Wise 1978, Wise & Rompré 1989).

   That habit-forming brain stimulation and habit-forming drugs might activate the same brain mechanism is consistent with the fact that drugs of abuse generally increase responding for brain stimulation reward; they do so by increasing the rewarding potency of the stimulation, and they do so at the sites and by the actions that make them rewarding in their own right. The present paper reviews the evidence (*a*) that addictive drugs and rewarding brain stimulation have actions in the same reward circuitry and (*b*) that their actions in that circuitry are synergistic. Evidence suggests that the same reward circuitry also accounts for the habit-forming effects of natural rewards such as food or sexual contact; this evidence is much less fully developed, and the interested reader is referred to discussions elsewhere (Conover & Shizgal 1994; Pfaus & Phillips 1990; Wise 1982, 1985).

## BASIC FEATURES OF BRAIN STIMULATION REWARD

Early studies of the motivational effects of focal brain stimulation dealt with two primary questions: What are the brain sites at which stimulation is rewarding? and What are the drugs that influence responding for stimulation? Simple response rate was the most frequent dependent variable, and the stimulation parameters were usually fixed by the experimenter at arbitrary levels. Using such methods, Olds & Olds (1963) showed that stimulation of a broad range of limbic and diencephalic structures could be rewarding; other investigators have added substantially to the list. Stimulation can be rewarding not only when it activates diencephalic and telencephalic circuitry, but also when it activates a variety of brain stem sites, including several sites in the dorsal tegmentum (Simon et al 1975), sites in the region of the nucleus of the solitary tract (Carter & Phillips 1975), and even regions of the cerebellum (Corbett et al 1982). In addition to many areas traditionally classified as associational, several areas where stimulation is rewarding appear to be frankly sensory (Phillips 1970), and others are argued to be frankly motor (van der Kooy & Phillips 1977).

The behavioral characteristics of intracranial self-stimulation can vary substantially depending on the electrode site (Olds & Olds 1963, St-Laurent 1988). In many (but not all) cases, animals that work to initiate stimulation will also work to terminate it (Bower & Miller 1958, Roberts 1958). This and other forms of ambivalence vary with the site of stimulation and are generally thought to be due to the concurrent activation of antagonistic systems within the same stimulation field (Bielajew & Shizgal 1980, Kent & Grossman 1969, Skelton & Shizgal 1980). Aversive or other (Rompré 1987) side effects of stimulation may well account for much of the variety in self-stimulation patterns (Kent & Grossman 1969), but substantial between-site differences almost certainly exist in the rewarding impact of stimulation as well (Olds & Olds 1963, Prado-Alcala & Wise 1984, Prado-Alcala et al 1984).

Of the various reward sites thus far identified, sites along the length of the medial forebrain bundle (particularly as it courses through the lateral and posterior hypothalamus), the ventral tegmental area (Olds & Olds 1963), and on into the pons (Simon et al 1975) are associated with the strongest rewarding effects of stimulation and the lowest levels of ambivalence. These sites are usually used in studies of the effects of drugs on intracranial self-stimulation, particularly in the case of drugs of abuse.

## Independent Variables

In addition to the advantage that its effects can be localized within the deeper levels of the brain, brain stimulation reward can be controlled more precisely than can more natural rewards. The parameters of brain stimulation reward and the timing of its delivery and receipt can be precisely controlled and measured. Parametric studies of brain stimulation reward have given us important information as to the characteristics of the directly activated neurons that carry the reward signal at various levels of the circuitry (Gallistel et al 1981). The most significant parameters of stimulation are the stimulation intensity, the frequency and duration of the stimulation pulses, and the number of pulses given in each stimulation train. Also, the pattern of stimulation (stimulation can be administered in trains of paired-pulses or pulse-bursts), the delay between the response, and the onset of its reward can be varied.

The earliest studies of brain stimulation reward involved 60-Hz sinusoidal stimulation. In these cases reward magnitude was controlled as a function of stimulation intensity. Current intensity controls the size of the effective stimulation field; at threshold currents, the effective spread of current appears to be on the order of a few hundred microns (Fouriezos & Wise 1984). The fact that increased stimulation current causes increased levels of responding implies that the reward system is, in most regions, diffusely organized in systems larger than the effective stimulation field. Experimental manipulation of stimulation intensity is most useful in studies where anatomical localization is of interest; stimulation at high intensity is most likely to reach distal reward sites, whereas stimulation at low intensity is most useful in attempts to localize their boundaries (Wise 1972).

In studies where stimulation frequency can be controlled, rectangular cathodal pulse trains are usually given. The minimal stimulation frequency for rewarding effects tends to be on the order of 20 Hz; depending on the stimulation site, response rate can continue to increase, with increases in frequency up to 200 or even 400 Hz (Carter & Phillips 1975). Experimental manipulation of stimulation frequency is preferable in pharmacological studies, where stimulation intensity (and thus the size of the stimulation field) should be held constant.

Train duration can also be varied; longer train durations produce more vigorous responding, up to a point (Frank et al 1987). With train durations longer than 200 ms, animals begin to attempt to lever-press for additional stimulation before the previously earned train is over. With longer train lengths, response rate varies inversely with train lengths (Lepore & Franklin 1992). In most studies, train length is fixed at a value between 200 and 500 ms.

Finally, pulse duration can be varied. Although longer pulses are generally more rewarding, pulses with durations on the order of 100 μs are useful because such pulses depolarize axons near the tip of the electrode once and only once and make theoretical analyses less complicated (Gallistel et al 1981).

## Dependent Variables

Since the early demonstration of Olds & Milner (1954), lever-pressing has been the traditional response, and rate of responding has been the traditional dependent variable in studies of brain stimulation reward. The advantages are obvious: Spontaneous lever-pressing is infrequent; thus learned (reward-dependent) lever-pressing is easily identified. The criterion for the response is objective, and the measurement of the response and the recording of the response pattern are easily automated. Moreover, Skinner (1950, p. 200) argued, with respect to lever-pressing for food reinforcement, that "rate of responding appears to be the only datum which varies significantly and in the expected direction under conditions which are relevant to the 'learning process'." However, while Skinner's dictum may be valid to some extent for the period when a response habit is being acquired, rate of responding also varies with motivational conditions that can dominate the maintenance of an already-established habit. Because most brain stimulation reward specialists and drug abuse specialists studied the performance of already-established responses rather than the acquisition of new ones, rate of responding was soon identified as a contaminated variable (Valenstein 1964), and the attempt was made to find rate-free measures of the rewarding impact of the stimulation.

In point of fact, there are no measures of operant behavior that do not involve assessment of the number of responses made or the number of rewards earned over some arbitrary unit of time (or, conversely, the amount of time to reach some arbitrary criterion of responding); thus there is no measure of brain stimulation reward that is truly response-rate independent. Even the preference for one of several concurrently available choices involves the determination of relative rates of response. However, the simple response rate associated with a single arbitrary set of stimulation parameters is clearly not a useful indication of the reward value of the stimulation. This fact is made clear in a paradigm (Gallistel 1987, Miliaressis et al 1986) in which the strength of brain stimulation is varied much like the dose of drug in a traditional pharmacological study (Liebman 1983, Wise et al 1992); this paradigm is termed the curve-shift paradigm of brain stimulation reward. When other parameters are held constant and stimulation intensity, pulse duration, pulse frequency (Keesey 1962), or train duration (Frank et al 1987) is varied systematically, dose-

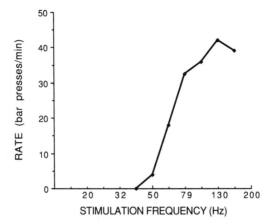

*Figure 2* Rate of responding for hypothalamic brain stimulation as a function of stimulation frequency (stimulation intensity, pulse width, and train duration held constant at arbitrary but moderate values). [Adapted from Gallistel & Karras (1984) with permission of the authors and publisher.]

response functions for brain stimulation reward can be determined.[4] Such functions resemble traditional pharmacological dose-response curves (Figure 2): Low doses of stimulation fail to sustain the lever-press habit at higher-than-chance levels; response rates rise progressively over some range of moderate stimulation doses; and finally, a high stimulation dose is reached, beyond which further increases in dose do not produce further increases in response rate. Similar rate-frequency functions are obtained when measuring the running speed of animals traversing a runway for the opportunity to self-stimulate (Edmonds & Gallistel 1974, Franklin 1978, Gallistel et al 1974).

From rate-frequency, rate-intensity, or rate-duration functions, we can draw inferences about the potency and efficacy of stimulation and about how the potency or efficacy, if either, are altered by drugs. As with a dose-response curve, shifts to the right or left reflect changes in potency, whereas shifts up or down reflect changes in efficacy. Drug-induced changes in the efficacy of stimulation (on the maximum response level) are assumed to reflect drug-induced changes in the performance capability of the animal, since they can be mimicked by changes in task difficulty (Edmonds & Gallistel 1974). Drug-induced changes in the potency of stimulation (leftward or rightward shifts of the curve) are assumed to reflect synergism or antagonism of the rewarding impact of the stimulation and are mimicked by concurrent manipulations of

[4]The most appropriate measure of the rewarding impact of brain stimulation is the total charge per reward (Gallistel 1978). Because the total charge is directly proportional to the stimulation intensity, pulse duration, pulse frequency, and train duration, any one of these is, within limits and with the others held constant, a valid reflection of the rewarding dose of the stimulation.

other stimulation parameters that contribute to the total charge per train of rewarding stimulation pulses (Edmonds & Gallistel 1974).

## DRUG ALTERATIONS OF BRAIN STIMULATION REWARD POTENCY

Lesion (Fibiger et al 1987, Lippa et al 1973, Phillips & Fibiger 1978), pharmacological (Fouriezos & Wise 1976, Franklin & McCoy 1979, Gallistel & Davis 1983, Liebman and Butcher 1974, Zarevics & Setler 1979), and anatomical mapping (Corbett & Wise 1979, 1980) studies have established reasonably well that the rewarding effects of medial forebrain bundle brain stimulation depend on the ability of the brain stimulation to activate (presumably transsynaptically) (Wise 1980a, Yeomans 1982) the mesocorticolimbic dopamine system. Not surprisingly, from this perspective, the drugs that alter the potency of brain stimulation reward are, almost exclusively, dopamine agonists or antagonists (Wise & Rompré 1989).

### Reward Antagonists

The drugs that cause rightward shifts of the rate-frequency or rate-intensity function, with one exception (Gallistel & Freyd 1987), are dopamine (Franklin 1978, Gallistel & Freyd 1987, Gallistel & Karras 1984, Wise 1985) or opiate (West & Wise 1988a) antagonists. When given in low doses, dopamine antagonists cause parallel rightward shifts of the rate-frequency function (e.g. see Gallistel & Karras 1984). When given in high doses they produce not only a rightward shift but also a downward shift (the rightward but not the downward component of the shift is reversed by the indirect dopamine agonist morphine) (Rompré & Wise 1989). The rightward shift is interpreted as reflecting antagonism of the rewarding action of the stimulation; by analogy to the dose-response analysis of pharmacological effects we would say that, to the degree that they cause rightward shifts in the rate-frequency curve with no shift in maximum response rate, dopamine antagonists are competitive antagonists of brain stimulation reward. Presumably, dopamine antagonists impair the synaptic transmission of the reward signal itself, either across the dopaminergic synapse (Wise 1980a,b; Yeomans 1982; Yeomans et al 1993) or across a synapse where dopamine plays a modulatory role (Gallistel 1986).

### Reward Synergists[5]

The drugs known to synergize with brain stimulation reward—drugs that shift the rate-frequency or rate-intensity function to the left—are, for the most part,

---

[5]The term synergist is used here in its nonspecialized meaning, simply to reflect a cooperative interaction and to provide a term for the class having opposite effects to those of antagonists. The synergy discussed here is additive, not multiplicative as is sometimes reflected in specialized use of the term.

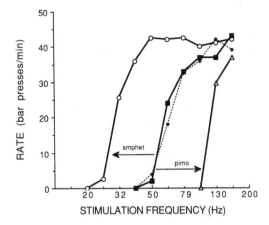

*Figure 3* Rate of responding for hypothalamic brain stimulation as a function of stimulation frequency and drug treatments. Dotted line and small diamonds indicate data from the no-drug condition; open circles indicate data taken during treatment with 2 mg kg$^{-1}$ of amphetamine; open triangles indicate data taken during treatment with 0.3 mg kg$^{-1}$ of pimozide; and filled squares indicate data taken during treatment with the combination of 2 mg amphetamine and 0.3 mg kg$^{-1}$ pimozide. [Adapted from Gallistel & Karras (1984) with permission of the authors and publisher.]

drugs of abuse. The best-studied shifts are those associated with amphetamine, a drug that causes impulse-independent (Carboni et al 1989, Hurd & Ungerstedt 1989a, Westerink et al 1987) release of dopamine from nerve terminals (Heikkila et al 1975b) and that blocks dopamine inactivation by blocking its reuptake (Heikkila 1975a,b). Amphetamine causes parallel leftward shifts of the rate-frequency function for medial forebrain bundle brain stimulation reward (Gallistel & Karras 1984, Wise & Munn 1993). A dose of amphetamine that causes a 0.3-log unit shift of the curve to the left just cancels the effects of a dose of dopamine antagonist that causes a 0.3-log unit shift to the right (Figure 3) (Gallistel & Karras 1984). Thus amphetamine acts as a synergist of brain stimulation reward as judged by two criteria: (*a*) by reducing the amount of stimulation needed to produce a given level of responding and (*b*) by nullifying the effects of a reward antagonist that has the opposite effect. Similarly, ventral tegmental injections of morphine—which elevate forebrain dopamine levels by disinhibiting the firing of dopaminergic neurons near the injection site (Johnson and North 1992)—also antagonize the rightward shift caused by pimozide in the rate-frequency paradigm (Rompré & Wise 1989).

Although they have not all been tested for their ability to counter the reward-antagonizing effects of dopamine antagonists, several other drugs of abuse have been shown to cause parallel leftward shifts in the brain stimulation reward rate-frequency function. In addition to amphetamine (Gallistel

& Freyd 1987, Gallistel & Karras 1984, Wise & Munn 1993) and morphine (Bauco et al 1993, Carlezon & Wise 1993a, Rompré & Wise 1989), which were already mentioned, these drugs include selective mu and delta opioids (Jenck et al 1987), cocaine (Bauco et al 1992, Wise et al 1992), nicotine (Bauco & Wise 1994), phencyclidine (Carlezon & Wise 1993b), and cannabis (EL Gardner, personal communication). Apomorphine, a drug not abused by humans but intravenously self-administered by rats (Baxter et al 1974), also causes leftward shifts in brain stimulation rate-intensify functions (Leith 1983).

Consistent with these findings are data from studies in which brain stimulation reward thresholds were determined by other methods; data from such methods identify a single point on the rate-frequency or rate-intensity function and thus indicate leftward or rightward shifts in the curve without revealing whether the shifts are parallel or whether a given treatment causes a change in the performance capability of the animal. Brain stimulation reward thresholds are decreased by amphetamines (Esposito et al 1980, Hubner et al 1988, Schaefer & Holtzman 1979, Schaefer & Michael 1988, Stein & Ray 1960), cocaine (Bain & Kornetsky 1987; Frank et al 1988, 1992; Kornetsky & Esposito 1981), heroin and morphine (Esposito & Kornetsky 1977, Esposito et al 1979, Hubner & Kornetsky 1992, Marcus & Kornetsky 1974, Nazzaro et al 1981), heroin (Bozarth et al 1980, Hubner & Kornetsky 1992), nicotine (Huston-Lyons et al 1992), phencyclidine (Kornetsky & Esposito 1979), cannabis (Gardner et al 1988, 1989), and possibly ethanol (Bain & Kornetsky 1989, Moolton & Kornetsky 1990; but see De Witte & Bada 1983, Schaefer & Michael 1987). As can be inferred from the curve-shift and threshold data, drugs of abuse generally tend to increase simple response rate when optimal stimulation parameters are chosen (Gardner 1992, Wise 1980b).

## Hypothesized Sites of Rewarding Action

If drugs of abuse are presumed to potentiate the rewarding effects of brain stimulation by actions at the same sites where the drugs are habit forming in their own right, we should, to some degree, be able to identify those sites of action. The data are very incomplete, but much of the literature is consistent with the hypothesis that drugs of abuse summate with brain stimulation reward and initiate their own habit-forming actions at sites intimately related to the mesolimbic dopamine system.

Like brain stimulation reward itself (Bauco et al 1994, Blaha & Phillips 1990, Gratton et al 1988), most drugs of abuse also elevate dopamine concentration at forebrain dopamine terminals, including nucleus accumbens: This is true for amphetamine (Carboni et al 1989, Hurd & Ungerstedt 1989a, Westerink et al 1987, Zetterström et al 1981), cocaine (Church et al 1987, Di Chiara & Imperato 1988, Hurd et al 1988), mu and delta opiates (Devine et al 1993, Di Chiara & Imperato 1988), nicotine (Di Chiara & Imperato 1988), phency-clidine (Bowyer et al 1984, Gerhardt et al 1987, Hernandez et al 1988), ethanol (Di Chiara & Imperato 1988), and cannabis (Gardner et al 1988, 1989).

Studies in which hypothalamic brain stimulation is given in the form of trains of paired pulses indicate that such stimulation probably does not directly depolarize a significant number of axons of the mesolimbic dopamine system. Variations of the inter-pulse interval indicate that the distribution of refractory periods for the directly activated, reward-relevant neurons at the electrode tip is generally too fast to reflect much contribution of the small, unmyelinated, dopaminergic fibers (Yeomans 1979). When paired-pulse stimulation is di-vided between two electrodes at different sites along the trajectory of the medial forebrain bundle (the conditioning pulse from each pair to one electrode and the test pulse to the other), the distribution of conduction velocities for the reward-relevant fibers can be estimated; again, the estimates suggest little contribution of directly activated dopaminergic fibers (Bielajew & Shizgal 1982, Shizgal et al 1980). Dual-electrode experiments can further suggest the direction of conduction of the reward pathway, and descending fibers have clearly been implicated (Bielajew & Shizgal 1986). Thus current evidence suggests that the bulk[6] of the reward signal triggered by depolarizing medial forebrain bundle axons is carried by descending myelinated axons. Because the trajectory of these axons conforms closely to the dorsal, ventral, medial, and lateral extent of the dopamine system (Corbett & Wise 1980, Wise 1981) and because pharmacological blockade of the dopamine system blocks the rewarding effects of the stimulation (Fouriezos & Wise 1976, Franklin 1978), some investigators have suggested that the descending myelinated reward fibers of the medial forebrain bundle may synapse directly on the dopaminergic cells of the ventral tegmental area and substantia nigra (Corbett & Wise 1980; Yeomans 1982; Wise 1980a,b); alternatively, the descending myelinated fibers may activate the dopamine system transsynaptically through actions on inter-vening neurons (Yeomans et al 1993). Rewarding hypothalamic stimulation usually fails to activate the dopamine fibers directly because these fibers have a very high threshold for direct activation (Yeomans et al 1988) and because quite low levels of stimulation are sufficient to establish reliable responding. Ultimately, however, rewarding hypothalamic brain stimulation does result,

[6]Because the current density falls off with the square of the distance from the electrode tip, some dopaminergic fibers are almost certainly activated at the electrode-tissue interface. With smaller electrode tips and higher stimulation intensities than are usually used, more small unmyelinated fibers can be activated (Yeomans et al 1985). However the number of such fibers is thought to be small under the traditional stimulation conditions (Yeomans 1989).

presumably transsynaptically, in strong activation of the mesolimbic dopamine system (Bauco et al 1994, Blaha & Phillips 1990, Gratton et al 1988).

The site of the rewarding and reward-enhancing actions of amphetamine appears to be at the dopamine transporter (Fisher & Cho 1979) on the terminals of the mesolimbic projection to nucleus accumbens. Rats learn to lever-press for amphetamine injections into this region (Hoebel et al 1983); amphetamine injections into this region establish conditioned place preferences (Carr & White 1983); and such injections potentiate brain stimulation reward as reflected in the curve-shift paradigm (Colle & Wise 1988). Systemic injections of dopamine antagonists (Davis & Smith 1975; Risner & Jones 1976; Yokel & Wise 1975, 1976) or dopamine-selective lesions of nucleus accumbens (Lyness et al 1979) block the rewarding effect of intravenous amphetamine. Amphetamine's ability to elevate extracellular dopamine levels can also be localized to dopamine terminal fields such as the nucleus accumbens (Zetterström et al 1988). Amphetamine actions at the level of the dopaminergic cell bodies are opposite in direction; at the soma, amphetamine causes dopamine release (Robertson et al 1991) that results in autoreceptor-mediated inhibition of dopaminergic cell firing (Bunney et al 1973).

The rewarding site of action of cocaine is not so clearly established. Although lesions of the mesolimbic dopamine system (Roberts et al 1977, 1980; Zito et al 1985) and neuroleptic injections into (Phillips et al 1983) nucleus accumbens disrupt the rewarding effects of intravenous cocaine, rats are not easily trained to work for nucleus accumbens injections of cocaine (Goeders & Smith 1983; but see Wise & Carlezon 1994). Rats do readily learn to work for cocaine injections into the medial frontal cortex, which has led to the suggestion of different sites of rewarding action for cocaine and amphetamine (Goeders & Smith 1983). Against this suggestion was the fact that 6-hydroxydopamine lesions of nucleus accumbens disrupt intravenous cocaine self-administration (Roberts et al 1977, 1980), but it could be argued that 6-hydroxy- dopamine lesions of accumbens also damage fibers of passage en route to the frontal cortex. However, nucleus accumbens injections of neuroleptics—which should spare the frontal cortex dopamine projections—also block cocaine self-administration (Phillips et al 1983). Similarly, kainic acid lesions of nucleus accumbens spare frontal cortex dopamine projections but disrupt intravenous cocaine self-administration (Zito et al 1985). Moreover, 6-hydrox- ydopamine lesions of frontal cortex fail to disrupt intravenous cocaine self-administration (Martin-Iverson et al 1986). Recent evidence suggests that even frontal cortex cocaine injections increase dopamine turnover in nucleus accumbens (Goeders & Smith 1993). Finally, although cocaine is not readily self-administered into nucleus accumbens, the more efficacious (Nomikos et al 1990) dopamine uptake inhibitor nomifensine is (Wise & Carlezon 1994). The bulk of the evidence thus appears to implicate nucleus accumbens in the rewarding effects of cocaine. Cocaine infused locally into nucleus accumbens is sufficient to elevate nucleus accumbens dopamine levels (Hurd & Ungerstedt 1989b); the action of cocaine at the dopamine cell body, like that of amphetamine, is to inhibit dopaminergic cell firing (Einhorn et al 1988).

Like the direct rewarding effects of amphetamine and cocaine and the reward-enhancing effects of amphetamine, the rewarding and reward-enhancing effects of opiates appear to be mediated via nucleus accumbens. Independent rewarding actions at two mesolimbic sites have been reported in this regard. First, rats work for morphine (Bozarth & Wise 1981, van Ree & de Wied 1980, Welzl et al 1989) and mu and delta (Devine & Wise 1994) opioid injections given locally into the region of dopaminergic cell bodies in the ventral tegmental area. Mu opioids injected here inhibit GABAergic neurons, increasing dopaminergic cell firing by releasing the adjacent dopaminergic neurons from tonic inhibition (Johnson & North 1992). Opiate injections into this region also cause conditioned place preferences (Bozarth 1987, Phillips & LePiane 1980); similarly, morphine and mu opioid injections into this region potentiate the rewarding effects of hypothalamic brain stimulation (Broekkamp et al 1976, Jenck et al 1987, Rompré & Wise 1989). Morphine and mu and delta opioid injections into this region increase nucleus accumbens dopamine levels, and the relative potencies of these agents in elevating nucleus accumbens dopamine is proportional to their relative potencies as rewards in their own right (Devine et al 1993).

Rats also work for morphine or enkephalin injections directly into the nucleus accumbens (Goeders et al 1984, Olds 1982). Morphine injections into this area can cause conditioned place preferences (van der Kooy et al 1982), and morphine and mu and delta opioid injections into this region facilitate brain stimulation reward (West & Wise 1988b). In this case the rewarding effects of opiates are thought to be independent of the dopamine system (Koob 1992), perhaps being triggered at the level of the medium spiny neurons that are the main synaptic targets of the meso-accumbens dopaminergic terminals (Sesack et al 1994).

Sites of action for the rewarding, reward-potentiating, and dopamine-elevating actions of other drugs of abuse are not yet so well characterized; however, a good deal of evidence suggests mechanisms of rewarding drug action that are linked to dopamine. The nicotinic agonist cytisine causes conditioned place preferences when injected into the ventral tegmental area (Museo & Wise 1994); systemically applied nicotine activates dopaminergic cells in this region (Clarke et al 1985, Grenhoff et al 1986, Mereu et al 1987). Nicotinic receptors are apparently localized to the dopaminergic neurons themselves (Clarke & Pert 1985). Phencyclidine blocks dopamine uptake (a nucleus accumbens action) (Gerhardt et al 1987) and increases dopaminergic impulse flow (a transsynaptic action expressed at the level of the ventral tegmental area) (French 1986); it has known rewarding actions in nucleus accumbens, but it shares them with MK-801 (Carlezon & Wise 1993c), which apparently does not share phencyclidine's ability to block dopamine uptake (Reid et al 1990).[7] Ethanol increases dopaminergic cell firing, but the mechanism is unknown (Gessa et al 1985). Cannabis elevates dopamine levels; it appears to do so by blocking dopamine uptake (Chen et al 1994).

Thus where the sites or mechanisms of action are known, common sites and mechanisms appear to subserve the ability of drugs of abuse to potentiate rewarding brain stimulation, to serve as rewards in their own right, and to activate the mesolimbic dopamine system or its efferents. Summation of effects within or at the synapses of the mesolimbic system is thus, at the present time, the most likely candidate for the mechanism underlying the synergism between rewarding drug injections and rewarding hypothalamic brain stimulation.

## CONCLUSIONS

One should not overgeneralize from the literature cited. Not all habit-forming drugs cause reliable reductions in reward thresholds or elevate extracellular dopamine levels. Although ethanol augments dopaminergic cell firing and elevates extracellular dopamine levels, there are as many reports that it fails to facilitate brain stimulation reward as there are that it succeeds; even based on the favorable reports, its facilitatory effects on brain stimulation reward are small and may be restricted to a subpopulation of responsive animals (Bain & Kornetsky 1989, Moolton & Kornetsky 1990). While caffeine stimulates dopamine turnover (Govoni et al 1984), it elevates brain stimulation reward thresholds (Mumford et al 1988). Although barbiturates and benzodiazepines are self-administered, modestly, by rats (Collins et al 1984, Szostak et al 1987) and although they have been reported to decrease brain stimulation reward thresholds (or at least potentiate response rate under certain circumstances) (Seeger et al 1981, Wise 1980b), these drugs are usually found to decrease rather than increase extracellular dopamine levels (Finlay et al 1992, Wood 1982; but see Di Chiara et al 1991).

Thus it does not seem to be the case that all habit-forming drugs activate the same reward mechanism in the brain. The evidence is good, however, that several of the more addictive substances—the psychomotor stimulants, the opiates, nicotine, phencyclidine, and cannabis—synergize with rewarding me-

---

[7]MK-801 does, however, share phencyclidine's ability to increase dopaminergic cell firing (French & Ceci 1990). Thus some of the NMDA receptor–mediated actions of phencyclidine and MK-801 may also be dopamine dependent.

dial forebrain bundle brain stimulation, and elevate—as does the stimulation itself (Bauco et al 1994, Blaha & Phillips 1990, Gratton et al 1988)—dopamine concentrations in the nucleus accumbens and other dopamine terminal fields. The degree to which other drugs of abuse activate the same system less dramatically or less directly remains to be more fully explored. For example, Rassnick et al (1992) and Samson et al (1993) have argued that the habit-forming effects of ethanol—on the strength of other forms of evidence—depend on the dopamine system (Rassnick et al 1992, Samson et al 1993). The effects of barbiturates and benzodiazepines are still a matter of debate (Di Chiara et al 1991) and may have ambiguous effects because of concurrent actions at different levels of the anatomical cascade of GABAergic neurons originating near the dopamine terminals of nucleus accumbens. The habit-forming effects of caffeine seem clearly not to fit within the present framework. It would be interesting to know if caffeine, like neuroleptics and clonidine (Gallistel & Freyd 1987), acts as a competitive antagonist to brain stimulation reward.

Although the first question of interest is whether two drugs or two treatments act by the same or a different mechanism, no two drugs will act by identical mechanisms. At the same time, drugs like the opiates and psychomotor stimulants, once argued to act through completely independent reward mechanisms (Ettenberg et al 1982), clearly have a common substrate of action, interfacing with the same circuitry at different synaptic levels (Hubner & Koob 1990, Wise & Bozarth 1987). Studies of brain stimulation reward contribute significantly to this emerging conclusion and may offer a powerful quantitative model (e.g. see Gallistel & Freyd 1987) for assessing correlates of the habit-forming properties of addictive drugs.

## Literature Cited

Bain GT, Kornetsky C. 1987. Naloxone attenuation of the effect of cocaine on rewarding brain stimulation. *Life Sci.* 40: 1119–25

Bain GT, Kornetsky C. 1989. Ethanol oral self-administration and rewarding brain stimulation. *Alcohol* 6:499–503

Bauco P, Rivest R, Wise RA. 1994. Extracellular nucleus accumbens dopamine and metabolite levels during earned and unearned lateral hypothalamic brain stimulation. *Soc. Neurosci. Abstr.* 20:823

Bauco P, Wang Y, Wise RA. 1992. Cocaine potentiation of lateral hypothalamic brain stimulation reward: a dose response and repeated treatment analysis. *Soc. Neurosci. Abstr.* 18:1573

Bauco P, Wang Y, Wise RA. 1993. Lack of sensitization or tolerance to the facilitating effect of ventral tegmental area morphine on lateral hypothalamic brain stimulation reward. *Brain Res.* 617:303–8

Bauco P, Wise R. 1994. Potentiation of lateral hypothalamic and midline mesencephalic brain stimulation reinforcement by nicotine: examination of repeated treatment. *J. Pharmacol. Exp. Ther.* 271:294–301

Baxter BL, Gluckman MI, Stein L, Scerni RA. 1974. Self-injection of apomorphine in the rat: positive reinforcement by a dopamine receptor stimulant. *Pharmacol. Biochem. Behav.* 2:387–91

Beninger RJ, Bellisle F, Milner PM. 1977. Schedule control of behavior reinforced by electrical stimulation of the brain. *Science* 196:547–49

Bielajew C, Shizgal P. 1980. Dissociation of the substrates for medial forebrain bundle self-stimulation and stimulation escape using a two-electrode stimulation technique. *Physiol. Psychol.* 25:707–11

Bielajew C, Shizgal P. 1982. Behaviorally derived measures of conduction velocity in the substrate for rewarding medial forebrain bundle stimulation. *Brain Res.* 237:107–19

Bielajew C, Shizgal P. 1986. Evidence implicating descending fibers in self-stimulation of the medial forebrain bundle. *J. Neurosci.* 6:919–29

Bindra D. 1974. A motivational view of learning, performance, and behavior modification. *Psychol. Rev.* 81:199–213

Blaha CD, Phillips AG. 1990. Application of in vivo electrochemistry to the measurement of changes in dopamine release during intracranial self-stimulation. *J. Neurosci. Meth.* 34:125–33

Bower GH, Miller NE. 1958. Rewarding and punishing effects from stimulating the same place in the rat's brain. *J. Comp. Physiol. Psychol.* 51:69–72

Bowyer JF, Spuhler KP, Weiner N. 1984. Effects of phencyclidine, amphetamine and related compounds on dopamine release from and uptake into striatal synaptosomes. *J. Pharmacol. Exp. Ther.* 229:671–80

Bozarth MA. 1987. Neuroanatomical boundaries of the reward-relevant opiate-receptor field in the ventral tegmental area as mapped by the conditioned place preference method in rats. *Brain Res.* 414:77–84

Bozarth MA, Gerber GJ, Wise RA. 1980. Intracranial self-stimulation as a technique to study the reward properties of drugs of abuse. *Pharmacol. Biochem. Behav.* 13 (Suppl. 1):245–47

Bozarth MA, Wise RA. 1981. Intracranial self-administration of morphine into the ventral tegmental area in rats. *Life Sci.* 28:551–55

Broekkamp CLE, Van den Bogaard JH, Heijnen HJ, Rops RH, Cools AR, Van Rossum JM. 1976. Separation of inhibiting and stimulating effects of morphine on self-stimulation behavior by intracerebral microinjections. *Eur. J. Pharmacol.* 36:443–46

Bunney BS, Walters JR, Roth RH, Aghajanian GK. 1973. Dopaminergic neurons: effect of antipsychotic drugs and amphetamine on single cell activity. *J. Pharmacol. Exp. Ther.* 185:560–71

Carboni E, Imperato A, Perezzani L, Di Chiara G. 1989. Amphetamine, cocaine, phencyclidine and nomifensine increase extracellular dopamine concentrations preferentially in the nucleus accumbens of freely moving rats. *Neuroscience* 28:653–61

Carlezon WA Jr, Wise RA. 1993a. Morphine-induced potentiation of brain stimulation reward is enhanced by MK-801. *Brain Res.* 620:339–42

Carlezon WA Jr, Wise RA. 1993b. Phencyclidine-induced potentiation of brain stimulation reward: Acute effects are not altered by repeated administration. *Psychopharmacology* 111:402–8

Carlezon WA Jr, Wise RA. 1993c. Rats self-administer the non-competitive NMDA receptor antagonists phencyclidine (PCP) and MK-801 directly into the nucleus accumbens. *Soc. Neurosci. Abstr.* 19:830

Carr GD, White NM. 1983. Conditioned place preference from intra-accumbens but not intra-caudate amphetamine injections. *Life Sci.* 33:2551–57

Carter DA, Phillips AG. 1975. Intracranial self-stimulation at sites in the dorsal medulla oblongata. *Brain Res.* 94:155–60

Chen J, Paredes W, Gardner EL. 1994. $\Delta^9$-Tetrahydrocannabinol's enhancement of nucleus accumbens dopamine resembles that of reuptake blockers rather than releasers—evidence from in vivo microdialysis experiments with 3-methoxytyramine. Presented at Annu. Meet. Coll. Prob. Drug Depend., 57th, Palm Beach

Church WH, Justice JB Jr, Byrd LD. 1987. Extracellular dopamine in rat striatum following uptake inhibition by cocaine, nomifensine and benztropine. *Eur. J. Pharmacol.* 139:345–48

Clarke PBS, Hommer DW, Pert A, Skirboll LR. 1985. Electrophysiological actions of nicotine on substantia nigra single units. *Br. J. Pharmacol.* 85:827–35

Clarke PBS, Pert A. 1985. Autoradiographic evidence for nicotine receptors on nigrostriatal and mesolimbic dopaminergic neurons. *Brain Res.* 348:355–58

Colle LM, Wise RA. 1988. Effects of nucleus accumbens amphetamine on lateral hypothalamic brain stimulation reward. *Brain Res.* 459:356–60

Collins RJ, Weeks JR, Cooper MM, Good PI, Russell RR. 1984. Prediction of abuse liability of drugs using IV self-administration by rats. *Psychopharmacology* 82:6–13

Conover KL, Shizgal P. 1994. Competition and summation between rewarding effects of sucrose and lateral hypothalamic stimulation in the rat. *Behav. Neurosci.* 108:537–48

Corbett D, Fox E, Milner PM. 1982. Fiber pathways associated with cerebellar self-stimulation in the rat: a retrograde and anterograde tracing study. *Behav. Brain Res.* 6:167–84

Corbett D, Wise RA. 1979. Intracranial self-stimulation in relation to the ascending noradrenergic fiber systems of the pontine tegmentum and caudal midbrain: a moveable electrode mapping study. *Brain Res.* 177: 423–36

Corbett D, Wise RA. 1980. Intracranial self-stimulation in relation to the ascending dopaminergic systems of the midbrain: a moveable electrode mapping study. *Brain Res.* 185:1–15

Davis WM, Smith SG. 1975. Effect of haloperidol on (+)-amphetamine self-administration. *J. Pharm. Pharmacol.* 27:540–42

Devine DP, Leone P, Pocock D, Wise RA. 1993. Differential involvement of ventral tegmental *mu, delta,* and *kappa* opioid receptors in modulation of basal mesolimbic dopamine release: in vivo microdialysis studies. *J. Pharmacol. Exp. Ther.* 266:1236–46

Devine DP, Wise RA. 1994. Self-administration of morphine, DAMGO, and DPDPE into the ventral tegmental area of rats. *J. Neurosci.* 14:1978–84

De Witte P, Bada MF. 1983. Self-stimulation and alcohol administered orally or intraperitoneally. *Exp. Neurol.* 82:675–82

Di Chiara G, Acquas E, Carboni E. 1991. Role of mesolimbic dopamine in the motivational effects of drugs: brain dialysis and place preference studies. In *The Mesolimbic Dopamine System: From Motivation to Action,* ed. P Willner, J Scheel-Kruger, pp. 367–84. New York: Wiley

Di Chiara G, Imperato A. 1988. Drugs of abuse preferentially stimulate dopamine release in the mesolimbic system of freely moving rats. *Proc. Natl. Acad. Sci. USA* 85:5274–78

Edmonds DE, Gallistel CR. 1974. Parametric analysis of brain stimulation reward in the rat: III. Effect of performance variables on the reward summation function. *J. Comp. Physiol. Psychol.* 87:876–83

Einhorn LC, Johansen PA, White FJ. 1988. Electrophysiological effects of cocaine in the mesoaccumbens dopamine system: studies in the ventral tegmental area. *J. Neurosci.* 8:100–12

Esposito R, Kornetsky C. 1977. Morphine lowering of self-stimulation thresholds: lack of tolerance with long-term administration. *Science* 195:189–91

Esposito RU, McLean S, Kornetsky C. 1979. Effects of morphine on intracranial self-stimulation to various brain stem loci. *Brain Res.* 168:425–29

Esposito RU, Perry W, Kornetsky C. 1980. Effects of d-amphetamine and naloxone on brain stimulation reward. *Psychopharmacology* 69:187–91

Ettenberg A, Pettit HO, Bloom FE, Koob GF. 1982. Heroin and cocaine intravenous self-administration in rats: mediation by separate neural systems. *Psychopharmacology* 78: 204–9

Fibiger HC. 1978. Drugs and reinforcement mechanisms: a critical review of the catecholamine theory. *Annu. Rev. Pharmacol. Toxicol.* 18:37–56

Fibiger HC, LePiane FG, Jakubovic A, Phillips AG. 1987. The role of dopamine in intracranial self-stimulation of the ventral tegmental area. *J. Neurosci.* 7:3888–96

Finlay JM, Damsma G, Fibiger HC. 1992. Benzodiazepine-induced decreases in extracellular concentrations of dopamine in the nucleus accumbens after acute and repeated administration. *Psychopharmacology* 106: 202–8

Fisher JF, Cho AK. 1979. Chemical release of dopamine from striatal homogenates: evidence for an exchange diffusion model. *J. Pharmacol. Exp. Ther.* 208:203–9

Fouriezos G, Wise RA. 1976. Pimozide-induced extinction of intracranial self-stimulation: response patterns rule out motor or performance deficits. *Brain Res.* 103:377–80

Fouriezos G, Wise RA. 1984. Current-distance relation for rewarding brain stimulation. *Behav. Brain Res.* 14:85–89

Frank RA, Manderscheid PZ, Panicker S, Williams HP, Kokoris D. 1992. Cocaine euphoria, dysphoria, and tolerance assessed using drug-induced changes in brain-stimulation reward. *Pharmacol. Biochem. Behav.* 42:771–79

Frank RA, Markou A, Wiggins LL. 1987. A systematic evaluation of the properties of train duration response functions. *Behav. Neurosci.* 101:546–59

Frank RA, Martz S, Pommering T. 1988. The effect of chronic cocaine on self-stimulation train-duration thresholds. *Pharmacol. Biochem. Behav.* 29:755–58

Franklin KBJ. 1978. Catecholamines and self-stimulation: reward and performance effects dissociated. *Pharmacol. Biochem. Behav.* 9: 813–20

Franklin KBJ, McCoy SN. 1979. Pimozide-induced extinction in rats: stimulus control of responding rules out motor deficit. *Pharmacol. Biochem. Behav.* 11:71–75

French ED. 1986. Effects of phencyclidine on ventral tegmental $A_{10}$ dopamine neurons in the rat. *Neuropharmacology* 25:241–48

French ED, Ceci A. 1990. Non-competitive N-methyl-D-aspartate antagonists are potent activators of ventral tegmental $A_{10}$ dopamine neurons. *Neurosci. Lett.* 119:159–62

Gallistel CR. 1978. Self-stimulation in the rat: quantitative characteristics of the reward pathway. *J. Comp. Physiol. Psychol.* 92: 977–98

Gallistel CR. 1986. The role of the dopaminergic projections in MFB self-stimulation. *Behav. Brain Res.* 20:313–21

Gallistel CR. 1987. Determining the quantitative characteristics of a reward pathway. In *Biological Determinants of Reinforcement*, ed. RM Church, ML Commons, JR Stellar, AR Wagner, pp. 1–30. Hillsdale, NJ: Erlbaum Assoc.

Gallistel CR, Davis AJ. 1983. Affinity for the dopamine D2 receptor predicts neuroleptic potency in blocking the reinforcing effect of MFB stimulation. *Pharmacol. Biochem. Behav.* 19:867–72

Gallistel CR, Freyd G. 1987. Quantitative determination of the effects of catecholaminergic agonists and antagonists on the rewarding efficacy of brain stimulation. *Pharmacol. Biochem. Behav.* 26:731–41

Gallistel CR, Karras D. 1984. Pimozide and amphetamine have opposing effects on the reward summation function. *Pharmacol. Biochem. Behav.* 20:73–77

Gallistel CR, Shizgal P, Yeomans J. 1981. A portrait of the substrate for self-stimulation. *Psychol. Rev.* 88:228–73

Gallistel CR, Stellar JR, Bubis E. 1974. Parametric analysis of brain stimulation reward in the rat: I. The transient process and the memory-containing process. *J. Comp. Physiol. Psychol.* 87:848–59

Gardner EL. 1992. Brain reward mechanisms. In *Substance Abuse: A Comprehensive Textbook*, ed. JH Lowinson, P Ruiz, RB Millman, pp. 70–99. Baltimore: Williams & Wilkins

Gardner EL, Paredes W, Smith D, Donner A, Milling C, et al. 1988. Facilitation of brain stimulation reward by Δ⁹-tetrahydrocannabinol. *Psychopharmacology* 96:142–44

Gardner EL, Paredes W, Smith D, Zukin RS. 1989. Facilitation of brain stimulation reward by D⁹tetrahydrocannabinol is mediated by an endogenous opioid mechanism. *Adv. Biosci.* 75:671–74

Gerhardt GA, Pang K, Rose GM. 1987. In vivo electrochemical demonstration of the presynaptic actions of phencyclidine in rat caudate nucleus. *J. Pharmacol. Exp. Ther.* 241:714–21

Gessa GL, Muntoni F, Collu M, Vargiu L, Mereu G. 1985. Low doses of ethanol activate dopaminergic neurons in the ventral tegmental area. *Brain Res.* 348:201–4

Goeders NE, Lane JD, Smith JE. 1984. Self-administration of methionine enkephalin into the nucleus accumbens. *Pharmacol. Biochem. Behav.* 20:451–55

Goeders NE, Smith JE. 1983. Cortical dopaminergic involvement in cocaine reinforcement. *Science* 221:773–75

Goeders NE, Smith JE. 1993. Intracranial cocaine self-administration into the medial prefrontal cortex increases dopamine turnover in the nucleus accumbens. *J. Pharmacol. Exp. Ther.* 265:592–600

Govoni S, Petkov VV, Montefusco O, Missale C, Battaini F, et al. 1984. Differential effects of caffeine on dihydroxyphenylacetic concentrations in various rat brain regions. *J. Pharm. Pharmacol.* 36:458–60

Gratton A, Hoffer BJ, Gerhardt GA. 1988. Effects of electrical stimulation of brain reward sites on release of dopamine in rat: an in vivo electrochemical study. *Brain Res. Bull.* 21:319–24

Grenhoff J, Aston-Jones G, Svensson TH. 1986. Nicotinic effects on the firing pattern of midbrain dopamine neurons. *Acta Physiol. Scand.* 128:351–58

Heath RG. 1963. Intracranial self-stimulation in man. *Science* 140:394–96

Heath RG. 1972. Pleasure and brain activity in man. *J. Nerv. Ment. Dis.* 154:3–18

Heath RG. 1975. Brain function and behavior. I. Emotion and sensory phenomena in psychotic patients and in experimental animals. *J. Nerv. Ment. Dis.* 160:159–75

Heikkila RE, Orlansky H, Cohen G. 1975a. Studies on the distinction between uptake inhibition and release of (³H)dopamine in rat brain tissue slices. *Biochem. Pharmacol.* 24:847–52

Heikkila RE, Orlansky H, Mytilineou C, Cohen G. 1975b. Amphetamine: evaluation of *d*- and *l*-isomers as releasing agents and uptake inhibitors for ³H-dopamine and ³H-norepinephrine in slices of rat neostriatum and cerebral cortex. *J. Pharmacol. Exp. Ther.* 194:47–56

Hernandez L, Auerbach S, Hoebel BG. 1988. Phencyclidine (PCP) injected in the nucleus accumbens increases extracellular dopamine and serotonin as measured by microdialysis. *Life Sci.* 42:1713–23

Hoebel BG, Monaco AP, Hernandez L, Aulisi EF, Stanley BG, Lenard L. 1983. Self-injection of amphetamine directly into the brain. *Psychopharmacology* 81:158–63

Hubner CB, Bird M, Rassnick S, Kornetsky C. 1988. The threshold lowering effects of MDMA (ecstasy) on brain-stimulation reward. *Psychopharmacology* 95:49–51

Hubner CB, Koob GF. 1990. The ventral striatum plays a role in mediating cocaine and heroin self-administration in the rat. *Brain Res.* 508:20–29

Hubner CB, Kornetsky C. 1992. Heroin, 6-acetylmorphine and morphine effects on threshold for rewarding and aversive brain stimulation. *J. Pharmacol. Exp. Ther.* 260:562–67

Hurd YL, Kehr J, Ungerstedt U. 1988. In vivo microdialysis as a technique to monitor drug transport: correlation of extracellular cocaine levels and dopamine overflow in the rat brain. *J. Neurochem.* 51:1314–16

Hurd YL, Ungerstedt U. 1989a. Ca²⁺ dependence of the amphetamine, nomifensine, and

Lu 19-005 effect on in vivo dopamine transmission. *Eur. J. Pharmacol.* 166:261–69

Hurd YL, Ungerstedt U. 1989b. Cocaine: an in vivo microdialysis evaluation of its acute action on dopamine transmission in rat striatum. *Synapse* 3:48–54

Huston-Lyons DJ, Bain GT, Kornetsky C. 1992. The effects of nicotine on the threshold for rewarding brain stimulation in rats. *Pharmacol. Biochem. Behav.* 41:755–59

Jenck F, Gratton A, Wise RA. 1987. Opioid receptor subtypes associated with ventral tegmental facilitation of lateral hypothalamic brain stimulation reward. *Brain Res.* 423:34–38

Johanson CE. 1978. Drugs as reinforcers. In *Contemporary Research in Behavioral Pharmacology*, ed. DE Blackman, DJ Sanger, pp. 325–90. New York: Plenum

Johnson SW, North RA. 1992. Opioids excite dopamine neurons by hyperpolarization of local interneurons. *J. Neurosci.* 12:483–88

Josefsson J-O, Johansson P. 1979. Naloxone-reversible effect of opioids on pinocytosis in *Amoeba proteus. Nature* 282:78–80

Kavaliers M, Hirst M. 1987. Slugs and snails and opiate tales: opioids and feeding behavior in invertebrates. *Fed. Proc.* 46:168–72

Keesey RE. 1962. The relation between pulse frequency, intensity, and duration and the rate of responding for intracranial stimulation. *J. Comp. Physiol. Psychol.* 55:671–78

Kent E, Grossman SP. 1969. Evidence for a conflict interpretation of anomalous effects of rewarding brain stimulation. *J. Comp. Physiol. Psychol.* 69:381–90

Koob GF. 1992. Drugs of abuse: anatomy, pharmacology and function of reward pathways. *Trends Pharmacol. Sci.* 13:177–84

Kornetsky C, Esposito RU. 1979. Euphorigenic drugs: effects on the reward pathways of the brain. *Fed. Proc.* 38:2473–76

Kornetsky C, Esposito RU. 1981. Reward and detection thresholds for brain stimulation: dissociative effects of cocaine. *Brain Res.* 209:496–500

Leith NJ. 1983. The effects of apomorphine on self-stimulation responding: Does the drug mimic the current? *Brain Res.* 277:129–36

Lepore M, Franklin KBJ. 1992. Modelling drug kinetics with brain stimulation: Dopamine antagonists increase self-stimulation. *Pharmacol. Biochem. Behav.* 41:489–96

Liebman JM. 1983. Discriminating between reward and performance: a critical review of intracranial self-stimulation methodology. *Neurosci. Biobehav. Rev.* 7:45–72

Liebman JM, Butcher LL. 1974. Comparative involvement of dopamine and noradrenaline in rate-free self-stimulation in substantia nigra, lateral hypothalamus, and mesencephalic central gray. *Naunyn-Schmiedeberg's Arch. Pharmacol.* 284:167–94

Lippa AS, Antelman SM, Fisher AE, Canfield DR. 1973. Neurochemical mediation of reward: a significant role for dopamine. *Pharmacol. Biochem. Behav.* 1:23–28

Lyness WH, Friedle NM, Moore KE. 1979. Destruction of dopaminergic nerve terminals in nucleus accumbens: effect on d-amphetamine self-administration. *Pharmacol. Biochem. Behav.* 11:553–56

Marcus R, Kornetsky C. 1974. Negative and positive intracranial reinforcement thresholds: effects of morphine. *Psychopharmacologia* 38:1–13

Martin-Iverson MT, Szostak C, Fibiger HC. 1986. 6-Hydroxydopamine lesions of the medial prefrontal cortex fail to influence intravenous self-administration of cocaine. *Psychopharmacology* 88:310–14

Mereu G, Yoon K-WP, Boi V, Gessa GL, Naes L, Westfall TC. 1987. Preferential stimulation of ventral tegmental area dopaminergic neurons by nicotine. *Eur. J. Pharmacol.* 141: 395–400

Miliaressis E, Rompré P-P, Laviolette LP, Philippe L, Coulombe D. 1986. The curve-shift paradigm in self-stimulation. *Physiol. Behav.* 37:85–91

Moolten M, Kornetsky C. 1990. Oral self-administration of ethanol and not experimenter-administered ethanol facilitates rewarding electrical brain stimulation. *Alcohol* 7:221–25

Mumford GK, Neill DB, Holtzman SG. 1988. Caffeine elevates reinforcement threshold for electrical brain stimulation: tolerance and withdrawal changes. *Brain Res.* 459: 163–67

Museo E, Wise RA. 1994. Place preference conditioning with ventral tegmental injections of cytisine. *Life Sci.* 55:1179–86

Nazzaro JM, Seeger TF, Gardner EL. 1981. Morphine differentially affects ventral tegmental and substantia nigra brain reward thresholds. *Pharmacol. Biochem. Behav.* 14: 325–31

Nomikos GG, Damsma G, Wenkstern BA, Fibiger HC. 1990. In vivo characterization of locally applied dopamine uptake inhibitors by striatal microdialysis. *Synapse* 6:106–12

Olds J. 1956. Pleasure centers in the brain. *Sci. Am.* 195:105–16

Olds J. 1958. Self-stimulation of the brain. *Science* 127:315–24

Olds J, Allan WS, Briese E. 1971. Differentiation of hypothalamic drive and reward centers. *Am. J. Physiol.* 221:368–75

Olds J, Milner PM. 1954. Positive reinforcement produced by electrical stimulation of septal area and other regions of rat brain. *J. Comp. Physiol. Psychol.* 47:419–27

Olds J, Olds ME. 1965. Drives, rewards, and the brain. In *New Directions in Psychology,* ed. TM Newcombe, pp. 327–410. New York: Holt, Rinehart & Winston

Olds ME. 1982. Reinforcing effects of morphine in the nucleus accumbens. *Brain Res.* 237:429–40

Olds ME, Olds J. 1963. Approach-avoidance analysis of rat diencephalon. *J. Comp. Neurol.* 120:259–95

Pavlov IP. 1927. *Conditioned Reflexes.* Oxford: Oxford Univ. Press

Pfaus JG, Phillips AG. 1990. Differential effects of dopamine receptor antagonists on the sexual behavior of male rats. *Psychopharmacology* 98:363–68

Phillips AG. 1970. Enhancement and inhibition of olfactory bulb self-stimulation by odours. *Physiol. Behav.* 5:1127–31

Phillips AG. 1984. Brain reward circuitry: a case for separate systems. *Brain. Res. Bull.* 12:195–201

Phillips AG, Broekkamp CLE, Fibiger HC. 1983. Strategies for studying the neurochemical substrates of drug reinforcement in rodents. *Prog. Neuro-Psychopharmacol. Biol. Psychiat.* 7:585–90

Phillips AG, Fibiger HC. 1978. The role of dopamine in maintaining intracranial self-stimulation in the ventral tegmentum, nucleus accumbens, and medial prefrontal cortex. *Can. J. Psychol.* 32:58–66

Phillips AG, LePiane FG. 1980. Reinforcing effects of morphine microinjection into the ventral tegmental area. *Pharmacol. Biochem. Behav.* 12:965–68

Prado-Alcala R, Streather A, Wise RA. 1984. Brain stimulation reward and dopamine terminal fields. II. Septal and cortical projections. *Brain Res.* 301:209–19

Prado-Alcala R, Wise RA. 1984. Brain stimulation reward and dopamine terminal fields. I. Caudate-putamen, nucleus accumbens and amygdala. *Brain Res.* 297:265–73

Rassnick S, Pulvirenti L, Koob GF. 1992. Oral ethanol self-administration in rats is reduced by the administration of dopamine and glutamate receptor antagonists into the nucleus accumbens. *Psychopharmacology* 109:92–98

Reid AA, Monn JA, Jacobson AE, Rice KC, Rothman RB. 1990. Pseudoallosteric modulation by (+)-MK-801 of NMDA-coupled phencyclidine binding sites. *Life Sci.* 46: PL77–PL82

Risner ME, Jones BE. 1976. Role of noradrenergic and dopaminergic processes in amphetamine self-administration. *Pharmacol. Biochem. Behav.* 5:477–82

Roberts DCS, Corcoran ME, Fibiger HC. 1977. On the role of ascending catecholaminergic systems in intravenous self-administration of cocaine. *Pharmacol. Biochem. Behav.* 6: 615–20

Roberts DCS, Koob GF, Klonoff P, Fibiger HC. 1980. Extinction and recovery of cocaine self-administration following 6-OHDA lesions of the nucleus accumbens. *Pharmacol. Biochem. Behav.* 12:781–87

Roberts WW. 1958. Both rewarding and punishing effects from stimulation of posterior hypothalamus of cat with same electrode at same intensity. *J. Comp. Physiol. Psychol.* 51:400–7

Robertson GS, Damsma G, Fibiger HC. 1991. Characterization of dopamine release in the substantia nigra by in vivo microdialysis in freely moving rats. *J. Neurosci.* 7:2209–16

Rompré P-P. 1987. Effects of concomitant motor reactions on the measurement of rewarding efficacy of brain stimulation. *Behav. Neurosci.* 101:827–31

Rompré P-P, Wise RA. 1989. Opioid-neuroleptic interaction in brain stem self-stimulation. *Brain Res.* 477:144–51

Samson HH, Hodge CW, Tolliver GA, Haraguchi M. 1993. Effect of dopamine agonists and antagonists on ethanol-reinforced behavior: the involvement of the nucleus accumbens. *Brain Res. Bull.* 30:133–41

Schaefer GJ, Holtzman SG. 1979. Free-operant and auto-titration brain self-stimulation procedures in the rat: a comparison of drug effects. *Pharmacol. Biochem. Behav.* 10: 127–35

Schaefer GJ, Michael RP. 1987. Ethanol and current thresholds for brain self-stimulation in the lateral hypothalamus of the rat. *Alcohol* 4:209–13

Schaefer GJ, Michael RP. 1988. An analysis of the effects of amphetamine on brain self-stimulation behavior. *Behav. Brain Res.* 29: 93–101

Seeger TF, Carlson KR, Nazarro JM. 1981. Pentobarbitol induces a naloxone-reversible decrease in mesolimbic self-stimulation threshold. *Pharmacol. Biochem. Behav.* 15: 583–86

Sesack SR, Aoke C, Pickel VM. 1994. Ultrastructural localization of D2 receptor-like immunoreactivity in midbrain dopamine neurons and their striatal targets. *J. Neurosci.* 14:88–106

Shizgal P, Bielajew C, Corbett D, Skelton R, Yeomans J. 1980. Behavioral methods for inferring anatomical linkage between rewarding brain stimulation sites. *J. Comp. Physiol. Psychol.* 94:227–37

Siegel R. 1989. *Intoxication: Life in Pursuit of Artificial Paradise.* New York: Dutton

Simon H, LeMoal M, Cardo B. 1975. Self-stimulation in the dorsal pontine tegmentum in the rat. *Behav. Biol.* 13:339–47

Skelton RW, Shizgal P. 1980. Parametric analysis of ON- and OFF- responding for hypothalamic stimulation. *Physiol. Behav.* 25:699–706

Skinner BF. 1937. Two types of conditioned

reflex: a reply to Konorski and Miller. *J. Gen. Psychol.* 16:272–79

Skinner BF. 1938. *The Behavior of Organisms.* New York: Appleton-Century-Crofts

Skinner BF. 1950. Are theories of learning necessary? *Psychol. Rev.* 57:193–216

Spealman RD, Goldberg SR. 1978. Drug self-administration by laboratory animals: control by schedules of reinforcement. *Annu. Rev. Pharmacol. Toxicol.* 18:313–39

St-Laurent J. 1988. Behavioral correlates of self-stimulation, flight and ambivalence. *Brain Res. Bull.* 21:61–77

Stein L. 1968. Chemistry of reward and punishment. In *Proceedings of the American College of Neuropsychopharmacology*, ed. DH Efron, pp. 105–23. Washington, DC: US Gov. Print. Off.

Stein L, Ray OS. 1960. Brain stimulation reward "thresholds" self-determined in rat. *Psychopharmacology* 1:251–56

Szostak C, Finlay JM, Fibiger HC. 1987. Intravenous self-administration of the short-acting benzodiazepine midazolam in the rat. *Neuropharmacology* 26:1673–76

Thompson T. 1968. Drugs as reinforcers: experimental addiction. *Int. J. Addict.* 3:199–206

Thorndike EL. 1898. Animal intelligence: an experimental study of the associative processes in animals. *Psychol. Monogr.* 8:1–109

Thorndike EL. 1933. A theory of the action of the after-effects of a connection upon it. *Psychol. Rev.* 40:434–39

Trowill JA, Panksepp J, Gandelman R. 1969. An incentive model of rewarding brain stimulation. *Psychol. Rev.* 76:264–81

Valenstein ES. 1964. Problems of measurement with reinforcing brain stimulation. *Psychol. Rev.* 71:415–37

van der Kooy D, Mucha RF, O'Shaughnessy M, Bucenieks P. 1982. Reinforcing effects of brain microinjections of morphine revealed by conditioned place preference. *Brain Res.* 243:107–17

van der Kooy D, Phillips AG. 1977. Trigeminal substrates of intracranial self-stimulation. *Science* 196:447–49

van Ree JM, de Wied D. 1980. Involvement of neurohypophyseal peptides in drug-mediated adaptive responses. *Pharmacol. Biochem. Behav.* 13(Suppl. 1):257–63

Welzl H, Kuhn G, Huston JP. 1989. Self-administration of small amounts of morphine through glass micropipettes into the ventral tegmental area of the rat. *Neuropharmacology* 28:1017–23

West TEG, Wise RA. 1988a. Effects of naltrexone on nucleus accumbens, lateral hypothalamic and ventral tegmental self-stimulation rate-frequency functions. *Brain Res.* 462:126–33

West TEG, Wise RA. 1988b. Nucleus accumbens opioids facilitate brain stimulation reward. *Soc. Neurosci. Abstr.* 14:1102

Westerink BHC, Tuntler J, Damsma G, Rollema H, De Vries JB. 1987. The use of tetrodotoxin for the characterization of drug-enhanced dopamine release in conscious rats studied by brain dialysis. *Naunyn-Schmiedeberg's Arch. Pharmacol.* 336:502–7

Wise RA. 1972. Spread of current from monopolar stimulation of the lateral hypothalamus. *Am. J. Physiol.* 223:545–48

Wise RA. 1978. Catecholamine theories of reward: a critical review. *Brain Res.* 152:215–47

Wise RA. 1980a. The dopamine synapse and the notion of "pleasure centers" in the brain. *Trends Neurosci.* 3:91–94

Wise RA. 1980b. Action of drugs of abuse on brain reward systems. *Pharmacol. Biochem. Behav.* 13:213–23

Wise RA. 1981. Intracranial self-stimulation: mapping against the lateral boundaries of the dopaminergic cells of the substantia nigra. *Brain Res.* 213:190–94

Wise RA. 1982. Neuroleptics and operant behavior: the anhedonia hypothesis. *Behav. Brain Sci.* 5:39–87

Wise RA. 1985. The anhedonia hypothesis: mark III. *Behav. Brain Sci.* 8:178–86

Wise RA. 1987. Intravenous drug self-administration: a special case of positive reinforcement. In *Methods of Assessing the Reinforcing Properties of Abused Drugs*, ed. MA Bozarth, pp. 117–41. New York: Springer-Verlag

Wise RA. 1989. The brain and reward. In *The Neuropharmacological Basis of Reward*, ed. JM Liebman, SJ Cooper, pp. 377–424. Oxford: Oxford Univ. Press

Wise RA, Bauco P, Carlezon WA Jr, Trojniar W. 1992. Self-stimulation and drug reward mechanisms. *Ann. NY Acad. Sci.* 654:192–98

Wise RA, Bozarth MA. 1984. Brain reward circuitry: four circuit elements "wired" in apparent series. *Brain Res. Bull.* 12:203–8

Wise RA, Bozarth MA. 1987. A psychomotor stimulant theory of addiction. *Psychol. Rev.* 94:469–92

Wise RA, Carlezon WA Jr. 1994. *Self-administration of the dopamine uptake inhibitor nomifensine into nucleus accumbens of rats.* Presented at Annu. Meet. Coll. Prob. Drug Depend., 57th, Palm Beach

Wise RA, Munn E. 1993. Effects of repeated amphetamine injections on lateral hypothalamic brain stimulation reward and subsequent locomotion. *Behav. Brain Res.* 55:195–201

Wise RA, Rompré P-P. 1989. Brain dopamine and reward. *Annu. Rev. Psychol.* 40:191–225

Wood PL. 1982. Actions of GABAergic agents on dopamine metabolism in the nigrostriatal pathway of the rat. *J. Pharmacol. Exp. Ther.* 222:674–79

Woods JH, Schuster CR. 1971. Opiates as reinforcing stimuli. In *Stimulus Properties of Drugs,* ed. T Thompson, R Pickens, pp. 163–75. New York: Appleton-Century-Crofts

Yeomans JS. 1979. Absolute refractory periods of self-stimulation neurons. *Physiol. Behav.* 22:911–19

Yeomans JS. 1982. The cells and axons mediating medial forebrain bundle reward. In *The Neural Basis of Feeding and Reward,* ed. BG Hoebel, D Novin, pp. 405–17. Brunswick, ME: Haer Inst.

Yeomans JS. 1989. Two substrates for medial forebrain bundle self-stimulation: myelinated axons and dopamine axons. *Neurosci. Biobehav. Rev.* 13:91–98

Yeomans JS, Maidment NT, Bunney BS. 1988. Excitability properties of medial forebrain bundle axons of A9 and A10 dopamine cells. *Brain Res.* 450:86–93

Yeomans JS, Mathur A, Tampakeras M. 1993. Rewarding brain stimulation: role of tegmental cholinergic neurons that activate dopamine neurons. *Behav. Neurosci.* 107: 1077–87

Yeomans JS, Mercouris N, Ellard C. 1985. Behaviorally measured refractory periods are lengthened by reducing electrode tip exposure or raising current. *Behav. Neurosci.* 99: 913–28

Yokel RA, Wise RA. 1975. Increased leverpressing for amphetamine after pimozide in rats: implications for a dopamine theory of reward. *Science* 187:547–49

Yokel RA, Wise RA. 1976. Attenuation of intravenous amphetamine reinforcement by central dopamine blockade in rats. *Psychopharmacology* 48:311–18

Zarevics P, Setler P. 1979. Simultaneous rate-independent and rate-dependent assessment of intracranial self-stimulation: evidence for the direct involvement of dopamine in brain reinforcement mechanisms. *Brain Res.* 169: 499–512

Zetterström T, Herrera-Marschitz U, Ungerstedt U. 1981. Simultaneous estimation of dopamine release and rotational behaviour induced by d-amphetamine in 6-OH-DA denervated rats. *Neurosci. Lett.* 7:27–32 (Suppl.)

Zetterström T, Sharp T, Collin AK, Ungerstedt U. 1988. In vivo measurement of extracellular dopamine and DOPAC in rat striatum after various dopamine-releasing drugs; implications for the origin of extracellular DOPAC. *Eur. J. Pharmacol.* 148:327–34

Zito KA, Vickers G, Roberts DCS. 1985. Disruption of cocaine and heroin self-administration following kainic acid lesions of the nucleus accumbens. *Pharmacol. Biochem. Behav.* 25:1029–36

*Annu. Rev. Neurosci. 1996. 19:341–77*

# MECHANISMS AND MOLECULES THAT CONTROL GROWTH CONE GUIDANCE

## Corey S. Goodman

Howard Hughes Medical Institute, Division of Neurobiology, Department of Molecular and Cell Biology, Life Science Addition Room 519, University of California, Berkeley, California 94720

KEY WORDS:  axon guidance, cell recognition, target recognition, cell adhesion, repulsion, attraction

### ABSTRACT

Neuronal growth cones traverse long distances along appropriate pathways to find their correct targets. This review presents an overview of the mechanisms and molecules that control these events. Secreted and cell surface ligands in the growth cone's environment bind to receptors on the growth cone's surface, trigger second-messenger signals, and lead to appropriate steering decisions. Growth cones appear to be guided by at least four different mechanisms: contact-mediated attraction, chemoattraction, contact-mediated repulsion, and chemorepulsion. These mechanisms are mediated by many different families of guidance molecules, including neural cell adhesion molecules of the immunoglobulin superfamily, netrins, and semaphorins, all of which appear to be highly conserved from worms and fruitflies to mice and humans. We are just beginning to gain insights into the functions of these and other molecules in the developing organism by the use of genetic analysis.

## INTRODUCTION

The problem is daunting: How do $10^{12}$ neurons make over $10^{15}$ specific synaptic connections to generate our functioning brain? Neurobiologists have been struggling with this question for decades, trying to uncover to what extent this precision of connections reflects instructions in the DNA, how these instructions are translated into meaningful patterns of guidance signals in the developing embryo, and how these signals are deciphered by the growing tips of neurons—the growth cones. Evolution has apparently endowed the genome with sufficient information to accomplish the task of building the initial scaffold of projections and connections. Growth cones express appropriate recep-

341

0147-006X/96/0301-0341$08.00

342    GOODMAN

tors to allow them to read the molecular landscape, traverse long distances along appropriate pathways, and find and recognize their correct targets. Patterns of neuronal activity then drive the refinement of these initial connections into highly tuned circuits, a process that continues throughout life.

The aim here is to present an overview of the activity-independent mechanisms that control the initial formation of specific synaptic connections. This review is not intended to be comprehensive, and thus only certain studies and molecules are used as examples. Moreover, the review reflects a bias: that functional analysis (by genetics or other specific perturbations) in the developing organism is the best way to elucidate the authentic function of a molecule. In vitro experiments using paradigms that mimic relevant in vivo contexts and functions are the most informative; many in vitro experiments do not pass this test. Some of these ideas have been discussed previously (e.g. Goodman & Shatz 1993, Goodman 1994).

## MECHANISMS

Growth cones appear to be guided by at least four different mechanisms: contact-mediated attraction, chemoattraction, contact-mediated repulsion, and chemorepulsion (Figure 1). Guidance signals can be either positive (permissive or attractive) or negative (inhibitory or repulsive), and each of these types of signals can either be attached to a physical cell surface or extracellular matrix (ECM), thus establishing a short-range step function of expression, or be diffusible, thereby establishing a long-range gradient. Although these definitions imply distinct mechanisms, in reality, the differences between these mechanisms are often blurred, as a secreted molecule may in certain contexts become immobilized by binding to the cell surface or ECM, and as a result,

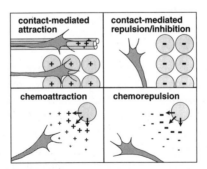

*Figure 1*  Growth cone guidance can be categorized into four different mechanisms: contact-mediated attraction, chemoattraction, contact-mediated repulsion, and chemorepulsion, where the word attraction includes a range of permissive and attractive responses and the word repulsion includes a range of inhibitory and repulsive responses.

form a rather sharp gradient that appears more like a boundary or short-range step function.

## Attraction

In 1963, Sperry (1963) proposed the chemoaffinity hypothesis, suggesting that the "homing behavior" of neuronal growth cones for their targets could be explained by "differential chemical attraction." In the 1970s, a variety of studies in tissue culture showed that differential adhesion can exert a major influence on growth cone guidance (e.g. Letourneau 1975). Thus, the combined focus on both attractive and contact-mediated guidance mechanisms led scientists to speculate that growth cone guidance and target recognition might be controlled by the differential expression of cell adhesion molecules (CAMs) on the surfaces of growth cones and the cells they contact (e.g. Edelman 1985). During the past decade, numerous CAMs from a variety of different gene families have been discovered. Many of these neural CAMs have been shown to promote neurite outgrowth in vitro. These CAMs and other cell surface signaling molecules are included in the broad category of contact-mediated attraction.

However, attractive signals need not be bound to the cell surface and need not mediate cell adhesion. Attractive signals can also be secreted. At the turn of the century, Ramon y Cajal (1893) proposed that growth cones might be guided by attractive gradients of diffusible factors emanating from their targets. In the late 1970s, Gunderson & Barrett (1979) showed that sensory neuron growth cones responded to a gradient of nerve growth factor (NGF). Although NGF is not likely to be a long-range chemoattractant in the developing organism, this study initiated a resurgence of interest in chemotropism. What followed were several studies using an in vitro assay in which tissues or cells are embedded in a collagen gel matrix, which allows for the establishment of stable gradients. These assays showed that different target tissues can attract appropriate growth cones from a distance, including the innervation of the maxillary whisker epithelium by the trigeminal nerve (Lumsden & Davies 1983, 1986), the attraction of commissural axons to the ventral floorplate of the spinal cord (Tessier-Lavigne et al 1988, Placzek et al 1990), and the projection of descending axons from the cerebral cortex to the basilar pons of the mammalian brain (Heffner et al 1990, O'Leary et al 1990, Sato et al 1994). In addition, in vivo perturbations suggest that dermamyotome may attract appropriate motoneurons from a distance (Tosney 1987).

## Repulsion

During the 1980s, evidence also began to accumulate that suggested the existence of negative influences on growth cone guidance (Kapfhammer et al

1986, Walter et al 1987a, Caroni & Schwab 1988), a notion similar to what cell biologists had been studying for many years in the form of contact inhibition (e.g. Abercrombie 1970). For example, growth cones and axons of two different types of vertebrate neurons (sympathetic vs retinal) avoid one another (Kapfhammer et al 1986); on contact with the inhibitory axons, the growth cones collapse, i.e. their filopodia and lamellipodia retract. The establishment of two in vitro assays—the growth cone collapse assay (Kapfhammer & Raper 1987a) and the stripe assay (Walter et al 1987b; see below)—led to numerous studies on the role of inhibition or repulsion in growth cone guidance.

Using these assays, examples of contact-mediated repulsive and inhibitory cues in growth cone guidance include the avoidance of CNS axons by PNS axons, and vice versa (Kapfhammer & Raper 1987b, Raper & Kapfhammer 1990); the avoidance of posterior optic tectum by temporal retinal axons (Walter et al 1987a, Cox et al 1990); the avoidance of nasal retinal axons by temporal retinal axons (Raper & Grunewald 1990); the avoidance of CNS oligodendrocytes and myelin by many types of growth cones (Schwab & Caroni 1988, Bandtlow et al 1990); the avoidance of posterior somite by motoneuron growth cones (Davies et al 1990, Oakley & Tosney 1993); the avoidance of sensory axons by preganglionic sympathetic growth cones (Moorman & Hume 1990); and the avoidance of optic chiasm cells by temporal retinal growth cones (Godement et al 1990, Sretavan 1990, Sretavan et al 1992, Wizenmann et al 1993, Godement et al 1994).

Repulsion was initially identified in assays where contact is required. Subsequent experiments, however, using the same sort of in vitro collagen gel matrix assay as had been used to demonstrate chemoattraction, revealed that repulsion can also function at a distance, leading to the notion of chemorepulsion (Pini 1993, Fitzgerald et al 1993).

The terms inhibition and repulsion signify two operationally different but functionally related types of growth cone behavior. Repulsion implies that the growth cone, upon contact with a particular factor, is repelled and continues to grow away from the source of the factor (i.e. either steering away from a contact-mediated signal or down a gradient). Inhibition implies that the growth cone is inhibited or prevented from taking some action. A particular factor might function as a growth or motility inhibitor, a branch inhibitor, a synapse inhibitor, etc. Whether these differences in inhibition vs repulsion are biologically significant or simply reflect operational differences in the types of assays used remains to be determined. The same molecule might appear to function as an inhibitor in one assay and as a repellent in another. For example, the growth cone collapse assay revealed an inhibitory function (Kapfhammer & Raper 1987a), yet this same factor can function as a repellent to influence growth cone steering in vitro when presented in a more localized fashion (Fan & Raper 1995).

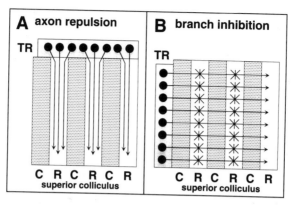

*Figure 2* In the mammalian visual system, axons from the temporal retina (TR) normally project and branch in the rostral (R) superior colliculus (SC; the mammalian homologue of the optic tectum of amphibians and birds) and not the caudal (C) SC. (*A*) When an explant of temporal retinal is oriented perpendicular to the SC stripes, retinal axons grow out parallel to the stripes. (*B*) Orienting the temporal retinal explant parallel to the SC membrane stripes allows the retinal axons to grow out perpendicular to the stripes. Temporal retinal axons extend across the alternating membrane stripes from the topographically correct rostral SC and incorrect caudal SC. They preferentially branch on the correct membranes and do not branch on the wrong ones (see text for further discussion) (Walter et al 1987b, Simon & O'Leary 1992, Roskies & O'Leary 1994).

There are certain contexts in vivo in which growth cones come to a complete stop. For example, growth cones stop when they reach their targets (e.g. Baird et al 1992, Bolz 1994) or when they reach transient targets or waiting zones (e.g. Allendoerfer & Shatz 1994). Target-derived stop signals do not repel the growth cone, but rather induce the shutdown in motility and its transformation into a presynaptic terminal arbor. Whether target-derived stop signals use mechanisms that share common components with inhibitory and repulsive guidance mechanisms is unknown.

An example of inhibition vs repulsion during growth cone guidance emerges from the experimental analysis of retinotopic specificity in chick and mammal (Figure 2). Bonhoeffer and colleagues devised an assay based on alternating stripes of membranes prepared from either rostral or caudal tectum in chick (Walter et al 1987b). A retinal explant strip is oriented perpendicular to the tectal stripes; thus retinal axons grow out parallel to the stripes. The results of these experiments are quite striking: The temporal retinal axons grow only on membranes from rostral tectum. This response is based on avoidance of the caudal tectum (Walter et al 1987a). In experiments using the growth cone collapse assay, membrane fragments from caudal tectum cause temporal growth cones to collapse (Cox et al 1990).

O'Leary and colleagues used similar assays to study the development of retinotopic projections to the superior colliculus (SC, the mammalian homologue of the tectum) in the rat. In one experiment (Simon & O'Leary 1992), they used the same in vitro stripe assay described above and obtained similar results (Figure 2A). In another experiment (Roskies & O'Leary 1994), they modified the assay by orienting the retinal explant parallel to the SC membrane stripes, thus allowing the retinal axons to grow out perpendicular to the stripes (Figure 2B). In this assay, temporal retinal axons extend across the alternating membrane stripes from the topographically correct rostral SC and the incorrect caudal SC of embryonic rats. They are not repelled by the incorrect membranes, but rather preferentially branch on the correct membranes and do not branch on the wrong ones. This branching preference is due to a molecule in the caudal SC that inhibits branching of temporal retinal axons.

Thus, depending upon the assay used, a membrane factor in the caudal tectum or superior colliculus can appear to be either an axon repellent or a branch inhibitor. These differences are observed in vivo as well. In the chick, temporal retinal axons avoid the caudal tectum, whereas in the rat, they enter caudal superior colliculus but only branch in rostral SC (Simon & O'Leary 1992). The same or a related molecule is likely involved in both systems, and these functional differences may reflect operational differences due to the in vitro assays used and contextual differences due to the spatial and temporal relations in vivo.

In summary, growth cone guidance can be categorized into four different mechanisms: contact-mediated attraction, chemoattraction, contact-mediated repulsion, and chemorepulsion (Figure 1); the word attraction includes a range of permissive and attractive responses, and the word repulsion includes a range of inhibitory and repulsive responses. What molecules mediate these mechanisms?

## MOLECULES

The list of guidance signals and their receptors is still far from complete. This section focuses on three families of guidance molecules: neural cell adhesion molecules of the immunoglobulin (Ig) superfamily, netrins, and semaphorins. While this review was in press, a new repulsive axon guidance signal (RAGS/ AL-1) was described that is a member of the family of ligands for the Eph receptor tyrosine kinases (Dresher et al 1995, Cheng et al 1995). This family of guidance signals and receptors is briefly described later in this review in the section on tyrosine kinases and phosphatases. Amongst the CAMs of the Ig superfamily, the section describes the functional analysis of two related proteins—vertebrate neural cell adhesion molecule (NCAM) and *Drosophila* fasciclin II. For brevity, studies on other guidance molecules (e.g. Ig CAMs, cadherins, integrins, laminins, tenascin/janusin) are not discussed here (e.g. Bixby & Harris 1991, Hynes & Lander 1992).

Invertebrates and vertebrates appear to use many of the same mechanisms and molecules for growth cone guidance. The phylogenetic conservation of molecular structure and apparent function in the CAMs, netrins, and semaphorins is quite striking and indicates that all three represent ancient families of axon guidance molecules. The combination of the in vivo analysis in invertebrates (particularly using genetic analysis in flies and worms) and in vitro analysis in vertebrates is providing complementary insights into the functions of these molecules.

Another important lesson is that just as a great variety of molecules can mediate attractive guidance, so too can a great variety of molecules inhibit or repel various aspects of growth and guidance. The semaphorins represent one family of repulsive or inhibitory guidance molecules (Kolodkin et al 1992, 1993; Luo et al 1993, 1995; Matthes et al 1995; Puschel et al 1995; Messersmith et al 1995). Other molecules have also been identified that can function in a repulsive or inhibitory fashion, including the cell surface proteins connectin (Nose et al 1994), myelin-associated glycoprotein (MAG) (McKerracher et al 1994, Mukhopadhyay et al 1994), and RAGS/AL-1 (Drescher et al 1995); the diffusible protein netrin-1 (Colamarino & Tessier-Lavigne 1995); the related ECM proteins tenascin and janusin/restrictin/J1-160-180 (Faissner & Kruse 1990, Pesheva et al 1993); and proteoglycans (see Snow et al 1990).

Finally, the data suggest that there is not always a precise one-to-one match between molecules and mechanisms, because individual molecules or families of molecules do not always fit into only one of the four mechanistic categories (Figure 3). Thus, an important insight is that some guidance molecules are not

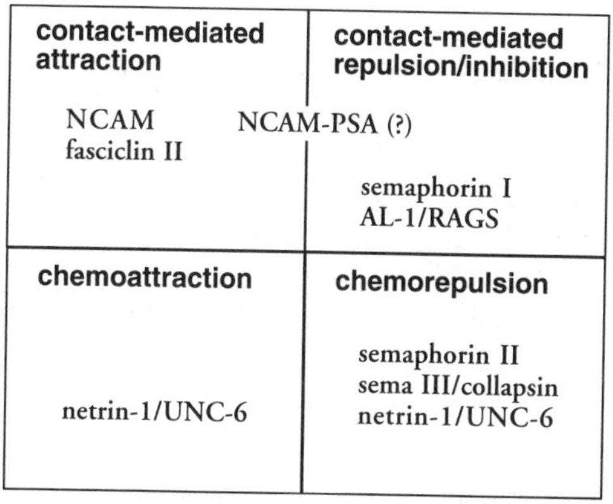

| contact-mediated attraction | contact-mediated repulsion/inhibition |
|---|---|
| NCAM          NCAM-PSA (?)<br>fasciclin II | semaphorin I<br>AL-1/RAGS |
| **chemoattraction** | **chemorepulsion** |
| netrin-1/UNC-6 | semaphorin II<br>sema III/collapsin<br>netrin-1/UNC-6 |

*Figure 3* Diagram showing some of the guidance molecules discussed in this review. See text for details and for references.

exclusively attractive or repulsive, but rather are bifunctional (e.g. Colamarino & Tessier-Lavigne 1995), playing different roles for different growth cones.

## Cell Adhesion Molecules

CAMs of a variety of different gene families are expressed on restricted subsets of axons during development (reviewed by Grenningloh & Goodman 1992). Many of these CAMs are members of the Ig superfamily (e.g. Grenningloh et al 1990, Rathjen & Jessell 1991), the focus of this section. Some of the best-characterized subfamilies of vertebrate neural Ig CAMs include (with their *Drosophila* relatives in parentheses) NCAM (fasciclin II); L1, NgCAM, NrCAM (neuroglian); SC1/DM-GRASP/BEN (irreC); and TAG-1/axonin-1/F11. Many of these CAMs have been implicated in the events of growth cone guidance because of their patterns of expression in vivo and their ability to promote neurite outgrowth in vitro. However, in only a few cases have their functions been examined in vivo.

In general, cell adhesion molecules are thought of as functioning as contact-mediated attractive signals during growth cone guidance. In vitro analysis suggests that some of these CAMs function as signal transduction molecules (Doherty & Walsh 1994, Bixby et al 1994). However, CAMs need not always function as attractants; two CAMs—MAG and connectin—may also function as contact-mediated inhibitors or repellents (Mukhopadhyay et al 1994, Mc-Kerracher et al 1994, Nose et al 1994) .

Loss-of-function genetic analysis of the genes encoding a few neural Ig CAMs in both *Drosophila* and mouse have been reported. In most cases, the loss-of-function mutations lead to more subtle defects in guidance and connectivity than were predicted based on their patterns of expression and functions in vitro. In each case, specific defects are observed, but overall the mutant nervous systems look quite normal. In mouse, this genetic analysis includes the *Ncam* (Tomasiewicz et al 1993, Cremer et al 1994, Ono et al 1994) and *Mag* (Li et al 1994, Montag et al 1994) genes. In *Drosophila,* this genetic analysis includes the *neuroglian* (Bieber et al 1989), *fasciclin II* (Grenningloh et al 1991, Lin & Goodman 1994, Lin et al 1994), and *irreC* (Ramos et al 1993) genes. This section focuses on the genetic analysis of *Ncam* in mouse and *fasciclin II* in *Drosophila*. Genetic analysis of other *Drosophila* CAMs are described in other sections, as are the phenotypes of mutations in the human *L1* gene (Jouet et al 1993, 1994; Vits et al 1994).

## NCAM and PSA

NCAM is expressed abundantly in the developing and mature nervous system. NCAM has five Ig and two fibronectin (FN) type-III domains in its extracellular region (Cunningham et al 1987) and comes in a variety of different isoforms, including alternatively spliced cytoplasmic domains, various membrane linkages, and extracellular microexons. The three major isoforms are NCAM-120 (phospholipid-anchored; 120 = 120 kDa), NCAM-140, and NCAM-180 (the last two being transmembrane forms with different-length cytoplasmic domains). Many of the different NCAM isoforms have characteristic and dynamic patterns of expression and functional properties. For example, one of the isoforms involves the 30-bp VASE microexon. Neurite outgrowth in vitro is high on NCAM-140 lacking the VASE exon and is much lower on NCAM-140 + VASE (Doherty et al 1994).

NCAM has been implicated in controlling a variety of events of neuronal development, including cell migration, neurite outgrowth, selective fasciculation and axon sorting, target recognition, and synaptic plasticity (e.g. Rutishauser 1993, Doherty & Walsh 1994, Bixby et al 1994, Luthl et al 1994). A related molecule in *Drosophila,* fasciclin II, has been shown to function in selective fasciculation and axon sorting (Lin & Goodman 1994, Lin et al 1994; see below), while a related molecule in *Aplysia,* apCAM, has been implicated in synaptic plasticity (Mayford et al 1992).

Amongst the most intriguing hypotheses put forward for NCAM function is the notion that a specific carbohydrate moiety on NCAM can control axon sorting by modulating axon-axon adhesion mediated by both NCAM and other neural CAMs. One form of NCAM carries an unusually large carbohydrate moiety of polysialic acid (PSA) (e.g. Rutishauser et al 1988, Rutishauser 1991). NCAM is evidently the major protein in the vertebrate embryo to carry such a large amount of PSA. During development, NCAM generally shifts from a PSA+ form to a PSA− form.

The removal of PSA from NCAM leads to an increase in NCAM-dependent adhesion and to a decrease in NCAM-dependent neurite outgrowth in vitro (reviewed by Doherty et al 1994). PSA on NCAM has been proposed to modulate cell-cell adhesion by not only decreasing the adhesivity of NCAM itself, but also by decreasing the adhesivity of other neighboring CAMs. In terms of growth cone guidance, NCAM has been thought of as a contact-mediated attractive signal. However, in the PSA model, NCAM-PSA is considered a modulator of and, in certain contexts, an inhibitor of other contact-mediated guidance molecules. The best support for this hypothesis comes from the analysis of avian motoneurons.

In the chick, as motoneuron growth cones exit the neural tube, the axons destined to innervate different muscles are intermingled within eight separate spinal nerves. Upon reaching the base of the limb bud, they converge to form the crural and sciatic plexuses. It is within these plexuses that specificity first unfolds, as motor axons sort out into motoneuron pool-specific groups and make specific pathway choices (Tosney & Landmesser 1985a,b). Changes in cell adhesion (as modulated by the PSA moiety associated with NCAM) have been shown to influence the ability of motoneuron axons to sort out in the plexus (Tang et al 1992). The higher the levels of PSA are, the lower the axon-axon adhesion mediated by NCAM and other CAMs is. During normal development, when motor axons reach the plexus region, the tightly fasciculated axons defasciculate and begin to sort themselves out (Tosney & Landmesser 1985a). At the same time, the levels of PSA go up dramatically on motoneuron growth cones. When PSA is enzymatically removed during this period, an increase in the number of projection errors is observed (Tang et al 1992). These results suggest that PSA can modulate the ability of motoneuron growth cones to respond to other guidance cues.

The analysis of loss-of-function mutations in the mouse *Ncam* gene has supported some but not all of the many hypotheses of NCAM function (Tomasiewicz et al 1993, Cremer et al 1994, Ono et al 1994). The *Ncam* mutant mice display more subtle defects in guidance and connectivity than had been expected. Cremer et al (1994) created a complete knockout in the mouse *Ncam* gene that eliminates all forms of NCAM, while Tomasiewicz et al (1993) generated a mutation that eliminates the largest form of the protein, the neural-specific NCAM-180 variant (although the expression of NCAM-140 is also affected). Both mutations lead to a nearly complete loss of PSA in the brains of these mice.

These NCAM-deficient mice appear surprisingly healthy and fertile. Their brains appear fairly normal, and they display normal activity and motor abilities. However, the animals show deficits in spatial learning when tested in the Morris water maze (Cremer et al 1994), suggesting a possible role for NCAM in synaptic plasticity (e.g. Luthl et al 1994). Alternatively, the hippocampus in the NCAM-180 mutant mice shows some slight morphological abnormalities, suggesting that the behavioral defect might be caused by a developmental defect in the underlying circuitry.

The *Ncam* mutant mice show a 10% reduction in their overall brain weight and a 36% decline in the size of the olfactory bulb. The most conspicuous mutant phenotype is observed in the olfactory bulb, where granule cells are reduced in number and appear disorganized. The precursors of these cells are found to be accumulated at their origin in the subependymal zone at the lateral ventricle, suggesting that the mutant defect is largely due to a lack of migration of the olfactory granule cell precursors. Interestingly, the PSA form of NCAM-180 is normally expressed along this migratory pathway. Subtle but distinct abnormalities are observed in other regions of the brain.

Ono et al (1994) analyzed the undersized olfactory bulb and accumulation of precursors in the subependymal layer in *Ncam* mutant mice and demonstrated that this defect can be duplicated by injection of an enzyme that specifically destroys the PSA moiety associated with NCAM. They interpret these results as supporting a role for PSA in specific cell-cell interactions—in this case, in specific cell migrations. They suggest that much of the *Ncam* mutant phenotype might be explained by the absence of PSA.

## Fasciclin II

Experiments in the grasshopper embryo showed that growth cones can distinguish one group of axons from another, leading to specific patterns of selective fasciculation (e.g. Raper et al 1984, Bastiani et al 1984, Goodman et al 1984). Fasciclin II was identified based on its dynamic pattern of expression on a subset of fasciculating axons in the grasshopper embryo (Bastiani et al 1987), making it a prime candidate to be a guidance molecule that controls selective fasciculation. The cloning of *fasII* in grasshopper (Snow et al 1988, Harrelson & Goodman 1988) and *Drosophila* (Grenningloh et al 1991) revealed that Fas II is a neural CAM of the Ig family and is related to vertebrate NCAM in both structure and sequence (~23% amino acid identity).

In *Drosophila,* Fas II is dynamically expressed on a subset of embryonic CNS axons, many of which selectively fasciculate in the vMP2, MP1, and FN3 pathways. Fas II is also expressed on all motor axons in the periphery and on other cell types and tissues as well. Lin et al (1994) examined the in vivo growth cone guidance function of Fas II by using genetic analysis. The complementary phenotypes produced by the loss-of-function and gain-of-function conditions define an in vivo function for Fas II as a guidance molecule that controls specific patterns of selective fasciculation (Figure 4).

When the levels of Fas II are decreased in *fasII* loss-of-function mutants, the axons in all three CNS pathways that normally express Fas II defasciculate and these axon fascicles do not form; as a result, the longitudinal connectives and neuropil regions are disorganized. Nevertheless, these growth cones extend in the normal direction at a normal rate. Two types of phenotypes are observed in *fasII* gain-of-function conditions. First, transgenic constructs that specifically drive Fas II expression on the axons in these same three pathways can rescue the defasciculation phenotype in a *fasII* loss-of-function background, thus creating a refasciculation of these three fascicles. Second, in both wild-type and *fasII* mutant backgrounds, these transgenic constructs can lead to a gain-of-function phenotype in which these axons fasciculate incorrectly. Pairs of pathways that should remain separate instead become abnormally joined together.

Fas II is normally expressed on all motoneuron growth cones and axons during the period of outgrowth and synapse formation. A separate study reported on the effects of increasing Fas II on these motor axons (Lin & Goodman 1994). Increased Fas II can block the defasciculation of motor axons at specific choice points. In other cases, increased Fas II can drive the fasciculation of axons that would normally not bundle together. The effects of increasing Fas

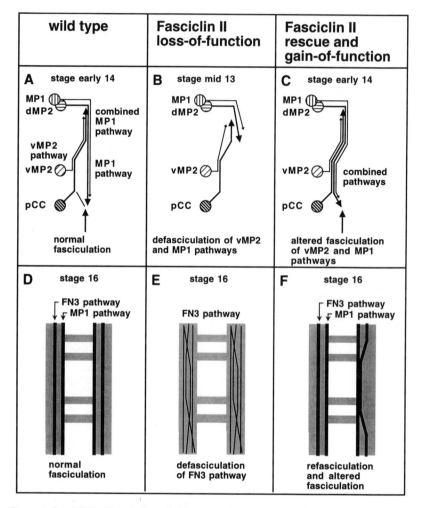

*Figure 4* Fasciclin II (Fas II) is dynamically expressed on a subset of embryonic CNS axons (shown schematically in *A* and *D* at two different stages of embryonic development). In *fasII* loss-of-function mutants (*B, E*), the axons in three CNS pathways that normally express Fas II defasciculate and these axon fascicles do not form. Two types of phenotypes are observed in *fasII* gain-of-function conditions (*C, F*). First, transgenic constructs that specifically drive Fas II expression on the axons in these same three pathways can rescue the defasciculation phenotype in a *fasII* loss-of-function background, thus creating a refasciculation of these three fascicles. Second, these transgenic constructs can lead to a gain-of-function phenotype in which these axons fasciculate incorrectly. Pairs of pathways that should remain separate instead become abnormally joined together. See text for further discussion (Lin et al 1994).

II on CNS axons (Lin et al 1994) are similar to the effects of increasing Fas II on motor axons (Lin & Goodman 1994); the common feature is that increased levels of Fas II lead to increased axon fasciculation.

Thus, by increasing and decreasing the levels of Fas II in the developing organism via genetic analysis, researchers have shown that Fas II functions in selective fasciculation and axon sorting. Moreover, the results also define other aspects of growth cone initiation, outgrowth, and guidance in which Fas II function is not required.

## Netrins

In vitro experiments have shown that the floor plate at the ventral midline of the developing mammalian spinal cord provides chemotropic guidance signals for the growth cones of commissural neurons whose axons extend towards the floor plate (Tessier-Lavigne et al 1988, Placzek et al 1990). When a dorsal spinal cord explant is placed within a few hundred microns of a floor plate explant in a collagen gel matrix, commissural growth cones turn and extend towards the floor plate from a distance. Tessier-Lavigne and colleagues reported on the molecular and functional characterization of two vertebrate netrins (Serafini et al 1994, Kennedy et al 1994), at least one of which appears to function as the chemoattractant emanating from the floor plate. The netrins are homologues of UNC-6, which plays a role in circumferential guidance in the nematode (Hedgecock et al 1990, Ishii et al 1992).

Netrins have a high degree of sequence similarity (~50%) to the nematode UNC-6 protein (Hedgecock et al 1990, Ishii et al 1992), a remarkable level of conservation considering the 600 million years separating worms and vertebrates. Two *netrin* genes have been cloned in *Drosophila* (Mitchell et al 1994; M Seeger, personal communication). UNC-6 and netrins are secreted proteins of ~600 amino acids. The N-termini of the netrins (~450 amino acids) are related to the N-termini of laminin subunits, in particular subunit B2, although B2 is much larger (>1600 amino acids). The C-terminal ~150–amino acid domain diverges from laminins but is highly conserved among the netrins and UNC-6. The netrins appear to be both diffusible and cell associated.

The present model is that netrins and UNC-6 are secreted proteins that can form a gradient in the extracellular environment. In chick, *netrin-1* is expressed by the floor plate and could form a ventral-to-dorsal gradient in the developing spinal cord (Serafini et al 1994, Kennedy et al 1994). In the nematode, *unc-6* is expressed in the ventral region (W Wadsworth & E Hedgecock, personal communication). In *Drosophila, netrin-A* is expressed along the ventral midline of the CNS (Mitchell et al 1994).

Experimental analysis suggests that netrins and UNC-6 are bifunctional, playing roles of both chemoattractants and chemorepellents. The initial evi-

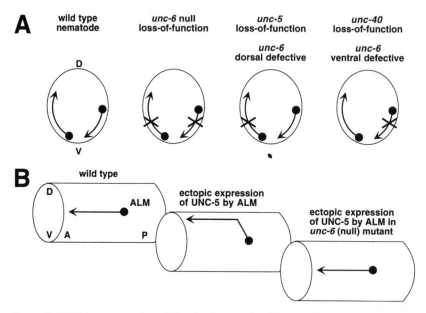

*Figure 5* UNC-6 appears to be a bifunctional secreted guidance molecule in the nematode. Vertebrate netrins are the homologues of nematode UNC-6. (*A*) Behavior of circumferential growth cones in wild-type, *unc-6* null loss-of-function, *unc-6* dorsal defective hypomorphic loss-of-function, *unc-6* ventral defective hypomorphic loss-of-function, *unc-5* loss-of-function, and *unc-40* loss-of-function mutant alleles. (*B*) Behavior of growth cones of ALM mechanosensory neurons in wild-type (in which they do not express UNC-5), when UNC-5 is ectopically expressed in the ALMs, and with UNC-5 ectopic expression in an *unc-6* null mutant. These results are compatible with a model in which UNC-6 is a bifunctional guidance molecule and (given that UNC-6 appears to be expressed along the ventral surface of the nematode) in which UNC-5 is a repulsive receptor for UNC-6 (Hedgecock et al 1990, Ishii et al 1992, Hamelin et al 1993)

dence proposing the bifunctionality of netrins and UNC-6 comes from the analysis of nematode mutants (Figure 5) (Hedgecock et al 1990, McIntire et al 1992, Hamelin et al 1993). First, there are three different classes of *unc-6* mutations. Null (complete loss-of-function) mutations disrupt both dorsal and ventral migrations, whereas certain partial loss-of-function mutations disrupt either ventral (ventral defective) or dorsal (dorsal defective) migrations, but not both, suggesting that different domains of the protein control dorsal vs ventral growth cone guidance.

Second, mutations in *unc-5* and *unc-40* also alter circumferential guidance. Mutations in *unc-5* disrupt dorsal but not ventral migrations, whereas mutations in *unc-40* disrupt ventral but not dorsal migrations. The *unc-5* gene encodes a transmembrane protein that has the structural features of a receptor (Leung-

Hagesteijn et al 1992). Mosaic analysis suggests that UNC-5 is required in the neurons that fail to extend properly in the *unc-5* mutant. The *unc-40* gene also encodes a transmembrane protein that has the structural features of a receptor (J Culotti, personal communication).

The third line of evidence comes from experiments in which UNC-5 is ectopically expressed by mechanosensory neurons whose growth cones normally extend either laterally (the ALMs) or ventrally; the growth cones of these neurons ectopically expressing UNC-5 now instead extend dorsally away from the ventral midline (Hamelin et al 1993). These altered dorsal trajectories depend upon the presence of UNC-6, because ectopic expression of UNC-5 in an *unc-6* mutant leads to a wild-type trajectory.

These results are compatible with a model in which axons can respond in two ways to a presumptive UNC-6 gradient, depending upon the particular UNC-6 receptor that they express. Neurons expressing UNC-5, which extend their axons dorsally away from the ventral midline, would move down the UNC-6 gradient, whereas neurons expressing a different receptor (perhaps the UNC-40 gene product), which extend their axons ventrally, would move up the UNC-6 gradient.

The conservation between netrin and UNC-6 sequences and the apparent conservation in the function of these proteins in attracting growth cones ventrally raised the question of whether netrins in vertebrates might also function to repel growth cones dorsally from the floor plate. To test this possibility, Colamarino & Tessier-Lavigne (1995) studied the development of the motor axons that normally exit the hindbrain in the trochlear nerve (cranial nerve IV). Trochlear motor axons originate near the floor plate and extend dorsally away from the floor plate. Colamarino & Tessier-Lavigne (1995) showed that the floor plate and COS cells secreting recombinant netrin-1 can repel trochlear motor axons at a distance in vitro, suggesting that netrin-1 secreted by floor plate cells functions as a chemorepellent to guide trochlear motor axons away from the ventral midline.

Thus, netrin-1 appears to be a bifunctional guidance signal in vertebrates (Figure 6). This is consistent with the dual function of UNC-6 in the nematode. In both cases, netrin and UNC-6 secreted by ventral cells appear to attract some growth cones toward the ventral midline but to repel other growth cones dorsally from the ventral midline. In the nematode, UNC-5 is a good candidate for UNC-6 repulsive receptor. The identities of other receptors are still unknown, although UNC-40 may also be a UNC-6 receptor.

## Semaphorins

The semaphorins are a family of cell surface and secreted proteins that are conserved from insects to humans (Kolodkin et al 1993). All available evidence

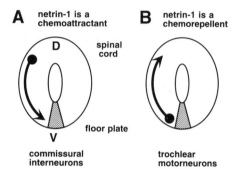

*Figure 6*   Netrin-1 appears to be a bifunctionai secreted guidance molecule in vertebrates. (*A*) In the spinal cord, commissural axons originate dorsally and extend ventrally towards the floor plate. Netrin-1 secreted by floor plate cells appears to function as a chemoattractant to guide commissural axons towards the ventral midline. (*B*) In the hindbrain, trochlear motor axons originate near the floor plate and extend dorsally away from the floor plate. Netrin-1 secreted by floor plate cells appears to function as a chemorepellent to guide trochlear motor axons away from the ventral midline (Serafini et al 1994, Kennedy et al 1994, Colamarino & Tessier-Lavigne 1995).

thus far suggests that different members of the semaphorin family function as chemorepellents or inhibitors of axon pathfinding, branching, or targeting (Kolodkin et al 1992, Luo et al 1993, Matthes et al 1995, Messersmith et al 1995). Semaphorins are ~750 amino acids in length (including signal sequence) and are defined by a conserved ~500–amino acid extracellular semaphorin domain (from about amino acid 50 to 550) containing 14–16 cysteines, many blocks of conserved residues, and no obvious repeats (Kolodkin et al 1992, 1993; Luo et al 1993).

How many semaphorins are encoded in any one genome is still unknown. There are at least two in *Drosophila* (Kolodkin et al 1993), two in nematode (P Roy & J Culotti, personnal communication), five in chick (Luo et al 1993, 1995), five in mouse (Puschel et al 1995), and at least four in human (Kolodkin et al 1993, Messersmith et al 1995), but this is probably an incomplete representation of any genome. In addition, two more divergent semaphorins are encoded in viral genomes (Kolodkin et al 1993).

The first member of the family was semaphorin I (formerly fasciclin IV) (Kolodkin et al 1992), a transmembrane protein. Sema I is expressed on subsets of fasciculating axons in the developing grasshopper CNS and on stripes of epithelial cells, particularly in the developing limb bud. Antibody-blocking experiments in the limb bud show that sema I functions to stall and then steer a pair of growth cones (from the Til pioneer neurons) as they encounter epithelial cells expressing it; sema I also prevents the axons that encounter it from defasciculating and branching (Figure 7A,B). However, sema I is not an absolute inhibitor of growth, because these growth cones normally extend on sema I–expressing cells. Thus, sema I can function to steer a pair of growth cones, prevent defasciculation, and inhibit branching.

Beginning with the grasshopper sema I sequence, other members of the family were subsequently identified in *Drosophila* (sema I and sema II) and

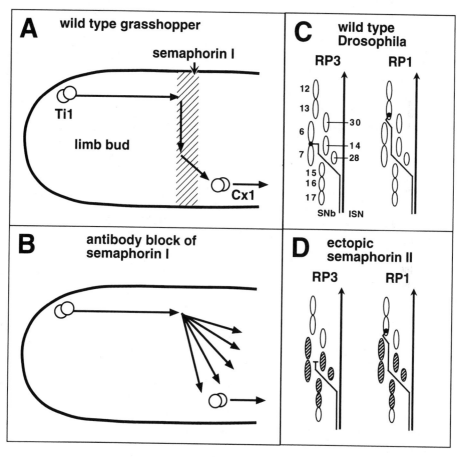

*Figure 7*  Schematic diagrams showing the functions of semaphorin I (sema I) and semaphorin II (sema II) in the guidance of sensory and motor axons in insect embryos. (*A, B*) Diagram of grasshopper embryo limb bud. Ti1 and Cx1 are pairs of early differentiating sensory neurons. (*C, D*) Diagram of *Drosophila* embryo ventral muscles and motor nerves as viewed in cross section. Numbers refer to muscles; RP3 and RP1 are identified motoneurons. Hatched muscles in *D* show those muscles that ectopically express semaphorin II (see text for details) (Kolodkin et al 1992, 1993; Matthes et al 1995).

human (sema III) (Kolodkin et al 1993). A member of this family, collapsin, was independently identified in chicken on the basis of its ability to cause collapse of sensory growth cones in vitro (Luo et al 1993). Collapsin and sema III appear to be homologues. Sema II, sema III, and several other semaphorins lack transmembrane domains and are secreted.

Semaphorin II, a secreted semaphorin in *Drosophila,* is transiently expressed by a subset of neurons in the CNS and by a single large muscle during motoneuron outgrowth and synapse formation. Loss-of-function mutations in *semaII* lead to behavioral abnormalities (Kolodkin et al 1993), but little is yet known about the cellular basis of these phenotypes (Matthes et al 1995). To test the in vivo function of sema II during growth cone guidance, transgenic *Drosophila* were created that generate ectopic semaphorin II expression by muscles that normally do not express it (Matthes et al 1995). The results of this gain-of-function analysis show that sema II inhibits certain identified motoneuron growth cones (e.g. RP3) from forming normal synaptic terminal arborizations on their target muscles, while other growth cones (e.g. RP1) appear unresponsive (Figure 7C,D). Thus, semaphorin II can function in vivo as a selective target-derived signal that inhibits the formation of specific synaptic terminal arbors.

Messersmith et al (1995) examined the effects of human semaphorin III on different classes of rat sensory axons. Distinct classes of primary sensory neurons in dorsal root ganglia (DRG) subserve different sensory modalities, terminate in different dorsoventral locations in the spinal cord, and display different neurotrophin response profiles. Large diameter muscle afferents that terminate in the ventral spinal cord are NT3 responsive, whereas small diameter afferents subserving pain and temperature are NGF responsive and terminate in the dorsal spinal cord.

Previous in vitro studies had shown that the developing ventral spinal cord secretes a long-range diffusible factor that inhibits the growth of sensory axons (Fitzgerald et al 1993). Messersmith et al (1995) showed that this diffusible factor repels NGF-responsive axons but has little effect on NT3-responsive axons. They then showed that sema III is expressed by ventral spinal cord cells (but not the floor plate) at the appropriate stages of development. Finally, they showed that sema III can mediate this effect; COS cells secreting sema III mimic the inhibitory effect of the ventral spinal cord on axons extending in response to NGF but show no effect on NT3-responsive axons (Figure 8). These results suggest that semaphorin III functions to pattern sensory projections by selectively repelling axons that normally terminate dorsally.

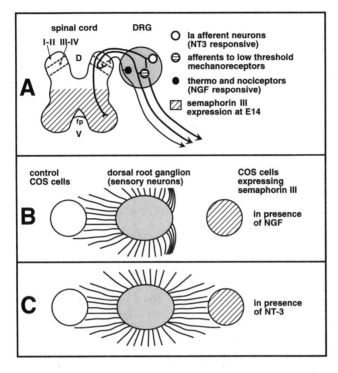

*Figure 8* Schematic diagrams showing the function of semaphorin III (sema III) in the guidance of sensory axons in the developing mammalian spinal cord. (*A*) DRG, dorsal root ganglion; fp, floor plate. Group Ia sensory afferents (NT3-responsive) (*open circle*) enter the spinal cord and project to the ventral region to synapse on motoneurons. Axons from NGF-responsive neurons (*solid circle*) project into the spinal cord starting around E16 and target the dorsal-most laminae I and II. *semaIII* transcripts are detected in the ventral spinal cord during this period. An additional class of afferents are the low-threshold mechanoreceptors (*hatched circle*) that enter the spinal cord, overshoot their targets, and then turn upward to terminate in laminae III and IV. The neurotrophin dependence of these cells is uncertain. (*B, C*) In the collagen gel matrix assay in vitro, COS cells secreting sema III repel the NGF-responsive DRG axons but show no effect on NT3-responsive axons. These results suggest that sema III functions to pattern sensory projections by selectively repelling axons that normally terminate dorsally (Messersmith et al 1995).

Collapsin causes rapid collapse of growth cones of chick DRG axons when added acutely to these axons (Luo et al 1993); recombinant human sema III (its mammalian homologue) has similar effects (Messersmith et al 1995). However, when growth cones are exposed to a localized but nondiffusible source of collapsing signal (in the form of chick brain extract enriched in collapsin that is immobilized on beads), these growth cones often turn away from the source without collapsing, following filopodial contact with the source (Fan & Raper 1995). Growth cone collapse may only occur when many filopodia are exposed simultaneously and rapidly to a step change in concentration of the factor, as is achieved in the collapse assay.

Thus, in the developing organism, semaphorins appear capable of inhibiting branching (sema I in grasshopper), influencing steering decisions (sema I in grasshopper and sema III in mammals), preventing axons from entering certain target regions (sema II in *Drosophila* and perhaps sema III in mammals), or inhibiting the formation of synaptic terminal arborizations (sema II in *Drosophila*). The full range of effects of different or even individual semaphorin family members on developing axons remain to be determined. In particular, given the evidence that some guidance cues can have both repulsive and attractive effects (e.g. Colamarino & Tessier-Lavigne 1995), it will be important to determine whether any semaphorins function as attractants. Nothing is yet known about the identity of the receptors for semaphorin family members.

## GENETIC ANALYSIS

Two methods of genetic analysis can be used to identify and functionally characterize molecules that control growth cone guidance. One approach is reverse genetics in which one begins with a molecule of interest and proceeds to generate mutations in the gene encoding it. This approach has been used in a variety of organisms to test the in vivo function of molecules that have been implicated in growth cone guidance by either their in vivo expression or their in vitro function.

An alternative approach is classical genetics in which one screens for a mutant phenotype and proceeds to identify the gene and encoded molecule. In principle, screening directly and systematically for mutations that perturb particular events is a powerful approach to identifying molecules involved in developmental processes, although the proteins discovered in this fashion could function at any step in the process (i.e. regulators, signals, receptors, effectors). One of the best examples of such an approach is found in the pioneering studies of Nüsslein-Volhard, Wieschaus, and their colleagues (Nüsslein-Volhard & Wieschaus 1980), who undertook a near-saturation screen for zygotic mutations in *Drosophila* affecting segment number and polarity.

A variation on this approach is an interactive genetic screen in which one begins with a mutation in a particular gene that creates a sensitized background (using either partial loss-of-function or gain-of-function mutants), and then screens for mutations in other genes that either enhance or suppress this phenotype. Such screens can be used to identify interacting genes and, when combined with epistasis, to identify upstream and downstream components in a genetic pathway (e.g. Simon et al 1991).

Classical genetic screens have been used to search for genes that control growth cone guidance in a variety of different organisms, most notably nematode and *Drosophila.* Up until a few years ago, most screens aimed at identifying guidance and connectivity mutants relied on either behavioral defects or alternatively on the anatomy of the adult nervous system, both as indirect measures of guidance and targeting events during development. However, the recent availability of anatomical probes that reveal specific growth cones and axons has made it possible to directly screen for anatomical defects in axon guidance during development (see below).

Similar large-scale systematic mutant screens have now been conducted in the *Zebrafish* by Nüsslein-Volhard and colleagues (Mullins et al 1994) and her former student Driever and colleagues (Solnica-Krezel et al 1994). Bonhoeffer and colleagues used anatomical probes to screen these mutants for defects in the retinotectal projection and have identified a number of mutants with interesting defects (reported in Kuwada 1995). Conducting similar types of systematic mutant screens in the mouse should also be possible (e.g. Joyner et al 1992).

In principle, a large-scale mutant screen should recover mutations in nearly all of the genes encoding the key signals and receptors involved in growth cone guidance and target recognition. However, there are three very important ways in which genes might be missed. First, if there is major redundancy or overlap in gene function, then such genes will largely go undetected in screens in which genes are mutated one at a time. Even if there is only partial overlap of function, mutations in certain genes might be overlooked because they lead to very subtle, partially penetrant, or variable phenotypes. Second, if a gene has a more general or early function such that mutations in it greatly disrupt development, then these genes may also go largely undetected in such screens. Third, the ability to identify mutations depends upon the resolution of the screen. The more subtle the phenotype, the more difficult it may be to detect.

Such primary screens typically uncover hundreds of potentially interesting mutations, only a few of which are ultimately shown to directly disrupt growth cone guidance. For example, three systematic large-scale screens have been conducted in *Drosophila* in search of mutations that disrupt guidance and connectivity (Seeger et al 1993, Van Vactor et al 1993, Martin et al 1995). Although each primary screen generated hundreds of mutations that disrupt the patterns of axon pathways and connections, after further analysis using more refined probes, most of these mutations were found to disrupt pathfinding and connectivity indirectly by perturbing cell fate and patterning. Nevertheless, all three screens yielded a number of new genes that appear to control important aspects of guidance (see below).

## Nematode

Hedgecock et al (1990) used one set of markers, and McIntire et al (1992) another, to anatomically screen the nervous systems of adult worms from an existing set of 110 behaviorally uncoordinated mutants (Brenner 1974). Within this collection, they identified genes required for pioneering circumferential growth of axons (*unc-6, unc-5, unc-40*) (Hedgecock et al 1990; see section on netrins), genes required for longitudinal growth and fasciculation (e.g. *unc-71, unc-76*, and *unc-34*), and genes required for more general aspects of axon elongation (McIntire et al 1992, Garriga et al 1993). The *unc-33, unc-53*, and *vab-8* genes appear to control certain aspects of axon guidance as well (Li et al 1992, Hekimi and Kershaw 1993; B Wightman & G Garriga, personal communication).

Among the initial set of behavioral mutants, *unc-4* encodes a homeodomain protein that controls synaptic connectivity (Miller et al 1992), while *unc-51* encodes a serine-threonine kinase that controls neurite outgrowth (Ogura et al 1994). Finally, *unc-7* encodes a protein related to *Drosophila Passover* (see below) and appears to play a role in the formation of certain kinds of synapses (Starich et al 1993).

## *Drosophila*

A behavioral screen for jumpless mutants in *Drosophila* (Thomas & Wyman 1982, 1984) led to the identification of the *Passover* gene (aka *shakB*) (Crompton et al 1992, Krishnan et al 1993), which is related to nematode UNC-7 and *Drosophila Ogre* and which may encode an invertebrate gap junction protein (Barnes 1994); and the *bendless* gene, which appears to encode a ubiquitin-conjugating enzyme (Muralidhar & Thomas 1993, Oh et al 1994). Another screen was based on grooming behavior of adult flies (Phillis et al 1993).

A different screen used anatomical methods to identify mutations that alter adult brain structure (Heisenberg et al 1985) and led to the identification of the *irreC* (*irregular chiasm C*) gene (Boschert et al 1990). Mutations in *irreC* lead to defects in the optic chiasm and cell death in the developing retina. The *irreC* gene encodes an Ig CAM that is related to vertebrate SC1/DM-GRASP/ BEN (Ramos et al 1993).

Three large-scale screens have been conducted in *Drosophila* by using anatomical probes to directly identify mutations that disrupt guidance and connectivity: CNS axon pathways, with a focus on the midline (Seeger et al 1993); neuromuscular connectivity (Van Vactor et al 1993, Sink and Goodman 1994); and retinotopic connectivity (Martin et al 1995). The CNS screen led to the identification of two genes with highly penetrant midline phenotypes: *commissureless* (*comm*) and *roundabout* (*robo*) (Seeger et al 1993). In *comm* mutants, growth cones that would normally cross the midline instead stay on their own side. In *robo* mutants, growth cones that would normally stay on their own side instead now cross the midline. The *comm* gene encodes a novel transmembrane protein expressed by midline cells (G Tear, R Harris, S Sutaria, K Kilomanski, CS Goodman & MA Seeger, unpublished results).

The CNS guidance screen also led to mutations in genes that disrupt the formation of longitudinal axon pathways, including *longitudinals lacking* (*lola*) and *longitudinals gone* (*logo*). *lola* encodes a nuclear protein that is expressed in many if not all neurons, and in other tissues as well (Giniger et al 1994).

The motoneuron screen of the second chromosome (Van Vactor et al 1993) led to the identification of five genes with highly penetrant phenotypes: *beaten path, stranded, short stop, walkabout,* and *clueless.* Another screen (D Van Vactor & D Lin, personal communication) led to the identification of another second chromosome gene: *bypass.* The screen of the third chromosome (Sink & Goodman 1994) led to the identification of several genes, including *sidestep.* Two of these genes (*stranded* and *short stop*) appear to control general aspects of guidance and growth, three others (*beaten path, bypass,* and *sidestep*) control the recognition of the correct target region, and two (*walkabout* and *clueless*) control specific aspects of muscle target recognition.

The retinal connectivity screen (Martin et al 1995) led to the identification of four genes with highly penetrant phenotypes: *diva, eddy, limbo,* and *nonstop.* One gene (*eddy*) appears to control axonal outgrowth, two others (*limbo* and *nonstop*) control target recognition, and one of them (*limbo*) controls retinotopy.

Reverse genetics has provided a complementary approach in *Drosophila* to identify guidance molecules. Such analysis has led to the identification of two proteins that have been shown, using genetic analysis and refined anatomical probes, to be involved in CNS growth cone guidance [fasciclin II (Lin et al 1994) and derailed (Callahan et al 1995)]. It has also led to the identification of several proteins (e.g. connectin, fasciclin III, and semaphorin II) that are expressed by subsets of muscles during the formation of neuromuscular connections (Nose et al 1992, 1994; Chiba et al 1995; Matthes et al 1995). However, loss-of-function mutations in the *connectin, fasIII,* and *semaII* genes do not lead to easily observable defects. Moreover, the systematic genetic screen for neuromuscular mutants described above did not recover mutations in any of these three genes.

Nevertheless, the ectopic expression of all three proteins shows that they can function in growth cone guidance—connectin as a bifunctional signal (Nose et al 1994), fasciclin III as an attractive signal (Chiba et al 1995), and semaphorin II as an inhibitory signal (Matthes et al 1995). This trend, in which the ectopic expression yields a stronger phenotype than does the loss-of-function, might reflect some inherent property in the way targeting systems are built (see Conclusions). Some guidance molecules may be in part refractory to loss-of-function genetic analysis in which gene functions are removed one at a time. Rather, methods that rely on ectopic expression may be required to reveal the function of some guidance molecules.

## Mouse

Reverse genetic approaches have been used in mouse to generate knockout mutations in the genes encoding putative guidance molecules [e.g. NCAM (Tomasiewicz et al 1993, Cremer et al 1994)]. Classical genetic approaches have also been used to study this problem by isolating mutations with behavioral defects. A number of mouse mutations have been collected based on their primary behavioral phenotypes and the secondary analysis of their underlying defects in cortical development (Caviness & Rakic 1978). One of these is the recessive mutation *reeler,* which leads to impaired motor coordination.

Neurons in the brains of *reeler* mutant mice fail to reach their proper locations, leading to a disrupted laminar organization in both the cerebellar and cerebral cortices. D'Arcangelo et al (1995) cloned the *reeler* gene and showed that the Reelin protein appears to be an ECM protein with domains similar to both F-spondin and epidermal growth factor (EGF)–like repeats. Reelin is expressed by subsets of neurons in target regions at appropriate times, and might be involved in the control of neuronal migration. Thus, Reelin is likely to mediate neuronal adhesion and migration at critical stages of cortical development.

## Human

Because humans are so sensitive in detecting and reporting on neurological and psychiatric diseases, there is a wealth of medical information on mutations in human genes that cause behavioral defects. In the case of neural CAMs of the Ig superfamily, much is known about the neurological consequences of mutations in the human gene encoding L1. Jouet et al (1993) reported that mutations in the *L1* gene lead to X-linked hydrocephalus resulting from stenosis of the aqueduct of Sylvius (HSAS). Other mutations in the *L1* gene lead to X-linked spastic paraplegia and MASA syndrome (mental retardation, aphasia, shuffling gait, and adducted thumbs) (Vits et al 1994, Jouet et al 1994). Defects in cell migrations or axon guidance might be the cause of some or all of these inherited neurological diseases.

Another example is Kallmann's syndrome, an X-linked disorder. Patients with Kallmann's disease suffer from a number of dysfunctions, including anosmia (inability to smell), apparently due to a defect in the embryonic migration of olfactory neurons to the olfactory bulb, and hypogonadism, apparently due to a defect in the embryonic migration of neurons that express gonadotropin-releasing hormone (Gn-RH) to the pituitary. Both the olfactory and Gn-RH neurons originate in the olfactory placode and share a common migratory pathway. Their migration is impaired by the mutation in the *Kall-*

*mann* (*KAL*) gene, suggesting that *KAL* encodes a protein that guides these migrations.

The KAL protein has a signal sequence but no transmembrane domain or apparent membrane linkage, contains several fibronectin (FN) type-III repeats and a four-disulfide-core domain, and appears to be a secreted protein (Franco et al 1991, Legouis et al 1991). The expression of KAL is highly restricted to neurons in certain regions of the developing and adult CNS; it is not expressed by glia (Legouis et al 1993, Rugarli et al 1993). In the developing and adult chick, for example, KAL is expressed by mitral cells in the olfactory bulb and by Purkinje cells in the cerebellar cortex.

KAL is a secreted protein that appears to function to guide the migration of a specific subset of CNS neurons. KAL might function as a chemoattractant for these migrations. Alternatively, KAL might be an ECM component that labels these pathways.

## SIGNAL TRANSDUCTION

During growth cone guidance, secreted and cell surface ligands bind to receptors on filopodia and lamellipodia and trigger second-messenger signals in appropriate regions of the growth cone. Evidence suggests that the most immediate target of these guidance signals is the cytoskeleton (e.g. Lin & Forscher 1995). Local changes in the regulation and translocation of F-actin appear to underlie the steering decisions of pioneer growth cones in the limb bud of the grasshopper embryo (O'Connor & Bentley 1993, Bentley & O'Connor 1994); turning towards the signal appears to be due to a local accumulation of F-actin, which leads to recruitment of microtubules. Repulsive signals lead to the loss of F-actin at the leading edge of the growth cone (Fan et al 1993). When DRG growth cones are exposed to a localized but nondiffusible source of collapsing signal (chick brain extract enriched in collapsin that is immobilized on beads), following filopodial contact, these growth cones often turn away from the source without collapsing (Fan & Raper 1995); turning away from the signal appears to be due to a local loss of F-actin.

Little is known about the second messengers that transduce the attractive and repulsive signals from the cell surface to the cytoskeleton during growth cone guidance in the developing organism. The reasons are threefold. First, although a number of new families of ligands have been discovered (e.g. netrins and semaphorins), we do not yet know the molecular identity of most of the receptors that mediate these responses (with the exception of the family of Eph receptor tyrosine kinases and their ligands, as discussed below). Second, most experiments on signal transduction mechanisms have been conducted in vitro and have relied on neurite outgrowth assays; only a few have used steering and turning assays that resemble in vivo guidance. Third, most in vitro studies rely on pharmacological reagents to dissect second-messenger pathways, an approach often fraught with problems of specificity. Specificity can be a major problem in cases where only a single reagent is available. For example, many studies have used pertussis toxin (PTX) as a specific reagent to inactivate G proteins. However, a recent study has shown that PTX can generate many of these results by a mechanism independent of direct G-protein inactivation (Kindt & Lander 1995).

In only a few cases has the function of signal transduction molecules during growth cone guidance been examined in the developing organism. These studies include the Abl tyrosine kinase in *Drosophila* (Elkins et al 1990); the UNC-51 serine-threonine kinase in nematode (Ogura et al 1994); calcium-calmodulin in *Drosophila* (VanBerkum and Goodman 1995); GAP-43 in mouse (Strittmatter et al 1995); the derailed receptor tyrosine kinase in *Drosophila* (Callahan et al 1995); and the DPTP69D, DPTP99A, and DLAR receptor tyrosine phophates in *Drosophila* (see below).

A large body of literature, based on the use of various in vitro assays or biochemical analysis of growth cone particles, has implicated almost every known second-messenger system in aspects of neurite outgrowth, neurite extension, growth cone collapse, and growth cone turning. These second-messenger systems include both receptor and nonreceptor tyrosine kinases (RTKs and NRTKs) and tyrosine phosphatases (RTPs and NRTPs), serine-threonine kinases, calcium, calmodulin, cAMP, G proteins, GAP-43, $IP_3$ and DAG, protein kinase C, and nitric oxide. In addition, the Rho, Rac, and Cdc42 GTPases have been shown to trigger the formation of actin-based structures in the filopodia of 3T3 cells and to regulate the motility of these cells (Nobes & Hall 1995); this family of GTPases are thus good candidates to mediate F-actin accumulation in growth cones (e.g. Chant & Stowers 1995). In *Drosophila,* a dominant-negative form of the small GTPase Drac1 has been expressed in embryonic neurons, and this elimination of Drac1 function has been shown to lead to defects in axon outgrowth (Luo et al 1994). A few of these signaling systems in growth cones are discussed below.

## Calcium and Calmodulin

Calcium has been implicated in the regulation of a wide range of growth cone behaviors (Kater & Mills 1991, Cypher & Letourneau 1992). For example, increases in internal $Ca^{2+}$ have been associated with axonogenesis (Silver et al 1990, Bentley et al 1991), axon turning up a gradient of acetylcholine in vitro (Zheng et al 1994), specific growth cone contacts in vivo (Bentley et al 1991), changes in filopodial morphology (Davenport & Kater 1992, Rehder & Kater 1992), certain kinds of growth cone collapse in vitro (Bandtlow et al 1993, Moorman & Hume 1993), and CAM-mediated neurite outgrowth in vitro

(Bixby et al 1994, Doherty & Walsh 1994). In fact, $Ca^{2+}$ signaling has been implicated in so many different aspects of growth cone behavior in vitro that it is difficult to predict which of these behaviors it controls in vivo.

Most $Ca^{2+}$ signaling is transduced by calmodulin (CaM). To test the role of calcium-CaM during growth cone guidance in the developing organism, calcium-CaM function was selectively disrupted in a specific subset of growth cones in transgenic *Drosophila* embryos (VanBerkum & Goodman 1995). A specific enhancer element was used to drive the expression of the kinesin motor domain fused to a CaM antagonist peptide (kinesin-antagonist, or KA, which blocks CaM binding to target proteins) or CaM itself (kinesin-CaM, or KC, which acts as a $Ca^{2+}$-binding protein). In both KA and KC mutant embryos, specific growth cones exhibit dosage-dependent stalls in axon extension and errors in axon guidance, including both defects in fasciculation and abnormal crossings of the midline. These results demonstrate an in vivo function for $Ca^{2+}$-CaM signaling in growth cone guidance, and suggest that $Ca^{2+}$-CaM may in part regulate specific growth cone decisions, including when to defasciculate and whether or not to cross the midline.

## *Tyrosine Kinases and Phosphatases*

NRTKs such as Src are localized at high levels in growth cones (Maness et al 1990). Bixby & Jhabvala (1993) showed that growth cone particles contain high levels of Src, Fyn, and Yes. Wu & Goldberg (1993) observed regulated tyrosine phosphorylation at the tips of growth cone filopodia. A number of RTPs have also been found at high levels in growth cones and developing axons (e.g. Zinn 1993).

Two studies show the function of both a NRTK (Ellkins et al 1990) and a RTK (Callahan et al 1995) in the events of growth cone guidance in the developing organism. Elkins et al (1990) showed that although embryos carrying mutations in either the Abl NRTK or the fasciclin I CAM have a relatively normal CNS, embryos doubly mutant for both fasciclin I and Abl have major abnormalities in the development of their axon commissures. Callahan et al (1995) showed that the derailed RTK in *Drosophila* (related to vertebrate Ryk) is expressed by a subset of neurons whose axons selectively fasciculate into two distinct CNS axon pathways. In *derailed* mutant embryos, these axons fail to properly fasciculate and make incorrect pathway choices.

Soriano and colleagues generated knockout mice for Src, Fyn, and Yes (Stein et al 1994). NCAM-dependent neurite outgrowth is inhibited in vitro in the neurons from Fyn-minus mice (Beggs et al 1994), whereas L1-dependent neurite outgrowth is impaired in vitro in neurons from Src-minus mutant mice (Ignelzi et al 1994). These studies suggest that individual NRTKs function downstream from specific CAMs.

Doherty, Walsh, and colleagues carried out a detailed analysis of the signal transduction events involved in the induction of neurite outgrowth by NCAM (e.g. Doherty & Walsh 1994). Their most intriguing result suggests that activation of the FGF receptor (a RTK) regulates neurite outgrowth in vitro as stimulated by NCAM, L1, and N-cadherin (Williams et al 1994a). They have also characterized the second-messenger pathway underlying this stimulation of neurite outgrowth (Williams et al 1994b).

The Eph family of RTKs comprises over a dozen members; the ligands for these receptors are all membrane anchored via either a phospholipid anchor or a transmembrane domain (reviewed by Tessier-Lavigne 1995). Both the Eph RTKs and their ligands are expressed in the developing nervous system. Winslow et al (1995) used an in vitro assay to show that AL-1, a human ligand for the Rek7 Eph RTK, can function in axon fasciculation (similar to the derailed RTK in *Drosophila*) (Callahan et al 1995).

Ligands of the Eph RTKs have been implicated as candidates for positional guidance signals in the developing retinotectal system. Drescher et al (1995) show that RAGS, the chick homologue of AL-1, is expressed in a graded fashion with its peak in the caudal (posterior) part of the developing optic tectum. The recombinant protein induces growth cone collapse and repulsion of retinal ganglion cell axons in vitro, although it displays little specificity for axons of nasal vs temporal origin. Cheng et al (1995) show that another ligand for Eph RTKs, ELF-1, is expressed in a similar gradient in the chick optic tectum, with its highest expression also in caudal tectum. Mek4 and Sek are RTKs that bind ELF-1, and Cheng et al (1995) show that while Sek is expressed in a uniform fashion in the retina, Mek4 is expressed in a gradient in the developing retina, with highest expression in the temporal retina, the region whose axons innervate the rostral tectum and are repelled by the caudal tectum. These results suggest that membrane-anchored ligands for the Eph family of RTKs are a potentially important family of axon guidance signals in the developing nervous system.

There is growing evidence that tyrosine phosphatases also play important roles in the control of growth cone guidance. Five RTPs have been identified in *Drosophila*, four of which are expressed predominantly on axons and growth cones in the developing nervous system (Zinn 1993). Genetic analysis reveals that three RTPs—DPTP69D, DPTP99A, and DLAR—function in motor axon guidance (C Desai, J Gindhart, LSB Goldstein, & K Zinn, unpublished results; N Krueger, D Van Vactor, H Wan, B Gelbart, CS Goodman, & H Saito, unpublished results). In these RTP mutants, motor axons in the periphery fail to defasciculate from the common motor nerve at specific choice points and enter their appropriate target regions; instead, they display a variety of stall, bypass, and detour phenotypes.

Thus, mutations in genes encoding CAMs (Lin et al 1995), RTKs (Callahan et al 1995), and RTPs (see above) all alter the events of fasciculation and defasciculation in the *Drosophila* embryo, suggesting that CAMs, RTKs, and RTPs might interact in common guidance decisions. Interestingly, Peles et al (1995) have shown that a vertebrate RTP, RPTPβ, binds the CAM contactin, further supporting this notion.

## GAP-43

GAP-43/neuromodulin is expressed at high levels in growth cones during both development and regeneration; many experiments have shown that GAP-43 is associated with growth and plasticity (reviewed by Skene 1989). GAP-43 is regulated by protein kinase C (PKC), GAP-43 regulates $G_0$ (a GTP binding protein) (Strittmatter et al 1990), and GAP-43 binds to CaM (e.g. Skene 1990); all of these signaling molecules are major components of growth cones.

Several studies have analyzed the behavior of neurons in the absence of GAP-43. Aigner & Caroni (1993, 1995) used antisense oligonucleotides to eliminate GAP-43 from DRG neurons in vitro. In the absence of GAP-43, growth cones extend but adhere poorly, display dynamic but unstable lamellar extensions, and are devoid of F-actin concentrations (Aigner & Caroni 1995).

Strittmatter et al (1995) generated GAP-43 knockout mice to determine the essential functions of GAP-43 in the developing organism. They found normal neurite outgrowth and initial extension by retinal axons in these mutant mice. However, the retinal axons remain trapped in the optic chiasm for six days, unable to navigate past this midline decision point. Over the subsequent weeks, most of the axons did enter the appropriate tracts, and the adult CNS is grossly normal. They conclude that GAP-43 is not essential for axon outgrowth, but rather is required at certain decision points, such as the optic chiasm, suggesting that GAP-43 functions to amplify pathfinding signals.

## CONCLUSIONS

In vitro studies have revealed a diversity of positive and negative mechanisms, including short-range (contact-mediated attraction and contact-mediated repulsion) and long-range (chemoattraction and chemorepulsion) signals. In the developing organism, however, the distinction between long-range vs short-range signals may not be so clear as they appear to function in a continuum. Moreover, although in vitro studies have led to the operational distinction between inhibition vs repulsion, it is unclear whether this distinction represents a true mechanistic difference in vivo.

In making the transition from mechanisms to molecules, the insight arises that particular families of molecules do not fit into distinct mechanistic categories. The functional analysis of three families of signals (CAMs of the Ig superfamily, netrins, and semaphorins) has shown that there is not a precise one-to-one match between molecule and mechanism, but rather some guidance molecules are bifunctional, playing different roles for different growth cones. Individual signals (e.g. netrins-1) can be either attractive or repulsive, suggesting the existence of multiple receptors. Moreover, members of the same family (e.g. semaphorins) can be both cell surface and secreted, further blurring the distinction between short-range and long-range signals.

Combining the results from genetic analysis of candidate guidance molecules (reverse genetics) with the results from genetic screens for mutations that disrupt guidance (classical genetics) leads to three important observations. First, for some guidance molecules, the loss-of-function mutant phenotypes are often more subtle than was predicted by in vivo expression and in vitro function. Second, for some of these molecules, ectopic expression in the developing organism at relevant times and places often leads to more dramatic gain-of-function phenotype than does the loss-of-function, suggesting that the molecule may indeed function in guidance in vivo. Third, large-scale genetic screens of large portions of the genome, although leading to the discovery of important new genes, have not revealed as many new genes as would be predicted if each growth cone decision was governed by a unique set of genes (encoding signal, receptor, downstream components) functioning in a single pathway.

These three observations can be explained by a number of different models. One explanation is to evoke redundancy. But there is little evidence for two or more putative guidance molecules having the identical pattern of expression and identical function. Moreover, with the exception of recent gene duplications (leading to pairs of genes that have not yet diverged in expression and function) (e.g. Li & Noll 1994), strict redundancy seems unlikely to be the cause for the surprisingly subtle phenotypes observed in so many knockouts of the genes encoding putative guidance molecules.

Several alternative explanations do not evoke strict redundancy. For example, all of the observations to date are consistent with a combinatorial model involving multiple targeting signals that are bifunctional, are aligned along common boundaries, have partially overlapping functions, and work together to guide growth cones either toward or away from specific targets. In such a model, the identity of a target is specified not by a single molecular label but rather by a combination of molecules expressed by that target and by neighboring targets. An individual growth cone is guided toward the target that (according to its receptors) appears attractive and away from neighboring targets that appear repulsive. Such an over-specified guidance system would be highly reliable and robust. It would also have the properties that removal of any one component would have a much less disruptive effect on the final outcome than would be the introduction of this same component in a novel location, thus creating a discontinuity in the pattern of signals.

If different guidance signals are normally aligned along common boundaries, then removing any one of them might leave enough other cues to define the boundary and might still allow a growth cone to ultimately make the correct choice, albeit in a bit less reliable or robust fashion. However, the ectopic expression of one of these components would place this cue out of alignment with the others, thus creating clashing signals, and in so doing disrupt the ability of the growth cone to correctly read the remaining cues.

Such a model represents just one alternative, and we do not yet know enough about the identity, function, and distribution of guidance and targeting signals and receptors to distinguish between various models. As more is learned, it should be possible to test and refine these ideas and derive a model that is consistent with normal development, and with functional (largely genetic) perturbations that add or delete individual molecules at specific times and places. In this way we hope to determine how the genome provides sufficient information to begin the life-long process of brain wiring.

ACKNOWLEDGMENTS

I thank David Van Vactor, Marc Tessier-Lavigne, Mark VanBerkum, Dennis O'Leary, and Sophie Petersen for thoughtful discussions and suggestions on the manuscript. Some of the experiments from my lab described here were supported by NIH grants NS18366 and HD21294. CSG is an Investigator with the Howard Hughes Medical Institute.

## Literature Cited

Abercrombie M. 1970. Contact inhibition in tissue culture. *In Vitro* 6:128–42

Aigner L, Caroni P. 1993. Depletion of 43-kD growth-associated protein in primary sensory neurons leads to diminished formation and spreading of growth cones. *J. Cell Biol.* 123:417–29

Aigner L, Caroni P. 1995. Absence of persistent spreading, branching, and adhesion in GAP-43-depleted growth cones. *J. Cell Biol.* 128: 647–60

Allendoerfer KL, Shatz CJ. 1994. The subplate, a transient neocortical structure: its role in the development of connections between thalamus and cortex. *Annu. Rev. Neurosci.* 17:185–218

Baird DH, Baptista CA, Wang LC, Mason CA. 1992. Specificity of a target cell-derived stop signal for afferent axonal growth. *J. Neurobiol.* 23:579–91

Bandtlow C, Zachleder T, Schwab ME. 1990. Oligodendrocytes arrest neurite growth by contact inhibition. *J. Neurosci.* 10:3837–48

Bandtlow CE, Schmidt MF, Hassinger TD, Schwab ME, Kater SB. 1993. Role of intracellular calcium in NI-35-evoked collapse of neuronal growth cones. *Science* 259:80–83

Barnes TM. 1994. OPUS: a growing family of gap junction proteins? *Trends Genet.* 10: 303–5

Bastiani MJ, Harrelson AL, Snow PM, Goodman CS. 1987. Expression of fasciclin I and II glycoproteins on subsets of axon pathways during neuronal development in the grasshopper. *Cell* 48:745–55

Bastiani MJ, Raper JA, Goodman CS. 1984. Pathfinding by neuronal growth cones in grasshopper embryos. III. Selective affinity of the G growth cone for the P cells within the A/P fascicle. *J. Neurosci.* 4:2311–28

Beggs HE, Soriano P, Maness PF. 1994. NCAM-dependent neurite outgrowth is inhibited in neurons from Fyn-minus mice. *J. Cell Biol.* 127:825–33

Bentley D, Guthrie PB, Kater SB. 1991. Calcium ion distribution in nascent pioneer axons and coupled preaxonogenesis neurons in situ. *J. Neurosci.* 11:1300–8

Bentley D, O'Connor TP. 1994. Cytoskeletal events in growth cone steering. *Curr. Opin. Neurobiol.* 4:43–48

Bieber AJ, Snow PM, Hortsch M, Patel NH, Jacobs JR, et al. 1989. *Drosophila* neuroglian: a member of the immunoglobulin superfamily with extensive homology to the vertebrate neural adhesion molecule L1. *Cell* 59:447–60

Bixby JL, Grunwald GB, Bookman RJ. 1994. $Ca^{2+}$ influx and neurite growth in response to purified N-cadherin and laminin. *J. Cell Biol.* 127:1461–75

Bixby JL, Harris WA. 1991. Molecular mechanisms of axon growth and guidance. *Annu. Rev. Cell Biol.* 7:117–59

Bixby JL, Jhabvala P. 1993. Tyrosine phosphorylation in early embryonic growth cones. *J. Neurosci.* 13:3421–32

Bolz J. 1994. Cortical circuitry in a dish. *Curr. Opin. Neurobiol.* 4:545–49

Boschert U, Ramos RG, Tix S, Technau GM, Fischbach KF. 1990. Genetic and developmental analysis of irreC, a genetic function required for optic chiasm formation in *Drosophila*. *J. Neurogenet.* 6:153–71

Brenner S. 1974. The genetics of *Caenorhabditis elegans*. *Genetics* 77:71–94

Callahan CA, Muralidhar MG, Lundgren SE, Scully AL, Thomas JB. 1995. Control of neuronal pathway selection by a *Drosophila* receptor protein-tyrosine kinase family member. *Nature* 376:171–74

Caroni P, Schwab ME. 1988. Antibody against myelin-associated inhibitor of neurite growth neutralizes nonpermissive substrate properties of CNS white matter. *Neuron* 1:85–96

Caviness VS Jr, Rakic P. 1978. Mechanisms of cortical development: a view from mutations in mice. *Annu. Rev. Neurosci.* 1:297–326

Chant J, Stowers L. 1995. GTPase cascades choreographing cellular behavior: movement, morphogenesis, and more. *Cell* 81:1–4

Cheng H-J, Nakamoto M, Bergemann AD, Flanagan JG. 1995. Complementary gradients in expression and binding of ELF-1 and Mek4 in development of the topographic retinotectal projction map. *Cell* 82:371–81

Chiba A, Snow P, Keshishian H, Hotta Y. 1995. Fasciclin III as a synaptic target recognition molecule in *Drosophila*. *Nature* 374:166–68

Colamarino SA, Tessier-Lavigne M. 1995. The axonal chemoattractant netrin-1 is also a chemorepellent for trochlear motor axons. *Cell.* In press

Cox EC, Muller B, Bonhoeffer F. 1990. Axonal guidance in the chick visual system: posterior tectal membranes induce collapse of growth cones from the temporal retina. *Neuron* 2:31–37

Cremer H, Lange R, Christoph A, Plomann M, Vopper G, et al. 1994. Inactivation of the N-CAM gene in mice results in size reduction of the olfactory bulb and deficits in spatial learning. *Nature* 367:455–59

Crompton DE, Griffin A, Davies JA, Miklos GL. 1992. Analysis of a cDNA from the neurologically active locus *shaking-B (Passover)* of *Drosophila melanogaster*. *Gene* 122:385–86

Cunningham BA, Hemperly JJ, Murray BA, Prediger EA, Brackenbury R, Edelman GM. 1987. Neural cell adhesion molecule: structure, immunoglobulin-like domains, cell surface modulation and alternative RNA splicing. *Science* 236:799–806

Cypher C, Letourneau PC. 1992. Growth cone motility. *Curr. Opin. Cell Biol.* 4:4–7

D'Arcangelo G, Miao GG, Chen S-C, Soares HD, Morgan JI, Curran T. 1995. A protein related to extracellular matrix proteins deleted in the mouse mutant *reeler*. *Nature* 374:719–23

Davenport RW, Kater SB. 1992. Local increases in intracellular calcium elicit local filopodial responses in Helisoma neuronal growth cones. *Neuron* 9:405–16

Davies JA, Cook GMW, Stern CD, Keynes RJ. 1990. Isolation from chick somites of a glycoprotein fraction that causes collapse of dorsal root ganglion growth cones. *Neuron* 2:11–20

Doherty P, Furness J, Williams EJ, Walsh FS. 1994. Neurite outgrowth stimulated by the tyrosine kinase inhibitor herbimycin A requires activation of tyrosine kinases and protein kinase C. *J. Neurochem.* 62:2124–31

Doherty P, Walsh FS. 1994. Signal transduction events underlying neurite outgrowth stimulated by cell adhesion molecules. *Curr. Opin. Neurobiol.* 4:49–55

Drescher U, Kremoser C, Handwerker C, Löschinger J, Noda M, Bonhoeffer F. 1995. In vitro guidance of retinal ganglion cell axons by RAGS, a 25 kDa tectal protein related to ligands for Eph receptor tyrosine kinases. *Cell* 82:359–70

Edelman GM. 1985. Cell adhesion molecule expression and the regulation of morphogenesis. *Cold Spring Harbor Symp. Quant. Biol.* 50:877–89

Elkins T, Zinn K, McAllister L, Hoffmann FM, Goodman CS. 1990. Genetic analysis of a *Drosophila* neural cell adhesion molecule: interaction of fasciclin I and Abelson tyrosine kinase mutations. *Cell* 60:565–75

Faissner A, Kruse J. 1990. J1/tenascin is a repulsive substrate for central nervous system neurons. *Neuron* 5:627–37

Fan J, Mansfield SG, Redmond T, Gordon-Weeks PR, Raper JA. 1993. The organization of F-actin and microtubules in growth cones exposed to a brain-derived collapsing factor. *J. Cell Biol.* 121:867–78

Fan J, Raper JA. 1995. Localized collapsing cues can steer growth cones without inducing their full collapse. *Neuron* 14:263–74

Fitzgerald M, Kwiat GC, Middleton J, Pini A. 1993. Ventral spinal cord inhibition of neurite outgrowth from embryonic rat dorsal root ganglia. *Development* 117:1377–84

Franco B, Guioli S, Pragliola A, Incerti B, Bardoni B, et al. 1991. A gene deleted in Kallmann's syndrome shares homology with neural cell adhesion and axonal path-finding molecules. *Nature* 353:529–36

Garriga G, Desai C, Horvitz HR. 1993. Cell interactions control the direction of outgrowth, branching and fasciculation of the HSN axons of *Caenorhabditis elegans*. *Development* 117:1071–87

Giniger E, Tietje K, Jan LY, Jan YN. 1994. *lola* encodes a putative transcription factor required for axon growth and guidance in *Drosophila*. *Development* 120:1385–98

Godement P, Salaün J, Mason CA. 1990. Retinal axon pathfinding in the optic chiasm: divergence of crossed and uncrossed fibers. *Neuron* 5:173–86

Godement P, Wang LC, Mason CA. 1994. Retinal axon divergence in the optic chiasm: dynamics of growth cone behavior at the midline. *J. Neurosci.* 14:7024–39

Goodman CS. 1994. The likeness of being: phylogenetically conserved molecular mechanisms of growth cone guidance. *Cell* 78:353–56

Goodman CS, Bastiani MJ, Doe CQ, du Lac S, Helfand SL, et al. 1984. Cell recognition during neuronal development. *Science* 225:1271–79

Goodman CS, Shatz CJ. 1993. Developmental mechanisms that generate precise patterns of neuronal connectivity. *Cell* 72:77–98

Grenningloh G, Bieber A, Rehm J, Snow P, Traquina Z, et al. 1990. Molecular genetics of neuronal recognition in *Drosophila*: evolution and function of immunoglobulin superfamily cell adhesion molecules. *Cold Spring Harbor Symp. Quant. Biol.* 55:327–40

Grenningloh G, Goodman CS. 1992. Pathway recognition by neuronal growth cones: genetic analysis of neural cell adhesion molecules in *Drosophila*. *Curr. Opin. Neurobiol.* 2:42–47

Grenningloh G, Rehm EJ, Goodman CS. 1991. Genetic analysis of growth cone guidance in

*Drosophila*: fasciclin II functions as a neuronal recognition molecule. *Cell* 67:45–57

Gunderson RW, Barrett JN. 1979. Neuronal chemotaxis: chick dorsal root axons turn toward high concentrations of nerve growth factor. *Science* 206:1079–80

Hamelin M, Zhou Y, Su MW, Scott IM, Culotti JG. 1993. Expression of the UNC-5 guidance receptor in the touch neurons of *C. elegans* steers their axons dorsally. *Nature* 364:327–30

Harrelson AL, Goodman CS. 1988. Growth cone guidance in insects: fasciclin II is a member of the immunoglobulin superfamily. *Science* 242:700–8

Hedgecock EM, Culotti JG, Hall DH. 1990. The *unc-5*, *unc-6*, and *unc-40* genes guide circumferential migrations of pioneer axons and mesodermal cells in the epidermis in *C. elegans*. *Neuron* 2:61–85

Heffner CD, Lumsden AGS, O'Leary DDM. 1990. Target control of collateral extensions and directional axon growth in the mammalian brain. *Science* 247:217–20

Heisenberg M, Borst A, Wagner S, Byers D. 1985. Drosophila mushroom body mutants are deficient in olfactory learning. *J. Neurogenet.* 2:1–30

Hekimi S, Kershaw D. 1993. Axonal guidance defects in *Caenorhabditis elegans* mutant reveal cell extrinsic determinants of neuronal morphology. *J. Neurosci.* 13:4254–71

Hynes RO, Lander AD. 1992. Contact and adhesive specificities in the associations, migrations, and targeting of cells and axons. *Cell* 68:303–22

Ignelzi MA Jr, Miller DR, Soriano P, Maness PF. 1994. Impaired neurite outgrowth of src-minus cerebellar neurons on the cell adhesion molecule L1. *Neuron* 12:873–84

Ishii N, Wadsworth WG, Stern BD, Culotti JG, Hedgecock EM. 1992. UNC-6, a laminin-related protein, guides cell and pioneer axon migrations in *C. elegans*. *Neuron* 9:873–81

Jouet M, Rosenthal A, Armstrong G, MacFarlane J, Stevenson R, et al. 1994. X-linked spastic paraplegia (SPG1), MASA syndrome and X-linked hydrocephalus result from mutations in the L1 gene. *Nature Genet.* 7:402–7

Jouet M, Rosenthal A, MacFarlane J, Donnai D, Kenwrick SA. 1993. Missense mutation confirms the *L1* defect in X-linked hydrocephalus (HSAS). *Nature Genet.* 4:331

Joyner AL, Auerbach A, Skarnes WC. 1992. The gene trap approach in embryonic stem cells: the potential for genetic screens in mice. *Ciba Found. Symp.* 165:277–88

Kapfhammer JP, Grunewald BE, Raper JA. 1986. The selective inhibition of growth cone extension by specific neurites in culture. *J. Neurosci.* 6:2527–34

Kapfhammer JP, Raper JA. 1987a. Collapse of

growth cone structure on contact with specific neurites in culture. *J. Neurosci.* 7:201–12

Kapfhammer JP, Raper JA. 1987b. Interactions between growth cones and neurites growing from different neural tissues in culture. *J. Neurosci.* 7:1595–600

Kater SB, Mills LR. 1991. Regulation of growth cone behavior by calcium. *J. Neurosci.* 11:891–99

Kennedy TE, Serafini T, de la Torre JR, Tessier-Lavigne M. 1994. Netrins are diffusible chemotropic factors for commissural axons in the embryonic spinal cord. *Cell* 78:425–35

Kindt RM, Lander AD. 1995. Pertussis toxin specifically inhibits growth cone guidance by a mechanism independent of direct G-protein inactivation. *Neuron* 15:79–88

Kolodkin AL, Matthes DJ, Goodman CS. 1993. The *semaphorin* genes encode a family of transmembrane and secreted growth cone guidance molecules. *Cell* 75:1389–99

Kolodkin AL, Matthes DJ, O'Connor TP, Patel NH, Admon A, et al. 1992. Fasciclin IV: sequence, expression, and function during growth cone guidance in the grasshopper embryo. *Neuron* 9:831–45

Krishnan SN, Frei E, Swain GP, Wyman RJ. 1993. *Passover:* a gene required for synaptic connectivity in the giant fiber system of *Drosophila. Cell* 73:967–77

Kuwada JY. 1995. Development of the zebrafish nervous system: genetic analysis and manipulation. *Curr. Opin. Neurobiol.* 5:50–54

Legouis R, Hardelin JP, Levilliers J, Claverie JM, Compain S, et al. 1991. The candidate gene for the X-linked Kallmann syndrome encodes a protein related to adhesion molecules. *Cell* 67:423–35

Legouis R, Lievre CA, Leibovici M, Lapointe F, Petit C. 1993. Expression of the KAL gene in multiple neuronal sites during chicken development. *Proc. Natl. Acad. Sci. USA* 90:2461–65

Letourneau PC. 1975. Cell-to-substratum adhesion and guidance of axonal elongation. *Dev. Biol.* 44:92–101

Leung-Hagesteijn C, Spence AM, Stern BD, Zhou Y, Su MW, et al. 1992. UNC-5, a transmembrane protein with immunoglobulin and thrombospondin type 1 domains, guides cell and pioneer axon migrations in *C. elegans. Cell* 71:289–99

Li C, Tropak MB, Gerlai R, Clapoff S, Abramow-Newerly W, et al. 1994. Myelination in the absence of myelin-associated glycoprotein. *Nature* 369:747–50

Li W, Herman RK, Shaw JE. 1992. Analysis of the *Caenorhabditis elegans* axonal guidance and outgrowth gene *unc-33. Genetics* 132:675–89

Li X, Noll M. 1994. Evolution of distinct developmental functions of three *Drosophila* genes by acquisition of different *cis*-regulatory regions. *Nature* 367:83–87

Lin C-H, Forscher P. 1995. Growth cone advance is inversely proportional to retrograde F-actin flow. *Neuron* 14:763–71

Lin DM, Fetter RD, Kopczynski C, Grenningloh G, Goodman CS. 1994. Genetic analysis of Fasciclin II in *Drosophila*: Defasciculation, refasciculation, and altered fasciculation. *Neuron* 13:1055–69

Lin DM, Goodman CS. 1994. Ectopic and increased expression of Fasciclin II alters motoneuron growth cone guidance. *Neuron* 13:507–23

Lumsden A, Davies AM. 1983. Earliest sensory nerve fibres are guided to peripheral targets by attractants other than nerve growth factor. *Nature* 306:786–88

Lumsden A, Davies AM. 1986. Chemotropic effect of specific target epithelium in the developing mammalian nervous system. *Nature* 323:538–39

Luo L, Liao YJ, Jan LY, Jan YN. 1994. Distinct morphogenetic functions of similar small GTPases: *Drosophila* Drac1 is involved in axonal outgrowth and myoblast fusion. *Genes Dev.* 8:1787–802

Luo Y, Raible D, Raper JA. 1993. Collapsin: a protein in brain that induces the collapse and paralysis of neuronal growth cones. *Cell* 75:217–27

Luo Y, Shepherd I, Li J, Renzi MJ, Chang S, Raper JA. 1995. A family of molecules related to collapsin in the embryonic chick nervous system. *Neuron* 14:1131–40

Luthl A, Laurent JP, Figurov A, Muller D, Schachner M. 1994. Hippocampal long-term potentiation and neural cell adhesion molecules L1 and NCAM. *Nature* 372:777–79

Maness PF, Shores CG, Ignelzi M. 1990. Localization of the normal cellular src protein to the growth cone of differentiating neurons in brain and retina. *Adv. Exp. Med. Biol.* 265:117–25

Martin KA, Poeck B, Roth H, Ebens AJ, Ballard L, Zipursky SL. 1995. Mutations disrupting neuronal connectivity in the *Drosophila* visual system. *Neuron* 14:229–40

Matthes DJ, Sink H, Kolodkin AK, Goodman CS. 1995. Semaphorin II can function as a selective inhibitor of specific synaptic arborizations. *Cell* 81:631–39

Mayford M, Barzilai A, Keller F, Schacher S, Kandel ER. 1992. Modulation of an NCAM-related adhesion molecule with long-term synaptic plasticity in *Aplysia. Science* 256:638–44

McIntire SL, Garriga G, White J, Jacobson D, Horvitz HR. 1992. Genes necessary for directed axonal elongation or fasciculation in *C. elegans. Neuron* 8:307–22

McKerracher L, David S, Jackson DL, Kottis

V, Dunn RJ, Braun PE. 1994. Identification of myelin-associated glycoprotein as a major myelin-derived inhibitor of neurite growth. *Neuron* 13:805–11

Messersmith EK, Leonardo ED, Shatz CJ, Tessier-Lavigne M, Goodman CS, Kolodkin AL. 1995. Semaphorin III can function as a selective chemorepellent to pattern sensory projections in the spinal cord. *Neuron* 14: 949–59

Miller DM, Shen MM, Shamu CE, Burglin TR, Ruvkun G, et al. 1992. *C. elegans unc-4* gene encodes a homeodomain protein that determines the pattern of synaptic input to specific motor neurons. *Nature* 355:841–45

Mitchell KJ, Doyle J, Tear G, Serafini T, Kennedy TE, et al. 1994. Genetic analysis of growth cone guidance at the midline in *Drosophila*: expression and function of D-netrin. *Soc. Neurosci. Abstr.* 20:1064

Montag D, Giese KP, Bartsch U, Martini R, Lang YJ, et al. 1994. Mice deficient for the myelin-associated glycoprotein show subtle abnormalities in myelin. *Neuron* 13: 229–46

Moorman SJ, Hume RI. 1990. Growth cones of chick sympathetic preganglionic neurons in vitro interact with other neurons in a cell-specific manner. *J. Neurosci.* 10:3158–63

Moorman SJ, Hume RI. 1993. Omega-conotoxin prevents myelin-evoked growth cone collapse in neonatal rat locus coeruleus neurons in vitro. *J. Neurosci.* 13:4727–36

Mukhopadhyay G, Doherty P, Walsh FS, Crocker PR, Filbin MT. 1994. A novel role for myelin-associated glycoprotein as an inhibitor of axonal regeneration. *Neuron* 13: 757–67

Mullins M, Hammerschmidt M, Haffter P, Nüsslein-Volhard C. 1994. Large-scale mutagenesis in the zebrafish: in search of genes controlling development in a vertebrate. *Curr. Opin. Biol.* 4:189–202

Muralidhar MG, Thomas JB. 1993. The *Drosophila bendless* gene encodes a neural protein related to ubiquitin-conjugating enzymes. *Neuron* 11:253–66

Nobes CD, Hall A. 1995. Rho, Rac, and Cdc42 GTPases regulate the assembly of multimolecular focal complexes associated with actin stress fibers, lamellipodia, and filopodia. *Cell* 81:53–62

Nose A, Mahajan VB, Goodman CS. 1992. Connectin: a homophilic cell adhesion molecule expressed on a subset of muscles and the motoneurons that innervate them in *Drosophila. Cell* 70:553–67

Nose A, Takeichi M, Goodman CS. 1994. Ectopic expression of connection reveals a repulsive function during growth cone guidance and synapse formation. *Neuron* 13: 525–39

Nüsslein-Volhard C, Wieschaus E. 1980. Mu-

tations affecting segment number and polarity in *Drosophila. Nature* 287:795–801

Oakley RA, Tosney KW. 1993. Contact-mediated mechanisms of motor axon segmentation. *J. Neurosci.* 13:3773–92

O'Connor TP, Bentley D. 1993. Accumulation of actin in subsets of pioneer growth cone filopodia in response to neural and epithelial guidance cues in situ. *J. Cell Biol.* 123:935–48

Ogura K, Wicky C, Magnenat L, Tobler H, Mori I, et al. 1994. *Caenorhabditis elegans unc-51* gene required for axonal elongation encodes a novel serine/threonine kinase. *Genes Dev.* 8:2389–400

Oh CE, McMahon R, Benzer S, Tanouye MA. 1994. *bendless*, a *Drosophila* gene affecting neuronal connectivity, encodes a ubiquitin-conjugating enzyme homolog. *J. Neurosci.* 14:3166–79

O'Leary DDM, Bicknese AR, De Carlos JA, Heffner CD, Koester SE, et al. 1990. Target selection by cortical axons: alternative mechanisms to establish axonal connections in the developing brain. *Cold Spring Harbor Symp. Quant. Biol.* 55:453–68

Ono K, Tomasiewicz H, Magnuson T, Rutishauser U. 1994. N-CAM mutation inhibits tangential neuronal migration and is phenocopied by enzymatic removal of polysialic acid. *Neuron* 13:595–609

Peles E, Nativ M, Campbell PL, Sakurai T, Martinez R, et al. 1995. The carbonic anhydrase domain of receptor tyrosine phosphatase β is a functional ligand for the axonal cell recognition molecule contactin. *Cell* 82: 251–60

Pesheva P, Gennarini G, Goridis C, Schachner M. 1993. The F3/11 cell adhesion molecule mediates the repulsion of neurons by the extracellular matrix glycoprotein J1–160/180. *Neuron* 10:69–82

Phillis RW, Bramlage AT, Wotus C, Whittaker A, Gramates S, et al. 1993. Isolation of mutations affecting neural circuitry required for grooming behavior in *Drosophila melanogaster. Genetics* 133:581–92

Pini A. 1993. Chemorepulsion of axons in the developing mammalian central nervous system. *Science* 261:95–98

Placzek M, Tessier-Lavigne M, Jessell T, Dodd J. 1990. Orientation of commissural axons in vitro to a floor plate-derived chemoattractant. *Development* 110:19–30

Puschel AW, Adams RH, Betz H. 1995. Murine semaphorin D/collapsin is a member of a diverse gene family and creates domains inhibitory for axonal extension. *Neuron* 14: 941–48

Ramon y Cajal S. 1893. La Retine des Vertebres. *La Cellule* 9:119–258

Ramos RG, Igloi GL, Lichte B, Baumann U, Maier D, et al. 1993. The irregular chiasm

C-roughest locus of *Drosophila*, which affects axonal projections and programmed cell death, encodes a novel immunoglobulin-like protein. *Genes Dev.* 7:2533–47

Raper J, Kapfhammer J. 1990. The enrichment of a neuronal growth cone collapsing activity from embryonic chick brain. *Neuron* 4:21–29

Raper JA, Bastiani MJ, Goodman CS. 1984. Pathfinding by neuronal growth cones in grasshopper embryos. IV. The effects of ablating the A and P axons upon the behavior of the G growth cone. *J. Neurosci.* 4:2329–45

Raper JA, Grunewald EB. 1990. Temporal retinal growth cones collapse on contact with nasal retinal axons. *J. Exp. Neurol.* 109:70–74

Rathjen FG, Jessell TM. 1991. Glycoproteins that regulate the growth and guidance of vertebrate axons: domains and dynamics of the immunoglobulin/fibronectin type III subfamily. *Semin. Neurosci.* 3:297–308

Rehder V, Kater SB. 1992. Regulation of neuronal growth cone filopodia by intracellular calcium. *J. Neurosci.* 12:3175–86

Roskies AL, O'Leary DD. 1994. Control of topographic retinal axon branching by inhibitory membrane-bound molecules. *Science* 265:799–803

Rugarli EI, Lutz B, Kuratani SC, Wawersik S, Borsani G, et al. 1993. Expression pattern of the Kallmann syndrome gene in the olfactory system suggests a role in neuronal targeting. *Nat. Genet.* 4:19–26

Rutishauser U. 1991. Pleiotropic biological effects of the neural cell adhesion molecule (NCAM). *Semin. Neurosci.* 3:265–70

Rutishauser U. 1993. Adhesion molecules of the nervous system. *Curr. Opin. Neurobiol.* 3:709–15

Rutishauser U, Acheson A, Hall AK, Mann DM, Sunshine J. 1988. The neural cell adhesion molecule (NCAM) as a regulator of cell-cell interactions. *Science* 240:53–57

Sato M, Lopez-Mascaraque L, Heffner CD, O'Leary DD. 1994. Action of a diffusible target-derived chemoattractant on cortical axon branch induction and directed growth. *Neuron* 13:791–803

Schnell L, Schwab ME. 1990. Axonal regeneration in the rat spinal cord produced by an antibody against myelin-associated neurite growth inhibitors. *Nature* 343:269–72

Schwab ME, Caroni P. 1988. Oligodendrocytes and CNS myelin are nonpermissive substrates for neurite growth and fibroblast spreading in vitro. *J. Neurosci.* 8:2381–93

Seeger M, Tear G, Ferres-Marco D, Goodman CS. 1993. Mutations affecting growth cone guidance in *Drosophila*: genes necessary for guidance toward or away from the midline. *Neuron* 10:409–26

Serafini T, Kennedy TE, Galko MJ, Mirzayan C, Jessell TM, Tessier-Lavigne M. 1994. The netrins define a family of axon outgrowth-promoting proteins homologous to *C. elegans* UNC-6. *Cell* 78:409–24

Silver RA, Lamb AG, Bolsover SR. 1990. Calcium hotspots caused by L-channel clustering promote morphological changes in neuronal growth cones. *Nature* 343:751–54

Simon DK, O'Leary DD. 1992. Responses of retinal axons in vivo and in vitro to position-encoding molecules in the embryonic superior colliculus. *Neuron* 9:977–89

Simon MA, Bowtell DD, Dodson GS, Laverty TR, Rubin GM. 1991. Ras1 and a putative guanine nucleotide exchange factor perform crucial steps in signaling by the sevenless protein tyrosine kinase. *Cell* 67:701–16

Sink H, Goodman CS. 1994. Mutations in *sidestep* lead to defects in pathfinding and synaptic specificity during the development of neuromuscular connectivity in *Drosophila*. *Soc. Neurosci. Abstr.* 20:1283

Skene JH. 1989. Axonal growth-associated proteins. *Annu. Rev. Neurosci.* 12:127–56

Skene JH. 1990. GAP-43 as a 'calmodulin sponge' and some implications for calcium signalling in axon terminals. *Neurosci. Res. Suppl.* 13:S112–25

Snow DM, Lemmon V, Carrino DA, Caplan AI, Silver J. 1990. Sulfated proteoglycans in astroglial barriers inhibit neurite outgrowth in vitro. *J. Exp. Neurol.* 109:111–30

Snow PM, Zinn K, Harrelson AL, McAllister L, Schilling J, et al. 1988. Characterization and cloning of fasciclin I and fasciclin II glycoproteins in the grasshopper. *Proc. Natl. Acad. Sci. USA* 85:5291–95

Solnica-Krezel L, Schier A, Driever W. 1994. Efficient recovery of ENU-induced mutations from the zebrafish germline. *Genetics* 136:1401–20

Sperry RW. 1963. Chemoaffinity in the orderly growth of nerve fiber patterns and connections. *Proc. Natl. Acad. Sci. USA* 50:703–10

Sretavan DW. 1990. Specific routing of retinal ganglion cell axons at the mammalian optic chiasm during embryonic development. *J. Neurosci.* 10:1995–2007

Sretavan DW, Siegel M, Reichardt L. 1992. Retinal ganglion cell axons fail to form an optic chiasm following embryonic ablation of ventral diencephalon neurons. *Soc. Neurosci. Abstr.* 18:1274

Starich TA, Herman RK, Shaw JE. 993. Molecular and genetic analysis of *unc-7*, a *Caenorhabditis elegans* gene required for coordinated locomotion. *Genetics* 133:527–41

Strittmatter SM, Fankhauser C, Huang PL, Mashimo H, Fishman MC. 1995. Neuronal pathfinding is abnormal in mice lacking the neuronal growth cone protein GAP-43. *Cell* 80:445–52

Strittmatter SM, Valenzuela D, Kennedy TE, Neer EJ, Fishman MC. 1990. G0 is a major growth cone protein subject to regulation by GAP-43. *Nature* 344:836–41

Tang J, Landmesser L, Rutishauser U. 1992. Polysialic acid influences specific pathfinding by avian motoneurons. *Neuron* 8:1031–44

Tessier-Lavigne M. 1995. Eph receptor tyrosine kinases, axon repulsion, and the development of topographic maps. *Cell* 82: 345–48

Tessier-Lavigne M, Placzek M, Lumsden AG, Dodd J, Jessell TM. 1988. Chemotropic guidance of developing axons in the mammalian central nervous system. *Nature* 336: 775–78

Thomas JB, Wyman RJ. 1982. A mutation in *Drosophila* alters normal connectivity between two identified neurons. *Nature* 298: 650–51

Thomas JB, Wyman RJ. 1984. Mutations altering synaptic connectivity between identified neurons in *Drosophila*. *J. Neurosci.* 4:530–38

Tomasiewicz H, Ono K, Yee D, Thompson C, Goridis C, et al. 1993. Genetic deletion of a neural cell adhesion molecule variant (N-CAM-180) produces distinct defects in the central nervous system. *Neuron* 11:1163–74

Tosney KW. 1987. Proximal tissues and patterned neurite outgrowth at the lumbosacral level of the chick embryo: deletion of the dermamyotome. *Dev. Biol.* 22:540–58

Tosney KW, Landmesser LT. 1985a. Growth cone morphology and trajectory in the lumbosacral region of the chick embryo. *J. Neurosci.* 5:2345–58

Tosney KW, Landmesser LT. 1985b. Specificity of early motoneuron growth cone outgrowth in the chick embryo. *J. Neurosci.* 5: 2336–44

Van Vactor D, Sink H, Fambrough D, Tsoo R, Goodman CS. 1993. Genes that control neuromuscular specificity in *Drosophila*. *Cell* 73:1137–53

VanBerkum MFA, Goodman CS. 1995. Targeted disruption of calcium-calmodulin signaling in *Drosophila* growth cones leads to stalls in axon extension and errors in axon guidance. *Neuron* 14:43–56

Vits L, Van Camp G, Coucke P, Fransen E, De Boulle K, et al. 1994. MASA syndrome is due to mutations in the neural cell adhesion gene L1CAM. *Nat. Genet.* 7:408–13

Walter J, Henke-Fahle S, Bonhoeffer F. 1987a. Avoidance of posterior tectal membranes by temporal retinal axons. *Development* 101: 909–13

Walter J, Kern-Veits R, Huf J, Stolze B, Bonhoeffer F. 1987b. Recognition of position-specific properties of tectal cell membranes by retinal axons in vitro. *Development* 101: 685–96

Williams EJ, Furness J, Walsh FS, Doherty P. 1994a. Activation of the FGF receptor underlies neurite outgrowth stimulated by *L1*, N-CAM, and N-cadherin. *Neuron* 13:583–94

Williams EJ, Furness J, Walsh FS, Doherty P. 1994b. Characterisation of the second messenger pathway underlying neurite outgrowth stimulated by FGF. *Development* 120:1685–93

Winslow JW, Moran P, Valverde J, Shih A, Yuan JQ, et al. 1995. Cloning of AL-1, a ligand for an Eph-related tyrosine kinase receptor involved in axon bundle formation. *Neuron* 14:973–81

Wizenmann A, Thanos S, von Boxberg Y, Bonhoeffer F. 1993. Differential reaction of crossing and non-crossing rat retinal axons on cell membrane preparations from the chiasm midline: an in vitro study. *Development* 117:725–35

Wu DY, Goldberg DJ. 1993. Regulated tyrosine phosphorylation at the tips of growth cone filopodia. *J. Cell Biol.* 123:653–64

Zheng JQ, Felder M, Connor JA, Poo MM. 1994. Turning of nerve growth cones induced by neurotransmitters. *Nature* 368: 140–44

Zinn K. 1993. *Drosophila* protein tyrosine phosphatases. *Semin. Cell Biol.* 4:397–401

*Annu. Rev. Neurosci. 1996. 19:379–404*

# LEARNING AND MEMORY IN HONEYBEES: From Behavior to Neural Substrates

*R. Menzel and U. Müller*

Institut für Neurobiologie, Freie Universität Berlin, Königin-Luise-Strasse 28/30, 14195 Berlin, Germany

KEY WORDS:   orientation, associative learning, olfactory conditioning, mushroom bodies, signaling cascades

## ABSTRACT

Learning and memory in honeybees is analyzed on five levels, using a top-down approach. (*a*) Observatory learning is applied during navigation and dance communication. (*b*) Local cues at the feeding site are learned associatively. (*c*) Classical conditioning of the proboscis extension response to olfactory stimuli provides insight into behavioral, neural, and neuropharmacological mechanisms of associative learning. (*d*) At the neural level, the pathways coding the conditioned and the unconditioned stimulus are identified. The reinforcing function of the unconditioned stimulus is traced to a particular neuron. (*e*) At the cellular level, the cAMP pathway is found to be critically involved. Nitric oxide is an essential mediator for the transfer from short- to long-term memory.

## INTRODUCTION

This article reviews the literature on learning and memory in honeybees, applying a top-down approach from behavioral to neural and cellular studies. We emphasize mechanistic analyses that try to relate behavioral phenomena to the neural substrate and its cellular components. Observations on freely behaving animals are also included, with the aim of relating the behavioral phenomena studied under tightly controlled conditions with those at work in the natural context. This neuroethological approach is particularly useful for studying an insect whose rich behavioral repertoire is strongly influenced by individual experience.

### A Historical Note

Throughout this century the honeybee has been used as a model organism for the discovery of unknown sensory capacities in animals and to unravel the

379

0147-006X/96/0301-0379$08.00

mysteries of social interactions in a community of 50,000 to 100,000 animals. The experimental tool used in all of these studies was the training technique of single and individually marked animals, as introduced by Karl von Frisch at the turn of the century. He studied the perception of colors, patterns, and odors, and developed a unique and effective method for handling bees in discrimination tasks (reviewed in von Frisch 1967). Rewarding a marked foraging bee with sucrose solution at an artificial feeding site allowed von Frisch to relate the experience of this particular bee with its choice behavior after it had systematically manipulated the signals attached to the feeding site. Using this technique, von Frisch and his colleagues discovered a whole range of sensory capacities in bees, e.g. sensitivity to ultraviolet light (Kühn 1927), trichromatic color vision (Daumer 1956), polarized light vision (von Frisch 1949), dance communication, sun compass orientation, time sense (Lindauer 1954, von Frisch 1967), sensitivity to the earth's magnetic field (Martin & Lindauer 1973), and many more (most of the work is described in von Frisch 1967). The process of learning, however, was not at the center of von Frisch's interest, and it became a research topic only much later (Lindauer 1970), although the question of which of the impressive cognitive capacities of bees are learned and which are innate was a matter of debate for decades (Buttel-Reepen 1900; Gould 1984; Lindauer 1959, 1963, 1967; von Frisch 1937).

## The Biological Context

von Frisch's training technique utilized the fast and effective learning abilities of a social insect, which needs to return reliably to its nest site for brood care, protection, and shelter. The colony as a whole is an organism with a potentially infinite life span, and its members are exposed to highly variable ecological conditions during the course of the year and in different habitats. Therefore, the individuals, although living only for a short period of time, cannot be prepared genetically for the ecological conditions of any particular habitat. Because bees search for food (nectar and pollen) at unpredictable sites, they have to learn the celestial and terrestrial cues guiding them on their foraging trips over long distances (in the range of several kilometers). The social bees benefit from a highly variable flower market by adjusting their search behavior continuously, and they cannot be tuned to any close flower-pollinator relationship (Heinrich 1983, Kevan & Baker 1983, Seeley 1985).

Bees perform these tasks with a rather small brain of approximately 1 mm$^3$ and fewer than 960,000 neurons (Witthöft 1967). As a result, some researchers suspect that learning is guided more closely by genetic predisposition (Gould 1984, Lindauer 1970, Menzel 1990). The search for the neural basis of the adaptive components of behavior is facilitated by the organization of the insect nervous system, which is well compartmentalized, with clear separations between multisensory higher-order neuropiles in the brain and neuropiles serving sensory-motor routines in the ventral cord. Many neurons in the insect CNS are individually recognizable and large enough to be recorded intracellularly during ongoing behavior, including learning.

## LEARNING IN THE NATURAL CONTEXT

### Biology of Foraging

Angiosperm plants provide food (nectar, pollen) at their reproductive organs, the flowers, to use insects as vectors for the transport of their genetic material (Darwin 1876, Sprengel 1793). Because outbreeding appears to be favored by evolution (Waser & Price 1983), flower fidelity on the part of the individual insect is one of the evolutionary traits developed in flower-pollinator interaction (Feinsinger 1983, Waser 1983). Flowers increase the specific pollen transfer by providing signals for localization and recognition and by offering only minute amounts of reward per flower, thus forcing the insect to visit many flowers on each foraging bout. Pollinators, for their part, gain from flower fidelity if flowers of the same species are abundant and if foraging becomes more effective and less risky by staying with one species (Heinrich 1983, Laverty 1994). Because flowers are often complicated structures that require special handling skills, flower fidelity may well be favored because of improved handling skills acquired by sticking to one kind of flower (Heinrich 1984, Laverty 1994). The costs and benefits of learning become apparent when species of bumble bees, which belong to pollinator specialists (innate preference for one plant species) and generalists (potential pollinator of many species), are compared (Laverty & Plowright 1988). The specialist always handles its flower better than the generalist, but the latter can learn new manipulatory movements and improve its performance with practice. Hymenopteran pollinators apply innate search images for selecting potential food sources (Giurfa et al 1995, Menzel 1985) or locations of food on flowers (Daumer 1958, Lunau 1991). However, learning flower cues and flower location in space is the most important factor in foraging. Honeybees acquire a knowledge base for effective foraging by two means: exploratory experience and the information transfer during the dance performance of recruiting bees (von Frisch 1967).

### Choice Performance in Foraging

After a patch of flowers has been found that provides enough food, the individual bee sticks to a particular kind of flower within a patch and may travel kilometers visiting up to hundreds of flowers during one foraging bout. Because the bee participates in the colony's information flow about alternative food sites (Seeley 1985, 1994), it is constantly informed about the effectiveness of its foraging performance. Out in the field, it focuses its efforts on the most productive flowers, indicating that it evaluates and learns the reward conditions of multiple feeding sites. Bees optimize their foraging efforts by choosing flowers or artificial feeders with high rewards, as opposed to those with lower rewards (Heinrich 1984). Two strategies are applied: maximizing (choice of only one feeder) or matching (choice frequency increases with the profitability of the feeder). In a patch with up to eight artificial feeders, each of which produces a different but constant flow rate of sucrose solution, bees maximize if one feeder provides a flow rate >1 µl/min; they match if the total flow rate in the patch lies between 0.4 and 1 µl/min; and they terminate foraging for total flow rates <0.4 µl/min (Greggers & Menzel 1993). In contrast to earlier model explanations of choice behavior based on random walk assumptions (Pyke 1984, Schmid-Hempel 1986), bees develop feeder-specific memories that relate feeder signals with measures of their profitability. Accordingly, incentive contrast is a robust and long-lasting phenomenon (Couvillon & Bitterman 1984). Interestingly, colors appear to signal unconditioned stimulus (US) strength less reliably than odors (Buchanan & Bitterman 1988, Couvillon & Bitterman 1988). The most important measure of profitability is the time spent licking per visit; energy content (sucrose solution concentration) is of secondary importance (Greggers et al 1993).

An analysis of the sequences of visits within a patch of two, four, or eight feeders and of feeding activities at each visit reveals learning rules and the existence of several incentive memories. Learning performance can be formalized according to the difference rule (Rescorla & Wagner 1972, Sutton & Barto 1981) and by assuming two forms of memory, a general patch memory and feeder-specific memories. The feeder-specific memories are highly dynamic and change their content over time and as a result of new experience (Greggers & Menzel 1993). Corresponding temporal dynamics of choice behavior were found for bumble bees foraging in a natural environment (Chittka et al 1995). The probability of the bumble bee switching to another kind of flower depends on the licking times during the last visits and the time interval between choices.

NAVIGATION    Central place foragers like the bee navigate in an egocentric reference system based on celestial compass information (e.g. sun, polarized light patterns) and route-specific landmark memories (Collett 1993, Wehner 1992). Celestial compass orientation requires the use of a time-compensated measure of the sun's azimuth. Bees learn the full solar azimuth/time function from temporary restricted segments (e.g. when they are exposed only to the late afternoon sun), indicating that they refer to an innate template that is adjusted by the learning process (Dyer & Dickinson 1994). Wehner (1984) first suggested this form of learning, which is used by the desert ant *Cataglyphis fortis*. The celestial compass provides the reference system for the rotatory component around the high body axis during flight, and this measure is used together with estimates of flight distances to continuously integrate the flight path (Wehner & Wehner 1990). Path integration (or dead reckoning) is a form of automatic observatory learning that enables the animal to return to the hive at any time along a straight line. Landmarks are learned within the framework of the celestial compass (von Frisch 1967) and can also be used to derive the compass direction under a fully overcast sky (Dyer & Gould 1981, von Frisch & Lindauer 1954). Landmark memories are stored as images arranged in sequences as they occur during outbound and homebound flights. Image memories are formed when the animal looks at the goal from a guide post at a selected distance and direction (Collett & Baron 1994, Collett & Cartwright 1983, Vollbehr 1975). Each image memory may be a retinotopi-cally stable flash memory as demonstrated recently for *Drosophila melanogaster* (Dill et al 1993), thus simplifying the procedure of establishing and re-trieving a picture memory of a location. Observations of the flight path and model calculations support this view (Cartwright & Collett 1983, 1987). Se-quences of landmark memories are established for multiple feeding places visited successively (Collett 1993) and en route when bees pass landmarks on a continuous flight towards a food source (Chittka & Geiger 1995). In the first case, bees expect the food source at a specific location relative to the surround-ing landmarks; in the second, bees expect a particular sequence of, and distance between, several landmarks en route.

Landmark and cue memories are linked to time (von Frisch 1967). Bees learn to fly toward a particular feeding place at a particular time of day and to expect reward at a particular set of cues (Bogdany 1978, Gould 1987, Koltermann 1971). The vector memories attached to each feeding place are retrieved specifically by the time of day and the landmarks surrounding the feeding place. For example, marathon dancers—bees that dance for a long time or can be stimulated to dance at any time, even at night—indicate the direction and distance of the place visited closest to the actual time (Lindauer 1954). Furthermore, bees trained to two different feeding places in the morning and afternoon fly in the correct sun compass direction according to the time of the day when released at an unfamiliar place. However, if they are released halfway between the morning and afternoon places, both vector memories are retrieved and a large proportion of bees fly straight toward the hive, indicating that the two vector memories are integrated (Menzel et al 1995a,b). Thus the navigational behavior of bees is guided by contextual and content addressable memories. Therefore, bees appear not to represent spatial memory in a geo-centric map-like organization, as suggested by Gallistel (1990) and Gould (1986b). No convincing evidence yet establishes that bees are able to determine their position at a new place relative to a desired place other than by dead reckoning, whereby they steer a novel flight route according to the geometric arrangement of landmarks (Wehner 1992, Wehner & Menzel 1990). Instead, they reach partial solutions to cope with problems arising from unexpected displacement by referring to different memories in a hierarchical order.

LEARNING DURING DANCE COMMUNICATION    Information transfer during dance communication has been proven beyond a doubt by using a robot bee to indicate distance and direction of a test location (Michelsen et al 1992). The learning process of the attending bees is not yet well understood. Because dancers do not always feed the attending bees—in particular when they are indicating the locations of pollen, water, resin, and new nest sites (Seeley 1985, von Frisch 1967)—learning cannot be a type of appetitive reward learning. The acoustic pulse emitted by the dancing bee at the moment of correct body position (Michelsen et al 1986) probably provides the necessary signal for the significant phase of the dance. In food foraging, olfactory and gustatory stimuli are exchanged between the dancing and the attending bees to reduce ambiguity regarding the source the dancer is recruiting. These stimuli are indeed learned, because a recruited bee expects, for example, a food source with the odor it had smelled during dance communication (von Frisch 1967). The probability that a recruited bee will be successful grows with the number of attended dances (Mautz 1971). The code for distance and direction need not be learned, but both the dancing and the attending bees must have learned the solar azimuth/time function and be familiar with the location of the hive. The level of complexity of cognitive functions involved in dance communication is controversial (Gould & Gould 1982). There is no convincing evidence to show that attending bees compare different dances before making a decision (Seeley & Towne 1992) or reject flying toward an impossible place (e.g. a feeding station in a lake), as claimed by Gould & Gould (1982).

## APPETITIVE CUE LEARNING AND MEMORY IN FREELY BEHAVING BEES

### Salient Stimuli in Operant Conditioning

Appetitive learning in freely flying bees follows some rules of operant conditioning (Grossmann 1973). Bees on fixed-ratio reinforcement perform twice as many choices as those trained on continuous reinforcement, and bees on fixed-interval schedules respond less than those on fixed-ratio schedules. The stimuli associated with reward differ with respect to their saliency. A few signals that are perceived in other contexts (e.g. polarized and flashing lights, rotating sectors) are not learned, but most signals, even those that do not resemble flower cues, are learned well, although at different rates (Menzel 1990). Color signals are a case in point. Intensity differences are not learned, although they are perceived in spatial vision and optomotor flight control (Lehrer 1994) and phototaxis (Menzel & Greggers 1985). Spectral lights around 400–420 nm (bee UV-blue) are learned fastest; those around 490 nm (bee blue-green) slowest (Menzel 1967). Hue salience reflects an innate predisposition and can be understood as an innate expectancy of reward probability (Giurfa et al 1995). Odor salience is also ranked according to biological significance (Koltermann 1973), but even repellent stimuli or the bee's own sting pheromone are readily learned (Menzel 1990).

## Contiguity

To learn local floral cues (e.g. odor, color) as reward predicting stimuli (conditioned stimuli, CSs), bees need to perceive them at arrival just before they experience the reward (unconditioned stimulus, US) (Gould 1987; Grossmann 1970, 1971; Menzel 1968; Opfinger 1931). The optimal time interval between CS and US is a few seconds in single-trial conditioning (Menzel 1968). However, this interval depends on the training conditions. For example, the CS-US interval can be extended by introducing a secondary reinforcer (e.g. odor) (Grossmann 1971). Bees also learn at departure when they perform characteristic circling flights (Couvillon et al 1991, Gould 1986a, Lehrer 1994, Lehrer & Collett 1994), which resemble those known from orientation flights of young bees at the hive entrance (Buttel-Reepen 1900, Vollbehr 1975) and from sand wasps at their nest sites (Baerends 1941, Tinbergen 1932). The extent of the orientation flights depends on the degree of novelty of the CSs involved and the US. Observatory learning taking place during these hovering and circling flights focuses on surrounding landmarks and increases with flight duration but appears to be independent of the US. Therefore, learning during departure is probably not a form of backward conditioning.

## The Effect of Other CSs

Stimuli predicting an important outcome do not appear in isolation; they appear usually in compounds of several stimuli. In the case of appetitive learning in bees, color and odor are most important and thus were studied intensively by applying paradigms developed in experimental psychology (e.g. blocking, overshadowing, within-compound association, CS pre-exposure) (Bitterman 1988). Experiments with freely behaving bees are, however, complicated by the fact that color is a far distance signal, and odor a close-up signal. Because the bee's settling on a target is usually taken as the behavioral criterion for learning, there is strong bias in favor of the odor effects. The influence of color is hardly quantifiable in color-odor compounds, owing to the uncertainty with which color is perceived, attended to, and chosen. Using color and odor as CSs, Couvillon et al (1983) found no blocking effects, and overshadowing (odor over color) appeared to be an unstable phenomenon. A particular color appears to be associated with the odor if the bee is rewarded with the compound of the two stimuli (indicating within-compound association) (Couvillon & Bitterman 1982, 1988), but it is as yet unknown whether the mechanism at work is second-order conditioning (odor being the primary CS because it is at closer contiguity to the US) or whether color is a context or occasion setting stimulus for the association of odor.

## The Effects of Unpaired USs and CSs

Additional USs not predicted by the CS have a profound effect on conditioning in vertebrates and lead to a reduction of CS acquisition due to a loss of predictive power by the CS (Rescorla 1967). Appetitive USs cannot be presented independently of CSs and contextual stimuli in freely behaving bees. In an attempt to study the effect of unsignaled USs, Abramson and colleagues (Abramson 1986, Abramson & Bitterman 1986) used an aversive conditioning procedure developed by Nunez & Denti (1970), in which the bees are taught to stop sucking sucrose solution to avoid an averse electric stimulus (see also Kirchner et al 1991, Towne & Kirchner 1989). They found stronger aversive learning for unsignaled USs than for USs signaled by vibration or air stream. Unfortunately, these experiments were not developed to the point where it was possible to scrutinize the role of the predictive relation between contextual stimuli, the CSs, and the aversive US. Latent inhibition, the effect of unreinforced CS preexposure, was tested with the same paradigm. Subsequent conditioning was found to be retarded, indicating the possibility that CS experience, uncorrelated with the US, leads to inhibitory learning.

## Aversive Conditioning

In this context one might ask whether aversive learning is of any biological relevance. Bees avoid, or learn to approach from the side, flowers that knock them off with a click of the petals designed for larger pollinators (e.g. Alfalfa flowers). Gould (1988) successfully trained bees to avoid strongly vibrating artificial petals. Spiders spin webs with patterns that appear to attract bees, because they might mistake them for flowers. Bees that manage to escape from them avoid them afterward (Craig 1994a,b; Craig & Ebert 1995). Spiders, for their part, try to compensate for avoidance learning by changing the web pattern (Craig 1994b).

# MEMORY DYNAMICS AND MEMORY CONTENTS

## Time Course of Retention and Memory Consolidation

Long-term memory (LTM) in bees can last for several months, surviving the winter rest of the colony (Lindauer 1963) and, in summer, lasting the life span of a forager (about 2 weeks)—even after only three learning trials (Menzel

1968). A single learning trial initiates time-dependent processes, leading to high retention immediately after the trial, low retention 2–4 min later, consolidation to a high level within the next 10–15 min, and vanishing retention over several days (Erber 1975a,b; Menzel 1968). The early phase depends on the strength of the US and is highly sensitive to extinction and reversal trials (Menzel 1968, 1969, 1979). Retrograde amnesia induced by weak electroconvulsive shocks, cooling, or narcosis is most effective during this early phase (Erber 1976, Menzel 1968). The dynamics of memory are interpreted to reflect memory formation rather than retrieval processes. The model assumes that a transitional short-term memory (STM), which is susceptible to experiential and experimental interference, is followed by a slowly consolidating middle-term memory (MTM), which is more stable in the face of new experience and unaffected by amnestic treatments (Menzel 1983, 1984). Multiple trials lead to LTM, which can be established within less than 2 min, if the trials follow each other quickly. Thus, the transfer from STM to MTM and LTM is both time and event dependent. Memory dynamics in bees resemble many of those features known from studies in mammals and humans (Squire 1987). Correspondence between the early dual time course and the consolidation phase is particularly striking. As in vertebrate studies, it is difficult to distinguish between retrieval and storage phenomena in the expression of memory. In contrast to the interference theory of consolidation (Keppel 1984), modern biological concepts refer to the notion that what is stored can change; thus consolidation is likely to reflect memory formation processes (McGaugh & Herz 1972). The same arguments have been applied to the bee on the basis of physiological data and more tightly controlled behavioral experiments (Hammer & Menzel 1995) (see also below).

# OLFACTORY CONDITIONING: BEHAVIORAL ANALYSIS

## The Olfactory Conditioning Paradigm

PROBOSCIS EXTENSION RESPONSE    Many insects extend their tongues (proboscis) reflexively when the sucrose receptors at the antennae, mouth parts, or tarsae are stimulated. Kuwabara (1957) and Takeda (1961) found that the proboscis extension response (PER) of bees can be conditioned to visual and olfactory stimuli if the bees are allowed to suck sucrose solution following the presentation of these CSs. Odors are associated with sucrose much faster (Menzel et al 1974, Vareschi 1971) than visual stimuli (Masuhr & Menzel 1972). PER strength is expressed by the probability of PER in a group of animals equally treated or in graded response values, such as the number of muscle spike of the muscle M17, the latency or duration of PER or muscle spikes, the number of licking movements of the glossa, and other measures (Rehder 1987, Smith & Menzel 1989a,b).

CLASSICAL CONDITIONING OF PER    Olfactory PER conditioning is a typical case of classical conditioning, with many features known from vertebrate literature. These aspects were reviewed more recently by Bitterman (1988), Menzel (1990), Menzel et al (1991, 1993b), and Hammer & Menzel (1995). A single pairing of the odor as the CS with sucrose as the US changes the PER probability from a spontaneous level of usually ≤10 to a level of ≥60% and multiple trials to an asymptotic level of ≥80%. Timing of CS and US presentation during the conditioning trial is an essential requirement. Forward pairing (CS precedes US by 1–3 s) is most effective. Backward pairing or unpaired CS and US presentations do not change PER probability (Bitterman et al 1983), and an inhibitory component is uncovered by the retardation of acquisition during subsequent forward pairing (Menzel 1990). The inhibitory effect is strongest for intermediate US-CS intervals and low for short and long US-CS intervals (Hellstern & Hammer 1994). Excitatory and inhibitory conditioning are combined in differential conditioning leading to a high (≥80%) PER probability to the CS$^+$ and zero to CS$^-$. Because bees generalize between the CSs, the response to CS$^-$ is initially high and decreases when learning progresses. Conditioned PER also develops in the absence of the US, e.g. in the case of second-order conditioning (Bitterman et al 1983, Menzel 1990). Furthermore, conditioning may be prevented or reduced even with CS-US forward pairing, e.g. in blocking and overshadowing paradigms (Rescorla & Holland 1982). Blocking between two odors was convincingly demonstrated by Smith & Cobey (1994), and overshadowing occurs in olfactory-tactile compound conditioning (Bitterman et al 1983).

NONASSOCIATIVE COMPONENTS OF PER    The appetitive stimulus, sucrose, arouses the animal and transiently sensitizes feeding-related responses, e.g. probability and strength of PER (Menzel et al 1991). Sensitization depends on the number and duration of sucrose stimulations and the site of stimulation (antennae, proboscis, or both as a compound) (Hammer et al 1994; Menzel 1990; Menzel et al 1989, 1991). For example, three compound stimulations are less effective than one. A long proboscis stimulation is more effective than a short one, but a long antennal stimulation is less effective than a short stimulation. These and other results were compiled in a model that assumes that during sucrose stimulation, bees acquire specific properties of the sensitizing stimulus by employing excitatory and inhibitory forms of learning, depending on the site of input, its frequency, and its duration (Hammer et al 1994).

Habituation to multiple sucrose stimulation at the antennae develops quickly for low concentrations of sucrose and slowly for high concentrations (Braun & Bicker 1992). Habituation is restricted to the input side, and stimulation of the contralateral antenna with high concentrations dishabituates the animal. A single session of habituation trials leads to a short-lived effect ($\leq 10$ min), multiple sessions to longer-lasting effects (about 24 h) (Bicker & Hähnlein 1994).

## Variations and Applications of PER Conditioning

AVERSIVE PER CONDITIONING    Bees quickly learn to discriminate between two odors; one of which is paired with sucrose, the other with sucrose plus electric shock (Smith et al 1991). PER probability to the shock-paired odor is not only much lower, but PER to sucrose is withheld or delayed in the context of this odor. The response strategies during PER conditioning appear to differ considerably between individuals, suggesting that aversive conditioning recruits different preparatory responses in different groups of animals, possibly with respect to their genetic background, age, and ethotype (actual duties in the colony).

OLFACTORY DISCRIMINATION AND KIN RECOGNITION BY CHEMICAL CUES    Olfactory PER conditioning has been successfully used to study olfactory discrimination (Smith & Getz 1994, Vareschi 1971) and kin recognition (Getz & Page 1991, Getz et al 1986). Bees can also be trained to nonvolatile chemicals (e.g. cuticular waxes) by touching the antennae with glass rods smeared with these substances. Bees readily learn to distinguish between substances collected from workers of different ages, workers or eggs from the same hive, and animals with a different genetic relationship.

ANOTHER CONDITIONING PARADIGM: ANTENNAL RESPONSE CONDITIONING    The coordinated movement of the antennae in response to visual stimuli, e.g. movement of a striped pattern, can be conditioned appetitively (Erber & Schildberger 1980). A strong bias was found for the direction of movement: Only ventral to dorsal movement simulating an approach flight is associated. Erber and coworkers (Erber et al 1993) developed another paradigm in which the spatial screening movement of the antenna is conditioned. The antenna more frequently probes an area where it experiences a flat obstacle. The memory for the localization of the obstacle lasts several minutes. This paradigm is particularly suitable for the study of modulatory actions of transmitters in the bee brain (Erber & Kloppenburg 1995, Kloppenburg & Erber 1995).

GENETICS OF LEARNING    Behavioral traits, including foraging behavior, have strong genetic components (Fewell & Page 1993, Moritz & Brandes 1987,

Robinson & Page 1989). Analysis of these traits is facilitated by the fact that all workers in a colony are sisters, and the male bees (drones) are haploid. Drones perform equally well in olfactory PER conditioning. Learning, as expressed in PER conditioning and color learning in freely flying bees, has a high heritable component (Bhagavan et al 1994, Brandes 1988), and lines of good or poor learners are selected within one or two generations (Benatar et al 1995, Brandes 1987). Improvement and reduction of learning in the selected lines is independent of the sensory modalities (olfaction, color vision) and of the learning set (PER conditioning, freely flying bees) (Brandes & Menzel 1990). A correlation analysis of the genetic effects of the nonassociative and associative components in PER conditioning reveals that lines of poor learners lack the initial high response rate indicative of US-induced sensitization (Brandes et al 1988). In a recent study, Bhagavan et al (1994) show that genotype influences learning more strongly than age or ethotype.

## OLFACTORY CONDITIONING: NEURAL SUBSTRATES

### The Olfactory Pathway

The axons of ~30,000 chemoreceptors on each antenna project to 156 glomeruli of the antennal lobe (Flanagan & Mercer 1989a), where they synapse with ~4700 local interneurons and ~1000 projection neurons. The projection neurons leave the antennal lobe in three main tracts: (a) the median antennoglomerularis tract (mAGT), which reaches the calyces of the mushroom bodies (mb) first and then the lateral protocerebrum; (b) the lateral AGT (lAGT), which innervates the lateral protocerebrum first and then the calyces; and (c) the medio-lateral AGT (mlAGT), which projects only to the lateral protocerebrum, a region of the brain where all three tracts terminate on descending neurons. A major portion of the mAGT contains acetylcholinesterase (AChE) and ACh-receptors, suggesting that the olfactory input to the mbs is partly cholinergic (Kreissl & Bicker 1989). The mb is formed by 170,000 densely packed local neurons, the Kenyon cells (Mobbs 1982, Witthöft 1967). The dendrites of the Kenyon cells receive inputs from the two projection neuron tracts in the upper part of the calyx, the lip region. Other parts of the calyx receive inputs from other modalities, the collar from visual neurons, the basal ring from visual and olfactory neurons. Output neurons connect the two lobes of the mb, $\alpha$ and $\beta$ lobes, with its own input region, the lateral protocerebrum, the contralateral mb, and many other brain regions (Menzel et al 1994, Rybak & Menzel 1993).

Neurons of the olfactory pathway were examined with respect to their olfactory-coding properties (Flanagan & Mercer 1989b, Gronenberg 1987, Homberg 1984, Sun et al 1993), and a single identified mb-extrinsic neuron, the PE1, was studied extensively in the context of nonassociative and associative learning (Mauelshagen 1993). The response of PE1 to olfactory stimuli does not change in a sensitization protocol with a compound sucrose stimulus applied to antennae and proboscis, but selectively, as a consequence of forward-pairing trials of the conditioned odor and this sucrose stimulus. A single forward pairing leads to a reduction, and multiple trials to an increase, in its CS-evoked response. Both the single- and multiple-trial effects are transient and disappear at longer (>10 min) intervals. Other forms of neural plasticity in the olfactory pathway were described by Erber (1981).

## The US Pathway

Sucrose receptors at the antennae and proboscis project to the subesophagal ganglion and terminate in close apposition to premotor and motorneurons involved in PER (Rehder 1989). In addition, a group of ventral unpaired median (VUM) neurons receive input from the sucrose receptors. One of the neurons, $VUM_{mx1}$, was found to be sufficient to serve the US-reinforcing function when its activity follows the CS at an optimal time interval for forward PER conditioning (Hammer 1993). The axonal arborizations of $VUM_{mx1}$ converge with the olfactory (CS) pathway at three sites, the antennal lobe, the lateral protocerebral lobe, and the mb calyces (lip and basal ring). $VUM_{mx1}$ was depolarized shortly after CS presentation in order to demonstrate that it mediates reinforcement; it thus substituted for the US in a single olfactory trial. A conditioned response to the CS was found after forward pairing, but not after backward pairing, of the CS with the depolarization of $VUM_{mx1}$. The transmitter of the VUM neurons is most likely octopamine, because these neurons stain with an antibody against octopamine (Kreissl et al 1994). Because VUM activity does not activate the reflex pathway, $VUM_{mx1}$ may specifically serve US reinforcement in olfactory PER conditioning via octopamine release; it executes this function in two neuropiles in parallel (antennal lobe, calyx of mb) (Hammer & Menzel 1995).

Knowledge of the neuron representing the US-reinforcing function in PER conditioning provides the opportunity to ask whether $VUM_{mx1}$ exerts properties that may allow the tracing of behaviorally unobservable variables, such as expectancy, attention, and certain forms of stimulus representations governing the associability of stimuli in inhibitory learning, blocking, and second-order conditioning. For example, $VUM_{mx1}$ develops a response to $CS^+$ but not to $CS^-$ in differential conditioning (Hammer 1993). The prolonged response to $CS^+$ may be a neural substrate for second-order conditioning, since it occurs in a neuron that has a reinforcing function. Other properties of $VUM_{mx1}$ have yet to be analyzed.

## *Localization and Dynamics of the Olfactory Memory Trace*

Local cooling of selected parts of the bee brain induces retrograde amnestic effects, which depend on the time interval between the single learning trial and the site of cooling (Erber et al 1980, Menzel et al 1974). The time course of growing resistance to amnesia induced by cooling the calyx region of the mbs resembles that for cooling the whole animal with a half-effect time interval of 3–4 min. Corresponding half-effect intervals are 1–2 min for the antennal lobes and 2-3 min for the α lobes. No amnestic effect occurs when the lateral protocerebral lobe is cooled. These results indicate that the mbs, in particular their input sites, the calyces, are essential structures for the formation of an amnesia-resistant memory trace but that other structures, such as the antennal lobes, participate at an early stage of the memory trace. The role of the mbs in insect olfactory learning is supported by the projection pattern of the $VUM_{mx1}$ neuron (see above) and the finding that structural mb mutations (Heisenberg et al 1985) and chemically ablated mbs in *D. melanogaster* (de Belle & Heisenberg 1994) lead to a total loss of olfactory learning, with negligible or no sensory and motor side effects.

One of the questions arising from these results is whether the mbs are the only substrate for the consolidation of associative olfactory learning. This question can be approached for the honey bee because the neural substrate of the US pathway is known, and its activity in the various neuropiles can be simulated by injecting the presumed transmitter, octopamine (OA), into these neuropiles as a substitute for the US in CS-OA pairing trials. Under these conditions, bees develop a conditioned response to the CS for both the injections into the calyx and the antennal lobe but not into the lateral protocerebrum, indicating that an associative trace is established in either of these two structures and independently from each other (Hammer & Menzel 1994). Another question relates to the problem of whether US-induced sensitization is a requirement for associative learning. Reserpine depletes biogenic amines in the bee brain, as in vertebrate brains, and causes a pronounced reduction in sensitization and conditioning (Braun & Bicker 1992, Menzel et al 1991). Injection of OA into the brain of reserpinized bees prior to conditioning selectively restores conditioning but not sensitization (Menzel et al 1993c), indicating that sensitization is not necessary for associative learning. A third general issue, which can be addressed using PER conditioning, relates to the problem of how the two components of memory, retrievability and memory formation, are connected. Dopamine injected into the brain blocks memory retrieval, but learning is not affected (Menzel et al 1988). Dopamine restores the motor components of PER in reserpinized animals but not the learning-related components (Menzel et al 1993c). Furthermore, satiated animals show neither the sucrose-released reflex nor the sucrose-induced sensitization, but olfactory conditioning is still possible, as indicated by the conditioned response when the animals are hungry (J Klein, M Hammer & V Steffen, personal communication). Thus retrieval and memory formation may be two separable processes in PER conditioning.

Repeated learning trials leads to amnesia resistant long-term memory (LTM) within less than a minute (Menzel & Sugawa 1986). The accelerated transfer into LTM requires associative trials and does not occur with CS-only or US-only trials. The question has been posed whether LTM in bees depends on protein synthesis, as it does in vertebrates (e.g. Squire & Davis 1975) and in long-term facilitation in *Aplysia* sensory neurons (Schacher et al 1988). Injection of cycloheximide directly into the bee brain blocks protein synthesis to a high degree (≥95%), but neither learning nor LTM is reduced. This finding applies to olfactory PER conditioning (Menzel et al 1993a, Wittstock et al 1993) and color learning in freely flying bees (Wittstock & Menzel 1994). LTM was tested up to three days after learning. These results do not exclude the possibility that bees, too, have protein synthesis–dependent forms of LTM, but if they exist, they do not limit long-term performance in conditioned behavior tested so far. In fact, two parallel forms of LTM were found in *D. melanogaster* recently, one lasting about four days and insensitive to protein synthesis inhibition or cooling, and the other lasting longer than four days and sensitive to protein synthesis inhibition and cooling (Tully et al 1994). The latter form of LTM requires de novo gene expression, probably mediated by CREB protein (Yin et al 1994).

Thus, at the circuit level, the neural substrate of olfactory PER conditioning is characterized by two neuropiles (antennal lobe, mb) and three tracts (the m-, ml- and lAGTs). The antennal lobes and mb calyces, but not the lateral protocerebrum, appear to develop an associative memory trace. The mb calyces receive inputs from other sensory modalities, and the mbs are the substrates for intense multisensory integration at the highest level of the insect nervous system (Erber et al 1987, Menzel et al 1994, Mobbs 1982, Rybak & Menzel 1993). Thus these memory traces in the antennal lobes and mbs may store different aspects with respect to the CSs involved—the nonassociative and associative components and the reinforcing effects (e.g. excitatory and inhibitory conditioning).

# CELLULAR AND MOLECULAR SUBSTRATES OF OLFACTORY LEARNING

A first step toward the elucidation of the cellular and molecular mechanisms of learning and memory relates to the identification of first and second transmitters involved and to the potential role of protein kinases and their substrates. The putative first transmitters were identified (CS pathway: ACh; US pathway:

OA), and protein kinases known to play a key role in cellular plasticity were characterized in the bee (Altfelder & Müller 1991, Altfelder et al 1991, Müller & Altfelder 1991).

## Second-Messenger Pathways

ANTENNAL LOBE: MODULATION OF PROTEIN KINASE A BY THE US AND CS   Second-messenger-regulated protein kinases in the antennal lobe (AL) were analyzed to investigate whether they are affected by the US and CS. Whereas application of sucrose to an antenna causes a rapid and transient elevation of PKA activity in the AL, mechanosensory or CS stimulation of an antenna does not affect protein kinase A (PKA) activity (Hildebrandt 1994; Hildebrandt & Müller 1995a,b). Because the transient modulation of the PKA is evoked by the OA-cAMP system but not by other monoamines detected in the AL (Mercer et al 1983) the VUM$_{mx1}$ neuron, a member of an OA cell group (Kreissl et al 1994), could be a mediator. These findings together with those reported above on reserpinized bees suggest that the US- or OA-evoked activation of PKA in the AL may be implicated in mechanisms of appetitive learning at the AL level (Hildebrandt 1994).

MUSHROOM BODIES: SIGNALING CASCADES   The predominant expression of gene products of *D. melanogaster* memory mutants *dunce* and *rutabaga* encoding for a cAMP phosphodiesterase and a $Ca^{2+}$-calmodulin–dependent adenylyl cyclase in the mbs indicates the crucial role played by the cAMP cascade during learning and memory at the mb level (Davis 1993). Although how and which synaptic mechanisms are affected by the cAMP cascade is still unknown, the cAMP pathway and its targets are of the utmost importance in mechanisms of learning in *D. melanogaster* (Drain et al 1991, Yin et al 1994). Biochemical determination of kinase activity in distinct areas of the bee's mbs reveals a two- to fourfold higher activity of PKA, protein kinase C, and $Ca^{2+}$-calmodulin–dependent kinase, compared to other neural tissues. This, taken together with the distinct localization of still uncharacterized substrate proteins of the kinases on the input and output sides of the mb, points to a complex network of different second-messenger cascades implicated in the processing of neuronal signals in the mbs (Hildebrandt 1994; U Müller, unpublished observations).

MODULATION OF CURRENTS IN KENYON CELLS   The intrinsic elements of the mb, the Kenyon cells, are the potential common postsynaptic sites of the CS and US pathway, and thus were examined with respect to their currents and receptors. Whole-cell patch measurements were performed on dissociated, short-term cultured Kenyon cells (Kreissl & Bicker 1992). Five voltage-dependent currents were characterized: a fast inactivating TTX-sensitive Na current, a typical insect Ca current that is insensitive to ω-conotoxin, a rapidly inactivating A-type K current, a charybdotoxin sensitive Ca-activated K current, and a delayed rectifier current (Schäfer et al 1994). Application of γ-amino-η-butyric acid (GABA) evokes a picrotoxin-sensitive chloride current, while ACh activates a nonselective cation current with a high proportion of calcium. The ACh receptor turned out to be very similar to the neural vertebrate receptor (Zhang & Peltz 1990).

Transmitters, known to be involved in learning and memory (Bicker & Menzel 1989), were tested for their modulatory effects on the ACh and the A currents. Histamine, serotonin, and glutamate increases, and noradrenaline decreases, the A current. The transmitters OA, tyramine, dopamine, were found to be ineffective on the A current, and serotonin reduces the ACh-mediated $Ca^{2+}$ current reversibly (Rosenboom et al 1994). In addition, high intracellular $Ca^{2+}$ reduces the ACh current. Modulation of the ACh-mediated $Ca^{2+}$ current by OA is of special interest, since OA is the putative transmitter of the US pathway, which may converge with the cholinergic CS pathway on the dendrites of the Kenyon cells.

Besides patch-clamp recordings, $Ca^{2+}$-fluorescence measurement on Kenyon cells reveals additional evidence of an ACh-mediated $Ca^{2+}$ influx (Bicker & Kreissl 1994, Rosenboom et al 1994). Moreover, the $Ca^{2+}$ measurements are currently used to elucidate the function of transmitter evoked changes in $Ca^{2+}$ level and its cross talk with components of other second-messenger cascades.

TOWARD AN UNDERSTANDING OF CELLULAR PATHWAYS The biochemical analysis of the coupling between transmitters, modulators, and second messengers turns out to be rather complex in the mbs, because (*a*) receptor subtypes exist that differ in their effects on second messengers, (*b*) multiple cross talk occurs between different second-messenger systems, and (*c*) distinct Kenyon cell subpopulations express different components of the second-messenger cascades. Initial evidence of the expression of special OA receptor types in the mbs is indicated by the distinct pharmacology of OA-binding sites in the mbs, compared to other brain areas (Erber et al 1993). In addition to a subclass of OA receptors that stimulates adenylyl cyclase (Evans & Robb 1993), a distinct, separate G-protein coupling has been demonstrated for a cloned *D. melanogaster* OA-tyramine receptor (Saudou et al 1990). While OA is more effective in the elevation of intracellular $Ca^{2+}$ levels, tyramine is more potent in inhibition of adenylate cyclase activity (Robb et al 1994). Although it is not yet possible to relate receptor subtypes and their coupling with G proteins to distinct functional areas within the mbs, biochemical evidence supports the idea of different OA receptor G-protein coupling in the mb's calyces. In contrast to the OA receptors in the AL, which activate adenylyl cyclase, OA has no direct effect on PKA activity in the mbs (Hildebrandt 1994).

No specific hypothesis for the processing of the US and CS in the mbs can yet be presented at the molecular level. Further studies need to take into account the functional differences between Kenyon cells, as indicated by biochemical and immunohistochemical findings (U Müller, unpublished observations).

## Nitric Oxide and Its Potential Functional Role

Nitric oxide (NO) was recently identified as a signaling molecule in the brain of bees (Müller 1994). In the mammalian nervous system the role of the free radical gas NO as a signaling molecule is well documented (Dawson & Snyder 1994, Schuman & Madison 1994a). Apart from its other functions, NO seems to play a role in the modulation of synaptic functions like long-term depression and potentiation and animal learning and in the development and regeneration of the nervous system (Dawson & Snyder 1994, Schuman & Madison 1994a,b). In contrast to conventional transmitters that are restricted to single synapses, NO easily diffuses from its site of production through membranes to act on neighboring targets. Therefore, the spatial distribution of NO within a cellular compartment provides a potential mechanism of modulation of neuropilar volumes, limited by half-life, diffusion constant, and diffusion barriers (Breer & Shepherd 1993).

NO synthase (NOS) in the bee brain has properties similar to those described for the vertebrate NOS. NOS-expressing cells and neuropiles can be reliably identified by NADPH-diaphorase histochemistry (Müller 1994). The localization of NOS in the AL and the lip of the mb calyces, sites of CS and US convergence, suggests that the NO system plays a role in the processing of chemosensory information and possibly also in mechanisms of learning and memory. In the AL, interneurons, which have an integrative function in chemosensory signal processing (see above), are responsible for the high NOS concentration within the glomeruli. Thus, the release of NO from AL interneurons (Müller & Bicker 1994) may modulate lateral mechanisms in neighboring cells within the circuitry of the glomeruli, which contribute to sensory adaptation and habituation (Breer & Shepherd 1993). Local inhibition of the NO-cGMP system in a single AL specifically affects habituation (Hildebrandt et al 1993, Müller & Hildebrand 1995). Repetitive chemosensory stimulation in vivo causes a gradual increase of PKA activity in the AL, mediated by NO activation of a soluble guanylate cyclase (Hildebrandt et al 1994). Thus, in the bee, the NO-cGMP system in the AL is a component of the machinery involved in nonassociative adaptive mechanisms during chemosensory information processing, as supposed by Breer & Shepherd (1993).

Recent findings indicate that inhibition of the NOS in the whole brain interferes with components of long-term associative odor memory (U Müller 1995). Inhibition of the NOS during the CS-US pairing (three trials, intertrial interval 2–6 min) alters neither US-induced sensitization, nor acquisition of a conditioned reponse (CR) to the odor during conditioning, nor memory retrieval up to 12 h after conditioning. However, memory tests 24 h after conditioning reveal that no LTM is formed. Interestingly, inhibition of the NOS during a single CS-US pairing does not affect the CR tested 24 h after conditioning, indicating that LTM requires multiple learning trials (see above). The molecular target of NO action, mediating LTM, is as yet unknown. In both vertebrates and insects, a major effector of NO is the soluble guanylate cyclase (Elphick et al 1993, Schuman & Madison 1994a). The cGMP produced via activation of the NOS can modulate different components of the cAMP cascade, e.g. it can activate PKA (Altfelder & Müller 1991), change the cAMP-levels by modulating phosphodiesterases, and interact with cyclic nucleotide–regulated ion channels (Dawson & Snyder 1994, Schuman & Madison 1994a). As a result, the release of the diffusible messenger NO during CS-US pairing may modulate cAMP-mediated signaling in adjacent neurons, possibly the Kenyon cells. Furthermore, NO represents a suitable transmitter for the synchronous coupling and/or strengthening of populations of neurons synaptically connected within a confined area (Schuman & Madison 1994b) that may also be required for distinct components of memory formation. Such networks oscillate, and NO-mediated changes in oscillations seem to play a role in the neuronal processing of chemosensory information, as recently reported for the mollusc *Limax maximus* (Gelperin 1994). In this context it is tempting to ask whether NO is also implicated in the odor evoked oscillation in subsets of Kenyon cells. Such oscillations were found recently in the locust brain (Laurent & Naraghi 1994, Laurent & Davidowitz 1994), but whether the oscillations are related to olfactory recognition and/or olfactory learning is still unknown.

## CONCLUSIONS

A top-down approach as applied to learning and memory in honeybees provides the opportunity of relating different levels of complexity to each other and of analyzing the rules and mechanisms from the viewpoint of the respective next higher level. Freely flying bees in a controlled environment learn in a way that can be interpreted with reference to the ecological conditions faced by a social animal that adapts its foraging efforts to fast and unpredictable changes in the food market. Olfactory conditioning of harnessed bees exemplifies essential elements of associative learning and, in general, forms a bridge between the systems and the cellular levels of analysis. Intracellular recordings of identified neurons during olfactory conditioning play a key role in this effort.

They allow testing of the assumptions made by modern behavioral theories of associative learning and provide access to cellular and molecular studies, owing to the identification of their transmitters and the peculiarities of the connectivities. Analysis of this intermediate level of complexity is particularly profitable in the bee, because essential neural elements of the associative network are known and can be tested during ongoing learning behavior. In this respect, the honeybee offers unique properties for the building of bridges between the molecular, cellular, neuronal, neuropilar, and behavioral levels of associative learning.

## Literature Cited

Abramson CI. 1986. Aversive conditioning in honeybees (*Apis mellifera*). *J. Comp. Psychol.* 100:108–16

Abramson CI, Bitterman ME. 1986. The US-preexposure effect in honeybees. *Anim. Learn. Behav.* 14:374–79

Altfelder K, Müller U. 1991. Cyclic nucleotide dependent protein kinases in the neural tissue of the honeybee *Apis mellifera*. *Insect Biochem.* 21:487–94

Altfelder K, Müller U, Menzel R. 1991. $Ca^{2+}$/calmodulin- and $Ca^{2+}$/phospholipid-dependent protein kinases in the neural tissue of the honeybee *Apis mellifera*. *Insect Biochem.* 21:479–86

Baerends GP. 1941. Fortpflanzungsverhalten und Orientierung der Grabwespe *Ammophila campestris*. *Tijdschr. Entomol.* 84:71–248

Benatar S, Cobey S, Smith BH. 1995. Selection on a haploid genotype for discrimination learning performance: correlation between drone honeybees (*Apis mellifera*) and their worker progeny. *Insect Behav.* In press

Bhagavan S, Benatar S, Cobey S, Smith BS. 1994. Effect of genotype but not of age or caste on olfactory learning performance in the honey bee, *Apis mellifera*. *Anim. Behav.* 48:1357–69

Bicker G, Hähnlein I. 1994. Long-term habituation of an appetitive reflex in the honey bee. *NeuroReport.* 6:54–56

Bicker G, Kreissl S. 1994. Calcium imaging reveals nicotinic acetylcholine receptors on cultured mushroom body neurons. *J. Neurophysiol.* 71:808–10

Bicker G, Menzel R. 1989. Chemical codes for the control of behaviour in arthropods. *Nature* 337:33–39

Bitterman ME. 1988. Vertebrate-invertebrate comparisons. *NATO ASI Series-Intell. Evol. Biol.* 17:251–75

Bitterman ME, Menzel R, Fietz A, Schäfer S. 1983. Classical conditioning of proboscis extension in honeybees (*Apis mellifera*). *J. Comp. Psychol.* 97:107–19

Bogdany FJ. 1978. Linking of learning signals in honey bee orientation. *Behav. Ecol. Sociobiol.* 3:323–36

Brandes C. 1987. Effects of bidirectional selection on learning behavior in honeybees (*Apis mellifera capensis*). In *Chemistry and Biology of Social Insects*, ed. J Eder, H Rembold, pp. 192–93. München: Peperny

Brandes C. 1988. Estimation of heritability of learning behaviour in honeybees (*Apis mellifera capensis*). *Behav. Genet.* 18:119–32

Brandes C, Frisch B, Menzel R. 1988. Time-course of memory formation differs in honey bee lines selected for good and poor learning. *Anim. Behav.* 36:981–85

Brandes C, Menzel R. 1990. Common mechanisms in proboscis extension conditioning and visual learning revealed by genetic selection in honey bees (*Apis mellifera capensis*). *J. Comp. Physiol. A* 166:545–52

Braun G, Bicker G. 1992. Habituation of an appetitive reflex in the honeybee. *J. Neurophysiol.* 67:588–98

Breer H, Shepherd GM. 1993. Implications of the NO/cGMP system for olfaction. *Trends Neurosci.* 16:5–9

Buchanan GM, Bitterman ME. 1988. Learning in honeybees as a function of amount and frequency of reward. *Anim. Learn. Behav.* 16:247–55

Buttel-Reepen H. 1900. *Sind die Bienen Reflexmaschinen? Experimentelle Beiträge zur Biologie der Honigbiene.* Leipzig: Verlag

Cartwright BA, Collett TS. 1983. Landmark learning in bees—experiments and models. *J. Comp. Physiol.* 151:521–43

Cartwright BA, Collett TS. 1987. Landmark maps for honeybees. *Biol. Cybern.* 57:85–93

Chittka L, Geiger K. 1995. Honeybee long distance orientation in a controlled environment. *Ethology.* 99:117–26

Chittka L, Gumbert A, Kunze J. 1995. Flower constancy in bumble bees and its relationship to memory dynamics and flower colour. *Behav. Ecol. Sociobiol.* Submitted

Collett TS. 1993. Route following and the retrieval of memories in insects. *Comp. Biochem. Physiol. A* 104:709–16

Collett TS, Baron J. 1994. Biological compasses and the coordinate frame of landmark memories in honeybees. *Nature* 368:137–40

Collett TS, Cartwright BA. 1983. Eidetic images in insects: their role in navigation. *Trends Neurosci.* 6:101–5

Couvillon PA, Bitterman ME. 1982. Compound conditioning in honeybees. *J. Comp. Physiol. Psychol.* 96:192–99

Couvillon PA, Bitterman ME. 1984. The overlearning-extinction effect and successive negative contrast in honeybees (*Apis mellifera*). *J. Comp. Psychol.* 98:100–9

Couvillon PA, Bitterman ME. 1988. Compound-component and conditional discrimination of colors and odors by honeybees: further tests of a continuity model. *Anim. Learn. Behav.* 16:67–74

Couvillon PA, Klosterhalfen S, Bitterman ME. 1983. Analysis of overshadowing in honeybees. *J. Comp. Psychol.* 97:154–66

Couvillon PA, Leiato TG, Bitterman ME. 1991. Learning by honeybees (*Apis mellifera*) on arrival at and departure from a feeding place. *J. Comp. Psychol.* 105(2):177–84

Craig CL. 1994a. Limits to learning: effects of predator pattern and colour on perception and avoidance-learning by prey. *Anim. Behav.* 47:1087–99

Craig CL. 1994b. Predator foraging behavior in response to perception and learning by its prey: interactions between orb-spinning spiders and stingless bees. *Behav. Ecol. Sociobiol.* 35:45–52

Craig CL, Ebert K. 1995. Color and pattern in predator-prey interactions: the bright body colors and patterns of tropical orb-spinning spiders attract flower-seeking prey. *Funct. Ecol.* In press

Darwin C. 1876. *Cross and Self Fertilization In The Vegetable Kingdom.* London: Murray

Daumer K. 1956. Reizmetrische Untersuchung des Farbensehens der Bienen. *Z. Vergl. Physiol.* 38:413–78

Daumer K. 1958. Blumenfarben wie sie die Bienen sehen. *Z. Vergl. Physiol.* 41:49–110

Davis RL 1993. Mushroom bodies and Drosophila learning. *Neuron* 11:1–14

Dawson TM, Snyder SH. 1994. Gases as biological messengers: nitric oxide and carbon monoxide in the brain. *J. Neurosci.* 14:5147–59

de Belle JS, Heisenberg M. 1994. Associative odor learning in *Drosophila* abolished by chemical ablation of mushroom bodies. *Science* 263:692–95

Dill M, Wolf R, Heisenberg M. 1993. Visual pattern recognition in *Drosophila* involves retinotopic matching. *Nature* 365:751–53

Drain P, Folkers E, Quinn WG. 1991. cAMP-dependent protein kinase and the disruption of learning in transgenic flies. *Neuron* 6:71–82

Dyer FC, Dickinson JA. 1994. Development of sun compensation by honeybees: how partially experienced bees estimate the sun's course. *Proc. Natl. Acad. Sci. USA* 91:4471–74

Dyer FC, Gould JL. 1981. Honey bee orientation: a backup system for cloudy days. *Science* 214:1041–42

Elphick MR, Green IC, O'Shea M. 1993. Nitric oxide synthesis and action in an invertebrate brain. *Brain Res.* 619:344–46

Erber J. 1975a. The dynamics of learning in the honeybee (*Apis mellifica carnica*). I. The time dependence of the choice reaction. *J. Comp. Physiol.* 99:231–42

Erber J. 1975b. The dynamics of learning in the honeybee (*Apis mellifica carnica*). II. Principles of information processing. *J. Comp. Physiol.* 99:243–55

Erber J. 1976. Retrograde amnesia in honeybees (*Apis mellifera carnica*). *J. Comp. Physiol. Psychol.* 90:41–46

Erber J. 1981. Neural correlates of learning in the honeybee. *Trends Neurosci.* 4:270–72

Erber J, Homberg U, Gronenberg W. 1987. Functional roles of the mushroom bodies in insects. In *Arthropod Brain: Its Evolution, Development, Structure, and Functions,* ed. AP Gupta, pp. 485–511. New York: Wiley

Erber J, Kloppenburg P. 1995. The modulatory effects of serotonin and octopamine in the visual system of the honey bee (*Apis mellifera* L.) I. Behavioral analysis of the motion-sensitive antennal reflex. *J. Comp. Physiol. A* 176:111–18

Erber J, Kloppenburg P, Scheidler A. 1993. Neuromodulation by serotonin and octopamine in the honeybee: behaviour, neuroanatomy and electrophysiology. *Experientia* 49:1073–83

Erber J, Masuhr T, Menzel R. 1980. Localization of short-term memory in the brain of the bee, *Apis mellifera. Physiol. Entomol.* 5:343–58

Erber J, Schildberger K. 1980. Conditioning of an antennal reflex to visual stimuli in bees (*Apis mellifera* L.). *J. Comp. Physiol.* 135:217–25

Evans PD, Robb S. 1993. Octopamine receptor subtypes and their modes of action. *Neurochem. Res.* 18:869–74

Feinsinger P. 1983. Coevolution and pollination. In *Coevolution*, ed. DJ Futuyma, M Slatkin. pp. 282–310. Sunderland, MA: Sinauer

Fewell JH, Page RE Jr. 1993. Genotypic variation in foraging responses to environmental stimuli by honey bees, *Apis mellifera*. *Experientia* 49:1106–12

Flanagan D, Mercer AR. 1989a. An atlas and 3-D reconstruction of the antennal lobes in the worker honey bee, *Apis mellifera* L. (Hymenoptera: *Apidae*). *Int. J. Insect Morphol. Embryol.* 18:145–59

Flanagan D, Mercer AR. 1989b. Morphology and response characteristics of neurones in the deutocerebrum of the brain in the honeybee *Apis mellifera*. *J. Comp. Physiol. A* 164:483–94

Gallistel CR. 1990. *The Organization of Learning*. Cambridge, MA: MIT Press

Gelperin A. 1994. Nitric oxide mediates network oscillations of olfactory interneurons in a terrestrial mollusc. *Nature* 369:61–63

Getz WM, Brückner D, Smith KB. 1986. Conditioning honeybees to discriminate between heritable odors from full and half sisters. *J. Comp. Physiol. A* 159:251–56

Getz WM, Page REJ. 1991. Chemosensory kin-communication systems and kin recognition in honey bees. *Ethology* 87:298–315

Giurfa M, Nunez JA, Chittka L, Menzel R. 1995. Colour preferences of flower-naive honeybees. *J. Comp. Physiol. A.* 177:247–59

Gould JL. 1984. The natural history of honey bee learning. See Marler & Terrace, pp. 149–80

Gould JL. 1986a. Landmark learning by honey bees. *Anim. Behav.* 35:26–34

Gould JL. 1986b. The locale map of honey bees: Do insects have cognitive maps? *Science* 232:861–63

Gould JL. 1987. Timing of landmark learning by honey bees. *Am. Nat.* 130:1–6

Gould JL. 1988. Honey bees store learned flower-landing behaviour according to time of day. *Anim. Behav.* 36:1–5

Gould JL, Gould CG. 1982. The insect mind: physics or metaphysics? In *Animal Mind-Human Mind*, ed. DR Griffin. New York: Springer

Greggers U, Küttner A, Mauelshagen J, Menzel R. 1993. *Proc. Göttingen Neurobiol. Conf. 21st, Göttingen*, p. 841. Stuttgart: Thieme

Greggers U, Menzel R. 1993. Memory dynamics and foraging strategies of honeybees. *Behav. Ecol. Sociobiol.* 32:17–29

Gronenberg W. 1987. Anatomical and physiological properties of feedback neurons of the mushroom bodies in the bee brain. *Exp. Biol.* 46:115–25

Grossmann K. 1970. Erlernen von Farbreizen an der Futterquelle durch Honigbienen während des Anfluges und während des Saugens. *Z. Tierpsychol.* 27:553–62

Grossmann KE. 1971. Belohnungsverzögerung beim Erlernen einer Farbe an einer künstlichen Futterstelle durch Honigbienen. *Z. Tierpsychol.* 29:28–41

Grossmann KE. 1973. Continuous, fixed-ratio, and fixed-interval reinforcement in honey bees. *J. Exp. Anal. Behav.* 20:105–9

Hammer M. 1993. An identified neuron mediates the unconditioned stimulus in associative olfactory learning in honeybees. *Nature* 366:59–63

Hammer M, Braun G, Mauelshagen J. 1994. Food induced arousal and nonassociative learning in honeybees: dependence of sensitization on the application site and duration of food stimulation. *Behav. Neural Biol.* 62:210–23

Hammer M, Menzel R. 1994. Octopamine local injections into the mushroom body calyces and the antennal lobe substitute for the unconditioned stimulus (US) in honeybee olfactory conditioning. *Soc. Neurosci. Abstr.* 20:582

Hammer M, Menzel R. 1995. Learning and memory in the honeybee. *J. Neurosci.* 15(3):1617–30

Heinrich B. 1983. Insect foraging energetics. See Jones & Little 1983, pp. 187–214

Heinrich B. 1984. Learning in invertebrates. See Marler & Terrace, pp. 135–47

Heisenberg M, Borst A, Wagner S, Byers D. 1985. *Drosophila* mushroom body mutants are deficient in olfactory learning. *J. Neurogenetics* 2:1–30

Hellstern F, Hammer M. 1994. *Proc. Göttingen Neurobiol. Conf. 22nd, Göttingen*, 2:827. Stuttgart: Thieme

Hildebrandt H. 1994. *cAMP-abhängige Proteinkinase im Antennallobus der Honigbiene: Regulation bei der Verarbeitung chemosensorischer Signale*. PhD thesis. Freie Univ., Berlin

Hildebrandt H, Müller U. 1995a. Octopamine mediates rapid stimulation of protein kinase A in the antennal lobe of honeybees. *J. Neurobiol.* 27(1):44–50

Hildebrandt H, Müller U. 1995b. PKA activity in the antennal lobe of honeybees is regulated by chemosensory stimulation in vivo. *Brain Res.* 679:281–88

Hildebrandt H, Müller U, Menzel R. 1993. *Proc. Göttingen Neurobiol. Conf. 21st. Göttingen*, 21:8. Stuttgart: Thieme

Hildebrandt H, Müller U, Menzel R. 1994. *Proc. Göttingen Neurobiol. Conf. 22nd, Göttingen*, 22:653. Stuttgart: Thieme

Homberg U. 1984. Processing of antennal information in extrinsic mushroom body neurons of the bee brain. *J. Comp. Physiol. A* 154:825–36

Jones CE, Little RJ, eds. 1983. *Handbook of Experimental Pollination Biology.* New York: Van Nostrand Reinhold

Keppel G. 1984. Consolidation and forgetting theory. In *Memory Consolidation: Psychobiology of Cognition,* ed. H Weingartner, ES Parker, pp. 149–61. Hillsdale, NJ: Erlbaum

Kevan PG, Baker HG. 1983. Insects as flower visitors and pollinators. *Annu. Rev. Entomol.* 28:407–53

Kirchner WH, Dreller C, Towne WF. 1991. Hearing in honeybees: operant conditioning and spontaneous reactions to airborne sound. *J. Comp. Physiol. A* 168:85–89

Kloppenburg P, Erber J. 1995. The modulatory effects of serotonin and octopamine in the visual system of the honey bee (*Apis mellifera* L.). II. Electrophysiological analysis of motion-sensitive neurons in the lobula. *J. Comp. Physiol. A* 176:119–29

Koltermann R. 1971. 24-Std-Periodik in der Langzeiterinnerung an Duft- und Farbsignale bei der Honigbiene. *Z. Vergl. Physiol.* 75: 49–68

Koltermann R. 1973. Rassen-bzw. artspezifische Duftbewertung bei der Honigbiene und ökologische [Adaptation]. *J. Comp. Physiol.* 85:327–60

Kreissl S, Bicker G. 1989. Histochemistry of acetylcholinesterase and immunocytochemistry of an acetylcholine receptor-like antigen in the brain of the honey bee. *J. Comp. Neurol.* 286:71–84

Kreissl S, Bicker G. 1992. Dissociated neurons of the pupal honeybee brain in cell culture. *Neurocytol.* 21:545–56

Kreissl S, Eichmüller S, Bicker G, Rapus J, Eckert M. 1994. Octopamine-like immunoreactivity in the brain and suboesophageal ganglion of the honeybee. *J. Comp. Neurol.* 348:583–95

Kühn A. 1927. Über den Farbensinn der Bienen. *Z. Vergl. Physiol.* 5:762–800

Kuwabara M. 1957. Bildung des bedingten Reflexes von Pavlovs Typus bei der Honigbiene, *Apis mellifica. J. Fac. Sci. Hokkaido Univ. Ser. VI Zool.* 13:458–64

Laurent G, Davidowitz H. 1994. Encoding of olfactory information with oscillating neural assemblies. *Science* 265:1872–75

Laurent G, Naraghi M. 1994. Odorant-induced oscillations in the mushroom bodies of the locust. *J. Neurosci.* 14:2993–3004

Laverty TM. 1994. Costs to foraging bumble bees of switching plant species. *Can. J. Zool.* 72:43–47

Laverty TM, Plowright RC. 1988. Flower handling by bumblebees: a comparison of specialists and generalists. *Anim. Behav.* 36: 733–40

Lehrer M. 1994. Spatial vision in the honeybee: the use of different cues in different tasks. *Vision Res.* 34:2363–85

Lehrer M, Collett TS. 1994. Approaching and departing bees learn different cues to the distance of a landmark. *J. Comp. Physiol. A* 175:171–77

Lindauer M. 1954. Dauertänze im Bienenstock und ihre Beziehung zur Sonnenbahn. *Naturwissenschaften* 41:506–7

Lindauer M. 1959. Angeborene und erlernte Komponenten in der Sonnenorientierung der Bienen. *Z. Vergl. Physiol.* 42:43–62

Lindauer M. 1963. Allgemeine Sinnesphysiologie. Orientierung im Raum. *Fortschr. Zool.* 16:58–140

Lindauer M. 1967. Recent advances in bee communication and orientation. *Annu. Rev. Entomol.* 12:439–70

Lindauer M. 1970. Lernen und Gedächtnis—Versuche an der Honigbiene. *Naturwissenschaften* 57: 463–67

Lunau K. 1991. Innate flower recognition in bumblebees (*Bombus terrestris, B. lucorum; Apidae*): Optical signals from stamens as landing reaction releasers. *Ethology* 88:203–14

Marler P, Terrace H, eds. 1984. *The Biology Of Learning.* Berlin: Springer

Martin H, Lindauer M. 1973. Orientierung im Erdmagnetfeld. *Fortschr. Zool.* 21:211–28

Masuhr T, Menzel R. 1972. Learning experiments on the use of sidespecific information in the olfactory and visual system in the honeybee of (*Apis mellifica*). In *Information Processing in the Visual Systems of Arthropods,* ed. R Wehner. pp. 315–22. Berlin: Springer

Mauelshagen J. 1993. Neural correlates of olfactory learning in an identified neuron in the honey bee brain. *J. Neurophysiol.* 69:609–25

Mautz D. 1971. Der Kommunikationseffekt der Schwänzeltänze bei *Apis mellifica carnica* (Pollm.). *Z. Vergl. Physiol.* 72:197–220

McGaugh JL, Herz MJ. 1972. *Memory Consolidation.* San Francisco: Albion

Menzel R. 1967. Untersuchungen zum Erlernen von Spektralfarben durch die Honigbiene (*Apis mellifica*). *Z. Vergl. Physiol.* 56:22–62

Menzel R. 1968. Das Gedächtnis der Honigbiene für Spektralfarben. I. Kurzzeitiges und langzeitiges Behalten. *Z. Vergl. Physiol.* 60: 82–102

Menzel R. 1969. Das Gedächtnis der Honigbiene für Spektralfarben. II. Umlernen und Mehrfachlernen. *Z. Vergl. Physiol.* 63:290–309

Menzel R. 1979. Behavioral access to short-term memory in bees. *Nature* 281:368–369

Menzel R. 1983. Neurobiology of learning and memory: the honey bee as a model system. *Naturwissenschaften* 70:504–11

Menzel R. 1984. Short-term memory in bees.

In *Primary Neural Substrates of Learning and Behavioral Change,* ed. DL Alkon, I Farley, pp. 259–74. Cambridge: Cambridge Univ. Press

Menzel R. 1985. Learning in honey bees in an ecological and behavioral context. In *Experimental Behavioral Ecology,* ed. B Hölldobler, M Lindauer. pp. 55–74. Stuttgart: Fischer

Menzel R. 1990. Learning, memory, and "cognition" in honey bees. In *Neurobiology of Comparative Cognition,* ed. RP Kesner, DS Olten. pp. 237–92. Hillsdale, NJ: Erlbaum

Menzel R, Durst C, Erber J, Eichmüller S, Hammer M, et al. 1994. The mushroom bodies in the honeybee: from molecules to behavior. In *Neural Basis of Behavioral Adaptations.* Vol. 39. *Fortschritte der Zoologie,* ed. K Schildberger, N Elsner. pp. 81–102. Stuttgart: Fischer

Menzel R, Erber J, Masuhr T. 1974. Learning and memory in the honeybee. In *Experimental Analysis of Insect Behaviour,* ed. L Barton-Browne. pp. 195–217. Berlin: Springer

Menzel R, Gaio UC, Gerberding M, Nemrava EA, Wittstock S. 1993a. Formation of long-term olfactory memory in honeybees does not require protein synthesis. *Naturwissenschaften* 80:380–82

Menzel R, Geiger K, Chittka L, Joerges J, Kunze J, Müller U. 1995a. The knowledge base of bee navigation. *J. Exp. Biol.* In press

Menzel R, Greggers U. 1985. Natural phototaxis and its relationship to colour vision in honeybees. *J. Comp. Physiol.* 157:311–21

Menzel R, Greggers U, Hammer M. 1993b. Functional organization of appetitive learning and memory in a generalist pollinator, the Honey Bee. In *Insect Learning: Ecological and Evolutionary Perspectives,* ed. D Papaj, AC Lewis. pp. 79–125. New York: Chapman & Hall

Menzel R, Hammer M, Braun G, Mauelshagen J, Sugawa M. 1991. Neurobiology of learning and memory in honeybees. In *The Behaviour and Physiology of Bees,* ed. LJ Goodman, RC Fisher. pp. 323–53. Wallingford, UK: CAB Int.

Menzel R, Hammer M, Schneider U, Heyne M, Durst C. 1993c. *Proc. Göttingen Neurobiol. Conf. 21st, Göttingen,* p. 842. Stuttgart: Georg Thieme

Menzel R, Hammer M, Sugawa M. 1989 *Proc. Int. Congr. Neuroethol., 2nd,* p. 221. Stuttgart: Thieme

Menzel R, Joerges J, Müller U. 1995b. *Proc. Göttingen Neurobiol. Conf. 23rd, Göttingen.* p. 31. Stuttgart: Thieme

Menzel R, Michelsen B, Rüffer P, Sugawa M. 1988. Neuropharmacology of learning and memory in honey bees. In *Synaptic Transmission and Plasticity in Nervous Systems,* ed. G Herting, HC Spatz. pp. 335–50. Berlin: Springer

Menzel R, Sugawa M. 1986. Time course of short-term memory depends on associative events. *Naturwissenschaften* 73:564–65

Mercer A, Mobbs PG, Evans PD, Davenport A. 1983. Biogenic amines in the brain of the honey bee, *Apis mellifera. Cell Tissue Res.* 234:655–77

Michelsen A, Andersen BB, Storm J, Kirchner WH, Lindauer M. 1992. How honeybees perceive communication dances, studied by means of a mechanical model. *Behav. Ecol. Sociobiol.* 30:143–50

Michelsen A, Kirchner WH, Lindauer M. 1986. Sound and vibrational signals in the dance language of the honeybee, *Apis mellifera. Behav. Ecol. Sociobiol.* 18:207–12

Mobbs PG. 1982. The brain of the honeybee *Apis mellifera.* I. The connections and spatial organization of the mushroom bodies. *Philos. Trans. R. Soc. London Ser. B.* 298:309–54

Moritz RF, Brandes C. 1987. Behavior genetics of honeybees (*Apis mellifera* L.). In *Neurobiology and Behavior of Honeybees,* ed. R Menzel, A Mercer. pp. 22–35. Berlin: Springer

Müller U. 1994. $Ca^{2+}$/calmodulin-dependent nitric oxide synthase in *Apis mellifera* and *Drosophila melanogaster. Eur. J. Neurosci.* 6:1362–70

Müller U. 1995. Inhibition of NO-synthase significantly impairs long-term memory in the honeybee, *Apis mellifera. Soc. Neurosci. Abstr.* 21:In press

Müller U, Altfelder K. 1991. The $Ca^{2+}$-dependent proteolytic system—calpain-calpastatin—in the neural tissue of the honeybee *Apis mellifera. Insect Biochem.* 21:473–77

Müller U, Bicker G. 1994 Calcium activated release of nitric oxide and cellular distribution of nitric oxide synthesizing neurons in the nervous system of the locust. *J. Neurosci.* 14:7521–28

Müller U, Hildebrandt H. 1995. The nitric oxide/cGMP system in the antennal lobe of *Apis mellifera* is implicated in integrative processing of chemosensory stimuli. *Eur. J. Neurosci.* In press

Nunez JA, Denti A. 1970. Respuesta de abejas recolectoras a un estimulo nociceptivo: habituation and learning in foraging bees. *Acta Physiol. Latinoam.* 20:140–46

Opfinger E. 1931. Über die Orientierung der Biene an der Futterquelle. *Z. Vergl. Physiol.* 15:432–87

Pyke GH. 1984. Optimal foraging theory: a critical review. *Annu. Rev. Ecol. Syst.* 15:523–75

Rehder V. 1987. Quantification of the honeybee's proboscis reflex by electromyographic recordings. *J. Insect Physiol.* 33:501–7

Rehder V. 1989. Sensory pathways and motoneurons of the proboscis reflex in the suboesophageal ganglion of the honey bee. *J. Comp. Neurol.* 279:499–513

Rescorla RA. 1967. Pavlovian conditioning and its proper control procedures. *Psychol. Rev.* 74:71–80

Rescorla RA, Holland PC. 1982. Behavioral studies of associative learning in animals. *Annu. Rev. Psychol.* 33:265–308

Rescorla RA, Wagner AR. 1972. A theory of classical conditioning: variations in the effectiveness of reinforcement and non-reinforcement. In *Classical Conditioning II: Current Research and Theory*, ed. AH Black, WF Prokasy. pp. 64–99. New York: Appleton-Century-Crofts

Robb S, Cheek TR, Hannan FL, Hall LM, Midgley JM, Evans PD. 1994. Agonist-specific coupling of a cloned *Drosophila* octopamine/tyramine receptor to multiple second messenger systems. *EMBO J.* 13: 1325–30

Robinson GE, Page REJ. 1989. Genetic determination of nectar foraging, pollen foraging, and nest-site scouting in honey bee colonies. *Behav. Ecol. Sociobiol.* 24:317–23

Rosenboom H, Goldberg F, Schäfer S, Menzel R. 1994. Ionic currents of Kenyon cells possibly involved in olfactory learning in insects. *Soc. Neurosci. Abstr.* 20:803

Rybak J, Menzel R. 1993. Anatomy of the mushroom bodies in the honey bee brain: the neuronal connections of the alpha-lobe. *J. Comp. Neurol.* 334:444–65

Saudou F, Amlaiky N, Plassat JL, Borrelli E, Hen R. 1990. Cloning and characterization of a *Drosophila* tyramine receptor. *EMBO J.* 9:3611–17

Schacher S, Castellucci VF, Kandel ER. 1988. cAMP evokes long-term facilitation in *Aplysia* sensory neurons that requires new protein synthesis. *Science* 240:1667–69

Schäfer S, Rosenboom H, Menzel R. 1994. Ionic currents of Kenyon cells from the mushroom body of the honeybee. *J. Neurosci.* 14:4600–12

Schmid-Hempel P. 1986. The influence of reward sequence on flight directionality in bees. *Anim. Behav.* 34:831–37

Schuman EM, Madison DV. 1994a. Nitric oxide and synaptic function. *Annu. Rev. Neurosci.* 17:153–83

Schuman EM, Madison DV. 1994b. Locally distributed synaptic potentiation in the hippocampus. *Science* 263:532–36

Seeley TD. 1985. *Honeybee Ecology. A Study of Adaptation in Social Life.* Princeton: Princeton Univ. Press

Seeley TD. 1994. Honey bee foragers as sensory units of their colonies. *Behav. Ecol. Sociobiol.* 34:51–62

Seeley T, Towne WF. 1992. Tactics of dance choice in honey bees: do foragers compare dances? *Behav. Ecol. Sociobiol.* 30:59–69

Smith B, Menzel R. 1989a. The use of electromygram recordings to quantify odourant discrimination in the honey bee, *Apis mellifera*. *J. Insect Physiol.* 35:369–75

Smith B, Menzel R. 1989b. An analysis of variability in the feeding motor program of the honey bee: the role of learning in releasing a modal action pattern. *Ethology* 82:68–81

Smith BH, Abramson CI, Tobin TR. 1991. Conditional withholding of proboscis extension in honeybees (*Apis mellifera*) during discriminative punishment. *J. Comp. Psychol.* 105(4):345–56

Smith BH, Cobey S. 1994. The olfactory memory of the honey bee, *Apis mellifera*. II: Blocking between odorants in binary mixtures. *J. Exp. Biol.* 195:91–108

Smith BH, Getz WM. 1994. Nonpheromonal olfactory processing in insects. *Annu. Rev. Entomol.* 39:351–75

Sprengel CK. 1793. *Das entdeckte Geheimnis der Natur im Bau und in der Befruchtung der Blumen.* Berlin: Vieweg

Squire LR. 1987. *Memory and Brain.* New York/Oxford: Oxford Univ. Press

Squire LR, Davis HP. 1975. Cerebral protein synthesis inhibition and discrimination training: effects of extent and duration of inhibition. *Behav. Biol.* 13:1379–86

Sun XJ, Fonta C, Masson C. 1993. Odour quality processing by bee antennal lobe interneurones. *Chem. Senses* 18:355–77

Sutton RS, Barto AG. 1981. Toward a modern theory of adaptive networks: expectation and prediction. *Psychol. Rev.* 88:135–70

Takeda K. 1961. Classical conditioned response in the honey bee. *J. Insect Physiol.* 6:168–79

Tinbergen N. 1932. Über die Orientierung des Bienenwolfes. *Z. Vergl. Physiol.* 16:305–34

Towne WF, Kirchner WH. 1989. Hearing in honey bees: detection of air-particle oscillations. *Science* 244:686–88

Tully T, Preat T, Boynton SC, Del Vecchio M. 1994. Genetic dissection of consolidated memory in *Drosophila*. *Cell* 79:35–47

Vareschi E. 1971. Duftunterscheidung bei der Honigbiene—Einzelzell-Ableitungen und Verhaltensreaktionen. *Z. Vergl. Physiol.* 75: 143–73

Vollbehr J. 1975. Zur Orientierung junger Honigbienen bei ihrem 1. Orientierungsflug. *Zool. Jahrb. Abt. Allg. Zool. Physiol.* 79:33–69

von Frisch K. 1937. Psychologie der Bienen. *Z. Tierpsychol.* 1:9–21

von Frisch K. 1949. Die Polarisation des Himmelslichtes als orientierender Faktor bei den Tänzen der Bienen. *Experientia* 5:142–48

von Frisch K. 1967. *The Dance Language and*

*Orientation of Bees*. Cambridge, MA: Harvard Univ. Press

von Frisch K, Lindauer M. 1954. Himmel und Erde in Konkurrenz bei der Orientierung der Bienen. *Naturwissenschaften* 41:245–53

Waser NM. 1983. The adaptive nature of floral traits: Ideas and evidence. In *Pollination Biology*, ed. LA Real. pp. 241–85. New York: Academic

Waser NM, Price MV. 1983. Optimal and actual outcrossing in plants, and the nature of plant-pollinator interaction. See Jones & Little, pp. 341–59

Wehner R. 1984. Astronavigation in insects. *Annu. Rev. Entomol.* 29:277–98

Wehner R. 1992. Arthropods. In *Animal Homing*, ed. F Papi. pp. 45–144. London: Chapman & Hall

Wehner R, Menzel R. 1990. Do insects have cognitive maps? *Annu. Rev. Neurosci.* 13:403–14

Wehner R, Wehner S. 1990. Insect navigation: use of maps or Ariadne's thread? *Ethol. Ecol. Evol.* 2:27–48

Witthöft W. 1967. Absolute Anzahl und Verteilung der Zellen im Hirn der Honigbiene. *Z. Morph. Tiere.* 61:160–84

Wittstock S, Kaatz HH, Menzel R. 1993. Inhibition of protein synthesis by cyclohemimide does not affect formation of long-term memory in honey bees after olfactory conditioning. *J. Neurosci.* 13(4):1379–86

Wittstock S, Menzel R. 1994. Color learning and memory in honey bees are not affected by protein synthesis inhibition. *Behav. Neural Biol.* 62:224–29

Yin JCP, Wallach JS, Del Vecchio M, Wilder EL, Zhou H, et al. 1994. Induction of a dominant negative CREB transgene specifically blocks long-term memory in *Drosophila*. *Cell* 79:49–58

Zhang ZW, Peltz P. 1990. Nicotinic acetylcholine receptors in porcine hypophyseal intermediate lobe cells. *J. Physiol.* 422:83–101

Annu. Rev. Neurosci. 1996. 19:405–36

# SYNAPTIC REGULATION OF MESOCORTICOLIMBIC DOPAMINE NEURONS

*Francis J. White*

Neuropsychopharmacology Laboratory, Department of Neuroscience, Finch University of Health Sciences, The Chicago Medical School, 3333 Green Bay Road, North Chicago, Illinois 60063-3095

KEY WORDS:    dopamine neurons, autoreceptors, behavioral sensitization, antipsychotic drugs, drug addiction

## ABSTRACT

The mesocorticolimbic dopamine system is thought to be involved in psychosis and drug addiction. Many intrinsic membrane properties and synaptic regulators have been identified and characterized in the midbrain ventral tegmental dopamine neurons that give rise to this system. Other neuronal properties remain subject to considerable debate. The present review brings together a large literature regarding neurophysiological studies of the synaptic mechanisms thought to regulate the activity of ventral tegmental dopamine neurons. This work is placed within the context of the behavioral responsiveness of midbrain dopamine neurons and the alterations produced in these neurons by repeated treatment with and withdrawal from antipsychotic drugs and drugs of abuse.

## INTRODUCTION

Historically, the mesolimbic dopamine (DA) system was defined as originating in the A10 DA cells of the mesencephalic ventral tegmental area (VTA) and projecting to structures closely associated with the limbic system, most prominently the ventromedial portion of the striatal complex referred to as the nucleus accumbens, but also the olfactory tubercle, septal nucleus, and other limbic-related areas (Dahlström & Fuxe 1964, Ungerstedt 1971). This system was considered to be separate from the nigrostriatal DA system, which originates in the more lateral substantia nigra (A9 DA cell group) and projects to the striatum proper. Identification of the mesocortical DA systems, as well as other quantitatively minor but perhaps qualitatively important ascending DA projections systems (e.g. to the hippocampus, habenula, and amygdala) expanded the concept of the mesolimbic system and led to modified terminolo-

405

gies describing the ascending DA pathways. The mesencephalic dopamine systems were together rechristened the mesotelencephalic system (Moore & Bloom 1978), but subdivisions of the DA system differed, depending upon whether the author(s) considered the nucleus accumbens and olfactory tubercle to be a part of the striatal complex and hence within the mesostriatal DA system (e.g. Lindvall & Björklund 1983), or to be part of the limbic system and thus within the mesocortical (also referred to as mesocorticolimbic or mesolimbocortical) DA system (e.g. Moore & Bloom 1978).

Despite their somewhat overlapping anatomy, the nigrostriatal and mesocorticolimbic DA systems subserve qualitatively distinct aspects of behavior. Current formulations are based upon a wealth of information indicating that the nigrostriatal DA system facilitates stimulus-response coupling through effects on motor preparatory processes ("response set") whereas the mesocorticolimbic (in particular, the mesoaccumbens) DA system facilitates the impact of stimulus-reward associations on behavior—i.e. incentive motivational processes (Robbins & Brown 1990, Robbins & Everitt 1992, Salamone 1994). The mesocorticolimbic DA system has received considerable attention owing to its speculated roles in incentive motivational processes, drug abuse, and schizophrenia (see below).

In this review, I focus on the mesocorticolimbic DA system while attempting to (a) coalesce in vivo and in vitro findings regarding synaptic mechanisms controlling (or modulating) VTA DA neurons, (b) suggest how such mechanisms might be related to motivationally relevant events and their encoded signals within the nervous system, and (c) describe alterations in VTA DA neuronal activity resulting from behaviorally relevant chronic drug treatments. Because detailed historical accounts of this topic and related aspects of mesocorticolimbic function have recently been provided in various other sources (see texts by Kalivas & Nemeroff 1988, Willner & Scheel-Krüger 1991), I concentrate my review primarily on recent major advances in the neurophysiology of synaptic regulation. Similar advances in the neurochemistry of such regulation have also been achieved and comprehensively reviewed elsewhere (Kalivas 1993). I also limit discussion primarily to studies of the VTA DA neuron population. However, the reader should be aware that similar mechanisms and relationships often exist within the substantia nigra.

## ELECTROPHYSIOLOGICAL IDENTIFICATION OF DA NEURONS

Over 20 years ago, Bunney and colleagues demonstrated that rat VTA DA neurons, like their counterparts within the substantia nigra, could be identified by their distinctive electrophysiological "signature" when monitored with extracellular single-unit recording techniques (Bunney et al 1973a,b). Midbrain

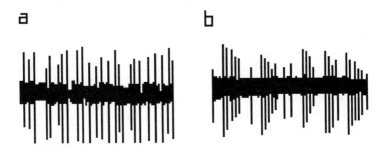

*Figure 1*  Oscilloscope tracings from in vivo extracellular recordings of mouse VTA DA neurons. *a*: Irregular firing pattern; *b*: burst-firing pattern with decreased spike amplitudes within each burst. Calibration bar 0.1 mv, 0.5 s.

DA neurons were distinguished from surrounding non-DA neurons by (*a*) long-duration action potentials (>2.5 ms); (*b*) slow (1–10 Hz) irregular or burst-firing patterns (Figure 1); (*c*) inhibition by low doses of DA agonists (Bunney et al 1973a,b; Wang 1981a) or by microiontophoretic administration of DA (Aghajanian & Bunney 1977); and (*d*) slow conduction velocities (0.5 m/s) due to thin, unmyelinated axons (Guyenet & Aghajanian 1978, Wang 1981a). Pioneering in vivo intracellular recordings combined with histochemical methods later confirmed that midbrain cells with these special characteristics were, in fact, DA-containing neurons (Grace & Bunney 1980, 1983).

In 1980, many investigators turned their attention to detailing the specific characteristics of VTA DA neurons. Three separate investigations demonstrated that rat mesoaccumbens DA neurons, identified by antidromic stimulation, possessed physiological and pharmacological properties identical to those of nigrostriatal A9 DA cells (Deniau et al 1980, German et al 1980, Maeda & Mogenson 1980). In 1981, Wang confirmed and extended these findings, providing the most comprehensive characterization of identified mesoaccumbens and mesocortical DA neurons within the VTA (Wang 1981a,b,c). From this solid foundation, many investigators have built our present knowledge of both intrinsic and synaptic mechanisms regulating and modulating the activity of VTA DA neurons. Before reviewing the mechanisms of intrinsic and synaptic regulation, I briefly describe what is currently known about the relationship between VTA DA cell firing and behavior.

## BEHAVIORAL RELEVANCE OF VTA DA NEURON DISCHARGE

Our understanding of the relationship between the activity of DA neurons and behavior is in its infancy. But as any new parent will attest, infancy is an exciting period of development. Our excitement stems from studies of DA cell activity in unanesthetized rats, cats, and, in particular, monkeys. Midbrain DA neurons are not activated in relation to specific motor acts or general changes in sensory input, but instead respond to stimuli only in specific behavioral contexts (see Schultz 1992 for review). DA neurons in behaving rats are capable of switching firing pattern between irregular and bursting modes (Freeman & Bunney 1987); the latter mode appears to be related to the presentation of behaviorally salient stimuli requiring a response on the part of the animal (Miller et al 1981).

Recent studies by Wolfram Schultz and collaborators, using monkeys required to learn and perform specific responses, have clarified the nature of DA neuron reactivity to the environment. Midbrain DA neurons respond to unconditioned stimuli (UCS), such as novel, unexpected events, with an increase in firing and sometimes with bursts of activity (Schultz 1992). Among the most efficacious unexpected stimuli are primary rewards (food, water), particularly when presented in unpredictable ways (Mirenowicz & Schultz 1994). When previously neutral stimuli (lights, tones, etc) are repeatedly associated with primary rewards such that they become conditioned stimuli (CS), DA neurons now respond to the CS, but not the UCS (Schultz 1992). Stimuli associated with a CS, and to which the animal must pay attention to detect the CS, can also come to activate DA neurons (Schultz & Romo 1990). With repeated performance of specific tasks (overtraining), DA neuron responsiveness to the CS wains in a task-specific manner. Responses to the UCS are not lost, per se, because they are again observed when the UCS is presented outside of the learning task (Mirenowicz & Schultz 1994). These responses are seen with most midbrain DA neurons, although those within the medial substantia nigra and VTA respond more frequently to primary rewards.

Taken together, the available data indicate that VTA DA neurons are particularly involved in reward-driven learning. They respond specifically to salient stimuli that have alerting properties, perhaps allowing the animal to respond to the most significant stimuli in a given context. The coordinated activation of DA neurons would signal to their postsynaptic targets the need for processing the most salient features of a given situation, thereby "focusing" postsynaptic structures on the most behaviorally relevant inputs and "biasing" the selection of appropriate responses to meet the demands of that situation.

## CHARACTERIZATION OF VTA DA NEURONS

Dopamine neurons of the VTA exhibit fusiform or multipolar somata (Figure 2a) with radially projecting dendrites, many of which are particularly enriched in spine-like protrusions; a thin axon typically arises from a major dendrite or somatic appendage (Grace & Onn 1989). When recorded from in vitro brain

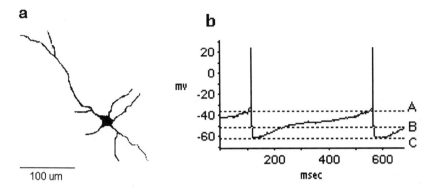

*Figure 2* Morphology and action potentials of mouse DA neurons visualized and recorded from a VTA slice preparation. *a*: Computer reconstruction of a DA neuron stained by Lucifer yellow dye. This cell has a medium-sized, multipolar soma with five clearly identifiable dendrites extending in a radial manner from the cell. *b*: Action potentials recorded in a mouse VTA DA neuron are driven by a slow depolarization from the resting membrane potential (−52 mV) indicated by dashed line B) to the spike threshold of about −36 mV (line A). Action potentials were followed by a prominent afterhyperpolarization that plateaued at −63 mV (line C).

slices or primary cultures, VTA DA neurons exhibit many of the characteristics observed in vivo, including: long-duration action potentials; hyperpolarization and inhibition by DA and DA agonists; lack of response to μ-opioid receptor agonists; a large-amplitude, slow depolarizing potential driving rhythmic spike activity; a relatively depolarized spike threshold; and substantial afterhyperpolarization (Grace & Onn 1989, Johnson & North 1992a, Rayport et al 1992). Unlike the in vivo situation, VTA DA neurons fail to exhibit a burst-firing pattern in vitro; instead they fire in pacemaker fashion (Grace & Onn 1989, Johnson et al 1992b, Rayport et al 1992, Wang & French 1993b). Similar findings have recently been obtained for mouse VTA DA neurons in vitro (see Figure 2*b*). The lack of burst firing in vitro clearly demonstrates the important role of afferent regulation of VTA DA cell firing.

Before reviewing the synaptic mechanisms regulating VTA DA cell activity, I briefly describe the intrinsic physiological properties of DA neurons. For a more detailed account, the interested reader should consult other recent reviews (e.g. Grace & Bunney 1995). Action potential generation in VTA DA neurons in vitro appears to involve a number of events: a voltage-dependent, pacemaker-like slow depolarization apparently involving both $Na^+$ and $Ca^{2+}$ conductances; a low-threshold $Ca^{2+}$-dependent depolarization that is activated during rebound from brief membrane hyperpolarizations; high-threshold $Ca^{2+}$ spikes that appear to be dendritic in origin, and that may trigger a $Ca^{2+}$-activated afterhyperpolarization; and a high-threshold initial segment $Na^+$ spike (Grace & Onn 1989, Rayport et al 1992). VTA DA neuron depolarizations are modulated by several other processes, including a delayed repolarization, in- stantaneous and time-dependent anomalous rectification, afterhyperpolarization, and an electrogenic $Na^+$ pump (Grace & Onn 1989, Johnson & North 1992a, Johnson et al 1992b, Rayport et al 1992, Wang & French 1993b). Modulation of these various conductances may be a primary means by which synaptic inputs alter the firing rates and patterns of VTA DA neurons in vivo. Because of their distinctive physiological properties, as well as important pharmacologic responses (see below), DA neurons of the VTA (principal cells) are readily distinguishable from non-DA neurons (secondary cells).

## AUTOREGULATION OF VTA DA NEURONS

Perhaps the most frequently studied aspect of midbrain DA neurophysiology is the autoregulatory capacity of these cells provided by receptors for their own transmitter, i.e. autoreceptors. Early studies indicated that local administration of DA, and both systemic and local administration of DA agonists, could inhibit the firing of DA neurons, including those in the VTA. This effect was reversed by DA antagonists, indicating the presence of impulse-regulating somatodendritic autoreceptors (Aghajanian & Bunney 1977). Intracellular recordings from VTA DA neurons in vitro demonstrated similar suppression of firing, as well as hyperpolarization, by DA (Mueller & Brodie 1989, Lacey et al 1990b, Johnson & North 1992a). Most VTA DA neurons appear to possess functional DA autoreceptors capable of regulating impulse activity, although mesocortical DA neurons may be an exception (see below). Ultrastructural studies have indicated the presence of DA autoreceptors primarily in distal dendrites of VTA DA neurons (Sesack et al 1994), as well as a relative lack of recurrent axon collaterals (Bayer & Pickel 1990). Thus, autoreceptors are more likely to be activated by DA released from dendrites. By means of both in vitro and in vivo procedures, VTA DA neurons have been demonstrated to release DA from dendrites (Beart et al 1979, Kalivas et al 1989). In addition to autoregulation through DA receptors, VTA DA neurons may also regulate one another electrically via gap junctions (Grace & Onn 1989). This mechanism might provide the basis for coordinated activity of DA neuronal ensembles to amplify normal influences on specific sets of neurons within their terminal domains.

### DA Autoreceptor Subtypes?

Current molecular evidence indicates that CNS DA receptors can best be classified into two subfamilies, $D_1$ and $D_2$ (Civelli et al 1993, Gingrich &

Caron 1993), each made up of multiple subtypes that are included on the basis of sequence homology and ligand selectivity. VTA DA autoreceptors were first classified as the $D_2$ subtype based upon in vivo recordings combined with microiontophoretic administration of receptor-selective compounds (White & Wang 1984b). Intracellular recordings conducted either in brain slices or in culture confirmed this conclusion (Lacey et al 1987, Johnson & North 1992a, Rayport et al 1992). Inhibition of DA neurons via $D_2$ autoreceptor stimulation occurs through pertussis toxin–sensitive G proteins (Innis & Aghajanian 1987, Lacey et al 1988, Liu et al 1994). The relevant action is an increase in whole-cell $K^+$ conductance (Lacey et al 1987, 1988; Chiodo & Kapatos 1992), involving at least three different $K^+$ currents, the transient A current ($I_A$), the delayed rectifier ($I_K$), and an anomolous rectifier, two of which ($I_A$ and $I_K$) utilize a common transduction system involving $G_o$ (Liu et al 1994, Chiodo et al 1995). Preliminary evidence suggests that DA $D_2$ autoreceptors might also decrease specific calcium currents via pertussis toxin–sensitive mechanisms (Chiodo et al 1995).

The discovery of the DA $D_3$ receptor, and its suggested expression by midbrain DA neurons (Bouthenet et al 1991), fueled subsequent claims of autoreceptor function for this $D_2$-class DA receptor. Indeed, reports indicate that the DA $D_3$ receptor can regulate mesolimbic DA synthesis (Nissbrandt et al 1995), release (Rivet et al 1994), and neuronal activity (Lejeune & Millan 1995). All of these positive findings are based on the selectivity of certain drugs for DA $D_3$ vs $D_2$ receptors, as determined by binding affinities in transfected cell lines. But such selectivity may not exist in vivo (Liu et al 1994, Potenza et al 1994), when DA $D_3$ receptors may be "occluded" as a result of tight receptor occupancy by DA (Schotte et al 1992). Functional comparisons between DA $D_2$ and $D_3$ receptors in transfected cells have also failed to support the degree of ligand selectivity suggested by binding affinities (Chio et al 1994, Potenza et al 1994, Gonzalez & Sibley 1995).

More recent in situ hybridization histochemistry studies have found little to no evidence for DA $D_3$ receptor expression by DA-containing (tyrosine hydroxylase positive) neurons within the ventral mesencephalon (Meador-Woodruff et al 1994, Diaz et al 1995). This finding is consistent with our own recent results questioning an autoreceptor role for DA $D_3$ receptors. Using homologous recombination of embryonic mouse stem cells, we have generated mice lacking the gene that encodes the DA $D_3$ receptor (Koeltzow et al 1995). The ability of the putative DA $D_3$ receptor-selective agonist PD128907 to inhibit the firing of VTA DA neurons was identical in the DA $D_3$ receptor "knockout" mice and their wild-type litter mates, clearly indicating that DA $D_2$ receptors are responsible for this effect (Figure 3).

Despite a wealth of evidence to the contrary (see White & Hu 1993 for review), some investigators have suggested that DA $D_1$ receptors might serve autoregu-

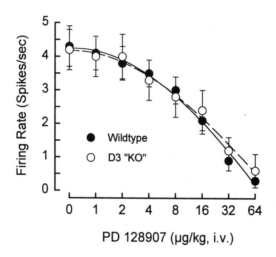

*Figure 3* Effects of the putative DA D3 receptor-selective agonist PD 128907 [(+)-(4aR, 10bR)-3,4,4a,10b-Tetrahydro-4-propyl-2H,5H-[1]benzopyrano-[4,3-b]-1,4-oxazin-9-ol Hcl] on the firing rates of VTA DA neurons in DA D3 receptor "knockout" (KO) mice and their wild-type litter mates. PD 128907 was administered i.v. on a cumulative dose regimen in which each injection equaled the prior total dose (1,1,2,4, etc μg/kg). Injections were separated by 60–90 s.

latory functions (Momiyama et al 1993a,b). However, the responses on which such claims are based more likely reflect DA $D_1$ receptor modulation of other afferent inputs either presynaptically in the VTA or via long-loop feedback mechanisms. DA $D_1$ receptors are quite sparse in the VTA and mRNA for the DA $D_2$ (but not $D_1$) receptor has been identified in VTA DA neurons (e.g. Meador-Woodruff et al 1991, Mansour et al 1995). However, recent recordings of postsynaptic potentials (PSPs) evoked in VTA DA neurons have identified DA $D_1$ receptor regulation of γ-aminobutyric acid (GABA) type B receptor–mediated inhibitory PSPs, apparently occurring via enhanced presynaptic modulation of GABA release (Cameron & Williams 1993). Accordingly, dendritic release of DA would cause stimulation of DA $D_1$ receptors and thereby increase GABA release to exert an inhibitory influence on VTA DA neurons. The extent to which such regulation occurs under normal conditions is questionable because drugs with selectivity for DA $D_1$ receptors fail to influence the activity of VTA DA neurons (White & Wang 1984b; Wachtel et al 1989).

## MESOCORTICAL DA AUTORECEPTORS?

During the early 1980s, a series of studies found evidence for a regional heterogeneity in autoregulation of DA neurons (Bannon & Roth 1983, Chiodo et al 1984). Antidromically identified mesocortical DA neurons were reported to exhibit extremely fast basal firing rates, considerable bursting activity, and complete insensitivity to the inhibitory effects of DA and direct-acting DA receptor agonists (Chiodo et al 1984). We also found significantly faster basal firing rates for mesocortical DA cells, as compared to mesoaccumbens DA cells, and a reduced sensitivity to DA and DA agonists (White & Wang 1984a). We concluded that the mesocortical DA neuron population differed from other VTA DA neurons in that it contained a higher proportion of cells characterized by a relatively low density (or sensitivity) of somatodendritic autoreceptors. However, other studies failed to support these findings, instead demonstrating similar firing rates of different DA neuron subpopulations and similar agonist sensitivities (Shepard & German 1984, Gariano et al 1989). The reasons for these discrepancies need elucidation with more definitive tools. Given the preferential sensitivity of mesocortical DA neurons to both stress and anesthetics (Roth & Elsworth 1995), in vivo electrophysiological studies in anesthetized animals may not provide adequate conditions for resolution of this important issue. General anesthesia can dramatically reduce extracellular glutamate levels (Moghaddam & Bolinao 1994) and thereby decrease firing rates and perhaps alter DA autoreceptor function via heteroregulation (see below), conceivably in such a way as to preferentially alter mesocortical DA neurons. Recent in situ hybridization studies have suggested that VTA DA neurons in humans may express few, if any, DA $D_2$ autoreceptors (Meador-Woodruff et al 1994), suggesting similarities to the mesocortical DA population observed (by some investigators) in rats.

## Physiological Relevance of DA Autoreceptors?

Autoregulation of DA neuron activity is certainly relevant from a pharmacological perspective. But do autoreceptors normally play a role in regulation of neuronal activity? Let us address the issue from the perspective of past DA receptor antagonism studies. If autoreceptors are tonically activated by dendritically released DA, and are thus of normal physiological relevance to spike activity, then autoreceptor antagonism should increase the firing of DA neurons. Early studies of VTA (and nigral) DA neurons found that, unlike systemic administration of DA receptor antagonists, iontophoretic application of such drugs failed to increase firing rates (Aghajanian & Bunney 1977, Wang 1981b). Such findings led to the conclusion that the increase in activity produced by systemic administration of DA receptor antagonists resulted from activation of a long-loop, forebrain feedback pathway. Lesion studies supported this notion for nigral, but not VTA, DA neurons (Iwatsubo & Clouet 1977, Hand et al 1987). The findings regarding nigral DA neurons have recently been questioned because acute transection of forebrain pathways to the nigra failed to diminish the excitatory effects of intravenous DA receptor antagonists (Pucak & Grace 1994). Close scrutiny of previous ionotophoretic work shows that certain DA receptor antagonists (e.g. sulpiride, haloperidol, clozapine) increased the firing of many VTA DA neurons, particularly those with slower basal firing rates (White & Wang 1984b; Hand et al 1987). These and related findings led to our conclusion that the sensitivity of DA neurons to autoreceptor stimulation was not determined by the level of spontaneous activity, but that the level of spontaneous activity was determined by the basal level of autoreceptor activation (White & Wang 1984a). From this perspective, autoreceptor modulation is a critical determinant of DA neuronal activity.

Although mesocortical DA neurons appear relatively deficient in autoregulatory capacity (see above), antidromically identified mesoaccumbens DA neurons might be more tightly regulated by somatodendritic DA autoreceptors than are other midbrain DA neurons. Such speculation has been based on comparisons of the effects of both DA agonists and antagonists on mesoaccumbens and nigrostriatal DA neurons (see White 1991 for review). However, recent studies have found little difference between these DA neuron populations with respect to autoreceptor-mediated effects (Clark & Chiodo 1988) and, in fact, have demonstrated potentially greater density or coupling efficiency of autoreceptors within the nigra (Cox & Waszczak 1990). In vitro studies also suggest little difference between the two groups of DA neurons with respect to autoregulation (Lacey et al 1988, Bowery et al 1994).

## Neuropeptide Autoreceptors?

Many VTA DA neurons in rat brain coexpress the neuropeptides cholecystokinin octapeptide (CCK) and neurotensin (see Deutch & Bean 1995 for recent review). Because these peptides may also be released from dendrites of VTA DA neurons, and because DA neurons express receptors for these peptides, we must consider the possibility of CCK and neurotensin autoreceptors. Indeed, it appears that these neuropeptides may modulate DA autoreceptor function.

CCK excites a subpopulation of midbrain DA neurons, presumably those that coexpress the peptide (Skirboll et al 1981). Such excitation appears to be due to activation of a nonselective cation current (Wu & Wang 1994). Paradoxically, CCK can also potentiate the inhibitory effects of DA and DA agonists on VTA DA neurons through an as yet unidentified mechanism (Hommer & Skirboll 1983, Stittsworth & Mueller 1990). It is currently unclear whether these two effects of CCK occur on the same or different subsets of DA neurons. The excitatory effects of CCK on VTA DA neurons appear to be mediated by the $CCK_A$ receptor (Kelland et al 1991, Wu & Wang 1994), but there is disagreement regarding the relative roles of $CCK_A$ and $CCK_B$

receptors in the potentiation of DA agonist effects (Kelland et al 1991, Meltzer et al 1993).

Like CCK, neurotensin depolarizes and increases the firing rates of VTA DA neurons in vitro (Seutin et al 1989) by inducing a TTX-insensitive, voltage-dependent sodium current (Mercuri et al 1993a). But unlike CCK, neurotensin decreases the inhibitory efficacy of DA autoreceptor stimulation apparently by downregulating the sensitivity of DA autoreceptors through a cAMP/protein kinase A–dependent process (Shi & Bunney 1990,1991,1992).

The potential modulation of VTA DA neuronal activity in general, and of DA autoreceptor function in particular, by coexisting neuropeptides adds further complexity to sites and mechanisms of regulatory control on DA system functioning. However, the extent to which such mechanisms operate under normal conditions is unknown. It is often the case that neuropeptide cotransmitters are released from nerve terminals only under conditions of high neuronal activity. If a similar relationship exists for dendrites, then modulation of DA firing by peptides might occur during burst-firing episodes, which are associated with the introduction of novel, motivationally salient external stimulation. But caution is warranted with respect to the generality of this work because, as compared to rats, humans exhibit considerably less coexpression of either CCK or neurotensin in mesencephalic DA neurons (Palacios et al 1989, Gaspar et al 1990). Thus, autoreceptor modulation by peptide autoreceptors may reflect a species-specific phenomenon of limited relevance to human behavior. Alternatively, it is possible that human DA neurons express receptors for neurotensin and CCK that are innervated by other synaptic or "paracrine" sources of the peptides, leading to similar modulation of transmission.

## AFFERENT REGULATION OF VTA DA NEURONS

Although there have been many studies aimed at identifying the different afferents that project into the VTA and, more specifically, synapse onto DA neurons (see Pickel & Sesack 1995 for recent review), only recently have we begun to understand the synaptic mechanisms that control or modulate the firing of VTA DA neurons. Because VTA DA neurons have only recently become the subject of current- and voltage-clamp analyses, the work is best described as ongoing, and there are many differences of opinion regarding the receptor populations involved in modulation of certain identified membrane currents.

### Excitatory Amino Acids

Anatomical studies have identified projections to the VTA originating primarily in prefrontal cortex (Beckstead 1979, Phillipson 1979). Electrophysiological studies of prefrontal cortex neurons in vivo have identified antidromic responses during stimulation of the VTA (Thierry et al 1983), indicating a direct projection. Lesion studies suggested that the cortical input to VTA might utilize aspartate as a transmitter (Christie et al 1985). An additional excitatory input from the amygdala has been suggested (Wallace et al 1989, Gonzales & Chesselet 1990), although the transmitter may be a neuropeptide rather than an excitatory amino acid (EAA) (Cassell et al 1986). Electrophysiological studies in a slice preparation have also identified an apparent EAA projection from the habenula to the VTA (Matsuda & Fujimura 1992). Although subthalamic nucleus and the pedunculopontine region send EAA projections to the substantia nigra, the extent to which these regions provide excitatory inputs to VTA DA neurons has not been thoroughly investigated. One recent study has described excitation of mesoaccumbens DA neurons following stimulation of the pedunculopontine nucleus, but the responsible transmitter(s) and receptor(s) were not examined (Kelland et al 1993).

One of the most influential lines of recent investigation deals with the likelihood that cortical afferents to the VTA regulate burst firing of DA neurons. Electrical stimulation of the medial prefrontal cortex/anterior cingulate cortex region induces burst firing of DA neurons (Gariano & Groves 1988, Tong et al 1995). Inactivation of this projection by cooling of the cortex or by acute transection reduces burst firing, but not firing rates, of VTA DA neurons, replacing it with pacemaker-like activity (Svensson & Tung 1989, Zhang et al 1992). Similar effects are observed after administration of nonselective glutamate receptor antagonists or selective antagonists of the $N$-methyl-D-aspartate (NMDA) subtype (Charlety et al 1991, Chergui et al 1993). Although local in vivo administration of NMDA to VTA DA neurons induces rate increases and burst firing (Suaud-Chagny et al 1992, White et al 1995), administration of NMDA in vitro was initially reported to increase firing rate without induction of burst firing (Seutin et al 1990, Mercuri et al 1992, Wang & French 1993a). More recent studies have identified NMDA-induced burst firing in vitro, particularly when the bee venom toxin apamin is used to block a $Ca^{2+}$-activated $K^+$ conductance (Shepard & Bunney 1988, 1991; Johnson et al 1992b; Seutin et al 1993a). One recent report also demonstrated burst firing of VTA DA neurons when NMDA was administered for prolonged periods in the absence of apamin (Wang et al 1994), but no such phenomenon was observed in another study (Mercuri et al 1992).

The mechanism underlying NMDA-induced burst firing in vitro has been proposed to involve $Na^+$ influx via NMDA-gated channels. These bursts are terminated by activation of a ouabain-sensitive electrogenic $Na^+$ pump that hyperpolarizes the neuron while extruding the intracellular $Na^+$ (Johnson et al 1992b). One puzzling feature of this work is that the action potentials within bursts did not exhibit clear decreases in amplitude, as are seen in vivo and in other in vitro reports (Shepard & Bunney 1991, Wang et al 1994). Moreover, the proposed mechanism appears quite unlike that reported for induction of burst firing in vivo, which depends upon increased $Ca^{2+}$ conductance (Grace & Bunney 1984). Further work is needed to resolve the mechanisms underlying burst firing of VTA DA neurons, particularly given the important role of this pattern for DA release and for responsiveness to environmental stimuli (see below).

Non-NMDA ionotropic receptors (AMPA/kainate) also contribute to the effects of EAAs on VTA DA neurons. Agonists of non-NMDA receptors increase the firing rates of VTA DA neurons but do not readily induce burst firing (Johnson et al 1992b, Suaud-Chagny et al 1992, White et al 1995). Unlike most CNS neurons, NMDA receptors on VTA DA neurons appear to be selectively activated by low levels of glutamate, whereas non-NMDA receptors appear to mediate excitation of VTA DA neurons only at high concentrations of glutamate (Wang & French 1993b). EAA-mediated EPSPs in VTA DA neurons in vitro involve both NMDA and non-NMDA components. The underlying current declines from peak levels biphasically with two separable time constants, a slow component mediated by NMDA receptors and a fast component mediated by non-NMDA receptors (Mereu et al 1991, Johnson & North 1992a).

VTA DA neurons also appear to express metabotropic glutamate receptors (mGluR). Thus, the mGluR agonist *trans*-1-amino-cyclopentane-1,3-dicarboxylate (*t*-ACPD) depolarizes and increases the firing of midbrain DA neurons in vitro and induces an inward current that is predominantly dependent upon external $Na^+$ (Mercuri et al 1993b). During in vivo recordings, microiontophoretic administration of the active form of *t*-ACPD ([1S,3R]-*t*-ACPD) causes only a slight increase in VTA DA neuron activity (X-F Zhang, X-T Hu, ME Wolf & FJ White, in preparation).

## γ-Aminobutyric Acid

Both biochemical and electrophysiological evidence suggests the existence of a GABAergic pathway from the nucleus accumbens to the VTA (e.g. Walaas & Fonnum 1980, Yim & Mogenson 1980), which might terminate on both DA and non-DA neurons. Although it has not yet been characterized electrophysiologically, a GABAergic projection from the ventral pallidum to the VTA has been described (Zahm 1989). Molecular anatomical studies also suggest that there are both extrinsic and intrinsic sources of GABA within the VTA (see Kalivas 1993, Pickel & Sesack 1995 for recent reviews). GABA-positive nerve terminals primarily synapse onto DAergic dendrites of neurons expressing relatively low levels of tyrosine hydroxylase (Bayer & Pickel 1991). Although the density of GABAergic intrinsic neurons appears relatively low (Bayer &

Pickel 1991), such neurons might provide the anatomical substrate for "paradoxical" excitatory responses of VTA DA neurons to electrical stimulation of GABAergic afferents as well as to systemic administration of agonists (muscimol) and positive modulators (benzodiazepines) of ionotropic $GABA_A$ receptors (Waszczak & Walters 1980, O'Brien & White 1987).

Initial intracellular recordings indicated that midbrain DA neurons possess $GABA_B$ receptors. The selective $GABA_B$ receptor agonist baclofen hyperpolarized and inhibited DA neurons in the VTA slice preparation (Mueller & Brodie 1989). The $GABA_B$ receptors on midbrain DA neurons appear to be linked to the same $K^+$ conductance that is regulated by DA $D_2$ autoreceptors (Innis & Aghajanian 1987, Lacey et al 1988). More recent recordings have studied PSPs generated in VTA DA neurons. In addition to the PSPs elicited by focal stimulation of the slice, spontaneous GABA-mediated PSPs were observed (Johnson & North 1992a; Sugita et al 1992). The spontaneous PSPs occurred in a subpopulation of VTA DA neurons and were mediated by ionotropic $GABA_A$ receptors regulating $Cl^-$ channels. These IPSPs appeared to result from action potentials generated by GABAergic neurons intrinsic to the slice because they were abolished by TTX, reduced by selective suppression of non-DA neuronal activity via $\mu$-opioid receptor activation, and were more readily observable under conditions favorable to action potential generation in the non-DA neurons—i.e. increased extracellular $K^+$ (Johnson & North 1992a; Seutin et al 1993b).

The IPSPs elicited in VTA DA neurons by focal stimulation of the slice exhibited both $GABA_A$ and $GABA_B$ receptor–mediated components (Johnson & North 1992a; Sugita et al 1992). The $GABA_A$ IPSP was isolated by preventing activation of EPSPs mediated by EAA receptors. When the $GABA_A$ potential was also masked by appropriate antagonists, a late hyperpolarizing potential was observed. This slowly rising, late IPSP was reversibly blocked by low-$Ca^{2+}$, high-$Mg^{2+}$ superfusate, and by the $GABA_B$ receptor antagonist phaclofen. This potential resulted from a $K^+$ conductance increase, reversing polarity at about $-105$ mV (Johnson & North 1992a; Sugita et al 1992). This component of the GABA response was not observed spontaneously and was therefore proposed to emanate from extrinsic GABA sources. Indeed, it was hypothesized that synaptic inputs to $GABA_A$ and $GABA_B$ receptors may originate from discrete afferent neurons (Sugita et al 1992).

## Acetylcholine

Surprisingly little attention has been paid to a possible physiological role of acetylcholine (ACh) in regulating VTA DA neurons. Both muscarinic ($M_1$ and $M_2$) and nicotinic ACh receptors are found within the VTA, and the source of cholinergic innervation appears to be the Ch5 pedunculopontine tegmental cholinergic neurons (Woolf 1991), which may colocalize glutamate with ACh (Lavoie & Parent 1994). Electrophysiological studies regarding the effects of locally applied ACh on DA neurons have been confined to the substantia nigra where the results are variable, with some investigators reporting excitation and others no effect (see White 1991 for review). Systemic administration of nicotine, however, excites VTA DA neurons and induces burst firing. The nicotinic ACh receptor (nAChR) antagonist mecamylamine generally produces opposite effects, suggesting a tonic ACh influence (Grenhoff et al 1986, Mereu et al 1987).

Recent in vitro intracellular recordings have demonstrated that ACh and nicotine depolarize and increase the firing of VTA DA neurons (Calabresi et al 1989). The ACh-induced depolarization differed from that caused by nicotine in that it exhibited both transient and slower component phases. The slower component phase was reduced or abolished by scopolamine and mimicked by muscarine, apparently through $M_1$ receptors (Lacey et al 1990a). The inward current modulated by the $M_1$ receptor remains to be identified. The nicotinic response of VTA DA neurons showed marked desensitization and voltage dependence and reversed polarity at about −4 mV; it was reduced by nicotinic antagonists.

## 5-Hydroxytryptamine (Serotonin)

Anatomical evidence indicates that 5-hydroxytryptamine (5-HT) neurons within the raphe nuclei are the source of a dense projection to the VTA (e.g. Vertes 1991) and that 5-HT terminals form primarily symmetric (presumably inhibitory) synaptic contacts with non-DA neurons as well as asymmetric (presumably excitatory) synapses with dendrites of DA neurons (Herve et al 1987, Van Bockstaele et al 1994). In addition, the VTA is relatively rich in 5-HT receptors, particularly those of the $5\text{-HT}_{1B}$ subtype (Pazos & Palacios 1985). Despite the presence of 5-HT terminals and receptors in the VTA, the nature of 5-HT modulation of midbrain DA neurons has been difficult to define and has generated considerable debate.

Early studies of identified DA neurons reported that 5-HT failed to affect DA neuronal activity directly but attenuated glutamate-induced excitation (see White 1991 for review). Electrical stimulation of the dorsal raphe nucleus inhibits mesoaccumbens DA neurons through as yet uncharacterized receptors (Kelland et al 1993). Systemic administration of 5-HT agonists and antagonists produces inconsistent effects on VTA DA neurons. The $5\text{-HT}_{1A}$ selective agent 8-hydroxy-2-(di-n-propylamino)tetralin (8-OH-DPAT) causes a modest increase in firing, and bursting, of a subpopulation of VTA DA neurons (Arborelius et al 1993, Prisco et al 1994), effects that are mimicked by local pressure injection (Arborelius et al 1993), but not by iontophoresis, of the agonist (Prisco et al 1994). In the latter study, the excitatory effects of systemic 8-OH-DPAT were abolished by lesions of the dorsal raphe, implicating disinhibition subsequent to suppression of dorsal raphe 5-HT neurons and suggesting tonic inhibitory serotonergic control over a subpopulation of VTA DA neurons (Prisco et al 1994).

Agonists with selectivity for $5\text{-HT}_{1B}$ receptors fail to influence VTA DA activity, whereas those with affinity for both $5\text{-HT}_{1B}$ and $5\text{-HT}_{1C}$ (now called $5\text{-HT}_{2C}$ (see Hoyer et al 1994) receptors partially suppress firing and bursting of VTA DA neurons (Prisco et al 1994). Antagonists with affinity for both $5\text{-HT}_2$ (now called $5\text{-HT}_{2A}$) and $5\text{-HT}_{2C}$ receptors caused the opposite effect—an observation attributed to the $5\text{-HT}_{2C}$ receptor because the $5\text{-HT}_{2A}$ receptor antagonist ritanserin failed to increase VTA DA firing rates (Prisco et al 1994). However, the selectivity of ritanserin for $5\text{-HT}_{2A}$ over $5\text{-HT}_{2C}$ receptors is questionable (Hoyer et al 1994), and an earlier study showed ritanserin to activate VTA DA neurons and induce burst firing (Ugedo et al 1989), albeit only at relatively high doses. The involvement of $5\text{-HT}_{2C}$ receptors in a tonic inhibitory control of a subpopulation of VTA DA neurons would fit in situ hybridization histochemistry results demonstrating moderate levels of $5\text{-HT}_{2C}$ mRNA in VTA neurons (Molineaux et al 1989).

A different story emerges from studies of 5-HT regulation of VTA DA neurons in vitro. When EAA and $GABA_A$ receptors were blocked, 5-HT decreased the amplitude of the isolated $GABA_B$ synaptic potential (Johnson et al 1992a; Sugita et al 1992). The effect appeared to result from activation of presynaptic $5\text{-HT}_{1B}$ receptors that reduce GABA release (Johnson et al 1992a). 5-HT also caused a slight depolarization in a subset of DA neurons since hyperpolarizing current had to be injected to maintain the resting membrane potential. Given the lack of evidence for synaptic specializations at 5-HT terminals apposed to axon terminals (Van Bockstaele et al 1994), such presynaptic regulation may occur through nonsynaptic means.

## Norepinephrine

Original studies suggested that norepinephrine (NE) hyperpolarizes and inhibits VTA DA cells by stimulating the DA $D_2$ receptor (White & Wang 1984b; Lacey et al 1987). That NE modulates the firing patterns of VTA DA neurons is suggested by more recent studies in which systemic administration of the NE $\alpha_2$ receptor-selective agonist clonidine failed to alter the firing rates of VTA DA neurons but regularized their firing patterns, converting irregular and burst-firing patterns to regular, pacemaker-like activity (Grenhoff & Svensson 1989). Similar results were obtained with the NE $\alpha_1$ receptor–selective antagonist prazosin, suggesting tonic regulation. Thus, the effect of clonidine would result from inhibition of the discharge of NE neurons within the locus coeruleus

(Grenhoff & Svensson 1993). Electrical stimulation studies indicate NE-me-diated excitatory modulation of VTA DA neurons by single-pulse stimulation of locus coeruleus. Unfortunately, the prevalence of burst firing was not addressed in this paper (Grenhoff et al 1993). Ultrastructural evidence indicating the presence of a quantitatively minor tyrosine hydroylase–positive, presumably noradrenergic, termination on tyrosine hydroxylase–positive, presumably dopaminergic, dendrites has been reported, supporting the possible regulatory role of NE afferents on VTA DA neuronal activity (Bayer & Pickel 1990).

## SYNTHESIS

The intrinsic membrane properties of VTA DA  neurons regulate tonic activity and limit the range of firing rates that these cells are capable of exhibiting. In the absence of normal synaptic inputs, most VTA DA neurons are pacemakers (some are not spontaneously active). Accordingly, synaptic inputs to VTA DA neurons act to modify tonic activity, in terms of both firing rates and firing patterns. Although many potential regulatory influences have been suggested for VTA DA neurons, in vitro studies of PSPs suggest that the major inputs use an EAA or GABA. However, the horizontal VTA DA slice preparation may limit which afferents can be effectively stimulated to produce detectable components of the compound PSP. Time will tell whether other components, perhaps mediated by 5-HT, NE, and ACh, might be demonstrable under different in vitro experimental conditions, as they are in vivo.

A primary source of excitatory synaptic influence on VTA DA neurons descends from the prefrontal cortex. The evidence from in vivo and in vitro preparations is complementary, indicating that this EAA projection induces NMDA receptor–mediated burst firing, through mechanisms that are being debated. But NMDA receptor stimulation alone appears insufficient to produce burst firing. Apparently, certain $K^+$ conductance decreases must also occur. Could this be the role of NE inputs from the locus coeruleus? In vivo studies suggest that tonic NE activation of $\alpha_1$ receptors is involved in maintaining the capacity of VTA DA neurons to fire in bursts. Inasmuch as NE $\alpha_1$ receptors have been shown to mediate excitatory effects on neurons by decreasing different $K^+$ conductances (e.g. see Aghajanian 1985), perhaps this is a relevant mechanism. A role for ascending NE regulation would be consistent with behavioral studies indicating responsiveness of VTA DA neurons to incentive stimuli, particularly when presented in novel situations, because the locus coeruleus is activated in states of heightened vigilance (see Valentino & Aston-Jones 1995 for recent review). Although less well understood in terms of behavioral significance and anatomical source, tonic ACh receptor stimulation as a possible modulator of the induction of burst firing should also be considered. Further characterization of extrinsic regulatory mechanisms underlying the induction of burst activity is of considerable practical importance because this firing pattern leads to significantly greater DA release from nerve terminals than would occur with the same number of action potentials in a regularly spaced pattern (Gonon 1988).

It is difficult to visualize clearly how 5-HT modulates VTA DA cell activity, primarily because the many discrepancies in the current literature are difficult to reconcile. Although extracellular recordings suggest a direct, perhaps tonic, inhibitory 5-HT regulation of a subpopulation of VTA DA neurons via 5-HT$_{2C}$ receptors, ultrastructural evidence suggests that direct regulation of mesoaccumbens DA neurons in the VTA occurs at asymmetric synapses typically considered to be excitatory. In addition, neurochemical studies suggest an excitatory influence of 5-HT on mesoaccumbens DA neurons in the VTA (Guan & McBride 1989). Intracellular recordings suggest that 5-HT acts primarily through a presynaptic 5-HT$_{1B}$ receptor to dampen GABA release onto GABA$_B$ receptors expressed by VTA DA neurons. Obviously, these and other forms of synaptic regulation may all occur, perhaps to different extents on different VTA DA neurons. What we need now are answers to questions regarding how and when these different forms of modulation are operable and to what extent they regulate behaviorally meaningful alterations in mesocorticolimbic DA transmission.

Inhibitory afferent input to VTA DA neurons appears to derive primarily from GABA neurons. GABAergic influences on the activity of VTA DA neurons may be either excitatory or inhibitory, depending upon whether they are direct from the forebrain (ventral striatal and pallidal regions) or indirect through forebrain GABAergic influences on other GABAergic (inter)neurons within the VTA. These two sources probably both impinge upon VTA DA neurons, but they may act through separate receptor populations, GABA$_A$ from local neurons and GABA$_B$ from forebrain sources (Sugita et al 1992). It has been proposed that GABAergic influences at B-type receptors oppose EAA induction of burst firing by increasing K$^+$ conductance and hyperpolarizing dendrites (Seutin et al 1994). Supporting this possibility are findings that systemic administration of the GABA$_B$ receptor agonist baclofen markedly regularizes the firing patterns of nigral DA neurons (Engberg et al 1993).

Yet the extent to which GABAergic forebrain feedback to the VTA is operative under normal conditions has not been determined. In anesthetized rats, lesions of the accumbens-VTA pathway fail to alter the firing rates of VTA DA neurons (Einhorn et al 1988). Similarly, systemic administration of GABA$_B$ receptor antagonists fail to alter the firing rates and patterns of DA neurons (Engberg et al 1993). Although the presence of the general anesthetic might contribute to this lack of effect, the influence of the inhibitory accumbens-VTA feedback pathway is clearly demonstrable in anesthetized rats when they are challenged with cocaine (Einhorn et al 1988). Based upon these findings, we previously suggested that forebrain feedback GABAergic mechanisms were not likely to exert significant effects of DA neuronal activity under normal conditions but might be activated during various states of perturbation that require unusual compensatory responses of the mesoaccumbens DA system (Einhorn et al 1988, White 1991).

If long-loop GABAergic regulation via $GABA_B$ receptors is not involved in tonic regulation of VTA DA neurons, what mechanisms are present to offset EAA-driven burst firing? Autoregulation via $D_2$ receptors is the most likely candidate. Activation of NMDA receptors on DA neurons by cortical EAA inputs would induce a period of burst firing. The switching of burst activity back to an irregular pattern, which does occur in unanesthetized rats, would likely result in stimulation of $D_2$ autoreceptors by increased dendritic release of DA. Indeed, NMDA receptor–mediated increases in dendritic DA release have been observed (Westerink et al 1992), as has prevention of NMDA-induced burst firing by DA autoreceptor–mediated hyperpolarization of VTA DA neuronal dendrites (Seutin et al 1994).

## ALTERATIONS IN VTA DA NEURONS PRODUCED BY CHRONIC NEUROLEPTIC TREATMENT

One of the most influential findings regarding DA neuron electrophysiology is the inactivation of DA neuron firing during the course of repeated antipsychotic drug treatments in rats (Bunney & Grace 1978, White & Wang 1983a). Antipyschotic drugs block DA receptors and increase DA turnover (Carlsson & Lindqvist 1963). After rats are treated repeatedly with antipsychotic drugs, the number of spontaneously active DA neurons, detected by conventional extracellular recording techniques, is markedly reduced. With classical neuroleptics, the reduction of active DA neurons occurs in both the nigra and the VTA, whereas with atypical antipsychotic drugs the effect is only observed in the VTA (Chiodo & Bunney 1983, White & Wang 1983b). This effect requires repeated drug administration and demonstrates a clear time-dependency (White & Wang 1983a), indicating a potential role in the delayed onset of therapeutic efficacy observed clinically. This inactivation of DA neurons has also been widely used in an animal model for the prediction of clinical efficacy (VTA effect) and extrapyramidal side-effect liability (nigra effect) of potential antipsychotic drugs.

The mechanism underlying neuroleptic-induced inactivation of DA neurons' firing was proposed to be "depolarization block" due to excess excitation. This conclusion was based on the reversal of inactivity by agents that hyperpolarized the neurons (Bunney & Grace 1978; White & Wang 1983a,b; Chiodo & Bunney 1983). Intracellular recordings conducted in vivo in rats treated chronically with the neuroleptic haloperidol confirmed that DA neurons were "over-depolarized" such that the resting membrane potential was above the threshold for action potential generation, and that it could be "reset" to fire action potentials by direct membrane hyperpolarization (Grace & Bunney 1986). The depolarizing influence appears to involve primarily forebrain feedback pathways because inactivation of nigral and VTA DA neurons during repeated neuroleptic treatment was prevented by lesions of the dorsal and ventral striatum, respectively (Bunney & Grace 1978, White & Wang 1983a).

Despite the widespread use of DA neuron inactivation as a model of chronic antipsychotic drug action, some investigators have questioned the existence of the phenomenon. Their doubt has been based primarily upon neurochemical evidence indicating no decrease in basal striatal DA levels following repeated neuroleptic treatment (e.g. Andén et al 1988). But given that only 50–80% of DA neurons undergo depolarization inactivation, it is not surprising that no decrease in basal DA levels is observed. Compensatory mechanisms are amazingly successful in maintaining normal extracellular levels of DA following marked reductions (up to 95%) in the number of DA neurons—reductions caused, for example, by 6-OHDA lesions (Castañeda et al 1990). This compensation clearly involves increases in firing rate (Hollerman & Grace 1989) and tyrosine hydroxylase activity (Zigmond et al 1986) in surviving DA neurons. Certainly, the most overlooked aspect of our original study on depolarization block was the finding of increased firing rates and markedly altered firing patterns in most of the remaining active DA neurons following repeated administration of antipsychotic drugs (White & Wang 1983a,b). Although burst firing was not precisely quantified at the time, inspection of the interspike interval histogram reported in the original paper, along with retrospective analysis of those neurons exhibiting irregular firing neurons, reveals obvious increases in burst firing, which would contribute to maintenance of normal DA levels despite reduced numbers of active DA neurons. Alternatively, normal extracellular levels of DA might be maintained despite depolarization block by NMDA receptor–mediated control of tonic, impulse-independent DA release at the level of DA nerve terminals, a suggestion that has formed the basis for a new hypothesis of antipsychotic drug action (Grace 1992).

In addition to neurochemical challenges to the depolarization block hypothesis of neuroleptic drug action, recent electrophysiological work has questioned the extent to which the phenomenon is an artifact of general anesthesia (Mereu et al 1995). Unlike Bunney & Grace (1978) and ourselves (FJ White, unpublished findings), Mereu and colleagues observed depolarization block in anesthetized, but not in unanesthetized, rats. They argued that the combination of depolarizing influences engendered by the neuroleptic and the anesthetic led to inactivation. Unfortunately, the experiment crucial to proving this argument was not conducted—namely, would the additional administration of a general anesthetic to a haloperidol-treated, unanesthetized rat put the active DA neurons in a state of quiescence? For now, one could argue that the lack of depolarization block in the Mereu et al (1995) experiments was an artifact of the unanesthetized preparation, perhaps created by the stress of stereotaxic confinement and surgical wounds in the absence of general anesthesia. Although the authors argued that immobilization stress should not have counteracted depolarization block, but instead should have favored depolarization of DA neurons (Mereu et al 1995), their argument was based upon results of neurochemical studies during mild immobilization stress. As was made clear above, assumptions regarding changes in DA neuronal activity that are based solely on DA neurochemistry at nerve terminals are not necessarily accurate. It is certainly possible that severe stress might activate (or inactivate) afferent systems and thus lead to hyperpolarization of DA neurons, thereby counteracting chronic neuroleptic-induced depolarization block. Until more definitive proof is offered one way or the other, the field would be better served if we exercised caution in assigning artifactual status to a phenomenon that has been influential in explaining a previously unknown aspect of antipsychotic drug action (delayed onset), driving drug development strategies, and generating new neurobiological conceptualizations of schizophrenic disorders and their treatment.

## ALTERATIONS IN VTA DA NEURONS PRODUCED BY REPEATED EXPOSURE TO DRUGS OF ABUSE

The positive reinforcing (rewarding) effects of most drugs of abuse, as well as those of natural rewards, are related to activation of the mesocorticolimbic DA system (Koob 1992). An important issue is how such drugs alter the activity of VTA DA neurons during repeated administration and withdrawal. During repeated administration of d-amphetamine or cocaine, DA $D_2$ VTA autoreceptors become subsensitive, which leads to elevations in both the number of spontaneously active VTA DA neurons and their basal firing rates (White & Wang 1984c, Henry et al 1989). These effects last up to a few days after the cessation of treatment, depending upon drug dose and the number and frequency of injections. Transient downregulation of DA $D_2$ VTA autoreceptors, lasting minutes to hours, occurs in vitro following acute administration of d-amphetamine (Seutin et al 1991), suggesting that autoreceptor subsensitivity observed shortly after withdrawal from repeated psychomotor stimulant administration may be an accentuation and prolongation of this phenomenon. The subsensitive DA receptor response is selective inasmuch as sensitivity to GABA is not altered (Henry et al 1989, Seutin et al 1991). The latter finding suggests an alteration at the receptor level. However, repeated cocaine administration also transiently decreases levels of the $\alpha$-subunits of inhibitory G proteins within the VTA (Nestler et al 1990), suggesting that $GABA_B$ and DA $D_2$ receptors may utilize separate G protein pools to regulate $K^+$ conductance. Whatever the case, the transient induction of functional VTA DA autoreceptor subsensitivity and the resulting increase in basal activity of the VTA DA neuron population are important initiators of more persistent neuroadaptations, including enhanced releasability of DA and DA $D_1$ receptor supersensitivity, that are involved in the expression of behavioral sensitization (reverse tolerance) to psychomotor stimulants (Henry et al 1989, Ackerman & White 1990). Indeed, repeated administration of such drugs into the VTA sensitizes rats to behaviors elicited by subsequent systemic administration (Kalivas & Stewart 1991) and causes an important neurochemical correlate of that sensitization within the nucleus accumbens, i.e. increased DA releasability (Vezina 1993).

VTA DA autoreceptor downregulation is not the only means of inducing behavioral sensitization. Repeated morphine injections, administered either systemically or intra-VTA, sensitize rats to the locomotor stimulant effect of this opioid without causing VTA DA autoreceptor subsensitivity (see Jeziorski & White 1995 for discussion). Yet, as with chronic amphetamine and cocaine, repeated morphine causes an increase in the basal activity of the VTA DA neuronal population (Figure 4). Because opioids alter the activity of DA neurons indirectly via inhibition of GABAergic (secondary) cells in the VTA (Johnson & North 1992b), it is not surprising that autoreceptor sensitivity would be unaffected by repeated morphine injections. An increase in the basal activity of VTA DA neurons, whether induced by opioid-dependent mechanisms (morphine) or DA autoreceptor–dependent mechanisms (cocaine, amphetamine), may be a key neuroadaptation leading to the development of behavioral sensitization. It is important to recognize that the increase in DA cell firing rate observed after short withdrawal is the only neuroadaptation identified to date that is common to these three psychomotor stimulants. This might also explain why a wide variety of pharmacological manipulations appear to prevent the development of psychomotor stimulant sensitization (Kalivas & Stewart 1991), because any treatment that prevents the increase in basal dopamine neuronal activity would be expected to prevent sensitization.

Glutamate receptors, including those within the VTA, are also involved in the induction of behavioral sensitization to psychomotor stimulants (Karler et al 1989, Wolf & Jeziorski 1993). Antagonists of NMDA receptor antagonists prevent both behavioral sensitization and the development of VTA DA autoreceptor subsensitivity (Wolf et al 1994), implying a possible heteroregulation of autoreceptor function by EAA receptors. A primary role for EAA neurotransmission in the effects of repeated psychomotor stimulant effects on DA systems is indicated by several recent findings including: prevention of cocaine sensitization by intra-VTA administration of the NMDA receptor antagonist MK-801 (Kalivas & Alesdatter 1993), enhanced responsiveness to glutamate

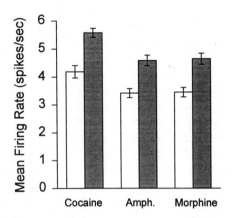

*Figure 4* Short-term withdrawal (24–48 h) from repeated administration of cocaine (14 days, 2 × 10 mg/kg/day), amphetamine (5 days, 5 mg/kg/day) and morphine (14 days, 10 mg/kg/day) increases the basal firing rates of VTA DA neurons in rats. All measures of neuronal rates were taken during population sampling and each measure is from more than 75 DA neurons. Each drug treatment group shows significantly higher firing rates than its appropriate control (repeated saline), as determined with Student's $t$-test ($p < 0.01$). Cocaine results are from Henry et al (1989).

following chronic cocaine or amphetamine treatments (White et al 1995), increased burst firing of VTA DA neurons in response to electrical stimulation of the prefrontal cortex after repeated amphetamine treatment (Tong et al 1995), and prevention of the horizontal locomotion component of amphetamine sensitization by prior lesions of the prefrontal cortex (Wolf et al 1995). Our working model suggests an essential role of cortico-VTA EAA pathways in the development of psychomotor stimulant sensitization. The relevant actions appear to be increased activity of VTA DA neurons as a direct result of enhanced EAA receptor sensitivity and an indirect heteroregulatory modulation of DA autoreceptor sensitivity. The relevant changes in the cortico-VTA system await elucidation.

As detailed above, VTA DA neurons exhibit increased basal activity shortly after the cessation of repeated psychomotor stimulant treatment. During more prolonged withdrawal from repeated exposure to drugs of abuse or, in some cases, shortly after withdrawal from more aggressive treatment regimens, there is a marked decrease in the basal activity of VTA DA neurons and resulting DA release in the nucleus accumbens (Weiss et al 1992, Diana et al 1993). This decrease involves inactivation of a subset of VTA DA neurons and/or a general suppression of their basal firing rates; it has been observed during withdrawal from cocaine (Ackerman & White 1992), morphine (Diana et al 1995), and ethanol (Diana et al 1993, Shen & Chiodo 1993). The inactivation of VTA DA cells during ethanol withdrawal appeared to result from depolarization block because it was reversed by hyperpolarization (Shen & Chiodo 1993). However, others failed to observe this effect, reporting only decrements in basal firing rates and burst firing (Diana et al 1993). The inactivation of VTA DA neurons observed in our cocaine studies did not appear to result from depolarization block (FJ White, unpublished observations). The reduced basal activity of VTA DA neurons during drug withdrawal states is likely to be related to the anhedonia and behavioral depression reported by drug addicts undergoing drug withdrawal. As such, the mechanisms underlying these various drug-induced states of hypoactivity need careful delineation.

## HORIZONS

As the field of DA neurophysiology enters its third decade, expectations for refinements in our understanding of these important neurons are high. We know many of the regulatory inputs to VTA DA neurons, but several questions remain about the relevant receptors and transduction systems linked to such regulation. Considerable debate exists with respect to the nature of tonic vs phasic regulation and, in some cases, to the source of the neurotransmitter afferents. From my perspective, two of the most exciting directions for discovery are 1. the nature and mechanisms of alterations produced by repeated exposure to—and withdrawal from—pharmacological agents, particularly antipsychotic drugs and drugs of abuse, and 2. the relationship between the activity of VTA DA neurons and distinct behavioral states associated with incentive motivation related to drug and other primary reinforcers. Clearly needed are further studies regarding the conditions under which VTA DA neurons exhibit plasticity, the forms of such plasticity, and their underlying mechanisms. Such advancements are on the horizon, and the means to reach them are, by necessity, many and varied. Indeed, a thorough understanding of the roles of these neurons in normal and abnormal behavior is more likely to be achieved if researchers appreciate the continued need for neurophysiological study at all levels of analysis, from single channels studied in isolated cell preparations to identified VTA DA neurons recorded from awake, intact animals behaving both in natural conditions and in environmental contexts requiring learning and execution of specific behavioral responses. We need to learn not only how environmental and pharmacological manipulations alter receptor, transducer,and effector mechanisms, but also how these mechanisms might change behavioral responses to salient stimuli. It seems certain that the increasing sophistication of applicable neurophysiological technologies will allow us to better understand VTA DA neurons and their roles in normal and aberrant behavior.

ACKNOWLEDGMENTS

I extend thanks to all my students and colleagues who participated in the work included in this review. I also thank Dr. Marina E. Wolf for many constructive criticisms of the manuscript. My research program has been generously supported by USPHS grants DA 04093 and MH 40832 and by a Research Scientist Development Award (DA 00207) from the National Institute on Drug Abuse.

## Literature Cited

Ackerman JM, White FJ. 1990. A10 somatodendritic dopamine autoreceptor sensitivity following withdrawal from repeated cocaine treatment. *Neurosci. Lett.* 117:181–87

Ackerman JM, White FJ. 1992. Decreased activity of rat A10 dopamine neurons following withdrawal from repeated cocaine. *Eur. J. Pharmacol.* 218:171–73

Aghajanian GK. 1985. Modulation of a transient outward current in serotonergic neurones by $\alpha_1$ adrenoceptors. *Nature* 315:501–3

Aghajanian GK, Bunney BS. 1977. Dopamine "autoreceptors": pharmacological characterization by microiontophoretic single unit recording studies. *Naunyn Schmiedebergs Arch. Pharmacol.* 297:1–7

Andén N-E, Grenhoff J, Svensson TH. 1988. Does treatment with haloperidol for 3 weeks produce depolarization block in midbrain dopamine neurons of anesthetized rats? *Psychopharmacology* 96:558–60

Arborelius L, Chergui K, Murase S, Nomikos GG, Höök BB, et al. 1993. The 5-HT1A receptor selective ligands, (*R*)-8-OH-DPAT and (*S*)-UH-301, differentially affect the activity of midbrain dopamine neurons. *Naunyn Schmiedebergs Arch. Pharmacol.* 347:353–62

Bannon MJ, Roth RH. 1983. Pharmacology of mesocortical dopamine neurons. *Pharmacol. Rev.* 35:53–68

Bayer VE, Pickel VM. 1990. Ultrastructural localization of tyrosine hydroxylase in the rat ventral tegmental area: relationship between immunolabeling density and neuronal associations. *J. Neurosci.* 10:2996–3013

Bayer VE, Pickel VM. 1991. GABA-labeled terminals form proportionally more synapses with dopaminergic neurons containing low densities of tyrosine hydroxylase-immunore-

activity in rat ventral tegmental area. *Brain Res.* 559:44–55

Beart PM, McDonald D, Gundlach AL. 1979. Mesolimbic dopaminergic neurones and somatodendritic mechanisms. *Neurosci. Lett.* 15:165–70

Beckstead RM. 1979. An autoradiographic examination of corticocortical and subcortical projections of the medial dorsal-projection (prefrontal) cortex in the rat. *J. Comp. Neurol.* 184:43–62

Bloom FE, Kupfer DJ, eds. 1995. *Psychopharmacology: The Fourth Generation of Progress.* New York: Raven

Bouthenet M-L, Souil E, Martres M-P, Sokoloff P, Giros B, Schwartz J-C. 1991. Localization of dopamine D3 receptor mRNA in the rat brain using in situ hybridization histochemistry: comparison with dopamine D2 receptor mRNA. *Brain Res.* 564:203–19

Bowery B, Rothwell LA, Seabrook GR. 1994. Comparison between the pharmacology of dopamine receptors mediating the inhibition of cell firing in rat brain slices through the substantia nigra pars compacta and ventral tegmental area. *Br. J. Pharmacol.* 112:873–80

Bunney BS, Aghajanian GK, Roth RH. 1973a. Comparison of effects of L-DOPA, amphetamine and apomorphine on the firing rate of rat dopaminergic neurons. *Nature* 245:123–25

Bunney BS, Grace AA. 1978. Acute and chronic haloperidol treatment: comparison of effects on nigral dopaminergic cell activity. *Life Sci.* 23:1715–28

Bunney BS, Walters JR, Roth RH, Aghajanian GK. 1973b. Dopaminergic neurons: effect of antipsychotic drugs and amphetamine on single cell activity. *J. Pharmacol. Exp. Ther.* 185:560–71

Calabresi P, Lacey MG, North RA. 1989. Nicotinic excitation of rat ventral tegmental neurones in vitro studied by intracellular recording. *Br. J. Pharmacol.* 98:135–40

Cameron DL, Williams JT. 1993. Dopamine D1 receptors facilitate transmitter release. *Nature* 366:344–47

Carlsson A, Lindqvist M. 1963. Effect of chlorpromazine or haloperidol on formation of 3-methoytyramine and normetanephrine in mouse brain. *Acta Pharmacol. Toxicol.* 20: 140–44

Cassell MD, Gray TS, Kiss JZ. 1986. Neuronal architecture in the rat central nucleus of the amygdala: a cytological, hodological and immunocytochemical study. *J. Comp. Neurol.* 246:478–99

Castañeda E, Whishaw IQ, Robinson TE. 1990. Changes in striatal dopamine neurotransmission assessed with microdialysis following recovery from a bilateral 6-OHDA lesion: variation as a function of lesion size. *J. Neurosci.* 10:1847–54

Charléty PJ, Grenhoff J, Chergui K, De La Chapelle B, Buda M, et al. 1991. Burst firing of mesencephalic dopamine neurons is inhibited by somatodendritic application of kynurenate. *Acta Physiol. Scand.* 142: 105–12

Chergui K, Charléty PJ, Akaoka H, Saunier CF, Brunet J-L, et al. 1993. Tonic activation of NMDA receptors causes spontaneous burst discharge of rat midbrain dopamine neurons in vivo. *Eur. J. Neurosci.* 5:137–44

Chio CL, Lajiness ME, Huff RM. 1994. Activation of heterologously expressed D3 dopamine receptors: comparison with D2 dopamine receptors. *Mol. Pharmacol.* 45: 51–60

Chiodo LA, Bannon MJ, Grace AA, Roth RH, Bunney BS. 1984. Evidence for the absence of impulse-regulating somatodendritic and synthesis-modulating nerve terminal autoreceptors on subpopulations of mesocortical dopamine neurons. *Neuroscience* 12:1–16

Chiodo LA, Bunney BS. 1983. Typical and atypical neuroleptics: differential effects of chronic administration on the activity of A9 and A10 midbrain dopaminergic neurons. *J. Neurosci.* 3:1607–19

Chiodo LA, Freeman AS, Bunney BS. 1995. Dopamine autoreceptor signal transduction and regulation. See Bloom & Kupfer 1995, pp. 221–26

Chiodo LA, Kapatos G. 1992. Membrane properties of identified mesencephalic dopamine neurons in primary dissociated cell culture. *Synapse* 11:294–309

Christie MJ, Bridge S, James LB, Beart PM. 1985. Excitotoxin lesions suggest an aspartatergic projection from rat medial prefrontal cortex to ventral tegmental area. *Brain Res.* 333:169–72

Civelli O, Bunzow JR, Grandy DK. 1993. Molecular diversity of the dopamine receptors. *Annu. Rev. Pharmacol. Toxicol.* 32:281–307

Clark D, Chiodo LA. 1988. Electrophysiological and pharmacological characterization of identified nigrostriatal and mesoaccumbens dopamine neurons in the rat. *Synapse* 2:474–85

Cox RF, Waszczak BL. 1990. Irreversible receptor inactivation reveals differences in dopamine receptor reserve between A9 and A10 dopamine systems: an electrophysiological analysis. *Brain Res.* 534:273–82

Dahlström A, Fuxe K. 1964. Evidence for the existence of monoamine-containing neurons in the central nervous system. I. Demonstration of monoamines in the cell bodies of brain stem neurones. *Acta Physiol. Scand.* 62:1–55

Deniau JM, Thierry AM, Feger J. 1980. Electrophysiological identification of mesencephalic ventromedial tegmental (VMT) neurons projecting to the frontal cortex, septum, and nucleus accumbens. *Brain Res.* 189:315–26

Deutch AY, Bean AJ. 1995. Colocalization in dopamine neurons. See Bloom & Kupfer 1995, pp. 197–206

Diana M, Pistis M, Carboni S, Gessa GL, Rossetti ZL. 1993. Profound decrement of mesolimbic dopaminergic neuronal activity during ethanol withdrawal syndrome in rats: electrophysiological and biochemical evidence. *Proc. Natl. Acad. Sci. USA* 90:5966–69

Diana M, Pistis M, Muntoni A, Gessa G. 1995. Profound decrease of mesolimbic dopaminergic neuronal activity in morphine withdrawn rats. *J. Pharmacol. Exp. Ther.* 272: 781–85

Diaz J, Lévesque D, Lammers CH, Griffon N, Martres M-P, et al. 1995. Phenotypical characterization of neurons expressing the dopamine D3 receptor in the rat brain. *Neuroscience* 65:731–45

Einhorn LC, Johansen PA, White FJ. 1988. Electrophysiological effects of cocaine in the mesoaccumbens dopamine system: studies in the ventral tegmental area. *J. Neurosci.* 8:100–12

Engberg G, Kling-Petersen T, Nissbrandt H. 1993. GABAB-receptor activation alters the firing pattern of dopamine neurons in the rat substantia nigra. *Synapse* 15:229–38

Freeman AS, Bunney BS. 1987. Activity of A9 and A10 dopaminergic neurons in unrestrained rats: further characterization and effects of apomorphine and cholecystokinin. *Brain Res.* 405:46–55

Gariano RF, Groves PM. 1988. Burst firing induced in midbrain dopamine neurons by stimulation of the medial prefrontal and anterior cingulate cortices. *Brain Res.* 462: 194–98

Gariano RF, Tepper JM, Sawyer SF, Young SJ, Groves PM. 1989. Mesocortical dopamine neurons. 1. Electrophysiological properties and evidence for soma-dendritic autoreceptors. *Brain Res. Bull.* 22:511–16

Gaspar P, Berger B, Febvret A. 1990. Neurotensin innervation of the human cerebral cortex: lack of colocalization with catecholamines. *Brain Res.* 530:181–95

German DC, Dalsass M, Kiser RS. 1980. Electrophysiological examination of the ventral tegmental (A10) area in the rat. *Brain Res.* 181:191–97

Gingrich JA, Caron MG. 1993. Recent advances in the moleclar biology of dopamine receptors. *Annu. Rev. Neurosci.* 16:299–321

Gonon FG. 1988. Nonlinear relationship between impulse flow and dopamine released by rat midbrain dopaminergic neurons as studied by in vivo electrochemistry. *Neuroscience* 24:19–28

Gonzales C, Chesselet MF. 1990. Amygdalonigral pathway: an anterograde study in the rat with *Phaseolus vulgaris* leucoagglutinin (PHA-L). *J. Comp. Neurol.* 297:182–200

Gonzalez AM, Sibley DR. 1995. [$^3$H]7-OH-DPAT is capable of labeling dopamine D$_2$ as well as D$_3$ receptors. *Eur. J. Pharmacol.* 272:R1–3

Grace AA. 1992. The depolarization block hypothesis of neuroleptic action: implications for the etiology and treatment of schizophrenia. *J. Neural Transm.* 36:91–131

Grace AA, Bunney BS. 1980. Nigral dopamine neurons: intracellular recording and identification with L-dopa injections and histofluorescence. *Science* 210:654–56

Grace AA, Bunney BS. 1983. Intracellular and extracellular electrophysiology of nigral dopaminergic neurons. 1. Identification and characterization. *Neuroscience* 10:301–15

Grace AA, Bunney BS. 1984. The control of firing pattern in nigral dopamine neurons: burst firing. *J. Neurosci.* 4:2877–90

Grace AA, Bunney BS. 1986. Induction of depolarization block in midbrain dopamine neurons by repeated administration of haloperidol: analysis using in vivo intracellular recording. *J. Pharmacol. Exp. Ther.* 238:1092–1100

Grace AA, Bunney BS. 1995. Electrophysiological properties of midbrain dopamine neurons. See Bloom & Kupfer 1995, pp. 163–77

Grace AA, Onn S-P. 1989. Morphology and electrophysiological properties of immunocytochemically identifed rat dopamine neurons recorded in vitro. *J. Neurosci.* 9:3463–81

Grenhoff J, Aston-Jones G, Svensson TH. 1986. Nicotinic effects of the firing pattern of midbrain dopamine neurons. *Acta Physiol. Scand.* 128:351–58

Grenhoff J, Nisell M, Ferré S, Aston-Jones G, Svensson TH. 1993. Noradrenergic modulation of midbrain dopamine cell firing elicited by stimulation of the locus coeruleus in the rat. *J. Neural Transm.* 93:11–25

Grenhoff J, Svensson TH. 1989. Clonidine modulates dopamine cell firing in rat ventral tegmental area. *Eur. J. Pharmacol.* 165:11–18

Grenhoff J, Svensson TH. 1993. Prazosin modulates the firing pattern of dopamine neurons in rat ventral tegmental area. *Eur. J. Pharmacol.* 233:79–84

Guan XM, McBride WJ. 1989. Serotonin microinfusion into the ventral tegmental increases accumbens dopamine release. *Brain Res. Bull.* 23:541–47

Guyenet PG, Aghajanian GK. 1978. Antidromic identification of dopaminergic and other ouput neurons of the rat substantia nigra. *Brain Res.* 150:69–84

Hand TH, Hu X-T, Wang RY. 1987. Differential effects of acute clozapine and haloperidol on the activity of ventral tegmental (A10) and nigrostriatal (A9) dopamine neurons. *Brain Res.* 415:257–69

Henry DJ, Greene MA, White FJ. 1989. Electrophysiological effects of cocaine in the mesoaccumbens dopamine system: repeated administration. *J. Pharmacol. Exp. Ther.* 251:833–39

Herve D, Pickel VM, Joh TH, Beaudet A. 1987. Serotonin axon terminals in the ventral tegmental area of the rat: fine structure and synaptic input to dopaminergic neurons. *Brain Res.* 435:71–83

Hollerman JR, Grace AA. 1989. Acute haloperidol administration induces depolarization block of nigral dopamine neurons in rats after partial dopamine lesions. *Neurosci. Lett.* 96:82–88

Hommer DW, Skirboll LR. 1983. Cholecystokinin-like peptides potentiate apomorphine-induced inhibition of dopamine neurons. *Eur. J. Pharmacol.* 91:151–52

Hoyer D, Clarke DE, Fozard JR, Hartig PR, Martin GR, et al. 1994. VII. International Union of Pharmacology classification of receptors for 5-hydroxytryptamine (serotonin). *Pharmacol. Rev.* 46:157–203

Innis RB, Aghajanian GK. 1987. Pertussis toxin blocks autoreceptor-mediated inhibition of dopaminergic neurons in rat substantia nigra. *Brain Res.* 411:139–43

Iwatsubo K, Clouet DH. 1977. Effects of morphine and haloperidol on the electrical activity of rat nigrostriatal neurons. *J. Pharmacol. Exp. Ther.* 202:429–36

Jeziorski M, White FJ. 1995. Dopamine receptor antagonists prevent expression, but not development, of morphine sensitization. *Eur. J. Pharmacol.* 275:235–44

Johnson SW, Mercuri NB, North RA. 1992a.

5-Hydroxytryptamine$_{1B}$ receptors block the GABA$_B$ synaptic potential in rat dopamine neurons. *J. Neurosci.* 15:2000–6

Johnson SW, North RA. 1992a. Two types of neurone in the rat ventral tegmental area and their synaptic inputs. *J. Physiol. (London)* 450:455–68

Johnson SW, North RA. 1992b. Opioids excite dopamine neurons by hyperpolarization of local interneurons. *J. Neurosci.* 12:483–88

Johnson SW, Seutin V, North RA. 1992b. Burst firing in dopamine neurons induced by N-methyl-D-aspartate: role of electrogenic sodium pump. *Science* 258:665–67

Kalivas PW. 1993. Neurotransmitter regulation of dopamine neurons in the ventral tegmental area. *Brain Res. Rev.* 18:75–113

Kalivas PW, Alesdatter JE. 1993. Involvement of N-methyl-D-aspartate receptor stimulation in the ventral tegmental area and amygdala in behavioral sensitization to cocaine. *J. Pharmacol. Exp. Ther.* 267:486–95

Kalivas PW, Bourdelais A, Abhold R, Abbott L. 1989. Somatodendritic release of endogenous dopamine: in vivo dialysis in the A10 dopamine region. *Neurosci. Lett.* 100:215–20

Kalivas PW, Nemeroff CB. 1988. *The Mesocorticolimbic Dopamine System.* New York: NY Acad. Sci.

Kalivas PW, Stewart J. 1991. Dopamine transmission in the initiation and expression of drug- and stress-induced sensitization of motor activity. *Brain Res. Rev.* 16:223–24

Karler R, Calder LD, Chaudhry IA, Turkanis SA. 1989. Blockade of "reverse tolerance" to cocaine and amphetamine by MK-801. *Life Sci.* 45:599–606

Kelland MD, Freeman AS, Rubin J, Chiodo LA. 1993. Ascending afferent regulation of rat midbrain dopamine neurons. *Brain Res. Bull.* 31:539–46

Kelland MD, Zhang J, Chiodo LA, Freeman AS. 1991. Receptor selectivity of cholecystokinin effects on mesoaccumbens dopamine neurons. *Synapse* 8:137–43

Koeltzow TE, Cooper DC, Hu X-T, Xu M, Tonegawa S, White FJ. 1995. In vivo effects of dopaminergic ligands in dopamine D$_3$ receptor deficient mice. *Soc. Neurosci. Abstr.* 21:364

Koob GF. 1992. Drugs of abuse: anatomy, pharmacology and function of reward pathways. *Trends Pharmacol. Sci.* 13:177–84

Lacey MG, Calabresi P, North RA. 1990a. Muscarine depolarizes rat substantia nigra zona compacta and ventral tegmental neurons in vitro through M$_1$-like receptors. *J. Pharmacol. Exp. Ther.* 253:395–40

Lacey MG, Mercuri NB, North RA. 1987. Dopamine acts on D$_2$ receptors to increase potassium conductance in neurones of the rat substantia nigra zona compacta. *J. Physiol. (London )* 410:437–53

Lacey MG, Mercuri NB, North RA. 1988. On the potassium conductance increase activated by GABA$_B$ and dopamine receptors in rat substantia nigra neurones. *J. Physiol. (London )* 401:437–54

Lacey MG, Mercuri NB, North RA. 1990b. Actions of cocaine on rat dopaminergic neurones in vitro. *Br. J. Pharmacol.* 99:731–35

Lavoie B, Parent A. 1994. Pedunculopontine nucleus in the squirrel monkey: distribution of cholinergic and monoaminergic neurons in the mesopontine tegmentum with evidence for the presence of glutamate in cholinergic neurons. *J. Comp. Neurol.* 344:190–209

Lejeune F, Millan MJ. 1995. Activation of dopamine D$_3$ autoreceptors inhibits firing of ventral tegmental dopaminergic neurones in vivo. *Eur. J. Pharmacol.* 275:R7–9

Lindvall O, Björklund A. 1983. Dopamine- and norepinephrine-containing neuron systems: their anatomy in the rat brain. In *Chemical Neuroanatomy*, ed. PC Emson, pp 229–55. New York: Raven

Liu J-C, Cox RF, Greif GJ, Freedman JE, Waszczak BL. 1994. The putative dopamine D$_3$ receptor agonist 7-OH-DPAT: lack of mesolimbic selectivity. *Eur. J. Pharmacol.* 264:269–78

Liu L, Shen R-Y, Kapatos G, Chiodo LA. 1994. Dopamine neuron membrane physiology: characterization of the transient outward current ($I_A$) and demonstration of a common signal transduction pathway for $I_A$ and $I_K$. *Synapse* 17:230–40

Maeda H, Mogenson GJ. 1980. An electrophysiological study of inputs to neurons of the ventral tegmental area from the nucleus accumbens and medial preoptic-anterior hypothalamic areas. *Brain Res.* 197:365–77

Mansour A, Meador-Woodruff JH, Bunzow JR, Civelli O, Akil H, Watson SJ. 1995. Localization of dopamine D$_2$ receptor mRNA and D$_1$ and D$_2$ receptor binding in the rat brain and pituitary: an in situ hybridization-receptor autoradiographic analysis. *J. Neurosci.* 10:2587–600

Matsuda Y, Fujimura K. 1992. Action of habenular efferents on ventral tegmental area neurons studied in vitro. *Brain Res. Bull.* 28:743–49

Meador-Woodruff JH, Damask SP, Watson SJ Jr. 1994. Differential expression of autoreceptors in the ascending dopamine systems of the human brain. *Proc. Natl. Acad. Sci. USA* 91:8297–301

Meador-Woodruff JH, Mansour A, Healy DJ, Kuehn R, Zhou Q-Y, et al. 1991. Comparison of the distributions of D$_1$ and D$_2$ dopamine receptor mRNAs in rat brain. *Neuropsychopharmacology* 5:231–42

Meltzer LT, Christoffersen CL, Serpa KA, Razmpour A. 1993. Comparison of the effects of the cholecystokinin-B receptor antagonist, PD 134308, and the cholecystokinin-A receptor antagonist, L-364,718, on dopamine neuronal activity in the substantia nigra and ventral tegmental area. *Synapse* 13:117–22

Mercuri NB, Stratta F, Calabresi P, Bernardi G. 1992. A voltage-clamp analysis of NMDA-induced responses on dopaminergic neurons of the rat substantia nigra zona compacta and ventral tegmental area. *Brain Res.* 593:51–56

Mercuri NB, Stratta F, Calabresi P, Bernardi G. 1993a. Neurotensin induces an inward current in rat mesencephalic dopaminergic neurons. *Neurosci. Lett.* 153:192–96

Mercuri NB, Stratta F, Calabresi P, Bonci A, Bernardi G. 1993b. Activation of metabotropic glutamate receptors induces an inward current in rat dopamine mesencephalic neurons. *Neuroscience* 56:399–407

Mereu G, Costa E, Armstrong DM, Vicini S. 1991. Glutamate receptor subtypes mediate excitatory synaptic currents of dopamine neurons in midbrain slices. *J. Neurosci.* 11:1359–66

Mereu G, Lilliu V, Vargiu P, Muntoni AL, Diana M, Gessa GL. 1995. Depolarization inactivation of dopamine neurons: an artifact. *J. Neurosci.* 15:1144–49

Mereu G, Yoon K-WP, Boi V, Gessa GL, Naes L, Westfall TC. 1987. Preferential stimulation of ventral tegmental area dopaminergic neurons by nicotine. *Eur. J. Pharmacol.* 141:395–99

Miller JD, Sanghera MK, German DC. 1981. Mesencephalic dopaminergic activity in the behaviorally conditioned rat. *Life Sci.* 29:1255–65

Mirenowicz J, Schultz W. 1994. Importance of unpredictability for reward responses in primate dopamine neurons. *J. Neurophysiol.* 72:1024–27

Moghaddam B, Bolinao ML. 1994. Glutamatergic antagonists attenuate ability of dopamine uptake blockers to increase extracellular levels of dopamine: implications for tonic influence of glutamate on dopamine release. *Synapse* 18:337–42

Molineaux SM, Jessel TM, Axel R, Julius D. 1989. 5-HT$_{1C}$ receptor is a prominent serotonin receptor subtype in the central nervous system. *Proc. Natl. Acad. Sci. USA* 86:6793–97

Momiyama T, Sasa M, Takaori S. 1993a. Enhancement of D$_2$ receptor agonist–induced inhibition by D$_1$ receptor agonist in the ventral tegmental area. *Br. J. Pharmacol.* 110:713–18

Momiyama T, Todo N, Sasa M. 1993b. A mechanism underlying dopamine D$_1$ and D$_2$ receptor–mediated inhibition of dopaminergic neurones in the ventral tegmental area in vitro. *Br. J. Pharmacol.* 109:933–40

Moore RY, Bloom FE. 1978. Central catecholamine neuron systems: anatomy and physiology of the dopamine systems. *Annu. Rev. Neurosci.* 1:129–69

Mueller AL, Brodie MS. 1989. Intracellular recording from putative dopamine-containing neurons in the ventral tegmental area of Tsai in a brain slice preparation. *J. Neurosci. Methods* 28:15–22

Nestler EJ, Terwilliger RZ, Walker JR, Sevarino KA, Duman RS. 1990. Chronic cocaine treatment decreases levels of the G protein subunits G$_{i\alpha}$ and G$_{o\alpha}$ in discrete regions of rat brain. *J. Neurochem.* 55:1079–82

Nissbrandt H, Ekman A, Eriksson E, Heilig M. 1995. Dopamine D$_3$ receptor antisense influences dopamine synthesis in rat brain. *NeuroReport* 6:573–76

O'Brien DP, White FJ. 1987. Inhibition of non-dopamine cells in the ventral tegmental area by benzodiazepines: relationship to A10 dopamine cell activity. *Eur. J. Pharmacol.* 142:343–54

Palacios JM, Savasta M, Mengod G. 1989. Does cholecystokinin colocalize with dopamine in the human substantia nigra? *Brain Res.* 488:369–75

Pazos A, Palacios JM. 1985. Quantitative autoradiographic mapping of serotonin receptors in rat brain. I. Serotonin-1 receptors. *Brain Res.* 346:205–30

Phillipson OT. 1979. Afferent projections to the ventral tegmental area of Tsai and interfascicular nucleus: a horseradish peroxidase study in the rat. *J. Comp. Neurol.* 187:117–44

Pickel VM, Sesack SR. 1995. Electron microscopy of central dopamine systems. See Bloom & Kupfer 1995, pp. 257–68

Potenza MN, Graminski GF, Schmauss C, Lerner MR. 1994. Functional expression and characterization of human D$_2$ and D$_3$ dopamine receptors. *J. Neurosci.* 14:1463–76

Prisco S, Pagannone S, Esposito E. 1994. Serotonin-dopamine interaction in the rat ventral tegmental area: an electrophysiological study in vivo. *J. Pharmacol. Exp. Ther.* 271:83–90

Pucak ML, Grace AA. 1994. Evidence that systemically administered dopamine antagonists activate dopamine neuron firing primarily by blockade of somatodendritic autoreceptors. *J. Pharmacol. Exp. Ther.* 271:1181–92

Rayport S, Sulzer D, Shi W-X, Sawasdikosol S, Monaco J, et al. 1992. Identified postnatal mesolimbic dopamine neurons in culture: morphology and electrophysiology. *J. Neurosci.* 12:4264–80

Rivet J-M, Audinot V, Gobert A, Peglion J-L, Millan MJ. 1994. Modulation of mesolimbic dopamine release by the selective dopamine D3 receptor antagonist, (+)-S 14297. *Eur. J. Pharmacol.* 265:175–77

Robbins TW, Brown VJ. 1990. The role of the striatum in the mental chronometry of action: a theoretical review. *Rev. Neurosci.* 2:181–213

Robbins TW, Everitt BJ. 1992. Functions of dopamine in the dorsal and ventral striatum. *Semin. Neurosci.* 4:119–27

Roth RH, Elsworth JD. 1995. Biochemical pharmacology of midbrain dopamine neurons. See Bloom & Kupfer 1995, pp. 227–43

Salamone JD. 1994. The involvement of nucleus accumbens dopamine in appetitive and aversive motivation. *Behav. Brain Res.* 61: 117–33

Schotte A, Janssen PFM, Gommeren W, Luyten WHLM, Leysen JE. 1992. Autoradiographic evidence for the occlusion of rat brain dopamine D3 receptors in vivo. *Eur. J. Pharmacol.* 218:373–75

Schultz W. 1992. Activity of dopamine neurons in the behaving primate. *Semin. Neurosci.* 4:129–38

Schultz W, Romo R. 1990. Dopamine neurons of the monkey midbrain: contingencies of responses to stimuli eliciting behavioral reactions. *J. Neurophysiol.* 63:607–24

Sesack SR, Aoki C, Pickel VM. 1994. Ultrastructural localization of D2 receptor–like immunoreactivity in midbrain dopamine neurons and their striatal targets. *J. Neurosci.* 14:88–106

Seutin V, Johnson SW, North RA. 1993a. Apamin increases NMDA-induced burst-firing of rat mesencephalic dopamine neurons. *Brain Res.* 630:341–44

Seutin V, Johnson SW, North RA. 1994. Effect of dopamine and baclofen on N-methyl-D-aspartate-induced burst firing in rat ventral tegmental neurons. *Neuroscience* 58:201–6

Seutin V, Messotte L, Dress A. 1989. Electrophysiological effects of neurotensin on dopamine neurons of the ventral tegmental area of the rat in vitro. *Neuropharmacology* 28:949–54

Seutin V, North RA, Johnson SW. 1993b. Transmitter regulation of mesencephalic dopamine cells. In *Limbic Motor Circuits and Neuropsychiatry*, ed. PW Kalivas, CD Barnes, pp. 89–100. Boca Raton: CRC

Seutin V, Verbanck P, Massotte L, Dresse A. 1990. Evidence for the presence of N-methyl-D-aspartate receptors in rat ventral tegmental area of the rat: an electrophysiological in vitro study. *Brain Res.* 514:147–50

Seutin V, Verbanck P, Massotte L, Dresse A. 1991. Acute amphetamine-induced subsensitivity of A10 dopamine autoreceptors in vitro. *Brain Res.* 558:141–44

Shen R-Y, Chiodo LA. 1993. Acute withdrawal after repeated ethanol treatment reduces the number of spontaneously active dopaminergic neurons in the ventral tegmental area. *Brain Res.* 622:289–93

Shepard PD, Bunney BS. 1988. Effects of apamin on the discharge properties of putative dopamine-containing neurons in vitro. *Brain Res.* 463:380–84

Shepard PD, Bunney BS. 1991. Repetitive firing properties of putative dopamine-containing neurons in vitro: regulation by an apamin-sensitive $Ca^{2+}$-activated $K^+$ conductance. *Exp. Brain Res.* 86:141–50

Shepard PD, German DC. 1984. A subpopulation of mesocortical dopamine neurons possess autoreceptors. *Eur. J. Pharmacol.* 114: 401–2

Shi W-S, Bunney BS. 1990. Neurotensin attenuates dopamine D2 agonist quinpirole–induced inhibition of midbrain dopamine neurons. *Neuropharmacology* 29:1095–97

Shi W-X, Bunney BS. 1991. Neurotensin modulates autoreceptor mediated dopamine effects on midbrain dopamine cell activity. *Brain Res.* 543:315–21

Shi W-X, Bunney BS. 1992. Roles of intracellular cAMP and protein kinase A in the actions of dopamine and neurotensin on midbrain dopamine neurons. *J. Neurosci.* 12:2433–38

Skirboll LR, Grace AA, Hommer DW, Rehfeld J, Goldstein M, et al. 1981. Peptide-monoamine coexistence: studies of the actions of cholecystokinin-like peptides on the electrical activity of midbrain dopamine neurons. *Neuroscience* 6:2111–24

Stittsworth JD, Mueller AL. 1990. Cholecystokinin octapeptide potentiates the inhibitory response mediated by D2 dopamine receptors in slices of the ventral tegmental area of the brain of the rat. *Neuropharmacology* 29:119–27

Suaud-Chagny MF, Chergui K, Chouvet G, Gonon F. 1992. Relationship between dopamine release in the rat nucleus accumbens and the discharge activity of dopaminergic neurons during local in vivo application of amino acids in the ventral tegmental area. *Neuroscience* 49:63–72

Sugita S, Johnson SW, North RA. 1992. Synaptic inputs to GABAA and GABAB receptors originate from discrete afferent neurons. *Neurosci. Lett.* 134:207–11

Svensson TH, Tung C-S. 1989. Local cooling of pre-frontal cortex induces pacemaker-like firing of dopamine neurons in rat ventral tegmental area in vivo. *Acta Physiol. Scand.* 136:135–36

Thierry AM, Chevalier G, Ferron A, Glowinski J. 1983. Diencephalic and mesencephalic efferents of the medial prefrontal cortex in the rat: electrophysiological evidence for the existence of branched axons. *Exp. Brain Res.* 50:275–82

Tong Z-Y, Overton PG, Clark D. 1995. Chronic administration of (+) amphetamine alters the reactivity of midbrain dopaminergic neurons to prefrontal cortex stimulation in the rat. *Brain Res.* 674:63–74

Ugedo L, Grenhoff J, Svensson TH. 1989. Ritanserin, a 5-HT$_2$ receptor antagonist, activates midbrain dopamine neurons by blocking serotonergic inhibition. *Psychopharmacology* 98:45–50

Ungerstedt U. 1971. Stereotaxic mapping of the monoamine pathways in the rat brain. *Acta Physiol. Scand. Suppl.* 367:1–48

Valentino RJ, Aston-Jones GS. 1995. Physiological and antomical determinants of locus coeruleus discharge: behavioral and clinical implications. See Bloom & Kupfer 1995, pp. 373–85

Van Bockstaele EJ, Cestari DM, Pickel VM. 1994. Synaptic structure and connectivity of serotonin terminals in the ventral tegmental area: potential sites for modulation of mesolimbic dopamine neurons. *Brain Res.* 647: 307–22

Vertes RP. 1991. A PHA-L analysis of ascending projections of the dorsal raphe nucleus in the rat. *J. Comp. Neurol.* 313:643–68

Vezina P. 1993. Amphetamine injected into the ventral tegmental area sensitizes the nucleus accumbens dopaminergic response to systemic amphetamine: an in vivo microdialysis study in the rat. *Brain Res.* 605:332–37

Wachtel SR, Hu X-T, Galloway MP, White FJ. 1989. D1 dopamine receptor stimulation enables the postsynaptic, but not autoreceptor, effects of D2 dopamine agonists in nigrostriatal and mesoaccumbens dopamine systems. *Synapse* 4:327–46

Walaas I, Fonnum F. 1980. Biochemical evidence for gamma-aminobutyrate containing fibers from the nucleus accumbens to the substantia nigra and ventral tegmental area in the rat. *Neuroscience* 5:63–72

Wallace DM, Magnuson DJ, Gray TS. 1989. The amygdalo-brainstem pathway: selective innervation of dopaminergic, noradrenergic and adrenergic cells in the rat. *Neurosci. Lett.* 97:252–58

Wang RY. 1981a. Dopaminergic neurons in the rat ventral tegmental area. I. Identification and characterization. *Brain Res. Rev.* 3:123–40

Wang RY. 1981b. Dopaminergic neurons in the rat ventral tegmental area. II. Evidence for autoregulation. *Brain Res. Rev.* 3:141–51

Wang RY. 1981c. Dopaminergic neurons in the rat ventral tegmental area. III. Effects of d- and l-amphetamine. *Brain Res. Rev.* 3:153–65

Wang T, French ED. 1993a. Electrophysiological evidence for the existence of NMDA and non-NMDA receptors on rat ventral tegmental dopamine neurons. *Synapse* 13:270–7

Wang T, French ED. 1993b. L-Glutamate excitation of A10 dopamine neurons is preferentially mediated by activation of NMDA receptors: extra- and intracellular electrophysiological studies in brain slices. *Brain Res.* 627:299–306

Wang T, O'Connor WT, Ungerstedt U, French ED. 1994. N-methyl-D-aspartic acid biphasically regulates the biochemical and electrophysiological response of A10 dopamine neurons in the ventral tegmental area: in vivo microdialysis and in vitro electrophysiological studies. *Brain Res.* 666:255–62

Waszczak BL, Walters JR. 1980. Intravenous GABA agonist administration stimulates firing of A10 dopaminergic neurons. *Eur. J. Pharmacol.* 66:141–44

Weiss F, Markou A, Lorang MT, Koob GF. 1992. Basal extracellular dopamine levels in the nucleus accumbens are decreased during cocaine withdrawal after unlimited-access self-administration. *Brain Res.* 593:314–18

Westerink BHC, Santiago M, De Vries JB. 1992. The release of dopamine from nerve terminals and dendrites of nigrostriatal neurons induced by excitatory amino acids in the conscious rat. *Naunyn Schmiedebergs Arch. Pharmacol.* 345:523–29

White FJ. 1991. Neurotransmission in the mesoaccumbens dopamine system. See Willner & Scheel-Krüger 1991, pp. 61–103

White FJ, Hu X-T. 1993. Electrophysiological correlates of D$_1$:D$_2$ interactions. In *D$_1$:D$_2$ Dopamine Receptor Interactions*, ed. JL Waddington, pp 79–114. San Diego: Academic

White FJ, Hu X-T, Zhang X-F, Wolf ME. 1995. Repeated administration of cocaine or amphetamine alters neuronal responses to glutamate in the mesoaccumbens dopamine system. *J. Pharmacol. Exp. Ther.* 273:445–54

White FJ, Wang RY. 1983a. Comparison of the effects of chronic haloperidol treatment on A9 and A10 dopamine neurons in the rat. *Life Sci.* 32:983–93

White FJ, Wang RY. 1983b. Differential effects of classical and atypical antipsychotic drugs on A9 and A10 dopamine neurons. *Science* 221:1054–57

White FJ, Wang RY. 1984a. A10 dopamine neurons: role of autoreceptors in determining firing rate and sensitivity to dopamine agonists. *Life Sci.* 34:1161–70

White FJ, Wang RY. 1984b. Pharmacological characterization of dopamine autoreceptors in the rat ventral tegmental area: microiontophoretic studies. *J. Pharmacol. Exp. Ther.* 231:275–80

White FJ, Wang RY. 1984c. Electrophysiological evidence for A10 dopamine autoreceptor subsensitivity following chronic *d*-amphetamine treatment. *Brain Res.* 309:283–92

Willner P, Scheel-Krüger J. 1991. *The Mesolimbic Dopamine System: From Motivation To Action.* New York: Wiley

Wolf ME, Dahlin SL, Hu X-T, Xue C-J, White K. 1995. Effects of lesions of prefrontal cortex, amygdala, or fornix on behavioral sensitization to amphetamine: comparison with *N*-methyl-D-aspartate antagonists. *Neuroscience* In press

Wolf ME, Jeziorski M. 1993. Coadministration of MK-801 with amphetamine, cocaine or morphine prevents rather than transiently masks the development of behavioral sensitization. *Brain Res.* 613:291–94

Wolf ME, White FJ, Hu X-T. 1994. MK-801 prevents alterations in the mesoaccumbens dopamine system associated with behavioral sensitization to amphetamine. *J. Neurosci.* 14:1735–45

Woolf NJ. 1991. Cholinergic systems in mammalian brain and spinal cord. *Prog. Neurobiol.* 37:475–524

Wu T, Wang H-L. 1994. CCK-8 excites substantia nigra dopaminergic neurons by increasing a cationic conductance. *Neurosci. Lett.* 170:229–32

Yim CY, Mogenson GJ. 1980. Effect of picrotoxin and nipecotic acid on inhibitory response of dopaminergic neurons in the ventral tegmental area to stimulation of the nucleus accumbens. *Brain Res.* 199: 466–72

Zahm DS. 1989. The ventral striatopallidal parts of the basal ganglia in the rat. II. Compartmentation of ventral pallidal efferents. *Neuroscience* 30:33–50

Zhang J, Chiodo LA, Freeman AS. 1992. Electrophysiological effects of MK-801 on rat nigrostriatal and mesoaccumbal dopaminergic neurons. *Brain Res.* 590:153–63

Zigmond MJ, Stachowiak MK, Berger TW, Stricker EM. 1986. Neurochemical events underlying continued function despite injury to monoaminergic systems. *Exp. Brain Res.* 13:119–28

*Annu. Rev. Neurosci. 1996. 19:437–62*

# LONG-TERM DEPRESSION IN HIPPOCAMPUS

## Mark F. Bear

Department of Neuroscience and Howard Hughes Medical Institute, Brown University, Providence, Rhode Island 02912

## Wickliffe C. Abraham

Department of Psychology and the Neuroscience Research Centre, University of Otago, Dunedin, New Zealand

KEY WORDS:    long-term potentiation, synaptic plasticity, learning and memory, calcium

### ABSTRACT

Long-term depression (LTD) is a lasting decrease in synaptic effectiveness that follows some types of electrical stimulation in the hippocampus. Two broad types of LTD may be distinguished. Heterosynaptic LTD can occur at synapses that are inactive, normally during high-frequency stimulation of a converging synaptic input. Homosynaptic LTD can occur at synapses that are activated, normally at low frequencies. Here we discuss the mechanisms of LTD and their possible relevance to hippocampal function.

## INTRODUCTION

It was not too many years ago that the mere existence of long-term depression (LTD) in hippocampus was seriously questioned. Skepticism stemmed mainly from the fact that LTD proved to be difficult to demonstrate in the commonly used slice preparations of hippocampus, contrasting sharply with the phenomenal success of the in vitro approach to the study of long-term potentiation (LTP). Fortunately, the introduction of new paradigms to study LTD in vitro have now put the phenomenon on firm ground, and this has also had the effect of bolstering confidence in earlier in vivo studies of LTD. This is not to say that the field today is without controversy. Although LTD is widely acknowledged as a genuine form of activity-dependent synaptic plasticity, a number of very basic questions remain largely unsettled. What are the mechanisms of LTD, and how do they relate to those of LTP? To what extent are the different forms of LTD related? Do all forms of LTD occur in vivo, and are they

437

0147-006X/96/0301-0437$08.00

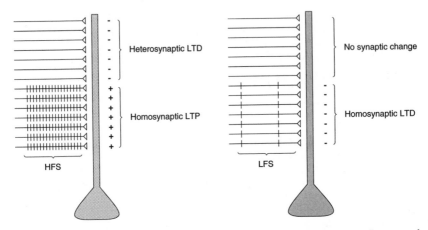

*Figure 1* Conditions for the induction of heterosynaptic and homosynaptic LTD. Represented schematically are hippocampal pyramidal cells receiving an array of synaptic inputs. Heterosynaptic LTD can occur at synapses that are inactive during high-frequency stimulation (HFS) of a converging synaptic input. Homosynaptic LTD can occur at synapses that are given low-frequency stimulation (LFS). Heterosynaptic LTD is most robust in dentate gyrus; homosynaptic LTD is most robust in CA1.

expressed at all ages? What is the function of LTD? In this review of hippocampal LTD, we focus on these questions.

Two broad types of LTD may be distinguished (Figure 1). In the case of heterosynaptic LTD, only the strengths of inactive synapses are depressed; in the case of homosynaptic LTD, only the strengths of active synapses are depressed. In either case, however, the level of postsynaptic activation during induction is a critical variable; by extension, therefore, so is the activity of converging synaptic inputs from other sources. Heterosynaptic LTD was discovered in the dentate gyrus in vivo over 15 years ago, while convincing evidence for homosynaptic LTD in CA1 has come only within the past several years. Therefore, in our review we recapitulate the history of hippocampal LTD by focusing first on heterosynaptic LTD of the perforant path inputs to dentate gyrus and then discussing homosynaptic LTD of the Schaffer collateral inputs to CA1.

## LTD IN DENTATE GYRUS

### Heterosynaptic LTD in the Dentate Gyrus

The excitatory monosynaptic pathway running from layer 2 cells of the entorhinal cortex via the perforant path to the granule cells of the dentate gyrus has proven over the years to be a reliable preparation for the study of hetero-synaptic LTD. Such LTD was first demonstrated in the crossed perforant path synapses following tetanization of the ipsilateral perforant path by Levy & Steward (1979). Since then, heterosynaptic LTD has been demonstrated at nearly all of the excitatory synaptic connections onto the granule cells, includ-ing the ipsilateral medial and lateral perforant path synapses (Abraham & Goddard 1983), the contralateral medial and lateral perforant path synapses (White et al 1988), and the commissural synapses arising from cells in the contralateral dentate hilus (Krug et al 1985, Abraham et al 1994). Although dentate heterosynaptic LTD has primarily been studied in anesthetized animals, it occurs readily, if not more so, in awake animals (Krug et al 1985, Abraham et al 1994). What remains perplexing is that heterosynaptic LTD in the dentate gyrus has not yet been observed in vitro (Wigstrom & Gustafsson 1983, Hanse & Gustafsson 1992). The loss of the entorhinal cortex during the slicing procedure does not appear to account for this difference, since neither tran-section of the perforant path fibers posterior to the stimulating site (Levy & Steward 1979) nor tetrodotoxin injections into the entorhinal cortex (Lopez et al 1990) appear to disrupt LTD in vivo. Thus the critical differences between these two experimental preparations remain to be identified.

Like long-term potentiation (LTP) in the hippocampus, heterosynaptic LTD is saturable (Levy & Steward 1983; Christie & Abraham 1992a,b) and revers-ible (Levy & Steward 1979, 1983; Abraham & Goddard 1983) and, depending on the tetanization parameters, can last for days or weeks (Krug et al 1985, Abraham et al 1994) (Figure 2). Reversal of LTP (depotentiation) by tetani-zation of a neighboring, converging pathway can occur at least as readily as heterosynaptic LTD of naive synapses (Levy & Steward 1979; Christie & Abraham 1992a,b), although the occasional exception to this principle has been reported when there is an interval of several hours between LTP and the heterosynaptic tetanization protocol (Abraham & Goddard 1983). In most experiments, the induction of heterosynaptic LTD in one pathway is accom-panied by LTP of the path that was tetanized. However, LTP is not required for the induction of LTD (Abraham & Goddard 1983). Nonetheless, no ex-perimental protocols or manipulations have yet been devised that routinely permit the induction of heterosynaptic LTD in the absence of LTP in neigh-boring synapses. This fact, plus the failure so far to observe this form of LTD in vitro, has restricted investigations of the mechanisms specifically underlying heterosynaptic LTD.

Some evidence is available, nonetheless, suggesting that strong depolariza-tion postsynaptically is an important signal for the induction of heterosynaptic LTD. First, the $GABA_A$ receptor antagonist bicuculline, which enhances post-synaptic depolarization during the tetanus, facilitates heterosynaptic LTD in-duction (Zhang & Levy 1993, Tomasulo et al 1993). Furthermore, LTD is

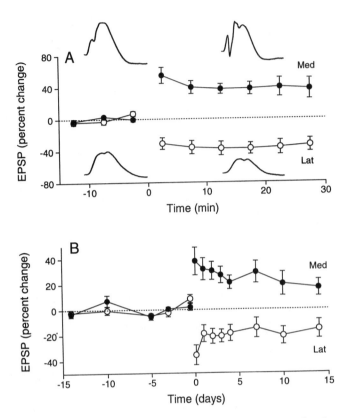

*Figure 2*  The induction and persistence of heterosynaptic LTD in the lateral perforant path of awake rats. (*A*) Mean (±SEM) LTP induction in the medial path (Med) and LTD induction in the lateral path (Lat) following high-frequency stimulation of the medial path. Waveforms are an average of responses recorded 10 min before and 30 min after tetanization. Calibration bars: 2 mV, 3 ms. (*B*) Persistence of the medial path LTP and lateral path LTD for the same animals as in *A*. Note that the LTD decays across days on average at about the same rate as LTP. Medial path (*closed circles*); lateral path (*open circles*). [Data taken from Abraham et al (1994).]

blocked by both competitive and noncompetitive antagonists of the *N*-methyl-D-aspartate (NMDA) receptor (Desmond et al 1991; Christie & Abraham 1992a,b). The block of heterosynaptic LTD is not dependent on the simultaneous block of LTP in the tetanized pathway because LTP remains intact when LTD is inhibited by L-type voltage-gated calcium channel antagonists (Christie & Abraham 1994). These data suggest that as for LTP, postsynaptic depolarization and calcium entry are important for induction of heterosynaptic LTD. However, for the induction of heterosynaptic LTD, the role of the NMDA receptor may be to facilitate the membrane depolarization responsible for activating voltage-gated calcium channels, rather than to provide the calcium signal directly. More direct measures and manipulations of membrane potential and calcium levels are necessary, however, to verify this hypothesis.

There is virtually no information regarding the downstream induction and expression mechanisms underlying heterosynaptic LTD. LTD, unlike LTP, appears to be insensitive to protein kinase C (PKC) activators (Lovinger & Routtenberg 1988), but other kinase or phosphatase manipulations have yet to be performed in the dentate gyrus. Desmond & Levy (1983) have suggested that the expression of heterosynaptic LTD may involve decreases in the size and complexity of synaptic profiles morphologically, but other candidate expression mechanisms, such as changes in transmitter release or glutamate receptors, still need to be investigated.

## Extrinsic Modulators of Heterosynaptic LTD

Scant attention has been paid to the question of how extrinsic afferents to the dentate gyrus may modulate the induction of heterosynaptic LTD. One study, however, has reported that chemical stimulation of the cholinergic medial septal neurons facilitates LTD induction in the medial perforant path of aged animals (Pang et al 1993). Conversely, the muscarinic receptor antagonist atropine blocks the relatively mild heterosynaptic LTD seen in pentobarbital anesthetized rats but does not prevent its induction in awake animals (WC Abraham, BR Christie & B Logan, unpublished). Therefore, acetylcholine released by medial septal afferents may be a modulator of LTD. Whether it acts by promoting membrane depolarization or by facilitating second-messenger pathways is not yet known.

Exogenous application of the neurotransmitter norepinephrine (NE) exerts complex effects on perforant path synaptic responses in dentate slices. NE potentiates medial perforant path responses yet depresses lateral path responses (Dahl & Sarvey 1989). These effects are mediated by β-adrenergic receptors because isoproterenol effectively substitutes for NE and the effects are blocked by propranolol (Dahl & Sarvey 1989). NE acting alone on β-receptors is probably insufficient to induce the LTD, however, because the LTD elicited in this manner is also blocked by the NMDA receptor antagonist 2-amino-5-phosphonovaleric acid (AP5) (Stanton et al 1989). Thus a specific interaction between NE and NMDA receptor activation by glutamate appears to be involved. This interaction may be two way, since NE can up-regulate NMDA receptor-mediated responses, and NMDA receptor activation is necessary for the NE-induced membrane depolarization and increase in input resistance to persist following NE washout (Stanton et al 1989). One important question that remains to be answered for this paradigm is whether the LTD of lateral path responses indeed represents a decrease in synaptic efficacy or whether it is secondary to the persistent alterations of the postsynaptic membrane parameters. Furthermore, it remains a mystery why the medial and lateral paths simultaneously respond in different directions to a single application of a β-receptor agonist.

## LTD of Synapses on Inhibitory Interneurons in the Dentate Gyrus

There is some evidence to suggest that perforant path synapses onto inhibitory interneurons in the dentate gyrus can also undergo LTD. Repeated tetanization of the lateral perforant path, for example, can cause an increase in the population spike amplitude recorded heterosynaptically for the medial path, even though the medial path EPSP is unchanged or decreased (Abraham et al 1985). One explanation for these results is that activation of the lateral path synapses on the interneurons causes heterosynaptic LTD of the medial path innervation of the interneurons, thus leading to reduced feed-forward inhibition elicited by medial path afferents. This idea has received recent support from a preliminary study showing heterosynaptic LTD of afferent driving of individually recorded interneurons in the dentate gyrus (Tomasulo 1994).

## Homosynaptic LTD in the Dentate Gyrus

On the whole, homosynaptic LTD has been much harder to demonstrate in the dentate gyrus in vivo than heterosynaptic LTD. Low-frequency stimulation (LFS) around 1–3 Hz readily generates short-term depression (or habituation) at lateral path synapses (Teyler & Alger 1976, McNaughton 1980, Abraham & Bliss 1985), but so far long-lasting changes (>15–30 min) have been small or nonexistent (Dragunow et al 1989, Errington et al 1994; WC Abraham, SE Mason-Parker & B Logan, submitted). Under certain conditions, however, the induction of homosynaptic LTD can be observed. For example, brief bursts of high-frequency stimulation (HFS) of the perforant path on the negative phase of the theta rhythm can cause LTD (Pavlides et al 1988). Similarly, LTD will result in lateral path synapses if single-pulse stimulation to the lateral path is given between short bursts of HFS applied at a theta rhythm frequency to the medial path (Christie & Abraham 1992b). In this latter case, the appearance of LTD also requires that prior priming stimulation, again preferentially at a theta rhythm frequency, has been given to the lateral path. The LTD arising from alternating HFS to the medial path and single pulses to the lateral path has been termed associative LTD (Stanton & Sejnowski 1989; Christie & Abraham 1992b), but we believe that this protocol simply provides another way of inducing homosynaptic LTD.

A critical feature of both the stimulation during the associative protocol and the stimulation during the negative phase of the theta rhythm may be that the activity occurs during a period of postsynaptic hyperpolarization. Support for this idea comes from an experiment showing that isolated NMDA synaptic potentials depress if they are generated while the postsynaptic membrane is hyperpolarized (Xie et al 1992). Furthermore, Zhang & Levy (1994) observed that LTD induced by LFS is facilitated by the GABA agonist muscimol in dentate slices. It is by no means clear what induction mechanisms are engaged by such activity, however. One possibility is that T-type calcium channels are involved, since they are generally inactivated at resting membrane potentials but become deinactivated by hyperpolarization and then activated during subsequent synaptic activity. Some evidence for T-channel involvement in cortical plasticity has already been reported (Komatsu & Iwakiri 1992), and preliminary evidence suggests that T channels may indeed contribute to LTD in the hippocampus (Schexnayder et al 1995).

## LTD IN CA1

### Induction of Homosynaptic LTD In Vitro

Until very recently, the existence of LTD in CA1 has been disputed. An important breakthrough came in 1992 with the introduction of a paradigm that can reliably produce homosynaptic LTD in vitro (Figure 3). Dudek & Bear (1992, 1993) found that several hundred stimuli delivered to the Schaffer collaterals at low frequencies produce a sustained depression of modest but significant magnitude. Although several points of contention remain, which are discussed further below, the basic observations of Dudek & Bear have been widely replicated (Mulkey & Malenka 1992; Abeliovich et al 1993; Izumi & Zorumski 1993; Kirkwood et al 1993; Mulkey et al 1993, 1994; Aiba et al 1994; Bolshakov & Siegelbaum 1994; Cummings et al 1994; Fitzpatrick & Baudry 1994; Maccaferri et al 1994; Stevens & Wang 1994; Stevens et al 1994; Xiao et al 1994; Hrabetova et al 1995; Kerr & Abraham 1995; Muller et al 1995; Selig et al 1995; Wagner & Alger 1995).

The key properties of LTD produced by LFS may be summarized as follows:

1. LFS-induced LTD is input specific. Only synapses receiving the low-frequency conditioning stimulation (LFS) show LTD; other synaptic inputs converging onto the same population of postsynaptic neurons are unaffected (hence the name homosynaptic LTD) (Dudek & Bear 1992, Mulkey & Malenka 1992, Wagner & Alger 1995, Muller et al 1995, Hrabetova & Sacktor 1994).
2. LFS-induced LTD is frequency dependent. LTD depends on the frequency of conditioning stimulation. LTD usually results from 0.5–3 Hz stimulation,

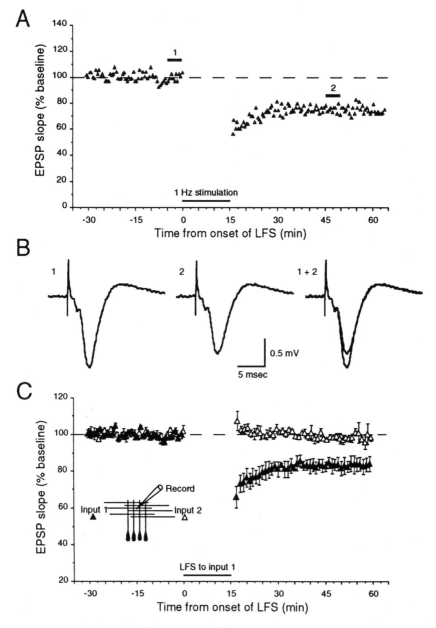

*Figure 3* Homosynaptic LTD in CA1 in vitro. (*A*) Record of a representative experiment in which LTD was induced by LFS. Each point represents a single measure of the initial slope of the population EPSP evoked by stimulation of the Schaffer collaterals at 0.03 Hz. The horizontal bar represents the period of 1 Hz conditioning stimulation. (*B*) Averages of 10 consecutive sweeps before and after LFS conditioning at the times indicated by the numbers in *A*. (*C*) Normalized averages (±SEM) of five experiments in which the response to two independent inputs was monitored (inset shows the stimulating and recording configuration). LFS (1 Hz) conditioning stimulation given to input 1 produced a depression of the response to input 1 only (*filled triangles*). This figure is from Dudek & Bear, 1992 (used with permission).

but the same number of pulses at higher frequencies can produce LTP instead (Dudek & Bear 1992, Mayford et al 1995).

3. LFS-induced LTD usually requires NMDA receptor activation. Under most circumstances, homosynaptic LTD is reversibly blocked by NMDA receptor antagonists (Dudek & Bear 1992, Mulkey & Malenka 1992, Kirkwood et al 1993, Izumi & Zorumski 1993, Xiao et al 1994, Stevens et al 1994, Hrabetova & Sacktor 1994, Kerr & Abraham 1995). However, there are also reports of NMDA-receptor–independent LTD (discussed further below).

4. LFS-induced LTD decreases with increasing age. The relative magnitude of homosynaptic LTD is age dependent. For example, the same LFS in CA1 of two-week-old rats produces a relative decrease in the EPSP that is double that seen in CA1 of five-week-old rats (Dudek & Bear 1992; see also Mulkey & Malenka 1992, Wagner & Alger 1995, Mayford et al 1995).

5. LFS-induced LTD is saturable and reversible. Repeated episodes of LFS in CA1 saturate LTD (Mulkey & Malenka 1992, Dudek & Bear 1993, Mulkey et al 1993). Furthermore, synapses that are depressed can be potentiated, and vice versa, indicating that LTD is not a result of lasting damage to the stimulated synapses (Mulkey & Malenka 1992, Dudek & Bear 1993, Mulkey et al 1993, Hrabetova & Sacktor 1994, Wagner & Alger 1995).

The observation that LFS can reverse LTP is not new. Many years ago Barrionuevo et al (1980) showed that LFS could partly reverse LTP in CA1. This effect, called depotentiation, was later shown to occur in awake animals (Staubli & Lynch 1990) and slices (Fujii et al 1991). However, the same stimulation that caused depotentiation failed to cause depression of naive synapses in these experiments. Two plausible explanations for these findings, not involving LTD, are (a) LFS disrupts the mechanisms of LTP consolidation or (b) LTP renders the conditioned synapses susceptible to excitotoxic damage caused by LFS. The first explanation is unlikely, as it has now been established that LFS can produce LTD in slices de novo. Dudek & Bear (1993) and Mulkey & Malenka (1992) addressed the second possibility by saturating LTP with repeated episodes of HFS before inducing LTD (depotentiation) with LFS. A subsequent episode of HFS restored the response to the fully saturated level, indicating (a) that LFS had unsaturated LTP and (b) that the same synapses whose effectiveness had been reduced by LFS could subsequently exhibit LTP. Mulkey et al (1993) performed the inverse experiment, showing that LTD could be reversibly unsaturated by HFS. The data, taken together, suggest that LTD and LTP reversibly affect synaptic effectiveness by acting at a common site. The relationship of LTD to depotentiation and deconsolidation of LTP is discussed below.

More recently, other stimulation protocols have been introduced that produce homosynaptic LTD in vitro with properties similar to those described above. In area CA1 of slice cultures, for example, Debanne et al (1994) report that if the Schaffer collaterals are repeatedly stimulated shortly after strong intracellular depolarizing pulses are delivered to a CA1 neuron, then LTD of the stimulated synapses will result. Like the homosynaptic depression described above, this effect is blocked by antagonists of the NMDA receptor. The authors ascribe the effect of the depolarizing pulse to $Ca^{2+}$ entry through voltage-gated calcium channels. Thus, like LFS, this protocol repetitively delivers $Ca^{2+}$ pulses to the postsynaptic neuron. Furthermore, like LFS-induced LTD, the interpulse interval is a critical variable. Strikingly, the interpulse interval that was optimal for LTD in this study (0.2–1.6 s) is in exactly the same range as that previously found to be optimal for LFS-induced LTD (Dudek & Bear 1992). Thus, it seems quite likely that both stimulation procedures activate the same intracellular mechanisms to induce LTD.

## Regulation of Homosynaptic LTD in Adult CA1

There appears to be widespread consensus that LFS of the Schaffer collaterals induces homosynaptic LTD in CA1 of immature animals (<four weeks old). However, in spite of the fact that homosynaptic LTD actually was first characterized in slices from young adults (Dudek & Bear 1992), there are now several reports that LFS fails to induce LTD in mature animals (Fujii et al 1991, O'Dell & Kandel 1994, Bortolotto et al 1994, Bashir & Collingridge 1994, Otani & Connor 1995). These results have raised the possibility that homosynaptic LTD is only of developmental significance.

Dudek & Bear (1992) showed that the magnitude of LTD varies according to stimulation frequency, but it is also apparent that it depends on other parameters, such as stimulation intensity and the state of inhibition. For example, using a stimulation intensity in slices of CA1 from adult animals that yielded an extracellular response $\leq0.5$ mV, Kerr & Abraham (1995) found that no LTD results, unless $GABA_A$ receptors are also blocked with picrotoxin. LFS at higher stimulation intensities did yield NMDA receptor–dependent LTD without picrotoxin, however, supporting the idea that induction of homosynaptic LTD, like that of LTP, requires a certain level of postsynaptic depolarization that is achieved when a critical number of afferents are coactivated. On the other hand, there may be an upper limit on the beneficial effect of increasing stimulation intensity for producing LTD, because at very high intensities, synaptic potentiation can override depression (C Dal Piaz & MF Bear, unpublished observations; see also Abraham et al 1986). Although they did not investigate the effects of varying stimulation intensity, Wagner & Alger (1995) also found that $GABA_A$ receptor antagonists significantly enhance

LFS-induced, NMDA receptor–dependent LTD in slices of CA1 from adult rats. Again, however, reducing inhibition too much may actually block expression of LTD by promoting LTP instead (Steele & Mauk 1994).

Ultimately, perhaps the most relevant question about homosynaptic LTD in adult CA1 is whether it occurs in vivo. Although more work needs to be done, the answer appears to be yes. Input-specific LTD follows conditioning stimulation at 1 Hz in adult, sodium pentobarbital–anesthetized rats, if care is taken to stimulate the ipsilateral Schaffer collaterals at the appropriate intensities (Heynen et al 1995) (Figure 4). Like the situation in vitro, LTD produced in this way is saturable and reversible. On the other hand, two groups using LFS (400–900 pulses at 0.5–3 Hz) of the commissural afferents to CA1 failed to observe significant LTD in adult anesthetized rats (Thiels et al 1994, Errington et al 1994). However, Thiels et al (1994) did show that paired-pulse stimulation (25 ms ISI) delivered every 2 seconds for 400 seconds was sufficient to produce LTD of the evoked population spike. Although input specificity was not assessed, this LTD was blocked by the NMDA receptor antagonist AP5, suggesting that it may be similar to the homosynaptic LTD evoked using LFS. Thiels et al propose that the paired-pulse requirement is explained by the fact that the second of the pair of EPSPs coincides with the disynaptic IPSP evoked by the first pulse. Indeed, administration of the $GABA_A$ receptor antagonist bicuculline blocked induction of LTD with paired-pulse stimulation. Again, an interpretation consistent with the results of this study is that LTD results if synaptic stimulation coincides with a critical range of postsynaptic membrane potentials. And, the depolarization achieved during conditioning is highly sensitive to inhibition, stimulation intensity, and stimulation frequency.

Repeated, asynchronous pairing of activity in two independent Schaffer collateral pathways has also been reported to produce LTD in adult CA1 in vitro (Stanton & Sejnowski 1989, Otani & Connor 1995; but see Kerr & Abraham 1993, Paulsen et al 1993), an effect also ascribed to recruiting IPSPs during the conditioning stimulation. Unlike the LTD discussed above, depression produced in this manner was not blocked by the NMDA receptor antagonist AP5. However, in one study, Otani & Connor (1995) blocked depression by coapplying AP5 and the metabotropic glutamate receptor (mGluR) antagonist (+)-α-methyl-4-carboxylphenyl-glycine (MCPG); in another experiment Stanton & Sejnowski (1989) blocked depression with 2-amino-3-phosphono-propionate (AP3), another mGluR antagonist. Similarly, in a study using slices from three-week-old animals, AP3 and MCPG (but not AP5) blocked LTD induced by pairing synaptic stimulation with bath application of GABA (Yang et al 1994). Below we discuss additional evidence for the involvement of metabotropic glutamate receptors in CA1 LTD and the possibility that some forms of LTD do not require NMDA receptor activation.

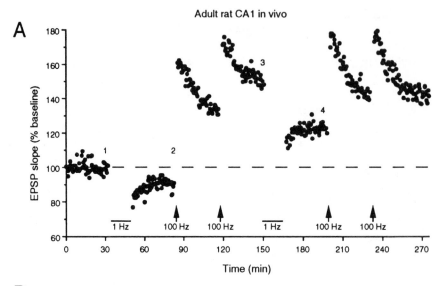

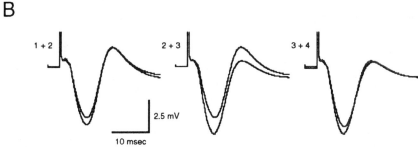

*Figure 4* An example of homosynaptic LTD and depotentiation in adult CA1 in vivo. (*A*) Each point represents a single measure of the initial slope of the CA1 population EPSP evoked by stimulation of the Schaffer collaterals at 0.03 Hz in a sodium pentobarbital–anesthetized Sprague-Dawley rat. 900 stimuli at 1 Hz produced LTD of the EPSP that could be reversed by a high-frequency tetanus (100 Hz for 1 s). Two consecutive tetani were given to saturate LTP, followed by 900 pulses at 1 Hz to produce depotentiation. Subsequent HFS restored the response to the saturated level of synaptic potentiation. (*B*) Superimposed averages of five consecutive sweeps at the times indicated by the numbers in *A* showing LTD, LTP, and depotentiation. [Data replotted with permission from Heynen et al (1995).]

## Homosynaptic Depression and Depotentiation

Homosynaptic LTD in adult CA1 is variable, and as we argued above, this variability may be explained by subtle differences in experimental protocol. However, even under conditions where LFS fails to produce LTD de novo, a consistent observation has been that LFS can reliably reverse previously established LTP (Barrionuevo et al 1980, Staubli & Lynch 1990, Fujii et al 1991, O'Dell & Kandel 1994, Bashir & Collingridge 1994). An important question is whether depotentiation and LTD are the same phenomenon.

Depotentiation and LTD have many features in common. Like homosynaptic LTD, depotentiation is induced by LFS (Barrionuevo et al 1980, Staubli & Lynch 1990, Fujii et al 1991, O'Dell & Kandel 1994) in a frequency-dependent manner (Fujii et al 1991, O'Dell & Kandel 1994) and depends on activation of NMDA receptors (Fujii et al 1991, O'Dell & Kandel 1994; but see Bashir & Collingridge 1994). LFS unsaturates previously established LTP (Dudek & Bear 1993, O'Dell & Kandel 1994), just as HFS unsaturates previously established LTD (Mulkey et al 1993). These results indicate that the change in synaptic responses following LFS occurs at a site in common with LTP, regardless of the initial state of the synapse.

Considering that depotentiation and LTD appear to be mechanistically identical, we are left with the question of why LTD occurs so much more readily after induction of LTP. When LFS fails to produce LTD, the synapses may already be in a fully depressed state, since there appears to be a limited dynamic range over which synaptic strengths can be altered (Dudek & Bear 1993, Mulkey et al 1993). According to this hypothesis, LTD follows LTP more readily because the high-frequency tetanus raises synaptic strength off the "floor" of the available dynamic range. Arguing against this possibility, however, is the evidence suggesting that depotentiation is restricted to a limited time window after establishment of LTP (Fujii et al 1991, O'Dell & Kandel 1994).

A more likely explanation is that the HFS primes the synapses so that they are now susceptible to LTD. Researchers have now shown in vivo (Christie et al 1993, Heynen et al 1995) and in vitro (Wexler & Stanton 1993, Wagner & Alger 1995) that prior stimulation, which alone has no lasting effect, can render synapses more susceptible to LFS-induced LTD. The mechanism for priming may include activation of certain protein kinases and, perhaps as a consequence, a transient reduction in the effectiveness of GABAergic inhibition. For example, Stanton (1994) has reported that brief application of the PKC activator phorbol 12,13-diacetate leads to a lasting change, such that LFS produces far greater LTD than under control conditions. Mayford et al (1995) have found that LFS-induced LTD is greatly enhanced in adult CA1 of transgenic mice carrying a point mutation that increases the $Ca^{2+}$-independent activity of $Ca^{2+}$- and calmodulin-dependent protein kinase II (CaMKII). Interestingly, $Ca^{2+}$-independent activity of CaMKII also increases after the types of HFS in CA1 that prime LTD (Ocorr & Schulman 1991). Adult mice lacking the gene for $\alpha$CaMKII also show less homosynaptic LTD in CA1 than do wild-type controls (Stevens et al 1994). The kinases may influence LTD induction by transiently altering inhibition. Wagner & Alger (1995) present data suggesting that the $GABA_B$ receptor–mediated modulation of GABA release is altered by prior stimulation in a way that makes LTD more likely.

Before leaving the topic of depotentiation, another activity-dependent modification of LTP that may be mechanistically distinct from LTD should be recognized. Larson et al (1993) report that 5 Hz (but not 1 Hz) stimulation reduces the magnitude of LTP attained after theta-burst stimulation, but only if it is delivered ≤5 min after the tetanus. This effect is promoted by application of NE and is blocked by antagonists of the A1 adenosine receptor. These data suggest that LTP may be consolidated over a period of approximately 5 min and that this can be disrupted by 5 Hz stimulation via A1 receptor activation. The temporal constraints and frequency dependence of such deconsolidation clearly differ from homosynaptic LTD.

## Mechanism of Homosynaptic LTD Induction

Mulkey & Malenka (1992) examined the intracellular mechanisms that lead to induction of NMDA receptor–dependent homosynaptic LTD in CA1. They showed that a necessary step in the induction by LFS of homosynaptic LTD was postsynaptic $Ca^{2+}$ entry through the NMDA receptor. Thus, suppressing postsynaptic NMDA currents with intracellular hyperpolarization or buffering intracellular rises in $[Ca^{2+}]$ by intracellular injection of the calcium chelator bis-(o-aminophenoxy)-$N,N,N',N'$-tetraacetic acid (BAPTA) prevented induction of LTD by LFS. The finding that BAPTA blocks LTD induction has been widely reproduced, even under experimental conditions where LTD is reportedly insensitive to AP5 (Brocher et al 1992, Xie et al 1992, Bolshakov & Siegelbaum 1994, Debanne et al 1994, Yang et al 1994, Otani & Connor 1995).

It is well established that $Ca^{2+}$ entry through the NMDA receptor is normally required for induction of LTP (Bliss & Collingridge 1993). How can the same signal—$Ca^{2+}$ entry through the NMDA receptor channel—be used to trigger both LTD and LTP? A large surge in intracellular $[Ca^{2+}]$ is widely believed to trigger LTP by activating $Ca^{2+}$-dependent protein kinases (Schwartz 1993). Lisman (1989) presented a model in which modest elevations of postsynaptic $Ca^{2+}$ (<1 μM) caused LTD by selectively activating protein phosphatases. Mulkey et al (1993) have provided data that support this model in CA1. They report that LFS-induced LTD can be completely blocked by bath application of either okadaic acid or calyculin A, both selective inhibitors of serine-threonine protein phosphatases (PP) 1 and 2A. In addition, intracellular injection of a third inhibitor—microcystin-L,R—also blocked LTD. Similar results were obtained for LFS-induced depotentiation in CA1 (O'Dell & Kandel 1994) and for a similar form of homosynaptic LTD in visual cortex (Kirkwood & Bear 1994).

If synaptic effectiveness were regulated by a balance of kinases and phosphatases acting on the same substrates, then one might expect that a disruption of this balance with an inhibitor by itself would be sufficient to alter the response magnitude. Indeed, Figurov et al (1993) have presented evidence that application of calyculin A causes an ~40% increase in the magnitude of the EPSP in naive slices. Neither the pharmacologically isolated NMDA component of the response nor the paired-pulse potentiation were altered, suggesting a selective action on the postsynaptic AMPA receptor. Calyculin-induced potentiation reduced (but did not fully occlude) subsequent tetanus-induced LTP. A complementary result was obtained by Hrabetova & Sacktor (1994) who found that inhibitors of PKC (chelerythrine and H7) cause an ~40% decrease in the magnitude of the EPSP in naive slices, which fully occludes LFS-induced LTD.

Not all investigators have observed the increase in synaptic responses during phosphatase inhibition, however. Mulkey et al (1993) reported that calyculin A had no effect on the baseline synaptic response but did reverse previously established homosynaptic LTD. In visual cortex, where a similar form of homosynaptic LTD has been described (Kirkwood et al 1993), okadaic acid was found to have no effect on baseline synaptic transmission either before or after induction of LTD (Kirkwood & Bear 1994). The reasons for these discrepancies remain to be elucidated, but one potentially confounding variable is that bath-applied calyculin A (but not okadaic acid) can increase presynaptic fiber excitability and transmitter release in an activity-dependent manner (Herron & Malenka 1994).

These unresolved issues notwithstanding, the data presently suggest that induction of LTD requires activation of PP1 and/or PP2a. However, neither of these enzymes is activated directly by $Ca^{2+}$, so the question remains how phosphatase activity is coupled to NMDA receptor activation and a rise in intracellular $[Ca^{2+}]$. According to Lisman's model, $Ca^{2+}$ activates PP2b (calcineurin), which then leads to activation of PP1 (and LTD) by dephosphorylating the PP1-regulatory protein I1 (inhibitor 1). Recent work supports this scheme, as selective calcineurin inhibitors prevent induction of LTD, both when bath-applied and injected postsynaptically (Mulkey et al 1994, Muller et al 1995). A caveat for interpreting this result comes from the recent finding that calcineurin inhibitors also alter the desensitization of NMDA receptors during repetitive activation (Tong et al 1995). Nonetheless, in accordance with the Lisman model, manipulations that prevent or reduce dephosphorylation of I1 (application of membrane-permeable cAMP analogues or intracellular injection of thiophosphorylated I1) also inhibit LTD induction (Mulkey et al 1994).

NMDA receptors are not the exclusive route for $Ca^{2+}$ entry into the cytosol. Activation of mGluRs 1 and 5, for example, can cause a rise in intracellular $[Ca^{2+}]$ by stimulating phosphoinositide hydrolysis and the production of $IP_3$

(Conn et al 1994). Indeed, the proposal was made many years ago that metabotropic glutamate receptors might play a key role in the induction of homosynaptic LTD, particularly during early postnatal life (Bear 1988, Dudek & Bear 1989). Support for this hypothesis has mainly come from studies using MCPG, a phenylglycine derivative that blocks several mGluR-mediated effects, including $IP_3$ formation associated with type 1 mGluRs (Watkins & Collingridge 1994). Bolshakov & Siegelbaum (1994) report that LFS-induced LTD in neonatal CA1 (three to seven days old) is prevented by MCPG (but not by AP5). Similarly, Bashir & Collingridge (1994) find that MCPG (but not AP5) blocks LFS-induced depotentiation in adult CA1. MCPG (but not AP5) has also been shown to block the LTD that results when synaptic stimulation (0.1–0.2 Hz) is applied during GABA application (Yang et al 1994). On the other hand, independent investigations in the Malenka and Bear labs (published together in Selig et al 1995b) uncovered no deficit in the induction of LTD or depotentiation during MCPG application, even in neonates (although AP5 effectively blocked LTD at all ages). Moreover, mice lacking the gene for mGluR1 apparently have normal homosynaptic LTD in CA1 (Aiba et al 1994). Tests for the involvement of type 5 mGluRs, which are highly expressed by CA1 pyramidal neurons but unaffected by MCPG (Joly et al 1995), await development of new pharmacological or genetic tools. The hypothesis therefore remains viable, but untested, that induction of homosynaptic LTD can be triggered by postsynaptic mGluR5 activation.

When it is effective in blocking LTD, MCPG may be acting presynaptically. The selective, but broad-spectrum, mGluR agonist *trans*-1-aminocyclopentane-1,3-dicarboxylic acid (*trans*-ACPD) depresses glutamate release by the Schaffer collaterals (Baskys & Malenka 1991) and GABA release from CA1 interneurons (Desai et al 1994). MCPG partially reverses these effects (Watkins & Collingridge 1994). Considering how sensitive LTD induction is to changes in excitation and inhibition during LFS (Kerr & Abraham 1995, Wagner & Alger 1995, Steele & Mauk 1994), it is not surprising that a drug that interferes with the regulation of neurotransmitter release also affects LTD. Comparing the postsynaptic responses during LFS in the presence and absence of MCPG might yield interesting results.

Besides $Ca^{2+}$ entry through NMDA-gated channels and intracellular $Ca^{2+}$ release from $IP_3$-sensitive pools, another means to elevate cytosolic $[Ca^{2+}]$ is voltage-gated calcium channels. However, there is conflicting evidence about the involvement of these channels in LTD induction. Bolshakov & Siegelbaum (1994) found that an L-type calcium channel antagonist, but not AP5, was effective in blocking LFS-induced LTD in slices from neonatal rats (P3–7). On the other hand, Selig et al (1995b) report that AP5, but not an L-type calcium channel antagonist, blocks homosynaptic LTD in newborn animals of the same age range. Calcium channel antagonists also have been reported to be ineffective in blocking homosynaptic LTD and depotentiation in older animals (Mulkey & Malenka 1992, Bashir & Collingridge 1994, Kerr & Abraham 1995).

If $Ca^{2+}$ is the trigger for LTD, elevations of postsynaptic $[Ca^{2+}]$ by any route may be sufficient to cause synaptic depression. The major route may vary depending on experimental conditions. While this idea is appealing and could help to resolve the differences in the literature, we note that no direct evidence demonstrates that an elevation in intracellular calcium concentration is, by itself, sufficient to trigger LTD. In fact, elevation of $[Ca^{2+}]_i$ through photolysis of the $Ca^{2+}$ chelator DM-nitrophen was found to be insufficient to cause synaptic depression (Bolshakov & Siegelbaum 1994). However, the timing as well as amplitude of $Ca^{2+}$ pulses appears to be critically important for triggering LTD. Considerable insight into this problem could be gained by measuring dendritic calcium transients during LFS at different frequencies.

## Expression and Maintenance of Homosynaptic LTD

As we mentioned above, induction of LTD unsaturates previously established LTP (Dudek & Bear 1993, O'Dell & Kandel 1994), and induction of LTP unsaturates previously established LTD (Mulkey et al 1993). These results strongly suggest that LTD and LTP are reversible modifications at a common site (Dudek & Bear 1993, Mulkey et al 1993). A site where LTP and LTD apparently converge is the regulation of protein kinase M$\zeta$ levels in CA1. PKM$\zeta$, a constituitively active fragment of PKC, increases in CA1 following induction of LTP (Sacktor et al 1993). Hrabetova & Sacktor (1994) have reported that LFS-induced LTD is associated with a decrease in PKM$\zeta$ and that this decrease can be reversed by subsequent induction of LTP. PKM$\zeta$ therefore is bidirectionally regulated, increasing in LTP and decreasing in LTD. As discussed previously, pharmacological inhibition of kinase activity also was reported to depress synaptic transmission and occlude LTD. Thus, PKM$\zeta$ may play a key role in the maintenance of synaptic strength following both LTD and LTP.

Consistent with the idea that LTD is the functional inverse of LTP, Stevens & Wang (1994) have presented data showing that both forms of plasticity are expressed as changes in the reliability of synaptic transmission in CA1. These and other data (Bolshakov & Siegelbaum 1994) suggest a presynaptic site of LTD (and LTP) expression. However, the assumptions underlying this interpretation may be flawed (Liao et al 1995, Issac et al 1995). In any case, a presynaptic site of expression does not preclude the possibility of a postsynaptic modification as well. For LTP, there is evidence that expression is both pre- and postsynaptic (Kullman & Nicoll 1992, Larkman et al 1992, Liao et al 1992).

The view that LTD and LTP affect synaptic efficacy at a common site of expression is complicated by the finding that LTD is associated with an equal decrease in the AMPA and NMDA components of the field EPSP (Xiao et al 1994). This contrasts with reports that LTP is associated with a larger increase in the AMPA component than in the NMDA component (Muller & Lynch 1988, Kauer et al 1988, Asztely et al 1992). However, this difference may be accounted for by the fact that NMDA receptors are subject to activity-dependent regulation in addition to that produced by LTD and LTP (Ben-Ari et al 1992, Selig et al 1995a).

To the extent that it is expressed presynaptically, LTD would seem to require a retrograde messenger—a factor that is released from the postsynaptic neuron to modify neurotransmitter release presynaptically—because induction of homosynaptic LTD occurs postsynaptically. Considering the confusion that persists regarding retrograde messengers and LTP (Williams et al 1993), it comes as no surprise that the few data available for LTD already are contradictory. Thus, inhibitors of the putative retrograde messenger nitric oxide have been found by some (Izumi & Zorumski 1993) to block LFS-induced LTD and by others (Cummings et al 1994, Bashir & Collingridge 1994) to have no effect. We anticipate that as the roles of the putative retrograde messengers in LTP are clarified, so will their roles in LTD.

## Heterosynaptic LTD in CA1

Establishing a model for heterosynaptic LTD in CA1 has been difficult. In 1977, Lynch et al (1977) reported that a consequence of producing LTP in one set of synapses was a generalized (heterosynaptic) depression of the effectiveness of the other synapses that converge on the same population of postsynaptic neurons. However, this was followed by numerous reports that indicated that heterosynaptic LTD in CA1 is not usually a consequence of tetanic stimulation of the Schaffer collaterals under normal physiological conditions, either in vivo or in vitro. Nonetheless, under special conditions in vitro, such as reduced inhibition or in the presence of BAYK8644, an agonist of the L-type voltage-gated calcium channel, a slight but significant heterosynaptic LTD can be observed (Abraham & Wickens 1991, Wickens & Abraham 1991). Pockett et al (1990) recently have identified conditions in CA1 under which LTD of large magnitude may be elicited. Thus, in the presence of high concentrations of $Mg^{2+}$ to block synaptic transmission, either bursts of antidromic stimulation or intracellular depolarizing pulses yield a large depression of the synaptic responses to orthodromic synaptic stimulation (upon wash out of the high $Mg^{2+}$ solution). Although the physiological relevance of these findings might be questioned for CA1, this approach may prove useful for dissecting the mechanisms of depression in locations such as the dentate gyrus, where heterosynaptic LTD is a robust phenomenon in vivo but has not been amenable to study in vitro.

## DISCUSSION

As this review of the literature clearly shows, the phenomenon of long-term depression in hippocampus is now on firm ground. The available data demonstrate the existence of two forms of LTD, heterosynaptic and homosynaptic, that are differentially expressed in dentate gyrus and CA1. We shall now address two final questions: (a) How closely related are homosynaptic and heterosynaptic LTD, and (b) what are the possible contributions of these forms of synaptic plasticity to hippocampal function?

### Do Homo- and Heterosynaptic Forms of LTD Share Common Expression Mechanisms?

The modes of induction for homosynaptic and heterosynaptic LTD clearly differ. Homosynaptic LTD requires activity at the synapses that are to be depressed, NMDA receptors appear to play a crucial role in providing the calcium necessary for triggering the downstream biochemical cascades, and the LTD can be facilitated by prior priming stimulation. Heterosynaptic LTD, on the other hand, involves no apparent presynaptic activity, it is unaffected by priming stimulation, and voltage-gated calcium channels are probably the critical source of calcium. These differences in induction mechanisms raise the critical question of whether each leads to a unique set of expression mechanisms or both funnel to a common trigger—for example, raised intracellular calcium levels—and then share downstream expression mechanisms. Answering this question has not been easy, since preparations favorable to obtaining one kind of LTD have generally not been favorable to obtaining the other.

Recent experiments using saturation and occlusion tests have suggested that homo- and heterosynaptic LTD do indeed share largely overlapping expression mechanisms. In the dentate gyrus in vivo, saturation of homosynaptic LTD in the lateral path by using the primed-associative protocol described above completely occluded the induction of more LTD by a heterosynaptic protocol involving medial path tetanization. Conversely, initial saturation of heterosynaptic LTD resulted in a 65% occlusion of homosynaptic LTD (Christie et al 1995). This latter result leaves open the possibility that homosynaptic LTD involves one or more mechanisms in addition to those underlying heterosynaptic LTD, but these results could also be explained by the homosynaptic protocol depressing a greater population of synapses, given that the spatial spread of heterosynaptic LTD is restricted by synaptic inhibition (White et al 1990, Abraham & Wickens 1991, Zhang & Levy 1993, Tomasulo et al 1993). These data, taken together with the fact that both forms of LTD can reverse established LTP, and are reversible by LTP, suggest that the two classes of induction protocols converge onto common expression mechanisms. A caveat regarding the saturation and occlusion experiments, however, is that induction of LTD by one protocol might cause a down-regulation of the induction mechanisms of the other, irrespective of any commonalty of the expression mechanisms. Additional approaches should help settle this question. For example, it will be interesting to see whether heterosynaptic LTD shows the same developmental regulation as homosynaptic LTD and the same dependence on phosphatase activation.

## What is The Function of LTD?

Assigning a precise functional role to hippocampal LTD is difficult because the precise contribution of the hippocampus to brain function is very poorly understood. Of course, hippocampus traditionally has been associated with certain forms of memory. Although the debate continues to rage about what types of information are stored in the hippocampus and for how long, it is agreed that a likely basis for information storage there is synaptic modification. LTD must therefore be considered as a candidate mechanism for memory formation in hippocampus.

Historically, increases in synaptic weights have received somewhat more attention as a learning mechanism. Although learning may occur through LTP-like processes alone, having LTD-like modifications clearly adds flexibility and information storage capacity to the system (Willshaw & Dayan 1990). Certainly most neural network models now include both up- and down-regulation of synaptic efficacy as storage mechanisms. From this general point of view, viewing LTP as a learning mechanism and LTD as a forgetting mechanism, as has sometimes been suggested, is probably unwise.

In the dentate gyrus of awake animals, a brief tetanus reliably causes heterosynaptic LTD, which is as persistent, robust, and physiological as LTP (Abraham et al, 1994). However, the lengthy induction protocols that are usually used to induce homosynaptic LTD in CA1 could make this form of plasticity an unlikely memory mechanism. Nine hundred pulses delivered over 15 min evoke a modification that is easily studied, but fewer stimuli over periods of less than 2 min can also evoke LTD that, although of smaller magnitude, is still significant (cf Mulkey & Malenka 1992, Kirkwood & Bear 1994). Furthermore, homosynaptic LTD can be induced rapidly with fewer pulses under the appropriate experimental conditions. In a particularly dramatic example, Huerta & Lisman (1995) have shown that a brief burst, lasting only

40 ms, can produce NMDA receptor–dependent LTD if the burst is delivered at the same time as a negative peak in the theta frequency oscillation that is induced in slices by application of carbachol. Neural network models store memories by assuming that small modifications, often occurring over many iterations, are distributed over a large number of synapses, so homosynaptic LTD should not be disqualified as a candidate memory mechanism on these grounds alone.

Still, the difficulty some have had in inducing LTD in adult CA1 has led to the suggestion that it is less robust than LTP and is therefore less likely to contribute to information storage. However, we physiologists have relatively crude tools—stimulating and recording electrodes—that are not optimal for studying the subtle synaptic modifications that underlie memory. The fact that under the appropriate conditions, LFS (and other stimulation) can reliably produce homosynaptic LTD in adult CA1 may be taken as occurrence proof that a mechanism for decreasing synaptic strengths exists. Similarly, LTP proves that a mechanism for increasing synaptic strength exists. But whether natural patterns of synaptic activation actually cause either LTP or LTD anywhere in the hippocampus remains to be determined.

We suggest that the mechanisms of LTP and LTD make complementary contributions to hippocampal function. Both forms of plasticity, acting in concert, seem ideally suited to contribute to the experience-dependent modification of stimulus selectivity. For example, neurons in CA1 respond when the animal is placed in specific regions of space, and this stimulus selectivity can be modified by experience (Wilson & McNaughton 1993). Experience-dependent shifts in selectivity (and the hippocampal representation of space) require (*a*) that neurons acquire responsiveness to new stimuli and (*b*) that neurons lose responsiveness to previously effective stimuli. LTP may reflect the mechanism underlying the first process, and homo- and heterosynaptic LTD may reflect the mechanisms of the second. Similarly, the mechanisms of LTP and LTD may play an important role in the activity-dependent acquisition of stimulus selectivity during development (Bear et al 1987). Indeed, LTD may be more robust early in life because it represents the initial stage of the widespread process that culminates in the pruning and physical retraction of axon terminals.

Directly testing these hypotheses will be a significant challenge. One way to assess the relative contributions and importance of LTP and LTD to learning is to selectively block each of the types of plasticity and compare the animals' performance with those of untreated controls. In the past, blockade of hippocampal LTP by NMDA antagonists has provided the most convincing evidence for a role of LTP in certain hippocampally dependent learning paradigms (Morris et al 1986), but the interpretation of such experiments is now clouded by the evidence that NMDA receptor blockade also prevents LTD induction.

Better success in dissecting independent contributions of LTP and LTD may be obtained in gene knockout experiments. In the end, however, the important goal remains to demonstrate directly that LTD and/or LTP do in fact occur at hippocampal synapses during learning and are necessary steps in hippocampus-dependent memory formation.

ACKNOWLEDGMENTS

The authors wish to acknowledge the support of the Fogarty International Center, the University of Otago, and the Health Research Council of New Zealand. We also thank the numerous colleagues who provided us with reprints and copies of manuscripts prior to publication, including BE Alger, NT Slater, RF Thompson, Y Frégnac, G Collingridge, ER Kandel, CF Stevens, WB Levy, JL Martinez, PK Stanton, JE Lisman, L Bindman, and T Sacktor. We apologize to those whose work was not discussed, but space limitations necessitated that we restrict the scope of the review.

*Literature Cited*

Abeliovich A, Chen C, Goda Y, Silva AJ, Stevens CF, Tonegawa S. 1993. Modified hippocampal long-term potentiation in PKC gamma-mutant mice. *Cell* 75:1253–62

Abraham W, Bliss T. 1985. An analysis of the increase in granule cell excitability accompanying habituation in the dentate gyrus of the anesthetized rat. *Brain Res.* 331:303–13

Abraham W, Bliss T, Goddard G. 1985. Heterosynaptic effects accompany long-term but not short-term potentiation in the rat perforant path. *J. Physiol.* 363:335–49

Abraham W, Gustafsson B, Wigstrom H. 1986. Single high strength afferent volleys can produce long-term potentiation in the hippocampus. *Neurosci. Lett.* 70:217–22

Abraham WC, Christie BR, Logan B, Lawlor P, Dragunow M. 1994. Immediate early gene expression associated with the persistence of heterosynaptic long-term depression in the hippocampus. *Proc. Natl. Acad. Sci. USA* 91:10049–53

Abraham WC, Goddard GV. 1983. Asymmetric relations between homosynaptic long-term potentiation and heterosynaptic long-term depression. *Nature* 305:717–19

Abraham WC, Wickens JR. 1991. Heterosynaptic long-term depression is facilitated by blockade of inhibition in area CA1 of the hippocampus. *Brain Res.* 546:336–40

Aiba A, Chen C, Herrup K, Rosenmund C, Stevens CF, Tonegawa S. 1994. Reduced hippocampal long-term potentiation and context-specific deficit in associative learning in mGluR1 mutant mice. *Cell* 79:365–75

Asztely F, Wigström H, Gustafsson B. 1992. The relative contribution of NMDA receptor channels in the expression of long-term potentiation in the hippocampal CA1 region. *Eur. J. Neurosci.* 4:681–90

Barrionuevo G, Schottler F, Lynch G. 1980. The effects of low frequency stimulation on control and "potentiated" synaptic responses in the hippocampus. *Life Sci.* 27: 2385–391

Bashir ZI, Collingridge GL. 1994. An investigation of depotentiation of long-term potentiation in the CA1 region of the hippocampus. *Exp. Brain Res.* 100:437–43

Baskys A, Malenka RC. 1991. Agonists at metabotropic glutamate receptors presynaptically inhibit EPSCs in neonatal rat hippocampus. *J. Physiol.* 444:687–701

Bear MF. 1988. Involvement of excitatory amino acid receptor mechanisms in the experience-dependent development of visual cortex. In *Frontiers in Excitatory Amino Acid Research*, ed. EA Cavalheiro, J Lehmann, L Turski, 46:393–401. New York: Liss

Bear MF, Cooper LN, Ebner FF. 1987. A physiological basis for a theory of synaptic modification. *Science* 237:42–48

Ben-Ari Y, Aniksztejn L, Bregestovski P. 1992. Protein kinase C modulation of NMDA currents: an important link for LTP induction. *Trends Neurosci.* 15:333–39

Bliss TVP, Collingridge GL. 1993. A synaptic model of memory: long-term potentiation in the hippocampus. *Nature* 361:31–39

Bolshakov VY, Siegelbaum SA. 1994. Postsynaptic induction and presynaptic expression of hippocampal long-term depression. *Science* 264:1148–52

Bortolotto ZA, Bashir ZI, Davis CH, Collingridge GL. 1994. A molecular switch activated by metabotropic glutamate receptors regulates induction of long-term potentiation. *Nature* 368:740–43

Brocher S, Artola A, Singer W. 1992. Intracellular injection of Ca²⁺ chelators blocks induction of long-term depression in rat visual cortex. *Proc. Natl. Acad. Sci. USA* 89:123–7

Christie B, Abraham W, Bear M. 1993. A possible "priming" requirement for low-frequency stimulation-induced homosynaptic depression (LTD) in area CA1 in vivo. *Soc. Neurosci. Abstr.* 19:1324

Christie BR, Abraham WC. 1992. NMDA-dependent heterosynaptic long-term depression in the dentate gyrus of anaesthetized rats. *Synapse* 10:1–6

Christie BR, Abraham WC. 1992. Priming of associative long-term depression in the dentate gyrus by theta frequency synaptic activity. *Neuron* 9:79–84

Christie BR, Abraham WC. 1994. L-type voltage-sensitive calcium channel antagonists block heterosynaptic long-term depression in the dentate gyrus of anaesthetized rats. *Neurosci. Lett.* 167:41–45

Christie BR, Stellwagen D, Abraham WC. 1995. Evidence for common expression mechanisms underlying heterosynaptic and associative long-term depression in the dentate gyrus. *J. Neurophysiol.* 74:1244–47

Conn PJ, Boss V, Chung DS. 1994. Second-messenger systems coupled to metabotropic glutamate receptors. In *The Metabotropic Glutamate Receptors*, ed. PJ Conn, J Patel, pp. 59–98. Totowa, NJ: Humana

Cummings JA, Nicola SM, Malenka RC. 1994. Induction in the rat hippocampus of long-term potentiation (LTP) and long-term depression (LTD) in the presence of a nitric oxide synthase inhibitor. *Neurosci. Lett.* 176:110–14

Dahl D, Sarvey J. 1989. Norepinephrine induces pathway-specific long-lasting potentiation and depression in the hippocampal dentate gyrus. *Proc. Natl. Acad. Sci. USA* 86:4776–80

Debanne D, Gahwiler BH, Thompson SM. 1994. Asynchronous pre- and postsynaptic

activity induces associative long-term depression in area CA1 of the rat hippocampus in vitro. *Proc. Natl. Acad. Sci. USA* 91:1148–52

Desai MA, McBain CJ, Kauer JA, Conn PJ. 1994. Metabotropic glutamate receptor-induced disinhibition is mediated by reduced transmission at excitatory synapses onto interneurons and inhibitory synapses onto pyramidal cells. *Neurosci. Lett.* 181:78–82

Desmond N, Levy W. 1983. Synaptic correlates of associative potentiation/depression: an ultrastructural study in the hippocampus. *Brain Res.* 265:21–30

Desmond NL, Colbert CM, Zhang DX, Levy WB. 1991. NMDA receptor antagonists block the induction of long-term depression in the hippocampal dentate gyrus of the anesthetized rat. *Brain Res.* 552:93–98

Dragunow D, Abraham W, Goulding M, Mason S, Robertson H, Faull R. 1989. Longterm potentiation and the induction of c-fos mRNA and proteins in the dentate gyrus of unanesthetized rats. *Neurosci. Lett.* 101:274–80

Dudek SM, Bear MF. 1989. A biochemical correlate of the critical period for synaptic modification in the visual cortex. *Science* 246:673–75

Dudek SM, Bear MF. 1992. Homosynaptic long-term depression in area CA1 of hippocampus and effects of N-methyl-D-aspartate receptor blockade. *Proc. Natl. Acad. Sci. USA* 89:4363–67

Dudek SM, Bear MF. 1993. Bidirectional longterm modification of synaptic effectiveness in the adult and immature hippocampus. *J. Neurosci.* 13:2910–18

Errington ML, Richter-Levin G, Bliss TVP. 1994. Low-frequency stimulation in the dentate gyrus in vivo does not produce long-term depression or depotentiaion. *Soc. Neurosci. Abstr.* 20:895

Figurov A, Boddeke H, Muller D. 1993. Enhancement of AMPA-mediated synaptic transmission by the protein phosphatase inhibitor calyculin A in rat hippocampal slices. *Eur. J. Neurosci.* 5:1035–41

Fitzpatrick JS, Baudry M. 1994. Blockade of long-term depression in neonatal hippocampal slices by a phospholipase A2 inhibitor. *Dev. Brain Res.* 78:81–86

Fujii S, Saito K, Miyakawa H, Ito K-i, Kato H. 1991. Reversal of long-term potentiation (depotentiation) induced by tetanus stimulation of the input to CA1 neurons of guinea pig hippocampal slices. *Brain Res.* 555:112–22

Hanse E, Gustafsson B. 1992. Postsynaptic, but not presynaptic, activity controls the early time course of long-term potentiation in the dentate gyrus. *J. Neurosci.* 12:3226–40

Herron C, Malenka RC. 1994. Activity dependent enhancement of synaptic transmission in hippocampal slices treated with the phosphatase inhibitor calyculin A. *J. Neurosci.* 14:6013–20

Heynen A, Bear MF, Abraham WC. 1995. Low-frequency stimulation of the Schaffer collaterals produces homosynaptic LTD in area CA1 of the adult rat hippocampus in vivo. *Soc. Neurosci. Abstr.* 21:2005

Hrabetova S, Sacktor TC. 1994. Down-regulation of protein kinase M$\zeta$ in maintenance of homosynaptic long-term depression in hippocampal CA1. *Soc. Neurosci. Abstr.* 20:446

Huerta P, Lisman JE. 1995. Single bursts may be the natural stimulus for the induction of LTP and LTD. *Neuron.* In press

Issac JTR, Nicoll RA, Malenka RC. 1995. Evidence for silent synapses: implications for the expression of LTP. *Neuron* 15:427–34

Izumi Y, Zorumski CF. 1993. Nitric oxide and long-term synaptic depression in the rat hippocampus. *Neuroreport* 4: 1131–34

Joly C, Gomeza J, Brabet I, Curry K, Bockaert J, Pin J-P. 1995. Molecular, functional and pharmacological characterization of the metabotropic glutamate receptor type 5 splice variants: comparison with mGluR1. *J. Neurosci.* 15:3970–81

Kauer JA, Malenka RC, Nicoll RA. 1988. A persistent postsynaptic modification mediates long-term potentiation in the hippocampus. *Neuron* 1:911–17

Kerr DS, Abraham WC. 1993. Comparison of associative and non-associative conditioning procedures in the induction of LTD in CA1 of the hippocampus. *Synapse* 14: 305–13

Kerr DS, Abraham WC. 1995. Cooperative interactions among afferents govern the induction of homosynaptic LTD. *Proc. Natl. Acad. Sci. USA.* In press

Kirkwood A, Bear MF. 1994. Homosynaptic long-term depression in the visual cortex. *J. Neurosci.* 14:3404–12

Kirkwood A, Dudek SM, Gold JT, Aizenman CD, Bear MF. 1993. Common forms of synaptic plasticity in the hippocampus and neocortex in vitro. *Science* 260:1518–21

Komatsu Y, Iwakiri M. 1992. Low-threshold $Ca^{2+}$ channels mediate induction of long-term potentiation in kitten visual cortex. *J. Neurophysiol.* 67:401–10

Krug M, Muller-Welde P, Wagner M, Ott T, Mathies H. 1985. Functional plasticity in two afferent systems of the granule cells in the rat dentate area: frequency related changes, long-term potentiation and heterosynaptic depression. *Brain Res.* 260:264–72

Kullman DM, Nicoll RA. 1992. Long-term potentiation is associated with increases in quantal content and quantal amplitude. *Nature* 357:240–44

Larkman A, Hannay T, Stratford K, Jack J. 1992. Presynaptic release probability influences the locus of long-term potentiation. *Nature* 360:70–73

Larson J, Xiao P, Lynch G. 1993. Reversal of LTP by theta frequency stimulation. *Brain Res.* 600:97–102

Levy WB, Steward O. 1979. Synapses as associative memory elements in the hippocampal formation. *Brain. Res.* 175:233–45

Levy WB, Steward O. 1983. Temporal contiguity requirements for long-term associative potentiation/depression in the hippocampus. *Neuroscience* 8:791–97

Liao D, Hessler NA, Malinow R. 1995. Activation of postsynaptically silent synapses during pairing-induced LTP in CA1 region of hippocampal slice. *Nature.* 375:400–4

Liao D, Jones A, Malinow R. 1992. Direct measurements of quantal changes underlying long-term potentiation in CA1 hippocampus. *Neuron* 9:1089–97

Lisman J. 1989. A mechanism for the Hebb and the anti-Hebb processes underlying learning and memory. *Proc. Natl. Acad. Sci. USA* 86: 9574–78

Lopez H, Burger B, Dickstein R, Desmond NL, Levy WB. 1990. Long-term potentiation and long-term depression in the dentate gyrus: quantitation of dissociable synaptic modification. *Synapse* 5:33–47

Lovinger D, Routtenberg A. 1988. Synapse-specific protein kinase C activation enhances maintenance of long-term potentiation in rat hippocampus. *J. Physiol.* 400:321–33

Lynch GS, Dunwiddie T, Gribkoff V. 1977. Heterosynaptic depression: a postsynaptic correlate of long term potentiation. *Nature* 266:737–39

Maccaferri G, Janigro D, Lazzari A, DiFrancesco D. 1994. Cesium prevents maintenance of long-term depression in rat hippocampal CA1 neurons. *NeuroReport* 5:1813–16

Mayford M, Wang J, Kandel ER, O'Dell T. 1995. CaMKII regulates the asociativity function of hipocampal synapses for the production of both LTD and LTP. *Cell* 81:891–904

McNaughton B. 1980. Evidence for two physiologically distinct perforant pathways to the fascia dentata. *Brain Res.* 199:1–20

Morris RGM, Anderson E, Lynch GS, Baudry M. 1986. Selective impairment of learning and blockade of long-term potentiation by an N-methyl-D-aspartate receptor antagonist, APV. *Nature* 319:774–76

Mulkey RM, Endo S, Shenolikar S, Malenka RC. 1994. Involvement of a calcineurin/inhibitor-1 phosphatase cascade in hippocampal long-term depression. *Nature* 369: 486–88

Mulkey RM, Herron CE, Malenka RC. 1993.

An essential role for protein phosphatases in hippocampal long-term depression. *Science* 261:1051–55

Mulkey RM, Malenka RC. 1992. Mechanisms underlying induction of homosynaptic long-term depression in area CA1 of the hippocampus. *Neuron* 9:967–75

Muller D, Heftt S, Figurov A. 1995. Heterosynaptic interactions between LTP and LTD in CA1 hippocampal slices. *Neuron* 14:599–605

Muller D, Lynch G. 1988. Long-term potentiation differentially affects two components of synaptic responses in hippocampus. *Proc. Natl. Acad. Sci. USA* 85:9346–50

Ocorr K, Schulman H. 1991. Activation of multifunctional Ca²⁺/calmodulin-dependent kinase in intact hippocampal slices. *Neuron* 6:907–14

O'Dell TJ, Kandel ER. 1994. Low-frequency stimulation erases LTP through an NMDA receptor-mediated activation of protein phosphatases. *Learning Memory* 1:129–39

Otani S, Connor JA. 1995. Long-term depression of naive synapses in adult hippocampus induced by asynchonous synaptic activity. *J. Neurophys.* 73:2596–601

Pang K, Williams M, Olton D. 1993. Activation of the medial septal area attenuates LTP of the lateral perforant path and enhances heterosynaptic LTD of the medial perforant path in aged rats. *Brain Res.* 632:150–60

Paulsen O, Li YG, Hvalby O, Anderson P, Bliss TV. 1993. Failure to induce long-term depression by an anti-correlation procedure in area CA1 of the rat hippocampal slice. *Eur. J. Neurosci.* 5:1241–46

Pavlides C, Greenstein Y, Grudman M, Winson J. 1988. Long-term potentiation in the dentate gyrus is induced preferentially on the positive phase of the theta rhythm. *Brain Res.* 439:383–87

Pockett S, Brookes NH, Bindman LJ. 1990. Long-term depression at synapses in slices of rat hippocampus can be induced by bursts of postsynaptic activity. *Exp. Brain Res.* 80:196–200

Sacktor TC, Osten P, Valsamis H, Jiang X, Naik MU, Sublette E. 1993. Persistent activation of the zeta isoform of protein kinase C in the maintenance of long-term potentiation. *Proc. Natl. Acad. Sci. USA* 90:8342–46

Schexnayder L, Christie B, Johnston D. 1995. Involvement of Ni²⁺-sensitive calcium channels in homosynaptic LTD in hippocampus. *Soc. Neurosci. Abstr.* 21:1808

Schwartz JH. 1993. Cognitive kinases. *Proc. Natl. Acad. Sci. USA* 90:8310–13

Selig DK, Hjelmstad GO, Herron CO, Nicoll RA, Malenka RC. 1995. Independent mechanisms for long-term depression of AMPA and NMDA responses. *Neuron* 15:417–26

Selig DK, Lee H-K, Bear MF, Malenka RC.

1995. Reexamination of the effects of MCPG on hippocampal LTP, LTD, and depotentiation. *J. Neurophys.* 74:1075–82

Stanton P, Mody I, Heinemann U. 1989. A role for N-methyl-D-aspartate receptors in norepinephrine-induced long-lasting potentiation in the dentate gyrus. *Exp. Brain Res.* 77:???

Stanton PK. 1994. Transient protein kinase C activation primes long-term depression and suppresses long-term potentiation of synaptic transmission in hippocampus. *Proc. Natl. Acad. Sci. USA* 92:1724–28

Stanton PK, Sejnowski TJ. 1989. Associative long-term depression in the hippocampus induced by hebbian covariance. *Nature* 229:215–18

Staubli U, Lynch G. 1990. Stable depression of potentiated synaptic responses in the hippocampus with 1–5 Hz stimulation. *Brain Res.* 513:113–18

Steele PM, Mauk MD. 1994. Inhibitory synaptic transmission modulates the induction of long-term depression in CA1 region of hippocampus. *Soc. Neurosci. Abstr.* 20:1514

Stevens CF, Tonegawa S, Wang Y. 994. The role of calcium-calmodulin kinase II in three forms of synaptic plasticity. *Curr. Opin. Biol.* 4:687–93

Stevens CF, Wang Y. 1994. Changes in reliability of synaptic function as a mechanism for plasticity. *Nature* 371:704–7

Teyler T, Alger B. 1976. Monosynaptic habituation in the vertebrate forebrain: the dentate gyrus examined in vitro. *Brain Res.* 115:413–26

Thiels E, Barrionuevo G, Berger TW. 1994. Excitatory stimulation during postsynaptic inhibition induces long-term depression in hippocampus in vivo. *J. Neurophysiol.* 71:3009–16

Tomasulo R, Ramirez J, Steward O. 1993. Synaptic inhibition regulates associative interactions between afferents during the induction of long-term potentiation and long-term depression. *Proc. Natl. Acad. Sci. USA* 90:11578–82

Tomasulo RA. 1994. Homosynaptic and heterosynaptic changes in driving of dentate gyrus interneurons following brief tetanic stimulation. *Soc. Neurosci. Abstr.* 20:1341

Tong G, Shaperd D, Jahr CE. 995. Synaptic desensitization of NMDA receptors by calcineurin. *Science* 267:1510–12

Wagner JJ, Alger BE. 1995. GABAergic and developmental influences on homosynaptic LTD and depotentiation in rat hippocamus. *J. Neurosci.* 15:1577–86

Watkins J, Collingridge G. 1994. Phenylglycine derivatives as antagonists of metabotropic glutamate receptors. *Trends Pharmacol. Sci.* 15:333–42

Wexler EM, Stanton PK. 1993. Priming of homosynaptic long-term depression in hippocampus by previous synaptic activity. *Neuro-Report* 4:591–94

White G, Levy WB, Steward O. 1988. Evidence that associative interactions between synapses during the induction of long-term potentiation occur within local dendritic domains. *Proc. Natl. Acad. Sci. USA* 85:2368–72

White G, Levy WB, Steward O. 1990. Spatial overlap between populations of synapses determines the extent of their associative interaction during the induction of long-term potentiation and depression. *J. Neurophysiol.* 64:1186–98

Wickens JR, Abraham WC. 1991. The involvement of L-type calcium channels in heterosynaptic long-term depression in the hippocampus. *Neurosci. Lett.* 130:128–32

Wigstrom H, Gustafsson B. 1983. Large long-lasting potentiation in the dentate gyrus in vitro during blockade of inhibition. *Brain Res.* 275:153–58

Williams JH, Errington ML, Li Y-G, Lynch MA, Bliss TVP. 1993. The search for retrograde messengers in long-term potentiation. *Semin. Neurosci.* 5:149–58

Willshaw D, Dayan P. 1990. Optimal plasticity from matrix memories: What goes up must come down. *Neural Comput.* 2:85–93

Wilson M, McNaughton B. 1993. Dynamics of the hippocampal ensemble that codes for space. *Science* 261:1055–58

Xiao MY, Wigstrom H, Gustafsson B. 1994. Long-term depression in the hippocampal CA1 region is associated with equal changes in AMPA and NMDA receptor-mediated synaptic potentials. *Eur. J. Neurosci.* 6:1055–57

Xie X, Berger TW, Barrionuevo G. 1992. Isolated NMDA receptor-mediated synaptic responses express both LTP and LTD. *J. Neurophysiol.* 67:1009–13

Yang X-D, Connor JA, Faber DS. 1994. Weak excitation and simultaneous inhibition induce long-term depression in hippocampal CA1 neurons. *J. Neurophysiol.* 71:1586–90

Zhang DX, Levy WB. 1993. Bicuculline permits the induction of long-term depression by heterosynaptic, translaminar conditioning in the hippocampal dentate gyrus. *Brain Res.* 613:309–12

Zhang DX, Levy WB. 1994. Muscimol permits homosynaptic LTD in the dentate gyrus in vitro. *Soc. Neurosci. Abstr.* 20:897

*Annu. Rev. Neurosci. 1996. 19:463–89*

# INTRACELLULAR SIGNALING PATHWAYS ACTIVATED BY NEUROTROPHIC FACTORS

*Rosalind A. Segal*

Department of Neurology, Beth Israel Hospital and Harvard Medical School, Boston, Massachusetts 02215

*Michael E. Greenberg*

Division of Neuroscience, Department of Neurology, Children's Hospital and Department of Neurobiology, Harvard Medical School, Boston Massachusetts 02115

KEY WORDS:    neurotrophins, CNTF, Trk, MAPK, c-*fos*

## ABSTRACT

Soluble and membrane embedded neurotrophic factors bind to specific receptors on responsive neurons and thereby initiate dramatic changes in the proliferation, differentiation, and survival of their target cells. Recent studies have elucidated many of the intracellular pathways by which neurotrophins and ciliary neurotrophic factor (CNTF) function to regulate gene expression and thereby achieve diverse biological responses. In this review we have focused particular attention on the importance of the Ras-MAP kinase pathway for neurotrophin signaling, and the role of the Jak-STAT pathway for CNTF signaling. Characterization of the enzymes, linker proteins, and transcription factors that are sequentially activated in response to neurotrophic factors has provided significant insight into the mechanisms by which these agents elicit specific biological responses during normal development and into the adaptive responses of mature neurons.

## INTRODUCTION

Extracellular stimuli play critical roles in the development and maintenance of the functioning nervous system in all animals. In vertebrates a variety of soluble neurotrophic factors and neurotransmitters have been identified that have profound effects on neuronal cell physiology. Such factors can stimulate the proliferation and differentiation of neuroblasts during development and trigger adaptive plasticity responses in the mature nervous system. The effects of extracellular stimuli are mediated via the activation of specific plasma

463

0147-006X/96/0301-0463$08.00

membrane–embedded receptors. In the case of neurotrophic factors, receptor activation typically leads to the activation of cellular tyrosine kinases that propagate the effects of the neurotrophic factors within cells (Barbacid et al 1991; Bothwell 1991, 1995; Chao 1992; Barbacid 1993; Raffioni et al 1993). This then leads to changes in gene expression that are critical for the overall cellular response. Neurotransmitters released at the synapse bind to their receptors on the postsynaptic cell and, via a variety of different intracellular signaling mechanisms, can trigger both transient and long-lasting cellular responses in developing and mature neurons (Greenberg et al 1986; Pincus et al 1990; Schwartz 1992; Cohen-Cory et al 1993; Ghosh & Greenberg 1995a).

In recent years there has been tremendous progress in the identification and characterization of the intracellular signaling pathways that mediate the neuronal response to neurotrophic factors and neurotransmitters. The characterization of these signaling mechanisms is beginning to elucidate how such diverse cellular responses as proliferation, differentiation, and death can be initiated in a single cell by varying the nature of the initial stimulating agents. Analysis of the mechanisms of signal transduction is also beginning to explain how events that are initiated at a synapse can be propagated over long distances to the nucleus to effect changes in gene expression. Finally, the characterization of signaling pathways that are regulated by neuronal activity in mature neurons is giving new information regarding the molecular basis of information storage in the brain. In this review we summarize recent progress in defining the intracellular signaling pathways that mediate the neuronal response to specific neurotrophic factors that signal via the activation of cellular tyrosine kinases. The intracellular signaling pathways that mediate the effects of neurotransmitters are the subject of several recent reviews and therefore are not discussed further here (Clapham 1995, Ghosh & Greenberg 1995a,b).

## NEUROTROPHINS

During development, many of the same peptide growth factors that regulate the proliferation of nonneuronal cells also stimulate the proliferation of neuroblasts. Insulin-like growth factors (LeRoith et al 1993), fibroblast growth factors (Unsicker et al 1993), the epidermal growth factor family, and the neurotrophins (Morrison 1993) have all been shown to enhance the proliferation of neuronal precursors cells. Under other circumstances a number of these growth factors can promote the survival and differentiation of distinct but overlapping subsets of neurons. In all cases these growth factors function by binding and activating specific receptor tyrosine kinases. This then triggers a cascade of events that culminates in specific programs of gene transcription and particular cellular responses (Heldin 1995).

Nerve growth factor (NGF) was the first growth factor to be identified and characterized. Over 40 years ago the experiments of Levi-Montalcini and Hamburger established the critical importance of NGF for the survival and development of sympathetic neurons and a subset of sensory neurons (Hamburger & Levi-Montalcini 1949, Levi-Montalcini & Angeletti 1968, Levi-Montalcini 1987). More recent experiments in which the NGF gene was disrupted by homologous recombination have corroborated the initial findings of Levi-Montalcini and Hamburger (Crowley et al 1994). NGF is a member of a family of closely related peptide factors, termed the neurotrophins. This family includes brain-derived neurotrophic factor (BDNF), neurotrophin-3 (NT-3), and neurotrophin-4/5 (NT-4/5) (Snider & Johnson 1989, Eide et al 1993, Korsching 1993, Snider 1994). A related factor, neurotrophin-6 (NT-6), has been identified in Teleost fish (Gotz et al 1994). There may also be additional family members in mammals; Berkemeier et al (1992) have identified several human DNA sequences encoding putative neurotrophic factors. The neurotrophins bind to and activate specific receptor tyrosine kinases of the Trk family. TrkA, TrkB, and TrkC are respectively the receptors for NGF, BDNF, and NT3 (Berkemeier et al 1991; Cordon-Cardo et al 1991; Glass et al 1991; Hempstead et al 1991; Kaplan et al 1991a,b; Klein et al 1991a,b; Lamballe et al 1991; Nebreda et al 1991; Soppet et al 1991; Squinto et al 1991). A second neurotrophin, NT-4/5, also binds to the TrkB receptor (Bothwell 1991, 1995; Chao 1992; Parada et al 1992; Eide et al 1993; Raffioni et al 1993).

The neurotrophins trigger a variety of biological responses—including proliferation (Confort et al 1991, Dicicco-Bloom et al 1993), differentiation, and survival of neuroblasts—and survival and adaptive responses of mature neurons. Among the differentiation responses that neurotrophins elicit are enhanced neurite outgrowth (Koo and Liebl 1992, Wright et al 1992, Lindholm et al 1993, Cohen et al 1994, Schnell et al 1994, Gao et al 1995, Segal et al 1995), changes in electrophyisological properties of the neurons (Kalman et al 1989, Levine et al 1995), and alterations in neuronal cell fate (Sieber-Blum 1991).

The mechanisms by which the neurotrophins generate such a diverse array of cellular responses have been studied extensively using the rat pheochromocytoma cell line PC12 (Green & Tischler 1976). PC12 cells are an actively dividing tumor cell line that express the NGF receptor, Trk A (Kaplan et al 1991a,b), and a p75 receptor (Radeke et al 1987). When exposed to NGF, PC12 cells traverse the cell cycle several times and then differentiate into postmitotic cells that in many ways resemble sympathetic neurons (Greene & Tischler 1982). Accumulating evidence indicates that NGF induction of PC12 cell differentiation is mediated via the TrkA receptor kinase and requires the activation of specific programs of immediate early gene (IEG) and delayed response gene (DRG) expression. Many IEGs encode transcription factors that help regulate the expression of NGF-specific DRGs (Curran & Franza 1988, Sheng & Greenberg 1990, Treisman 1994). In NGF-treated PC12 cells the DRGs encode proteins that contribute in a variety of ways to the differentiated PC12 cell phenotype (Arenander & Herschman 1993).

In contrast to NGF, epidermal growth factor (EGF) activates the EGF-receptor tyrosine kinase and is a mitogen for PC12 cells (Huff et al 1981). Although EGF and NGF initially induce very similar intracellular signaling pathways within PC12 cells, these two growth factors ultimately elicit very different cellular responses. Somehow NGF and EGF are able to induce similar programs of IEG transcription but different DRG programs. The detailed analyses and comparisons of the signaling pathways that regulate the PC12 cell response to NGF and EGF are beginning to provide insight into how these two agents generate such distinct biological responses (Marshall 1995).

## Trk Receptor Signaling

The primary event in neurotrophin signaling is the specific binding of the ligand to its receptor. The binding of NGF to TrkA on the surface of PC12 cells causes receptor dimerization and leads to activation of the intrinsic tyrosine kinase activity of TrkA (Jing et al 1992). The initial substrate of the Trk A kinase is the Trk molecule itself (Kaplan et al 1991a,b). Within the TrkA dimer, the individual TrkA subunits each catalyze the phosphorylation of the other subunit (Heldin 1995). The sites of phosphorylation include three tyrosines that are located within the kinase domain and two tyrosines that lie outside this domain (Middlemas et al 1994, Stephens et al 1994). Phosphorylation of the tyrosines within the kinase domain occurs first and appears to enhance the catalytic activity of the tyrosine kinase (Mitra 1991). Subsequently, the tyrosines that are located outside of the kinase domain are phosphorylated in *trans*. Tyrosine-phosphorylated TrkA then serves as a scaffolding for the recruitment of a variety of adapter proteins and enzymes that ultimately propagate the NGF signal (Vetter et al 1991; Ohmichi et al 1992; Soltolf et al 1992; Knipper et al 1993; Obermeier et al 1993a,b, 1994; Stephens 1994). Within the activated TrkA molecule the phosphotyrosines and their surrounding amino acid residues serve as specific recognition sites for effector molecules that contain a structural motif termed a Src homology 2 (SH2) domain. SH2 domains were initially characterized as a group of roughly 100–amino acid long sequences that are present in a variety of signaling molecules and bear striking similarity to an evolutionary conserved noncatalytic region of the pp60src tyrosine kinase (Schlessinger 1994, Cohen et al 1995). A combination of mutagenesis and peptide competition experiments has led to the characterization of the SH2 domain–containing proteins that interact with TrkA and have allowed the identification of their specific sites of interaction (Obermeier

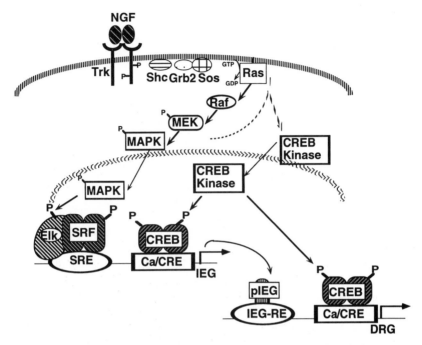

*Figure 1*  The Ras-MAP kinase pathways activated by neurotrophins regulates gene expression. Neurotrophins bind to Trk receptors, leading to dimerization and autophosphorylation. The linker Shc binds to tyrosine phosphorylated Trk and to a Grb2-Sos complex. Sos is a nucleotide exchange factor that activates Ras by replacing GDP with GTP. Activated Ras interacts directly with the serine-threonine kinase Raf. The activated Raf leads to the sequential activation of MEK, the MAP kinase–Erk kinase, and the mitogen activated protein kinase (MAPK). MAPK translocates to the nucleus, where it phosphorylates transcription factors such as Elk-1. Elk binds to a complex consisting of a dimer of the serum response factor (SRF) bound to the serum response element (SRE). A second pathway emanates from the Ras-MAP kinase pathway and leads to the activation of a distinct enzyme, CREB kinase. The CREB kinase is hypothesized to translocate to the nucleus, where it phosphorylates the transcription factor cAMP response element–binding protein (CREB). CREB binds to the cAMP-Ca response element (Ca-CRE) and activates transcription of immediate early genes (IEGs) in cooperation with the SRE complex. The protein products of several IEGs (pIEGs) are transcription factors that bind to immediate early gene response elements (IEG-REs) present within the regulatory regions of delayed response genes (DRGs). The pIEGs are proposed to activate DRG transcription in cooperation with CREB bound to the DRG promoter.

et al 1993a,b, 1994; Middlemas et al 1994; Stephens et al 1994). Among the proteins that interact with tyrosine phosphorylated TrkA are the enzymes phospholipase C$\gamma$ (PLC$\gamma$) (Vetter et al 1991, Obermeier et al 1993a, Stephens et al 1994) and PI3 kinase (Ohmichi et al 1992, Obermeier et al 1993b), and the adapter protein Shc (Obermeier et al 1993b, Stephens et al 1994). Recent evidence suggests that each of these molecules activates distinct signaling pathways that may have different functions. The Ras-MAP kinase pathways initiated by Shc, or by PLCγ, may be involved in differentiation, while the PI3 kinase pathway may be important for survival (Yao & Cooper 1995).

## Ras-MAP Kinase Pathway

SHC, GRB2, AND SOS    The binding of the adapter protein Shc to its recognition site on activated TrkA sets in motion a signaling pathway, termed the Ras-MAP kinase pathway, that is critical for NGF induction of PC12 cell differentiation (Figure 1). The Shc protein has no catalytic function of its own but rather serves as an adapter protein that mediates the association of additional proteins with activated TrkA. Shc binds to TrkA tyrosine 490 (Y490), which is located in the juxtamembrane region of TrkA (Obermeier et al 1993b, Stephens et al 1994). Following its association with TrkA, Shc becomes a substrate for the receptor tyrosine kinase. Phosphorylated Shc can then associate through its phosphotyrosine with another SH2 domain–containing protein, Grb2 (Rozakis-Adcock et al 1992). The sequence of the Grb2 protein reveals that in addition to a single SH2 domain, Grb2 possesses a second structural motif, the Src homology 3 (SH3) domain (Lowenstein et al 1992). SH3 domains also mediate specific protein-protein interactions, since they form recognition sites for proteins that contain proline-rich regions (Cohen et al 1995). The two SH3 domains in Grb2 mediate Grb2 association with the Ras GTP exchange factor Sos. Thus, as summarized in Figure 1, activated TrkA associates with and phosphorylates Shc. This promotes Shc's association with Grb2 and Sos and translocates these cytoplasmic proteins to the plasma membrane where Sos then activates the small G protein Ras.

RAS    Ras is a farnesylated protein that is associated with the inner surface of the plasma membrane through its lipid tail (Casey 1995). Ras is inactive when bound to GDP and active when bound to GTP. Sos activates Ras by promoting the exchange of GDP for GTP (McCormick 1994). It is presently unclear whether the activated TrkA-Shc-Grb2-Sos signaling complex functions solely to bring the GTP-GDP exchange factor Sos to the plasma membrane, in close proximity to Ras, or whether this signaling complex also has additional activating functions.

Ras activation is a critical event in neurotrophin-induced differentiation. When microinjected into PC12 cells, a mutated constitutively active form of H-Ras induces PC12 cell differentiation and thus mimics the action of NGF (Bar-Sagi & Feramisco 1985). Conversely, the introduction into PC12 cells of a dominant interfering form of Ras or of neutralizing antibodies to Ras inhibits NGF induction of PC12 cell differentiation (Hagag et al 1986, Szeberenyi et al 1990). Activated Ras functions at the inner surface of the plasma membrane by activating a specific kinase cascade that has recently been the subject of intense study (Marshall 1994)

RAS-TRIGGERED KINASE CASCADE  The GTP-bound, active form of Ras present in NGF-treated PC12 cells interacts with and activates the serine-threonine kinase Raf (Moodie et al 1992, Wood et al 1992, VanAelst et al 1993, Voltek et al 1993, Marshall 1994). The Ras-Raf interaction is believed to primarily bring Raf to the plasma membrane. This was demonstrated by experiments in which the Ras membrane–localizing sequence was fused to Raf, resulting in constitutive activation of Raf. However, growth factor addition further increases the activity of this chimeric kinase, suggesting that additional modifications of Raf may be required for full activation of this enzyme (Leevers et al 1994). When Raf, Ras, and a receptor tyrosine kinase are coexpressed in a baculovirus system, Raf becomes tyrosine phosphorylated. These tyrosine phosphorylation events increase Raf activity and may be important for Raf activation in vivo (Fabian et al 1993). However, to date, the phosphorylation of Raf on tyrosine residues has not been detected in NGF-stimulated PC12 cells. In PC12 cells, Raf-B, another member of the Raf family, appears to mediate the NGF signal (Oshima et al 1993, Jaiswal et al 1994, Traverse & Cohen 1994). The role tyrosine phosphorylation plays in Raf-B function remains to be determined.

Once activated, Raf-B phosphorylates and stimulates the activity of the dual specificity protein kinase MEK1 (Jaiswal et al 1994, Lange-Carter & Johnson 1994, Vaillancourt et al 1994). The phosphorylation of MEK1 at two sites (Ser217 and Ser221) is catalyzed by Raf-B (Alessi et al 1994). While MEK1 may be one of several Raf-B targets, it appears to be the substrate that plays a central role in NGF-induced PC12 cell differentiation. The activation of MEK1 is necessary and sufficient for certain aspects of PC12 differentiation: Expression of a constitutive form of MEK1 was sufficient to induce neurite outgrowth in PC12 cells, whereas expression of a dominant negative form of MEK1 blocked NGF-induced outgrowth (Cowley et al 1994). The only known substrates of MEK1 are the mitogen-activated protein kinases 1 and 2 (MAPK1 and 2), otherwise known as extracellular signal–regulated kinases (Erks) (Crews et al 1992). In NGF-treated PC12 cells, activated MEK1 catalyzes the phosphorylation of MAPK at a threonine and a tyrosine residue (Payne et al 1991). This then leads to MAPK activation and translocation of the MAPK to the nucleus (Chen et al 1992). Like cyclin-dependent kinases, the MAP kinases are proline-directed kinases; they phosphorylate the consensus site P-LXT-SP (Mansour et al 1994, Marshall 1994). Among the substrates of MAPKs are several transcription factors, as well as another family of kinases, the ribosomal S6 kinases (pp90rsks) (summarized in Chen et al 1993; Marshall 1994, 1995; Hill & Treisman 1995).

Three members of the Rsk family have been characterized, and all are expressed in PC12 cells. Although pp90Rsk1 was originally defined as an enzyme that catalyzes the phosphorylation of ribosomal protein S6 in vitro, this appears not to be the function of this family of enzymes in vivo. Instead, the Rsks apparently phosphorylate transcription factors and thereby control the expression of IEGs and, possibly, DRGs as well (Blenis 1993, Chen et al 1993).

GENE EXPRESSION: IMMEDIATE EARLY GENES    A variety of experiments indicate that a primary function of the Ras-MAP kinase signaling pathway is to induce differentiation by regulating gene expression (Hill & Treisman 1995). Although this pathway may play a major role in controlling the expression of a variety of genes, the evidence that the Ras-MAPK pathway regulates transcription of the c-*fos* proto-oncogene is most compelling. The c-*fos* gene is the best-characterized IEG. In response to a wide variety of extracellular stimuli, transcription of c-*fos* is rapidly and transiently induced, even in the absence of new protein synthesis (Greenberg et al 1986a,b; Morgan & Curran 1986; Curran & Morgan 1987; Sheng & Greenberg 1990). Several key regulatory elements have been identified within the c-*fos* promoter that mediate the c-*fos* response to NGF and other growth factors. One of these elements, the c-*fos* SRE, is a 20-bp region of dyad symmetry located approximately 310 nucleotides 5′ of the site of initiation of c-*fos* mRNA synthesis. The inner core of the SRE, the CArG box, is composed of the sequence $CC(A/T)_6GG$ and binds a transcription factor, serum response factor (SRF) (Treisman 1992, 1994). SRF binding appears to be critical for NGF stimulation of c-*fos* transcription because mutations within the SRE that prevent SRF binding also inhibit NGF-induced transcription (Sheng et al 1988).

Adjacent to the CArG box within the c-*fos* promoter, there is a second regulatory element, CAGGAT. Mutations within this sequence lead to a dramatic reduction in the ability of NGF to induce SRE-dependent transcription (Miranti et al 1995). The CAGGAT sequence is the binding site for a second transcription factor, Elk-1 (Treisman et al 1992). Elk-1 contacts both SRF and the CAGGAT sequence and only interacts with the SRE when SRF is already bound. Thus, a critical function of SRF is to serve as a docking site for Elk-1 (Mueller & Nordheim 1991, Hill et al 1993). Elk-1 is a member of a family of transcription factors that in addition to Elk-1, includes Sap1a, Sap1b, and Net (Dalton & Treisman 1992, Giovane et al 1994). Each of these Elk family members can form a ternary complex with SRF at the SRE, and it is not necessarily clear in a given cell type which of these Elk-related proteins mediates the c-*fos* response. Transfection studies in PC12 cells revealed that SRF and Elk-1 act together to mediate NGF induction of SRE-dependent transcription. In response to NGF, Elk-1 becomes newly phosphorylated at several MAPK consensus sites (e.g. Ser-383 and Ser-389) (Miranti et al 1995). Phosphorylation at these sites stimulates Elk-1's ability to activate transcription, perhaps by promoting interactions among Elk-1 and component(s) of the basic polII transcription machinery.

Several findings suggest that the phosphorylation of Elk-1 (or a related family member) in NGF-stimulated PC12 cells is mediated by the Ras-MAPK pathway. The expression of a dominant interfering form of Ras in PC12 cells blocks Elk-1 phosphorylation and SRE-dependent transcription. In addition, MAPK phosphorylates Elk-1 in vitro at the same sites that become inducibly phosphorylated in vivo (Janknecht et al 1993, Marais et al 1993). Moreover, mutations at these phosphorylation sites block the ability of Elk-1 to activate transcription in PC12 cells (Miranti et al 1995).

Although the ternary complex bound at the c-*fos* SRE plays a critical role in NGF-induced transcription, other transcription factors also contribute to the NGF response. The cAMP regulatory element–binding protein (CREB) can bind to three separate sequences within the c-*fos* promoter (Berkowitz et al 1989). Mutations within these sequence elements can abolish NGF induction of c-*fos* transcription under conditions in which the SRE is functionally intact (Bonni et al 1995). Ginty et al (1994) demonstrated that NGF and other growth factors induce CREB phosphorylation at a critical amino acid, Ser-133, by a Ras-dependent mechanism. Phosphorylation of CREB at Ser-133 appears to allow this factor to cooperate with factors such as SRF and Elk-1 to activate transcription (Bonni et al 1995). As illustrated in Figure 1, the transcription factors that mediate NGF induction of c-*fos* expression are also likely to control the transcriptional response of other IEGs. A number of IEGs have binding sites for SRF, Elk, and CREB within their regulatory regions, suggesting that the signaling pathways that regulate c-*fos* transcription may have a more general function.

GENE EXPRESSION: DELAYED RESPONSE GENES    Interestingly, several NGF-specific DRGs also have CREB-binding sites within their promoters, leading to speculation that CREB may be a regulator of DRG as well as IEG transcription. In two studies mutations within the CREB-binding site of one DRG, the *VGF* gene, significantly reduced the level of *VGF* transcription (Hawley et al 1992; Bonni et al 1995). It has been proposed that transcriptional activation of *VGF* requires cooperation between a transcription factor that is the product of an IEG and CREB. In NGF-stimulated cells, CREB is still phosphorylated at Ser-133 many hours after the initial stimulus, at which time the IEG protein products have accumulated. Under such circumstances CREB might cooperate with an IEG protein product to selectively induce DRG transcription (see Figure 1). In contrast, EGF only transiently stimulates CREB Ser-133 phosphorylation and so may not be able to induce *VGF* transcription.

By the time the IEG protein product is made in EGF-stimulated cells, CREB will no longer be phosphorylated and will not be capable of stimulating transcription (Bonni et al 1995). Whether this proposed mechanism contributes to the selective induction of DRGs in NGF-treated PC12 cells remains to be determined.

GENE EXPRESSION: SPECIFICITY    By controlling the phosphorylation of transcription factors such as Elk-1 and CREB, and thereby regulating IEG and possibly DRG expression, the Ras-MAPK pathway mediates NGF-induced differentiation. However, in other systems, the Ras-MAPK pathway has quite different functions. Related Ras-MAPK pathways have been implicated in innumerable responses to extracellular stimuli ranging from the yeast response to changes in osmolarity and mating factors to mitogen-mediated prolif- eration responses (Herskowitz 1995). How the Ras-MAPK pathway(s) are adapted in the nervous system to generate a variety of distinct proliferation and cellular differentiation responses remains to be clarified. Marshall (1995) has summarized evidence that NGF and EGF induce the Ras-MAPK pathway with different time courses and that the difference in kinetics accounts for the differential response of PC12 cells to these two agents. In PC12 cells exposed to NGF, sustained activation of the Ras-MAPK pathway lasts for several hours. In contrast, treatment with the mitogen EGF leads to transient activation of the Ras-MAPK signaling pathway, lasting only minutes after the initial EGF stimulus (Qiu & Green 1992, Traverse et al 1992). When EGF receptors are overexpressed in PC12 cells, the time course of EGF-induced Ras-MAPK activation is prolonged, and the overexpressing PC12 cells differentiate along a neuronal pathway in response to EGF (Traverse et al 1994). This provides evidence that the kinetics of activation may be critical in determining the biological response. The difference between the duration of the NGF- and EGF-triggered Ras-MAPK responses may reflect differences at the level of the receptor tyrosine kinases. One possibility is that specific phosphotyrosines within the EGF receptor that initiate the Ras-MAP kinase pathways are rapidly dephosphorylated, while the corresponding phosphotyrosines in the TrkA receptor are more slowly dephosphorylated.

The ability of NGF to activate the Ras-MAPK pathway for several hours may provide at least a partial explanation of how this factor induces neuronal differentiation. As discussed above, sustained activation of the Ras signaling pathway in NGF-treated PC12 cells can result in the sustained phosphorylation of transcription factors such as CREB. This may allow CREB to selectively activate DRGs that have CREB-binding sites within their regulatory regions. Such DRGs would be activated in response to NGF, but not EGF, and might encode proteins that contribute to the acquisition of a neuronal phenotype (Bonni et al 1995).

PHOSPHOLIPASE C PATHWAY    Given the obvious importance of the Ras-MAP kinase signaling pathway, it was surprising that a mutation within TrkA that abrogates its interaction with Shc had very little effect on the ability of NGF to induce MAPK activity (Stephens et al 1994). This was demonstrated through the use of a mutant PC12 cell line (nnr5 cells) that does not express TrkA and so is not responsive to NGF (Loeb et al 1991). When nnr5 cells were transfected with wild-type TrkA, the NGF response was restored. NGF addition to the transfected cells induced Ras, MAPK, and neuronal differentiation. Surprisingly, when the TrkA mutant that is incapable of interacting with Shc was introduced into nnr5 cells, NGF treatment still led to the induction of Ras and MAPK. The most likely explanation for this finding is that there are other routes to Ras.

NGF induction of PLCγ appears to be an alternative mechanism for Ras activation. PLCγ binds to activated TrkA at phosphorylated tyrosine 785 (Obermeier et al 1993a, Middlemas et al 1994, Stephens et al 1994). Only when both Y785 and the Shc binding site (Y490) of TrkA are mutated to phenylalanines is the receptor no longer capable of inducing Ras-MAPK signaling in response to NGF (Stephens et al 1994). This suggests that when activated TrkA interacts with PLCγ, a signal is transduced that leads to Ras activation. The association of PLCγ with TrkA results in the tyrosine phosphorylation and activation of PLCγ (Vetter et al 1991, Obermeier et al 1993a). The active enzyme cleaves phosphatidylinositol 4,5-bisphosphate to generate two signaling molecules, diacylglycerol (DAG) and inositoltrisphosphate ($IP_3$). $IP_3$ binds to its receptor on the endoplasmic reticulum and triggers the release into the cytoplasm of $Ca^{2+}$ from internal stores (Berninger et al 1993). The second lipid signaling molecule, DAG, stimulates protein kinase C (PKC), either alone or together with $Ca^{2+}$ (Berridge 1993). However, it is not yet clear how PKC and $Ca^{2+}$ lead to the activation of the Ras-MAP kinase pathway in NGF-treated PC12 cells.

## Ras-Independent Pathways

Several phenotypic responses to neurotrophins occur independently of Ras. Recent studies have begun to elucidate novel pathways by which NGF promotes electrical excitability, enhances survival, and induces a cessation of proliferation. NGF enhances the electrical excitability of PC12 cells by regulating transcription of sodium channel genes (Mandel et al 1988, Darcangelo et al 1993). This transcriptional response occurs approximately 5 h after NGF stimulation by a Ras-independent mechanism (Darcangelo & Halegoua 1993). A recent study has highlighted the fact that this transcriptional response is triggered by a very brief pulse of NGF and is also induced by γ interferon (Toledo-Aral et al 1995). These findings raise the possibility that $Na^+$ channel gene induction may be mediated by a Jak-STAT (signal transducers and activators of transcription) pathway, similar to that activated in response to ciliary neurotrophic factor (CNTF) and to cytokines (see below).

Other Ras-independent pathways have been implicated in neurotrophin-induced survival and cessation of proliferation. PI3 kinase is a lipid-protein kinase that is activated in response to neurotrophin binding to the Trk receptor and that has recently been implicated in the neurotrophin-induced survival response (Yao & Cooper 1995). A molecule termed SNT (suc-associated neurotrophic factor–induced tyrosine–phosphorylated target) has been implicated in the pathway whereby NGF induces PC12 cells to exit the cell cycle (Rabin et al 1993). Distinct signaling pathways mediate activation of SNT and the PI3 kinase response (Peng et al 1995).

PI3 KINASE   NGF binding to TrkA promotes TrkA association with PI3 kinase and leads to PI3 kinase activation (Ohmichi et al 1992). PI3 kinase is a heterodimer composed of an 85-kD SH2 domain containing regulatory subunit and a 110-kD catalytic subunit (Carpenter et al 1990). PI3 kinase was first characterized as an activity that catalyzes the phosphorylation of the 3' position of an inositol ring. Researchers now know that PI3 kinase can phosphorylate the 3' position on a variety of inositol lipids and serines on protein substrates (Auger et al 1989). The best-characterized protein substrates of the PI3 kinase are the PI3 kinase 85-kD subunit and the insulin receptor substrate 1 (IRS1) (Carpenter et al 1990, Dhand et al 1993, Lam et al 1994).

Whether the PI3 kinase interacts directly or indirectly with TrkA is a matter of some controversy (Ohmichi et al 1992, Obermeier et al 1993b, Kodaki et al 1994, Rodriguez-Viciana et al 1994, Wang et al 1995). However, regardless of the mechanism by which TrkA activates PI3 kinase, evidence is accumulating that suggests that the PI3 kinase may play an important role as a mediator of NGF responses. Recent evidence suggests that the PI3 kinase may mediate NGF's effects on cell survival. Inhibitors of the PI3 kinase promote PC12 cell apoptosis in the presence of NGF, suggesting that one function of NGF is to activate PI3 kinase, which then promotes cell survival. In PC12 cells engineered to express the platelet-derived growth factor (PDGF) receptor, PDGF was capable of substituting for NGF in promoting cell survival in serum-free medium. However, when the PDGF receptor was mutated so that it no longer could interact with or activate the PI3 kinase, PDGF was no longer effective at promoting cell survival (Yao & Cooper 1995). Following PDGF stimulation, the 3' phosphorylated lipids generated by PI3 kinase activity stimulate the Akt kinase (also known as protein kinase B) (Burgering & Coffer 1995, Franke et al 1995). It is unknown whether NGF-stimulated PI3 activates Akt and thereby mediates a survival signal.

Other downstream responses to PI3 kinase signaling may also be critical for neuronal survival. Several recent studies suggest that one downstream consequence of PI3 kinase activation is the activation of the pp70 ribosomal S6 kinase, a serine-threonine protein kinase (Cheatham et al 1994, Chung et al 1994). Determining whether this enzyme is also part of the survival signaling pathway will be important to future research. It may also prove useful to consider the possibility that PI3 kinase functions as part of the neuronal retrograde signaling pathway, since PI3 kinase plays a role in membrane vesicle trafficking in several systems (Schu et al 1993, Cheatham et al 1994, Chung et al 1994, Kundra et al 1994). Notably PDGF receptors that fail to interact with the PI3 kinase are not internalized properly (Joly et al 1994). The retrograde signaling by neurotrophins is initiated by binding the ligand to its receptor at the axon terminus, followed by internalizing the ligand and its receptor (Raivich et al 1991; Bothwell 1995). The NGF-TrkA complex may need to be internalized to promote survival, and appropriate internalization may require PI3 kinase activation.

SNT    One of the remarkable effects of NGF that distinguishes it from other growth factors, such as EGF, is that NGF can induce an actively dividing cell to withdraw from the cell cycle (Huff et al 1981, Rudkin et al 1989). Likewise, NGF's relatives, BDNF and NT-3, can induce certain types of neural precursors to exit from the cell cycle. The particular features of the neurotrophins and their receptors that endow them with this unique property have not yet been identified, but the protein SNT may be involved. SNT is an 85- to 100-kD protein that is inducibly phosphorylated, in a Trk-dependent and Ras-independent manner, within a minute of NGF addition to PC12 cells. Deletion of a sequence conserved in the Trks (KFG) eliminates SNT activation, without affecting Trk tyrosine phosphorylation or Ras, PLCγ, or PI3 kinase stimulation.Thus SNT defines a novel signaling pathway that is critical for the antimitogenic and neuritogenic activities of NGF (Peng et al 1995). Once phosphorylated, SNT translocates to the nucleus, where it may function as a transcription factor. SNT tyrosine phosphorylation is not triggered by agents like EGF, which fails to induce the withdrawal of PC12 cells from the cell cycle. An intriguing possibility is that SNT functions as a transcriptional regulator of genes that control cell cycle withdrawal (Rabin et al 1993).

p75 LOW-AFFINITY NEUROTROPHIN RECEPTOR    The neurotrophins are distinctive in that they can bind to more than one receptor. In addition to binding to a specific Trk, each of the four known neurotrophins (NGF, BDNF, NT-3 and NT-4/5) interacts with a common receptor, p75 (Rodriguez-Tebar et al 1990, Rodriguez et al 1992). p75 is structurally related to cytokine receptors that are characterized by four cysteine residues and tryptophan-serine (WS) repeats (Smith et al 1994). This receptor is primarily expressed on neurotrophin-responsive cells and has been implicated in apoptosis (Rabizadeh et al 1993) and cell migration (Anton et al 1994). Despite its ability to interact with each of the neurotrophins with a high degree of specificity, p75 appears to be neither necessary nor sufficient for many aspects of neurotrophin signaling. This is suggested by several lines of evidence. First, some neurotrophin responsive cells express a Trk but do not express p75 (Marsh et al 1993, Knusel et al 1994). Second, mutant PC12 cells that do not express Trk and only express p75 do not differentiate in response to NGF (Loeb et al 1991). Third, a mutant NGF that can only bind to TrkA receptors and does not bind to p75 can initiate PC12 cell neuronal differentiation (Ibáñez et al 1992).

Although p75 is not required for neurotrophin signaling, the presence of this receptor may modulate the cellular response to NGF and other neurotrophins. For example, the presence of p75 appears to enhance the sensitivity of TrkA's response to NGF. This has been demonstrated in several ways. First, sensory neurons from mice with targeted gene deletion of p75 require higher concentrations of NGF for survival (Davies et al 1993). This shift in the dose-response curve indicates that p75 can potentiate a response to NGF. Second, coexpression of p75 and Trk in a neuronal precursor cell line led to increased Trk phosphorylation and enhanced biological response to NGF when compared to cells that expressed only Trk (Verdi et al 1994). Furthermore, when NGF binding to p75 was selectively inhibited, NGF binding to TrkA was also reduced, as were other biological responses that are typically elicited when PC12 cells are exposed to NGF (Barker & Shooter 1994). In another study using 3T3 fibroblast cells transfected with Trk receptors and mutant p75 molecules, the extracellular domain of p75 was sufficient to increase Trk sensitivity (Hantzopoulos et al 1994). However, researchers still do not know whether p75 modulates Trk function by affecting receptor function directly (Hempstead et al 1991) or by acting on a downstream signaling pathway. The latter possibility is compatible with the recent finding that p75 itself is endowed with signaling capabilities. NGF binding to p75, in the absence of Trks, was found to activate the sphingomyelinase pathway (Dobrowsky et al 1994) and also to generate specific biological responses, such as a change in cell-cell adhesion (Itoh et al 1995).

The sphingomyelinase pathway is a lipid signaling pathway that was originally characterized in studies of the signaling pathways triggered by tumor necrosis factor $\alpha$ (TNF$\alpha$). Because p75 is related structurally to the p55 TNF$\alpha$ receptor (Smith et al 1994), it is not surprising that activation of p75 triggers similar signaling pathways. Details of these signaling pathways, as elucidated for TNF$\alpha$, are described in several recent reviews (Heller & Kronke 1994, Kolesnick & Golde 1994). Briefly, TNF$\alpha$ binding to its receptor results in the activation of a neutral, membrane-associated sphingomyelinase, which cleaves sphingomyelin to produce ceramide and phosphocholine. The ceramide then binds and activates a 100-kD protein kinase. This ceramide-activated kinase is a proline-directed serine-threonine kinase that has not yet been cloned (Joseph et al 1994).

## Are All Neurotrophins Equal?

Because each of the neurotrophins signals by interacting with p75 and a member of the Trk family, one might expect that the neurotrophins would each have very similar effects on cells. However, studies of the effects of the neurotrophins on the growth, differentiation, and survival of central nervous system neurons have revealed that BDNF and NT-3 can have very different effects even on the same target neuron. For example, cortical precursor cells express both the BDNF and the NT-3 receptors, TrkB and TrkC (Tessarollo et al 1993, Allendoerfer et al 1994, Lamballe et al 1994, Smith et al 1994). However, the biological response to these two factors is quite different: NT-3 enhances neuronal differentiation of the cortical precursors, while BDNF promotes their survival (Ghosh & Greenberg 1995b). Similarly, cerebellar granule precursor cells express both TrkB and TrkC, but in these cells, BDNF promotes survival and enhances axonal elongation, while NT-3 increases neurite fasciculation (Segal et al 1995). So far, no differences in the Trk-mediated signaling pathways have been reported when cortical or cerebellar precursors are exposed to BDNF and NT-3. However, such differences are likely to exist. For example, BDNF may selectively activate the PI3 kinase and thereby promote neuronal survival.

# CILIARY NEUROTROPHIC FACTOR

The neurotrophins have served as the prototype for understanding how extracellular stimuli can elicit biological responses in neurons. However, a variety of additional factors have been identified that are unrelated to the neurotrophins but are capable of inducing profound changes in neurons through the activation of cytoplasmic tyrosine kinases. Among these factors are members of the CNTF family, the integrin family, and the ligands for the Eph family of receptor tyrosine kinases .

CNTF promotes the differentiation and survival of a diverse array of neurons and glia: For example, it can inhibit growth and promote the differentiation of sympathetic precursor cells and induce cholinergic differentiation of mature sympathetic neurons. CNTF also triggers the differentiation of type-II astrocytes from glial progenitor cells and promotes the survival of spinal motor neurons and dopaminergic neurons of the substantia nigra (Hughes et al 1988, Ernsberger et al 1989, Hagg & Varon 1993, Burnham et al 1994, Patterson 1993). In many cases the same cells are responsive to both CNTF and neurotro-

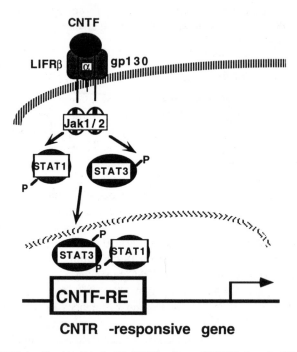

*Figure 2* CNTF signaling involves the Jak-STAT pathway. CNTF binds to the CNTF-specific α receptor and to the leukemia inhibitory factor (LIF)–receptor β subunit and the related gp130, forming a trimeric receptor. This activates the associated Janus tyrosine kinase (Jak 1 or Jak2), which phosphorylates STAT-1 and/or STAT3. The phosphorylated proteins translocate to the nucleus, where they bind to CNTF-responsive elements (CNTF-RE) on CNTF-responsive genes and thereby initiate transcription.

phins. Because CNTF and the neurotrophins are structurally unrelated and their effects are often additive or complementary, CNTF and the neurotrophins might activate distinct signaling pathways. Recent characterization of the CNTF signaling mechanism, depicted in Figure 2, has provided support for this idea.

CNTF is a member of a family of distantly related cytokines that includes leukemia inhibitory factor (LIF), oncostatin M (OSM), interleukin-6 (IL-6), interleukin-11 (IL-11), and cardiotrophin-1 (Bazan 1991). In the nervous system LIF and CNTF have overlapping effects (Ip & Yancopoulos 1992). The reason for this became clear when the LIF and CNTF receptors were identified. The CNTF receptor is a trimer composed of the structurally related gp130 and LIF Rβ subunits that form the LIF receptor, and a unique glycosylphosphatidylinositol-anchored α-receptor subunit (Davis et al 1991, Ip et al 1992). The α receptor confers specificity to the CNTF receptor. In its absence, the β subunits will bind LIF but not CNTF. None of the CNTF receptor subunits contain a tyrosine kinase domain. Instead, as is the case for LIF and IL-6, the CNTF receptor β components are constitutively associated with members of Jak-Tyk family of cytoplasmic tyrosine kinases (Stahl et al 1994).

The first step in CNTF receptor activation involves the binding of CNTF to the α-receptor subunit, which then promotes association with and dimerization of the two β-receptor subunits. The dimerization of the two β subunits leads to the activation of the associated Jak tyrosine kinases and propagation of the CNTF signal (Figure 2). Several distinct Jak-Tyk isoforms have been identified and three members of this family—Jak1, Jak2, and Jak3—can associate with the CNTF receptor β subunits (Stahl et al 1994). This provides a potential mechanism for specificity. By associating with different members of the Jak family, the activated CNTF receptor may be capable of phosphorylating different substrates (Stahl et al 1994).

CNTF activation of the Jak kinases induces the phosphorylation and activation of two members of the STAT family of transcription factors, STAT 1 and STAT 3 (Darnell et al 1994). These STAT proteins are SH2 domain–containing proteins that are located within the cytoplasm prior to CNTF addition. However, when phosphorylated by a Jak kinase, the tyrosine phosphorylated STATs form homo- or heterodimers via phosphotyrosine–SH2 domain interactions. The STAT dimers then rapidly translocate to the nucleus, where they bind to specific sequence determinants (consensus sequence 5'TTCCCCGA-A3') (Hirano et al 1994) within the regulatory regions of their target gene and activate transcription (Bonni et al 1993). STAT1 and STAT3 also can be phosphorylated on serine residues that lie within consensus sites for phosphorylation by MAP kinases. Although the kinase(s) responsible for serine phosphorylation of the STATs have not yet been identified, phosphorylation of STATs on serine residues may potentiate their transcriptional activating potential (Wen et al 1995, Zhang et al 1995). Some of the likely target genes for STATs are IEGs, because a number of these genes have STAT-binding sites, as well as SRE and CRE sites, within their promoters (Hirano et al 1994). Consistent with the finding that CNTF and the neurotrophins utilize different signaling mechanisms, CNTF and the neurotrophins induce distinct subsets of IEG and DRGs (Mulderry 1994). The differential effects of CNTF and neurotrophins on IEG expression may at least in part explain how these two groups of factors elicit distinct biological responses in their target cells.

In some cases, CNTF and neurotrophins can elicit similar biological responses; for example, both BDNF and CNTF are survival factors for motor neurons. The finding that neurotrophins and CNTF have synergistic effects on motor neuron survival (Mitsumoto et al 1994) suggests that neurotrophin and CNTF signaling pathways may interact in vivo. One intriguing possibility is that neurotrophin stimulation of MAP kinases increases serine phosphorylation of STATs and so potentiates their ability to mediate a CNTF response.

## CELL CONTACT–INITIATED SIGNALING

Recently, considerable attention has focused on the signaling pathways whereby cell-cell contact can regulate gene expression. The integrins and the Eph family of receptor tyrosine kinases are two types of receptors involved in such pathways. The integrins were initially identified based on their role in adhesion; however, they can also initiate a complex array of signaling events. Integrins are expressed in the developing and mature nervous system (Bronner-Fraser et al 1992, Duband et al 1992, Letourneau et al 1994) and have been implicated in processes such as migration and neurite outgrowth guidance (Arcangeli et al 1993, Fishman & Hatten 1993, Kuhn et al 1995). Clustering of integrins in focal adhesion areas activates several cytoplasmic tyrosine kinases that feed into the Ras-MAP kinase and PI3 kinase pathways discussed for neurotrophin signaling (Clark & Brugge 1995). Understanding how integrin-mediated contacts play a role in regulating migration, neurite outgrowth, and fasciculation during neuronal development and in adaptive responses will be an important area for future research (Kuhn et al 1995).

The Eph-related receptors constitute a large family of receptor tyrosine kinases, many of which are expressed at high levels in the nervous system (Lhotak et al 1991, Pasquale 1991, Sajjadi et al 1991, Pasquale et al 1992, Maisonpierre et al 1993, Sajjadi & Pasquale 1993, Ruiz & Robertson 1994). A family of ligands for the Eph receptors has recently been identified (Davis et al 1994). These ligands must be membrane embedded to activate their receptors on an adjacent cell surface (Cheng & Flanagan 1994, Davis et al 1994). Several such ligands associate with the outer surface of the plasma membrane by means of a glycosylphosphatidylinositol linkage (Cheng & Flanagan 1994, Kozlosky et al 1995). Glycosylphosphatidylinositol–linked ligands for the Eph receptors, including ELF-1 and RAGS (repulsive axon guidance signal), have been implicated in the topographical organization of the developing retinotectal system (Cheng et al 1995, Drescher et al 1995, Tessier-Lavigne 1995). The signal transduction pathways used by these novel membrane-bound ligands are as yet unknown. To date, the PI3 kinase pathway is the only signaling mechanism that has been demonstrated to be activated by the Eph receptor tyrosine kinases (Pandey et al 1994). In the future, understanding by which mechanisms these novel ligands determine cellular responses will be important.

## CONCLUSIONS

Extracellular stimuli play a major role in the development of the nervous system and in its adaptive responses. Remarkable progress has been made in the past few years in elucidating the intracellular pathways that enable neurotrophic factors to regulate neuronal proliferation, differentiation, and survival. Because many of the basic mechanisms involved in intracellular signaling are defined, studies on neuronal signal transduction can begin to focus on how conventional signaling pathways have been adapted to the requirements of the nervous system. Questions for the future are how these pathways have been modified to allow specificity of signaling and how they manage to convey a signal the length of a nerve process. Both the kinetics and location of the pathways may be important aspects that have been modified for neuronal signaling.

The most dramatic progress has been made in characterizing the signaling pathways used by neurotrophins. The Ras-MAP kinase pathway is critical in neurotrophin-induced differentiation responses, particularly in neurite outgrowth. This multicomponent pathway has been defined; it leads from the neurotrophin Trk receptors to nuclear transcription factors. The time course of Ras-MAP kinase activation may be important for specifying a differentiation rather than a proliferation response and may also enable a signal to be carried over a long distance. Ras-independent pathways are less well understood but appear to be critical in the effects of neurotrophins on electrical excitability, survival, and the cell cycle. The PI3 kinase signaling pathway has been implicated in intracellular targeting and the survival response. The kinetics and location of the components of the PI3 kinase pathway may help explain how target-derived growth factors, presented to nerve endings that can be meters away from the cell nucleus, can act as survival factors. In contrast to the neurotrophins, CNTF induces the Jak-STAT pathway, which is used by several cytokines. Unlike the Ras-MAP kinase pathway, this is a very direct path to transcriptional responses. Whether this difference is important in understanding the biological responses to CNTF remains to be determined.

The intracellular signaling pathways constitute the mechanisms by which all the neurotrophic factors execute their biological responses. Therefore, determining the pathways activated by distinct factors provides insight into the molecular basis by which specific and distinct biological effects can be elicited.

## Literature Cited

Alessi DR, Saito Y, Campbell DG Cohen P, Sithanandam G, et al. 1994. Identification of the sites in MAP kinase kinase-1 phosphorylated by p74raf-1. *EMBO J.* 13:1610–19

Allendoerfer K, Cabelli R, Escandon E, Kaplan D, Nikolics K, Shatz CJ. 1994. Regulation of neurotrophin receptors during the maturation of the mammalian visual system. *J. Neurosci.* 14:1795–811

Anton E, Weskamp G, Reichardt L, Matthew W. 1994. Nerve growth factor and its low affinity receptor promote Schwann cell migration. *Proc. Natl. Acad. Sci. USA* 91:2795–99

Arcangeli A, Becchetti A, Mannini A, Mugnai G, De Filippi P, et al. 1993. Integrin-mediated neurite outgrowth in neuroblastoma cells depends on the activation of potassium channels. *J. Cell. Biol.* 122:1131–43

Arenander A, Herschman H. 1993. Primary response gene expression in the nervous system. See Loughlin & Fallon 1993, pp. 89–128

Auger K, Serunian L, Soltoff S, Libby P, Cantley L. 1989. PDGF-dependent tyrosine phosphorylation stimulates production of novel phosphoinositides in intact cells. *Cell* 57:167–75

Barbacid M. 1993. Nerve growth factor: a tale of two receptors. *Oncogene* 8:2033–41

Barbacid M, Lamballe F, Pulido D, Klein R. 1991. The trk family of tyrosine protein kinase receptors. *Biochim. Biophys. Acta* 1072:115–27

Barker P, Shooter E. 1994. Disruption of NGF binding to the low affinity neurotrophin receptor p75LNTR reduces NGF binding to TrkA on PC12 cells. *Neuron* 13:203–15

Bar-Sagi D, Feramisco JR. 1985. Microinjection of the ras oncogene protein into PC12 induces morphologic differentiation. *Cell* 42:841–48

Bazan J. 1991. Neuropoietic cytokines in the hematopoietic fold. *Neuron* 7:197–208

Berkemeier LR, Ozcelik T, Francke U, Rosenthal A. 1992. Human chromosome 19 contains the neurotrophin-5 gene locus and three related genes that may encode novel acidic neurotrophins. *Somatic Cell Mol. Genet.* 18:233–45

Berkemeier LR, Winslow JW, Kaplan DR, Nikolics K, Goeddel DV, Rosenthal A. 1991. Neurotrophin-5: a novel neurotrophic factor that activates trk and trkB. *Neuron* 7:857–66

Berkowitz L, Riabowal K, Gilman M. 1989. Multiple sequence elements of a single functional class are required for cyclin AMP responsiveness of the mouse c-*fos* promoter. *Mol. Cell. Biol.* 9:4272–81

Berninger B, Garcia DE, Inagaki N, Hahnel C, Lindholm D. 1993. BDNF and NT-3 induce intracellular $Ca^{2+}$ elevation in hippocampal neurons. *NeuroReport* 4:1303–6

Berridge M. 1993. Inositol triphosphate and calcium signaling. *Nature* 361:315–25

Blenis J. 1993. Signal transduction via the MAP kinases: proceed at your own RSK. *Proc. Natl. Acad. Sci. USA* 90:5889–92

Bonni A, Frank E, Schindler C, Greenberg ME. 1993. Characterization of a pathway for ciliary neurotrophic factor signaling to the nucleus. *Science* 262:1575–79

Bonni A, Ginty D, Dudek H, Greenberg M. 1995. Serine 133-phosphorylated CREB induces transcription via a cooperative mechanism that may confer specificity to neurotrophin signals. *Mol. Cell. Neurosci.* 6:168–83

Bothwell M. 1991. Keeping track of neurotrophin receptors. *Cell* 65:915–18

Bothwell M. 1995. Functional interactions of neurotrophins and neurotrophin receptors. *Annu. Rev. Neurosci.* 18:223–53

Bronner-Fraser M, Artinger M, Muschler J, Horwitz AF. 1992. Developmentally regulated expression of alpha 6 integrin in avian embryos. *Development* 115:197–211

Burgering B, Coffer P. 1995. Protein kinase B (Akt) in phosphatidylinositol-3-OH kinase signal transduction. *Nature* 376:599–602

Burnham P, Louis J, Magal E, Varon S. 1994. Effects of ciliary neurotrophic factor on the survival and response to nerve growth factor of cultured rat sympathetic neurons. *Dev. Biol.* 161:96–106

Carpenter C, Auger KK, Duckworth B, Hou W, Schaffhausen B, Cantley L. 1990. Purification and characterization of phosphoinositide 3-kinase from rat liver. *J. Biol. Chem.* 265:19704–11

Casey P. 1995. Protein lipidation in cell signaling. *Science* 268:221–24

Chao M. 1992. Neurotrophin receptors: a window into neuronal differentiation. *Neuron* 9:583–93

Cheatham B, Vlahos C, Cheatham L, Wang L, Blenis J, Kahn C. 1994. Phosphatidylinositol 3-kinase activation is required for insulin stimulation of pp70 S6 kinase, DNA synthesis, and glucose transporter translocation. *Mol. Cell. Biol.* 14:4902–11

Chen RH, Sarnecki C, Blenis J. 1992. Nuclear localization and regulation of erk- and rsk-encoded protein kinases. *Mol. Cell. Biol.* 12:915–27

Chen RH, Tung RR, Abate C, Blenis J. 1993. Cytoplasmic to nuclear signal transduction by mitogen-activated protein kinase and 90 kDa ribosomal S6 kinase. *Biochem. Soc. Trans.* 21:895–900

Cheng HJ, Flanagan JG. 1994. Identification and cloning of ELF-1, a developmentally expressed ligand for the Mek4 and Sek receptor tyrosine kinases. *Cell* 79:157–68

Cheng HJ, Nakamoto M, Bergemann A, Flanagan J. 1995. Complementary gradients in expression and binding of ELF-1 and Mek-4 in development of the topographical retinotectal projection map. *Cell* 83:371–82

Chung J, Grammer T, Lemon K, Kazlauskas A, Blenis J. 1994. PDGF and insulin-dependent pp70$^{S6K}$ activation mediated by phosphatidylinositol-3-OH kinase. *Nature* 370:71–75

Clapham D. 1995. Calcium signaling. *Cell* 80: 259–68

Clark E, Brugge K. 1995. Integrins and signal transduction pathways: the road taken. *Science* 268:233–39

Cohen A, Bray G, Aguayo A. 1994. Neurotrophin4/5 (NT4/5) increases adult rat retinal ganglion cell survival and neurite outgrowth in vitro. *J. Neurobiol.* 25:953–59

Cohen G, Ren R, Baltimore D. 1995. Modular binding domains in signal transduction proteins. *Cell* 80:237–48

Cohen-Cory S, Elliott R, Dreyfus C, Black I. 1993. Depolarizing influences increase low-affinity NGF receptor gene expression in cultured Purkinje neurons. *Exp. Neurol.* 119: 165–75

Confort C, Charrasse S, Clos J. 1991. Nerve growth factor enhances DNA synthesis in cultured cerebellar neuroblasts. *NeuroReport* 2:566–68

Cordon-Cardo C, Tapley P, Jing S, Nanduri V, O'Rourke E, et al. 1991. The trk tyrosine protein kinase mediates the mitogenic properties of nerve growth factor and NT3. *Cell* 66:173–83

Cowley S, Paterson H, Kemp P, Marshall C. 1994. Activation of MAP kinase kinase is necessary and sufficient for PC12 differentiation and for transformation of NIH 3T3 cells. *Cell* 77:841–52

Crews C, Alessandrini A, Erikson E. 1992. The primary structure of MEK, a protein kinase that phosphorylates the ERK gene product. *Science* 258:478–80

Crowley C, Spencer S, Nishimura M, Chen K, Pitts-Meek S, et al. 1994. Mice lacking nerve growth factor display perinatal loss of sensory and sympathetic neurons yet develop basal forebrain cholinergic neurons. *Cell* 76: 1001–11

Curran T, Franza B. 1988. Fos and Jun: the AP-1 connection. *Cell* 55:395–97

Curran T, Morgan J. 1987. Memories of fos. *Bioessays* 7:255–58

Dalton S, Treisman R. 1992. Characterization of SAP-1, a protein recruited by serum response factor to the c-*fos* serum response element. *Cell* 68:597–612

Darcangelo G, Halegoua S. 1993. A branched signaling pathway for nerve growth factor is revealed by src-mediated, ras-mediated, and raf-mediated gene inductions. *Mol. Cell. Biol.* 13:3146–55

Darcangelo G, Paradiso K, Shepherd D, Brehm P, Halegoua S, Mandel G. 1993. Neuronal growth factor regulation of 2 different sodium channel types through distinct signal transduction pathways. *J. Cell. Biol.* 122: 915–21

Darnell JE Jr, Kerr IM, Stark GR. 1994. Jak-STAT pathways and transcriptional activation in response to IFNs and other extracellular signaling proteins. *Science* 264: 1415–21

Davies AM, Lee KF, Jaenisch R. 1993. p75-deficient trigeminal sensory neurons have an altered response to NGF but not to other neurotrophins. *Neuron* 11:565–74

Davis S, Aldrich TH, Valenzuela DM, Wong V, Furth ME, et al. 1991. The receptor for ciliary neurotrophic factor. *Science* 253:59–63

Davis S, Gale NW, Aldrich TH, Maisonpierre PC, Lhotak V, et al. 1994. Ligands for EPH-related receptor tyrosine kinases that require membrane attachment or clustering for activity. *Science* 266:816–19

Dhand R, Hiles I, Panayotou G, Roche S, Fry M, et al. 1993. PI3-kinase is a dual specificity enzyme: autoregulation by an intrinsic protein serine kinase activity. *EMBO J.* 13: 522–33

Dicicco-Bloom E, Friedman WJ, Black IB. 1993. NT-3 stimulates sympathetic neuroblast proliferation by promoting precursor survival. *Neuron* 11:1101–11

Dobrowsky R, Werner M, Castellino A, Chao M, Hannun Y. 1994. Activation of the sphingomyelin cycle through the low affinity neurotrophin receptor. *Science* 265:1596–99

Drescher U, Kremoser C, Handwerker C, Loschinger J, Moda M, Bonhoeffer F. 1995. In vitro guidance of retinal ganglion cell axons by RAGS, a 25 kDa tectal protein related to ligands for Eph receptor tyrosine kinases. *Cell* 82:359–70

Duband JL, Belkin AM, Syfrig J, Thiery JP, Koteliansky VE. 1992. Expression of alpha 1 integrin, a laminin-collagen receptor, during myogenesis and neurogenesis in the avian embryo. *Development* 116:585–600

Eide FF, Lowenstein DH, Reichardt LF. 1993. Neurotrophins and their receptors—current concepts and implication for neurologic disease. *Exp. Neurol.* 121:200–14

Ernsberger U, Sendtner M, Rohrer H. 1989. Proliferation and differentiation of embryonic chick sympathetic neurons: effects of ciliary neurotrophic factor. *Neuron* 2:1275–84

Fabian J, Daar I, Morrison D. 1993. Critical tyrosine residues regulate the enzymatic and biological activity of raf-1 kinase. *Mol. Cell. Biol.* 13:7170–79

Fishman RB, Hatten ME. 1993. Multiple receptor systems promote CNS neural migration. *J. Neurosci.* 13:3485–95

Franke T, Yang S, Chan T, Datta K, Kazlauskas A, et al. 1995. The protein kinase encoded by the Akt proto-oncogene is a target of the PDGF-activated phosphatidylinositol 3-kinase. *Cell* 81:727–36

Gao W, Zheng J, Karihaloo M. 1995. Neurotrophin-4/5 (NT-4/5) and brain derived neurotrophic factor (BDNF) act at later stages of cerebellar granule cell differentiation. *J. Neurosci.* 15:2656–67

Ghosh A, Greenberg ME. 1995a. Calcium signaling in neurons: molecular mechanisms and cellular consequences. *Science* 268:239–47

Ghosh A, Greenberg ME. 1995b. Distinct roles for bFGF and NT3 in the regulation of cortical neurogenesis. *Neuron* 15:1–20

Ginty DD, Bonni A, Greenberg ME. 1994. Nerve growth factor activates a ras-dependent protein kinase that stimulates c-*fos* transcription via phosphorylation of CREB. *Cell* 77:713–25

Giovane A, Pintzas A, Maira S, Sobieszczuk P, Wasylyk B. 1994. Net, a new ets transcription factor that is activated by Ras. *Genes Dev.* 8:1502–13

Glass DJ, Nye SH, Hantzopoulos P, Macchi MJ, Squinto SP, et al. 1991. TrkB mediates BDNF/NT3 dependent survival and proliferation in fibroblasts lacking the low-affinity NGF receptor. *Cell* 66:405–13

Gotz R, Koster R, Winkler C, Raulf F, Lattspeich F, et al 1994. Neurotrophin-6 is a new member of the nerve growth factor family. *Nature* 372:266–69

Greenberg ME, Hermanowski A, Ziff E. 1986a. Effect of protein synthesis inhibitor on growth factor activation of c-*fos*, c-*myc* and actin gene transcription. *Mol. Cell. Biol.* 6:1050–57

Greenberg ME, Ziff E, Greene L. 1986b. Stimulation of neuronal acetylcholine receptors induces rapid gene transcription. *Science* 234:80–83

Greene L, Tischler A. 1976. Establishment of a noradrenergic clonal line of rat adrenal pheochromocytoma cells which respond to nerve growth factor. *Proc. Natl. Acad. Sci. USA* 73:2424–28

Greene L, Tischler A. 1982. PC12 pheochromocytoma cultures in neurobiological research. *Adv. Cell. Neurobiol.* 3:373–414

Hagag N, Halegoua S, Viola M. 1986. Inhibition of growth factor induced differentiation of PC12 cells by microinjection of antibodies to ras p21. *Nature* 319:680–82

Hagg T, Varon S. 1993. Ciliary neurotrophic factor prevents degeneration of adult rat substantia nigra dopaminergic neurons in vivo. *Proc. Natl. Acad. Sci. USA* 90:6315–19

Hamburger V, Levi-Montalcini R. 1949. Proliferation, differentiation and degeneration in the spinal ganglia of the chick embryo under normal and experimental conditions. *J. Exp. Zool.* 111:457–501

Hantzopoulos P, Suri C, Glass D, Goldfarb M, Yancopoulos G. 1994. The low affinity NGF receptor, p75, can collaborate with each of the Trks to potentiate functional response to the neurotrophins. *Neuron* 13:187–201

Hawley R, Scheibe R, Wagner J. 1992. Nerve growth factor induces the expression of the *VGF* gene through a cAMP response element. *J. Neurosci.* 12:2573–81

Heldin C. 1995. Dimerization of cell surface receptors in signal transduction. *Cell* 80:213–23

Heller R, Kronke M. 1994. Tumor-necrosis factor receptor mediated signaling pathways. *J. Cell. Biol.* 126:5–9

Hempstead BL, Martin ZD, Kaplan DR, Parada LR, Chao MV. 1991. High-affinity NGF binding requires coexpression of the trk proto-oncogene and the low-affinity NGF receptor. *Nature* 350:678–83

Herskowitz I. 1995. MAP kinase pathways in yeast: for mating and more. *Cell* 80:187–97

Hill C, Treisman R. 1995. Transcriptional regulation by extracellular signals: mechanisms and specificity. *Cell* 80:199–211

Hill CS, Marais RR, John S, Wynne J, Dalton S, Treisman R. 1993. Functional analysis of a growth factor-responsive transcription factor complex. *Cell* 73:395–406

Hirano T, Matsuda T, Nakajima K. 1994. Signal transduction through gp130 that is shared among the receptor for the interleukin 6 related cytokine subfamily. *Stem Cells* 12:262–77

Huff K, End D, Guroff G. 1981. Nerve-growth factor induced alteration in the response of PC12 pheochromocytoma cells to epidermal growth factor. *J. Cell. Biol.* 88:189–98

Hughes S, Lillien L, Raff M, Rohrer H, Sendtner M. 1988. Ciliary neurotrophic factor (CNTF) as an inducer of type–2 astrocyte differentiation in rat optic nerve. *Nature* 335:70–73

Ibáñez CF, Ebendal T, Barbany G, Murray RJ, Blundell TL, Persson H. 1992. Disruption of the low affinity receptor-binding site in NGF allows neuronal survival and differentiation by binding to the trk gene product. *Cell* 69:329–41

Ip NY, Nye SH, Boulton TG, Davis S, Taga T, et al. 1992. CNTF and LIF act on neuronal cells via shared signaling pathways that involve the IL-6 signal transducing receptor component gp130. *Cell* 69:1121–32

Ip NY, Yancopoulos G. 1992. Ciliary neurotrophic factor and its receptor complex. *Prog. Growth Factor Res.* 4:139–55

Itoh K, Brackenbury R, Akeson R. 1995. Induction of L1 mRNA in PC12 cells by NGF is modulated by cell-cell contact and does not require the high affinity NGF receptor. *J. Neurosci.* 15:2504–12

Jaiswal RK, Moodie SA, Wolfman A, Landreth GE. 1994. The mitogen-activated protein kinase cascade is activated by B-Raf in response to nerve growth factor through interaction with p21ras. *Mol. Cell. Biol.* 14:6944–53

Janknecht R, Ernst W, Pingoud V, Nordheim A. 1993. Activation of ternary complex factor Elk-1 by MAP kinases. *EMBO J.* 12:5097–104

Jing S, Tapley P, Barbacid M. 1992. Nerve growth factor mediates signal transduction through trk homodimer receptors. *Neuron* 9:1067–79

Joly M, Kazlauskas A, Fay F, Corvera S. 1994. Disruption of PDGF receptor trafficking by mutation of its PI-3 kinase binding sites. *Science* 263:684–87

Joseph C, Byun H, Bittman R, Kolesnick R. 1994. Substrate recognition by ceramide-activated protein kinase. *J. Biol. Chem.* 268:20002–6

Kalman D, Wong B, Horvai A, Cline M, O'Lague P. 1989. Nerve growth factor acts through cAMP-dependent protein kinase to increase the number of sodium channels on PC12 cells. *Neuron* 2:355–66

Kaplan DR, Hempstead BL, Martin-Zanca D, Chao MV, Parada LF. 1991. The trk proto-oncogene product: a signal transducing receptor for nerve growth factor. *Science* 252:558–61

Kaplan DR, Martin-Zanca D, Parada LF. 1991. Tyrosine phosphorylation and tyrosine kinase activity of the trk proto-oncogene product induced by NGF. *Nature* 350:158–60

Klein R, Jing SQ, Nanduri V, O'Rourke E, Barbacid M. 1991. The trk proto-oncogene encodes a receptor for nerve growth factor. *Cell* 65:189–97

Klein R, Nanduri V, Jing SA, Lamballe F, Tapley P, et al. 1991. The trkB tyrosine protein kinase is a receptor for brain-derived neurotrophic factor and neurotrophin-3. *Cell* 66:395–403

Knipper M, Beck A, Rylett J, Breer H. 1993. Neurotrophin induced cAMP and IP(3) responses in PC12 cells—different pathways. *FEBS Lett.* 324:147–52

Knusel B, Rabin S, Hefti F, Kaplan D. 1994. Regulated neurotrophin receptor responsiveness during neuronal migration and early differentiation. *J. Neurosci.* 14(3):1542–54

Kodaki T, Woscholski R, Hallberg R, Rodriguez-Viciana P, Downward J, Parker P.

1994. The activation of phosphatidylinositol 3-kinase by Ras. *Curr. Biol.* 4:798–806

Kolesnick R, Golde DW. 1994. The sphingomyelin pathway in tumor necrosis factor and interleukin-1 signaling. *Cell* 77:325–28

Koo PH, Liebl DJ. 1992. Inhibition of nerve growth factor-stimulated neurite outgrowth by methylamine-modified alpha 2-macroglobulin. *J. Neurosci. Res.* 31:678–92

Korsching S. 1993. The neurotrophic factor concept: a reexamination. *J. Neurosci.* 13:2739–48

Kozlosky CJ, Maraskovsky E, McGrew JT, VandenBos T, Teepe M, et al. 1995. Ligands for the receptor tyrosine kinases hek and elk: isolation of cDNAs encoding a family of proteins. *Oncogene* 10:299–306

Kuhn TB, Schmidt MF, Kater SB. 1995. Laminin and fibronectin guideposts signal sustained but opposite effects to passing growth cones. *Neuron* 14:275–85

Kundra V, Escobedo J, Kazlauskas A, Kim H, Rhee S, et al. 1994. Regulation of chemotaxis by the platelet-derived growth factor receptor-β. *Nature* 367:474–76

Lam K, Carpenter C, Ruderman N, Friel J, Kelly K. 1994. The phosphotidylinositol 3-kinase serine kinase phosphorylates IRS-1. Stimulation by insulin and inhibition by wortmannin. *J. Biol. Chem.* 269:20648–52

Lamballe F, Klein R, Barbacid M. 1991. TrkC, a new member of the trk family of tyrosine protein kinases, is a receptor for neurotrophin 3. *Cell* 66:967–79

Lamballe F, Smeyne RJ, Barbacid M. 1994. Developmental expression of trkC, the neurotrophin-3 receptor, in the mammalian nervous system. *J. Neurosci.* 14:14–28

Lange-Carter, CA, Johnson GL. 1994. Ras-dependent growth factor regulation of MEK kinase in PC12 cells. *Science* 265:1458–61

Leevers SJ, Paterson HF, Marshall CJ. 1994. Requirement for Ras in Raf activation is overcome by targeting Raf to the plasma membrane. *Nature* 369:411–14

LeRoith D, Roberts C, Werner H, Bondy C, Raizada M, Adamo M. 1993. Insulin-like growth factors in the brain. See Loughlin & Fallon 1993, pp. 391–442

Letourneau PC, Condic ML, Snow DM. 1994. Interactions of developing neurons with the extracellular matrix. *J. Neurosci.* 14:915–28

Levi-Montalcini R. 1987. The nerve growth factor 35 years later. *Science* 237:1154–62

Levi-Montalcini R, Angeletti PU. 1968. Nerve growth factor. *Physiol. Rev.* 48:534–69

Levine E, Dreyfus C, Black I, Plummer M. 1995. Differential effects of NGF and BDNF on voltage-gated calcium currents in embryonic basal forebrain neurons. *J. Neurosci.* 15:3084–91

Lhotak V, Greer P, Letwin K, Pawson T. 1991. Characterization of elk, a brain-specific receptor tyrosine kinase. *Mol. Cell. Biol.* 11: 2496–502

Lindholm D, Castren E, Tsoulfas P, Kolbeck R, Berzaghi M, et al. 1993. Neurotrophin-3 induced by tri-iodothyronine in cerebellar granule cells promotes Purkinje cell differentiation. *J. Cell. Biol.* 122:443–50

Loeb D, Maragos J, Martin-Zanca D, Chao M, Parada L, Greene L. 1991. The trk proto-oncogne rescues NGF responsiveness in mutant NGF-responsive PC12 cell lines. *Cell* 66: 961–66

Loughlin S, Fallon J, eds. 1993. *Neurotrophic Factors.* Boston: Academic

Lowenstein E, Daly R, Batzer A, Li W, Margolis B, et al. 1992. The SH2 and SH3 domain-containing protein GRB2 links receptor tyrosine kinases to ras signaling. *Cell* 70:431–42

Maisonpierre PC, Barrezueta NX, Yancopoulos GD. 1993. Ehk-1 and Ehk-2: two novel members of the Eph receptor-like tyrosine kinase family with distinctive structures and neuronal expression. *Oncogene* 8:3277–88

Mandel G, Cooperman S, Maue R, Goodman R, Brehm P. 1988. Selective induction of brain type II $Na^+$ channels by nerve growth factor. *Proc. Natl. Acad. Sci. USA* 85:924–28

Mansour SJ, Resing KA, Candi JM, Hermann AS, Gloor JW, et al. 1994. Mitogen-activated protein (MAP) kinase phosphorylation of MAP kinase kinase: determination of phosphorylation sites by mass spectrometry and site-directed mutagenesis. *J. Biochem.* 116:304–14

Marais R, Wynne J, Treisman R. 1993. The SRF accessory protein Elk-1 contains a growth factor-regulated transcriptional activation domain. *Cell* 73:381–93

Marsh HN, Scholz WK, Lamballe F, Klein R, Nanduri V, et al. 1993. Signal transduction events mediated by the BDNF receptor gp145(trkB) in primary hippocampal pyramidal cell culture. *J. Neurosci.* 13:4281–92

Marshall CJ. 1994. MAP kinase kinase kinase, MAP kinase kinase and MAP kinase. *Curr. Opin. Genet. Devel.* 4:82–89

Marshall CJ. 1995. Specificity of receptor tyrosine kinase signaling: transient versus sustained extracellular signal-regulated kinase activation. *Cell* 80:179–85

McCormick F. 1994. Activators and effectors of ras p21 proteins. *Curr. Opin. Genet. Devel.* 4:71–76

Middlemas DS, Meisenhelder J, Hunter T. 1994. Identification of TrkB autophosphorylation sites and evidence that phospholipase C-gamma 1 is a substrate of the TrkB receptor. *J. Biol. Chem.* 269:5458–66

Miranti C, Ginty D, Huang G, Chatila T, Greenberg ME. 1995. Calcium activates serum response factor-dependent transcription by a Ras- and Elk-1-independent mechanism that involves a $Ca^{2+}$/Calmodulin-dependent kinase. *Mol. Cell. Biol.* 15:3672–84

Mitra G. 1991. Mutational analysis of conserved residues in the tyrosine kinase domain of the human trk oncogene. *Oncogene* 6: 2237–41

Mitsumoto H, Ikeda K, Klinkosz B, Cederbaum JM, Wong V, Lindsay RM. 1994. Arrest of motor neuron disease in wobbler mice cotreated with CNTF and BDNF. *Science* 265:1107–10

Moodie S, Willumsen B, Weber M, Wolfman A. 1992. Complexes of Ras-GTP with Raf-1 and mitogen-activated protein kinase. *Science* 260:1658–61

Morgan J, Curran T. 1986. Role of ion flux in the control of c-*fos* expression. *Nature* 322: 552–55

Morrison R. 1993. Epidermal growth factor: structure, expressin and functions in the central nervous system. See Loughlin & Fallon 1993, pp. 339–58

Mueller C, Nordheim A. 1991. A protein domain conserved between yeast MCM1 and human SRF directs ternary complex formation. *EMBO J.* 10:4219–29

Mulderry P. 1994. Neuropeptide expression by newborn and adult rat sensory neurons in culture: effects of nerve growth factor and other neurotrophic factors. *Neuroscience* 59: 673–88

Nebreda AR, Martin ZD, Kaplan DR, Parada LF, Santos E. 1991. Induction by NGF of meiotic maturation of *Xenopus* oocytes expressing the trk proto-oncogene product. *Science* 252:558–61

Obermeier A, Bradshaw RA, Seedorf K, Choidas A, Schlessinger J, Ullrich A. 1994. Neuronal differentiation signals are controlled by nerve growth factor receptor/trk binding sites for SHC and PLC. *EMBO J.* 13: 1585–90

Obermeier A, Halfter H, Wiesmuller KH, Jung G, Schlessinger J, Ullrich A. 1993a. Tyrosine 785 is a major determinant of Trk-substrate interaction. *EMBO J.* 12:933–41

Obermeier A, Lammers R, Wiesmuller K, Jung G, Schlessinger J, Ullrich A. 1993b. Identification of trk binding sites for SHC and phosphatidylinositol 3′-kinase and formation of a multimeric signaling complex. *J. Biol. Chem.* 268:22963–66

Ohmichi M, Decker S, Saltiel A. 1992. Activation of phosphatidylinositol 3-kinase by nerve growth factor involves indirect coupling of the trk proto-oncogne with src Homology 2 domains. *Neuron* 9:769–77

Oshima M, Sithanandam G, Rapp U, Guroff G. 1993. The phosphorylation and activation of B-raf in PC12 cells stimulated by nerve growth factor. *J. Biol. Chem.* 266: 23753–60

Pandey A, Lazar DF, Saltiel AR, Dixit VM. 1994. Activation of the Eck receptor protein tyrosine kinase stimulates phosphatidylinositol 3-kinase activity. *J. Biol. Chem.* 269: 30154–57

Parada LF, Tsoulfas P, Tessarollo L, Blair J, Reid SW, Soppet D. 1992. The trk family of tyrosine kinases—receptors for NGF-related neurotrophins. *Cold Spring Harbor Symp. Quant. Biol.* 57:43–51

Pasquale EB. 1991. Identification of chicken embryo kinase 5, a developmentally regulated receptor-type tyrosine kinase of the Eph family. *Cell Regul.* 2:523–34

Pasquale EB, Deerinck TJ, Singer SJ, Ellisman MH. 1992. Cek5, a membrane receptor-type tyrosine kinase, is in neurons of the embryonic and postnatal avian brain. *J. Neurosci.* 12:3956–67

Patterson PH. 1993. Cytokines and the function of the mature nervous system. *C. R. Acad. Sci. Ser. III* 316:1150–57

Payne D, Rossamondo A, Martino P, Erickson A, Her J, et al. 1991. Identification of the regulatory phosphorylation sites in pp42/Mitogen-activated protein kinase (MAP Kinase). *EMBO J.* 10:885–92

Peng X, Greene L, Kaplan D, Stephens R. 1995. Deletion of a conserved juxtamembrane sequence in Trk abolishes NGF-promoted neuritogenesis. *Neuron* 15:395–406

Pincus DW, DiCicco BE, Black IB. 1990. Vasoactive intestinal peptide regulates mitosis, differentiation and survival of cultured sympathetic neuroblasts. *Nature* 343:564–67

Qiu M, Green S. 1992. PC12 cell neuronal differentiation is associated with prolonged p21 ras activity and consequent prolonged ERK activity. *Neuron* 9:705–17

Rabin S, Cleghorn V, Kaplan D. 1993. SNT, a differentiation-specific target of neurotrophic factor-induced tyrosine kinase activity in neurons and PC12 cells. *Mol. Cell. Biol.* 13:2203–13

Rabizadeh S, Oh J, Zhong LT, Yang J, Bitler CM, et al. 1993. Induction of apoptosis by the low-affinity NGF receptor. *Science* 261: 345–48

Radeke MJ, Miisko TP, Hsu C, Herzenberg LA, Shooter EM. 1987. Gene transfer and molecular cloning of the rat nerve growth factor receptor. *Nature* 325:593–97

Raffioni S, Bradshaw RA, Buxser SE. 1993. The receptors for nerve growth factor and other neurotrophins. *Annu. Rev. Biochem.* 62:823–50

Raivich G, Hellweg R, Kreutzberger G. 1991. NGF receptor-mediated reduction in axonal NGF uptake and retrograde transport following sciatic nerve injury and during regeneration. *Neuron* 7:151–64

Rodriguez-Tebar A, Dechant G, Barde YA. 1990. Binding of brain-derived neurotrophic factor to the nerve growth factor receptor. *Neuron* 4:487–92

Rodriguez-Tebar A, Dechant G, Gotz R, Barde YA. 1992. Binding of neurotrophin-3 to its neuronal receptors and interactions with nerve growth factor and brain-derived neurotrophic factor. *EMBO J.* 11:917–22

Rodriguez-Viciana P, Warne P, Hand RD, Vanhaesebroeck B, Gout I, et al. 1994. Phosphatidylinositol-3-OH kinase as a direct target of Ras. *Nature* 370:527–32

Rozakis-Adcock A, McGlade J, Mbamalu G, Pelicci G, Daly R, et al. 1992. Association of the SHC and Grb2/sem-5 SH2 containing proteins is implicated in activation of the ras pathway by tyrosine kinases. *Nature* 360: 689–92

Rudkin B, Lazarovici P, Levi B, Abe Y, Fujit K, Guroff G. 1989. Cell-cycle specific action of nerve growth factor in PC12 cells: differentiation without proliferation. *EMBO J.* 8: 3319–25

Ruiz JC, Robertson EJ. 1994. The expression of the receptor-protein tyrosine kinase gene, eck, is highly restricted during early mouse development. *Mech. Devel.* 46:87–100

Sajjadi FG, Pasquale EB. 1993. Five novel avian Eph-related tyrosine kinases are differentially expressed. *Oncogene* 8:1807–13

Sajjadi FG, Pasquale EB, Subramani S. 1991. Identification of a new eph-related receptor tyrosine kinase gene from mouse and chicken that is developmentally regulated and encodes at least two forms of the receptor. *New Biol.* 3:769–78

Schlessinger J. 1994. SH2/SH3 signaling proteins. *Curr. Opin. Genet. Dev.* 4:25–30

Schnell L, Schneider R, Kolbeck R, Barde YA, Schwab ME. 1994. Neurotrophin-3 enhances sprouting of corticospinal tract during development and after adult spinal cord lesion. *Nature* 367:170–73

Schu P, Takegawa K, Fry M, Stack J, Waterfield M, Emr S. 1993. Phosphatidylinositol 3-kinase encoded by yeast VPS34 gene essential for protein sorting. *Science* 260:88–91

Schwartz JP. 1992. Neurotransmitters as neurotrophic factors: a new set of functions. *Int. Rev. Neurobiol.* 34:1–23

Segal RA, Pomeroy S, Stiles C. 1995. Axonal growth and fasciculation linked to differential expression of BDNF and NT3 receptors in developing cerebellar granule cells. *J. Neurosci.* 15(7):4970–81

Sheng M, Dougan S, McFadden G, Greenberg ME. 1988. Calcium and growth factor pathways of c-*fos* transcriptional activation require distinct upstream regulatory sequences. *Mol. Cell. Biol.* 8:2787–96

Sheng M, Greenberg ME. 1990. The regulation and function of c-*fos* and other immediate early genes in the nervous system. *Neuron* 4:477–85

Sieber-Blum M. 1991. Role of the neurotrophic factor BDNF and NGF in the commitment of pluripotent neural crest cells. *Neuron* 6: 949–55

Smith C, Farrah T, Goodwin R. 1994. The TNF receptor superfamily of cellular and viral proteins: activation, costimulation and death. *Cell* 76:959–62

Snider WD. 1994. Functions of neurotrophins during nervous system development: What are knockouts teaching us? *Cell* 77:627–38

Snider WD, Johnson EM. 1989. Neurotrophic molecules. *Ann. Neurol.* 26:489–506

Soltoff SP, Rabin SL, Cantley LC, Kaplan DR. 1992. Nerve growth factor promotes the activation of phosphatidylinositol-3 kinase and its association with the trk tyrosine kinase. *J. Biol. Chem.* 267:17472–77

Soppet D, Escandon E, Maragos J, Middlemas DS, Reid SW, et al. 1991. The neurotrophic factors brain-derived neurotrophic factor and neurotrophin-3 are ligands for the trkB tyrosine kinase receptor. *Cell* 65:895–903

Squinto SP, Snitt TN, Aldrich TH, Davis S, Bianco SM, et al. 1991. TrkB encodes a functional receptor for brain-derived neurotrophic factor and neurotrophin 3 but not nerve growth factor. *Cell* 65:1–20

Stahl N, Boulton T, Farruggella T, Ip N, Davis S, et al. 1994. Association and activation of JAK-Tyk kinases by CNTF-LIF-OSM-IL-6 β receptor components. *Science* 263:92–95

Stephens RD, Loeb D, Copeland T, Pawson T, Greene L, Kaplan D. 1994. Trk receptors use redundant signal transduction pathways involving SHC and PLC gamma 1 to mediate NGF responses. *Neuron* 12:691–705

Szeberenyi J, Cai H, Cooper G. 1990. Effect of a dominant inhibitory Ha-ras mutation on neuronal differentiation of PC12 cells. *Mol. Cell. Biol.* 10:5324–32

Tessarollo L, Tsoulfas P, Martin-Zanca D, Gilbert DJ, Jenkins NA, et al. 1993. Trkc, a receptor for neurotrophin-3, is widely expressed in the developing nervous system and in non-neuronal tissues. *Development* 118:463–75

Tessier-Lavigne M. 1995. Eph receptor tyrosine kinases, axon repulsion, and the development of topographical maps. *Cell* 82: 345–48

Toledo-Aral J, Brehm P, Halegoua S, Mandel G. 1995. A single pulse of nerve growth factor triggers long-term neuronal excitability through sodium channel gene induction. *Neuron* 14:607–11

Traverse S, Cohen P. 1994. Identification of a latent MAP kinase kinase kinase in PC12 cells as B-raf. *FEBS Lett.* 350:13–18

Traverse S, Gomez N, Paterson H, Marshall C, Cohen P. 1992. Sustained activation of the mitogen-activated protein (MAP) kinase cascade may be required for differentiation of PC12 cells. Comparison of the effects of nerve growth factor and epidermal growth factor. *Biochem. J.* 288:351–55

Traverse S, Seedorf K, Paterson H, Marshall CJ, Cohen P, Ullrich A. 1994. EGF triggers neuronal differentiation of PC12 cells that overexpress the EGF receptor. *Curr. Biol.* 4:694–701

Treisman R. 1992. The serum response element. *Trends Biochem. Sci.* 17:423–26

Treisman R. 1994. Ternary complex factors: growth regulated transcriptional activators. *Curr. Opin. Genet. Dev.* 4:694–701

Treisman R, Marais R, Wynne J. 1992. Spatial flexibility in ternary complexes between SRF and its accessory proteins. *EMBO J.* 11:4631–40

Unsicker K, Grothe G, Ludecke G, Otto D, Westerman R. 1993. Fibroblast growth factors: their roles in the central and peripheral nervous system. See Loughlin & Fallon 1993, pp. 313–38

Vaillancourt RR, Gardner AM, Johnson GL. 1994. B-Raf-dependent regulation of the MEK-1/mitogen-activated protein kinase pathway in PC12 cells and regulation by cyclic AMP. *Mol. Cell. Biol.* 14:6522–30

VanAelst L, Barr M, Marcus S, Polverino P, Wigler M. 1993. Complex formation between Ras and Raf and other protein kinases. *Proc. Natl. Acad. Sci. USA* 90:6213–17

Verdi J, Birren S, Ibáñez CI, Persson H, Kaplan D, et al. 1994. p75LNGFR regulates Trk signal transduction and NGF-induced neuronal differentiation in MAH cells. *Neuron* 12: 733–45

Vetter ML, Martin-Zanca D, Parada LF, Bishop JM, Kaplan DR. 1991. Nerve growth factor rapidly stimulates tyrosine phosphorylation of phospholipaseC by a kinase activity associated with the product of the trk protooncogene. *Proc. Natl. Acad. Sci. USA* 88:5650–54

Voltek A, Gollenberg S, Cooper J. 1993. Mammalian Ras interacts directly with the serine/threonine kinase Raf. *Cell* 74:205–14

Wang J, Auger K, Jarvis L, Shi Y, Roberts T. 1995. Direct association of Grb2 with the p85 subunit of phosphatidylinositol 3-kinase. *J. Biol. Chem.* 270:12774–80

Wen Z, Zhong Z, Darnell JE. 1995. Maximal activation of transcription by Stat1 and Stat3 requires both tyrosine and serine phosphorylation. *Cell* 82:241–50

Wood KW, Sarnecki C, Roberts TM, Blenis J. 1992. Ras mediates nerve growth factor receptor modulation of three signal-transducing protein kinases: MAP kinase, Raf-1, and RSK. *Cell* 68:1041–50

Wright EM, Vogel KS, Davies AM. 1992. Neurotrophic factors promote the maturation of developing sensory neurons before they become dependent on these factors for survival. *Neuron* 9:139–50

Yao R, Cooper G. 1995. Requirement for phosphatidylinositol-3 kinase in the prevention of apoptosis by nerve growth factor. *Science* 267:2003–6

Zhang X, Blenis J, Li H, Schindler C, Chen-Kiang S. 1995. Requirement of serine phosphorylation for formation of Stat-promoter complexes. *Science* 267:1990–94

*Annu. Rev. Neurosci. 1996. 19:491–515*

# THE NEUROTROPHINS AND CNTF: Two Families of Collaborative Neurotrophic Factors

*Nancy Y. Ip*[1] *and George D. Yancopoulos*

Regeneron Pharmaceuticals, Inc., 777 Old Saw Mill River Road, Tarrytown, New York 10591

KEY WORDS:   nerve growth factor, neurotrophin, ciliary neurotrophic factor, leukemia inhibitory factor, receptor tyrosine kinase

## ABSTRACT

Because the actions of neurotrophic factors appear distinct from those of traditional growth factors and cytokines, it was long assumed that the neurotrophic factors utilized receptors and signaling systems fundamentally different from those used by growth factors operating elsewhere in the body. Recent advances in the understanding of the structure of the receptors for neurotrophic factors have unexpectedly revealed that they are in fact similar to the receptors used by the traditional growth factors and cytokines. The expression of the receptors for the neurotrophic factors is exclusively or predominantly in the nervous system; activation of these receptors in the context of the neuron allows these factors to display distinctive actions. While the precise roles of the neurotrophic factors and their therapeutic potential in various disease states still remain to be elucidated, this review describes studies on their receptor systems, their notable biological activities in the nervous system, and recent insights provided by targeted gene disruptions.

## INTRODUCTION

Cell-cell communication is often made possible by the production of protein ligands in one cell that are recognized by cell-surface receptors on other cells. This review focuses on two families of protein ligands that use receptors that are largely limited to neuronal cells in their distributions and can be classified as either receptor tyrosine kinases or cytokine receptors. Because of the restricted distributions of their receptors, the actions of these protein ligands are

[1]Current address: Department of Biology, Hong Kong University of Science and Technology, Hong Kong

491

largely specific to neuronal cells. Several of these factors were indeed discovered for their ability to support the survival of neuronal cells, based on the concept that the survival of neurons during development is influenced by signals provided by their target tissues, and were thus termed neuronal survival factors or neurotrophic factors. In addition to regulating neuronal survival, neurotrophic factors can exert other effects that are critical for the differentiation and maintenance of the nervous system and that in some cases do not appear to be provided by the neuronal target.

The first family of neurotrophic factors discussed in this review is collectively known as the neurotrophins and is comprised of several members that are related to nerve growth factor (NGF). The second family is represented by ciliary neurotrophic factor (CNTF), which shares receptor components with a number of distantly related cytokines, such as leukemia inhibitory factor (LIF) and interleukin-6 (IL-6); these cytokines can mimic the actions of CNTF on neurons in some cases but have much broader actions throughout the rest of the body.

Initially discovered based on their ability to support neuronal survival, neurotrophic factors were originally thought to be fundamentally distinct from traditional growth factors and cytokines operating elsewhere in the body. However, recent progress has unexpectedly revealed similarities between the receptors and signaling systems used by neurotrophic factors and other growth factors (Ip & Yancopoulos 1993, 1994). For example, the neurotrophins utilize a family of receptor tyrosine kinases, known as the Trk family, that appear unexpectedly similar to the receptor tyrosine kinases used by traditional growth factors such as fibroblast growth factor (FGF) or platelet-derived growth factor (PDGF). CNTF, on the other hand, utilizes what is referred to as a cytokine receptor system; this multicomponent receptor system includes a specificity-conferring α-receptor component that functions in concert with signal transducing receptor components shared with its distant cytokine relatives.

In addition to their distinct effects on target neuronal cells, the different classes of neurotrophic factors can collaborate and in some cases act synergistically. In this review, we focus on the receptor systems utilized by the neurotrophic factors, their notable biological activities in the nervous system, and recent insights provided by targeted gene disruptions that have generated mice lacking neurotrophic factors or their receptors.

## THE NEUROTROPHINS

Five members of the neurotrophin family have thus far been isolated, with nerve growth factor (NGF) serving as the prototype for this family. The other members include brain-derived neurotrophic factor (BDNF), neurotrophin-3 (NT3), neurotrophin-4 (NT4, alternatively neurotrophin-5 or NT4/5) and neurotrophin-6 (NT6) (Götz et al 1994), although there is no evidence that NT6, which was discovered in fish, has a mammalian counterpart. An excellent recent review by Bothwell (1995) has summarized the discoveries of the neurotrophins and their receptors as well as their interactions. In this review, we focus on some aspects that were not covered in the previous review.

## The Trks as Functional Neurotrophin Receptors

A major breakthrough in understanding how the neurotrophins initiate signaling on target neurons came with the initial identification of the TrkA protooncogene as a functional receptor for NGF (Kaplan et al 1991, Klein et al 1991); subsequent studies revealed that other neurotrophins utilize other members of this Trk family of receptor tyrosine kinases (reviewed in Barbacid 1993, Glass & Yancopoulos 1993, Bothwell 1995). The conclusion of a number of studies is that the three known Trk receptors display overlapping specificities for the neurotrophins: NGF is the ligand specific for the TrkA receptor; BDNF and NT4/5 are the preferred ligands for the TrkB receptor; and NT3 primarily acts on the TrkC receptor, although it can also activate the TrkB receptor in certain cell types (Ip et al 1993c, Barbacid 1993, Bothwell 1995). In most respects the Trk receptors appear quite similar to other receptor tyrosine kinases, such as those used by traditional growth factors such as FGF. Thus, the neurotrophins activate the intrinsic tyrosine kinase activity of the Trk receptors via ligand-induced dimerization (Jing et al 1992), and the intracellular signaling substrates activated by the Trks overlap with those utilized by other receptor tyrosine kinases (Table 1) (Rasouly et al 1995).

**Table 1**  Intracellular signaling substrates for the neurotrophic factors

| Substrates | Activated in response to | |
| | Neurotrophins | CNTF |
| --- | --- | --- |
| FAK | + | ? |
| MAPK (ERK) | + | + |
| MEK | + | + |
| PI-3 kinase | + | + |
| PLC-$\gamma$ | + | + |
| Raf | + | + |
| Ras | + | + |
| RSK | + | ? |
| SHC | + | + |
| SNT | + | ? |
| pp120 src substrate | ? | + |
| PTP1D | ? | + |
| JAK1 | − | + |
| JAK2 | − | + |
| TYK2 | − | + |
| STAT1 | − | + |
| STAT3 | − | + |

## Responses Mediated by the Trk Receptors are Dependent on the Cellular Context

The major feature that distinguishes the Trks from other receptor tyrosine kinases is their limited distributions to specific neuronal subpopulations. Thus the neurotrophins were discovered as neuronal survival factors, instead of as traditional growth factors, not because of unique signaling capabilites of their Trk receptors, but because of the restricted expression of the Trks to neurons. While the Trk receptors mediate survival and differentiation when expressed in a neuronal cell, ectopic expression of Trk receptors in nonneuronal cells, such as fibroblasts, allows them to mediate mitogenic effects in a manner indistinguishable from those elicited by other growth factors such as FGF (Glass et al 1991). This finding is important because it points out that there may well be cell types in which the Trks are normally expressed, such as in mitotically-competent neuronal precursors, where the neurotrophins indeed act as proliferative factors, as opposed to survival molecules. Because the response elicited by Trk activation is so context dependent, different types of neuron will likely respond quite differently to Trk activation, e.g. by regulating distinct sets of neuropeptides, neurotransmitters, or other gene products (reviewed in Snider 1994).

In addition to determining the types of responses elicited, the cellular context in which the Trk receptors are expressed can also alter their specificity for their ligands (Ip et al 1993c). For example, certain neuronal environments (such as PC12 cells, or primary neurons from dorsal root ganglion) can restrict the ability of TrkB to interact with its nonpreferred ligand, NT3. Although the contributing factors for such restrictive environments are still unclear, some cell type–specific accessory molecules (such as variant forms of the Trk receptors) may play an important role.

## The Role of p75

In addition to the Trk receptors, all of the neurotrophins can also bind with low affinity to a cell-surface receptor known as the low-affinity NGF receptor, or p75 (reviewed in Meakin & Shooter 1992, Bothwell 1995). Although earlier studies proposed that p75 was required for the formation of a functional neurotrophin receptor (Hempstead et al 1989), several lines of evidence now argue against this possibility. Among these are the findings that the Trk receptors can mediate functional responses to physiologically relevant concentrations of the neurotrophins even in the absence of p75 (Glass et al 1991, Verdi et al 1994b). Furthermore, a mutant NGF that cannot bind p75 but retains its ability to bind to TrkA is still capable of eliciting biological responses (Ibáñez et al 1992). Thus, current evidence indicates that the Trk receptors can directly bind to and mediate responses to the neurotrophins in the absence of p75. However, p75 may still act as an accessory molecule to modulate the responses of the Trk receptors to the neurotrophins. Indeed, a recent study indicated that a truncated form of p75 displayed a remarkably dramatic ability to functionally interact with each of the Trk receptors, allowing the Trk receptors to respond to lower concentrations of the neurotrophins (Hantzopoulos et al 1994); similar, though less striking, results were also seen with full-length p75 (Verdi et al 1994a, Barker & Shooter 1994, Hantzopoulos et al 1994). The mechanism of potentiation of Trk function by p75 may be analogous to that seen with the tumor necrosis factor receptor (Tartaglia et al 1993), in which the accessory receptor component appears to aid in presentation of the ligand to the signal-transducing receptor component. Because the amounts of the neurotrophins may be limiting in vivo, such a potentiating function of p75 could play an important role.

Some studies suggest that p75 alone may also play a signaling role (Rabiadeh et al 1993, Anton 1994, Dobrowsky et al 1994). These studies should be viewed with caution, however, because the specificity of responsiveness to the neurotrophins correlates with Trk expression, not with p75 expression. Thus far, no neuron has convincingly been shown to respond to neurotrophins in the absence of a Trk receptor, thereby arguing against a signaling role for p75 alone. However, p75 may somehow serve to modify the intracellular signaling capabilities of a Trk receptor.

## Biological Activities of the Neurotrophins and Insights Into Physiological Roles as Demonstrated by Gene Disruptions in Mice

A variety of in vitro and in vivo studies, in which the effects of exogenously provided neurotrophins on a particular neuronal population are examined, have revealed that the neurotrophins can act on both overlapping and discrete sets of neurons. It should be emphasized, however, that just because a particular neurotrophin can elicit a response from a particular neuronal population in vitro or when exogenously provided in vivo, it does not necessarily imply that this response is physiologically relevant. The early use of neutralizing antibodies to NGF and the more recent use of homologous recombination technology to generate mice disrupted at each of the neurotrophin or Trk alleles have been quite useful for determining which actions of exogenously provided neurotrophins are physiologically relevant, as well as for identifying previously unsuspected roles of the neurotrophins (Table 2). Examples are provided below in which in vitro studies have correctly predicted a physiological role of a particular neurotrophin for particular neurons, as well as examples in which this has not been the case. It should also be noted that just because an effect of an exogenously provided neurotrophin is not physiologically relevant, as

**Table 2**  Phenotypes of mice mutant for neurotrophic factors and receptors[a]

| Gene disrupted: | Percentage of Control | | | | | | | | | |
|---|---|---|---|---|---|---|---|---|---|---|
| | NGF | TrkA | NT3 | TrkC | BDNF | NT4 | BDNF + NT4 | TrkB | CNTF | CNTFRα |
| **PNS** | | | | | | | | | | |
| SCG[b] | 20 | 10 | 50 | ND | 100 | 100 | 100 | 100 | ND | 100 |
| DRG (lumbar) | 30 | 20 | 40 | 80 | 70 | 100 | ND | 70 | ND | 100 |
| Nodose and petrosal | 100 | 100 | 50 | 50 | 50[c] | 50[d] | 10 | 10 | ND | ND |
| TH[+] subset | | | | | 40[c] | 90[d] | 20 | 20 | | |
| Geniculate | ND | ND | 75 | ND | 50 | 50 | 10 | ND | ND | ND |
| Trigeminal | 30 | 20 | 40 | ND | 60 | 100 | 60 | 40 | 100 | 100 |
| Vestibular | ND | ND | 70 | ND | 10[c] | 100 | 10 | ND | ND | ND |
| Cochlear | ND | ND | 15 | ND | +[e] | + | + | ND | ND | ND |
| **CNS** | | | | | | | | | | |
| Facial MN | 100 | 100 | 100 | 100 | 100 | 100 | 100 | 100 | 100 | 60 |
| Lumbar MN | 100 | 100 | 100 | 100 | 100 | 100 | 100 | 100 | 100 | 70 |
| Substantia nigra[f] | ND | ND | ND | + | + | + | ND | + | ND | ND |
| Basal forebrain[f] | + | + | ND | + | + | ND | ND | + | ND | ND |

[a] Values represent approximation of numbers from manuscripts cited in text as follows: NGF (Crowley et al 1994), TrkA (Smeyne et al 1994), BDNF (Jones et al 1994, Farinas et al 1994, Ernfors et al 1994a, Conover et al 1995, Liu et al 1995; Erickson et al, submitted; Bianchi et al, submitted), NT4 and BDNF + NT4 double knockout (Conover et al 1995, Liu et al 1995; Erickson et al, submitted; Bianchi et al, submitted),; TrkB (Klein et al 1993), differences from published data (e.g. reflecting no loss of motor neurons) depends on recent analysis by Barbacid and colleagues (personal communication), NT3 (Ernfors et al 1994b, Farinas et al 1994), TrkC (Klein et al 1994), CNTF (Masu et al 1993), CNTFRα (DeChiara et al 1995).
[b] SCG, superior cervial ganglia; DRG, dorsal root ganglia; TH, tyrosine hydroxylase; MN, neurons; ND, not determined.
[c] Extent of loss depends on gene dosage, indicating BDNF is limiting.
[d] Extent of loss does not depend on gene dosage, indicating NT4 is in excess.
[e] Indicates not qualitatively different from controls.
[f] No quantitative data available for these neuronal populations.

judged by the observation that the normal biology of the responding neurons is not discernably altered in mice lacking that neurotrophin, it does not necessarily mean that this effect of the neurotrophin cannot be exploited for therapeutic purposes, as will also be discussed below.

NGF AND TRKA    Numerous in vitro studies predicted that each of the neurotrophins would be critical for the survival of different populations of peripheral neurons (reviewed in Lindsay 1994). Early findings that NGF had in vitro actions on sympathetic neurons and a large proportion of dorsal root ganglion (DRG) sensory neurons (in particular, the small-sized neurons in the DRG) were found to reflect a physiological role of NGF for these neuronal populations when the activities of NGF were neutralized in vivo by using NGF-specific antibodies (reviewed in Lindsay 1994). More recent examinations of mice lacking either NGF or TrkA (Smeyne et al 1994, Crowley et al 1994) found that both these neuronal populations were ablated. Rather surprisingly, however, basal forebrain cholinergic neurons, which are one of the few populations of CNS neurons that have been shown to respond to NGF and express TrkA, are not notably affected (Smeyne et al 1994, Crowley et al 1994). This finding does not rule out the possibility that more subtle aspects of neuronal function, other than survival per se, have been affected in these cells. Furthermore, it does not exclude the use of NGF as having therapeutic value for these neurons in diseases in which they appear to be particularly affected, such as Alzheimer's Disease.

NT3 AND TRKC    In vitro and in vivo experimentation using exogenously provided BDNF, NT3, and NT4 has not been as straightforwardly useful in predicting the physiological roles of these factors. For example, all were found to act as motor neuron survival molecules (reviewed in Lindsay 1994), although α motor neuron numbers appear quite normal in animals lacking each of these factors individually (Jones et al 1994; Farinas et al 1994; Ernfors et al 1994a,b; Conover et al 1995), in a reanalysis of animals lacking either the TrkB or TrkC receptors (M Barbacid, personal communication), and even in animals lacking both BDNF and NT4 (Conover et al 1995, Liu et al 1995). However, it was correctly suggested that NT3 might be a specific survival factor for large proprioceptive neurons found in the DRG (reviewed in Lindsay 1994). Analysis of mice lacking either NT3 (Ernfors et al 1994b, Farinas et al 1994) or TrkC (Klein et al 1994) have verified the absolute dependence of these neurons on NT3 for survival. Not only are the proprioceptive neurons ablated in mice lacking NT3, but so are their muscle sensory organs, and the mice exhibit a profound neurological abnormality that is consistent with the loss of positional sensing ability. Mice lacking NT3 and TrkC have also been noteworthy for their deficits in their cochlear neurons, as well as in nodose and sympathetic ganglia (which is discussed further, below).

BDNF AND NT4    Understanding the physiological roles of BDNF and NT4 has been particularly dependent on analysis of mice lacking these factors, individually and in combination. This is because the actions of exogenously provided BDNF and NT4 are quite similar, in vitro (Ip et al 1993a) and in vivo, since they act via the same receptor. Whereas mice lacking BDNF have revealed a critical role for BDNF for a subpopulation of DRG and trigeminal ganglion sensory neurons, for most vestibular neurons, and for the visceral sensory neurons found in the nodose-petrosal and geniculate ganglia, mice lacking NT4 have only been noted to have deficits in their nodose-petrosal and geniculate ganglia. The loss of vestibular neurons in mice lacking BDNF explains the dominant neurological abnormality in these mice—the inability to maintain their sense of balance. However, whereas the BDNF mutant mice fail to thrive and die within the first few weeks of life, the NT4 mutant mice thrive and are long lived. Detailed analysis of the visceral sensory neurons in the nodose and petrosal ganglia of these mice has shed light on the differences in survival between the BDNF and NT4 mutant mice. The BDNF-dependent neurons appear to comprise a subset of the ganglia that is different from that of NT4, including a subset [the tyrosine hydroxylase-positive (TH⁺) subpopulation] that innervates the carotid body and appears critical for maintaining cardiorespiratory homeostasis. Thus about 60% of the critical TH⁺ subpopulation is lost in BDNF mutant mice, whereas only about 10% of these critical neurons are lost in the NT4 mutant mice, apparently correlating with the early lethality of BDNF mutant mice (JT Erickson, JC Conover, RJ Smeyne, TM DeChiara, M Barbacid, GD Yancopoulos & DM Katz, submitted).

Further analysis of mice lacking different combinations of BDNF and NT4 alleles has revealed that neurons can display different types of neurotrophin requirements. For example, losses in the nodose and petrosal ganglia are almost precisely proportional to the numbers of functional BDNF alleles (i.e. about 25% of the neurons are lost in mice heterozygous for the disrupted BDNF allele, while about 50% are lost in the homozygote) (JT Erickson, JC Conover, RJ Smeyne, TM DeChiara, M Barbacid, GD Yancopoulos & DM Katz, submitted). Similarly, vestibular neurons display a strict dose dependence on the number of functional BDNF alleles (LM Bianchi, J Conover, RM Lindsay & GD Yancopoulos, submitted). These data directly support a central tenet of the neurotrophic hypothesis, namely that neurotrophic factor concentration is limiting for neuronal survival in vivo. On the other hand, NT4 appears to be present in excess for the subset of the nodose and petrosal neurons it supports: Neuronal losses require the mice to be homozygous for the NT4 mutation (JT Erickson, JC Conover, RJ Smeyne, TM DeChiara, M Barbacid, GD Yancopoulos & DM Katz, submitted).

SEQUENTIAL REQUIREMENT OF NEUROTROPHINS?    Comparison of the various neurotrophin mutant mice largely reveals that these factors are required for different populations of neurons (Table 2). Notable exceptions include sympathetic neurons and nodose-petrosal neurons. While almost all sympathetic neurons are lost in mice lacking NGF or TrkA, about 50% of these neurons are also lost in mice lacking NT3. Similarly, while almost all nodose-petrosal neurons are lost in mice lacking both BDNF and NT4, about 50% of these neurons are also lost in mice lacking NT3. Thus at least a subpopulation of sympathetic and nodose-petrosal neurons appear to require NT3 in addition to either NGF or BDNF and NT4. The most likely explanation seems to be that this requirement for NT3 might occur at a different point in development (probably at an earlier point) than the requirement for the other neurotrophins. This explanation would be consistent with in vitro evidence that NT3 might act as a mitogen on early neural crest cells and sympathetic neuroblasts, might be required as a survival factor before sympathetic neuroblasts become NGF dependent, and might in fact induce NGF responsiveness (Kalchem et al 1992, Dicicco-Bloom et al 1993, Birren et al 1993, Verdi & Anderson 1994).

CNS REQUIREMENT OF THE NEUROTROPHINS    One of the surprising aspects of the analysis of neurotrophin and Trk mutants has been the relative paucity of deficits noted within the CNS. As was noted above, basal forebrain cholinergic neurons were present in NGF and TrkA mutant mice, and these neurons were also not absent in mice mutant for the other neurotrophins. Similarly, numbers of α motor neurons were not altered in mutants of any of the neurotrophins or Trks. Furthermore, dopaminergic and retinal ganglion neurons were present in BDNF and NT4 mutant mice, although these factors have impressive survival effects on these neurons in vitro or when provided during injury in vivo (Hyman et al 1994, LaVail et al 1992). The roles of the neurotrophins in the CNS may be much less obvious than in the PNS because CNS neurons do not require them for survival or because CNS neurons receive much more redundant trophic support due to their far more numerous connections. The CNS may also require injury to expose a requirement for neurotrophin support. In any event, the neurotrophins, in addition to any potential survival roles, also have subtler differentiative effects on CNS neurons that can be revealed during analysis of mutant mice, as noted by alterations in neuropeptides and calcium-binding proteins in the brains of mice deficient in BDNF (Jones et al 1994). Other physiologically relevant in vivo functions, such as mediating the activity-dependent control of axonal branching during development of the CNS (Cabelli et al 1995) and regulating synaptic strength at developing as well as mature synapses (Lohof et al 1993, Kang & Schuman 1995) may be revealed by further analysis of the mutant mice. In any case, the ability of the neurotrophins to positively intervene in models of CNS injury and disease (Lindsay 1994) suggests that they might indeed have important therapeutic applications regardless of their physiologic roles in the CNS.

MICE DISRUPTED FOR P75    In contrast to the severe phenotypes displayed by the mutant mice lacking Trk receptors, mice carrying a targeted p75 gene were able to proceed normally during their development; only relatively mild neuronal deficiencies were observed (Lee et al 1992, 1994). Consistent with the in vitro data mentioned above, these findings strongly argue against the proposal that p75 is required for signaling responses to the neurotrophins and are more consistent with an accessory role for this receptor, as discussed above.

## CILIARY NEUROTROPHIC FACTOR

### Biological Activities of CNTF

Early studies have identified CNTF as a trophic factor that supports the survival of embryonic chick ciliary ganglion neurons in vitro (Adler et al 1979, Lin et al 1989, Stöckli et al 1989). Subsequent cloning and sequencing of CNTF revealed that it is unrelated to the neurotrophins; as discussed below, it is part of a cytokine family that includes more generally acting cytokines such as LIF and IL-6. It is now well-established that in addition to ciliary neurons, a wide variety of peripheral and central neurons also respond to CNTF (reviewed in Ip et al 1991b, Manthorpe et al 1992). For example, CNTF inhibits the proliferation of sympathetic precursors (Ernsberger et al 1989), affects the differentiation of developing sympathetic neurons, and induces the expression of neuropeptide genes in neuronal cell lines (Saadat et al 1989, Rao et al 1992, Symes et al 1993, Fann & Patterson 1993). CNTF also exhibits survival effects on cultured embryonic motor neurons (Arakawa et al 1990, Oppenheim et al 1991, Martinou et al 1992), preganglionic sympathetic neurons (Blottner et al 1989), and sensory neurons (Barbin et al 1984, Thaler et al 1994). In the CNS, CNTF has been shown to increase the survival of cultured hippocampal neurons (Ip et al 1991a); degeneration of specific neuronal populations in the CNS could be prevented in the presence of CNTF (Hagg et al 1992, Hagg & Varon 1993). The ability of CNTF to affect excitability of its target neurons was recently investigated; such studies demonstrated for the first time that CNTF could regulate voltage-gated ion channels in neuroblastoma cells and potentiate the release of transmitters (Lesser & Lo 1995, Stoop & Poo 1995).

Although the known actions of CNTF appear to be restricted to neuronal cells, there are nonneuronal cells that also respond to CNTF. For example, differentiation of glial progenitor cells into type-2 astrocytes (Hughes et al 1988, Lillien et al 1988) and the survival and maturation of oligodendrocytes (Louis et al 1993, Barres et al 1993) also appear to depend on CNTF. CNTF can also induce acute-phase protein expression in hepatocytes (Schooltinck et al 1992) and can inhibit the differentiation of pluripotent embryonic stem cells in culture in a manner similiar to that observed with LIF (Conover et al 1994).

Of all the known biological actions of CNTF, those on the motor neurons have attracted the most attention. Aside from its well-characterized actions on cultured motor neurons (Arawaka et al 1990, Wong et al 1993), CNTF can prevent the degeneration of axotomized facial motor neurons as well as markedly ameliorate the motor deficits in mice with neuromuscular dysfunction (Sendtner et al 1990, 1992a). Curtis et al (1993) demonstrated retrograde axonal transport of CNTF to motor neurons, which was greatly increased following peripheral nerve injury. This finding is consistent with the proposal that the retrograde transport of trophic factors such as CNTF is important in the regenerative response of neuronal cells following axotomy.

## Evidence for CNTF as an Injury Factor in the CNS and PNS

Several lines of evidence indicate that CNTF may be a key player in the injury response in the nervous system. Dramatic changes in the level of expression of CNTF occur following instances of neural trauma in the CNS and PNS. For example, mechanical lesions in the brain resulted in a dramatic increase in CNTF mRNA and protein bordering the wound site (Ip et al 1993d, Asada et al 1995). Such an increase in CNTF production could be localized to reactive astrocytes in the resulting glial scar (Ip et al 1993d, Rudge et al 1992). While it is reasonable to assume that CNTF may act as a trophic factor for damaged neurons at the sites of CNS injury, it is also conceivable that astrocytes potentially represent an important target site for CNTF following injury. Rudge et al (1994) demonstrated that the expression of CNTF receptor (see below) switched from a purely neuronal localization to cells in the glial scar at the edge of the wound, suggesting that astrocytes as well as fibroblasts, which are prevalent at the wound site, possess functional CNTF receptor complexes. Thus these cell types have the ability to respond to CNTF made available following injury as part of a repertoire of complex cellular interactions that occur after trauma.

In the PNS, high levels of CNTF mRNA and protein were localized within the Schwann cells of the sciatic nerve (Stöckli et al 1991, Rende et al 1992, Friedman et al 1992). The CNTF expressed by the Schwann cells is sequestered within the cytoplasm presumably because CNTF lacks a signal sequence for secretion. Following nerve injury, the abundant level of CNTF mRNA decreases dramatically in the distal nerve (Friedman et al 1992). Such a decrease in gene expression is coupled with detection of CNTF protein in the extracellular fluid (Sendtner et al 1992b). Although the process by which CNTF is released from its site of synthesis is unknown, Schwann cells may release CNTF as a consequence of neural injury. A hypothesis for possible interactions between the released CNTF and its receptor is discussed below.

Recent evidence obtained using the light-damage model with retina (LaVail et al 1992) also indicated that changes in CNTF expression is an early marker of neural injury. A dramatic increase in the level of CNTF expression was observed coincidental with degenerations of photoreceptor cells (R Steinberg, NY Ip, GD Yancopoulos & MM LaVail, unpublished observations). Administration of BDNF into the rat eye not only rescued the photoreceptors from degeneration, but also prevented the upregulation of CNTF expression following constant light damage of the retina (R Steinberg, NY Ip, GD Yancopoulos & MM LaVail, unpublished observations). Taken together, these findings suggest that CNTF is a key player in the injury response in the retina.

## Molecular Cloning and Expression of a CNTF-Binding Protein CNTFRα

In contrast with the wealth of knowledge generated concerning the biological actions of CNTF, the signal transduction mechanisms utilized by CNTF to act on responsive cells were not understood until recently. One of the major clues came from the analysis of a cloned CNTF-binding protein. A novel scheme using a genetically engineered CNTF molecule tagged with an epitope allowed an antibody against the tag to detect the CNTF receptor on neuronal cells (Squinto et al 1990). Such a receptor detection scheme, together with the use of a cDNA expression library derived from a neuronal cell line, allowed for the cloning of a CNTF-binding protein (Davis et al 1991), referred to as CNTFRα.

Consistent with previous observations for neurotrophic factor receptors, the expression of CNTFRα was largely restricted to neural tissues, including all known peripheral targets of CNTF, such as sympathetic, sensory, and parasympathetic ganglia (Ip et al 1993b). In addition, prominent expression of CNTFRα in both upper and lower motor neurons such as in the cortex and facial nucleus, as well as in motor-related brain areas and spinal cord, is consistent with a distinctive role for CNTF in maintaining motor system function. The broad expression of CNTFRα throughout the central nervous system also suggests that additional potential cellular targets of CNTF action exist. In the periphery, CNTFRα was unexpectedly found to be expressed in skeletal muscle; direct myotrophic action for CNTF in reducing atrophy in denervated rat muscle is consistent with this expression pattern (Helgren et al 1994).

The cloned CNTFRα exhibits several unexpected novel features (Davis & Yancopoulos 1993). One surprising feature is that CNTFRα lacks transmembrane and cytoplasmic domains, and is instead anchored to the cell surface via a glycosyl phosphatidylinositol (GPI) linkage (Davis et al 1991). This characteristic is quite distinct from other neurotrophic factor receptors. Furthermore, CNTFRα was unexpectedly found to display sequence homology with IL-6Rα, one of the receptor components utilized by a hematopoietic cytokine, IL-6. IL-6Rα by itself is not capable of initiating signaling, but after binding to IL-6, it can recruit a second receptor component, gp130, to initiate signaling (Taga et al 1989, Hibi et al 1990, Kishimoto et al 1992).

Concurrent findings from a theoretical analysis revealed that CNTF and IL-6, together with a number of hematopoietic cytokines such as LIF, granulocyte colony stimulating factor (G-CSF), and oncostatin M, in fact share structural homologies and constitute a newly defined superfamily of hematopoietic cytokines (Bazan 1991, Rose & Bruce 1991). Although LIF displays broad actions outside the nervous system similar to its distantly related cytokines, the actions of LIF on neuronal cells often mimic those of CNTF (Hilton & Gough 1991, Murphy et al 1991, Hall & Rao 1992). LIF is identical to the cholinergic differentiation factor that was characterized and cloned based on its ability to switch the phenotype of cultured sympathetic neurons from adrenergic to cholinergic (Yamamori et al 1989). In contrast to the limited actions of CNTF on the nervous system, its cytokine relatives are known to act on a variety of cell types throughout the body. The structural insights obtained from the observed homologies for these cytokines and their receptors have raised a number of intriguing questions. Given the homology between CNTFRα and IL-6Rα and the lack of a cytoplasmic domain in CNTFRα, does CNTF require a second receptor component, similar to the signal transducer gp130 used by IL-6? Can overlapping actions exhibited by members of this cytokine family be explained by sharing of receptor subunits and/or intracellular signaling molecules? What is the basis for the relatively restricted actions of CNTF on neuronal cells? As discussed below, efforts in understanding the CNTF receptor structure have provided answers to these questions.

## CNTF Shares Receptor Components, LIFRβ and Gp130, With Other Hematopoietic Cytokines

Subsequent to the cloning of CNTFRα, progress in understanding the structure of the receptor system utilized by CNTF was greatly facilitated by the wealth of knowledge available on the IL-6 receptor system (Kishimoto et al 1992). The homology between CNTFRα and IL-6Rα, together with the lack of a cytoplasmic domain in CNTFRα, raised the possibility that CNTF might require a second receptor component, similar to the signal transducer gp130 used by IL-6, to initiate intracellular signaling. The ability to identify neuronal cell lines responsive to CNTF and LIF allowed for the characterization of CNTF- and LIF-induced signaling responses, and of the receptor systems initiating these responses (Ip et al 1992). A series of experiments revealed that CNTF uses the same signal transducer as IL-6, namely gp130, as well as an additional signal transducer designated LIFRβ because it was initially cloned as a LIF-binding protein (Ip et al 1992; Ip & Yancopoulos 1992; Gearing et al 1991, 1992, 1994; Stahl et al 1993; Davis et al 1993b; Baumann et al 1993). Insights into the receptor system used by CNTF led to the proposal of uniform receptor systems for all the CNTF-related cytokines (Ip et al 1992). The important feature of this model is that all the receptors include dimers of their signal transducing β receptor components, whether they use only gp130 (in which case the receptors contain gp130 homodimers) or gp130 in conjunction with LIFRβ (in which case they contain gp130-LIFRβ heterodimers); some of the factors also require specificity-conferring α components that restrict their sites of action (Figure 1). These receptor models account for the cell-type specificity of the various cytokines while simultaneously explaining why some of the cytokines appear to have identical actions on some targets. In particular, the neuronal specificity of CNTF can be explained by the restricted expression of the CNTF receptor LIFRβ com-

**A.**

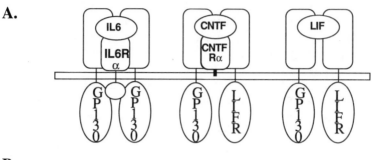

**B.**

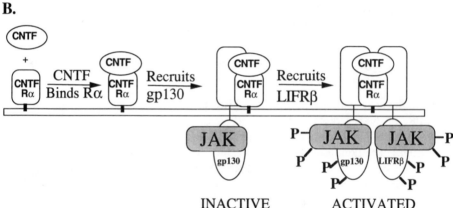

INACTIVE              ACTIVATED

*Figure 1* (A) A unified model for the receptor complexes used by CNTF, LIF, and IL-6. CNTF is a cytokine that shares β receptor components, gp130 and/or LIFRα, with LIF and IL-6. (B) Stepwise formation of an activated CNTF receptor complex. Sequential assembly of the CNTF receptor complex involves the binding of CNTF to CNTFRα followed by the recruitment of each of the β signal transducing components. Activation of the Jak and Tyk kinases that are preassociated with the β components results in tyrosine phosphorylation of the β components as well as other cellular proteins.

ponent. In contrast, the variety of actions of LIF are due to the widespread distributions of its receptor components, gp130 and LIFRβ, and to the fact that it does not require a restricted LIFRβ component. The fact that CNTF and LIF share their β components explains why CNTF and LIF appear to have identical actions on their shared neuronal targets. A series of experiments subsequently verified the receptor structures that had been proposed (Gearing et al 1992, Stahl et al 1993, Davis et al 1993b, Baumann et al 1993, Murakami et al 1993).

## Dimerization of β Components Initiates Signaling for CNTF

The realization that the CNTF-related cytokines all utilize receptor systems containing dimers of their α components suggested a mechanism by which receptor activation might occur. Analogous to the receptor tyrosine kinases that are activated by ligand-induced dimerization, it was suggested that the various cytokine receptor components are initially unassociated on the cell surface, and that ligand induces sequential association of these components in a stepwise fashion (Ip et al 1992). Subsequent studies verified this notion, and revealed that the first step in the formation of the receptor complex involves binding of the ligand to the α component, followed by the dimerization of the β components and signal activation (Figure 1; Davis et al 1993b, Ip & Yancopoulos 1992, Murakami et al 1993, Davis & Yanopoulos 1993, Stahl & Yancopoulos 1993). In the case of CNTF or LIF, it is the heterodimerization of gp130 and LIFRβ, while for IL-6, homodimerization of gp130 by IL-6 initiates subsequent signaling events (Davis et al 1993b, Murakami et al 1993).

## CNTF Utilizes the Jak-Tyk Family of Cytoplasmic Tyrosine Kinases

Ligand stimulation of the CNTF receptor results in tyrosine phosphorylation of the two β receptor components and other cellular proteins, which are required for subsequent downstream biological responses to the cytokine (Table 1). Unlike the Trk receptors or other receptor tyrosine kinases, however, the receptor components for CNTF lack intrinsic kinase domains or regions with recognizable catalytic function. These findings suggested that activation of nonreceptor tyrosine kinase(s) associated with the cytoplasmic domains of the β receptor components might be required in order to transduce signals subsequent to ligand stimulation. Indeed, important clues as to how CNTF or its distant relatives can initiate signal transduction came from a recently discovered family of cytoplasmic tyrosine kinases, Jak-Tyk. The known members of this nonreceptor kinase family include Jak1, Jak2, and Tyk2, which were originally cloned by homology-based strategies (Firmbach-Kraft et al 1990, Wilks et al 1991) and recently found to be crucial in the signaling pathways for several cytokines. For example, recent studies have demonstrated that Jak1 and Tyk2 were required for interferon α signaling (Muller et al 1993, Velazquez et al 1992), Jak1 and Jak2 were required for responses to interferon γ (Muller et al 1993, Watling et al 1993), and Jak2 could be activated in response to growth hormone and erythropoietin (Argetsinger et al 1993, Witthuhn et al 1993). Studies initiated to examine whether the CNTF family of cytokines also utilize the Jak-Tyk family of tyrosine kinases demonstrated the crucial involvement of these kinases (Stahl et al 1994, Stahl & Yancopoulos 1994). In particular, unlike other cytokines that activate only a certain subset of the

Jak-Tyk kinases, the CNTF-related cytokines are unique in being capable of activating all the known members of the Jak-Tyk family of kinases. Furthermore, Stahl et al (1994) demonstrated that these kinases were constitutively preassociated with the β receptor components and that they became activated during ligand-mediated dimerization of the β components.

## Modular Motifs in the β Receptor Components Modulates Jak Choice of Substrates

The convergence of many cytokines on the utilization of the same Jak-Tyk kinases for signaling raise the important issue of how these cytokines activate distinct signaling pathways. Recent studies show that the recruitment of specific downstream signaling substrates is not determined by the Jak-Tyk kinases but rather by specific modular domains on the β receptor components (Stahl et al 1995). That is, specific "docking sites" on the receptor, upon tyrosine phoshorylation by the Jak-Tyk kinases, can recruit particular substrates via their SH2 domains (Figure 2). Indeed, analysis of intracellular signaling pathways activated by the CNTF and its relatives has identified a set of proteins overlapping with those induced by other cytokines and growth factors, many of which contain SH2 domains (Table 1) (Boulton et al 1994). Included among these are the STATs (signal transducer and activator of transcription), which are important in mediating the actions of CNTF and its relatives (Lutticken et al 1994, Shuai et al 1994, Boulton et al 1995, Heim et al 1995). STAT3, in particular, is a well-characterized example of a signaling substrate preferentially activated by the CNTF-related cytokines because it is recruited to the receptor complex by modular motifs in the β components (Stahl et al 1995). Upon activation of the CNTF receptor, the STATs became tyrosine phosphorylated via Jak-Tyk kinases, then translocated to the nucleus to participate in the transcriptional responses to CNTF (Bonni et al 1993, Lutticken et al 1994, Boulton et al 1995).

## Activity of a Soluble Form of CNTFRα

As discussed above, one of the unique features of the CNTF receptor complex is that the α component lacks a transmembrane domain and is anchored to the cell surface via a GPI linkage (Davis et al 1991). This unusual linkage of CNTFRα, which is cleavable by phospholipases, suggests that CNTFRα could be released from the cell surface under certain circumstances and act as a soluble protein in a manner similar to that of IL-6Rα (Taga et al 1989). Both CNTFRα and IL-6Rα are analogous to one of the two subunits of a heterodimeric cytokine, natural killer stimulatory factor (NKSF) (Gearing & Cosman 1991); the other subunit resembles the cytokines themselves. These findings led to the suggestion that soluble CNTFRα can bind to its ligand and then

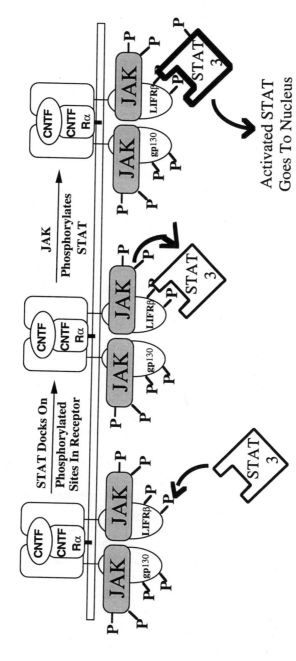

*Figure 2*    Motifs on the β components of the CNTF receptor complex modulate JAK choice of substrates. Activation of a CNTF receptor complex results in tyrosine phosphorylation of specific motifs on the β components. These motifs on the β components act as docking sites to recruit specific SH2 domain–containing substrates (e.g. STAT3), thus determining substrates that are acted upon by the rather generic JAK kinases. Note that some substrates may dock directly to phosphorylated motifs on the JAKs and that they are not designated by the β components.

activate the β signaling components. Indeed, Davis et al (1993a) demonstrated that a soluble form of CNTFRα can form a heterodimeric complex with CNTF to activate signaling in cells that express only the two β components (gp130 and LIFRβ) and normally respond only to LIF but not CNTF. For example, addition of CNTFRα either in a soluble or membrane-bound form to he-matopoietic cells resulted in functional responses to CNTF (Davis et al 1993a, Ip et al 1993b).

Findings of the collaboration between CNTF and soluble CNTFRα in vitro raised the possibility that a similar function might exist in vivo through the regulated release of soluble CNTFRα. Indeed, a physiological role for soluble CNTFRα is suggested by its presence in cerebrospinal fluid in vivo (Davis et al 1993a). Transection of the sciatic nerve also resulted in a transient dramatic increase in CNTFRα mRNA expression coincidental with the release of CNTFRα from skeletal muscle (Davis et al 1993a). CNTF released from the nerve together with soluble CNTFRα released from skeletal muscle might collaborate and act on diverse cell types (such as blood-derived monocytes) at the site of injury that normally do not respond to CNTF due to the absence of CNTFRα (Ip & Yancopoulos 1992). Such interaction might play an impor-tant role in the regeneration response following injury. Taken together, our findings suggest that in the presence of soluble CNTFRα, CNTF has the ability to act via the LIF receptors expressed in various cell types in vivo.

## Gene Disruptions Reveal Physiological Roles of CNTF and CNTFRα

As discussed above for the neurotrophins, the wealth of knowledge concerning the many observed activities of exogenously provided CNTF in vitro and in vivo was not predictive of the normal physiological role of CNTF. Because of the many observed actions of exogenously provided CNTF, it was surprising to find that mice lacking CNTF developed in a normal manner and displayed only mild motor neuron problems later in adulthood without any other major neurological abnormalities (Masu et al 1993). These findings, together with the striking observation that a substantial proportion (2.5%) of the Japanese population is homozygous for a null mutation of CNTF and yet appears quite normal even in old age (Takahashi et al 1994), suggested that CNTF is not crucial for development and that it may not even be absolutely required later in life in response to nerve injury or other trauma.

In contrast to the findings with mice lacking CNTF, recent ablation of the CNTFRα in mice resulted in a dramatic phenotype (Table 2). The newborn mutant pups were clearly abnormal; they did not initiate the feeding process, and all died shortly after birth. They also exhibited dramatic loss in all motor neuron populations examined (DeChiara et al 1995), as well as in a number of other PNS and CNS populations (TM DeChiara & GD Yancopoulos, un-published observations). Taken together with the results of null CNTF muta-tions in mice and humans, the phenotype of mice lacking CNTFRα strongly predicts that there exists an as yet undiscovered CNTF-related factor that also utilizes the CNTFRα in vivo, in much the same way that the TrkB receptor is utilized by both BDNF and NT4. This second CNTFRα ligand appears much more critical for normal development of the nervous system and for motor neurons, in particular. In fact, despite the plethora of factors known to support motor neurons when exogenously provided, mice mutant for CNTFRα were used to identify the only neurotrophic factor system shown to be physiologi-cally relevant for normal motor neuron development in vivo; thus the putative second ligand for CNTFRα may well correspond to the long sought after muscle-derived motor neuron trophic factor. Known members of the CNTF cytokine family, including the recently identified cardiotrophin 1 (Pennica et al 1995), do not appear to require CNTFRα. Similarly, a CNTF relative recently cloned from chicken, called growth promoting activity (GPA) (Leung et al 1992), appears to be the chicken counterpart to CNTF rather than a candidate for the CNTF relative. Thus the identity of the alternative ligand that shares the CNTFRα as predicted from the genetic evidence remains to be determined.

## Collaboration Between Different Classes of Neurotrophic Factors

Cytokines of the CNTF family, such as IL-6 and LIF, have been shown to affect various lineages of hematopoiesis (Miyajima et al 1992). Prominent expression of CNTFRα by neuronal progenitors in the neuroepithelial layer, together with the presence of CNTF transcript in rat embryos (Ip et al 1993b), suggested that CNTF might also be a key signaling molecule during neuro-poiesis in a manner similar to its cytokine relatives during hematopoiesis. The availability of a CNTF responsive cell line, the MAH cell line, allowed for examination of such a role for CNTF in the development of sympathetic neurons from sympathoadrenal progenitor cells. CNTF or LIF together with FGF was found to act collaboratively on these immortalized neuronal precur-sors to influence their differentiation program and to dramatically increase the proportion of these progenitor cells to become NGF-dependent postmitotic neurons (Ip et al 1994). Synergism at the signaling level was also observed in that CNTF could synergize with FGF to result in an increased and prolonged activation of the ERKs (Ip et al 1994). Furthermore, direct synergy between CNTF and NGF could be demonstrated if the Trk receptor was introduced into the MAH cell progenitors (Verdi 1994a). Similarly, Murphy et al (1993) showed that LIF collaborates with NGF in the development of sensory neurons from neural crest precursors. Taken together, these findings suggest that multiple factors (such as CNTF, FGF, and members of the neurotrophin family), utilizing distinct receptor systems and signaling mechanisms, can interact synergistically to promote differentiation of progenitors into postmitotic neurons. Such synergistic interaction could play an important role in modulating the differentiation of a wide assortment of neuronal precursors during embryonic development in vivo.

Recent data suggest that similar synergistic interactions may occur on mature neurons in the adult and may be therapeutically relevant. In vivo synergy has recently been demonstrated in the Wobbler mouse model of degenerative motor neuron disease: Treatment with both CNTF and BDNF in this model was shown to be substantially better than treatment with each factor individually (Mitsumoto et al 1994).

## CONCLUSION

As discussed above, the receptor system utilized by CNTF is quite different from the one used by the neurotrophins. However, there are many parallels. Both CNTF and the neurotrophins are primarily known for their neurotrophic actions not because of unusual signaling capabilities of their receptor systems—each of which is quite similar to receptors used by traditional growth factors and cytokines operating outside of the nervous system—but because their receptor systems are largely restricted to the nervous system in their expression. Also in both cases, ligand activates either an intrinsic or an associated tyrosine kinase activity by inducing dimerization of signal transducing receptor components. Multiple classes of neurotrophic factors may exist so that they may in some cases signal interactively or collaboratively; certainly there are striking examples of synergies between CNTF and the neurotrophins. Recent studies with mutant mice disrupted for the various neurotrophic factors and their receptors have provided great insight into the physiological role of these factors, much of which could not have been predicted by simply studying the effects of exogenously provided factors in vitro and even in vivo.

*Literature Cited*

Adler R, Landa KB, Manthorpe M, Varon S. 1979. Cholinergic neuronotrophic factors: intraocular distribution of soluble trophic activity for ciliary neurons. *Science* 204:1434–36

Anton ES, Westkamp G, Reichardt LF, Matthew WD. 1994. Nerve growth factor and its low affinity receptor promote Schwann cell migration. *Proc. Natl. Acad. Sci. USA* 91: 275–99

Arakawa Y, Sendtner M, Thoenen H. 1990. Survival effect of ciliary neurotrophic factor (CNTF) on chick embryonic motoneurons in culture: comparison with other neurotrophic factors and cytokines. *J. Neurosci.* 10:3507–15

Argetsinger LS, Campbell GS, Yang X, Witthuhn BA, Silvennoinen O, et al. 1993. Identification of JAK2 as a growth hormone receptor-associated tyrosine kinase. *Cell* 74:237–44

Asada H, Ip NY, Pan L, Razack N, Parfitt MM, Plunkett RJ. 1995. Time course of ciliary neurotrophic factor mRNA expression is coincident with the presence of protoplasmic astrocytes in traumatized rat striatum. *J. Neurosci. Res.* 40:22–30

Barbacid M. 1993. Nerve growth factor: a tale of two receptors. *Oncogene* 8:2033–42

Barbin G, Manthorpe M, Varon S. 1984. Purification of chick eye ciliary neuronotrophic factor. *J. Neurochem.* 43:1468–78

Barker PA, Shooter EM. 1994. Disruption of NGF binding to the low affinity neurotrophin receptor p75$^{LNTR}$ reduces NGF binding to Trk A on PC12 cells. *Neuron* 13:203–15

Barres BA, Schmid R, Sendtner M, Raff MC. 1993. Multiple extracellular signals are required for long-term oligodendrocyte survival. *Development* 118:283–95

Baumann H, Ziegler SF, Mosley B, Morella KK, Pajovic S, Gearing DP. 1993. Reconstitution of the response to leukemia inhibitory factor, oncostatin M, and ciliary neurotrophic factor in hepatoma cells. *J. Biol. Chem.* 268:8414–17

Bazan JF. 1991. Neuropoietic cytokines in the hematopoietic fold. *Neuron* 7:197–208

Birren SJ, Lo L, Anderson DJ. 1993. Sympathetic neuroblasts undergo a developmental switch in trophic dependence. *Development* 119:597–610

Blottner D, Bruggemann W, Unsicker K. 1989. Ciliary neurotrophic factor supports target-deprived preganglionic sympathetic spinal cord neurons. *Neurosci. Lett.* 105:316–20

Bonni A, Frank DA, Schindler C, Greenberg ME. 1993. Characterization of a pathway for ciliary neurotrophic factor signaling to the nucleus. *Science* 262:1575–79

Bothwell M. 1995. Functional interactions of neurotrophins and neurotrophin receptors. *Annu. Rev. Neurosci.* 18:223–53

Boulton TG, Stahl N, Yancopoulos GD. 1994. Ciliary neurotrophic factor/leukemia inhibitory factor/interleukin 6/oncostatin M family of cytokines induces tyrosine phosphorylation of a common set of proteins overlapping those induced by other cytokines and growth factors. *J. Biol. Chem.* 269:11648–55

Boulton TG, Zhong Z, Wen Z, Darnell JE Jr, Stahl N, Yancopoulos GD. 1995. STAT3 activation by cytokines utilizing gp130 and re-

lated transducers involves a secondary modification requiring an H7-sensitive kinase. *Proc. Natl. Acad. Sci. USA.* 92:6915–19

Cabelli RJ, Hohn A, Shatz CJ. 1995. Inhibition of ocular dominance column formation by infusion of NT-4/5 or BDNF. *Science* 267:1662–66

Conover JC, Erickson JT, Katz DM, Bianchi LM, Poueymirou WT, et al. 1995. Neuronal deficits, not involving motor neurons, in mice lacking BDNF and/or NT4. *Nature* 375:235–38

Conover JC, Ip NY, Poueymirou WT, Bates B, Goldfarb MP, et al. 1994. Ciliary neurotrophic factor maintains the pluripotentiality of embryonic stem cells. *Development* 119:559–65

Crowley C, Spencer SD, Nishimura MC, Chen KS, Pitts-Meek S, et al. 1994. Mice lacking nerve growth factor display perinatal loss of sensory and sympathetic neurons yet develop basal forebrain cholinergic neurons. *Cell* 76:1001–11

Curtis R, Adryan KM, Zhu Y, Harkness PJ, Lindsay RM, DiStefano PS. 1993. Retrograde axonal transport of ciliary neurotrophic factor is increased by peripheral nerve injury. *Nature* 365: 253–55

Davis S, Aldrich TH, Ip NY, Stahl N, Scherer S, et al. 1993a. Released form of CNTF receptor α component as a soluble mediator of CNTF responses. *Science* 259:1736–39

Davis S, Aldrich TH, Stahl N, Pan L, Taga T, Kishimoto T, Ip NY, Yancopoulos GD. 1993b. LIFRβ and gp130 as heterodimerizing signal transducers of the tripartite CNTF receptor. *Science* 260:1805–8

Davis S, Aldrich TH, Valenzuela DM, Wong V, Furth ME, et al. 1991. The receptor for ciliary neurotrophic factor. *Science* 253:59–63

Davis S, Yancopoulos GD. 1993. The molecular biology of the CNTF receptor. *Curr. Opin. Neurobiol.* 3:20–24

DeChiara TM, Vejsada R, Pouyemirou WT, Acheson A, Suri C, et al. 1995. Mice lacking the CNTF receptor, unlike mice lacking CNTF, exhibit profound motor neuron deficits at birth. *Cell.* 83:313–22

Dicicco-Bloom E, Friedman WJ, Black IB. 1993. NT-3 stimulates sympathetic neuroblast proliferation by promoting precursor survival. *Neuron* 11:1101–11

Dobrowsky RT, Werner MH, Castellino AM, Chao MV, Hannun YA. 1994. Activation of sphingomyelin cycle through the low-affinity neurotrophin receptor. *Science* 265:1596–99

Ernfors P, Lee K-F, Jaenisch R. 1994a. Mice lacking brain-derived neurotrophic factor develop with sensory deficits. *Nature* 368:147–50

Ernfors P, Lee K-F, Kucera J, Jaenisch R.

1994b. Lack of neurotrophin-3 leads to deficiencies in the periphral nervous system and loss of limb proprioceptive afferents. *Cell* 77:503–12

Ernsberger U, Sendtner M, Rohrer H. 1989. Proliferation and differentiation of embryonic chick sympathetic neurons: effects of ciliary neurotrophic factor. *Neuron* 2:1275–84

Fann M-J, Patterson PH. 1993. A novel approach to screen for cytokine effects on neuronal gene expression. *J. Neurochem.* 61: 1349–55

Farinas I, Jones KR, Backus C, Wang X-Y, Reichardt LF. 1994. Severe sensory deficits in mice lacking neurotrophin-3. *Nature* 369: 658–61

Firmbach-Kraft I, Byers M, Shows T, Dalla-Favera R, Krolewski JJ. 1990. Tyk 2, a prototype of a novel class of non-receptor tyrosine kinase genes. *Oncogene* 5:1329–36

Friedman B, Scherer S, Rudge JS, Helgren M, Morrisey D, et al. 1992. Regulation of ciliary neurotrophic factor expression in myelin-related Schwann cells in vivo. *Neuron* 9:295–305

Gearing DP, Comeau MR, Friend DJ, Gimpel SD, Thut CJ, et al. 1992. The IL-6 signal transducer, gp130: an oncostatin M receptor and affinity converter for the LIF receptor. *Science* 255:1434–37

Gearing DP, Cosman D. 1991. Homology of the p40 subunit of natural killer cell stimulatory factor (NKSF) with the extracellular domain of the interleukin-6 receptor. *Cell* 66:9–10

Gearing DP, Thut CJ, VandenBos T, Gimpel SD, Delaney PB, et al. 1991. Leukemia inhibitory factor receptor is structurally related to the IL-6 signal transducer, gp130. *EMBO J.* 10: 2839–48

Gearing DP, Ziegler SF, Comeau MR, Friend D, Thoma B, et al. 1994. Proliferative responses and binding properties of hematopoietic cells transfected with low-affinity receptors for leukemia inhibitory factor, oncostatin M, and ciliary neurotrophic factor. *Proc. Natl. Acad. Sci. USA* 91:1119–23

Glass DJ, Nye SH, Hantzopoulos P, Macchi MJ, Squinto SP, et al. 1991. TrkB mediates BDNF/NT-3-dependent survival and proliferation in fibroblasts lacking the low affinity NGF receptor. *Cell* 66:405–13

Glass DJ, Yancopoulos GD. 1993. The neurotrophins and their receptors. *Trends Cell Biol.* 3:262–68

Götz R, Köster R, Winkler C, Raulf F, Lottspeich F, et al. 1994. Neurotrophin-6 is a new member of the nerve growth factor family. *Nature* 372:266–69

Hagg T, Quon D, Higaki J, Varon S. 1992. Ciliary neurotrophic factor prevents neuronal degeneration and promotes low affinity

NGF receptor expression in the adult rat CNS. *Neuron* 8:145–58

Hagg T, Varon S. 1993. Ciliary neurotrophic factor prevents degeneration of adult rat substantia nigra dopaminergic neurons in vivo. *Proc. Natl. Acad. Sci. USA.* 90:6315–19

Hall AK, Rao MS. 1992. Cytokines and neurokines: related ligands and related receptors. *Trends Neurosci.* 15:35–37

Hantzopoulos PA, Suri C, Glass DJ, Goldfarb MP, Yancopoulos GD. 1994. The low-affinity NGF receptor, p75, can collaborate with each of the Trks to potentiate functional responses to the neurotrophins. *Neuron* 13: 187–201

Heim MH, Kerr IM, Stark GR, Darnell JE Jr. 1995. Contribution of STAT SH2 groups to specific interferon signaling by the Jak-STAT pathway. *Science* 267:1347–49

Helgren ME, Squinto SP, Davis HL, Parry DJ, Boulton TG, et al. 1994. Trophic effect of ciliary neurotrophic factor on denervated skeletal muscle. *Cell* 76:493–504

Hempstead BL, Schleifer LS, Chao MV. 1989. Expression of functional nerve growth factor receptors after gene transfer. *Science* 243: 373–75

Hibi M, Murakami M, Saito M, Hirano T, Taga T, Kishimoto T. 1990. Molecular cloning and expression of an IL6 signal transducer, gp130. *Cell* 63:1149–57

Hilton DJ, Gough NM. 1991. Leukemia inhibitory factor: a biological perspective. *J. Cell. Biochem.* 46:21–26

Hughes SM, Lillien LE, Raff MC, Rohrer H, Sendtner M. 1988. Ciliary neurotrophic factor induces type-2 astrocyte differentiation in culture. *Nature* 335:70–73

Hyman C, Juhasz M, Jackson C, Wright P, Ip NY, Lindsay RM. 1994. Overlapping and distinct actions of the neurotrophins BDNF, NT-3 and NT-4/5, on cultured dopaminergic and GABAergic neurons of the ventral mesencephalon. *J. Neurosci.* 14:335–47

Ibáñez CF, Ebendal T, Barbany G, Murray-Rust J, Blundell TL, Persson H. 1992. Disruption of the low affinity receptor-binding site in NGF allows neuronal survival and differentiation by binding to the trk gene product. *Cell* 69:329–41

Ip NY, Boulton TG, Li Y, Verdi JM, Birren SJ, et al. 1994. CNTF, FGF and NGF collaborate to drive the terminal differentiation of MAH cells into postmitotic neurons. *Neuron* 13: 443–55

Ip NY, Li Y, van de Stadt I, Panayotatos N, Alderson RF, Lindsay RM. 1991a. Ciliary neurotrophic factor enhances neuronal survival in embryonic rat hippocampal cultures. *J. Neurosci.* 11:3124–34

Ip NY, Li Y, Yancopoulos GD, Lindsay RM. 1993a. Cultured hippocampal neurons show responses to BDNF, NT-3, and NT-4, but not NGF. *J. Neurosci.* 13:3394–405

Ip NY, Maisonpierre PC, Alderson R, Friedman B, Furth ME, et al. 1991b. The neurotrophins and CNTF: specificity of action towards PNS and CNS neurons. *J. Physiol.* 85:123–30

Ip NY, McClain J, Barrezueta NX, Aldrich TH, Pan L, et al. 1993b. The α component of the CNTF receptor is required for signaling and defines potential CNTF targets in the adult and during development. *Neuron* 10:89–102

Ip NY, Nye SN, Boulton TG, Davis S, Taga T, et al. 1992. CNTF and LIF act on neuronal cells via shared signalling pathways that involve the IL-6 signal transducing receptor component gp130. *Cell* 69:1121–32

Ip NY, Stitt TN, Tapley P, Klein R, Glass DJ, et al. 1993c. Similarities and differences in the way neurotrophins interact with the Trk receptors in neuronal and nonneuronal cells. *Neuron* 10:137–49

Ip NY, Wiegand SJ, Morse J, Rudge JS. 1993d. Injury-induced regulation of ciliary neurotrophic factor mRNA in the adult rat brain. *Eur. J. Neurosci.* 5:25–33

Ip NY, Yancopoulos GD. 1992. Ciliary Neurotrophic Factor and its receptor complex. *Prog. Growth Factor Res.* 4:139–55

Ip NY, Yancopoulos GD. 1993. Receptors and signaling pathways of ciliary neurotrophic factor and the neurotrophins. *Semin. Neurosci.* 5:249–57

Ip NY, Yancopoulos GD. 1994. Neurotrophic factor receptors: just like other growth factor and cytokine receptors? *Curr. Opin. Neurobiol.* 4:400–5

Jing S, Tapley P, Barbacid M. 1992. Nerve growth factor mediates signal transduction through trk homodimer receptors. *Neuron* 9:1067–79

Jones KR, Farinas I, Backus C, Reichardt LF. 1994. Targeted disruption of the BDNF gene perturbs brain and sensory neuron development but not motor neuron development. *Cell* 76:989–99

Kalchem C, Carmeli C, Rosenthal A. 1992. NT-3 is a mitogen for cultured neural crest cells. *Proc. Natl. Acad. Sci. USA* 89:1661–65

Kang H, Schuman EM. 1995. Long-lasting neurotrophin-induced enhancement of synaptic transmission in the adult hippocampus. *Science* 267:1658–62

Kaplan DR, Hempstead BL, Martin-Zanca D, Chao MV, Parada LF. 1991. The trk proto-oncogene product: a signal transducing receptor for nerve growth factor. *Science* 252:554–56

Kishimoto T, Akira S, Taga T. 1992. Interleukin-6 and its receptor: a paradigm for cytokines. *Science* 258:593–97

Klein R, Jing SQ, Nanduri V, O'Rourke E,

Barbacid M. 1991. The trk proto-oncogene encodes a receptor for nerve growth factor. *Cell* 65:189–97

Klein R, Silos-Santiago I, Smeyne RJ, Lira SA, Brambilla R, et al. 1994. Disruption of the neurotrophin-3 receptor gene trkC eliminates Ia muscle afferents and results in abnormal movements. *Nature* 368:249–51

Klein R, Smeyne RJ, Wurst W, Long LK, Auerbach BA, et al. 1993. Targeted disruption of the trkB neurotrophin receptor gene results in nervous system lesions and neonatal death. *Cell* 75: 113–22

LaVail MM, Unoki K, Yasumura D, Matthes MT, Yancopoulos GD, Steinberg RH. 1992. Multiple growth factors, cytokines and neurotrophins rescue photoreceptors from the damaging effects of constant light. *Proc. Natl. Acad. Sci. USA* 89:11249–53

Lee K-F, Bachman K, Landis S, Jaenisch R. 1994. Dependence on p75 for innervation of some sympathetic targets. *Science* 263: 1447–49

Lee K-F, Li E, Huber J, Landis SC, Sharpe AH, et al. 1992. Targeted mutation of the gene encoding the low affinity NGF receptor p75 leads to deficits in the peripheral sensory nervous system. *Cell* 69:737–49

Lesser SS, Lo DC. 1995. Regulation of voltage-gated ion channels by NGF and ciliary neurotrophic factor in SK-N-SH neuroblastoma cells. *J. Neurosci.* 15:253–61

Leung DW, Parent AS, Cachianes G, Esch F, Coulombe JN, et al. 1992. Cloning, expression during development and evidence for release of a trophic factor for ciliary ganglion neurons. *Neuron* 8:1045–53

Lillien LE, Sendtner M, Rohrer H, Huges SM, Raff MC. 1988. Type-2 astrocyte development in rat brain cultures is initiated by a CNTF-like protein produced by type 1 astrocytes. *Neuron* 1:485–94

Lin L-FH, Mismer D, Lile JD, Armes LG, Butler ETIII, et al. 1989. Purification, cloning and expression of ciliary neurotrophic factor (CNTF). *Science* 246:1023–25

Lindsay RM. 1994. Neurotrophins and receptors. *Prog. Brain Res.* 103:3–14

Liu X, Ernfors P, Wu H, Jaenisch R. 1995. Sensory but not motor neuron deficits in mice lacking NT-4 and BDNF. *Nature* 375: 238–41

Lohof AM, Ip NY, Poo M-m. 1993. Potentiation of developing neuromuscular synapses by the neurotrophins NT-3 and BDNF. *Nature* 363:350–53

Louis J-C, Magal E, Takayama S, Varon S. 1993. CNTF protection of oligodendrocytes against natural and tumor necrosis factor-induced death. *Science* 259:689–92

Lutticken C, Wegenka UM, Yuan J, Buschmann J, Schindler C, et al. 1994. Association of transcription factor APRF and protein kinase Jak 1 with the interleukin-6 signal transducer gp130. *Science* 263:89–92

Manthorpe M, Hagg T, Varon S. 1992. Ciliary neurotrophic factor. In *Neurotrophic Factors*, ed. JA Fallon, SE Loughlin, pp. 443–73. New York: Academic

Martinou JC, Martinou I, Kato AC. 1992. Cholinergic differentiation factor (CDF/LIF) promotes survival of isolated rat embryonic motoneurons in vitro. *Neuron* 8:737–44

Masu Y, Wolf E, Holtmann B, Sendtner M, Brem G, Thoenen H. 1993. Disruption of the CNTF gene results in motor neuron degeneration. *Nature* 365:27–32

Meakin SO, Shooter EM. 1992. The nerve growth factor family of receptors. *Trends Neurosci.* 15:323–31

Mitsumoto H, Ikeda K, Klinkosz B, Cedarbaum JM, Wong V, Lindsay RM. 1994. Arrest of motor neuron disease in wobbler mice cotreated with CNTF and BDNF. *Science* 265:1107–10

Miyajima A, Kitamura T, Harada N, Yokota T, Arai K. 1992. Cytokine receptors and signal transduction. *Annu. Rev. Immunol.* 10:295–331

Muller M, Briscoe J, Laxton C, Guschin D, Ziemiecki A, et al. 1993. The protein tyrosine kinase JAK1 complements defects in interferon-$\alpha/\beta$ and -$\gamma$ signal transduction. *Nature* 366:129–35

Murakami M, Hibi M, Nakagawa N, Nakagawa T, Yasukawa K, et al. 1993. IL-6 induced homodimerization of gp130 and associated activation of a tyrosine kinase. *Science* 260:1808–11

Murphy M, Reid K, Brown MA, Barlett PF. 1993. Involvement of leukemia inhibitory factor and nerve growth factor in the development of dorsal root ganglion neurons. *Development* 117:1173–82

Murphy M, Reid K, Hilton DJ, Barlett PF. 1991. Generation of sensory neurons is stimulated by leukemia inhibitory factor. *Proc. Natl. Acad. Sci. USA* 88:3498–501

Oppenheim RW, Prevette D, Qin-Wei Y, Collins F, MacDonald J. 1991. Control of embryonic motoneuron survival in vivo by ciliary neurotrophic factor. *Science* 251:1616–18

Pennica D, King KL, Shaw KJ, Luis E, Rullamas J, et al. 1995. Expression cloning of cardiotrophin 1, a cytokine that induces cardiac myocyte hypertrophy. *Proc. Natl. Acad. Sci. USA* 92:1142–46

Rabizadeh S, Oh J, Zhong LT, Yang J, Bitler CM, et al. 1993. Induction of apoptosis by the low-affinity NGF receptor. *Science* 261:345–48

Rao MS, Tyrrell S, Landis SC, Patterson PH. 1992. Effects of ciliary neurotrophic factor (CNTF) and depolarization of neuropeptide

expression in cultured sympathetic neurons. *Dev. Biol.* 150:281–93

Rasouly D, Lazarovici P, Stephens RM, Kaplan DR. 1995. Neurotrophin receptors. *Promega Neural Notes Spring.* pp. 3–6

Rende M, Muir D, Ruoslahti E, Hagg T, Varon S, Manthorpe M. 1992. Immunolocalization of ciliary neuronotropic factor in adult rat sciatic nerve. *Glia* 5:25–32

Rose TM, Bruce G. 1991. Oncostatin M is a member of a cytokine family that includes leukemia-inhibitory factor, granulocyte colony-stimulating factor, and interleukin 6. *Proc. Natl. Acad. Sci. USA.* 88:8641–45

Rudge JS, Alderson RF, Pasnikowski E, McClain J, Ip NY, Lindsay RM. 1992. Expression of ciliary neurotrophic factor and the neurotrophins—nerve growth factor, brain-derived neurotrophic factor and neurotropin3—in cultured rat hippocampal astrocytes. *Eur. J. Neurosci.* 4:459–71

Rudge JS, Li Y, Pasnikowski EM, Mattsson K, Pan L, et al. 1994. Neurotrophic factor receptors and their signal transduction capabilities in astrocytes. *Eur. J. Neurosci.* 6: 693–705

Saadat S, Sendtner M, Rohrer H. 1989. Ciliary neurotrophic factor induces cholinergic differentiation of rat sympathetic neurons in culture. *J. Cell Biol.* 108:1807–16

Schooltinck H, Stoyan T, Roeb E, Heinrich PC, Rose-John S. 1992. Ciliary neurotrophic factor induces acute-phase protein expression in hepatocytes. *FEBS Lett.* 314:280–84

Sendtner M, Kreutzberg GW, Thoenen H. 1990. Ciliary neurotrophic factor prevents the degeneration of motor neurons after axotomy. *Nature* 345:4401

Sendtner M, Schmalbruch H, Stockli KA, Carroll P, Kreutzberg GW, Thoenen H. 1992a. Ciliary neurotrophic factor prevents degeneration of motor neurons in mouse mutant progressive motor neuronopathy. *Nature* 358:502–4

Sendtner M, Stöckli KA, Thoenen H. 1992b. Synthesis and localization of ciliary neurotrophic factor in the sciatic nerve of adult rat after lesion and in regeneration. *J. Cell Biol.* 118:139–48

Shuai K, Horvath CM, Huang LHT, Quareshi SA, Cowburn D, Darnell JE Jr. 1994. Interferon activation of the transcription factor Stat91 involves dimerization through SH2-phosphotyrosyl peptide interactions. *Cell* 76: 821–28

Smeyne RJ, Klein R, Schnapp A, Long LK, Bryant S, et al. 1994. Severe sensory and sympathetic neuropathies in mice carrying a disrupted Trk/NGF receptor gene. *Nature* 368: 246–49

Snider WD. 1994. Functions of the neurotrophins during nervous system development:

what the knockouts are teaching us. *Cell* 77: 627–38

Squinto SP, Aldrich TH, Lindsay RM, Morrissey DM, Panayotatos N, et al. 1990. Identification of functional receptors for ciliary neurotrophic factor on neuronal cell lines and primary neurons. *Neuron* 5:757–66

Stahl N, Boulton TG, Farruggella T, Ip NY, Davis S, et al. 1994. Association and activation of Jak-Tyk kinases by CNTF-LIF-OSM-IL-6 β receptor components. *Science* 263: 92–95

Stahl N, Davis S, Wong V, Taga T, Kishimoto T, et al. 1993. Cross-linking identifies leukemia inhibitory factor-binding protein as a ciliary neurotrophic factor receptor component. *J. Biol. Chem.* 268:7628–31

Stahl N, Farruggella TJ, Boulton TG, Zhong Z, Darnell JE Jr, Yancopoulos GD. 1995. Choice of STATs and other substrates specified by modular tyrosine-based motifs in cytokine receptors. *Science* 267:1349–53

Stahl N, Yancopoulos GD. 1993. The alphas, betas, and kinases of cytokine receptor complexes. *Cell* 74:587–90

Stahl N, Yancopoulos GD. 1994. The tripartite CNTF receptor complex: activation and signaling involves components shared with other cytokines. *J. Neurobiol.* 25:1454–66

Stöckli KA, Lillien LE, Naher-Noe M, Brietfeld G, Hughes RA, et al. 1991. Regional distribution, developmental changes, and cellular localization of CNTF-mRNA and protein in the rat brain. *J. Cell Biol.* 115:447–59

Stöckli KA, Lottspeich F, Sendtner M, Masiakowski P, Carroll P, et al. 1989. Molecular cloning, expression and regional distribution of rat ciliary neurotrophic factor. *Nature* 342: 920–23

Stoop R, Poo M-m. 1995. Potentiation of transmitter release by ciliary neurotrophic factor requires somatic signaling. *Science* 267:695–99

Symes A, Rao MS, Lewis SE, Landis SC, Hyman SE, Fink JS. 1993. Ciliary neurotrophic factor coordinately activates transcription of neuropeptide genes in a neuroblastoma cell line. *Proc. Natl. Acad. Sci. USA* 90:572–76

Taga T, Hibi M, Hirata Y, Yamasaki K, Yasukawa K, et al. 1989. Interleukin-6 triggers the association of its receptor with a possible signal transducer, gp130. *Cell* 58:573–81

Takahashi R, Yokoji H, Misawa H, Hayashi M, Hu J, Deguchi T. 1994. A null mutation in the human CNTF gene is not causally related to neurological diseases. *Nat. Genet.* 7:79–84

Tartaglia LA, Pennica D, Goeddel DV. 1993. Ligand passing: the 75-kDa tumor necrosis factor (TNF) receptor recruits TNF for signaling by the 55-kDa TNF receptor. *J. Biol. Chem.* 268:18542–48

Thaler CD, Suhr L, Ip NY, Katz DM. 1994. Leukemia inhibitory factor and neurotrophins support overlapping populations of rat nodose sensory neurons in culture. *Dev. Biol.* 161:338–44

Velazquez L, Fellous M, Stark GR, Pellegrini S. 1992. A protein kinase in the interferon α/β pathway. *Cell* 70:313–22

Verdi JM, Anderson DJ. 1994. Neurotrophins regulate sequential changes in neurotrophin receptor expression by sympathetic neuroblasts. *Neuron* 13:1359–172

Verdi JM, Birren SJ, Ibanez CF, Persson H, Kaplan DR, et al. 1994a. p75$^{LNGFR}$ regulates Trk signal transduction and NGF-induced neuronal differentiation in MAH cells. *Neuron* 12:733–45

Verdi JM, Ip NY, Yancopoulos GD, Anderson DJ. 1994b. Expression of p140 trk in MAH cells lacking p75$^{LNGFR}$ is sufficient to permit NGF-induced differentiation to post-mitotic neurons. *Proc. Natl. Acad. Sci. USA* 91: 3949–53

Watling D, Guschin D, Muller M, Silvennoinen O, Witthuhn BA, et al. 1993. Complementation by the protein tyrosine kinase JAK2 of a mutant cell line defective in the interferon-g signal transduction pathway. *Nature* 366: 166–70

Wilks AF, Harpur AG, Kurban RR, Ralph SJ, Zurcher G, Ziemiecki A. 1991. Two novel protein-tyrosine kinases, each with a second phosphotransferase-related catalytic domain, define a new class of protein kinase. *Mol. Cell Biol.* 11:2057–65

Witthuhn BA, Quelle FW, Silvennoinen O, Yi T, Tang B, et al. 1993. JAK2 associates with the erythropoietin receptor and is tyrosine phosphorylated and activated following stimulation with erythropoietin. *Cell* 74: 227–36

Wong V, Arriaga R, Ip NY, Lindsay RM. 1993. The neurotrophins BDNF, NT-3, and NT-4, but not NGF, upregulate the cholinergic phenotype of developing motor neurons. *Eur. J. Neurosci.* 5:466–74

Yamamori T, Fukada K, Aebersold R, Korsching S, Fann MJ, Patterson PH. 1989. The cholinergic neuronal differentiation factor from heart cells is identical to leukemia inhibitory factor. *Science* 246:1412–16

*Annu. Rev. Neurosci. 1996. 19:517–44*

# INFORMATION CODING IN THE VERTEBRATE OLFACTORY SYSTEM

## Linda B. Buck

Howard Hughes Medical Institute and Department of Neurobiology, Harvard
Medical School, 220 Longwood Avenue, Boston, Massachusetts 02115

KEY WORDS:    olfaction, odorant receptors, sensory, olfactory epithelium, olfactory bulb

### ABSTRACT

The olfactory systems of vertebrates are able to discriminate a vast array of
structurally diverse odorants. This perceptual acuity derives from a series of
information-processing events that occur within distinct neural structures through
which olfactory sensory information flows. This review discusses current knowledge concerning the mechanisms by which olfactory stimuli are initially detected
and transduced into electrical signals that are transmitted to the olfactory bulb
of the brain. It also reviews how information may initially be organized, or
encoded, and then reorganized as it flows through the system.

## INTRODUCTION

The vertebrate olfactory system is exquisitely tuned to discriminate an immense variety of odorous molecules of differing shapes and sizes that may be
present in vanishingly small quantities in the environment. The discriminatory
capacity of this system derives from a series of information-processing steps
that occur at distinct anatomical structures through which olfactory sensory
information progresses: the olfactory epithelium of the nose, where olfactory
sensory neurons detect odorous molecules (odorants); the olfactory bulb of the
brain, to which these neurons transmit signals; and higher-order structures of
the brain, such as the piriform cortex, which receive information from the
olfactory bulb and distribute it to other regions of the brain.

This review discusses current knowledge concerning the molecular mechanisms and organizational strategies used by the olfactory system to transduce,
encode, and process information at various levels in the neural pathway followed by olfactory sensory information. The discussion focuses exclusively
on the olfactory systems of vertebrates, primarily those of mammals. However,

517

it should be emphasized that considerable information has been gained about the olfactory systems of several invertebrate species and that there are some striking structural similarities between vertebrate and invertebrate olfactory systems that suggest that the strategies used for processing olfactory information in vertebrates and invertebrates may, in some respects, be similar (Kaissling 1986, Boeckh & Tolbert 1993, Sengupta et al 1993, Stocker 1994, Hildebrand 1995). In addition, genetic approaches to the study of several invertebrate systems promise to yield further insight into mechanisms underlying olfactory information coding (Siddiqi 1987, Carlson 1991, Bargmann et al 1993).

## SENSORY TRANSDUCTION IN THE OLFACTORY EPITHELIUM

### The Olfactory Neuron

The initial events in olfactory perception occur in olfactory sensory neurons (or olfactory neurons or receptor neurons), which are embedded in a pseudo-stratified columnar epithelium (the olfactory epithelium) that, in mammals, lines the posterior nasal cavity. This epithelium contains three major cell types: olfactory neurons, supporting cells (sustentacular cells), and olfactory stem cells (Graziadei & Monti-Graziadei 1979) (Figure 1). Olfactory neurons are unusual in that they are short-lived cells that generally live for only 30–60 days and are continuously replaced from the basal layer of stem cells (Graziadei & Monti-Graziadei 1979, Calof & Chikaraishi 1989, Caggiano et al 1994).

The olfactory neuron is a bipolar nerve cell. From its apical pole, it extends a single dendrite to the epithelial surface. Numerous cilia protrude from this dendrite into the mucus lining the nasal cavity, providing an extensive receptive surface for the detection of odorants. From its opposite pole, the neuron sends a single axon to the olfactory bulb of the brain. Odorants that dissolve in the nasal mucus bind to specific odorant receptors on the cilia of olfactory neurons and thereby induce a cascade of transduction events that culminate in the generation of action potentials in the sensory axon and the transmission of signals to the olfactory bulb of the brain.

### Perireceptor Events

Odorants are small, usually lipophilic, molecules that must partition into and traverse the nasal mucus in order to be recognized by odorant receptors on the cilia of olfactory neurons. The mucus coating of the olfactory epithelium is believed to be specialized to accommodate this necessity and to provide the appropriate ionic environment for changes in ciliary membrane potential that are critical to signal generation in the olfactory neuron (reviewed in Getchell et al 1984, Carr et al 1990, Pevsner & Snyder 1990, Anholt 1993). The quality

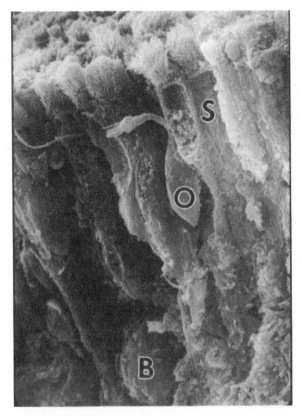

*Figure 1*   The olfactory epithelium. In this scanning electron micrograph, the three major cell types of the epithelium can be seen: olfactory sensory neurons (O), supporting cells (S), and basal stem cells (B). The dendrite and terminal cilia of one neuron can be seen clearly. [From Morrison & Constanzo (1990).]

of the mucus layer and the molecules secreted into it by supporting cells and by Bowman's glands underlying the olfactory epithelium are thought to be important in creating an optimal perireceptor environment.

Perhaps the most interesting and important molecules found in olfactory mucus are the soluble odorant-binding proteins (OBPs), which are small proteins that belong to a large family of ligand carrier proteins called lipocalins (Bignetti et al 1985, Pevsner et al 1986, Lee et al 1987, Pevsner & Snyder 1990). Small families of OBPs have been found in rat and in several invertebrate species, suggesting that different OBPs may be specialized to bind to different structural classes of odorants (Dear et al 1991, Krieger et al 1991, Vogt et al 1991, McKenna et al 1994, Pikielny et al 1994). The function of

OBPs is not yet clear. However, OBPs might enhance detection by concentrating odorants in mucus, by "presenting" odorants to their cognate receptors, or by removing odorants, thereby permitting rapid recovery and renewed sensitivity (Pevsner & Snyder 1990, Vogt et al 1991, Breer 1994).

## Odorant Receptors

The immense variety of odors perceived by mammals has long been presumed to derive from the existence of mechanisms that allow different olfactory neurons to respond to different odors (Allison & Warwick 1949, Le Gros Clark 1951). In the 1960s, Amoore proposed that selective deficiencies in the ability of individual humans to perceive various odorants (specific anosmias) derive from genetic variations in genes encoding specific odorant receptor (OR) proteins (Amoore et al 1964, Amoore 1977). Direct evidence for the existence of such receptors was subsequently provided by studies that showed the binding of radiolabeled odorants (amino acids) to olfactory cilia in fish (Rhein & Cagan 1980, 1983).

More recently, a large multigene family was identified in rat that appeared to code for hundreds of different ORs expressed by olfactory neurons (Buck & Axel 1991). The ORs belong to a large superfamily of G protein–coupled receptors (reviewed in Dohlman et al 1991); members of this family, including the ORs, exhibit seven hydrophobic domains that are likely to serve as membrane-spanning α-helices (Figure 2). Homologous families of OR genes have now been identified in a variety of vertebrate species, including human (Parmentier et al 1992, Selbie et al 1992, Schurmans et al 1993, Ben-Arie et al 1994), mouse (Nef et al 1992, Ressler et al 1993), salamander, catfish (Ngai et al 1993b), and zebrafish. Raming et al (1993) have shown that expression of at least one member of this gene family in heterologous cells endows those cells with the ability to selectively respond to a particular subset of odorants.

Three unusual features of the OR family are consistent with an ability to interact with a wide variety of structurally diverse odorous ligands. 1. The OR family is extremely large. Human and rodent OR gene families appear to contain 500–1000 genes (Ressler et al 1994b), whereas the gene family of catfish contains about 100 genes (Ngai et al 1993b). 2. The OR family is extraordinarily diverse. Although the ORs share sequence motifs, as a group they are extremely variable in amino acid sequence (Buck & Axel 1991, Lancet & Ben-Arie 1993, Ngai et al 1993b, Ressler et al 1994b). 3. Although sequence diversity in the OR family is not confined to any particular region of ORs, it is accentuated in several of the transmembrane domains (Buck & Axel 1991, Ngai et al 1993b) (Figure 2). Mutational analyses of other G protein–coupled receptors indicate that a ligand-binding pocket may be formed in the plane of

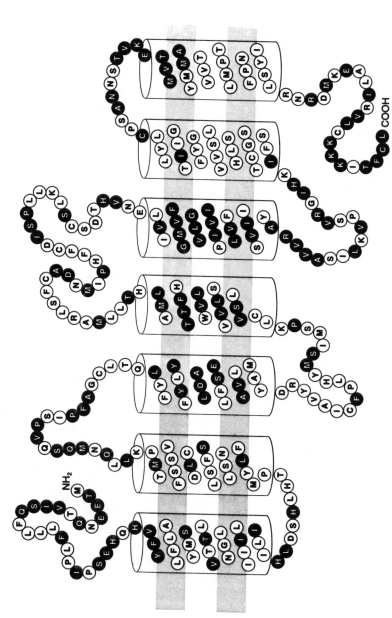

*Figure 2*   An odorant receptor. A typical odorant receptor (I15) is shown schematically in the membrane (*horizontal bars*) with its N-terminus located extracellularly. Seven hydrophobic domains (*within cylinders*) are likely to serve as membrane-spanning α-helices, which together form a ligand-binding pocket in the membrane. Black balls indicate residues that are especially diverse among odorant receptors initially examined. The abundance of variable residues in the transmembrane domains is consistent with an ability of the receptor family to interact with a variety of odorous ligands. [From Buck & Axel (1991).]

the membrane by a combination of the membrane-spanning α-helices (O'Dowd et al 1989, Strader et al 1989). This suggests that the transmembrane domains that show exceptional diversity among odorant receptors may be directly involved in ligand binding (Buck & Axel 1991, Ngai et al 1993b). Interestingly, sequence comparisons of OR genes suggest that there has been a positive selection for diversification of these transmembrane domains via point mutation (Hughes & Hughes 1993, Ngai et al 1993b). Comparison of OR protein sequences has suggested that diversity in the OR family has also been enhanced by the transfer of small segments of coding sequence from one OR gene to another via gene conversion mechanisms (Buck & Axel 1991).

Despite their diversity, OR genes can be grouped into subfamilies on the basis of nucleotide sequence similarity and the consequent ability of subsets of OR genes to hybridize to one another (Buck & Axel 1991, Ngai et al 1993b). Southern blot analyses indicate that most mouse OR genes (~90%) belong to such subfamilies, with the average size of a subfamily being about 7–10 genes (Ressler et al 1994b). Members of the same subfamily encode receptors that are highly related in amino acid sequence and therefore might recognize structurally related ligands (Buck & Axel 1991, Ngai et al 1993b).

## Olfactory Signal Transduction

The binding of odorants to ORs induces a transduction cascade that ultimately leads to depolarization of the neuron and action potential generation in the olfactory axon (reviewed in Firestein 1992, Reed 1992, Ronnett & Snyder 1992, Breer 1994, Dionne & Dubin 1994, Shepherd 1994) (Figures 3 and 4). In the early 1970s, Kurihara & Koyama reported the presence of high levels of adenylyl cyclase in the olfactory epithelium and proposed that cAMP plays a central role in olfactory signal transduction (Kurihara & Koyama 1972). The first direct evidence for this came from ground-breaking studies that showed that odors induce increases in adenylyl cyclase activity in olfactory cilia (Pace et al 1985, Sklar et al 1986). Subsequent studies showed that the cAMP increase is rapid, peaking after only 50 ms, and transient, consistent with an ability to mediate rapid signaling (Boekhoff et al 1990, Breer et al 1990) (Figure 3). A means by which increases in cAMP could result in the changes in membrane potential necessary for signal transmission was provided by the discovery of a cyclic nucleotide–gated cation conductance in olfactory cilia (Nakamura & Gold 1987), which was shown to be activated by exposure to odorants (Bruch & Teeter 1990, Firestein et al 1991).

Presumed components of the cAMP transduction pathway were subsequently identified by molecular cloning (Figure 4). These include the G protein–coupled odorant receptors already discussed and isoforms of a stimulatory G-protein α subunit ($G_{\alpha olf}$) (Jones & Reed 1989); an adenylyl cyclase (adenylyl

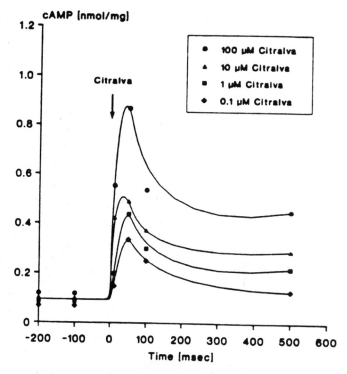

*Figure 3*  Odorants induce a cAMP increase in olfactory cilia. This figure shows the effect of citralva, a fruity odorant, on cAMP concentrations in rat olfactory cilia membranes. Odor-induced elevations in cAMP are rapid, transient, and dose-dependent, consistent with a role in sensory signaling. [From Boekhoff et al (1990).]

cyclase type III) (Bakalyar & Reed 1990); and two cyclic nucleotide gated (CNG) channel subunits, oCNC1 (Dhallan et al 1990, Ludwig et al 1990, Goulding et al 1992) and oCNC2 (Bradley et al 1994, Liman & Buck 1994)— all of which are highly expressed in olfactory neurons and at least several of which (all of those tested) have been localized ultrastructurally to olfactory cilia (Menco et al 1992).

In several vertebrate species, odorants have been shown to elicit rapid increases in IP$_3$ as well as in cAMP, suggesting the possible existence of two separate pathways of olfactory signal transduction (Huque & Bruch 1986, Boekhoff et al 1990, Ronnett et al 1993). Conceivably, increases in IP$_3$ could be generated by interaction of ORs with a G-protein $\alpha$ subunit of the $G_{\alpha q}$ class. Alternatively, the interaction of a single receptor with $G_{\alpha olf}$ could give rise to both cAMP and IP$_3$ via the activation of adenylyl cyclase by the G-protein $\alpha$

## OLFACTORY SIGNAL TRANSDUCTION

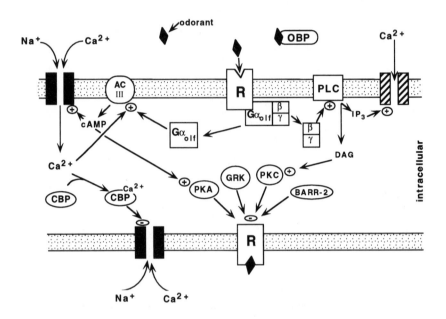

## OLFACTORY SIGNAL TERMINATION

*Figure 4* Models of olfactory signal transduction (*upper*) and termination (*lower*). Molecules implicated in these processes and their proposed interactions are shown. Odorants bind to odorant receptors (R), which interact with a G protein, inducing the release of its $\alpha$ and $\beta\gamma$ subunits. The $\alpha$ subunit ($G_{\alpha olf}$) stimulates an adenylyl cyclase (AC III), causing an increase in cAMP. The cAMP opens a cyclic nucleotide–gated (CNG) cation channel (*paired vertical black bars*), leading to a change in membrane potential that culminates in action potential generation in the sensory axon. A second transduction pathway involving $IP_3$ may be initiated by the $\beta\gamma$ subunits (or an unidentified $G_\alpha$), which stimulate phospholipase C (PLC), causing increases in $IP_3$ and diacylglycerol (DAG). $IP_3$ opens an $IP_3$-gated $Ca^{2+}$ channel, leading to $Ca^{2+}$ influx, which might cause either excitation or inhibition. Signal termination (*below*) may result from the phosphorylation of odorant receptors by one or more protein kinases (PKA, PKC, and GRK) and by interaction with $\beta$-arrestin-2 (BARR-2). $Ca^{2+}$ flux through CNG channels may lead to channel closure mediated by a $Ca^{2+}$-binding (or -regulated) protein (CBP). Odorant-binding proteins (OBPs) in the nasal mucus may aid in signal termination by removing odorants, or they may enhance stimulation.

subunit and the activation of phospholipase C (PLC) by its $\beta\gamma$ subunits (Sternweis 1994). Odorant-induced increases in $IP_3$ are thought to result in an elevation in cytosolic $Ca^{2+}$ due to influx of $Ca^{2+}$ through $IP_3$-gated $Ca^{2+}$ channels in the ciliary plasma membrane (Restrepo et al 1990, Kalinoski et al 1992).

At present, the role of the $IP_3$ pathway in vertebrate olfactory signal transduction is not understood. In the lobster, cAMP and $IP_3$ transduction pathways coexist in individual olfactory neurons but have opposite effects, with $IP_3$ mediating excitation and cAMP mediating inhibition (reviewed in Ache 1994). Evidence for the coexistence of cAMP and $IP_3$ pathways in catfish neurons has raised speculation that the two pathways might also have opposing functions in vertebrates (with cAMP mediating excitation and $IP_3$ mediating inhibition) (reviewed in Ache 1994). Although indirect evidence supports this idea (Bacigalupo et al 1993), direct analyses of $IP_3$ effects have revealed either no effects (salamander) (Firestein et al 1991, Lowe & Gold 1993) or an excitatory effect (catfish) (Miyamoto et al 1992). One interesting speculation is that $IP_3$ acts indirectly to dampen the cAMP pathway (Ache 1994). In this model, $IP_3$-mediated $Ca^{2+}$ increases would have inhibitory effects on CNG channel function similar to those resulting from influx of $Ca^{2+}$ through CNG channels, which have been proposed to play an important role in signal termination (see below).

Recent studies have implicated additional molecules in olfactory sensory transduction, including cGMP and the gas carbon monoxide (Verma et al 1993), as well as a number of different ionic conductances that may be modulated by odorants ($Na^+$, $Ca^{2+}$, $Cl^-$, $K^+$) (reviewed in Dionne & Dubin 1994). However, the precise roles of these molecules in sensory transduction have not yet been defined.

## Olfactory Signal Termination

One important feature of the olfactory system is its ability to adapt, which is best illustrated by the well-known failure to continue smelling a foul odor over extended periods of exposure. Another important feature is the ability of the olfactory system to rapidly recover the ability to sense an odorant, which may be important in sensing concentration changes that may be critical to the tracking of moving prey. As already discussed, OBPs may play a role in signal termination by serving as scavengers to remove odorants from the mucus. In addition, recent studies have identified several different mechanisms of signal termination that may act inside the neuron to affect different steps of the transduction pathway (Figure 4).

The first element in the transduction pathway that may be a target for signal termination mechanisms is the odorant receptor. Previous studies indicate that the desensitization of a G protein–coupled receptor can be mediated by internalization of the receptor or by phosphorylation of the receptor by protein kinases with broad specificities [e.g. protein kinase A (PKA) or protein kinase C (PKC)] or by specialized G protein–coupled receptor kinases (GRKs) [e.g. rhodopsin kinase or β-adrenergic receptor kinase (BARK)] (Lefkowitz et al 1990, Collins et al 1992, Lefkowitz 1993). Consistent with a role for kinase-mediated receptor phosphorylation in olfactory signal termination, odorants have been shown to induce phosphorylation of immunoprecipitable odorant receptors (Kreiger et al 1993). Current evidence suggests that both broad specificity protein kinases and a GRK that is related to BARK II are involved, although it is not yet known whether both types of kinase act by directly phosphorylating odorant receptors (Boekhoff & Breer 1992, Dawson et al 1993, Schleicher et al 1993). In the case of the β- adrenergic receptor, further quenching of signal transduction requires the binding of the protein β-arrestin (BARR) to the phosphorylated receptor (Lefkowitz 1993). Immunohistochemical studies indicate that one isoform of BARR, BARR-2, is present in olfactory cilia, suggesting that this molecule may be involved in olfactory signal termination(Dawson et al 1993).

Another component of the transduction pathway that appears to be a target for signal termination mechanisms is the CNG channel. It appears that $Ca^{2+}$ flux through open CNG channels results in elevations in intracellular $Ca^{2+}$ that then inhibit these channels. At present, the mechanisms underlying the $Ca^{2+}$-mediated inhibition of CNG channels are unclear. Direct effects of $Ca^{2+}$ ions on the channel are controversial (Zufall et al 1991, Kleene 1993). Several studies indicate that a $Ca^{2+}$-regulated protein, possibly calmodulin, directly inhibits CNG channels (Kramer & Siegelbaum 1992, Lin et al 1994). Borisy et al (1992) have also proposed that elevations in cytosolic $Ca^{2+}$ might stimulate a $Ca^{2+}$-activated calmodulin-dependent phosphodiesterase, which might reduce cAMP and lead to channel closure.

## INFORMATION CODING IN THE OLFACTORY EPITHELIUM

A major goal of olfactory research over the past 50 years has been to determine how the olfactory system organizes sensory information. One obvious possibility is that, like other sensory systems, the olfactory system uses physical space within the nervous system to organize, or encode, information. This possibility has been investigated over the years in studies employing a variety of anatomical, biochemical, and more recently, molecular biological techniques. The results of these studies suggest that the olfactory system does indeed use physical space to encode information, but it does so in a manner distinctly different from that used in other sensory systems.

### Functional Responses of Single Neurons

Exposure to odorants generally elicits a depolarizing, excitatory response in the olfactory neuron, although hyperpolarizing, inhibitory responses to odorants can also be observed (Gesteland et al 1965, O'Connell & Mozell 1969,

Getchell & Shepherd 1978, Revial et al 1978, Sicard & Holley 1984, Dionne 1992). The frequency of action potentials generated in the neuron by odorants depends upon odorant concentration, with higher concentrations of odorant leading to increases in the number of action potentials generated. Numerous studies, performed over the past 30 years with increasingly refined electro-physiological techniques, have examined the odorant specificities of individual olfactory neurons (Gesteland et al 1965, Getchell 1974, Revial et al 1978, Sicard & Holley 1984, Firestein et al 1993). These studies have uniformly indicated that each olfactory neuron can respond to a variety of different odors but that different neurons respond to different sets of odorants (Figure 5). The broad tuning of the olfactory neuron indicated by these findings indicates that different odorants are encoded by overlapping sets of olfactory neurons. Observations that responses of single neurons to different odors can vary in temporal patterning (e.g. duration or latency) suggest the possibility that these differences may also contribute to the encoding of an odorant (Getchell & Shepherd 1978, Revial et al 1982, Trotier & MacLeod 1983, Firestein et al 1993, Ivanova & Caprio 1993).

## Odorant Receptor Expression in Individual Neurons

In situ hybridization studies indicate that each OR gene is expressed in only a small fraction of olfactory neurons (Nef et al 1992; Strotmann et al 1992; Ngai et al 1993a, 1993b; Ressler et al 1993; Vassar et al 1993; Chess et al 1994; Strotmann et al 1994). Quantitative analyses in the mouse indicate that individual OR genes (or gene subfamilies) are expressed in only ~0.1 to 0.2% of neurons, although some are expressed at up to 10 times higher or lower frequencies (Ressler et al 1993, 1994b; Chess et al 1994). In the catfish, each OR gene appears to be expressed in ~1% of olfactory neurons (Ngai et al 1993a). Given the estimated number of mouse and catfish OR genes (500–1000 and ~100, respectively), this suggests that each neuron may express only a single receptor gene. These observations suggest that the information transmitted to the olfactory bulb by each neuron may directly reflect the ligand-binding capabilities of a single type of OR.

## Spatial Patterns of Odorant Responses in the Olfactory Epithelium

A variety of techniques have been employed to determine whether responsiveness to an odorant is confined to a particular region of the olfactory epithelium. Early studies of the summed activity [the electro-olfactogram (EOG)] elicited by odorants at different sites on the surface of the frog olfactory epithelium suggested that the magnitude of an odor response could differ by as much as

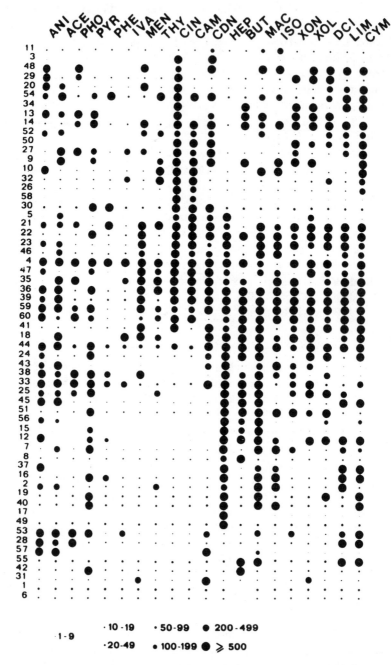

*Figure 5* Responses of single olfactory neurons to odorants. The recorded activity of individual frog olfactory neurons (*numbered at left*) in response to different odorants (*top*) are diagrammed here. Black spots indicate excitatory responses, spot size being proportional to response intensity (*bottom*). Note that most neurons respond to multiple odorants. [From Sicard & Holley (1984).]

a factor of two in different regions but that responses to a single odorant could be found in all regions of the epithelium (Mustaparta 1971). Finer-grained maps of EOG responses to several odorants obtained in the salamander (MacKay-Sim et al 1982) gave similar results and further showed that although each odorant could stimulate all regions of the epithelium, the regions of peak responsiveness differed among odorants. Results obtained with EOG recordings and with 2-deoxyglucose and voltage-sensitive dyes in several different species uniformly indicate that, in both amphibians and mammals, maximal responsiveness to different odors occurs in different regions of the epithelium but that information about each odorant is provided by the responses of neurons in many regions of the olfactory epithelium (Thommesen & Doving 1977, MacKay-Sim et al 1982, Edwards et al 1988, Kent & Mozell 1992, MacKay-Sim & Kesteven 1994).

## Spatial Patterns of Odorant Receptor Expression in the Olfactory Epithelium

As discussed above, each OR gene is expressed in only a small fraction of olfactory neurons. In situ hybridization studies show that neurons that express the same OR gene are not localized to one small patch in the epithelium; rather, they are broadly scattered over a large area of the epithelial sheet. In the catfish, neurons that express the same receptor gene appear to be randomly distributed throughout the entire olfactory epithelium (Ngai et al 1993a). In mouse and rat, neurons expressing the same OR gene are also broadly distributed, but only in certain regions of the olfactory epithelium (Nef et al 1992, Strotmann et al 1992, Ressler et al 1993, Vassar et al 1993, Strotmann et al 1994).

In mouse and rat, the olfactory epithelium is divided into at least four distinct spatial zones in which different sets of OR genes are expressed (Ressler et al 1993, Vassar et al 1993, Strotmann et al 1994) (Figure 6). Each zone consists of a series of elongated bands or stripes that extend along the anterior-posterior axis of the nasal cavity. These stripes are organized into sets in which there is a consistent dorsal-ventral organization. Sets of the four zonal stripes are found on various structures in the nasal cavity, including the nasal septum and individual turbinates that protrude into the nasal cavity. The zonal patterns are bilaterally symmetrical in the two nasal cavities and are virtually identical in different individuals. Although a few exceptions to this characteristic patterning have been observed (e.g. Strotmann et al 1992), the vast majority of the OR genes examined thus far are expressed in one of the four zones (Sullivan et al 1995). Neurons that express the same receptor gene, and therefore presumably recognize the same odors, are confined to the same zone. Moreover, neurons that express members of the same OR subfamily, which are likely to recognize the same (or similar) odors, are also generally located in the same zone (Ressler et al 1993).

The existence of four distinct spatial zones expressing different sets of OR genes suggests that sensory information is broadly organized in the nose prior to its transmission to the brain. Le Gros Clark's early studies showed that the axonal projection from the olfactory epithelium to the olfactory bulb is organized along the dorsal-ventral axis (Le Gros Clark 1951). More recent studies have defined the patterning of this projection in considerable detail with retrograde tracers and with antibodies that recognize subsets of olfactory neurons and their axons (Saucier & Astic 1986, Schwob & Gottlieb 1986, Stewart & Pedersen 1987, Schwarting & Crandall 1991, Schoenfeld et al 1994). Comparison of the OR spatial zones with the patterns of labeled neurons seen in those studies indicates that different OR expression zones project axons to different domains in the olfactory bulb (Ressler et al 1993). This suggests that the initial zonal organization of sensory information in the nose is maintained when that information is transmitted to the olfactory bulb.

## INFORMATION CODING IN THE OLFACTORY BULB

### Synaptic Input from the Olfactory Epithelium

The axons of olfactory neurons in each nasal cavity project to the ipsilateral olfactory bulb, which lies just above, and posterior to, the cavity. Both the anatomy and physiology of the olfactory bulb have been analyzed extensively in a variety of studies employing different approaches (for reviews, see Harrison & Scott 1986, Mori 1987, Shepherd 1990, Kauer & Cinelli 1993, Scott et al 1993, Shepherd 1993, Trombley & Shepherd 1993, Shipley et al 1995). These studies have shown that, in the olfactory bulb, the axons of olfactory neurons form synapses within specialized bundles of neuropil, called glomeruli, which are arrayed over the surface of the bulb. In the glomerulus, the sensory axon forms synapses with the dendrites of periglomerular interneurons, which surround the glomerulus, and with the dendrites of more deeply located mitral and tufted secondary neurons, which relay information to the olfactory cortex. The single axon of each olfactory neuron, and the primary dendrite of each mitral and tufted relay neuron, innervates only a single glomerulus, indicating that glomeruli are discrete functional units. In rodents, there are about 2000 glomeruli, each of which receives input from 1000–3000 olfactory neurons and is innervated by about 25 mitral cells and 50 tufted cells, providing a convergence ratio of about 100:1.

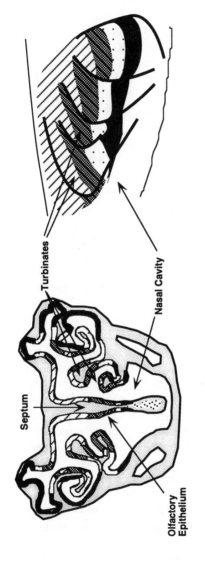

**Coronal View**

**Lateral View**

*Figure 6*   Odorant receptor expression zones. The four spatial zones in which different sets of OR genes are expressed are shown schematically in a coronal section (*left*) and a view of the lateral turbinates (*right*).Each zone consists of a series of anterior-posterior stripes. Sets of four zonal stripes are evident on either side of each turbinate and on the nasal septum. Neurons expressing the same receptor are confined to one zone,  but are randomly distributed in that zone. [Adapted from Ressler et al (1994b).]

## Anatomical and Functional Patterning of the Olfactory Epithelium to Bulb Projection

One of the two most salient features of the axonal projection from the olfactory epithelium to the olfactory bulb is that individual glomeruli receive input from olfactory neurons that are found in many regions of the olfactory epithelium. The first evidence for this came from the early experiments of Le Gros Clark & Warwick, who examined the patterning of connections between the epithelium and bulb by making small lesions in the bulb and then looking at the patterns of degenerating neurons in the epithelium (Le Gros Clark & Warwick 1946). They found that even small lesions resulted in the degeneration of neurons widely distributed in the olfactory epithelium. This indicated that point-to-point connections like those seen in the visual system (the retinal-tectal projection) do not exist in the epithelium bulb projection. Instead, olfactory neurons that are dispersed in the epithelium converge on small regions of the bulb. Concrete evidence for this was subsequently provided by detailed studies using anatomical tracers (Dubois-Dauphin et al 1981, Kauer 1981, Jastreboff et al 1984) and by experiments that showed that odorants applied to many different regions of the epithelium can stimulate a single mitral cell (Kauer & Moulton 1974).

A second important feature of the epithelium to bulb axonal projection is that glomeruli appear to be functional units that respond differentially to different odorants. The possibility that glomeruli are functional units and that different odors might be mapped onto different glomeruli was first suggested by Le Gros Clark and Warwick on the basis of their anatomical studies of the olfactory bulb (Le Gros Clark & Warwick 1946, Allison & Warwick 1949, Le Gros Clark 1951). The earliest support for this idea came from the pioneering studies of Adrian in the early 1950s, which suggested that different mitral cells (which innervate different glomeruli) are stimulated by different odorants or different sets of odorants (Adrian 1950, 1951). More definitive evidence for this was provided by studies that examined responses at the level of individual glomeruli (Leveteau & MacLeod 1966). Studies that monitored odor-induced changes in 2-deoxyglucose uptake and c-*fos* expression later took these analyses further by examining odor responses over the entire olfactory bulb (Stewart et al 1979, Jourdan et al 1980, Astic & Saucier 1983, Coopersmith et al 1986, Onoda 1992, Guthrie et al 1993, Kauer & Cinelli 1993, Sallaz & Jourdan 1993). These studies indicate that each odorant typically elicits activity in a number of different glomeruli (Figure 7). However, the number of glomeruli involved can vary among odorants and is highly dependent upon odor concentration, with higher concentrations leading to the activation of more glomeruli.

## Patterning of Odorant Receptor Input to the Olfactory Bulb

The patterns of functional activity observed in the olfactory bulb could be accounted for in several ways (Scott et al 1993, Ressler et al 1994a). Different glomeruli that respond to the same odorant could conceivably receive input either from neurons expressing the same OR or from neurons expressing different ORs that recognize that odorant. Similarly, a single glomerulus might respond to multiple odorants because it receives input from one OR that recognizes all of those odorants or, alternatively, because it is innervated by different types of ORs that interact with those different odorants.

These questions have recently been addressed by studies that examined the patterns of connections formed in the olfactory bulb by olfactory neurons expressing different types of ORs (Ressler et al 1994a, Vassar et al 1994). In these studies, the presence of OR mRNAs in the axons of olfactory neurons allowed for a visualization (by in situ hybridization) of glomeruli that receive input from neurons expressing different ORs (Figure 7). Each OR probe hybridized to only a small number of glomeruli that were clustered at only a

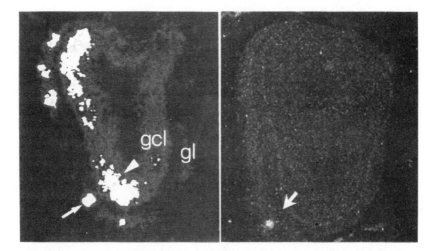

*Figure 7* Spatial patterning of functional activity versus odorant receptor input in the olfactory bulb. Inhalation of a single odorant activates many olfactory bulb glomeruli. Some of the glomeruli activated by peppermint odor are evident in the olfactory bulb section on the left as regions of focal increases of c-*fos* expression in the glomerular layer (gl) (*arrow*). Activity is also seen in the granule cell layer (gcl) (*arrowhead*). In contrast, olfactory neurons expressing the same odorant receptor converge on only a few olfactory bulb glomeruli, and each glomerulus may receive input from only a single receptor type. One mouse glomerulus that receives input from the A16 receptor is shown in the in situ hybridized section on the right. [From Guthrie et al (1993) (*left*) and Ressler et al (1994a) (*right*).]

few sites in the olfactory bulb. This finding indicates that neurons expressing the same OR converge on only a small subset of glomeruli. Importantly, the percentage of glomeruli that hybridized to individual OR probes was found to be similar to the percentage of OR genes that hybridized to those probes. This correspondence further suggested the possibility that each glomerulus might be dedicated to only a single OR and receive input only from neurons expressing that OR. Surprisingly, the glomeruli that hybridized to each OR probe had approximately the same locations in different individuals. Thus there appears to be a stereotyped mapping of input from different ORs onto the glomerular array.

These findings suggest that information that is highly distributed within individual spatial zones in the nose is transformed in the bulb into a highly organized and spatially stereotyped information map, which is, in essence, a map of information provided by different ORs (Ressler et al 1994a, Vassar et al 1994). They also suggest that the many glomeruli activated by a single odorant in functional studies receive input from different ORs rather than from a single OR that sends information to many glomeruli. Furthermore, these findings suggest that the ability of a single glomerulus to respond to a number of different odorants derives not from the innervation of that glomerulus by olfactory neurons expressing multiple ORs responsive to those odorants, but rather from the ability of a single OR to recognize a number of different odorants.

If each glomerulus truly receives input from a single OR type, as these studies suggest, the implication would be that each odorant is recognized by multiple different ORs and that each OR recognizes multiple different odorants. In fact, this is consistent with the ability of individual olfactory neurons, which may express only a single OR gene, to respond to multiple different odorants (see above). The tremendous structural diversity of ORs suggests that different ORs that interact with the same odorant might well recognize different structural features of that odorant, just as different antibodies can recognize different structural features, or epitopes, of an antigen. Different odorants might share one epitope but differ in others.

From this point of view, the map in the bulb might be viewed as an epitope map, or a map of individual structural features of odorants (Beets 1970, Ham & Jurs 1985, Kauer 1991, Katoh et al 1993, Mori & Shepherd 1994, Ressler et al 1994a, Vassar et al 1994, Hildebrand 1995). In this model, each glomerulus might respond to an epitope shared by a variety of different odorants, including those that are perceived as having different odor qualities (i.e. odors that smell very different). Each odorant would be recognized by a variety of ORs, many of which recognize different structural features, or epitopes, of the odorant. As a consequence, each odorant would be represented spatially in the bulb by a unique combination, or ensemble, of glomeruli. Each glomerulus could serve as part of the code for numerous different odorants, thus providing for the discrimination of a vast variety of different odors.

One can envision several distinct advantages of such a coding mechanism. One, of course, is the ability to discriminate many more odorants than there are individual ORs. A second potential advantage is that an epitope map might allow for the recognition of odorants never before encountered and also ensure the ability to recognize odorants not encountered over long periods of time (Ressler et al 1994b). Occasional axonal activity may be required to maintain the integrity of synapses between olfactory neurons and the olfactory bulb (reviewed in Brunjes 1994). Thus, if an OR were to recognize only a single odorant, the neural connections necessary for its detection might degenerate in the odorant's absence. In contrast, with an epitope map, the individual components of the code for the absent odorant would be maintained by the presence of other odorants whose codes include different components of that code.

## Processing of Sensory Information in the Olfactory Bulb

Sensory information received by the olfactory bulb may be processed and refined prior to transmission to the olfactory cortex via intrabulbar circuits involving two classes of interneurons, the periglomerular cells and the granule cells. In addition, centrifugal inputs to the olfactory bulb from several other brain regions may modulate intrabulbar circuits and thereby influence signal transmission from the olfactory bulb. Several excellent reviews should be consulted for detailed discussions of the anatomical and physiological bases of these interactions (Scott & Harrison 1987, Shepherd & Greer 1990, Kauer & Cinelli 1993, Scott et al 1993, Trombley & Shepherd 1993, Shipley et al 1995). Here, only two proposed modes of signal modification in the olfactory bulb are mentioned.

The first potential site of signal processing in the olfactory bulb is the glomerulus. Here, periglomerular cells, which receive excitatory input from sensory axons and form dendrodendritic synapses with mitral and tufted (M/T) neurons, may mediate two types of inhibitory effects on signal transmission. First, periglomerular cells that express the inhibitory neurotransmitter GABA are proposed to inhibit M/T cells through their dendrodendritic synapses (Shepherd 1971, 1972; White 1972; Getchell & Shepherd 1975; Halasz et al 1981; Mugnaini et al 1984; Mori 1987; Baker 1988). Second, the presence of dopamine in periglomerular cells and the expression of D2 dopamine receptors on olfactory sensory axons has suggested that periglomerular cells might have a modulatory presynaptic effect (Halasz et al 1981, Mugnaini et al 1984, Baker 1988, Nickell & Shipley 1992, Sallaz & Jourdan 1993).

A second proposed site of signal modulation is deeper in the bulb at dendrodendritic synapses formed between M/T cells and granule cell interneurons. In addition to their primary dendrites, which (in mammals) are confined to a single glomerulus, individual M/T cells extend long basal dendrites laterally within the external plexiform layer underlying the glomeruli (Rall et al 1966, Price & Powell 1970, Shepherd 1972, Orona et al 1984). These basal dendrites form reciprocal synapses with the dendrites of granule cells; excitation of the granule cell by the M/T cell leads to inhibition of the M/T cell by the granule cell, the proposed neurotransmitters involved being glutamate and GABA, respectively (Jacobson et al 1986, Trombley & Westbrook 1990, Nicoll 1971, Jahr & Nicoll 1982, Yokoi et al 1995). Because each granule cell contacts the dendrites of numerous M/T cells, it has been proposed that these connections might mediate lateral inhibition between M/T cells that innervate different glomeruli.

Both anatomical and functional analyses support the existence of lateral inhibitory mechanisms by which activity in M/T cells innervating one glomerulus may lead to suppression of activity in M/T cells innervating neighboring glomeruli (Meredith 1986, Mori 1987, Wilson & Leon 1987, Scott et al 1993). Recent studies have provided physiological evidence for the involvement of M/T cell–granule cell dendrodendritic synapses in lateral inhibition (Yokoi et al 1995). Interestingly, these studies also suggest a role for these interactions in the refinement of sensory information (Figure 8). Examination of responses of individual M/T cells to inhalation of a series of normal aliphatic aldehydes (i.e. aldehydes with different carbon chain lengths) revealed that many individual cells were excited by one subset of these odorants, inhibited by another subset, and unaffected by yet another subset. The inhibitory responses were shown to be suppressed by agents that blocked reciprocal synapses between M/T cells and granule cells. Remarkably, the odorants that excited an individual M/T cell were found to have numerically consecutive carbon chain lengths, whereas those that inhibited the cell were one to several carbons longer and/or shorter in length. Molecules much longer or shorter had no effect. On the basis of these observations, it was postulated (see also Mori & Shepherd 1994) that each glomerular unit (the glomerulus and the M/T cells that innervate it) has the potential to respond to a wide range of related odorants but also receives inhibitory inputs from neighboring glomerular units through lateral inhibitory mechanisms (Yokoi et al 1995). Weak responses to a given odorant in one glomerular unit may be thereby suppressed by strong responses in another unit. The predicted outcome is that although the glomerular unit receives information about many related odorants, it relays information to the cortex only about a subset of those odorants.

## INFORMATION CODING IN THE OLFACTORY CORTEX

From the olfactory bulb, sensory information is transmitted to the olfactory cortex via the axons of M/T cells, which travel in the lateral olfactory tract

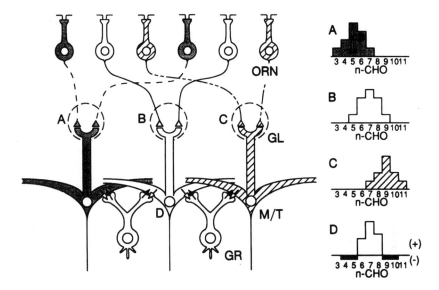

*Figure 8*  A model of information refinement in the olfactory bulb. In this model, glomerular (GL) units (*A,B,C*) receive input from olfactory neurons (ORN) that respond to a relatively wide range of *n*-aliphatic aldehydes as shown in *A, B,* and *C* on the right. The mitral or tufted cell (M/T) D exhibits excitatory responses (+) to a narrower range of aldehydes as well as inhibitory responses (–) to molecules related to the excitatory aldehydes (those with somewhat longer or shorter hydrocarbon chains). The narrowness of this response relative to the input to the glomerular unit (B) is postulated to result from lateral inhibitory mechanisms mediated by reciprocal den-drodendritic synapses between M/T and granule (GR) cells. [From Yokoi et al (1995).]

(LOT) that extends along the olfactory cortex (for detailed discussions of cortical olfactory pathways, see Scott 1986, Price 1987, Haberly 1990, Scott et al 1993). The olfactory cortex is divided into five main areas: the anterior olfactory nucleus (AON), which appears to mediate communication among bilaterally symmetrical regions of the two olfactory bulbs (Schoenfeld & Macrides 1984, Scott et al 1985); the piriform cortex; the olfactory tubercle; the amygdala, which projects to the hypothalamus; and the entorhinal area, which projects to the hippocampus. Information relayed from the latter four areas to the orbitofrontal cortex [either directly, or indirectly through the thalamus (Price 1987)] is thought to give rise to the conscious perception and discrimination of odors. In contrast, the limbic pathways are thought to mediate the affective aspects of olfactory perception.

At present, little is known about how sensory information about different odorants might be organized in the cortex. Individual mitral cell axons can extend for long distances in the LOT and give off multiple collaterals that can innervate different regions of the piriform cortex and the olfactory tubercle (Scott et al 1980, Scott 1981, Luskin & Price 1982, Schneider & Scott 1983, Ojima et al 1984). Recordings from single pyramidal cells in the piriform cortex indicate that, similar to M/T cells in the olfactory bulb, individual pyramidal cells can generally be excited and/or inhibited by at least several different odorants (Takagi 1986). However, in contrast to the bulb, functional studies of the piriform cortex using 2-deoxyglucose (Sharp et al 1977, Astic & Cattarelli 1982) have failed to reveal any discernible functional organization. Nonetheless, the projection from the bulb to the AON is topographically organized (Haberly & Price 1977, Scott et al 1980, Ojima et al 1984, Schoenfeld & Macrides 1984), and the projection to the olfactory tubercle appears to be at least roughly organized along the dorsal-ventral axis (Price 1970). In addition, there is some evidence to suggest that neighboring mitral cells (which are likely to innervate the same glomerulus) project to the same sites in the piriform cortex (Buonviso et al 1991). These findings raise the interesting possibility that there is an underlying organization of inputs in the cortex in which the organization of inputs to olfactory bulb glomeruli from the olfactory epithelium is maintained in the projection to the cortex but is represented in multiple copies.

## Literature Cited

Ache B. 1994. Towards a common strategy for transducing olfactory information. *Semin. Cell. Biol.* 5:55–63

Adrian ED. 1950. Sensory discrimination with some recent evidence from the olfactory organ. *Br. Med. Bull.* 6:330–33

Adrian ED. 1951. Olfactory discrimination. *L'Annee Psychol.* 50:107–13

Allison AC, Warwick RTT. 1949. Quantitative observations on the olfactory system of the rabbit. *Brain* 72:186–97

Amoore JE. 1977. Specific anosmia and the concept of primary odors. *Chem. Senses Flavor* 2:267–81

Amoore JE, Johnston JW, Rubin M. 1964. The stereochemical theory of odor. *Sci. Am.* 210: 42–49

Anholt R. 1993. Molecular neurobiology of olfaction. *Crit. Rev. Neurobiol.* 7:1–22

Astic L, Cattarelli M. 1982. Metabolic mapping of functional activity in the rat olfactory system after a bilateral transection of the lateral olfactory tract. *Brain Res.* 245:17–23

Astic L, Saucier D. 1983. Ontogenesis of the functional activity of guinea-pig olfactory bulb: autoradiographic study with the 2-deoxyglucose method. *Dev. Brain Res.* 10:257–63

Bacigalupo J, Morales B, Ugarte G, Delgado R, Jorquera O, Labarca P. 1993. Electrophysiological studies in toad olfactory receptor neurons. *Chem. Senses* 18:525

Bakalyar HA, Reed RR. 1990. Identification of a specialized adenylyl cyclase that may mediate odorant detection. *Science* 250:1403–6

Baker H. 1988. Neurotransmitter plasticity in the juxtaglomerular cells of the olfactory bulb. In *Molecular Neurobiology of the Olfactory System,* ed. F Margolis, T Getchell, pp. 185–216. New York: Plenum

Bargmann C, Hartwieg E, Horvitz H. 1993. Odorant-selective genes and neurons mediate olfaction in *C. elegans. Cell* 74:515–27

Beets M. 1970. The molecular parameters of olfactory response. *Pharm. Rev.* 22:1–34

Ben-Arie N, Lancet D, Taylor C, Khen M, Walker N, et al. 1994. Olfactory receptor gene cluster on human chromosome 17: possible duplication of an ancestral receptor repertoire. *Hum. Mol. Genet.* 3:229–35

Bignetti E, Cavaggioni A, Pelosi P, Persaud K, Sorbi R, Tirindelli R. 1985. Purification and characterization of an odorant-binding protein from cow nasal tissue. *Eur. J. Biochem.* 149:227–31

Boeckh J, Tolbert LP. 1993. Synaptic organization and development of the antennal lobe in insects. *Microsc. Res. Tech.* 24:260–80

Boekhoff I, Breer H. 1992. Termination of second messenger signaling in olfaction. *Proc. Natl. Acad. Sci. USA* 89:471–74

Boekhoff I, Tareilus E, Strotmann J, Breer H. 1990. Rapid activation of alternative second messenger pathways in olfactory cilia from rats by different odorants. *EMBO J.* 9:2453–58

Borisy F, Ronnet G, Cunningham A, Juilfs D, Beavo J, Snyder S. 1992. Calcium/calmodulin-activated phosphodiasterase expressed in olfactory receptor neurons. *J. Neurosci.* 12:915–23

Bradley J, Li J, Davidson N, Lester HA, Zinn K. 1994. Heteromeric olfactory cyclic nucleotide-gated channels: a subunit confers increased sensitivity to cAMP. *Proc. Natl. Acad. Sci. USA* 91:8890–94

Breer H. 1994. Odor recognition and second messenger signaling in olfactory receptor neurons. *Semin. Cell Biol.* 5:25–32

Breer H, Boekhoff I, Tareilus E. 1990. Rapid kinetics of second messenger formation in olfactory transduction. *Nature* 345:65–68

Bruch RC, Teeter JH. 1990. Cyclic AMP links amino acid chemoreceptors to ion channels in olfactory cilia. *Chem. Senses* 15:419–30

Brunjes PC. 1994. Unilateral naris closure and olfactory system development. *Brain. Res. Rev.* 10:257–63

Buck L, Axel R. 1991. A novel multigene family may encode odorant receptors: a molecular basis for odor recognition. *Cell.* 65:175–87

Buonviso N, Revial M, Jourdan F. 1991. The projections of mitral cells from small regions of the olfactory bulb: an anterograde tracing study using PHA-L (*Phaseolus vulgaris* leucoagglutinin). *Eur. J. Neurosci.* 1991:493–501

Caggiano M, Kauer J, Hunter D. 1994. Globose basal cells are neuronal progenitors in the olfactory epithelium: a lineage analysis using a replication-incompetent retrovirus. *Neuron* 13:339–52

Calof AL, Chikaraishi DM. 1989. Analysis of neurogenesis in a mammalian neuroepithelium: proliferation and differentiation of an olfactory neuron precursor in vitro. *Neuron* 3:115–27

Carlson J. 1991. Olfaction in *Drosophila*: genetic and molecular analysis. *Trends Neurosci.* 14:520–24

Carr W, Gleeson R, Trapido-Rosenthal H. 1990. The role of perireceptor events in chemosensory processes. *Trends Neurosci.* 13:212–15

Chess A, Simon I, Cedar H, Axel R. 1994. Allelic inactivation regulates olfactory receptor gene expression. *Cell* 78:823–34

Collins S, Caron M, Lefkowitz R. 1992. From ligand binding to gene expression: new insights into the regulation of G-protein-coupled receptors. *Trends Biochem. Sci.* 17:37–39

Coopersmith R, Henderson SR, Leon M. 1986. Odor specificity of the enhanced neural response following early odor experience in rats. *Dev. Brain Res.* 27:191–97

Dawson T, Arriza J, Jaworksy D, Borisy F, Attramadal H, et al. 1993. Beta-adrenergic receptor kinase-2 and beta-arrestin-2 as mediators of odorant-induced desensitization. *Science* 259:825–29

Dear T, Boehm T, Keverne E, Rabbitts T. 1991. Novel genes for potential ligand binding proteins in subregions of the olfactory mucosa. *EMBO J.* 10:2813–19

Dhallan RS, Yau KW, Schrader KA, Reed RR. 1990. Primary structure and functional expression of a cyclic nucleotide-activated channel from olfactory neurons. *Nature* 347:184–87

Dionne V. 1992. Chemosensory responses in isolated olfactory receptor neurons from Necturus maculosus. *J. Gen. Physiol.* 99:415–33

Dionne V, Dubin A. 1994. Transduction diversity in olfaction. *J. Exp. Biol.* 194:1–21

Dohlman HG, Thorner J, Caron MG, Lefkowitz RJ. 1991. Model systems for the study of seven-transmembrane-segment receptors. *Annu. Rev. Biochem.* 60:653–88

Dubois-Dauphin M, Tribollet E, Driefuss JJ. 1981. Relations somatotopiques entre la muqueuse olfactive et le bulbe olfactif chez les triton. *Brain Res.* 219:269–87

Edwards DA, Mather RA, Dodd GH. 1988. Spatial variation in response to odorants on the rat olfactory epithelium. *Experientia* 44:208–11

Finger TE, Silver WL, eds. 1987. *The Neurobiology of Taste and Smell.* New York: Wiley

Firestein S. 1992. Electrical signals in olfactory transduction. *Curr. Opin. Neurobiol.* 2:444–48

Firestein S, Darrow B, Shepherd GM. 1991. Activation of the sensory current in salamander olfactory receptor neurons depends on a G protein-mediated cAMP second messenger system. *Neuron* 6:825–35

Firestein S, Picco C, Menini A. 1993. The relation between stimulus and response in olfactory receptor cells of the tiger salamander. *J. Physiol.* 468:1–10

Firestein S, Zufall F, Shepherd G. 1991. Single odorant sensitive channels in olfactory receptor neurons are also gated by cyclic nucleotides. *J. Neurosci.* 11:3535–72

Gesteland R, Lettvin J, Pitts W. 1965. Chemical transmission in the nose of the frog. *J. Physiol.* 181:525–59

Getchell T. 1974. Unitary responses in frog olfactory epithelium to sterically related molecules at low concentrations. *J. Gen. Physiol.* 64:241–61

Getchell T, Margolis F, Getchell M. 1984. Perireceptor and receptor events in vertebrate olfaction. *Prog. Neurobiol.* 23:317–45

Getchell T, Shepherd G. 1975. Short-axon cells in the olfactory bulb: dendrodendritic synaptic interactions. *J. Physiol.* 251:523–48

Getchell T, Shepherd G. 1978. Responses of olfactory receptor cells to step pulses of odour at different concentrations in the salamander. *J. Physiol.* 282:512–40

Goulding EH, Ngai J, Kramer R, Colicos S, Axel R, et al. 1992. Molecular cloning and single-channel properties of the cyclic nucleotide-gated channel from catfish olfactory neurons. *Neuron* 8:45–58

Graziadei PPC, Monti-Graziadei GA. 1979. Neurogenesis and neuron regeneration in the olfactory system of mammals. I. Morphological aspects of differentiation and structural organization of the olfactory sensory neurons. *J. Neurocytology* 8:1–18

Guthrie KM, Anderson AJ, Leon M, Gall C. 1993. Odor-induced increases in c-*fos* mRNA expression reveal an anatomical "unit" for odor processing in olfactory bulb. *Proc. Natl. Acad. Sci. USA* 90:3329–33

Haberly L. 1990. Olfactory cortex. In *The Synaptic Organization of the Brain*, ed. G Shepherd, pp. 317–45. New York: Oxford. 3rd ed.

Haberly L, Price J. 1977. The axonal projection patterns of mitral and tufted cells of the olfactory bulb in the rat. *Brain Res.* 129:152–57

Halasz N, Johansson O, Hokfelt T, Ljungdahl, Goldstein M. 1981. Immunohistochemical identification of two types of dopamine neurons in the rat olfactory bulb as seen by serial sectioning. *J. Neurocytol.* 10:251–59

Ham C, Jurs P. 1985. Structure-activity studies of musk odorants using pattern recognition: monocyclic nitrobenzenes. *Chem. Senses* 10:491–505

Harrison T, Scott J. 1986. Olfactory bulb responses to odor stimulation: analysis of response pattern and intensity relationships. *J. Neurophysiol.* 56:1571–89

Hildebrand J. 1995. Analysis of chemical signals by nervous systems. *Proc. Natl. Acad. Sci. USA* 92:67–74

Hughes A, Hughes M. 1993. Adaptive evolution in the rat olfactory receptor gene family. *J. Mol. Evol.* 36:249–54

Huque T, Bruch R. 1986. Odorant and guanine nucleotide-stimulated phosphoinositide turnover in the olfactory cilia. *Biochem. Biophys. Res. Commun.* 137:37–42

Ivanova T, Caprio J. 1993. Odorant receptors activated by amino acids in sensory neurons to the channel catfish *Ictalurus punctatus*. *J. Gen. Physiol.* 102:1085–105

Jacobson I, Butcher S, Hamberger A. 1986. An analysis of the effects of excitatory amino acid receptor antagonists on evoked field potentials in the olfactory bulb. *Neuroscience* 19:267–73

Jahr C, Nicoll R. 1982. An intracellular analysis of dendrodendritic inhibition in the turtle in vitro olfactory bulb. *J. Physiol.* 326:213–24

Jastreboff PJ, Pedersen PE, Greer CA, Stewart WB, Kauer JS, et al. 1984. Specific olfactory receptor populations projecting to identified glomeruli in the rat olfactory bulb. *Proc. Natl. Acad. Sci. USA* 81:5250–54

Jones DT, Reed RR. 1989. Golf: an olfactory neuron-specific G-protein involved in odorant signal transduction. *Science* 244:790–95

Jourdan F, Duveau A, Astic L, Holley A. 1980. Spatial distribution of 2-deoxyglucose uptake in the olfactory bulb of rats stimulated with two different odors. *Brain Res.* 188:139–54

Kaissling K. 1986. Chemo-electrical transduction in insect olfactory receptors. *Annu. Rev. Neurosci.* 9:121–45

Kalinoski D, Aldinger S, Boyle A, Huque T, Marecek J, et al. 1992. Characterization of a novel inositol 1,4,5-triphosphate receptor in isolated olfactory cilia. *Biochem. J.* 281:449–56

Katoh K, Koshimoto H, Tani A, Mori K. 1993. Coding of odor molecules by mitral/tufted cells in rabbit olfactory bulb. II. Aromatic compounds. *J. Neurophysiol.* 70:2161–75

Kauer JS. 1981. Olfactory receptor cell staining using horseradish peroxidase. *Anat. Rec.* 200:331–36

Kauer JS. 1987. Coding in the olfactory system. See Finger & Silver 1987, pp. 205–31

Kauer J. 1991. Contributions of topography and parallel processing to odor coding in the vertebrate olfactory pathway. *Trends Neurosci.* 14:79–85

Kauer JS, Cinelli AR. 1993. Are there structural and functional modules in the vertebrate olfactory bulb? *Microsc. Res. Tech.* 24:157–67

Kauer JS, Moulton DG. 1974. Responses of olfactory bulb neurones to odour stimulation of small nasal areas in the salamander. *J. Physiol.* 243:717–37

Kent PF, Mozell MM. 1992. The recording of odorant-induced mucosal activity patterns with a voltage-sensitive dye. *J. Neurophysiol.* 68:1804–19

Kleene S. 1993. The cyclic nucleotide-activated conductance in olfactory cilia: effects of cytoplasmic $Mg^{2+}$ and $Ca^{2+}$. *J. Membr. Biol.* 131:237–43

Kleene SJ, Gesteland RC. 1991. Calcium-activated chloride conductance in frog olfactory cilia. *J. Neurosci.* 11:3624–29

Kramer RH, Siegelbaum SA. 1992. Intracellular $Ca^{2+}$ regulates the sensitivity of cyclic nucleotide-gated channels in olfactory receptor neurons. *Neuron* 9:897–906

Krieger J, Raming K, Breer H. 1991. Cloning of genomic and complementary DNA encoding insect pheromone binding proteins: evidence for microdiversity. *Biochim. Biophys. Acta* 1088:277–84

Krieger J, Scheicher S, Strotmann J, Wanner I, Boekhoff I, Raming K, de Gues P, Breer H. 1994. Probing olfactory receptors with sequence-specific antibodies. *Eur. J. Biochem.* 219:829–35

Kurihara K, Koyama N. 1972. High activity of adenyl cyclase in olfactory and gustatory organs. *Biochem. Biophys. Res. Commun.* 48: 30–34

Lancet D, Ben-Arie N. 1993. Olfactory receptors. *Curr. Biol.* 3:668–74

Lee K, Wells R, Reed R. 1987. Isolation of an olfactory cDNA: similarity to retinol-binding protein suggests a role in olfaction. *Science* 235:1053–56

Lefkowitz R. 1993. G protein-coupled receptor kinases. *Cell* 74:409–12

Lefkowitz R, Hausdorff W, Caron M. 1990. Role of phosphorylation in desensitization of the beta-adrenoreceptor. *Trends Pharmacol. Sci.* 11:190–94

Le Gros Clark WE. 1951. The projection of the olfactory epithelium on the olfactory bulb in the rabbit. *J. Neurol. Neurosurg. Psychiatry* 14:1–10

Le Gros Clark WE, Warwick RT. 1946. The pattern of olfactory innervation. *J. Neurol. Neurosurg. Psychiatry* 9:101–11

Leveteau J, MacLeod P. 1966. Olfactory discrimination in the rabbit olfactory glomerulus. *Science* 153:175–76

Liman E, Buck L. 1994. A second subunit of the olfactory cyclic nucleotide-gated channel confers high sensitivity to cAMP. *Neuron* 13:611–21

Lin M, Chen T, Ahamed B, Li J, Yau K. 1994. Calcium-calmodulin modulation of the olfactory cyclic nucleotide-gated cation channel. *Science* 266:1348–54

Lowe G, Gold GH. 1993. Contribution of the ciliary cyclic nucleotide-gated conductance to olfactory transduction in the salamander. *J. Physiol.* 462:175–96

Ludwig J, Margalit T, Eismann E, Lancet D, Kaupp UB. 1990. Primary structure of cAMP-gated channel from bovine olfactory epithelium. *FEBS Lett.* 270:24–29

Luskin M, Price J. 1982. The distribution of axon collaterals from the olfactory bulb and the nucleus of the horizontal limb of the diagonal band to the olfactory cortex, demonstrated by double retrograde labeling techniques. *J. Comp. Neurol.* 209:249–63

MacKay-Sim A, Kesteven S. 1994. Topographic patterns of responsiveness to odorants in the rat olfactory epithelium. *J. Neurophysiol.* 71:150–60

MacKay-Sim A, Shaman P, Moulton DG. 1982. Topographic coding of olfactory quality: odorant-specific patterns of epithelial responsivity in the salamander. *J. Neurophysiol.* 48:584–96

McKenna M, Hekmat-Scafe D, Gaines P, Carlson J. 1994. Putative *Drosophila* pheromone-binding proteins expressed in a subregion of the olfactory system. *J. Biol. Chem.* 269:16340–47

Menco BPM, Bruch RC, Dau B, Danho W. 1992. Ultrastructural localization of olfactory transduction components: the G protein subunit $G_{olf}$ and Type III Adenylyl Cyclase. *Neuron* 8:441–53

Meredith M. 1986. Patterned response to odor in mammalian olfactory bulb: the influence of intensity. *J. Neurophysiol.* 56:572–97

Miyamoto T, Restropo D, Cragoe E Jr, Teeter J. 1992. IP3- and cAMP-induced responses in isolated olfactory receptor neurons from the channel catfish. *J. Membr. Biol.* 127: 173–83

Mori K. 1987. Membrane and synaptic properties of identified neurons in the olfactory bulb. *Prog. Neurobiol.* 29:274–320

Mori K, Shepherd GM. 1994. Emerging principles of molecular signal processing by mitral/tufted cells in the olfactory bulb. *Semin. Cell Biol.* 13:771–90

Morrison E, Constanzo R. 1990. Morphology of the human olfactory epithelium. *J. Comp. Neurol.* 297:1–13

Mugnaini E, Oertel W, Wouterlood F. 1984. Immunocytochemical localization of GABA neurons and dopamine neurons in the rat main and accessory olfactory bulbs. *Neurosci. Lett.* 47:221–26

Mustaparta H. 1971. Spatial distribution of receptor responses to stimulation with different odours. *Acta Physiol. Scand.* 82:154–66

Nakamura T, Gold G. 1987. A cyclic nucleotide-gated conductance in olfactory receptor cilia. *Nature* 325:442–44

Nef P, Hermans-Borgmeyer I, Artieres-Pin H, Beasley L, Dionne VE, Heinemann SF. 1992. Spatial pattern of receptor expression in the olfactory epithelium. *Proc. Natl. Acad. Sci. USA* 89:8948–52

Ngai J, Chess A, Dowling MM, Necles N, Macagno ER, Axel R. 1993a. Coding of olfactory information: topography of odorant receptor expression in the catfish epithelium. *Cell* 72:667–80

Ngai J, Dowling MM, Buck L, Axel R, Chess A. 1993b. The family of genes encoding odorant receptors in the channel catfish. *Cell* 72:657–66

Nickell WT, Shipley MT. 1992. Neurophysiology of the olfactory bulb. In *Science of Olfaction*, ed. MJ Serby, KL Chobor, 1: 172–212. New York: Springer-Verlag

Nicoll R. 1971. Pharmacological evidence for GABA as the transmitter in granule cell inhibition in the olfactory bulb. *Brain Res.* 35: 137–49

O'Connell R, Mozell M. 1969. Quantitative stimulation of frog olfactory receptors. *J. Neurophysiol.* 32:51–63

O'Dowd BF, Lefkowitz RJ, Caron MG. 1989. Structure of the adrenergic and related receptors. *Annu. Rev. Neurosci.* 12:67–83

Ojima H, Mori K, Kishi K. 1984. The trajectory of mitral cell axons in the rabbit olfactory cortex revealed by intracellular HRP injection. *J. Comp. Neurol.* 230:77–87

Onoda N. 1992. Odor-induced fos-like immunoreactivity in the rat olfactory bulb. *Neurosci. Lett.* 137:157–60

Orona E, Rainer E, Scott J. 1984. Dendritic and axonal organization of mitral and tufted cells in the rat olfactory bulb. *J. Comp. Neurol.* 217:227–37

Pace U, Hanski E, Salomon Y, Lancet D. 1985. Odorant-sensitive adenylate cyclase may mediate olfactory reception. *Nature.* 316: 255–58

Parmentier M, Libert F, Schurmans F, Schiffmann S, Lefort A, et al. 1992. Expression of members of the putative olfactory receptor gene family in mammalian germ cells. *Nature* 355:453–55

Pevsner J, Sklar P, Snyder S. 1986. Odorant-binding protein: localization to nasal glands and secretions. *Proc. Natl. Acad. Sci. USA* 83:4942–46

Pevsner J, Snyder S. 1990. Odorant binding protein: odorant transport function in the nasal epithelium. *Chem. Senses* 15:217–22

Pikielny C, Hasan G, Rouyer P, Rosbash M. 1994. Members of a family of *Drosophila* putative odorant-binding proteins are expressed in different subsets of olfactory hairs. *Neuron* 12:35–49

Price J, Powell T. 1970. The synaptology of the granule cells of the olfactory bulb. *J. Cell. Sci.* 7:125–55

Price JL. 1987. The central olfactory and accessory olfactory systems. See Finger & Silver 1987, pp. 179–203

Rall W, Shepherd G, Reese T, Brightman M. 1966. Dendrodendritic synaptic pathway for inhibition in the olfactory bulb. *Exp. Neurol.* 14:44–56

Raming K, Krieger J, Strotmann J, Boekhoff I, Kubick S, et al. 1993. Cloning and expression of odorant receptors. *Nature* 361:353–56

Reed RR. 1992. Signaling pathways in odorant detection. *Neuron* 8:205–9

Ressler KJ, Sullivan SL, Buck LB. 1993. A zonal organization of odorant receptor gene expression in the olfactory epithelium. *Cell* 73:597–609

Ressler KJ, Sullivan SL, Buck LB. 1994a. Information coding in the olfactory system: evidence for a stereotyped and highly organized epitope map in the olfactory bulb. *Cell* 79:1245–55

Ressler KJ, Sullivan SL, Buck LB. 1994b. A molecular dissection of spatial patterning in the olfactory system. *Curr. Opin. Neurobiol.* 4:588–96

Restrepo D, Miyamoto T, Bryant B, Teeter J. 1990. Odor stimuli trigger influx of calcium into olfactory neurons of the channel catfish. *Science* 249:1166–68

Revial M, Duchamp A, Holley A, MacLeod P. 1978. Frog olfaction: odour groups, acceptor distribution, and receptor categories. *Chem. Senses* 3:23–33

Revial M, Sicard G, Duchamp A, Holley A. 1982. New studies on odour discrimination in the frog's olfactory receptor cells. II. Mathematical analysis of electrophysiological responses. *Chem. Senses* 8:179–90

Rhein LD, Cagan RH. 1980. Biochemical studies of olfaction: isolation, characterization, and odorant binding activity of cilia from rainbow trout olfactory rosettes. *Proc. Natl. Acad. Sci. USA* 77:4412–16

Rhein LD, Cagan RH. 1983. Biochemical studies of olfaction: binding specificity of odorants to a cilia preparation from rainbow trout olfactory rosettes. *J. Neurochem.* 41:569–77

Ronnett GV, Cho H, Lester LD, Wood S, Snyder S. 1993. Odorants differentially enhance phosphoinositide turnover and adenylyl cylase in olfactory receptor neuronal cultures. *J. Neurosci.* 13:1751–58

Ronnett GV, Snyder SH. 1992. Molecular messengers of olfaction. *Trends Neurosci.* 15: 508–13

Sallaz M, Jourdan R. 1993. C-fos expression and 2-deoxyglucose uptake in the olfactory bulb of odour-stimulated awake rats. *NeuroReport* 4:55–58

Saucier D, Astic L. 1986. Analysis of the topographical organization of olfactory epithelium projections in the rat. *Brain Res. Bull.* 16:455–62

Schleicher S, Boekhoff I, Arriza J, Lefkowitz R, Breer H. 1993. A beta-adrenergic receptor kinase-like enzyme is involved in olfactory signal termination. *Proc. Natl. Acad. Sci. USA* 90:1420–24

Schneider S, Scott J. 1983. Orthodromic response properties of rat olfactory bulb mitral and tufted cells correlate with their projection patterns. *J. Neurophysiol.* 50:358–78

Schoenfeld TA, Clancy AN, Forbes WB, Macrides F. 1994. The spatial organization of the peripheral olfactory system of the hamster 1. Receptor neuron projections to the main olfactory bulb. *Brain Res. Bull.* 34:183–210

Schoenfeld TA, Macrides F. 1984. Topographic organization of connections between the main olfactory bulb and pars externa of the anterior olfactory nucleus in the hamster. *J. Comp. Neurol.* 227:121–35

Schurmans S, Muscatelli F, Miot F, Mattei MG, Vassart G, Parmentier M. 1993. The OLFR1 gene encoding the HGMP07E putative olfactory receptor maps to the 17p13–p12 region of the human genome and reveals an MSPL restriction fragment length polymorphism. *Cytogenet. Cell Genet.* 6:200–4

Schwarting GA, Crandall JE. 1991. Subsets of olfactory and vomeronasal sensory epithelial cells and axons revealed by monoclonal antibodies to carbohydrate antigens. *Brain Res.* 547:239–48

Schwob JE, Gottlieb DI. 1986. The primary olfactory projection has two chemically distinct zones. *J. Neurosci.* 6:3393–404

Scott J. 1981. Electrophysiological identification of mitral and tufted cells and distribution of their axons in the olfactory system of the rat. *J. Neurophysiol.* 46:918–31

Scott J. 1986. The olfactory bulb and central pathways. *Experientia* 42:223–31

Scott J, McBride R, Schneider S. 1980. The organization of projections from the olfactory bulb to the piriform cortex and olfactory tubercle in the rat. *J. Comp. Neurol.* 194:519–34

Scott J, Rainer E, Pemberton J, Orena E, Mouradian L. 1985. Pattern of rat olfactory bulb mitral and tufted cell connections to the anterior olfactory bulb pars externa. *J. Comp. Neurol.* 242:415–24

Scott JW, Harrison TA. 1987. The olfactory bulb: anatomy and physiology. See Finger & Silver, pp. 151–77

Scott JW, Wellis DP, Riggott MJ, Buonviso N. 1993. Functional organization of the main olfactory bulb. *Microsc. Res. Tech.* 24:142–56

Selbie LA, Townsend-Nicholson A, Iismaa TP, Shine J. 1992. Novel G protein-coupled receptors: a gene family of putative human olfactory receptor sequences. *Brain Res.* 13:159–63

Sengupta P, Colbert H, Kimmel B, Dwyer N, Barmann C. 1993. The cellular and genetic basis of olfactory responses in *C. elegans*. *Ciba Found. Symp.* 179:235–44

Sharp FR, Kauer JS, Shepherd GM. 1977. Laminar analysis of 2-deoxyglucose uptake in olfactory bulb and olfactory cortex of rabbit and rat. *J. Neurophysiol.* 40:800–13

Shepherd G. 1971. Physiological evidence for dendrodendritic synaptic interactions in the rabbit's olfactory glomerulus. *Brain Res.* 32:212–17

Shepherd G. 1972. Synaptic organization of the mammalian olfactory bulb. *Physiol. Rev.* 52:864–917

Shepherd G. 1993. Principles of specificity and redundancy underlying the organization of the olfactory system. *Microsc. Res. Tech.* 24:106–12

Shepherd GM, ed. 1990. *The Synaptic Organization of the Brain.* New York: Oxford Univ. Press

Shepherd GM. 1994. Discrimination of molecular signals by the olfactory receptor neuron. *Neuron* 13:771–90

Shepherd GM, Greer CA. 1990. Olfactory bulb. See Shepherd 1990, pp. 133–69.

Shipley M, McLean J, Ennis M. 1995. Olfactory system. In *The Rat Nervous System*, ed. G Paxinos, pp. 899–926. New York: Academic

Sicard G, Holley A. 1984. Receptor cell responses to odorants: similarities and differences among odorants. *Brain Res.* 292:283–96

Siddiqi O. 1987. Neurogenetics of olfaction in *Drosophila melangoster. Trends Genet.* 3:137–42

Sklar PB, Anholt RRH, Snyder SH. 1986. The odorant-sensitive adenylate cyclase of olfactory receptor cells: differential stimulation by distinct classes of odorants. *J. Biol. Chem.* 261:15538–43

Sternweis P. 1994. The active role of βγ in signal transduction. *Curr. Opin. Cell Biol.* 6:198–203

Stewart WB, Kauer JS, Shepherd GM. 1979. Functional organization of rat olfactory bulb, analyzed by the 2-deoxyglucose method. *J. Comp. Neurol.* 185:715–34

Stewart WB, Pedersen PE. 1987. The spatial organization of olfactory nerve projections. *Brain Res.* 411:248–58

Stocker R. 1994. The organization of the chemosensory system in *Drosophila melanogaster:* a review. *Cell Tissue Res.* 275:3–26

Strader CD, Sigal IS, Dixon RAF. 1989. Structural basis of beta-adrenergic receptor function. *FASEB Lett.* 3:1825–32

Strotmann J, Wanner I, Helfrich T, Beck A, Breer H. 1994. Rostro-caudal patterning of receptor expressing neurones in the rat nasal cavity. *Cell Tissue Res.* 278:11–20

Strotmann J, Wanner I, Krieger J, Raming K, Breer J. 1992. Expression of odorant receptors in spatially restricted subsets of chemosensory neurons. *NeuroReport* 3:1053–56

Sullivan S, Ressler K, Buck L. 1995. Spatial patterning and information coding in the olfactory system. *Curr. Opin. Genet. Dev.* 5: 516–23

Takagi S. 1986. Studies on the olfactory nervous system in the Old World monkey. *Prog. Neurobiol.* 27:195–250

Thommesen G, Doving KB. 1977. Spatial distribution of the EOG in the rat; a variation with odour quality. *Acta Physiol. Scand.* 99: 270–80

Trombley P, Shepherd G. 1993. Synaptic transmission and modulation in the olfactory bulb. *Curr. Opin. Neurobiol.* 3:540–47

Trombley P, Westbrook G. 1990. Excitatory synaptic transmission of primary cultures of rat olfactory bulb. *J. Neurophys.* 1990:598–606

Trotier D, MacLeod P. 1983. Intracellular recordings from salamander olfactory receptor cells. *Brain Res.* 268:225–37

Vassar R, Chao SK, Sitchern R, Nunez JM, Vosshall LB, Axel R. 1994. Topographic organization of sensory projections to the olfactory bulb. *Cell* 79:981–91

Vassar R, Ngai J, Axel R. 1993. Spatial segregation of odorant receptor expression in the mammalian olfactory epithelium. *Cell* 74: 309–18

Verma A, Hirsch D, Glatt C, Ronnett G, Snyder S. 1993. Carbon monoxide: a putative neural messenger. *Science* 259:381–84

Vogt R, Prestwich G, Lerner M. 1991. Odorant binding protein subfamilies associate with distinct classes of olfactory receptor neurons in insects. *J. Neurobiol.* 22:74–84

White E. 1972. Synaptic organization in the olfactory glomerulus of the mouse. *Brain Res.* 37:69–80

Wilson D, Leon M. 1987. Evidence of lateral synaptic interactions in olfactory bulb output cell responses to odors. *Brain Res.* 417:175–80

Yokoi M, Mori K, Nakanishi S. 1995. Refinement of odor molecule tuning by dendrodendritic synaptic inhibition in the olfactory bulb. *Proc. Natl. Acad. Sci. USA* 92:3371–75

Zufall F, Shepherd GM, Firestein S. 1991. Inhibition of the olfactory cyclic nucleotide-gated ion channel by intracellular calcium. *Proc. R. Soc. London Ser. B* 246: 225–30

*Annu. Rev. Neurosci. 1996. 19:545–75*

# THE *DROSOPHILA* NEUROMUSCULAR JUNCTION: A Model System for Studying Synaptic Development and Function

*Haig Keshishian*

Department of Biology, Yale University, Box 208103, New Haven, Connecticut 06520-8103

*Kendal Broadie*

Department of Zoology, Cambridge University, Downing Street, Cambridge, England CB2 3EJ

*Akira Chiba*

Department of Cell and Structural Biology, University of Illinois, 505 South Goodwin Avenue, Urbana, Illinois 61801

*Michael Bate*

Department of Zoology, Cambridge University, Downing Street, Cambridge, England CB2 3EJ

KEY WORDS: axonogenesis, motoneurons, muscle, synaptic transmission, synaptogenesis

### ABSTRACT

The *Drosophila* neuromuscular junction has attracted widespread attention as an excellent model system for studying the cellular and molecular mechanisms of synaptic development and neurotransmission. In *Drosophila* the advantages of invertebrate small systems, where individual cells can be examined with single-cell resolution, are combined with the powerful techniques of patch-clamp analysis and molecular genetics. In this review we examine myogenesis and motoneuron development, the problems of axon outgrowth and target selection, the differentiation of the synapse, and the mechanisms of both synaptic function and plasticity in this model genetic system.

0147-006X/96/0301-0545$08.00

## INTRODUCTION

Understanding the cellular and molecular mechanisms governing synaptogenesis is a central issue in developmental neurobiology. Ideally, one wishes to examine these problems in an organism where the molecules involved can be identified and directly manipulated and where the choices of individual neurons can be examined and tested. In *Drosophila,* reverse-genetic methods can be used to identify the molecules expressed at developing synapses. It is also possible to screen for mutations that disrupt axonal guidance and target recognition. Furthermore, the *Drosophila* neuromuscular system can be directly manipulated. This can be accomplished by microsurgery, by mutations that alter the numbers or identities of motoneurons and muscle fibers, and by transgenes that alter the function or expression pattern of molecules of interest.

The *Drosophila* neuromuscular junction is a good model for studying fundamental questions about synapses. Many of the molecules involved in synaptic transmission are conserved between *Drosophila* and vertebrates. Using voltage clamp, one can examine genetic loss-of-function mutations of the vesicle-associated proteins involved in exocytosis to draw insights about function in higher systems. Also, the development of the *Drosophila* and vertebrate synapses is similar at both the cellular and molecular level. Several proteins with structural or functional similarity to vertebrate cell adhesion and cell repulsion molecules are expressed at developing *Drosophila* synapses. How the rich ensemble of adhesive and repulsive molecules combine to influence axon guidance and synaptogenesis may ultimately be best understood in *Drosophila,* where these proteins can be readily manipulated and their functions tested in a system of singly identified synapses.

## THE *DROSOPHILA* NEUROMUSCULAR JUNCTION

Virtually all of the studies of embryonic and larval neuromuscular junctions have focused on the abdominal segments, which have a simple and easily accessible array of overlapping, striated fibers that are attached to the inner surface of the body wall at specific sites. Each muscle is a single multinucleate cell, formed by the fusion of neighboring myoblasts, and differs therefore from a vertebrate muscle, which typically consists of many syncytial fibers bundled together as a functional unit. The overall pattern of muscles is made up of a series of segmental repeats, with a fixed set of 30 muscles in each hemisegment from A2 to A7 (Figure 1). The pattern in A1 is slightly different, and there are other, specialized muscle sets in the more anterior and posterior segments.

The muscle fibers can be considered individually specified cells, on the basis of both anatomical criteria—such as size, body wall insertion sites, and nuclear numbers (Johansen et al 1989a,b; Bate 1990; Budnik et al 1990; Chiba et al 1993)—and fiber-specific molecular expression patterns. Specific muscle fi-

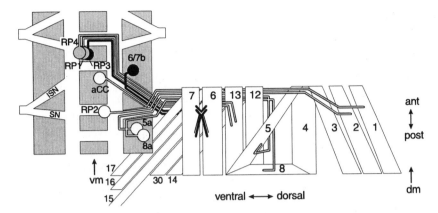

*Figure 1*  The *Drosophila* neuromuscular system. Several of the identified motoneurons and muscle fibers of an abdominal hemisegment are shown. The precision in synaptic wiring results from the recognition of individual muscle fibers by the motoneuron growth cones, a process that takes place during mid-to-late embryogenesis. On the basis of several features, including molecular expression patterns, each muscle fiber can be considered a singly specified cell. Abbreviations: ISN, intersegmental nerve; SN, segmental nerve; vm and dm, ventral and dorsal midline. [Adapted from Keshishian & Chiba (1993).]

bers express a variety of putative cell adhesion molecules (CAMs), including the Ig superfamily members Fasciclin III (Halpern et al 1991); Neuromusculin (Kania et al 1993, Kania & Bellen 1995); the LRR containing proteins Connectin (Nose et al 1992, 1994) and Toll (Nose et al 1992, Halfon et al 1995); and the putative signaling molecule Semaphorin II, which shares sequence similarity to the vertebrate Collapsin (Kolodkin et al 1993, Luo et al 1993, Messersmith et al 1995). In addition, there is a diversity of transcription factors expressed by the muscle fibers, which may be important in specifying their distinct cellular identities. These include LIM factor Apterous (Bourgouin et al 1992) and the homeodomain-containing transcription factors Even Skipped (Patel et al 1992) and S59 (Dohrmann et al 1990).

These features mark the muscle fibers as separate elements in a complex pattern. Yet despite their diversity, the larval body wall muscles are remarkably similar in terms of both their physiological (Jan & Jan 1976a,b) and structural properties (Crossley 1978), with apparently uniform patterns of structural gene expression generating the contractile apparatus (Epstein & Bernstein 1992). Thus, during development both a myogenic pathway responsible for the expression of gene products required in all muscles and mechanisms that lead to the local activation of genes endowing individual muscles with specific identities will be present. This suggests that during muscle development there will be (*a*) local controls over the numbers of cells fusing to form individual muscle precursors, (b) a regulatory mechanism allowing individual muscles to identify appropriate epidermal insertion sites, and (c) a diversification of surface molecules that allows the growth cones of innervating motoneurons to select between alternative target muscles (see section on Axon Outgrowth and Target Selection).

Extensive molecular and cellular studies have shown that the motoneurons in *Drosophila* embryos and larvae also are individually identified cells. Sink & Whitington (1991a) have estimated, on the basis of retrograde tracing methods, that about 35 motoneurons project onto each abdominal hemisegment. Intracellular dye-fills and cell-specific reporter constructs show that the neurons make stereotypic target choices, with each motoneuron projecting specifically to one or more muscle fibers, resulting in a precise and invariant wiring pattern (Halpern et al 1991; Sink & Whitington 1991a,b; Broadie & Bate 1993a; Callahan & Thomas 1994). Each body wall hemisegment is innervated by motoneurons from both its own and from the next anterior CNS segment, with cell bodies located on both the ipsi- and contralateral sides. There is no organized motoneuron topography in the CNS with respect to the locations of the innervated muscles (Sink & Whitington 1991a).

Electrophysiological studies by Jan & Jan (1976a,b) showed that the muscle fibers are multiply innervated, with glutamate serving as the excitatory transmitter. Immunohistochemical studies have since shown that the entire motoneuron population is glutamatergic (Johansen et al 1989a,b), with subsets expressing putative cotransmitters, including octopamine (Monastirioti et al 1995), the peptides proctolin (Anderson et al 1988) and PACAP (Feany & Quinn 1995, Zhong & Pena 1995, Zhong 1995), as well as peptides similar to leukokinin-1 (Cantera & Nässel 1992) and insulin (Gorczyca et al 1993). Retrograde tracing, dye fills of larval motoneuron boutons, and electron microscopical analysis have confirmed that *Drosophila* body wall muscle fibers are polyinnervated (Johansen et al 1989a, Budnik & Gorczyca 1992, Atwood et al 1993, Jia et al 1993, Kurdyak et al 1994). Whereas some motoneurons project exclusively to individual muscle fibers, others project to muscle fiber pairs or larger subsets, suggesting distinct motor control of both individual and groups of muscle fibers (Sink et al 1991a, Halpern et al 1991, Keshishian et al 1993). The larval motor endings fall into two classes (type I and II with further subdivisions), based on the size of the synaptic boutons and the anatomy of the arbors. These motor ending classes resemble those described for other arthropods and may correlate to slow and fast motoneurons (Budnik & Gorczyca 1992, Atwood et al 1993, Keshishian et al 1993, Kurdyak et al 1994).

## MYOGENESIS AND MOTONEURON DEVELOPMENT

### Myogenesis and Muscle Differentiation

All larval and adult muscles are derived from a population of ventral blastoderm cells that invaginate during embryonic gastrulation to form an internal layer of mesoderm (Bate 1993). These cells and their progeny give rise to the visceral muscles of the gut and heart as well as the somatic body wall muscles of the larva and adult. The entry of these cells into specific myogenic fates depends on features intrinsic to the mesodermal cells, established prior to or during gastrulation, and on inductive signals from the overlying ectoderm (Staehling-Hampton et al 1994, Frasch 1995, Baker & Schubiger, 1995).

Several genes have recently been identified in *Drosophila* that are thought to be involved in the decision of mesodermal cells to enter a myogenic pathway (Abmayr et al 1995). Among these is *nautilus (nau)* (Michelson et al 1990, Paterson et al 1991), the only MyoD family member identified thus far in the fly. This gene is apparently not the essential link in myogenesis, as it is expressed by only a subset of the larval muscles. However, the MADS box–containing gene *Dmef2,* the homologue of vertebrate *mef2* genes, may serve this function (Nguyen et al 1994, Lilly et al 1994, 1995). MEF2 is a transcription factor expressed in the progenitors of all visceral and somatic muscles of the embryo. In *Drosophila* the loss of the single *D-mef2* gene leads to a disruption in the development of all body wall muscles (as well as interference with the development of cardiac and visceral muscles). The role of genes such as *D-mef2* and *nau* in the later development of the neuromuscular junction has not been tested, although it is possible that they are responsible for activating genes required for normal postsynaptic development.

The first overt signs of muscle differentiation occur when neighboring myoblasts fuse to form the syncytial precursors of mature muscles during a period of *Drosophila* embryogenesis known as germ-band retraction [about 7.5 h after egg laying (AEL), or stage 12] [see Campos-Ortega & Hartenstein (1985) for staging]. Small syncytia, consisting of two or three fused myoblasts, appear in the most ventral mesoderm immediately over the inner face of the developing CNS (Bate 1990). These first syncytia are the precursors of ventral muscles that eventually lie on either side of the nervous system after germ-band retraction. As the germ band shortens, additional syncytia appear at dorsal, lateral, and ventral locations in the mesoderm, all of which are in close contact with the underlying epidermis. Thus, by the completion of germ-band shortening, each of the 30 muscles in A2–A7 is represented by a syncytial precursor at a specific site in the somatic mesoderm. Each syncytium enlarges by fusing with neighboring myoblasts, and as the precursors of the muscles grow, the pool of unfused myoblasts that surrounds them is steadily depleted.

Fusion is a continuous process that dictates the final size of individual muscles. Although the precursors of large and small muscles are initially the same size, fusion for a small muscle containing 3–4 nuclei is complete by the end of germ-band retraction, whereas for larger muscles, it continues for several hours. Several genes required for fusion have been identified in mutant screens, but the process of fusion has not been systematically analyzed. The enlarging muscle precursors put out growth cone–like processes, i.e. polar extensions of the syncytium, that migrate over the surface of the epidermis to the final sites where muscle attachments will be formed. Expression of DGluR, detected by in situ hybridization, begins before fusion is complete and may occur in single myoblasts before they have fused with a muscle precursor (Currie et al 1995). By 13 h AEL, the process of muscle assembly is essentially complete, and the growth cones of motoneurons are exploring the surfaces of the myotubes prior to the formation of functional neuromuscular junctions (Johansen et al 1989a,b; Halpern et al 1991; Sink & Whitington 1991a,b,c; Broadie & Bate 1993a). Because the peripheral nervous system (motor and sensory pathways) does not develop until after the elements of the larval muscle pattern have been laid down, innervation plays no part in the patterning of muscles in the embryo (Johansen et al 1989b, Bate 1990), and muscle differentiation proceeds normally in the absence of innervation (Broadie & Bate 1993c). However, there is good evidence that during pupal development, motoneuronal innervation is critical to the specification of at least one set of muscle fibers (Lawrence & Johnston 1986, Currie & Bate 1995).

## The Specification of Muscle Identity

The formation of the muscle pattern requires that groups of neighboring myoblasts that fuse to form one of the syncytial muscles gain access to the information required to make a particular muscle fiber. This can be envisaged as resulting from an information transfer process that occurs in two essential steps. First, the pattern is seeded by the specification and segregation in the mesoderm of a number of unique cells, each of which is the *founder* of a distinct muscle (Bate 1990, Dohrmann et al 1990). Second, in a nonspecific phase of fusion, these specialized founder cells recruit adjacent myoblasts to form the syncytial precursors of mature muscles.

A growing number of genes have been identified that are expressed in subsets of mesodermal cells and appear to mark muscle founder cells (Dohrmann et al 1990, Bourgouin et al 1992, Bate 1993). These genes encode putative transcription factors, and they are expressed first in small numbers of mesodermal cells and later in subsets of muscles. The role of these genes in muscle differentiation is unclear, although they are invaluable markers for the earliest phases of muscle development. Each muscle has its specific characteristics of size, shape, insertion sites, and innervation, and in principle, these could all be defined by the activation of downstream genes in muscle precursors expressing a particular transcription factor. However, the loss-of-function phenotype, which has been described so far for only one of these putative regulatory genes, *apterous* (*ap*), is relatively mild (Bourgouin et al 1992): *ap* is normally expressed in a subset of lateral and ventral muscles, and in *ap* mutant embryos, a variable loss of lateral muscles occurs. When *ap* is overexpressed under the control of a heat shock promoter, the phenotype is reversed, and a variable overproduction of lateral muscles occurs. For the moment, the question of the exact function of the founder cell genes in muscle differentiation awaits further genetic analysis of candidate loss-of-function alleles.

## Muscle Development Without Fusion

In grasshopper embryos, myogenesis is initiated by special large cells, the muscle pioneers (Ho et al 1983), that appear at specific locations in the embryonic mesoderm, span the territories of future muscles, and fuse with adjacent myoblasts to form syncytial muscles. Although no such pioneer muscle cells are detectable in wild-type *Drosophila* embryos, the underlying molecular processes of myogenesis in grasshoppers and flies may be identical. The expression of genes such as *S59* and *vestigial* (*vg*), which mark cells in the earliest phases of forming specific muscles, supports this view. As the initially expressing cells fuse with their neighbors, these cells are recruited to expression (Dohrmann et al 1990); in the absence of fusion, this recruitment does not take place (Rushton et al 1995). Moreover, in mutant embryos where cell fusion fails to occur, there appear to be of two kinds of prospective muscle forming cells: a relatively undifferentiated set of myoblasts and a smaller set of single cells that express markers such as *S59* and *vg*. In the mutant embryos, these cells migrate to appropriate positions, attach to the epidermis, and extend processes that span epidermal domains typical of the muscles that would express these genes in normal embryos (Rushton et al 1995). In many ways these special cells are like single-celled muscles: They express cell surface proteins such as Connectin (Nose et al 1992), they apparently receive appropriate innervation, and they succeed in forming neuromuscular junctions that appear normal at an ultrastructural level (A Prokop, personal communication). By contrast, the majority of the myoblasts in these mutant embryos remain rounded, and although they express muscle myosin, none of them has yet been shown to express any of the genes that are restricted to specific subsets of the muscles. Thus, in embryos where fusion fails, the myoblasts are shown to be of two classes: (*a*) a special class of cells, resembling the muscle pioneers in grasshoppers, that appears to have the information required to initiate the formation of particular muscles and (*b*) neighboring myogenic cells that would normally be recruited into syncytia by fusion. Myoblasts of the first class act as the founders of individual muscles in wild-type embryos, and therefore, the minimum requirement for the definition of the elements of the muscle pattern is the specification of a spaced pattern of 30 such specialized cells in the somatic mesoderm of each hemisegment.

## A Role for Wingless

The founder cells of particular muscles arise at highly reproducible locations in the developing mesoderm, in close contact with the overlying ectoderm.

The specification of founders depends on both a general mechanism that segregates single founder cells (see below) and a position-specific cue, or cues, that assign a characteristic identity to each founder. Founder cell identity could be defined by a point-to-point imposition of ectodermal information onto the mesoderm. Were this the case, all mutations affecting epidermal patterning would have predictable consequences for the underlying pattern of founders, which would correspond in location and severity to the epidermal phenotypes. However, only one member of the segment polarity class genes, *wingless* (*wg*), has so far been found to have extensive effects on the segregation of the muscle founder cells (Bate & Rushton 1993). *wg* encodes a secreted molecule (Nusse & Varmus 1992) that has been shown to be capable of operating across germ layers (Immergluck et al 1990, Lawrence et al 1994). Mutations in *wg* lead to a complete loss of the *S59*-expressing founder cells. Ectopic expression of *wg* throughout either the ectoderm or the mesoderm rescues the patterned expression of *S59* in the mesoderm of *wg* mutant embryos, which suggests that there is a restricted set of mesodermal cells capable of responding to the ectodermally derived *wg* by expressing *S59* (Baylies et al 1995). Because *wg* expression is initially dependent on pair rule gene expression and is then sustained by the expression of other segment polarity genes, notably *engrailed* (*en*) and *hedgehog* (*hh*) (for review, see Martinez Arias 1993), the fact that mutations in *en* and *hh* do not reproduce the *wg* phenotype in the mesoderm shows unequivocally that the requirement for *wg* is an early one, occurring during blastoderm to stage 10/11, which corresponds to the pair rule–dependent phase of its expression. Chu-LaGraff & Doe (1994) have shown that during this phase of ectodermal development, *wg* is required for the proper segregation of a subset of neuroblasts in the developing central nervous system of the embryo. Thus an equivalent system of positional cues may be involved in specifying both the progenitors of neurons and the progenitors of the muscles they will innervate.

## A Role for Neurogenic and Proneural Genes

The analogy in *Drosophila* between the formation of muscle founder cells in the embryonic mesoderm and the formation of neuroblasts in the neuroectoderm has been noted before (Bate 1990, Corbin et al 1991). In both cases, single cells are in some way selected from their neighbors and assigned unique properties. Genetic evidence shows that founder cells and neuroblasts are both selected from larger groups of equivalent cells by mechanisms that require the functions encoded by the neurogenic genes. In embryos that are mutant for neurogenic genes, the singling out of both neuroblasts in the ectoderm and founder cells in the mesoderm is deranged. In such mutant embryos, groups of neighboring ectodermal cells all form neuroblasts (for review, see Campos-

Ortega 1993) and enlarged clusters of cells in the mesoderm express founder cell marker genes (Corbin et al 1991, Bate et al 1993). Interestingly, at least some of the genetic pathways that are required for the spatial organization of neurogenesis seem to be replicated in the mesoderm and required for the proper segregation of muscle founder cells. In the neuroectoderm, the definition of the groups of equivalently competent cells from which neuroblasts will seg-regate depends on the activation of proneural genes, which include members of the *achaete-scute* complex (*AS*-C) (for review see Campos-Ortega 1993). Members of the *AS*-C, which encode transcription factors of the bHLH class, are expressed in clusters of ectodermal cells from which single neural progeni-tor cells will segregate. The absence of proneural gene functions leads to a loss of neural progenitors.

Recently, Carmena et al (1995) demonstrated that one member of the *AS*-C, *lethal of scute* (*l'sc*) is expressed in well-defined clusters of cells in the somatic mesoderm, from which muscle founder cells segregate (Carmena et al 1995). In each cluster *l'sc* expression is lost from all cells except one, which expresses *l'sc* at a higher level and forms a muscle progenitor. This progenitor cell goes on to divide and gives rise to two muscle founders. Identified founder cells are consistently derived from specific groups of *l'sc*-expressing cells. In addi-tion, loss of function in *l'sc* leads to a loss of specific muscles and muscle founder cells, while overexpression of *l'sc* causes muscles and founder cells to be duplicated. The origin of two distinct founders from one progenitor is highly reminiscent of events in the CNS, such as the formation of midline precursor neurons by the symmetrical division of a single progenitor, or the origin of pairs of distinct neurons from the division of ganglion mother cells in the neuroblast lineages. There may be a fundamental similarity between the mechanisms that underlie the specification of neurons and the specification of muscles, and this similarity may be central to the specificity with which nerve muscle connections are subsequently formed.

## *Regulation of Motoneuron Identity*

The *Drosophila* CNS originates from the neuroectoderm lying on either side of the ventral midline. Both neurogenic and proneural genes interplay to specify a set of neuroectodermal cells from their equivalent-group neighbors (for review, see Goodman & Doe 1993). These cells assume neuronal stem cell properties and delaminate internally. As a result, stereotyped arrays of neuroblasts form directly dorsal to the ventral ectoderm. Before the germ line is fully extended (embryonic stage 10), a full complement of neuroblasts consisting of 29 bilateral pairs of neuroblasts (NBs) and one median unpaired neuroblast (MNB) becomes visible in each segment. Each neuroblast is unique in terms of its position and pattern of gene expression and undergoes largely invariant neuroblast-specific cell divisions (Doe 1992, Doe & Technau 1993). In each segment, by the end of embryogenesis, the neuroblasts produce about 70 motoneurons and some 400 interneurons, at which time they become quiescent. Later, during larval development, they resume their proliferation to generate adult-specific interneurons. However, there is good evidence that most, if not all, adult motoneurons are generated by the neuroblasts during embryogenesis (Truman & Bate 1988). Lineage-tracing studies are beginning to reveal the lineages for specific motoneurons (Doe & Technau 1993) (summarized in Table 1).

Even prior to axonogenesis, many motoneurons are each uniquely identifiable by their cell body positions within the CNS (Figure 1). In addition, the motoneurons have many common features. They share at least one putative cell adhesion protein, Fasciclin II (Van Vactor et al 1993), and the antigen detected by MAb 22C10 (Fujita et al 1982), which is rarely expressed by interneurons; in addition they use glutamate as their excitatory neurotransmitter (Johansen et al 1989a). Therefore, many of these motoneuronal features are conceivable under the control of commonly expressed regulatory genes. Furthermore, many of the genes involved in embryonic segmentation are reexpressed in subsets of motoneurons and interneurons in the embryonic CNS—including the pair rule and segmentation genes *fushi tarazu* (*ftz*), *even-skipped* (*eve*), *engrailed* (*en*), and the homeotic selector gene *ultrabithorax* (*Ubx*). Loss of function of *ftz* and *eve* during CNS development can alter the morphogenesis of identified motoneurons (Doe et al 1988a,b). There are, in addition, expression patterns characteristic to specific motoneurons for an ensemble of cell surface glycoproteins, which may be involved in the highly stereotypic and cell-specific pathway and target choices characteristic of each motoneuron (summarized in Table 1).

Each of the transcription factors shown thus far in this review to mark subsets of muscles and their founder cells is also expressed in a subset of CNS neurons. These expression patterns may represent the coordinate specification of motoneurons and their synaptic targets. However, in only one instance has it been shown that a muscle and its innervating motoneuron coexpress the same transcription factor (*eve,* which is expressed in muscles 1 and by motoneuron aCC, which innervates it) (Patel et al 1992, Bodmer 1993). In other cases either the cells in the CNS that express the marker are not motoneurons (as in *apterous*) or (as with *vestigial*) the muscles that express the gene are innervated by nonexpressing motoneurons (M Landgraf, personal communication). The underlying logic of neuron and muscle specification has yet to be discovered; so far we can say that the processes of establishing a motoneuron map in the ectoderm and a muscle map in the mesoderm are at least remarkably similar.

**Table 1** Features of selected embryonic motoneurons in *Drosophila*

| Montoneuron | RP1 | RP3 | RP4 | RP5 | RP2 | aCC |
|---|---|---|---|---|---|---|
| Precursor neuroblast[a] | NB3-1 | NB3-1 | NB3-1 | NB3-1 | NB4-2 | NB1-1 |
| CNS axon pathway[b] | ac → 1c → ISNt → lateral exit (contralateral & immediately posterior hemisegment) | | | | SNt → lateral exit (ipsilateral) | ISNt → lateral exit (ipsilateral) |
| Peripheral nerve[c] | SNb | SNb | SNb | SNb | ISN | ISN |
| Muscle target[c] | 13 | 6, 7 | 13 | 12, 13, 14, 30 | 2 | 1 |
| Molecular expression (nuclear proteins)[d] | Ubx, Antp? | Ubx, Antp? | Ubx, Antp? | Ubx, Antp? | eve, ftz, Ubx, Antp | eve, ftz, Ubx, Antp |
| Molecular expression (membrane proteins)[e] | Fas I, Fas II, Fas III | Fas I, Fas II, Fas III | Fas I, Fas II, Fas III | Fas I, Fas II? | Conn? | Conn? |
| Neurotransmitters[f] | Glutamate | Glutamate | Glutamate | Glutamate | Glutamate | Glutamate |

[a] CQ Doe, personal communication.
[b] Jacobs & Goodman 1989, Johansen et al 1989b, Halpern et al 1991, Sink & Whitington 1991a.
[c] Halpern et al 1991, Sink & Whitington 1991a.
[d] Doe 1992, Doe & Technau 1993.
[e] Halpern et al 1991; Nose et al 1992, 1994; Van Vactor et al 1993.
[f] Jan & Jan 1976a,b; Johansen et al 1989a,b.

# AXONAL OUTGROWTH AND TARGET SELECTION

## Selecting Pathways Within the CNS

The initial axon paths taken by motoneurons are distinctive. Some extend single axons ipsilaterally (e.g. aCC, RP2), while other project contralaterally (RP1, RP3, RP4, RP5) or bilaterally (as in the ventral unpaired medial cells). There are three major axon pathways for the central neurons. A pair of longitudinal connectives runs through the entire length of the CNS on both sides of the midline, while in each segment two anterior and posterior commissures connect the left and right sides. There seems to be no obvious difference between the ways the axonal growth cones of motoneurons and interneurons behave as they navigate through their central axon pathways within the CNS. Because of their well-known axon pathways and synaptic targets, the RP motoneurons are a favored model for experiments that attempt to dissect general guidance mechanisms. For example, many neurons, including RP1 and RP3, send their axons across the midline. In *commissureless* mutants, no axon succeeds in crossing the midline (Seeger et al 1993). Yet, despite this major guidance error at the beginning of growth, the RP axons nevertheless project correctly through the CNS, and in the periphery, they select their appropriate target muscles, albeit on the ipsilateral side. These observations suggest that growth cones pass though a series of guidance choice points along their trajectory, each of which operates independently of each other (Seeger et al 1993).

## Exiting the CNS

The axons of nearly all motoneurons in each segment (except those in the transverse nerve) (Gorczyca et al 1994) exit the CNS through a common lateral nerve exit point on each side. At each exit point is a group of cells collectively known as exit glia. These peripheral glia arise within the CNS, and migrate distally to the nerve exit points by stage 14 of embryogenesis (Klämbt and Goodman, 1991, Ito et al 1995). The mechanisms responsible for motoneuron growth cones exiting the CNS remain unknown. These growth cones may be responding to local intermediate cues, with the exit glia as candidates for a guidepost role. Genetic deletion of a set of exit glia at the dorsal nerve exit point results in neurosecretory axons failing to exit the CNS from that site (Gorczyca et al 1994). Alternatively, the neurons may be responding to either diffusible chemoattractants or repellants, as has been demonstrated for classes of spinal cord neurons in vertebrates (Serafini et al 1994, Messersmith et al 1995).

Three peripheral nerves innervate the body wall in each abdominal hemisegment of embryos and larvae. Dorsal and ventral regions of the body wall are innervated by the intersegmental (ISN) and segmental (SN) nerves, respectively (Johansen et al 1989a,b). The smaller transverse nerve (TN) is a mixed motor and sensory projection with efferent axons that innervate at least two muscle fibers in mid–body wall regions (Cantera & Nassel 1992, Gorczyca et al 1994). Motoneurons navigate through a territory consisting largely of post-mitotic and differentiating cells, which serve as potential guidance cues for the axons. The embryonic nervous system sends out its first axons when the muscle fibers have already formed and when tracheal development has already commenced (Johansen et al 1989b, Bate 1990, Manning & Krasnow, 1993). The first efferents pioneer the ISN at around 10 h AEL, or stage 13, with motoneuron aCC leading the way. Midway to their dorsal targets, the ISN axons contact afferent fibers from dorsal sensory cells and fasciculate with their axons (Johansen et al 1989b). The ISN trajectory is influenced by the trachea, with the axons aligning on the lateral trunk (Giniger et al 1993). The trachea may also serve as guidance cues for sensory axons elsewhere in the body wall (Younossi-Hartenstein & Hartenstein 1993, Manning & Kransow 1993). The SN axons first exit the CNS around 11 h AEL, late-stage 14; so as a rule, the farther axons must project, the earlier they leave the CNS (Johansen et al 1989b).

There is good evidence that motoneuron growth cones do not depend upon the presence of their target muscle fibers to grow to within filopodial reach of that site. *Drosophila* growth cones can sample a radius of about 10 to 15 µm, or approximately the width of 2 to 3 muscle fibers in the embryo (Halpern et al 1991, Keshishian et al 1993). Muscle fiber ablations have shown that target-deprived motoneurons still project to preferred body wall sites even when their target muscle fibers are absent. Cash et al (1992) observed that the neurons would in addition innervate neighboring muscle fibers with ectopically placed but other-wise functional synaptic connections, which were located on the muscle fiber at sites near to where the native fiber would have been located. Sink & Whitington (1991c) examined the behavior of individually identified motoneurons whose target muscle fibers were surgically removed in cultured embryos. Despite the absence of the target muscle fibers, the motoneurons were capable of growing to the general vicinity of their targets, where they could contact multiple neighbor-ing fibers. Thus, for most of their trajectory these axons must depend on nontarget intermediate cues to reach their destinations. The identity of the signals respon-sible for guiding axons to these regions remains undefined, but by analogy to the better understood system of pioneer neurons in the grasshopper limb bud, the *Drosophila* cues could involve gradients, alignment with molecular stripes expressed on the epidermis, cellular targets such as guideposts, and selective fasciculation (Bentley & O'Connor 1994). In *Drosophila* both mutagenesis screens and reverse genetic approaches have yielded several candidates for the molecular elements involved these guidance choices (see section on Synaptic Function and Plasticity).

In the periphery the growth cones freely contact both target and nearby nontarget muscle fibers. The details of growth cone morphology can be best revealed by intracellular dye injection into the motoneurons. For example, both the RP1 and RP3 growth cones contact more or less the same set of ventral muscle fibers (Sink & Whitington 1991b, Halpern et al 1991). During this exploratory period motoneuron growth cones, including those of RP1 and RP3 are morphologically indistinguishable (Halpern et al 1991). However, a first sign of morphological change is noticed as their filopodia contact their specific target muscle fibers (Halpern et al 1991). Soon after the RP1 growth cone contacts muscle fiber 13, its target, the filopodial patterns revealed by staged dye fills begin to modify. Instead of maintaining about a half dozen filopodia that extend up to 10 μm around the growth cone head, the RP1 growth cone begins to consolidate its filopodial processes onto muscle fiber 13. Then, a small number of varicosities form on processes maintained on the target. A similar series of morphological changes is seen for RP3 as it innervates its targets, muscle fibers 6 and 7 (Halpern et al 1991; Sink & Whitington 1991a,b; Broadie & Bate 1993a). These responses to the target muscle fiber by the innervating growth cone are suggestive of specialized recognition events.

## Target Mismatch

What are the underlying mechanisms for specific growth cone behavior and target selection? Sink & Whitington (1991c) and Chiba et al (1993) have used surgery, mutations, heat shock, and/or laser cell ablation in various target mismatch experiments to alter the body wall musculature, and have studied the responses of the RP growth cones to either deletion or duplication of their target muscle fibers. When both muscle fibers 6 and 7 are deleted surgically, the RP3 growth cone fails to form synapses on alternative muscle fibers (Sink & Whitington 1991c). Instead, it maintains a relatively large number of filopodia that are often abnormally long. Its appearance remains undifferentiated as synaptogenesis by other unaffected motoneurons proceeds normally. Similarly, when the RP1 growth cone loses its normal target in *numb* mutants, it often keeps advancing distally without forming synaptic structures on any other muscle fibers, even during very late embryogenesis (Chiba et al 1993). Therefore, the complete loss of target muscle fiber(s) results in the growth cones initially failing to form synapses. However, there is good evidence that target-deprived motoneurons do eventually form alternative, ectopically placed synapses on neighboring muscle fibers (Cash et al 1992). These results support the idea that each motoneuron growth cone has an ability to specifically recognize its normal target(s) upon initial contact. Furthermore, they suggest that cues necessary for this specific target recognition are tightly associated with the target and/or its immediately neighboring cells. This is also consistent with the fact that the morphologically visible synaptic differentiation begins only after a direct growth cone–target contact.

Partial target deletions have further revealed the specificity of the growth cone–target recognition. Deletion of either muscle fiber 6 or 7, i.e. removal of one of the two normal target cells, leads to the RP3 growth cone settling on the remaining target fiber in 80% of the cases (Sink & Whitington 1991c, Chiba et al 1993). At early stages the RP3 growth cone sometimes extends filopodia-like processes that are often abnormally long and reach multiple neighboring muscle fibers (Sink & Whitington 1991a–c). However, after the normal period of synaptogenesis, it confines its contact to the surviving target muscle fiber (Chiba et al 1993). In about 20% of the cases, when it fails to innervate the remaining muscle fiber, the RP3 growth cone extends further distally and does not form presynaptic structures. When RP3 innervates the single remaining target muscle fiber, it does so from either the lateral or medial side of the muscle fiber at roughly equal frequency (Chiba et al 1993). Thus, the exact synaptic site on the muscle fiber can vary, whereas the motoneuron-to–target fiber matching remains specific.

To further test the idea that synaptogenesis is initiated through the specific target contacts, Chiba et al (1993) examined situations where RP1's normal target, muscle fiber 13, is duplicated. The RP3 growth cone, when faced with duplicated muscle fibers, proceeds to innervate them both, whereas other motoneurons innervate their targets normally. Thus the number of fibers that a single motoneuron growth cone innervates depends at least partially on particular interactions between the growth cone and its microenvironment.

Although these studies point to a positive response by the growth cone for its target, the possibility remains that recognition also involves the avoidance of nontarget fibers. To test whether the muscle fibers immediately next to the normal target fiber of RP3 may provide negative cues to the growth cone, Chiba et al (1993) examined situations where all the ventral longitudinal muscle fibers except muscle fiber 6 were deleted, as occurs in *numb* mutants. In this case, the RP3 growth cone still innervated the remaining target about 80% of the time, the same rate as when only muscle fiber 7 was absent. In the remaining 20% of the cases, it failed to innervate the target and instead advanced distally without differentiating into presynaptic structures. In other words, the success rate for RP3 to select the remaining target cell is independent of the presence of neighboring muscle cell(s). This implies that during this final step of target selection, the major cues guiding the growth cone are positive signals from the target cells and not negative signals from the immediate neighbor muscle fibers. Consistent with this finding, in other experiments where another muscle fiber is deleted, the neuromuscular organization on the remaining neighbor muscle fibers remains normal (Cash et al 1992).

These target mismatch experiments have yielded important clues to the molecular nature of the growth cone–target recognition. At least two steps are involved. First, the motoneuron growth cones seeks out a target region, which is defined as a site in the body wall musculature within filopodial reach of the target muscle fiber(s). Second, the growth cones actively selects a specific target muscle fiber(s). This second step requires the physical contact of the target muscle fibers by the growth cone. The target deletion studies have clearly demonstrated that the first step does not require the target cells themselves. In contrast, for the second step the muscle fibers are necessary, as motoneuron growth cones do not initially form synapses in the embryos without their targets.

## MOLECULAR GENETIC ANALYSIS OF SYNAPTOGENESIS

The above discussion suggests that a form of molecular recognition is involved in matching motoneurons to muscle fibers. Two experimental approaches have been employed in *Drosophila* to test the in vivo functions of putative cell recognition molecules deployed during synaptogenesis. First, phenotypic characterizations are possible for many of the candidate molecules by using either null mutations or alleles with varying degrees of reduced expression. Second, these molecules can be misexpressed using P-element enhancer detectors, such as the GAL4-UAS system (Brand & Perrimon 1993), or through transgenes with mesoderm-specific regulatory sequences (Nose et al 1994, Matthes et al 1995,Chiba et al 1995).

### Mutagenesis and Reverse-Genetic Studies

Van Vactor et al (1993) performed a second chromosome mutagenesis screen for phenotypes that involved axonal guidance and targeting defects, in order to identify genes that are important for neuromuscular development. Whole-mount mutant embryos were screened using Fasciclin II immunocytochemistry, a particularly effective label for efferent axons. Care was taken to identify mutant lines where the gross morphology of the CNS appeared normal. The recovered mutants fell into several phenotypic classes, including cases where axons had defective outgrowth (*beaten path, shortstop,* and *stranded*) or made connections either nonspecifically or onto multiple incorrect targets (*clueless* and *walkabout*). A separate third chromosome screen has recovered mutant lines falling into similar phenotypic classes (Sink & Goodman 1994). Given their frank morphological phenotypes, these embryonic lethal lines are likely to include mutations of genes for proteins essential for growth cone function, including mechanoenzymes and signal transduction machinery.

Enhancer detector and monoclonal antibody screens have revealed a second group of putative cell-recognition molecules with intriguing expression patterns and in vitro adhesion functions. Unlike the proteins recovered in mutagenesis screens, mutants lacking these molecules have generally had meager guidance and targeting phenotypes. Limited morphological phenotypes have been found for Fasciclins I, II, III; Neuroglian; Connectin; and Semaphorin II (see below). One explanation for this observation is that the motoneurons and muscles express multiple molecules with shared or overlapping functions, where each molecule is by itself sufficient, but not essential, for target recognition. Testing this hypothesis requires the analysis of double or multiple mutants, where the redundancy is eliminated—an approach that has not yet been extensively employed for the neuromuscular system in *Drosophila*. Alternatively, an effective strategy for testing function in a system with extensive redundancy is to misexpress a molecule of interest on inappropriate cells. This generates muscle fibers or motoneurons with novel combinations of putative recognition molecules, which may lead to interpretable guidance or targeting phenotypes. Five studies have been published testing the roles of Fasciclin II, Connectin, Fasciclin III, and Semaphorin II using this paradigm (Lin & Goodman 1994, Lin et al 1994, Nose et al 1994, Chiba et al 1995, Matthes et al 1995).

## *Differential Cell Adhesion During Axonal Guidance*

Fasciclin II (Fas II) is a cell adhesion molecule belonging to the immunoglobulin superfamily with sequence similarity to the vertebrate adhesion molecule NCAM (Harrelson & Goodman 1988, Grenningloh et al 1991). At least three Fas II isoforms have been identified in *Drosophila*, including two transmembrane forms and one with a phosphatidyl-inositol membrane linkage (Lin et al 1994). Fas II is expressed in the periphery on all efferent axons (Van Vactor et al 1993), while null mutations have no apparent phenotype with respect to peripheral guidance and fasciculation (Lin & Goodman 1994). In the CNS, axonal fasciculation errors arise following either the loss or hyperexpression of Fas II (Lin et al 1994).

Lin & Goodman (1994) examined the effect of neuronal hyperexpression of Fas II by using a GAL4-mediated expression system (Brand & Perrimon 1993), where Fas II expression was induced in all neurons. Increased dosage of UAS-Fas II led to a corresponding increased fasciculation of axons in a subset of the nerves examined in stage 16 embryos. This resulted in the failure of motor axons to defasciculate at their targets. The Fas II hyperfasciculation caused motoneuron axons to grow to inappropriate body wall regions, often distal to their targets. As Fas II levels dropped in later stages of embryogenesis the misguided axons attempted to innervate their appropriate targets, albeit from incorrect trajectories.

This experiment provides three important insights into neuromuscular development. First, it gives the first in vivo evidence that Fasciclin II is actually involved in axonal fasciculation, a function hypothesized from in vitro cell adhesion assays, where it mediates homophilic adhesion. Second, it suggests that motoneurons leave the peripheral nerves for their targets through a dynamic balance between attraction for the nerve, mediated by CAMs such as Fas II, and attraction for the muscle fiber target, reminiscent of axonal defasciculation in the chick limb bud mediated by the balance between the adhesion molecules PSA-NCAM and L1 (Landmesser et al 1992, Tang et al 1994). Finally, it shows that motoneurons can project to their targets at incorrect times and from inappropriate trajectories, supporting the hypothesis that synaptic connectivity involves specific target recognition by motoneuron growth cones.

## Evidence for Cell Repulsion

In the *Drosophila* body wall, motoneurons show a variety of behaviors that may be consistent with surface repulsion, including the withdrawal of processes from inappropriate muscle fibers (Sink & Whitington 1991a,b; Halpern et al 1991). No detailed time-lapse studies have been performed in embryonic fillets, so it is unknown whether growth cone collapse occurs in this system. However, in vivo evidence for repulsion is derived from two misexpression studies, involving the proteins Connectin and Semaphorin II.

Connectin was found by Nose et al (1992) to be expressed by eight muscle fibers in the body wall and, furthermore, by the motoneurons that innervate them. The expression is predominantly among fibers in the mid–body wall, and during synaptogenesis Connectin localizes to the neuromuscular contacts. Connectin is a transmembrane glycoprotein with 10 extracellular leucine-rich repeats (LRRs) and demonstrates homophilic adhesion in in vitro assays (Nose et al 1992). Loss-of-function null mutations of Connectin were found to have no obvious morphological defects with respect to neuromuscular innervation (Nose et al 1992, 1994). By using a transgene with the muscle-specific regulatory sequences of Toll, Connectin was misexpressed on a subset of ventral muscle fibers. Surprisingly, this resulted in transient denervation of the fibers, with innervation delayed until the misexpression declined in later stages of embryogenesis. As with the Fasciclin II hyperexpression study, this experiment showed that muscles were competent for innervation at later stages of development and that delaying innervation does not disrupt target recognition. Equally striking, the results show that Connectin misexpression interfered in some fashion with the normal innervation of the muscle fibers. These results provide possible in vivo evidence for a repulsive or masking function of Connectin in normal development. According to this scenario, Connectin may function as a nonpermissive surface cue for ventral motor axons, preventing them from advancing into mid–body wall domains. Further evidence favoring repulsive signals in the body wall have come from studies of misexpression of *Drosophila* Semaphorin II (Matthes et al 1995).

The idea that muscle-specific cues might function to repel axons in the *Drosophila* body wall received an impetus from the discovery that Collapsin, the chick growth cone collapsing factor (Luo et al 1993), shared sequence similarity with the Fasciclin IV, a transmembrane glycoprotein characterized in the grasshopper embryo that was previously shown to be necessary for normal axonal guidance in the embryonic limb bud (Kolodkin et al 1992). These proteins have been renamed Semaphorins, of which two have been identified in *Drosophila* (Kolodkin et al 1993). One of these, Semaphorin II (Sema II) has no transmembrane domain and is likely secreted. Sema II has a very limited peripheral expression, confined to a single muscle fiber on either side of the embryo (muscle 33 of segment T3, Matthes et al 1995). Loss-of-function null mutations of Sema II have no reported effects on the innervation of this muscle fiber, nor do they affect the behavior of motoneurons innervating adjacent muscle fibers in T3.

Matthes et al (1995) examined the ectopic expression of Sema II in abdominal segments by using the same paradigm as Nose et al (1994), namely a transgene with expression under the control of the mesoderm-specific Toll regulatory sequence. This leads to expression in specific ventral longitudinal muscle fibers, including muscle fibers 6 and 7, the targets of the motoneuron RP3. They observed that Sema II misexpression by these fibers disrupted RP3's innervation of muscle fiber 6 and 7, resulting in targeting errors. This observation, like that for Connectin, suggests that RP3 is sensitive to a repulsive signal conferred by Sema II, which is particularly intriguing because Sema III (the chick Collapsin) is a potent growth cone chemorepulsive factor (Luo et al 1993, Messersmith et al 1995). The Matthes et al (1995) study shows that a motoneuron that normally never comes in contact with a Sema II–positive muscle fiber nevertheless responds to the protein's ectopic presence. This suggests that the motoneuron carries either receptors specific to Sema II or receptors for a structurally similar ligand that is in fact expressed by the ventral abdominal muscle fibers.

## *Cell Recognition Molecules*

Fasciclin III (Fas III) is a transmembrane glycoprotein that exists in two isoforms with sequence similarity to members of the immunoglobulin super-family (Snow et al 1989). The protein is expressed by subsets of muscle fibers (Halpern et al 1991) and motoneurons (Jacobs & Goodman 1989, Halpern et al 1991). These subsets include the ventral longitudinal muscle fibers 6 and 7 and several of the RP motoneurons. Thus, both the motoneuron RP3 and its muscle fiber targets coexpress Fas III. This observation was tantalizing, as Snow et al (1989) had previously shown that Fas III mediates a homophilic cell adhesion in in vitro assays by using transient expression of the molecule. Nose et al (1992) discovered a similar and more extensive coexpression by synaptic partners of a putative cell adhesion molecule for Connectin.

Null mutations for Fas III have no defects in peripheral innervation, with RP3's targeting of muscle fibers 6 and 7 indistinguishable from wild-type embryos (Chiba et al 1995). Chiba et al (1995) misexpressed Fas III by using a transgene with expression under the control of the regulatory sequence of the myosin heavy chain. This resulted in Fas III expression by all body wall muscle fibers; neuronal expression was unaltered. They observed that the RP3 motoneuron responded to the novel expression pattern for Fas III by establishing synapses with many of the muscle fibers in the ventral body wall, both proximal and distal to the appropriate target. The degree of errors were dose dependent, rising to over 80% with two copies of the transgene. By contrast, the motoneuron RP1, which normally innervates a non–Fas III positive fiber, did not show any changes to its targeting preferences when the CAM was expressed ubiquitously in the musculature.

These results show that Fas III is sufficient to induce novel synapses by the RP3 motoneuron when it is misexpressed by muscle fibers but that in its absence the motoneuron can still find its correct target, which suggests that there are multiple overlapping targeting cues available to RP3, of which Fas III is one. This experiment provided the first evidence that one of the *Drosophila* CAMs is a functional participant in establishing a neuromuscular synapse.

Further controls and, especially, direct time lapse–imaging studies of growth cone behavior are needed to better understand these misexpression experiments. Nevertheless, assuming that repulsive and attractive signals exist in *Drosophila,* target recognition may involve a balance of cues directing motoneurons away from inappropriate regions of the body wall and onto preferred targets. Thus mid–body wall muscle fibers that express Connectin would act as a negative, nonpermissive domain for ventral motoneurons. Fasciclin III, by contrast, would act as a positive attractor for specific motoneurons when expressed by subsets of ventral muscle fibers.

## DIFFERENTIATION OF THE SYNAPSE

Approximately 8 h elapse from the arrival of the motoneuron growth cones at the target site to the time of hatching, a period when many of the properties of the synapse differentiate (at 25°C embryonic development lasts 21 h). Initially, as the growth cones explore the region of the muscle fiber, small amplitude excitatory junctional currents (EJCs) can be recorded from the muscle by using voltage clamp. These EJCs occur spontaneously and can be evoked by stimulation of the nerve (Broadie & Bate 1993a, Kidikoro & Nishikawa 1994). Extensive gap junction–mediated electrical and dye coupling is also observed during early stages between the embryonic muscle fibers (Johansen et al 1989, Broadie & Bate 1993a, Gho 1994), a situation that may affect voltage-clamp analysis. Dye coupling is lost around the time of the initial growth cone contacts (Johansen et al 1989, Broadie & Bate 1993a), although some electrical coupling persists between the muscle fibers near their insertion sites (Kidikoro & Nishikawa 1994). Nevertheless, the fibers can be clamped throughout embryogenesis and into the larval stages. A true endplate current, with a fast rise time and decay, can be distinguished from a lower amplitude current with a much slower time course, reflecting current spread from adjacent muscle fibers (Kidikoro & Nishikawa 1994).

Several synaptic components are expressed by the motoneuron and muscle fiber prior to their contact. For example, the neurotransmitter glutamate is first detected immunocytochemically in motoneuron growth cones as they enter their target areas, although differentiated presynaptic boutons are not evident until several hours later (Broadie & Bate 1993a). Similarly, in muscles the glutamate receptor transcripts are first expressed several hours before the arrival of motoneuron growth cones, at a time when myogenesis is still taking place (Currie et al 1995). However, sensitivity to iontophoretically applied glutamate is not detected in muscle fibers until the first motoneuronal contacts are made (Broadie & Bate 1993a).

Most of the studies on embryonic synaptic development and neurotransmission have involved the motoneuron RP3 and the larger of its two targets, muscle fiber 6 (Figure 1). At early stages, the motoneuron establishes a rapidly fatiguing synapse with low-amplitude EJCs. Following an initial plateau, the EJC amplitude increases steadily through subsequent embryogenesis, with a corresponding decline in synaptic fatigueability (Broadie & Bate 1993a,b; Kidikoro & Nishikawa 1994). This apparently occurs as a consequence of the enrichment of functional GluRs at the postsynaptic site (Broadie and Bate 1993a).

An initial issue to be settled was the degree to which the developmental mechanisms of synaptogenesis, already partially characterized at the vertebrate endplate (Hall & Sanes 1993), were conserved in *Drosophila*. For example, to what extent is the development of synaptic properties cell-autonomous, or alternatively, dependent on signals between motoneuron and muscle? *Drosophila* is particularly well suited for addressing problems of this sort. There are mutations that specifically affect pre- or postsynaptic cells (Broadie & Bate 1993c), and furthermore, it is possible to target gene expression to one side of the junction or the other using heterologous promoters and the GAL4 system (Sweeney et al 1995). Using these approaches, it has been possible to test the effects of delayed innervation on postsynaptic differentiation. Early developmental events, involving the expression of functional glutamate receptors or the cell adhesion molecules Fasciclin III and Connectin, still take place in denervated muscle fibers (Broadie & Bate 1993c). Also, the electrical and contractile properties of the muscle can still develop following denervation (Broadie & Bate 1993d). By contrast, the differentiation of mature postsynaptic properties and receptor patterning requires the presence of a functional motoneuron (Broadie & Bate 1993c). Normally, in *Drosophila* transmitter receptors cluster beneath the presynaptic terminals. Following a delay or absence of innervation, the receptors remain diffusely distributed, and furthermore, the normally occuring second wave of receptor accumulation is blocked (Broadie & Bate 1993c). Thus, both receptor clustering and the expression of functional receptors are dependent on the neuron, as is the case in vertebrates (Hall & Sanes 1993).

Despite these similarities, some aspects of the regulation of receptor expression and patterning in *Drosophila* may be different from vertebrates. For example, there is no localization of receptor transcripts to the subsynaptic region (Currie et al 1995), as occurs at the vertebrate endplate (Merlie & Sanes 1985, Fontaine et al 1988). Also, the *Drosophila* homologues of vertebrate agrins (McMahan 1990) have remained elusive, and it is unknown whether either ARIA (Falls et al 1990) or CGRP-like molecules (New & Mudge 1986, Fontaine et al 1986) are involved in regulating receptor expression. Finally, the role of electrical activity in regulating receptor expression may be different.

A precise test of the requirements for presynaptic electrical activity on receptor expression can be made in *Drosophila* by using conditional mutations that block action potentials. In vertebrate muscle the clustering of the acetylcholine receptors at the endplate occurs independent of electrical activity, and the role of muscle activity is to down-regulate receptor expression in extrajunctional nuclei (reviewed by Hall & Sanes 1993). The unexpected observation in *Drosophila* is that motoneuron action potentials are required for both the clustering and later enrichment of glutamate receptors (Broadie & Bate 1993e). Neither synaptic release from the motoneuron nor evoked activity in the muscle seems to be directly involved (Broadie et al 1994). However, the observed differences between *Drosophila* synaptogenesis and vertebrates may be due to the fact that the mammalian twitch fiber, which is the basis for many of the vertebrate studies, is a specialized muscle. Thus, the localized expression of receptor transcripts to the vertebrate endplate may result from the fact that relatively few nuclei lie close to the synaptic site. Because *Drosophila* muscle fibers are much smaller than their vertebrate counterparts, and all the nuclei lie close to synaptic terminals, a class of transcriptionally specialized nuclei may not be necessary.

# SYNAPTIC FUNCTION AND PLASTICITY

## *Functional Analysis of the Synapse*

Homologues of known synaptic proteins are now being systematically uncovered in *Drosophila* by using reverse genetic approaches. Many of the presynaptic proteins are proposed to play a role in excitation-secretion coupling. Every synaptic protein implicated in vertebrate transmission that has been studied in *Drosophila* has been found to be strongly conserved (typically 70–80% relative to the mammalian homologues) and expressed at the neuromuscular junction. These observations support the view that the underlying machinery at the *Drosophila* neuromuscular junction and in vertebrate central synapses is fundamentally similar. So far five proteins with lethal loss-of-function phenotypes have been identified: Synaptotagmin, Synaptobrevin (also known as VAMP), Syntaxin, Rop (a *Drosophila* homologue of the *n*-sec 1 family of proteins), and Cysteine String Protein (CSP).

Rop and Syntaxin both appear to be essential for the general machinery of exocytosis as well as for synaptic transmission. Null mutations in either gene block secretion from many different cell types in the *Drosophila* embryo (Harrison et al 1994, Schulze et al 1994), and null mutations in *syntaxin*, at least, completely block evoked vesicle release at the embryonic neuromuscular junction (Schulze et al 1995). As might be predicted for such functions, *rop*, at least, is a cell lethal mutation (Harrison et al 1994), and embryonic survival in both *rop* and *syntaxin* mutants seems to depend on maternally contributed protein (K Broadie, unpublished observations). The other three proteins, Synaptobrevin, Synaptotagmin, and CSP, appear to be specific for neural transmission at the synapse. Synaptobrevin is essential for synaptic transmission, and in its absence, evoked release is completely eliminated, although spontaneous vesicle release still occurs (Sweeney et al 1995). Both Synaptotagmin and CSP appear to be specialized neuronal regulators of exocytosis during synaptic transmission, as the loss of either of them does not completely block transmission (Broadie et al 1994, Zinsmaier et al 1994). Instead, mutations in either gene rather dramatically reduce the fidelity of excitation secretion coupling at the neuromuscular junction. Synaptotagmin acts at least partly as a negative regulator of synaptic vesicle release (Broadie et al 1994) and could act to reserve a pool of vesicles ready for $Ca^{2+}$-dependent exocytosis. Two complementary lines of evidence suggest that Synaptotagmin may be involved in docking synaptic vesicles at presynaptic release sites and/or in activating the fusion of vesicles by directly sensing the voltage-dependent $Ca^{2+}$ influx (Littleton et al 1993,1994). The role of CSP is less clear, although it seems to regulate vesicle release through an interaction between vesicles and presynaptic $Ca^{2+}$ channels (Zinsmaier et al 1994).

## Synaptic Plasticity

Despite the precise target preferences exhibited by embryonic motoneurons, there is good evidence that synaptic choices are not hard wired but instead show significant plasticity. The plasticity involves both alterations in synaptic connectivity and an activity-dependent regulation of the size and complexity of the larval motor endings. Cash et al (1992) noted that motoneurons deprived of their targets during embryogenesis often establish functional connections with neighboring muscle fibers, which demonstrates that motoneurons do not have obligate commitments to specific fibers. The alternate synapses are located ectopically, persist through larval development, and are electrophysiologically functional. Similarly, loss of motoneurons causes neighboring motor endings to establish collateral connections onto denervated fibers. When the RP motoneurons are eliminated through either genetic or microsurgical methods, the resulting partial or complete denervation of muscle fibers 7 and 6 lead to ectopic contacts on between 60 to 80% of the fibers (Keshishian et al 1994, Halfon et al 1995). The ectopic synapses arise from motoneurons innervating adjacent muscle fibers or as collateral sprouts from axons of neighboring nerves.

The cellular mechanisms governing these changes in local connectivity remain incompletely understood. However, a strikingly similar change in neuromuscular connectivity occurs when neural activity is reduced genetically or pharmacologically (Keshishian et al 1994, Jarecki & Keshishian 1995). Motoneurons throughout the body wall establish collateral branches on their neighbors when action potentials are inhibited either by TTX or in mutant lines with reduced Na channel number or function. By using conditional mutations, the critical period for sprouting has been found to extend from late embryogenesis into the first larval instar (Jarecki & Keshishian 1995). These results show that motoneurons are not firmly or irreversibly committed to their preferred muscle fibers but that they can alter target preferences in response to activity-dependent mechanisms.

Activity also regulates the size and complexity of neuromuscular endings in *Drosophila*. Motoneuron arbors branch extensively as muscle fibers enlarge during larval development (Keshishian et al 1993). Budnik et al (1990) discovered that hyperexcitable mutants with defective K$^+$ channels (*ether-a-gogo (eag) Shaker (Sh)* double mutants) develop increased numbers of motoneuron boutons and branches when compared to wild-type larvae. The expansion occurs in the fine, type II motoneuron arbors, which have since been shown to express the neuromodulatory transmitter octopamine (Monastirioti et al 1995). The activity dependence of motoneuron growth is an observation of considerable importance, as it shows that the *Drosophila* neuromuscular junction may be used as a model system for studying morphological forms of synaptic plasticity. The motoneuron terminal expansion also occurs in flies that are mutant for the gene *dunce* (*dnc*) (Zhong et al 1992), a learning mutant that has elevated cAMP levels (Byers et al 1981) and that encodes phosphodiesterase II, an enzyme that hydrolyzes cAMP (Byers et al 1981). Mutations in *dnc* interact synergistically with those of *eag* or *Sh* to enhance motoneuron arbor sizes, suggesting that some of the consequences of hyperexcitability occur through a cAMP pathway (Zhong et al 1992). This idea is supported by the observation that *rutabaga* (*rut*) mutants suppress the terminal expansion seen in *dnc* mutants as well as the enhanced expansion observed in *dnc; Sh* double mutants (Zhong et al 1992). Mutations in *rut* depress the levels of cAMP (Livingstone et al 1984), as the gene encodes a subunit of the $Ca^{2+}$/calmodulin-dependent adenylate cyclase (Dudai & Zvi 1985). This suggests that the activity-dependent growth of the motoneuron is influenced by presynaptic $Ca^{2+}$ accumulation in the arbor, which would in turn act on the cAMP second-messenger cascade. Mutations in *dnc* and *rut* also alter neuromuscular transmission by disrupting both synaptic facilitation and potentiation (Zhong & Wu 1991).

Both *dnc* and *rut* were initally characterized by their learning and memory phenotypes in adult flies (Dudai et al 1976, Aceves-Pina et al 1983). Determining how the cAMP pathway is involved in the long-term synaptic changes that underlie learning would be challenging in the relatively inaccessible synapses of the adult CNS. By contrast, the anatomical and functional changes that occur at the neuromuscular junction in these mutants provide a direct way to examine some of the elements involved in synaptic plasticity. For example, Feany & Quinn (1995) have shown that *amnesiac*, a mutant with defective learning retention, encodes a *Drosophila* homologue of the mammalian neuropeptide PACAP (pituitary adenylyl cyclase activating peptide). A PACAP-like peptide is also expressed in terminals of the larval neuromuscular junction (Zhong & Pena 1995). This peptide is released by high-frequency motoneuron stimulation, causing a slow muscle depolarization with delayed activation of $K^+$ channels, along with a 100-fold enhancement of the outward $K^+$ current. Zhong (1995) has showed at the neuromuscular junction that the PACAP-like peptide operates through the cooperative activation of both a *rut*-encoded adenylyl cyclase and the Ras-Raf second-messenger pathway. This elegant dissection of a second-messenger system confirms the view that the neuromuscular junction is a valuable tool for analyzing synaptic function and plasticity, and gives us a deeper insight into the potential role of the cAMP pathway in learning and memory within the CNS (Yin et al 1994, 1995; Kandel & Abel 1995).

## SUMMARY AND DIRECTIONS FOR THE FUTURE

In *Drosophila* a clear picture is emerging about the molecular mechanisms involved in cell recognition, synaptogenesis, and synaptic function. The major developmental problem to be solved remains the molecular mechanism of cell recognition and target selection. So far, mutagenesis and reverse-genetic approaches have yielded an ensemble of candidate molecules, whose functions in guidance and recognition still need to be examined in detail. Multiple mechanisms, involving both cell attraction and repulsion, are proposed to guide axons both to specific regions of the body wall and onto individual muscle targets. Understanding how a growth cone balances these influences will require better imaging techniques to observe neuronal behavior in situ, and in vitro and semi-intact culturing methods to test neuron-substrate interactions. The roles of the second-messenger systems that are activated during these cellular interactions need to be better understood. Assuming that cell recognition in *Drosophila* involves multiple, potentially redundant or overlapping cues, increasingly sophisticated genetic and misexpression paradigms to test function are needed.

The study of synaptic function is a second area ripe for further exploration. The analysis of mutations of known synaptic proteins identified by homology, apart from indicating likely in vivo functions for these proteins, has underscored the fundamental similarity of synaptic transmission at the *Drosophila* neuromuscular junction and other well-known synapses. As with the developmental studies, the true worth of the *Drosophila* system is in the genetic analysis. We can screen for novel genes and proteins required for synaptic transmission. At the same time, we can use genetic techniques to study the ways in which these proteins interact. There are few others systems in which the methods of electrophysiology, genetics, and molecular biology can be so effectively deployed to address fundamental questions of synaptic development and function.

## Literature Cited

Abmayr SM, Erickson MS, Bour BA. 1995. Embryonic development of the larval body-wall musculature of *Drosophila melanogaster. Trends Genet.* 11:153–59

Aceves-Pina EO, Booker R, Duerr JS, Livingstone MS, Quinn WG, et al. 1983. Learning and memory in *Drosophila,* studied with mutants. *Cold Spring Harbor Symp. Quant. Biol.* 48:831–40

Atwood HL, Govind CK, Wu C-F. 1993. Differential ultrastructure of synaptic terminals on ventral longitudinal abdominal muscles in *Drosophila* larvae. *J. Neurobiol.* 24:1008–24

Anderson MDS, Halpern ME, Keshishian H. 1988. Identification of the neuropeptide transmitter proctolin in *Drosophila* larvae: characterization of muscle fiber-specific neuromuscular endings. *J. Neurosci.* 8: 242–55

Baker R, Schubiger G. 1995. Ectoderm induces muscle-specific gene expression in *Drosophila* embryos. *Development* 121:387–98

Bate M. 1990. The embryonic development of larval muscle fibers in *Drosophila. Development* 110:791–804

Bate M. 1993. The mesoderm and its derivatives. See Bate & Martinez Arias 1993, pp. 1013–90

Bate CM, Martinez Arias A, eds. 1993. *The Development of Drosophila melanogaster.* New York: Cold Spring Harbor Press

Bate M, Rushton E. 1993. Myogenesis and muscle patterning in *Drosophila. C. R. Acad. Sci. Ser. C* 316:1047–61

Bate M, Rushton E, Frasch M. 1993. A dual requirement for neurogenic genes in *Drosophila* myogenesis. *Development.* pp. 149–61 (Suppl.)

Baylies MK, Martinez Arias AM, Bate M. 1995. *wingless* is required for the proper specification of a subset of muscle founder cells during *Drosophila* embryogenesis. *Development.* In press

Bentley D, O'Connor TP. 1994. Cytoskeletal events in growth cone steering. *Curr. Opin. Neurobiol.* 4:43–48

Bodmer R. 1993. The gene *tinman* is required for specification of the heart and visceral muscles in *Drosophila. Development* 118: 719–29

Bourgouin C, Lundgren SE, Thomas JB. 1992. *apterous* is a *Drosophila* LIM domain gene required for the development of a subset of embryonic muscles. *Neuron* 9:549–61

Brand AH, Perrimon N. 1993. Targeted gene expression as a means of altering cell fates and generating dominant phenotypes. *Development* 118:401–15

Broadie KS, Bate M. 1993a. Development of the embryonic neuromuscular synapse of *Drosophila melanogaster. J. Neurosci.* 13: 144–66

Broadie KS, Bate M. 1993b. Development of larval muscle properties in the embryonic myotubes of *Drosophila melanogaster. J. Neurosci.* 13:167–80

Broadie KS, Bate M. 1993c. Innervation directs receptor synthesis and localization in *Drosophila* embryo synaptogenesis. *Nature* 361: 350–53

Broadie K, Bate M. 1993d. Muscle development is independent of innervation during *Drosophila* embryogenesis. *Development* 119:533–43

Broadie K, Bate M. 1993e. Activity-dependent development of the neuromuscular synapse during *Drosophila* embryogenesis. *Neuron* 11:607–19

Broadie K, Bellen HJ, DiAntonio A, Littleton JT, Schwarz TL. 1994. Absence of synaptotagmin disrupts excitation-secretion coupling during synaptic transmission. *Proc. Natl. Acad. Sci. USA* 91:10727–31

Broadie K, Sink H, Van Vactor D, Fambrough D, Whitington PM, et al. 1993. From growth cone to synapse, the life history of the RP3 motor neuron. *Development.* pp. 227–38 (Suppl.)

Budnik V, Gorczyca M. 1992. SSB, an antigen that selectively labels morphologically distinct synaptic boutons at the *Drosophila* larval neuromuscular junction. *J. Neurobiol.* 23:1054–66

Budnik V, Zhong Y, Wu C-F. 1990. Morphological plasticity of motor axons in *Drosophila* mutants with altered excitability. *J. Neurosci.* 10:3754–68

Byers D. Davis RL, Kiger JA. 1981. Defect in cyclic AMP phosphodiesterase due to the *dunce* mutation of learning in *Drosophila melanogaster. Nature* 289:79–81

Callahan CA, Thomas JB. 1994. Tau-beta galactosidase, an axon-targeted fusion protein. *Proc. Natl. Acad. Sci. USA* 91:5972–76

Campos-Ortega JA. 1993. Early neurogenesis in *Drosophila.* See Bate & Martinez Arias 1993, pp. 1091–129

Campos-Ortega JA, Hartenstein V. 1985. *The Embryonic Development of Drosophila melanogaster.* Berlin/New York: Springer-Verlag. 227 pp.

Cantera R, Nässel DR. 1992. Segmental peptidergic innervation of abdominal targets in larval and adult dipteran insects revealed with an antiserum against leukokinin I. *Cell. Tissue Res.* 269:459–71

Carmena A, Bate M, Jimenez F. 1995. *lethal of scute,* a proneural gene is required for the specification of muscle progenitors during *Drosophila* embryogenesis. *Genes Dev.* In press

Cash S, Chiba A, Keshishian H. 1992. Alternate neuromuscular target selection following the loss of single muscle fibers in *Drosophila. J. Neurosci.* 12:2051–64

Chiba A, Hing H, Cash S, Keshishian H. 1993. The growth cone choices of *Drosophila* motoneurons in response to muscle fiber mismatch. *J. Neurosci.* 13:714–32

Chiba A, Snow P, Keshishian H, Hotta Y. 1995. Fasciclin III and a synaptic recognition molecule in *Drosophila. Nature* 374:166–68

Chu-LaGraff Q, Doe CQ. 1994. Neuroblast specification and formation regulated by *wingless* in the *Drosophila* CNS. *Science* 261:1594–97

Corbin V, Michelson AM, Abmayr SM, Neel V, Alcamo E, et al. 1991. A role for the *Drosophila* neurogenic genes in mesoderm differentiation. *Cell* 67:311–23

Crossley CA. 1978. The morphology and development of the *Drosophila* muscular system. In *The Genetics and Biology of Drosophila,* ed. M Ashburner, TRF Wright, 2b:499–560. New York: Academic

Currie DA, Truman JW, Burden SJ. 1995. *Drosophila* glutamate receptor RNA expression in embryonic and larval muscle fibers. *Dev. Dyn.* 203:311–16

Doe CQ. 1992. Molecular markers for identified neuroblasts and ganglion mother cells in the *Drosophila* central nervous system. *Development* 116:855–63

Doe CQ, Hiromi Y, Gehring WJ, Goodman CS. 1988. Expression and function of the segmentation gene *fushi tarazu* during *Drosophila* neurogenesis. *Science* 239:170–75

Doe CQ, Smouse D, Goodman CS. 1988. Control of neuronal fate by the *Drosophila* segmentation gene *even-skipped. Nature* 333: 376–78

Doe CQ, Technau GM. 1993. Identification and cell lineage of individual neural precursors in the *Drosophila* CNS. *Trends Neurosci.* 16:510–14

Dohrmann C, Azpiazu N, Frasch M. 1990 A new *Drosophila* homeobox gene is expressed in mesodermal precursor cells of distinct muscle fibers during embryogenesis. *Genes Dev.* 4:2098–111

Dudai Y, Jan YN, Byers D, Quinn WG, Benzer S. 1976. *dunce,* a mutant of *Drosophila* deficient in learning. *Proc. Natl. Acad. Sci. USA* 73:1684–88

Dudai Y, Zvi S. 1985. Multiple defects in the activity of adenylate cyclase from the *Drosophila* memory mutant rutabaga. *J. Neurochem.* 45:355–64

Epstein HF, Bernstein SI. 1992. Genetic approaches to understanding muscle development. *Dev. Biol.* 154:231–44

Falls DL, Harris DA, Johnson FA, Morgan MM, Corfas G, Fischbach GD. 1990. Mr 42,000 ARIA: a protein that may regulate the accumulation of acetylcholine receptors at developing chick neuromuscular junctions. *Cold Spring Harbor Symp. Quant. Biol.* 55: 397–406

Feany MB, Quinn WG. 1995. A neuropeptide gene defined by the *Drosophila* memory mutant *amnesiac. Science* 268:869–73

Fontaine B, Klarsfeld A, Hokfelt T, Changeux JP. 1986. Calcitonin gene-related peptide, a peptide present in spinal cord motoneurons, increases the number of acetylcholine receptors in primary cultures of chick embryo myotubes. *Neurosci. Lett.* 71:59–65

Fontaine B, Sassoon D, Buckingham M, Changeux JP. 1988. Detection of the nicotinic acetylcholine receptor alpha-subunit mRNA by in situ hybridization at neuromuscular junctions of 15-day-old chick striated muscles. *EMBO J.* 7:603–39

Frasch M. 1995. Induction of visceral and cardiac mesoderm by ectodermal Dpp in the early *Drosophila* embryo. *Nature* 374:464–67

Fujita SC, Zipursky SL, Benzer S, Ferrus A, Shotwell SL. 1982. Monoclonal antibodies against the *Drosophila* nervous system. *Proc. Natl. Acad. Sci. USA* 79:7929–33

Gho M. 1994. Voltage-clamp analysis of gap junctions between embryonic muscles of *Drosophila. J. Physiol.* 481:371–83

Giniger E, Jan LY, Jan YN. 1993. Specifying the path of the intersegmental nerve of the *Drosophila* embryo: a role for Delta and Notch. *Development.* 117:431–40

Goodman CS, Doe CQ. 1993. Embryonic development of the *Drosophila* central nervous system. See Bate & Martinez Arias 1993, pp. 1131–206

Gorczyca M, Augart C, Budnik V. 1993. Insulin-like receptor and insulin-like peptide are localized at neuromuscular junctions in *Drosophila. J. Neurosci.* 13:3692–704

Gorczyca MG, Phillis RW, Budnik V. 1994. The role of *tinman,* a mesodermal cell fate gene, in axon pathfinding during the development of the transverse nerve in *Drosophila. Development* 120:2143–52

Grenningloh G, Rehm EJ, Goodman CS. 1991. Genetic analysis of growth cone function in *Drosophila:* fasciclin II functions as a neuronal recognition molecule. *Cell* 67:45–57

Halfon M, Hashimoto C, Keshishian H. 1995. The *Drosophila* Toll gene functions zygotically and is necessary for proper motoneuron and muscle development. *Dev. Biol.* 169: 151–67

Hall ZW, Sanes JR. 1993. Synaptic structure and development: the neuromuscular junction. *Cell* 72:99–121

Halpern ME, Chiba A, Johansen J, Keshishian H. 1991. Growth cone behavior underlying the development of stereotypic synaptic connections in *Drosophila* embryos. *J. Neurosci.* 11:3227–38

Harrelson AL, Goodman CS. 1988. Growth cone guidance in insects: fasciclin II is a member of the immunoglobulin superfamily. *Science* 242:700–8

Harrison SD, Broadie K, van de Goor J, Rubin GM. 1994. Mutations in the *Drosophila* Rop gene suggest a function in general secretion and synaptic transmission. *Neuron* 13:555–66

Ho RK, Ball EE, Goodman CS. 1983. Muscle pioneers: large mesodermal cells that erect a scaffold for developing muscles and motoneurones in grasshopper embryos. *Nature* 301:66–69

Immergluck K, Lawrence PA, Bienz M. 1990. Induction across germ layers in *Drosophila* mediated by a genetic cascade. *Cell* 62:261–68

Ito K, Urban J, Technau GM. 1995. Distribution, classification, and development of *Drosophila* glial cells in the late embryonic and early larval nerve cord. *Roux Arch. Dev. Biol.* 204:284–307

Jacobs JR, Goodman CS. 1989. Embryonic development of axon pathways in the *Drosophila* CNS: II. Behavior of pioneer growth cones. *J. Neurosci.* 9:2412–24

Jan LY, Jan YN. 1976a. Properties of the larval

neuromuscular junction in *Drosophila melanogaster. J. Physiol.* 262:189–214

Jan LY, Jan YN. 1976b. L-Glutamate as an excitatory transmitter at the *Drosophila* larval neuromuscular junction. *J. Physiol.* 262: 215–36

Jarecki J, Keshishian H. 1995. Role of neural activity during synaptogenesis in *Drosophila. J. Neurosci.* 15:8177–90

Jia X-X, Gorczyca M, Budnik V. 1993. Ultrastructure of neuromuscular junctions in *Drosophila:* Comparison of wild type and mutants with increased excitability. *J. Neurobiol.* 24:1025–44

Johansen J, Halpern ME, Johansen KM, Keshishian H. 1989a. Stereotypic morphology of glutamatergic synapses on identified muscle fiber cells of *Drosophila* larvae. *J. Neurosci.* 9:710–25

Johansen J, Halpern ME, Keshishian H. 1989b. Axonal guidance and the development of muscle fiber-specific innervation in *Drosophila* embryos. *J. Neurosci.* 9:4318–32

Kandel E, Abel T. 1995. Neuropeptides, adenylyl cyclase, and memory storage. *Science* 268:825–26

Kania A, Bellen HJ. 1995. Mutations in *neuromusculin,* a gene encoding a cell adhesion molecule, cause nervous system defects. *Roux Arch. Dev. Biol.* 204:259–270

Kania A, Han P-L, Kim Y-T, Bellen H. 1993. neuromusculin, a *Drosophila* gene expressed in peripheral neuronal precursors and muscles, encodes a cell adhesion molecule. *Neuron* 11:673–87

Keshishian H, Chang TN, Jarecki J. 1994. Precision and plasticity during *Drosophila* neuromuscular development. *FASEB J.* 8: 731–37

Keshishian H, Chiba A. 1993. Neuromuscular development in *Drosophila:* insights from single neurons and single genes. *Trends Neurosci.* 16:278–83

Keshishian H, Chiba A, Chang TN, Halfon M, Harkins EW, et al. 1993. The cellular mechanisms governing the development of synaptic connections in *Drosophila melanogaster. J. Neurobiol.* 24:767–87

Kidokoro Y, Nishikawa K. 1994. Miniature endplate currents at the newly formed neuromuscular junction in *Drosophila* embryos and larvae. *Neurosci. Res.* 19:143–54

Klämbt C, Goodman CS. 1991. The diversity and pattern of glia during axon pathway formation in the *Drosophila* embryo. *Glia* 4: 205–13

Kolodkin AL, Matthes DJ, Goodman CS. 1993. The *semaphorin* genes encode a family of transmembrane and secreted growth cone guidance molecules. *Cell* 75:1389–99

Kolodkin AL, Matthes DJ, O'Connor TP, Patel NH, Admon A, et al. 1992. Fasciclin IV: sequence, expression, and function during

growth cone guidance in the grasshopper embryo. *Neuron* 9:831–45

Kurdyak P, Atwood HL, Steward BA, Wu CF. 1994. Differential physiology and morphology of motor axons to ventral longtidunal muscle in larval *Drosophila. J. Comp. Neurol.* 350:463–72

Landmesser L, Dahm L, Tang J, Rutishauser U. 1992. Polysialic acid as a regulator of intramuscular nerve branching during embryonic development. *Neuron* 4:655–67

Lawrence PA, Johnston P. 1986. The muscle patterns of a segment of *Drosophila* may be determined by neurons and not by contributing muscles. *Cell* 45:505–13

Lawrence PA, Johnston P, Vincent JP. 1994. *Wingless* can bring about a mesoderm-to-ectoderm induction in *Drosophila* embryos. *Development* 120:3355–59

Lilly B, Galewsky S, Firulli AB, Schulz RA, Olson EN. 1994. D-MEF2: a MADS box transcription factor expressed in differentiating mesoderm and muscle cell lineages during *Drosophila* embryogenesis. *Proc. Natl. Acad. Sci. USA* 91:5662–66

Lilly B, Zhao B, Ranganayakulu G, Paterson BM, Schulz RA, Olson EN. 1995. Requirement of MADS domain transcription factor D-MEF2 for muscle formation in *Drosophila. Science* 267:688–93

Lin DM, Fetter RD, Kopczynski C, Grenningloh G, Goodman CS. 1994. Genetic analysis of Fasciclin II in *Drosophila:* defasciculation, refasciculation, and altered fasciculation. *Neuron* 13:1055–69

Lin DM, Goodman CS. 1994. Ectopic and increased expression of fasciclin II alters motoneuron growth cone guidance. *Neuron* 13: 673–87

Littleton JT, Stern M, Perin M, Bellen HJ. 1994. Calcium dependence of neurotransmitter release and rate of spontaneous vesicle fusions are altered in *Drosophila synaptotagmin* mutants. *Proc. Natl. Acad. Sci USA* 91: 10888–92

Littleton JT, Stern M, Shulze K, Perin M, Bellen HJ. 1993. Mutational analysis of *Drosophila* synaptotagmin demonstrates its essential role in $Ca^{++}$-activated neurotransmitter release. *Cell* 74:1125–34

Livingstone MS, Sziber PP, Quinn WG. 1984. Loss of calcium/calmodulin responsiveness in adenylate cyclase of *rutabaga,* a *Drosophila* learning mutant. *Cell* 37:205–15

Luo Y, Raible D, Raper JA. 1993. Collapsin: a protein in brain that induces the collapse and paralysis of neuronal growth cones. *Cell* 75: 217–27

Manning G, Krasnow M. 1993. Development of the *Drosophila* tracheal system. See Bate & Martinez Arias 1993, pp. 609–85

Martinez Arias A. 1993. Development and patterning of the larval epidermis of *Droso-*

*phila.* See Bate & Martinez Arias 1993, pp. 517–607

Matthes DJ, Sink H, Kolodkin AL, Goodman CS. 1995. Semaphorin II can function as a selective inhibitor of specific synaptic arborizations. *Cell* 81:631–39

McMahan U. 1990. The agrin hypothesis. *Cold Spring Harbor Symp. Quant. Biol.* 50:407–18

Merlie JP, Sanes JR. 1985. Concentration of acetylcholine receptor mRNA in synaptic regions of adult muscle fibres. *Nature* 317:66–68

Messersmith EK, Leonardo ED, Shatz CJ, Tessier-Lavigne M, Goodman CS, Kolodkin AL. 1995. Semaphorin III can function as a selective chemorepellant to pattern sensory projections in the spinal cord. *Neuron* 14:949–59

Michelson AM, Abmayr SM, Bate M, Martinez Arias A, Maniatis T. 1990. Expression of a MyoD family member prefigures muscle pattern in *Drosophila* embryos. *Genes Develop.* 4:2086–97

Monastirioti M, Gorczyca M, Rapus J, Eckert M, White K, Budnik V. 1995. Octopamine immunoreactivity in the fruit fly *Drosophila melanogaster*. *J. Comp. Neurol.* 356:275–87

New HV, Mudge AW. 1986. Calcitonin generelated peptide regulates muscle acetylcholine receptor synthesis. *Nature* 323:809–11

Nguyen HT, Bodmer R, Abmayr SM, McDermott JC, Spoerel NA. 1994. D-mef2: a *Drosophila* mesoderm-specific MADS box-containing gene with a biphasic expression profile during embryogenesis. *Proc. Natl. Acad. Sci. USA* 91:7520–24

Nose A, Mahajan VB, Goodman CS. 1992. Connectin: a homophilic cell adhesion molecule on a subset of muscles and motoneurons that innervate them in *Drosophila.* *Cell* 70:553–67

Nose A, Takeichi M, Goodman CS. 1994. Ectopic expression of connectin reveals a repulsive function during growth cone guidance and synapse formation. *Neuron* 13:525–39

Nusse R, Varmus HE. 1992. Wnt genes. *Cell* 69:1073–87

Patel NH, Ball EE, Goodman CS. 1992. Changing role of even-skipped during the evolution of insect pattern formation. *Nature* 357:339–42

Paterson BM, Walldorf U, Eldridge J, Dübendorfer A, Frasch M, Gehring, WJ. 1991. The *Drosophila* homologue of vertebrate myogenic-determination genes encodes a transiently expressed nuclear protein marking primary myogenic cells. *Proc. Natl. Acad. Sci. USA* 88:3782–86

Rushton E, Drysdale R, Abmayr SM, Michelson AM, Bate M. 1995. Mutations in a novel gene, *myoblast city*, provide evidence in support of the founder cell hypothesis for *Drosophila* muscle development. *Development* 121:1979–88

Schulze KL, Broadie K, Perin MS, Bellen HJ. 1995. Genetic and electrophysiological studies of *Drosophila* syntaxin-1A demonstrate its role in nonneuronal secretion and neurotransmission. *Cell* 80:311–20

Schulze KL, Littleton JT, Salzberg A, Halachmi N, Stern M, et al. 1994. rop, a *Drosophila* homolog of yeast Sec1 and vertebrate n-Sec1/Munc-18 proteins, is a negative regulator of neurotransmitter release in vivo. *Neuron* 13:1099–108

Seeger M, Tear G, Ferres-Marco D, Goodman CS. 1993. Mutations affecting growth cone guidance in *Drosophila:* genes necessary for guidance toward or away from the midline. *Neuron* 10:409–26

Serafini T, Kennedy TE, Galko MJ, Mirzayan C, Jessell TM, Tessier-Lavigne M. 1994. The netrins define a family of axon outgrowth-promoting proteins homologous to *C. elegans* UNC-6. *Cell* 78:409–24

Sink H, Goodman CS. 1994. Mutations in *sidestep* lead to defects in pathfinding and synaptic specificity during the development of neuromuscular connectivity in *Drosophila.* *Soc. Neurosci. Abstr.* 20:1283.

Sink H, Whitington PM. 1991a. Location and connectivity of abdominal motoneurons in the embryo and larva of *Drosophila melanogaster.* *J. Neurobiol.* 22:298–311

Sink H, Whitington PM. 1991b. Pathfinding in the central nervous system and periphery by identified embryonic *Drosophila* motor axons. *Development* 112:307–16

Sink H, Whitington PM. 1991c. Early ablation of target muscles modulates the arborisation pattern of an identified embryonic *Drosophila* motor axon. *Development* 113:701–7

Snow PM, Bieber AJ, Goodman CS. 989. Fasciclin III: a novel homophilic adhesion molecule in *Drosophila.* *Cell* 59:313–23

Staehling-Hampton K, Hoffmann FM, Baylies MK, Rushton E, Bate M. 1994. dpp induces mesodermal gene expression in *Drosophila.* *Nature* 372:783–86

Sweeney ST, Broadie K, Keane J, Niemann H, O'Kane CJ. 1995. Targeted expression of tetanus toxin light chain in *Drosophila* specifically eliminates synaptic transmission and causes behavioral defects. *Neuron* 14:341–51

Tang J, Landmesser L, Rutishauser U. 1992. Polysialic acid influences specific pathfinding by avian motoneurons. *Neuron* 8:1031–44

Taylor MV, Beatty KE, Hunter HK, Baylies MK. 1995. *Drosophila* MEF2 is regulated by *twist* and is expressed in both the primordia and differentiated cells of the embryonic somatic, visceral and heart musculature. *Mech. Dev.* 50:29–41

Truman JW, Bate M. 1988. Spatial and temporal patterns of neurogenesis in the central nervous system of *Drosophila melanogaster. Dev. Biol.* 125:145–57

Van Vactor D, Sink H, Fambrough D, Tsoo R, Goodman CS. 1993. Genes that control neuromuscular specificity in *Drosophila. Cell* 73:1137–54

Yin JC, Del Vecchio M, Zhou H, Tully T. 1995. CREB as a memory modulator: induced expression of a dCREB2 activator isoform enhances long-term memory in *Drosophila. Cell* 81:107–15

Yin JC, Wallach JS, Del Vecchio M, Wilder EL, Zhou H, et al. 1994. Induction of a dominant negative CREB transgene specifically blocks long-term memory in *Drosophila. Cell* 79:49–58

Younossi-Hartenstein A, Hartenstein V. 1993. The role of the tracheae and musculature during pathfinding of *Drosophila* embryonic sensory axons. *Dev. Biol.* 158:430–47

Zhong Y. 1995. Mediation of PACAP-like neuropeptide transmission by coactivation of Ras/Raf and cAMP signal transduction pathways in *Drosophila. Nature* 375:88–92

Zhong Y, Budnik V, Wu C-F. 1992. Synaptic plasticity in *Drosophila* memory and hyperexcitable mutants: role of cAMP cascade. *J. Neurosci.* 12:644–51

Zhong Y, Pena LA. 1995. A novel synaptic transmission mediated by a PACAP-like neuropeptide in *Drosophila. Neuron* 14:527–36

Zhong Y, Wu. 1991. Altered synaptic plasticity in *Drosophila* memory mutants with a defective cyclic AMP cascade. *Science* 251:198–201

Zinsmaier KE, Eberle KK, Buchner E, Walter N, Benzer S. 1994. Paralysis and early death in cysteine string protein mutants of *Drosophila. Science* 263:977–80

*Annu. Rev. Neurosci. 1996. 19:577–621*

# VISUAL OBJECT RECOGNITION

## *Nikos K. Logothetis and David L. Sheinberg*

Division of Neuroscience, Baylor College of Medicine, One Baylor Plaza, Houston, Texas 77030

KEY WORDS: object representation, perceptual categorization, monkey, electrophysiology, inferotemporal cortex

---

### ABSTRACT

Visual object recognition is of fundamental importance to most animals. The diversity of tasks that any biological recognition system must solve suggests that object recognition is not a single, general purpose process. In this review, we consider evidence from the fields of psychology, neuropsychology, and neurophysiology, all of which supports the idea that there are multiple systems for recognition. Data from normal adults, infants, animals, and brain-damaged patients reveal a major distinction between the classification of objects at a basic category level and the identification of individual objects from a homogeneous object class. An additional distinction between object representations used for visual perception and those used for visually guided movements provides further support for a multiplicity of visual recognition systems. Recent evidence from psychophysical and neurophysiological studies indicates that one system may represent objects by combinations of multiple views, or aspects, and another may represent objects by structural primitives and their spatial interrelationships.

---

## INTRODUCTION

An essential behavior of animals is the visual recognition of objects that are important for their survival. Human activity, for instance, relies heavily on the classification or identification of a large variety of visual objects. We rapidly and effortlessly recognize these objects even when they are encountered in unusual orientations, under different illumination conditions, or partially occluded by other objects in a visually complicated environment.

How is this performance accomplished by the brain? What kind of information does the visual system derive from the retinal image to construct

577

0147-006X/96/0301-0577$08.00

descriptions of sets of object features that capture the invariant properties of objects? How are such descriptions stored, and how are they activated by the viewed object? Are object representations general, or are they specific to an action or to a cognitive process, such as learning, planning, or reasoning?

These questions have historically been addressed by scientists in a variety of disciplines, including cognitive psychology (Pinker 1985, Biederman 1987, Banks & Krajicek 1991), neurobiology (Gross 1973, Gross et al 1993, Miyashita 1993, Rolls 1994), neuropsychology (Humphreys & Riddoch 1987a, 1987b; Damasio et al 1990; Farah 1990; Grüsser & Landis 1991), and computation and engineering (Marr 1982, Ullman 1989, Koenderink 1990, Aloimonos 1993). In this chapter we review selected work from each of the aforementioned fields that, in combination, shed increasing light on the internal workings of this system.

Our aim is to provide evidence that a multipurpose general recognition system does not actually exist. Instead, in the process of biological recognition, multiple representations of an object are formed, each specific to the transformations required by either perception or action. The reviewed literature suggests that the recognition of prototypical members of an object category, the encoding of dynamic and plastic transformations of objects or object parts, the identification of individual members of a homogeneous object class, and the planning of movements habitually made when interacting with familiar objects rely on different representations that are formed in different neural sites or by different interconnectivity patterns.

We start with an overview of the basic capacities and limitations of the primate recognition system. After a brief description of some general principles of object categorization, we discuss the performance of human and nonhuman primates in different recognition tasks and relate this performance to relevant theoretical models. We then survey a number of human neuropsychological and animal lesion studies showing that damage in different regions of the brain often results in a selective disruption of different recognition processes. In the final section, we discuss findings from psychophysical and electrophysiological experiments in the monkey that examine the role of single neurons of cortical areas thought to be essential in the formation of object representations.

## CATEGORIZATION

The world has an infinite number of stimuli that can be discriminated from one another, to an arbitrary degree of detail. Which discriminations are essential for a given recognition system, and what is the basis for organizing information into equivalence classes? In this section, we examine how humans classify and recognize objects, and then we provide evidence that the same principles most likely underlie categorization performed by other biological recognition systems. Specifically, we show that the generalizations about the world that allow us to categorize objects are not the product of the development of language, but are instead of perceptual origin. Perceptual categorizations, in turn, reflect the redundant, correlational structure of the environment and occur most often at the level at which individual members of categories are most similar to each other and maximally different from members of other categories.

## Object Classes and Taxonomies

In human cultures, object categories are usually designated by words that capture the common functional properties of the category's members. Brown (1958) considered the question of why everyday "things" (e.g. pineapples and dimes) are referred to by the same name by most members of a society. He concluded that "[t]he most common name for each of these categorizes them as they need to be categorized for the community's nonlinguistic purposes. The most common name is at the level of usual utility" (Brown 1958, p. 16). The idea that categories may actually reflect more than just linguistic constructs was examined systematically by Rosch and her colleagues (1976a). They showed that human conceptual categories have a perceptual basis and are determined by the high correlational structure of the real world, in which certain combinations of attributes are more probable than others. For instance, attributes such as "feathers" and "wings" co-occur often, while combinations such as "feathers" and "wheels" generally do not. Bundles of such co-occurring attributes form the basis of a natural classification for objects.

Rosch et al (1976a) argued that the world contains "intrinsically separate things," and that there exists a taxonomy for objects within which categories are related to each other by class inclusion. Such categories form natural groupings of stimuli with different perceivable characteristics, or "cues." Cues with high frequency within a given category and low frequency in all other categories are valid category predictors. For example, the cue "long neck" has extremely high validity for the category giraffe because it reliably predicts the presence of a giraffe. The cue "hoofed foot," on the other hand, has low validity for the same category because all ungulate mammals, in addition to giraffes, have hooves. The notion of cue validity extends to categories and is conceived of as the sum of the cue validities of each of the category's features (Reed 1972, Rosch et al 1976a).

General categories, such as mammals, are highly inclusive but have low cue validity, since few perceivable characteristics are shared among their members. Categories such as doberman, on the other hand, are very specific but also have low cue validity because many properties are shared with other categories at the same level of abstraction (e.g. setter, pointer, golden retriever). The most inclusive category within which attributes are common to most category members is what Rosch and colleagues called the basic-level category, e.g. dog, and it is the category that has the maximum cue validity. Classifications more general than the basic level are called superordinate categories, while those that are more specific are called subordinate categories (Rosch et al 1976a).

When human subjects are asked to list as many attributes that apply to certain objects as they can, they report the greatest increase in the number of characteristic features when describing objects at the basic level (Rosch et al 1976a, Tversky & Hemenway 1984). Objects in the same basic-level category are also manipulated using common motor sequences and share considerable shape similarity with each other but not with objects of most other groups. The similarity of basic objects is such that shape-based averaging of two members of the category will often yield a new object that can also be recognized as a category member. In fact, in the case of a highly homogeneous group of objects like faces, photographic averages from two separate 20-person groups, selected according to gender and age, have been shown to yield two "average" faces remarkably similar to each other (Katz 1953).

Empirically, recognition of objects at the basic level often occurs more rapidly and more accurately than the recognition of objects at any other taxonomic level. Exceptions to this rule are atypical exemplars of basic categories that have pronounced shape differences from the prototype, or central tendency, of the class. For example, humans usually identify "penguins" or "racing cars" faster as such than as "birds" or "cars," which are the basic-level classifications for these objects (Jolicoeur et al 1984, Murphy & Brownell 1985). To describe the level at which specific objects are first accessed irrespective of inclusiveness or cue validity, Jolicoeur et al (1984) coined the term "entry point" of recognition.

Interestingly, the entry point of individual objects, which usually coincides with the basic level of classification, can shift to the subordinate level when perceivers become especially sensitive to subtle differences between objects of the same class. For example, Rosch et al (1976a) noted major differences in the descriptions of object attributes between experts in a field and unspecialized subjects. An airplane mechanic, for instance, when asked to list airplane attributes, spontaneously reported a large number of attributes of airplanes that are potentially available to the casual observer but that are usually ignored. In a systematic study of this phenomenon, Tanaka & Taylor (1991) showed that for experts in a field (such as bird watching), subordinate categories become as differentiated as basic-level categories. This entry point change is also evident in naming latencies, which become as short as those of the basic classifications as expertise increases.

In summary, humans systematically categorize objects in the world based on natural groupings of attributes. Do, however, such categorization principles also apply in situations where the observer has no prior conceptual information about the objects to be classified? Most importantly, do they apply for the nonverbal observer? If natural categories do develop independently of preexisting conceptual or linguistic labels, then the same principles of categorization may underlie the recognition skills of other animals, in which the neural representations of objects can be studied directly using neurophysiological techniques. At least three lines of evidence, discussed briefly below, suggest that categorization may indeed rely on principles applying to any recognition system.

## Perceptual Categorization

LEARNING NOVEL STIMULI   Evidence suggesting some universal principles in the formation of categories comes from recognition experiments with visually novel objects that are unrelated to any previously experienced verbal codes or abstract concepts. Such experiments show that in the process of learning basic objects, humans can detect consistent features of minimum interindividual variability, e.g. features of high cue validity, that characterize most exemplars of an object class, thereby extracting class invariances. In their seminal work, Posner & Keele (1968) probed the representations stored by humans when learning to classify patterns with individual variance around a common abstract structure. They used dot patterns (Figure 1a) as prototypes, and created individual category instances (exemplars) at specified distances, or deviations, from the original pattern by applying statistical distortion rules (Posner et al 1967). Their subjects were taught to classify distorted patterns constructed from three different prototypes, and they were subsequently tested in a recognition task in which they were exposed to the previously viewed patterns, the prototype pattern, and new, distorted patterns. Interestingly, subjects recognized the prototype pattern almost as quickly and as accurately as they recognized previously memorized patterns, even though they had never been directly exposed to it, suggesting that the prototype is a main constituent in the category's memorial representation (see also Franks & Bransford 1971, Strange et al 1970). Moreover, information about the central tendency of such sets of exemplars was found to be extracted and stored during learning and not during the process of recognition (Posner & Keele 1970, Homa et al 1973, Strange et al 1970).

Categorizations are not based upon a recognition threshold that, once exceeded, definitively endows a stimulus with class membership. In other words, class boundaries are not formed by sharp transition hypersurfaces in a multidimensional feature space. Instead, a familiarity continuum exists, according to which the probability of a correct classification depends on the structural typicality of the stimulus, determined by the closeness to the class prototype

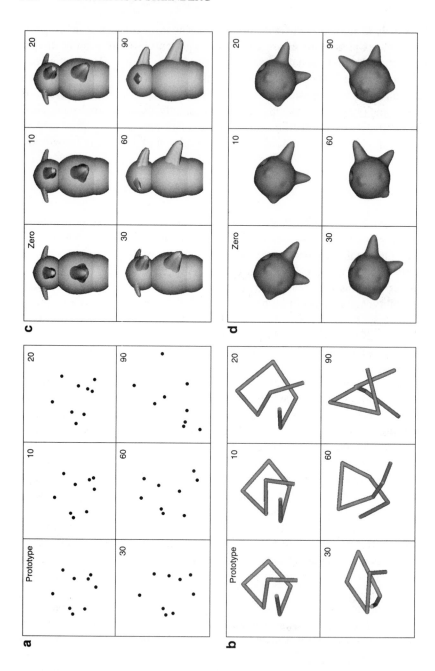

(Rosch et al 1976b). Atypical exemplars greatly differing from the prototype are recognized as individual entities rather than as class members, i.e. they themselves become the entry point of recognition.

What follows demonstrates that the principles emerging from experiments with novel objects appear to underlie the complex and sophisticated processes that infants possess for categorizing and representing their experiences with the environment.

CATEGORIZATION IN INFANTS    Piaget (1969), after carefully observing the development of perception and cognition in children, concluded that these capacities are rooted in prelinguistic constructs that are only later enriched through the use of language. A large number of studies using behavioral paradigms developed for research in preverbal observers—such as the preferential looking paradigm, a technique that capitalizes on infants' innate preference for novel stimuli (Fantz 1964)—have shown that infants as young as three or four months old can form categorical representations based on visual and auditory stimuli (for review, see Quinn & Eimas 1986).

Infants, for instance, can form categorical representations for animals from different basic-level categories that are sufficiently distinct (Quinn et al 1993, Eimas & Quinn 1994). Specifically, they categorized various horses as different from cats, zebras, and giraffes, and perceived cats as different from tigers and horses, but not female lions. The inclusion of female lions in the category of cats was found to disappear, however, by 6 to 7 months of age (Eimas & Quinn 1994), as more subordinate-level recognition skills develop with increasing demands for finer-level discriminations. On the other hand, it is remarkable that some types of subordinate recognition, such as the identification of familiar faces, appear to begin extremely early in life, as neonates can visually discriminate between their mother's face and the face of a stranger (Bushnell et al 1989). In contrast, superordinate-level classifications of object

---

*Figure 1*    Example stimuli used in object recognition experiments. (*a*) Random dot patterns formed by distorting a prototype (*upper left*) by increasing levels of dot-position perturbation. Each pattern consists of nine dots placed randomly in a 512 × 512 unit area. The number in the top right corner indicates the average distance each dot was displaced from its position in the prototype. Stimuli of this sort were first used by Posner et al (1967) to investigate how humans form abstract visual categories. [Figure adapted from Knapp & Anderson (1984).] (*b*) Prototype distortions of a three-dimensional (3D) wireframe object, similar to those used by Edelman & Bülthoff (1992). Distortions of the prototype were created by randomly displacing each of the vertices by a percentage of the original segment length. (*c*) Example of a "greeble" object (Gauthier 1995) used to study mechanisms underlying recognition performance of experts. The zero view of the object (*upper left*) is shown rotated around the vertical axis in five different poses. Degrees of rotation are indicated by the number in the upper right-hand corner. (*d*) Six views of a spheroidal object shown rotated in the image plane and similar to those used by Edelman & Bülthoff (1992) and Logothetis et al (1994).

pictures appear later and improve with age, usually in close relation to linguistic developments (Rosch et al 1976a).

Most interestingly, the processes involved in forming perceptual categories appear to be very similar to those involved in categorization in adults. By combining the preferential looking technique with the random dot stimuli introduced by Posner & Keele (1968), Bomba & Siqueland (1983) investigated the processes underlying the ability of infants to abstract a prototype from sets of novel stimuli.

The infants were first familiarized with distortions of one of three dot pattern prototypes: a square, a triangle, or a diamond. Once habituated to the exemplars, they were presented with the prototype of the learned exemplars paired with the prototype of one of the other categories. When the number of exemplars used was small (six stimuli) and the recognition test immediately followed the familiarization period, infants learned the individual examples but failed to extract a class representative prototype. In contrast, infants reliably associated the familiarized exemplars with their prototype when the number of examples was increased (12 stimuli) or when a delay of 3 min was introduced between training and testing (Posner & Keele 1970, Homa & Vosburgh 1976). In other words, infants, like human adults, tend to abstract the central tendency of a category when exposed to a sufficient number of exemplars, but they learn individual entities when presented with only a few exemplars. The reported "delay effects," whereby individual exemplars are remembered if testing immediately follows familiarization but the prototype is extracted when testing occurs after a delay, have also been observed in adult subjects and have been taken as evidence for different memory decay times for specific exemplars and for the category prototype.

Taken together these studies suggest that structure in the environment is more critical for categorization than are the linguistic labels assigned to stimulus classes. Although the transition between perceptual and conceptual categorization is by no means a settled issue, categorical representations of infants below 15 months are predominantly perceptual, and conceptual representations begin to slowly emerge only later in infancy (Eimas & Quinn 1994). In Lorenz's (1971) words, "a young child which is already capable of referring to all dogs as 'bow-wow' and all cats as 'miaow-miaow' has quite definitely not abstracted the zoological identification formula for *Canis familiaris* and *Felis ocreata*" (p. 306).

CATEGORIZATION IN ANIMALS   Finally, the generality of classification rules is perhaps best demonstrated in experiments examining concept formation and categorization performance in animals other than humans. Nonhuman primates clearly are capable of making various categorizations at different abstraction levels, of associating meaning or purpose to different objects, and of possessing natural concepts (e.g. see Lorenz 1971, Davis 1974). Monkeys, for instance, can learn to perform various types of complex classification tasks in the laboratory (e.g. see Davis 1974). They are obviously capable of making basic-level categorizations, but they can also easily learn to discriminate individual human or monkey faces (Rosenfeld & Van Hoesen 1979, CJ Bruce 1982) and novel artificial object classes, even generalizing learning across basic image transformations (Logothetis et al 1994, 1995).

Category formation is not exclusive to primates. For instance, the ability to discriminate between basic classes has been demonstrated in the goldfish (Bowman & Sutherland 1970) and in many different bird species, which have been shown to recognize even impoverished stimuli (Watanabe et al 1993). Herrnstein & Loveland (1964) showed that pigeons can easily learn to peck a key in the presence of a color slide containing people and to withhold pecks for slides not containing people. Similarly, Herrnstein et al (1976) found that pigeons can reliably classify novel photographs of either trees, water, or a particular woman.

Cerella (1979) used a similar procedure and found that pigeons could also learn to classify novel silhouettes of oak leaf patterns from other leaves, although he had great difficulty training the pigeons to respond selectively to a single, specific oak leaf. Based on this inability, he concluded that "the pigeon is most strongly disposed to code class (i.e. generic) descriptions of visual input. This tendency can be countered only to a limited extent, to achieve stimulus-specific descriptions" (Cerella 1979, p. 75).

However, certain subordinate-level discriminations are commonly performed by many animals. For instance, the development of personal recognition of specific individuals is essential in the closed societies of birds and mammals, as it allows recognition of nonmembers and of the internal rank order prevailing between group members. Ryan (1982) has shown that chickens can discriminate slides of one bird in a variety of poses from slides of other birds, and they can transfer this discrimination to novel sets of slides. A notable fact is that subordinate recognition in some species depends on the relevance of the objects to the animal. For example, although birds can recognize scrambled parts of the Charlie Brown cartoon figure as Charlie Brown, they fail to recognize a pigeon's head as such if it is not presented as a full, unscrambled face (Watanabe & Ito 1991).

## Task-Specific Representations

The studies reviewed above strongly suggest that categorizations made by humans and other animals have a general, perceptual basis, reflecting the structure of the world. Classifying objects at the basic level is a fundamental recognition task, and it is likely to be the only task that simple recognition systems perform. Nonetheless, as mentioned above, in primates and other mammals, subordinate-level recognition is also essential in various social and cognitive tasks. Do categorizations at different abstraction levels rely on the same type of stored representations? And, are similar representations used when the perceived object elicits a visually guided motor action?

REPRESENTING PROTOTYPES VS REPRESENTING EXEMPLARS    An interesting observation, pertinent to this question, was made by Homa and his colleagues, who investigated the abstraction of prototypical or exemplar information in categories having either uniform or mixed (low, medium, and high) distortion levels (Homa & Vosburgh 1976). They found that the breadth of a given category, in terms of mixing groups of patterns with different degrees of distortion, has a profound effect on both prototype abstraction and classification performance. In their experiments, recognition was little affected by retention delays (up to 10 weeks) as long as an adequate number of training exemplars were provided. However, when only a few training exemplars were learned, transfer to new stimuli was better if the original training set was not mixed but rather contained uniformly low distortion exemplars (Homa & Chambliss 1975, Homa & Vosburgh 1976). Importantly, training with a few, high distortion exemplars resulted in a form of generalization in which subjects appeared to store information about the individual exemplars and delay the extraction of any prototype information until the onset of the recognition testing. Furthermore, by systematically varying the similarity of old and new exemplars, Homa et al (1981) found that although in general the accuracy of classification of new instances depends on their similarity to old exemplars, the effect of old-new similarity is much greater for small categories (5 items) than for large ones (20 items), implying that for large categories, individual members are not specifically encoded.

At least two models can account for the effects of category breadth on categorization performance. One hypothesizes that all forms of categorization depend on the workings of a single system, based on distributed memory storage (e.g. Anderson et al 1977, Knapp & Anderson 1984, McClelland & Rumelhart 1985). In such a model, when the number of exemplars of the stimulus pattern is small, the new patterns are classified according to their similarity to the learned patterns, but as the size of the training set increases, accuracy depends on the new pattern's similarity to the category's prototype. Alternatively, category breadth–dependent performance can be explained by assuming the existence of two representation systems, the contribution of each of which depends on the level of classification. Support for the latter comes from a recent experiment by Marsolek (1995) that suggests that at least two separate visual form systems may exist in humans: (a) one that is used to classify different instances of an object as belonging to the same abstract category and that involves the left hemisphere and (b) another involving the right hemisphere that appears to preserve visual details of objects in order to distinguish specific exemplars of particular object classes (Marsolek 1995).

ACTION-RELATED REPRESENTATIONS    Recognition of a stimulus is often signaled by the ability of the subject to respond appropriately to that stimulus. In many instances such responses involve visually guided reaching and grasping. In humans and old-world monkeys, prehensile movements almost always require visually acquired shape information. Recent experiments with human subjects show that such action-relevant representations may differ from those used when performing various categorization tasks. For example, when normal human subjects reach for an object, they move their fingers into a certain spatial configuration appropriate for grasping the object. If, however, the perceived size of an object is different from its actual size, as may occur with some form illusions, then a dissociation is observed between the perceived size and the size of the object represented in the systems mediating the grasping behavior (Vishton & Cutting 1995). For instance, when subjects are asked to give verbal estimates of the size of a small horizontal line intercepted by a vertical line of the same length, and then close their eyes and reach for it, their verbal estimate reveals the expected biases from veridicality—in this case overestimating the length of the vertical segment—but their thumb-to-finger distance during grasping shows no such bias, providing reliable estimates for both the vertical and the horizontal lines.

Together, these findings indicate that multiple recognition systems may be employed during the categorization of stimuli at different levels of abstraction and during visuomotor activities. The following sections provide further evidence supporting this point by surveying studies on object constancy in the primate.

# RECOGNITION PERFORMANCE IN HUMANS AND MONKEYS

Introspection indicates that the recognition of familiar objects is largely insensitive to changes in their retinal image. Nonetheless, careful examination shows that image transformations, even simple scaling and translation, can sometimes affect recognition. Moreover, invariance to some transformations, such as rotations in depth, appears to depend strongly on familiarity, as well as on the nature of the object and task.

## Effects of Size and Position

The effects of scale changes on recognition have been examined in experiments in which subjects classify shapes as being the same or different, disregarding changes in size. In these tasks, recognition performance, typically assessed by measuring response latencies, varies as a function of the size ratio between the two stimuli, with increasing size discrepancy between the two shapes resulting in elevated reaction times (Jolicoeur 1987, Ellis et al 1989). Studies in which the viewing distance and the size of novel objects were manipulated have found that the perceived, and not the retinal, size of the objects determines the size ratio effects (Milliken & Jolicoeur 1992).

The partial dependency of recognition on object size has led to the belief that shapes are stored at a particular size and that their sensory representation has first to be scaled before recognition occurs (Ullman 1989). As both Jolicoeur (1987) and Biederman & Cooper (1992) have suggested, however, the effects of size obtained in these experiments may reflect the processes of memory-based comparisons rather than the perceptual representation of the objects. To isolate the perceptual effects from those of episodic discriminations, Biederman & Cooper (1992) used a picture-priming task in which objects viewed on one occasion are more quickly and accurately perceived when presented on a second occasion. In such priming experiments, reaction times were found to be independent of whether the primed object was presented at the same or a different size from when originally viewed (Biederman & Cooper 1992). In contrast, explicit memory tasks using the same stimuli showed clear size effects on recognition, suggesting that differential results obtained from priming and episodic memory experiments might reflect the differential functioning of two representation systems: one underlying the description of an object's shape, and the other its metric attributes, such as its size, orientation, or position (Biederman & Cooper 1992, see also Cooper et al 1992).

Somewhat less pronounced are the effects of stimulus position on recognition performance. Response latencies in visual priming tasks are affected very little by stimulus translation (Biederman & Cooper 1991). However, translational disparity between study and test has been shown to reduce recognition accuracy in a successive presentation same-different task (Foster & Kahn 1985) and in a memory task in which subjects were trained to recognize small novel dot stimuli and thin lines presented at only a single retinal location during the learning phase, and then tested with the same stimuli translated to two new positions (Nazir & O'Regan 1990). Although recognition accuracy initially decreased at the new positions, criterion performance was restored after only a few presentations. Interestingly, no effects of translation were found when the stimuli were either very simple or very complex patterns, both of which usually contain salient diagnostic features that are themselves translation invariant (O'Regan 1992), a finding that again suggests different recognition strategies for different tasks or object classes.

## *Effects of Rotation in the Picture Plane*

Studies specifically directed at assessing the effects of image-plane rotation on recognition of familiar shapes, such as letters and digits (Corballis et al 1978, Simion et al 1982), or of shapes with pronounced diagnostic features, such as line drawings of natural objects (Eley 1982, Jolicoeur 1985), have found relatively small costs—in terms of error rates or reaction times—associated with the misorientation of the stimuli. Moreover, after practice, even these small costs were found to generally disappear (Shinar & Owen 1973, Jolicoeur 1985, McMullen & Jolicoeur 1992). However, when the familiarity of test objects was more closely controlled by presenting subjects with novel, letter-like shapes possessing no diagnostic features, Tarr & Pinker (1989) found that responses to stimuli rotated away from the training view were slower and less accurate. Continued practice reduced the effects of rotation for the newly familiarized views, but this practice did not transfer to "surprise" views presented later. In contrast, transfer of practice to never-experienced views was found to occur for line drawings of everyday objects, which presumably possess rotation invariant, diagnostic features (Murray et al 1993).

Image plane rotations have been also studied in the monkey by using novel objects in exemplar identification tasks (Logothetis et al 1995). In the early phases of testing, the monkeys exhibited orientation dependency in their recognition performance. However, over time, their ability to generalize across rotations in the picture plane improved, even in the absence of feedback. Initial view-dependent performance often progressed rapidly, over the course of a few test sessions, to view-invariant performance.

Of particular interest are the effects of inversion on object recognition. In humans, inversion strongly affects the processing of distinct classes of objects, most notably faces (Valentine 1988). At present, whether faces, per se, are special or simply represent the most common class of objects that must be identified based on subtle shape differences is controversial (see below). Nevertheless, evidence showing an inversion effect for other overlearned but highly similar stimuli, such as different dog breeds (Diamond & Carey 1986) and artificial stimuli designed to mimic animate objects (Gauthier 1995; also see Figure 1c), indicates that recognition of individual members of homogeneous object classes relies predominately on the processing of configurational information, and very little, if at all, on the discrimination of features. This configuration-based recognition, on the other hand, reveals exactly the same inversion effects observed when recognizing faces. The importance of configurational processing for intra-class recognition has also been shown by asking subjects to identify the individual halves of composite face stimuli. Under these conditions, two unmatched halves interfere with the recognition of the upright but not the inverted composite face stimuli (Young et al 1987).

Interestingly, in monkeys, who commonly encounter faces from many viewpoints (not just the upright), the effects of inversion are not found (CJ Bruce 1982), highlighting the importance of experience in the development of configurational processing. Similarly, young children, who may not yet have developed appropriate sensitivity for holistic stimuli, are also less affected by face inversion than adults (Carey & Diamond 1977) and are likely to rely on individual facial features that can be recognized from any viewpoint.

## Effects of Rotations in Depth

One of the most active areas of recognition research in the past decade has concentrated on the effects of rotations in depth on the recognition of 3D objects. Experiments demonstrating viewpoint-invariant recognition have been interpreted as evidence that the visual system employs an object-centered reference frame for representing stimuli. Conversely, experiments demonstrating viewpoint-dependent recognition performance have been taken as evidence for viewer-centered representations. We propose here that the visual system actually uses both types of representation, depending on the classification task and the subject's familiarity with the test objects (see Ullman 1989, Tarr 1995).

DEPTH ROTATION OF COMMON OBJECTS    Similar to the effects of position and size changes, the effects of viewpoint changes on recognition are subtle for common familiar objects. Bartram (1974), using a visual priming paradigm, found that naming objects in pictures was facilitated most by a previous presentation of the identical view of the object, compared to presentations of either a different view of the same object or a view of a different object with the same name. In a sequential same-different matching task using photographs of objects with high- and low-frequency names, Bartram (1976) found little effect of viewpoint change for the pictures with high-frequency names, but significant effects on judgments of rotated low-frequency named objects and for different exemplars with the same name. These results suggest that frequently encountered everyday objects can be accessed equally well from multiple views, while the activation of less common object representations is subject to greater viewpoint dependence, perhaps because such objects are coded by storing specific object views (see below). Furthermore, the priming effects are at least partly visual, because the recognition of objects with the same name as the prime, but that are visually dissimilar, are not equally speeded. Biederman & Gerhardstein (1993) directly studied the effects of rotation in depth on recognition by using a naming task in which subjects identified line drawings of rapidly presented familiar objects, some of which had been seen during a prior priming block. During the test block, the naming latencies for objects that had appeared during the priming block were speeded, but the magnitude of the priming effect varied only slightly with changes in orientation, indicating again that at least under some circumstances, recognition of everyday objects seems invariant to rotations in depth.

Nevertheless, although recognition of highly familiar objects seems to be viewpoint invariant, it is still "view biased" in the sense that human subjects consistently label one or two views of common objects as subjectively better than all other views (Palmer et al 1981). Naming of objects occurs fastest when the stimulus is shown from such a better view, designated a canonical view, with response times increasing monotonically with increasing angular disparity between a test view and the canonical view.

In a recent experiment, Srinivas (1993) selected 42 common objects that were found to have both usual and unusual views for use in a name priming task. She found that recognition of all objects improved if they had been previously presented but also that the cost of switching viewpoints was greater for tests of unusual views. In other words, seeing the unusual view during study provided the same benefits for recognizing the usual view during test as did a previous encounter with the usual view. Seeing a usual view during study, however, did not facilitate recognition of unusual views as much as did the familiarization with the visually identical unusual view. Thus, while exposure to a usual view of an object may prime only nearby views of the same object, unusual views may activate both similar unusual views and automatically prime the object's canonical view (see also Warrington & Taylor 1973).

The largely viewpoint-invariant recognition of familiar objects can be explained by a number of different theoretical models (Ullman 1989). For example, recognition has been proposed to occur by detecting properties of objects that are invariant to all image transformations, such as the ratio between an object's apparent area and its volume, a compactness measure, or certain parametric shape descriptions, such as Fourier descriptors or object moments.

Alternatives to the invariant-properties approach include theories relying on the decomposition of objects into natural parts. A horse, for instance, can be thought of as a set of components, such as the torso, the legs, the head, and the tail, each of which can be recognized on its own. One can, therefore, assume that the process of recognition of a complex object can be reduced to the recognition of its parts and their relationships (Palmer 1977, Marr & Nishihara 1978). Recursive decomposition may lead to simple volumetric primitives, the combination of which can represent any complex object. Such descriptions obviously do not constitute a true theory of recognition, rather they simply provide one adequate type of representation with which a recognition theory could be constructed. The human body, for instance, can be described as a collection of points of different brightnesses, as a collection of lines and curves, as a group of planes, or as a set of 3D block structures.

Marr & Nishihara (1978) proposed a representational scheme for objects, using axis-based structural descriptions that can be decomposed into sets of generalized cones. The class includes simple shapes such as a pyramid or a sphere, as well as natural forms such as arms and legs. In a similar vein, Pentland (1986) suggested that most complex natural objects may be described by combinations of superquadric components, such as spheres, wedges, etc, which might be the basic components that the recognition system recovers while analyzing images.

A recent example of a structural description theory is the recognition by components (RBC) theory (Biederman 1987). The underlying assumption of this theory is that an object can be decomposed into volumetric parts called geons. These parts have simple spatial relationships to each other that remain invariant for all object views, and recognition involves the indexing of these parts and the detection of their structural relationships. A solution to the formidable inverse-optics problem for complex objects is reduced, in this approach, to an inverse-optics problem for simple volumes and their two-dimensional (2D) arrangements.

Biederman and colleagues suggest that recognition of most common objects can be accomplished by indexing structural descriptions based on geons, provided that the following three principal conditions are met: (a) Objects are decomposable, (b) they have different part descriptions, and (c) different viewpoints lead to the same configuration of geons. Such conditions are indeed often met when recognizing entry point objects. These conditions are clearly not met, however, when recognizing objects at the subordinate level, that is, when specific exemplars must be identified or when the objects to be recognized cannot be meaningfully decomposed into simpler parts. In these cases, recognition is often view dependent, and object constancy can only be achieved through perceptual learning.

Finally, viewpoint-invariant performance can also be explained by a viewer-centered recognition system that stores a limited number of object views or aspects and is capable of combining information from these views to recognize any view of the object (Seibert & Waxman 1990, Poggio 1990).

DEPTH ROTATION OF NOVEL OBJECTS    The study of recognition of unfamiliar, novel stimuli has provided important insights into the representation of visual objects. Familiar objects can often confound the issues of object constancy, since a recognition system based on 3D descriptions cannot easily be distinguished from a view-based system exposed to a sufficient number of object views.

The first demonstration of strong viewpoint dependence in the recognition of novel objects was that of Rock and his collaborators (Rock et al 1981, Rock & DiVita 1987). These investigators examined the ability of human subjects to recognize 3D, smoothly curved wire objects seen from one viewpoint, when encountered from a different attitude and thus having a different 2D projection on the retina. Although their stimuli were real objects (made from 2.5-mm wire) and provided the subject with full 3D information, there was a sharp drop in recognition for view disparities larger than approximately 30°. In fact, as subsequent investigations showed, subjects had difficulty even imagining how wire objects would appear when rotated, even when instructed to visualize the object from another viewpoint (Rock et al 1989).

A number of recent experiments have further studied subordinate-level recognition by using computer-rendered novel 3D stimuli, including wire or spheroidal objects (Bülthoff & Edelman 1992, Edelman & Bülthoff 1992), cube-composed stick figures (Tarr 1995), and novel clay shapes (Humphrey & Khan 1992). With these stimuli, recognition was again found to be strongly view dependent, and generalization could only be accomplished by familiarizing the subject with multiple views of target objects.

Such results may present an extreme case in terms of performance. Farah et al (1994) observed that when Rock's wire-forms were interpolated with a smooth clay surface (creating "potato chip" objects), subjects' recognition accuracy was less affected by the same changes in viewpoint tested by Rock. They concluded that object shape and structure plays a significant role in the ability of humans to compensate for variations in viewpoint. One possibility is that as the structure of objects becomes more regular (in terms of properties such as spatial relations and symmetries), the ability to compensate efficiently for changes in viewpoint is enhanced because the resultant image structure is more predictable (Vetter et al 1994, Liu et al 1995). Under these conditions, recognition may be faster than it is for less regular objects, although it is possible that mixed strategies or verification procedures will yield response times that are still dependent on viewpoint, even for familiar objects (Palmer et al 1981) and faces (V Bruce 1982).

The effects of rotation in depth on recognition were recently studied systematically in monkeys by using computer-rendered objects in both basic- and subordinate-level tasks (Logothetis et al 1994). In these experiments, the animals easily learned to generalize recognition to novel views of objects, such as those illustrated in Figures 1b and 1d, when the set of distractors included objects from other basic-level categories. Their ability to identify individual exemplars of either the wire or the spheroidal objects, however, was found to depend strongly on the viewpoint from which the object was encountered during the training phases. The monkeys were unable to recognize objects rotated more than approximately 40° from a familiar view. However, when two views of the target were presented in the training phase, 75 to 120° apart, the animals interpolated between them, often reaching perfect levels of performance for any novel view between the two trained views, as has been shown in human experiments (Bülthoff & Edelman 1992). For all the monkeys tested, training with a limited number of views (about 10 views for the entire viewing sphere) was usually sufficient to achieve view-independent performance. This ability of humans and monkeys to interpolate between different familiar views of a novel object is consistent with a recent viewer-centered theory of recognition based on regularization networks. In short this theory assumes the existence of units acting as blurred templates, the receptive field of which develops selectivity to the views seen in the training phase. Generalization occurs by linearly combining the output of such units (for details, see review by Poggio 1990).

## NEUROPSYCHOLOGICAL STUDIES

Behavioral studies of humans and animals with brain damage provide important insights about the organization of neural modules that participate in visual object recognition. While a detailed discussion of this literature is well beyond the scope of this review (see Farah 1990, Grüsser & Landis 1991), we present evidence here that dissociable clinical deficits in the visual processing of objects, for purposes of categorization, identification, and goal-directed action, strongly suggest a multiple systems architecture for recognition.

### Category-Specific Breakdowns

An essential subordinate classification task for both humans and nonhuman primates is the recognition of facial identity. Ischaemic infarcts in the inferomedial occipito-temporal region of the right hemisphere have been shown to disrupt the recognition of familiar faces (Landis et al 1988). The lesions typically involve the fusiform and lingual gyri or their interconnections and are caused by strokes of the right posterior cerebral artery that typically extends from the level of the splenium of the corpus callosum to the occipital pole (but see Damasio et al 1982).

Prosopagnosia, as this disorder was called by Bodamer (1947), was traditionally considered a specific agnosia, which renders human patients incapable of recognizing the faces of familiar or famous persons but spares their ability to recognize common objects. Prosopagnosic patients can recognize individuals by their voice or even by watching their gait but fail to do so by looking at their face (Damasio et al 1982). They can also recognize a face as the object "face" and name and point to its parts (Lhermitte et al 1972, Whiteley & Warrington 1977, Damasio et al 1982). Furthermore, configurational complexity does not appear to be the cause of this disorder, since the patients do recognize visual objects that may be structurally more complex than faces (Benton 1980, Damasio 1985).

The notion that prosopagnosia is the disruption of the function of a recognition system specialized for the processing of faces was questioned, however, by Faust (1955), who suggested, instead, that the major deficit of prosopagnosic patients is their inability to evaluate the structural signs in any figure that confers individuality. In support of this claim were later findings showing that prosopagnosics indeed have problems distinguishing individuals within other objects classes, such as fruits, playing cards, housefronts, and automobiles (De Renzi et al 1968, Macrae & Trolle 1956, Lhermitte & Pillon 1975). Bernstein et al (1969) reported a bird-watcher who had lost the ability to differentiate visually between birds and another patient who could no longer recognize his cows.

An appealing explanation of the various deficits observed in prosopagnosic patients was offered by Damasio and his collaborators (Damasio et al 1982). After systematically examining three such patients, these investigators noticed that, as previously described, the subjects were unable to identify various objects, including automobiles, articles of clothing, cooking utensils, and food ingredients. Recognition tests using a carefully chosen stimulus set—consisting of photographs of animals, abstract symbols, and motor vehicles—revealed that failure to identify any of these objects was caused primarily by visuo-structural similarity and was not due to complexity. The patients were able to recognize animals such as horses, owls, and elephants, but they confused simpler abstract objects such as a dollar sign, a British pound sign, and a musical clef. Like the infants described above, they also failed to discriminate animals such as cats, tigers, or panthers, which, despite their great size differences, do share many similarities in shape. Damasio et al (1982) suggested that focal brain damage, such as that observed in prosopagnosics, interferes with the patient's ability to perform within-category discriminations, or identifications, without affecting the recognition of the generic class to which the stimulus belongs (see also Gaffan & Heywood 1993).

The hypothesis that the focal brain damage causing agnosia of faces may be interfering with specific, subordinate-level recognition processes and that the preserved representations may be those general representations that are the easiest to access also receives support from the finding that a patient with car agnosia could still identify special-purpose vehicles, such as an ambulance or a fire engine (Lhermitte et al 1972). Given their very atypical appearance, such objects are likely to have their own entry point attributes.

Evidence against this hypothesis, on the other hand, comes from those clinical reports showing that prosopagnosia does occasionally occur without any other subordinate recognition deficits (De Renzi 1986). Moreover, Farah et al (1991) reported that even after factors of visual complexity, such as inter-item similarity or specificity of identification, were accounted for in the analysis of visual recognition performance of two visual agnosic patients, the recognition of "living things" was still disproportionately impaired compared to the recognition of "nonliving things." Finally, a recent study found that the well-known face-inversion effect found in normal subjects (see above) was reversed in a prosopagnosic patient, who could more accurately match inverted faces than upright ones (Farah et al 1995). Normal subjects in this task are especially proficient at matching upright faces and, unlike the prosopagnosic patient, perform much worse when the face stimuli are inverted. These clinical data support the view that specialized neural modules for recognition may coexist and may thus be selectively impaired.

Whether or not prosopagnosia is the result of failing to detect individuality, the aforementioned investigations show that at least two separate representation systems may be involved in the recognition of visual objects: one that represents prototypical objects and a second one that is employed when subordinate classifications are required. The latter system is often selectively affected by the infero-medial occipito-temporal lesions that also disturb the recognition of faces. It remains to be seen whether the capacity to perform subordinate-level recognition tasks can be selectively impaired in the absence of any deficits in face recognition, which would suggest that there are actually at least three different systems underlying recognition: one for basic-level classifications, one for subordinate-level identification, and one for recognizing the class of animate objects.

## Deficits in the Recognition of Facial Expressions

Another dissociation in the recognition of objects is that observed between the identification of faces and the recognition of facial expressions—an especially challenging task for a recognition system. Facial expressions, such as smiles and frowns, are nonrigid, stretching, or bulging nonaffine transforms that have to be "discarded" when recognizing the same face under different emotional conditions. At the same time, however, the dynamic configuration of a face is endowed with a number of different meanings that are essential for social interaction in many species. Recent evidence shows this type of facial expression analysis proceeds independently of face identification. Specifically, some prosopagnosics fail to recognize the identity of a face or another unique, individual object of a category, although they retain their ability to recognize facial expressions (Tranel et al 1988).

Recently, Damasio's group (Adolphs et al 1994) reported a case of a 30-year-old woman with a confined amygdala lesion. She suffered from Urbach-Wiethe disease, which led to almost complete bilateral destruction of the amygdala, while sparing her hippocampus and all neocortical structures. The patient was tested in rating facial expressions—such as happiness, surprise, fear, anger, disgust, sadness, as well as neutral faces—and was found to be much poorer at this task than age-matched controls. Her ability to identify individual faces, however, was completely preserved. A similar condition is found in monkeys with bilateral amygdalectomy, in which stimuli that would normally induce a fearful response fail to do so (Weiskrantz 1956, Blanchard & Blanchard 1972, Davis 1992).

Evidence for category-specific representations also comes from the description of a double dissociation between reported forms of metamorphopsia (Bodamer 1947). Metamorphopsias are severe pseudohallucinations in the perception of visual stimuli. One of three patients reported on by Bodamer (1947) experienced severe distortions while looking at faces [a condition later termed prosopo-metamorphopsia by Critchley (1953)] but did not lose the capacity to recognize the faces themselves. In contrast, there are also reports of prosopagnosic patients who are impaired in their ability to identify individual faces, but who also experience metamorphopsias with all visual stimuli except faces (Bodamer 1947).

## Selective Damage of Visuomotor Representations

The posterior parietal area is known to play an important role in sensory-motor integration, as lesions in this brain region in either humans or monkeys produce a variety of spatio-perceptual or spatio-behavioral disorders. The clinical neurologist Balint was the first to describe three characteristic deficits—now known together as Balint's syndrome—in a patient with bilateral posterior parietal lesions: (a) psychic paralysis of the gaze; (b) optic ataxia, i.e. impairment of object-bound movements of the hand performed under visual guidance; and (c) a form of simultanagnosia, which is the inability to perceive more than one object at a time, irrespective of the object's angular extent and despite preserved visual fields (for further discussion, see Farah 1990).

Optic ataxia, which was initially considered to be simply a deficit in reaching for objects (Damasio & Benton 1979), is now known to also be a disturbance of visually guided grasping, since patients appear not only to misreach for objects, but also to demonstrate impaired execution of finger movements and show a remarkable disturbance in the formation of finger grip and hand orientation before reaching for target objects (for review, see Jeannerod et al 1995). These deficits in "preshaping" occur despite the facts that the patients can correctly perceive the shape of the objects for which they attempt to reach and that movements that do not require visual guidance, such as those directed to the body, are executed correctly (Damasio & Benton 1979). Similarly, in the monkey, a unilateral parietal leukotomy, damaging a portion of the parieto-occipital white matter but sparing the optic radiations, causes a severe contralateral impairment of fine finger-movement control and misreaching (Haaxma & Kuypers 1975).

Recent studies demonstrated that the disturbance of preshaping can be dissociated from visuospatial perception, suggesting a double dissociation between the representations used for action and those that may underlie the perception of an object. For example, a recently reported agnosic patient was unable to perceive the orientation or size of objects, although she could still accurately use orientation and size information for visuomotor actions (Goodale et al 1991). The patient, who suffered an episode of carbon monoxide poisoning, had diffuse lesions in areas 18 and 19, with her primary visual cortex being largely intact. She was severely incapacitated in her ability to recognize visual objects based on shape information, being unable to perform elementary shape discrimination tasks (Milner & Heywood 1989). Nonetheless, she had adequate visual acuity, preserved central visual fields, and impaired but clear stereopsis, motion, and color vision (Milner & Heywood 1989).

When she was presented with a pair of rectangular plaques of the same or different dimensions, she was unable to distinguish between them. Similarly, her estimates of the width of a single plaque were only randomly related to the actual dimensions of the object, showing a considerable trial-to-trial variability (Goodale et al 1991). When the same patient, however, was asked to reach out and pick up the plaques, she did so with an index finger and thumb aperture that accurately reflected the dimensions of the objects. Based on these and similar findings, Goodale and colleagues suggested a dissociation between the representations used for apprehension and those used for action, in particular for prehensile hand movements that require accurate visual guidance.

## ELECTROPHYSIOLOGICAL STUDIES

### Inferotemporal Cortex

In order to discuss the physiological studies pertinent to object representation, it is important to explore, in some detail, the anatomical framework underlying visual processing in the primate brain. A survey of the anatomical organization of the visual system is also important, since the extreme diversity and complexity of areas and connections that make up the primate visual system themselves lend strong support to the idea that there are, indeed, multiple systems for recognition.

Sensory information from the primary visual cortex reaches the temporal and parietal lobes by a number of cortico-cortical stages that form two relatively separate pathways (Ungerleider & Mishkin 1982, Desimone & Ungerleider 1989). One pathway, roughly corresponding to the superior longitudinal fasciculus, passes dorsally in extrastriate cortex to end in the posterior parietal lobule and the frontal lobe; the other, corresponding to the inferior longitudinal fasciculus, passes ventrally in extrastriate cortex to reach the inferior temporal or inferotemporal cortex (IT).

IT is generally considered a large region of cortex extending approximately from just anterior to the inferior occipital sulcus to a couple of millimeters posterior to the temporal pole, and from the fundus of the superior temporal sulcus (STS) to the fundus of the occipito-temporal sulcus (Figure 2). It is roughly coextensive with Brodmann areas 20 and 21, or area TE of Von Bonin & Bailey (1947), which was later subdivided into the areas TE anteriorly and TEO posteriorly (Von Bonin & Bailey 1950, Iwai & Mishkin 1969). Area TEO forms a band extending from the lip of the STS to a few millimeters medial to the occipito-temporal sulcus. Its posterior border is close to the lip of the ascending portion of the inferior occipital sulcus, and its posterior-anterior extent is 10 to 15 mm (Boussaoud et al 1991). Area TE extends further anteriorly to about the sphenoid. Studies based on the deficits that follow focal lesions in IT—where TEO lesions lead to simple pattern deficits, while TE lesions result in associative and visual memory deficits—suggest two functional subdivisions—one posterior and one anterior—that are roughly coextensive with, but not identical to, the previously defined cytoarchitectonic TE and TEO subdivisions (Iwai 1978, 1981, 1985).

Based on topography and the laminar organization of projections, Felleman & Van Essen (1991) subdivided IT into PIT, CIT, and AIT (Figure 2), each having a ventral and dorsal portion. Based on cyto- and myeloarchitectonic criteria, as well as on the pattern of afferent cortical connections, the temporal cortex has been further subdivided into a large number of separate visual areas (Figure 3) (Seltzer & Pandya 1978, 1994), several of which have distinct physiological characteristics (Baylis et al 1987).

Area TEO receives feedforward, topographically organized cortical inputs from areas V2, V3, and V4 and has interhemispheric connections mediated mainly via the corpus callosum. Sparser inputs arise from areas V3A, V4t, and MT. Each of these areas receives a feedback connection from TEO (Distler et al 1993, Rockland et al 1994). TEO projects feedforwardly to the areas TEm, TEa, and IPa, all of which send feedback projections back to TEO. Feedback projections to this area also arise from the parahippocampal area TH and the areas TG and 36 (Distler et al 1993). Cortical projections of area TE include those to TH, TF, STP, frontal eye fields (FEF), and area 46 (Barbas & Mesulam 1981, Barbas 1988, Shiwa 1987). Area TE has both direct and indirect connections to the limbic structures. Direct connections have been reported to the amygdaloid complex (Amaral & Price 1984, Herzog & Van Hoesen 1976, Iwai & Yukie 1987, Turner et al 1980) and to the hippocampus (Yukie & Iwai 1988), which also receives an indirect projection via the parahippocampal gyrus (Van Hoesen 1982). TE does not project directly to entorhinal cortex (Insausti et al 1987), the cortical inputs of which arise primarily in the perirhinal and parahippocampal cortices (Suzuki & Amaral 1994).

Areas TEO and TE are also connected to a large number of subcortical

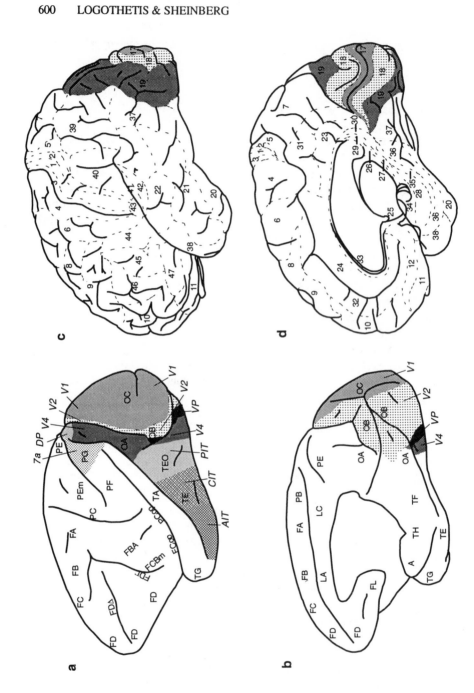

structures. Both areas receive nonreciprocal inputs from several nuclei of the thalamus and from the hypothalamus, locus coeruleus, reticular formation, basal nucleus of Meynert, and the dorsal and median raphe nuclei. Both are also reciprocally connected with the pulvinar and the ventral portion of the claustrum (Webster et al 1993). The main nonreciprocal output of both areas is a projection to the striatum, while TEO alone projects to superior colliculus and TE to the medial dorsal magnocellular nucleus of the thalamus (Webster et al 1993).

The pathway that begins in striate cortex, passes through the extrastriate and inferotemporal cortices, and reaches these subcortical areas is thought to underlie a variety of cognitive and visuomotor functions, such as recognition, habit formation, associative recall, and formation of visuomotor associations (Mishkin & Appenzeller 1987, Zola-Morgan & Squire 1993, Brothers & Ring 1993, Wilson & Rolls 1993). The diverse subcortical connections of many extrastriate areas, however, indicate that object-related information does not necessarily have to pass through IT to reach the striatum or the limbic structures. Different areas in the ventral, but also in the dorsal, pathway have reciprocal connections to structures such as the caudate, claustrum, amygdala, or hippocampal complex, areas which subserve different types of memory (Yeterian & Van Hoesen 1978, Webster et al 1993, Baizer et al 1993). Therefore some of the higher visual functions mentioned above may also be adequately accomplished by virtue of information derived from earlier processing levels. For example, categorizations based on the detection of some object features with high cue validity could, in principle, be possible even with damage to area TE, if these features are encoded in the activity of neurons in any of the earlier areas that project to limbic or striatal structures. Accordingly, the diversity and specificity of deficits observed in brain-damaged, agnosic patients may reflect an organization in which each processing stage copes with increasingly abstract representations and is capable, on its own, of supporting some types of categorization performance.

## Physiological Properties of Inferotemporal Neurons

Gross and colleagues were the first to obtain visually evoked responses in IT by using both macro- and microelectrodes in anesthetized and unanesthetized

---

*Figure 2*  Von Bonin & Bailey's (1947) map of the (*a*) lateral and (*b*) medial surface of the *Macaca mulatta* brain. Superimposed are the major visual areas as described by Felleman & Van Essen (1991). The names from the Von Bonin & Bailey (1947) parcellation are depicted on the brain, and the labels currently used by most investigators appear adjacent to the relevant areas: visual areas 1, 2, and 4 (V1, V2, V4), ventral posterior (VP), posterior inferotemporal (PIT), central inferotemporal (CIT), anterior inferotemporal (AIT), and dorsal parietal (DP). (*c*) Lateral and (*d*) medial surface of the human brain with Brodmann's areas numbered. [Adapted from Nieuwenhuys et al (1980).] Note that the relative sizes of the macaque and human brains are not to scale.

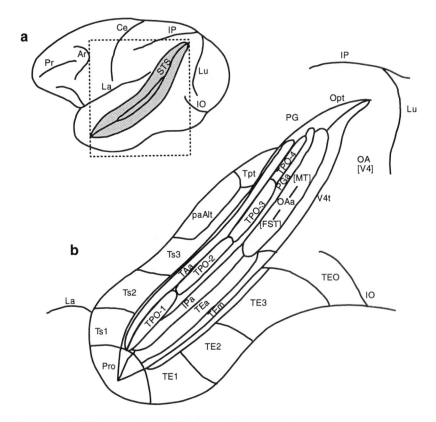

*Figure 3* Subdivision of monkey inferior temporal lobe centered around the superior temporal sulcus (STS). (*a*) Lateral view of the cortical surface with major visible sulci labeled: inferior occipital (IO), lunate (Lu), intraparietal (IP), central (Ce), lateral (Sylvian) fissure (La), arcuate (Ar), and principal (Pr). (*b*) Expanded view of the inferior temporal areas surrounding the STS. [Adapted from Seltzer & Pandya (1994).]

monkeys (for review, see Gross 1994). A large number of investigations confirmed and extended the initial findings, establishing the IT as the last exclusively visual area in the ventral pathway. More than 85% of the neurons in this area are excited or inhibited by different simple or complex visual patterns (Desimone et al 1984).

The observed properties of IT cells change significantly as one moves from the most posterior part of TEO, where cells have similar properties to those observed in area V4, to the most anterior part of TE, where neurons rarely respond to such simple stimuli. Among the changing characteristics are topography, receptive field size, and stimulus selectivity. Area TEO has a coarse visuotopic organization. It has an almost complete representation of the contralateral visual field, with receptive fields that are larger than those of the V4 neurons (Boussaoud et al 1991). In contrast, area TE is not visuotopically organized. Cells have large, ipsilateral, contralateral, or bilateral receptive fields, almost always including the fovea (Gross et al 1972). The response of the cells to stimuli presented at the fovea is usually more vigorous than elsewhere in the receptive field, possibly due to the fact that IT cortex receives a strong projection from those parts of the extrastriate cortex in which the central visual field is overrepresented, with a smaller contribution from those areas that process peripheral visual stimuli (Desimone et al 1980, Seltzer & Pandya 1978).

There is a systematic increase in the receptive field size along the posterior-anterior length of IT, with receptive field diameters in TEO as small as 1.5 to 2.5°, and those in TE reaching diameters of 30 to 50° (Boussaoud et al 1991, Tanaka 1993). The responsiveness of TE neurons to stimuli presented in the ipsilateral hemifield depends on the massive projections received from the opposite hemisphere via the splenium of the corpus callosum and the anterior commissure (Zeki 1973, Gross et al 1977). Sectioning of the splenium reduces the incidence of ipsilateral activation by about 50%, while combined splenium and anterior commissure sections entirely eliminate ipsilateral activation, suggesting that interhemispheric connections do play an essential role in the positional invariance observed in the response of many neurons in this area.

Many IT neurons are selective for a variety of stimulus attributes, such as color, orientation, texture, direction of movement, or shape (Gross et al 1972, Desimone & Gross 1979, Mikami & Kubota 1980). Of particular interest is the sensitivity of IT neurons to stimulus shape. Although shape selectivity has also been reported in earlier areas such as V4 (Desimone & Schein 1987, Tanaka et al 1991, Gallant et al 1993, Kobatake & Tanaka 1994), only in IT is this selectivity extensively encountered. Neurons in this area respond selectively to a variety of natural or synthetic objects (Desimone et al 1984), to parametric shape descriptors (Schwartz et al 1983), or to mathematically created 2D patterns, e.g. Walsh functions, that can be used to synthesize any arbitrary image with a given resolution (Richmond et al 1987). Groups of cells in IT have also been found that respond to the sight of biologically important objects such as faces or hands (see below). Face cells, which have been reported in monkeys as young as six weeks old (Rodman et al 1993), are two to ten times more sensitive to faces than to simple geometrical stimuli or 3D objects (Perrett et al 1979, 1982).

Interestingly, many IT neurons show various degrees of invariance to image transformations. The absolute response of the cells only rarely exhibits size or position constancy (e.g. see Logothetis et al 1995). However, their selectivity for shape, i.e. their relative preference for the optimal stimulus over several suboptimal stimuli, is preserved over large changes in stimulus size and position (Sato et al 1980, Schwartz et al 1983, Logothetis et al 1995, Ito et al 1995). In this sense, more than half of the IT neurons can be thought of as demonstrating size and position invariance. The response of the rest of the neurons indicates some degree of size specificity, suggesting that at least some object representations might be stored in a size-specific manner (Ito et al 1995). Selectivity for shape has also been found to be cue invariant, in the sense that cell responses to an optimal stimulus remain unchanged regardless of the cues (motion, texture, or luminance) determining the object's shape (Sáry et al 1993). Contrast polarity, on the other hand, appears to have large effects on the response of IT neurons (Ito et al 1994). The effects of contrast polarity corroborate the proposed role this area may play in shape processing, since the recovery of surface structure relies partly on shading information, which in turn depends on luminance contrast polarity (Cavanagh 1987, Ramachandran 1990).

In summary, IT appears to have all the machinery requisite for the formation of object descriptions. Cells respond selectively to stimulus attributes such as color and texture, to simple and complex patterns, and to complex natural objects such as faces. They also show a certain degree of translation and scale invariance. An obvious question, then, is, What is the encoding scheme used to represent visual objects? Are they represented explicitly by the firing of a few gnostic units? Are they represented by the firing of a small population of neurons, each encoding some features, aspects, or single views? Or, are they only implicitly represented by a large population of cells each acting as a specialized pattern filter that combines certain shapes with different surface properties of objects, such as their texture, color, or lightness?

Electrophysiological findings suggest the existence of at least two possible neural mechanisms for object representation. One system may code the prototypes of objects that can be decomposed into parts and recognized by indexing these parts and their metric or spatial relationships. A second, separate, system may be used when holistic configuration rather than individual features is important and may rely primarily on small populations of neurons with strong configurational selectivity.

## Combination Encoding

Recent careful studies of the properties of inferotemporal neurons have revealed a systematic organization in the temporal lobe, wherein neurons with relatively similar response properties are clustered in modules spanning the entire thickness of cortex (Fujita et al 1992, Tanaka 1993, Young 1993, Gawne & Richmond 1993, Kobatake & Tanaka 1994).

Columnar organization is a well-established cortical property in many different areas. In the early visual system, clustering is found for neurons responding selectively to simple stimulus attributes, such as position in the visual field, ocular preference, orientation, or direction of movement. In area TE, modular organization is less related to retinotopic organization and, instead, reflects similar preferences for combinations of shapes and other stimulus attributes. Details of this work are described elsewhere in this volume by Tanaka (1996). Of particular interest for this discussion are the "elaborate" cells reported by Tanaka et al (1991), which responded only to composite shapes. These cells were studied extensively by reducing the complexity of an effective visual stimulus in a systematic manner until the simplest pattern that would drive the cell maximally was determined. The degree of complexity required to drive an elaborate cell was found to increase, in general, from area TEO to area TE. In addition, cells of different modules showed greater differences in shape selectivity than cells within a single module. Based on these and related findings, it has been argued that the general class of an object could be represented by the activity across different IT modules, while detailed discriminations could, in principle, be accomplished by detecting small differences in the activity of neurons within single modules (Fujita et al 1992, Tanaka 1993, Young 1993, Gawne & Richmond 1993).

At present, it is unclear whether the critical features of the elaborate cells form a complete set of general shape descriptors that could represent any complex object or scene. Such a scheme would, in many ways, be similar to the RBC theory proposed by Biederman (1987), although there is not, as yet, evidence for cells in IT that code for the spatial relations between individual primitives. Nonetheless, the idea that prototypes may be represented by a relatively small number (estimated to be around 1300) of modules is both theoretically appealing and biologically plausible.

Is, however, such a system sufficient for representing individual exemplars of a given object category or when holistic configuration information is necessary to disambiguate individual objects? The perception of overall configuration is often crucial for subordinate-level discriminations, as in the case of face recognition mentioned earlier. Could the representation of holistic configuration be accomplished by the combination encoding scheme described above? While not excluding this possibility, two lines of evidence show that in the monkey visual system, alternative strategies are probably used when configuration or metric information is the determining factor for a categorization task. First, a large number of neurons in TE and STS seem to encode the overall shape of biologically important objects—not specific features or parts (Rolls 1994, Oram & Perrett 1994). Further, recordings from the IT of monkeys trained to identify individual objects from two novel object classes have shown that neurons can be found in TE that respond to a limited subset of views of the objects, as face-selective neurons do for views of faces (Logothetis & Pauls 1995).

## Selectivity for Biologically Important Stimuli

Responses of cells to biologically relevant stimuli—such as faces, face features, hands, and other body parts—have been reported by several investigators (Gross et al 1972; Perrett et al 1982, 1985, 1989; Rolls 1984; Desimone et al 1984; Yamane et al 1988; Hasselmo et al 1989; Young & Yamane 1992). Face cells, which seem to be the most prominent class, are found in the STS and IT (areas TPO, TEa, and TEm) (Baylis et al 1987); the amygdala (Sanghera et al 1979; Leonard et al 1985; Rolls 1992a, 1992b); the ventral striatum, which receives a projection from the amygdala (Williams et al 1993, Rolls 1994); and the inferior convexity of the prefrontal cortex (Wilson et al 1994). The large number of areas containing cells responsive to faces is consistent with the hypothesis that object representations are manifest in multiple parallel sites.

There is a considerable differentiation among face-selective neurons. One subgroup appears to be very similar to the elaborate neurons described above. Some cells are selective for particular features of the head and face, e.g. the eyes (Perrett et al 1982, 1992), whereas another population of cells can only be driven by simultaneous presentation of multiple parts of a face (Perrett & Oram 1993, Wachsmuth et al 1994). Yet other face cells require the entire face-view configuration, or even a combination of information such as eye gaze angle, head direction, and body posture (Perrett et al 1992), and in this sense, they encode holistic information about shape and not information about the existence of individual features.

The selectivity of these neurons for faces is maintained over changes in stimulus size and position but less over changes in orientation. Face cells are sensitive to rotations in the picture plane, with a strong bias for upright faces (Tanaka et al 1991), and most of them show selectivity for a specific vantage point. In particular, some cells are maximally sensitive to the front view of a face, and their response falls off as the head is rotated into the profile view, while some others are sensitive to the profile view with no response to the front view of the face (see also Figures 4c and 4d). A detailed investigation of these types of cells by Perrett et al (1985) reported a total of five types of cells in the STS, each maximally responsive to one view of the head: full face, profile, back of the head, head up, and head down. In addition, two subtypes have been discovered that respond only to left profile or only to right profile, confirming that these cells are involved in visual analysis rather than in representing specific behavioral or emotional responses.

Interestingly, a study using correlation analysis between quantified facial features and neural responses has shown that face neurons can detect combinations of the distances between facial parts, such as eyes, mouth, eyebrows, and hair (Yamane et al 1988). These cells show a remarkable redundancy of coding characteristics, as only two facial dimensions seem to be necessary for explaining most of the variance in the population. For example, all of the width measurements—such as the width of the eyes, the mouth, and the interocular distance—covary with the general width of the face. Young & Yamane (1992) also found that each face cell typically exhibits graded responses to a wide variety of face stimuli and therefore presumably participates in the representation of many different faces.

In general, face-selective neurons responsive explicitly to the identity of faces are found in IT, while cells that respond to facial expressions, gaze direction, and vantage point are mostly located in the STS (Hasselmo et al 1989, Perrett et al 1992). Such functional localization is in agreement with lesion experiments in monkeys (Heywood & Cowey 1992), showing that removal of the cortex in the banks and floor of the rostral STS of monkeys results in deficits in the perception of gaze directions and the facial expression, but not face identification (Heywood & Cowey 1992). It is also in agreement with preoperative electrophysiological recordings in epileptic patients that also suggest separate processing of facial identity and facial expressions (Ojemann et al 1993). In fact, PET studies suggest that the posterior fusiform gyrus is activated during face matching or gender discrimination testing (Haxby et al 1991, Sergent et al 1992), while the presentation of a unique face activates the mid-fusiform gyrus (Sergent et al 1992).

Cells in IT also respond to the sight of the entire human body or of body parts (Wachsmuth et al 1994). About 90% of these neurons responded to the human body in a viewer-centered fashion, whereas the rest responded equally well to any view of the stimulus. Of particular interest here is the observation that about one fifth of the neurons studied responded only to the entire body and not to the sight of any of the constituent body parts alone (Wachsmuth et al 1994).

In summary, the evidence presented here suggests that at least some objects are represented by neurons with a complex configurational selectivity that cannot be reduced to selectivity of individual features or even constellations of such features. One obvious question is whether such a configurational selectivity is specific for animate objects, such as faces and body forms. Clinical observations, as mentioned above, have shown that the recognition of living things can be selectively impaired (see e.g. Farah et al 1991). Thus one possibility is that the perception of these shapes is mediated by specialized neural populations. If so, then the complex-pattern selectivity for faces and body parts reported above may be unique to the representation of the class of living things, with different encoding mechanisms responsible for the recognition of other objects. In support of this hypothesis are the observations that face cells appear very early in the ontogenesis of monkeys (Rodman et al 1993) and that newborn human infants show a special affinity for the sight of faces (Goren et al 1975, Johnson et al 1991). Alternatively, a system based on neurons that are selective for complex configurations may provide a general mechanism for encoding any object that cannot undergo useful decomposition in the process of recognition.

## Configurational Selectivity for Novel Objects

Recently, Logothetis and his colleagues (Logothetis et al 1994, 1995; Logothetis & Pauls 1995) set out to determine whether the configurational selectivity found for IT neurons is specific for faces or body parts or whether it can be generated for any novel object as a result of extensive training. In these combined psychophysical and neurophysiological experiments, macaque monkeys were trained to become experts at identifying novel computer-generated wire and spheroidal objects, similar to those shown in Figures 1b and 1d. These objects had never been experienced by the monkeys, nor did they possess any inherent biological relevance. Nonetheless, after training, the animals learned to discriminate individual objects from a set of highly similar distractors (Logothetis et al 1994), a task not unlike the problem of identifying a specific face or a particular bird species. Because all of the objects used in testing were composed of the same basic parts, good performance in this task relied upon the detection of subtle shape differences. These experiments were thus directed at understanding how objects are neurally represented when they are encountered in the context of a subordinate-level recognition task.

Physiological recordings from individual neurons in the inferior temporal lobe, near the anterior medial temporal sulcus (AMTS), revealed a subpopulation of cells that were activated selectively by views of previously unfamiliar objects (Logothetis et al 1995, Logothetis & Pauls 1995). Many neurons fired selectively for a small set of views of a spheroidal or wire object that the monkey had learned to recognize from all viewpoints. The cells were most active when the target was presented from one particular view (Figure 4a), and their activity declined as the object was rotated in depth.

The neurons found in the temporal lobe of the expert monkeys bear interesting similarities to face-selective cells found in the banks of the rostral STS (NK Logothetis & DL Sheinberg, unpublished observations). Cells from both populations exhibit object-specific as well as view-specific selectivity. Figure 4 illustrates response profiles of four different cells selective for specific views of wire objects (Figures 4a and 4b) and faces (Figures 4c and 4d). The neurons depicted in Figures 4a and 4c show a marked preference for a single view of the test object and a steady decrease in mean activity for increasing object rotations. Such neurons seem to act like blurred templates, with their tolerance for small rotations in depth representing a form of limited generalization. The cells shown in Figures 4b and 4d reveal a broader form of generalization by their selective response for pseudo-reflected object views, i.e. object views 180° apart that appear as mirror images. Psychological studies have shown that enantiomorphic views of objects are not easily discriminated by children (Bernstein et al 1978, Corballis & McLaren 1984) and seem to be categorized equivalently by adults (Biederman & Cooper 1991, Cooper et al 1992). Generalization across enantiomorphs, at both the cellular and behavioral level, may be evidence of the primate recognition system's ability to extract a more global representation of a shape, ignoring local deviations in cases where  differentiating between the two views is almost always unnecessary for the purpose of object recognition.

The narrow tuning curves encountered for views of stimuli from the two novel object classes clearly shows that high stimulus selectivity is not limited to faces or other biological forms. Thus the ability to make subordinate-level judgments about novel objects may rely on some of the same mechanisms, or perhaps even the same population of cells, that are involved in the recognition of faces.

That the stimulus selectivity of cells in IT can be altered as a result of experience has also been suggested by Kobatake et al (1993). In their experiment, a monkey was first trained for more than two months in a discrimination task, using 28 shapes composed of two or three geometric primitives. Following training, cells in IT were isolated and tested using a battery of visual stimuli while the monkey was anesthetized and immobilized. Interestingly, a much higher proportion of cells in the trained monkey were strongly activated by stimuli in the test set, compared to untrained monkeys. These results imply that long-term changes in tuning characteristics can be induced by experience and that these changes can be observed in the anesthetized animal. These data lend support to the claim that the selectivity of the view-selective cells reported by Logothetis & Pauls (1995) was, in fact, tuned throughout the course of the animals' training. In addition, they further emphasize the importance of characterizing the properties of cells in IT in the context of a behaviorally relevant task.

## Action-Related Representations

As mentioned above, damage to the parietal cortex can cause a severe impairment in spatial perception (Lynch 1980, Andersen 1989). In the monkey, this cortical region covers Brodmann's areas 5 (superior parietal lobule) and 7 (inferior parietal lobule), or approximately, areas PE, PG, and PF, as defined by Von Bonin & Bailey (see Figure 2). Physiological investigations have supported the clinical and lesion studies. A major route from the occipital lobe into the parietal lobe is via the middle temporal area (MT) that is located on the posterior bank and floor of the caudal third of the STS (see references in Snowden 1994, Logothetis 1994). Area MT has neurons that are highly selective for binocular disparity, speed, and direction of stimulus motion. Informa-

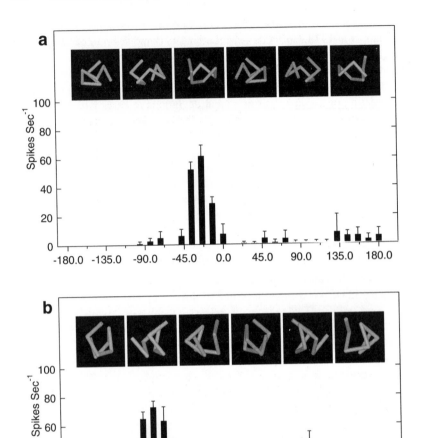

tion from MT is routed to areas MST and FST, which in turn project to parietal
areas, such as LIP, VIP, and 7a (Ungerleider & Desimone 1986, Felleman &
Van Essen 1991). The responsiveness of many parietal neurons is strongly
modulated by attention, and many neurons are related to visuomotor activity
(Robinson et al 1978, Lynch 1980, Mountcastle et al 1984).

Parietal cells are also sensitive to those visual qualities of an object that
determine the posture of the hand and fingers during a grasping movement. In

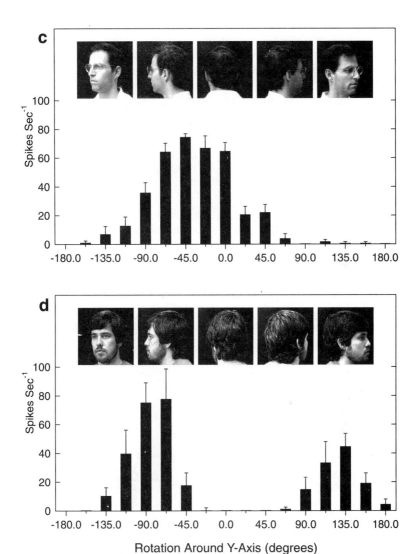

*Figure 4*  Four different IT neurons selective for views of wires and faces. (*a–d*) Two of the neurons shown here responded maximally for a single view of an object (*a, c*), and response magnitude decreased gradually as the object was rotated in depth away from the preferred view. Figure 4*b* shows an example of a cell responding to two views of a wire object separated by 180°, and Figure 4*d* shows data from a cell that exhibited its maximum response for the left-facing profile of a head and nearly the same response for the right-facing profile. (Error bars indicate standard deviations of mean response rates.)

particular, for reaching and grasping tasks, neurons in area 7 of the monkey have been found to be selectively activated depending upon the configuration and orientation of the target object (Taira et al 1990, Sakata & Taira 1994). Taira et al (1990) reported a class of motor-dominant neurons that fired during hand movements in either the light or the dark but not during the visual fixation of the manipulandum. Another class of visual-and-motor neurons were found to be active in all three conditions, and a third class of visual-dominant cells fired during hand-movement or visual fixations but not in the dark. The first two classes were closely related to hand manipulation and fired more consistently in the light than in the dark, suggesting that the visual-dominant cells recorded in this area may be providing the visual input to these neurons controlling the hand configuration.

Although these experiments clearly show that cells in parietal cortex are responsive to visual stimuli, they do not directly address the question of whether some parietal neurons are truly sensitive to an object's shape. Sereno & Maunsell (1995) recently reported such evidence, however, in a study of single-unit activity in the LIP of monkeys who were trained to perform a short-term memory task. About one third of the units recorded in these experiments did indeed show significant response differences, dependent upon stimulus shape, and about a third of the units showed significant differences in delay period activity, dependent on the shape of the sample. These results suggest that parietal cells may contribute to the memory of shape features, as well as participate in the execution of visually guided actions.

## CONCLUSIONS

The research reviewed in this paper suggests that recognition of visual objects relies on different types of stored representations, each employed according to the requirements of the task under study.

A fundamental problem is the classification of objects at the basic categorization level, a task that may encompass all that simple recognition systems are capable of performing. In humans, basic-level object names are the first to enter a child's vocabulary and are used to a much greater extent than any other term to describe categories. Recognition at the basic object level appears largely invariant to surface illumination, hue, or image transformations, such as scaling, translation, or rotation around any axis. The representations used in prototype recognition may rely on structural decomposition of the objects into parts and on indexing these parts and their relationships. In agreement with this notion are electrophysiological studies suggesting that basic object forms may be represented by the activity of neurons across different modules of the inferior temporal cortex, each encoding combinations of various complex forms with surface properties such as texture or color.

Categorization at the subordinate level appears to involve different types of representation, each of which may rely on different neural mechanisms than those used for the recognition of objects at the basic level. Subordinate-level recognition is initially strongly view dependent, with generalization accomplished through perceptual learning. The extent to which the different types of representations are used in recognition is likely to vary depending on the object type or familiarity. Recognition at the subordinate level is differentially affected by brain damage in the absence of any deficits in the recognition of objects at the basic level. Agnosic patients can occasionally recognize natural or synthetic objects with distinct shapes that belong to different classes, but they fail to do so when identification of individual entities is required, as when recognizing personal items such as their own wallets, cars, or articles of clothing. Identification of items of a structurally homogeneous class may rely on the activity of neurons with high configurational selectivity, such as the cells found in the experiments described above.

Neuropsychological evidence suggests a further dissociation between the representations used for the recognition of living things and nonliving things. We have discussed evidence regarding the specificity of prosopagnosia that can occur in the absence of any other associative failure. Similarly, facial memory, i.e. the matching of unfamiliar but previously presented faces, is also a somewhat different task than recognition of familiar faces. Even more striking, however, is the specific loss of humans' or monkeys' ability to recognize facial expressions and the associated emotions, despite the preserved capacity to recognize face identity. While the failure to recognize the identity of a face may simply be due to a general inability to detect individuality within a given homogeneous object class, the agnosia of facial expressions strongly suggests a specialized mechanism for the processing of biologically important configurations, as does the double dissociation between prosopo-metamorphopsia and the metamorphopsias for other visual stimuli. Evidence for brain mechanisms specialized for recognizing biological forms, especially faces, also comes from the electrophysiological findings reviewed in the previous section, showing that neurons in the temporal lobe respond selectively to faces, hands, or other body parts. Finally, both physiological and clinical work suggests a dissociation between the representations underlying perceptual categorization and those used for the action of object grasping.

In reviewing the current state of research in the field of visual object recognition, we were interested to find that the field of memory research has, in the last decade, been characterized in much the same way that we have suggested here (Tulving & Schacter 1990). Insofar as visual recognition necessarily requires memory, similar ideas should hold in both fields. In closing, then, we borrow a conclusion from a recent review on memory research that, after substituting the term object recognition for memory, seems to apply remarkably well to the field of visual object recognition:

> [I]t is becoming increasingly clear that there are no universal principles of memory and that facts discovered about one form of memory need not hold for other forms. This is why systematic classification of memory systems, both psychological and physiological, is an essential prerequisite for the successful pursuit of the empirical and theoretical understanding of memory processes and mechanisms. The systems approach combined with appropriate processing theories seems to provide the most direct route to the future (Tulving & Schacter 1990, p. 305).

ACKNOWLEDGMENTS

Special thanks to E Bricolo and D Murray for assistance in preparing the manuscript and figures. We also thank J Pauls, D Leopold, and Dr J Assad for reading the manuscript and making many useful suggestions. Nikos K Logothetis was supported by the Office of Naval Research (contract N000 14-93-1-0209, 1992), the National Institute of Health (grant NIH 1RO1EY10089-01), an AASERT award from the Office of Naval Research, and the McKnight Endowment Fund for Neuroscience (1993). David L Sheinberg was supported by the National Institute of Health (grant NRSA 1F32EY06624).

## Literature Cited

Adolphs R, Tranel D, Damasio H, Damasio A. 1994. Impaired recognition of emotion in facial expressions following bilateral damage to the human amygdala. *Nature* 372:669–72

Aloimonos Y, ed. 1993. *Active Perception.* Hillsdale, NJ: Erlbaum

Amaral DG, Price JL. 1984. Amygdalo-cortical projections in the monkey (*macaca fascicularis*). *J. Comp. Neurol.* 230:465–96

Andersen RA. 1989. Visual and eye movement functions of the posterior parietal cortex. *Annu. Rev. Neurosci.* 12:377–403

Anderson JA, Silverstein JW, Ritz SA, Jones RS. 1977. Distinctive features, categorical perception, and probability learning: some applications of a neural model. *Psychol. Rev.* 84:413–51

Baizer JS, Desimone R, Ungerleider LG. 1993. Comparison of subcortical connections of inferior temporal and posterior parietal cortex in monkeys. *Vis. Neurosci.* 10:59–72

Banks WP, Krajicek D. 1991. Perception. *Annu. Rev. Psychol.* 42:305–31

Barbas H. 1988. Anatomic organization of basoventral and mediodorsal visual recipient prefrontal regions in the rhesus monkey. *J. Comp. Neurol.* 276:313–42

Barbas H, Mesulam MM. 1981. Organization of afferent input to subdivisions of area 8 in the rhesus monkey. *J. Comp. Neurol.* 200:407–31

Bartram DJ. 1974. The role of visual and semantic codes in object naming. *Cogn. Psychol.* 6:325–56

Bartram DJ. 1976. Levels of coding in picture-picture comparison tasks. *Mem. Cogn.* 4:593–602

Baylis GC, Rolls ET, Leonard CM. 1987. Functional subdivisions of the temporal lobe neocortex. *J. Neurosci.* 7:330–42

Benton AL. 1980. The neuropsychology of facial recognition. *Am. Psychol.* 35:176–86

Biederman I. 1987. Recognition-by-components: a theory of human image understanding. *Psychol. Rev.* 94:115–47

Biederman I, Cooper EE. 1991. Evidence for

complete translational and reflectional invariance in visual object priming. *Perception* 20:585–95

Biederman I, Cooper EE. 1992. Size invariance in visual object priming. *J. Exp. Psychol.: Hum. Percept. Perform.* 18:121–33

Biederman I, Gerhardstein PC. 1993. Recognizing depth-rotated objects: evidence and conditions for three-dimensional viewpoint invariance. *J. Exp. Psychol.: Hum. Percept. Perform.* 19:1162–82

Blanchard DC, Blanchard RJ. 1972. Innate and conditioned reactions to threat in rats with amygdaloid lesions. *J. Comp. Physiol. Psychol.* 81:281–90

Bodamer J. 1947. Die Prosop-Agnosie (Die Agnosie des Physiognomieerkennens). *Arch. Psychiatr. Nervenkr.* 179:6–54

Bomba PC, Siqueland ER. 1983. The nature and structure of infant form categories. *J. Exp. Child. Psychol.* 35:294–328

Bornstein MH, Gross CG, Wolf JZ. 1978. Perceptual similarity of mirror images in infancy. *Cognition* 6:89–116

Bornstein MH, Sroka H, Munitz H. 1969. Prosopagnosia with animal face agnosia. *Cortex* 5:164–69

Boussaoud D, Desimone R, Ungerleider LG. 1991. Visual topography of area TEO in the macaque. *J. Comp. Neurol.* 306:554–75

Bowman RS, Sutherland NS. 1970. Shape discrimination by goldfish: coding of irregularities. *J. Comp. Physiol. Psychol.* 72:90–97

Brothers L, Ring B. 1993. Mesial temporal neurons in the macaque monkey with responses selective for aspects of social stimuli. *Behav. Brain Res.* 57:53–61

Brown R. 1958. How shall a thing be called? *Psychol. Rev.* 65:14–21

Bruce CJ. 1982. Face recognition by monkeys: absence of an inversion effect. *Neuropsychologia* 20:515–21

Bruce V. 1982. Changing faces: visual and nonvisual coding processes in face recognition. *Br. J. Psychol.* 73:105–16

Bülthoff HH, Edelman S. 1992. Psychophysical support for a two-dimensional view interpolation theory of object recognition. *Proc. Natl. Acad. Sci. USA* 89:60–64

Bushnell IWR, Sai F, Mullin JT. 1989. Neonatal recognition of the mother's face. *J. Dev. Psychol.* 7:3–15

Carey S, Diamond R. 1977. From piecemeal to configuration representation of faces. *Science* 195:312–13

Cavanagh P. 1987. Reconstructing the third dimension: interactions between color, texture, motion, binocular disparity and shape. *Comp. Vis. Graphics Image Proc.* 37:171–95

Cerella J. 1979. Visual classes and natural categories in the pigeon. *J. Exp. Psychol.: Hum. Percept. Perform.* 5:68–77

Cooper LA, Schacter DL, Delaney SM,

Ballesteros S, Moore C. 1992. Priming and recognition of transformed three-dimensional objects: effects of size and rotation. *J. Exp. Psychol.: Learn. Mem. Cogn.* 18:43–57

Corballis MC, McLaren R. 1984. Winding one's ps and qs: mental rotation and mirror-image discrimination. *J. Exp. Psychol.: Hum. Percept. Perform.* 10:318–27

Corballis MC, Zbrodoff NJ, Shetzer LI, Butler PB. 1978. Decisions about identity and orientation of rotated letters and digits. *Mem. Cogn.* 6:98–107

Critchley M, ed. 1953. *The Parietal Lobes.* New York: Hafner

Damasio AR. 1985. Disorders of complex visual processing: agnosia, achromatopsia, Balint's syndrome, and related difficulties of orientation and construction. In *Principles of Behavioural Neurology,* ed. MM Mesulam, pp. 259–88. Philadelphia: Davis Co.

Damasio AR, Benton AL. 1979. Impairment of hand movements under visual guidance. *Neurology* 29:170–78

Damasio AR, Damasio H, Van Hoesen GW. 1982. Prosopagnosia: anatomic basis and behavioral mechanisms. *Neurology* 32:331–41

Damasio AR, Tranel D, Damasio H. 1990. Face agnosia and the neural substrate of memory. *Annu. Rev. Neurosci.* 13:89–109

Davis M. 1992. The role of amygdala in fear and anxiety. *Annu. Rev. Neurosci.* 15:352–75

Davis RT. 1974. *Monkeys as Perceivers.* New York: Academic

De Renzi E. 1986. Slowly progressive visual agnosia or apraxia without dementia. *Cortex* 22:171–80

De Renzi E, Faglioni P, Spinnler H. 1968. The performance of patients with unilateral brain damage on face recognition tasks. *Cortex* 4:17–34

Desimone R, Albright TD, Gross CG, Bruce CJ. 1984. Stimulus-selective properties of inferior temporal neurons in the macaque. *J. Neurosci.* 4:2051–62

Desimone R, Fleming JFR, Gross CG. 1980. Prestriate afferents to inferior temporal cortex: an HRP study. *Brain Res.* 184:41–55

Desimone R, Gross CG. 1979. Visual areas in the temporal cortex of the macaque. *Brain Res.* 178:363–80

Desimone R, Schein SJ. 1987. Visual properties of neurons in area V4 of the macaque: sensitivity to stimulus form. *J. Neurophysiol.* 57:835–67

Desimone R, Ungerleider LG. 1989. Neural mechanisms of visual processing in monkeys. In *Handbook of Neuropsychology,* ed. F Boller, J Grafman, 2:267–99. Amsterdam: Elsevier

Diamond R, Carey S. 1986. Why faces are and are not special: an effect of expertise. *J. Exp. Psychol.* 115:107–17

Distler C, Boussaoud D, Desimone R, Ungerleider LG. 1993. Cortical connections of inferior temporal area TEO in macaque monkeys. *J. Comp. Neurol.* 334:125–50

Edelman S, Bülthoff HH. 1992. Orientation dependence in the recognition of familiar and novel views of 3D objects. *Vis. Res.* 32: 2385–400

Eimas PD, Quinn PC. 1994. Studies on the formation of perceptually based basic-level categories in young infants. *Child Devel.* 65: 903–17

Eley MG. 1982. Identifying rotated letter-like symbols. *Mem. Cogn.* 10:25–32

Ellis R, Allport DA, Humphreys GW, Collis J. 1989. Varieties of object constancy. *Q. J. Exp. Psychol.* 41A:775–96

Fantz RL. 1964. Visual experience in infants: decreased attention to familiar patterns relative to novel ones. *Science* 146:668–70

Farah MJ. 1990. *Visual Agnosia.* Cambridge, MA: MIT Press

Farah MJ, McMullen PA, Meyer MM. 1991. Can recognition of living things be selectively impaired? *Neuropsychologia* 29:185–93

Farah MJ, Rochlin R, Klein KL. 1994. Orientation invariance and geometric primitives in shape recognition. *Cogn. Sci.* 18:325–44

Farah MJ, Wilson KD, Drain HM, Tanaka JR. 1995. The inverted face inversion effect in prosopagnosia: evidence for mandatory, face-specific mechanisms. *Vis. Res.* 35: 2089–93

Faust C, ed. 1955. *Die zerebralen Herdstörungen nach Hinterhauptverletzungen und ihre Beurteilung.* Stuttgart: Thieme

Felleman DJ, Van Essen DC. 1991. Distributed hierarchical processing in primate cerebral cortex. *Cereb. Cortex* 1:1–47

Foster DH, Kahn JI. 1985. Internal representations and operations in the visual comparison of transformed patterns: effects of pattern point-inversion, positional symmetry, and separation. *Biol. Cybern.* 51:305–12

Franks JJ, Bransford JD. 1971. Abstraction of visual patterns. *J. Exp. Psychol.* 90:65–74

Fujita I, Tanaka K, Ito M, Cheng K. 1992. Columns for visual features of objects in monkey inferotemporal cortex. *Nature* 360: 343–46

Gaffan D, Heywood CA. 1993. A spurious category-specific visual agnosia for living things in human and nonhuman primates. *J. Cogn. Neurosci.* 5:118–28

Gallant JL, Braun J, Van Essen DC. 1993. Selectivity for polar, hyperbolic, and cartesian gratings in macaque visual cortex. *Science* 259:100–3

Gauthier I. 1995. *Becoming a "greeble" expert: exploring the face recognition mechanism.* MS thesis. Yale Univ.

Gawne TJ, Richmond BJ. 1993. How independent are the messages carried by adjacent inferior temporal cortical neurons? *J. Neurosci.* 13:2758–71

Goodale MA, Milner AD, Jakobson LS, Carey DP. 1991. A neurological dissociation between perceiving objects and grasping them. *Nature* 349:154–56

Goren CC, Sarty M, Wu RWK. 1975. Visual following and pattern discrimination of face-like stimuli by newborn infants. *Pediatrics* 56:544–49

Gross CG. 1973. Visual functions of inferotemporal cortex. In *Handbook of Sensory Physiology,* ed. R Jung, 7/3B:451–82. Berlin: Springer-Verlag

Gross CG. 1994. How inferior temporal cortex became a visual area. *Cereb. Cortex* 4:455–69

Gross CG, Bender DB, Mishkin M. 1977. Contributions of the corpus callosum and the anterior commissure to visual activation of inferior temporal neurons. *Brain Res.* 131: 227–39

Gross CG, Rocha-Miranda CE, Bender DB. 1972. Visual properties of neurons in inferotemporal cortex of the macaque. *J. Neurophysiol.* 35:96–111

Gross CG, Rodman HR, Gochin PM, Colombo MW. 1993. Inferior temporal cortex as a pattern recognition device. In *Computational Learning and Cognition: Proc. 3rd NEC Res. Symp.,* ed. E Baum, p. 44. Slam: NEC Res.

Grüsser OJ, Landis T, eds. 1991. *Visual Agnosias and Other Disturbances of Visual Perception and Cognition.* London: Macmillan

Haaxma R, Kuypers HGJM. 1975. Intrahemispheric cortical connexions and visual guidance of hand and finger movements in the rhesus monkey. *Brain* 98:239–60

Hasselmo ME, Rolls ET, Baylis GC, Nalwa V. 1989. Object-centered encoding by face-selective neurons in the cortex in the superior temporal sulcus of the monkey. *Exp. Brain Res.* 75:417–29

Haxby JV, Grady CL, Horwitz B, Ungerleider LG, Mishkin M, et al. 1991. Dissociation of object and spatial visual processing pathways in human extrastriate cortex. *Proc. Natl. Acad. Sci. USA* 88:1621–25

Herrnstein RJ, Loveland DH. 1964. Complex visual concept in the pigeon. *Science* 146: 549–51

Herrnstein RJ, Loveland DH, Cable C. 1976. Natural concepts in pigeons. *J. Exp. Psychol.: Anim. Behav. Proc.* 2:285–302

Herzog AG, Van Hoesen GW. 1976. Temporal neocortical afferent connections to the amygdala in the rhesus monkey. *Brain Res.* 115:57–69

Heywood CA, Cowey A. 1992. The role of the 'face-cell' area in the discrimination and recognition of faces by monkeys. *Philos. Trans. R. Soc. London Ser. B* 335:31–38

Homa D, Chambliss D. 1975. The relative contributions of common and distinctive information on the abstraction from ill-defined categories. *J. Exp. Psychol.: Hum. Learn. Mem.* 1:351–59

Homa D, Cross J, Cornell D, Goldman D, Shwartz S. 1973. Prototype abstraction and classification of new instances as a function of number of instances defining the prototype. *J. Exp. Psychol.* 101:116–22

Homa D, Sterling S, Trepel L. 1981. Limitations of exemplar-based generalization and the abstraction of categorical information. *J. Exp. Psychol.: Hum. Learn. Mem.* 7:418–39

Homa D, Vosburgh R. 1976. Category breadth and the abstraction of prototypical information. *J. Exp. Psychol.: Hum. Learn. Mem.* 2:322–30

Humphrey GK, Khan SC. 1992. Recognizing novel views of three-dimensional objects. *Can. J. Psychol.* 46:170–90

Humphreys GW, Riddoch MJ. 1987a. *To See But Not To See: A Case Study of Visual Agnosia.* Hillsdale, NJ.: Erlbaum

Humphreys GW, Riddoch MJ, eds. 1987b. *Visual Object Processing: A Cognitive Neuropsychological Approach.* Hillsdale, NJ.: Erlbaum

Insausti R, Amaral DG, Cowan WM. 1987. The entorhinal cortex of the monkey. II. Cortical afferents. *J. Comp. Neurol.* 264:356–95

Ito M, Fujita I, Tamura H, Tanaka K. 1994. Processing of contrast polarity of visual images in inferotemporal cortex of the macaque monkey. *Cereb. Cortex* 4:499–508

Ito M, Tamura H, Fujita I, Tanaka K. 1995. Size and position invariance of neuronal responses in monkey inferotemporal cortex. *J. Neurophysiol.* 73:218–26

Iwai E. 1978. The visual learning area in the inferotemporal cortex of monkeys. In *Integrative Control Functions of the Brain,* ed. M Ito, pp. 419–27. Tokyo: Kodansha

Iwai E. 1981. Visual mechanisms in the temporal and prestriate association cortices of the monkey. *Adv. Physiol. Sci.* 17:279–86

Iwai E. 1985. Neuropsychological basis of pattern vision in macaque monkeys. *Vis. Res.* 25:425–39

Iwai E, Mishkin M. 1969. Further evidence on the locus of the visual area in the temporal lobe of the monkey. *Exp. Neurol.* 25:585–94

Iwai E, Yukie M. 1987. Amygdalofugal and amygdalopetal connections with modality-specific visual cortical areas in macaques (*macaca fuscata, m. mulatta* and *m. fascicularis*). *J. Comp. Neurol.* 261:362–87

Jeannerod M, Arbib MA, Rizzolatti G, Sakata H. 1995. Grasping objects: the cortical mechanisms of visuomotor transformation. *Trends Neurosci.* 7:314–20

Johnson MH, Dziurawiec S, Ellis H, Morton J. 1991. Newborns' preferential tracking of face-like stimuli and its subsequent decline. *Cognition* 40:1–19

Jolicoeur P. 1985. The time to name disoriented natural objects. *Mem. Cogn.* 13:289–303

Jolicoeur P. 1987. A size-congruency effect in memory for visual shape. *Mem. Cogn.* 15:531–43

Jolicoeur P, Gluck MA, Kosslyn SM. 1984. Pictures and names: making the connection. *Cogn. Psychol.* 16:243–75

Katz D, ed. 1953. *Studien zur experimentellen Psychologie.* Basel: Schwabe

Knapp AG, Anderson JA. 1984. Theory of categorization based on distributed memory storage. *J. Exp. Psychol.* 10:616–37

Kobatake E, Tanaka K. 1994. Neuronal selectivities to complex object features in the ventral visual pathway of the macaque cerebral cortex. *J. Neurophysiol.* 71:856–67

Kobatake E, Tanaka K, Wang G, Tamori Y. 1993. Effects of adult learning on the stimulus selectivity of cells in the inferotemporal cortex. *Soc. Neurosci. Abstr.* 19:975

Koenderink JJ. 1990. *Solid Shape.* Cambridge, MA: MIT Press

Landis T, Regard M, Bliestle A, Kleihues P. 1988. Prosopagnosia and agnosia for noncanonical views. An autopsied case. *Brain* 111:1287–97

Leonard CM, Rolls ET, Wilson FA, Baylis GC. 1985. Neurons in the amygdala of the monkey with responses selective for faces. *Behav. Brain. Res.* 15:159–76

Lhermitte F, Chain F, Escourolle R, Ducarne B, Pillon B. 1972. Etude anotomo-clinique d'un cas de prosopagnosie. *Rev. Neurol.* 126:329–46

Lhermitte F, Pillon B. 1975. La prosopagnosie. Role de l'hemisphere droit dans la perception visuelle. *Rev. Neurol.* 131:791–812

Liu Z, Knill DC, Kersten D. 1995. Object classification for human and ideal observers. *Vis. Res.* 35:549–68

Logothetis NK. 1994. Physiological studies of motion inputs. In *Visual Detection of Motion.* ed. AT Smith, RJ Snowden, pp. 177–216. New York: Academic

Logothetis NK, Pauls J. 1995. Psychophysical and physiological evidence for viewer-centered representations in the primate. *Cereb. Cortex* 5:270–88

Logothetis NK, Pauls J, Bülthoff HH, Poggio T. 1994. View-dependent object recognition by monkeys. *Curr. Biol.* 4:401–14

Logothetis NK, Pauls J, Poggio T. 1995. Shape representation in the inferior temporal cortex of monkeys. *Curr. Biol.* 5:552–63

Lorenz K. 1971. *Studies in Animal and Human Behaviour.* Vol. 2. Cambridge, MA: Harvard Univ. Press

Lynch JC. 1980. The functional organization of posterior parietal association cortex. *Behav. Brain Sci.* 3:485–534

Macrae D, Trolle W. 1956. The defect of function in visual agnosia. *Brain* 79:94–110

Marr D. 1982. *Vision.* New York: Freeman

Marr D, Nishihara HK. 1978. Representation and recognition of the spatial organization of three-dimensional shapes. *Proc. R. Soc. London Ser. B* 200:269–94

Marsolek CJ. 1995. Abstract visual-form representation in the left cerebral hemisphere. *J. Exp. Psychol.: Hum. Percept. Perform.* 21:375–86

McClelland JL, Rumelhart DE. 1985. Distributed memory and the representation of general and specific information. *J. Exp. Psychol.* 114:159–88

McMullen B, Jolicoeur P. 1992. The reference frame and effects of orientation on finding the top of rotated objects. *J. Exp. Psychol.: Hum. Percept. Perform.* 18:802–20

Mikami A, Kubota K. 1980. Inferotemporal neuron activities and color discrimination with delay. *Brain Res.* 182:65–78

Milliken B, Jolicoeur P. 1992. Size effects in visual recognition memory are determined by perceived size. *Mem. Cogn.* 20:83–95

Milner AD, Heywood CA. 1989. A disorder of lightness discrimination in a case of visual form agnosia. *Cortex* 25:489–94

Mishkin M, Appenzeller T. 1987. The anatomy of memory. *Sci. Am.* 256:80–89

Miyashita Y. 1993. Inferior temporal cortex: where visual perception meets memory. *Annu. Rev. Neurosci.* 16:245–63

Mountcastle VB, Motter BC, Steinmetz MA, Duffy CJ. 1984. Looking and seeing: the visual functions of the parietal lobe. In *Dynamic Aspects of Neocortical Function.* ed. GM Edelman, WE Call, WM Cowan, pp. 159–93. New York: Wiley

Murphy GL, Brownell HH. 1985. Category differentiation in object recognition: typicality constraints on the basic category advantage. *J. Exp. Psychol.: Learn. Mem. Cogn.* 11:70–84

Murray JE, Jolicoeur P, McMullen PA, Ingleton M. 1993. Orientation-invariant transfer of training in the identification of rotated natural objects. *Mem. Cogn.* 21:604–10

Nazir TA, O'Regan JK. 1990. Some results on translation invariance in the human visual system. *Spatial Vis.* 5:81–100

Nieuwenhuys R, Voogd J, van Huijzen C. 1980. *Das Zentrainervensystem des Menschen.* Berlin: Springer-Verlag

Ojemann GA, Ojemann JG, Haglund M, Holmes M, Lettich E. 1993. Visually related activity in human temporal cortical neurons. In *Functional Organisation of the Human Cortex.* ed. B Gulyaas, D Ottoson, PE Roland, pp. 279–89. Oxford: Pergamon

Oram MW, Perrett DI. 1994. Modeling visual recognition from neurobiological constraints. *Neural Networks* 7:945–72

O'Regan JK. 1992. Solving the 'real' mysteries of visual perception: the world as an outside memory. *Can. J. Psych.* 46:461–88

Palmer SE. 1977. Hierarchical structure in perceptual representation. *Cogn. Psychol.* 9:441–74

Palmer SE, Rosch E, Chase P. 1981. Canonical perspective and the perception of objects. In *Attention and Performance IX,* ed. J Long, A Baddeley, pp. 135–51. Hillsdale: Erlbaum

Pentland AP. 1986. Perceptual organization and the representation of natural form. *Artif. Intell.* 28:293–331

Perrett DI, Harries MH, Bevan R, Thomas S, Benson PJ, et al. 1989. Frameworks of analysis for the neural representation of animate objects and actions. *J. Exp. Biol.* 146:87–113

Perrett DI, Hietanen JK, Oram MW, Benson PJ. 1992. Organization and functions of cells responsive to faces in the temporal cortex. *Philos. Trans. R. Soc. London Ser. B* 335:23–30

Perrett DI, Oram MW. 1993. Neurophysiology of shape processing. *Image Vis. Comput.* 11:317–33

Perrett DI, Rolls ET, Caan W. 1979. Temporal lobe cells of the monkey with visual responses selective for faces. *Neurosci. Lett. Suppl.* S3:S358

Perrett DI, Rolls ET, Caan W. 1982. Visual neurones responsive to faces in the monkey temporal cortex. *Exp. Brain Res.* 47:329–42

Perrett DI, Smith PAJ, Potter DD, Mistlin AJ, Head AS, et al. 1985. Visual cells in the temporal cortex sensitive to face view and gaze direction. *Proc. R. Soc. London Ser. B* 223:293–317

Piaget J. 1969. *The Mechanisms of Perception.* Transl. M. Cook. New York: Basic Books

Pinker S. 1985. Visual cognition: an introduction. In *Visual Cognition.* ed. S Pinker, pp. 1–63. Cambridge, MA: MIT Press

Poggio T. 1990. A theory of how the brain might work. *Cold Spring Harbor Symp. Quant. Biol.* 55:899–910

Posner MI, Goldsmith R, Welton KE. 1967. Perceived distance and the classification of distorted patterns. *J. Exp. Psychol.* 73:28–38

Posner MI, Keele SW. 1968. On the genesis of abstract ideas. *J. Exp. Psychol.* 77:353–63

Posner MI, Keele SW. 1970. Retention of abstract ideas. *J. Exp. Psychol.* 83:304–8

Quinn PC, Eimas PD. 1986. Categorization in early infancy. *Merrill-Palmer Q.* 32:331–63

Quinn PC, Eimas PD, Rosenkrantz SL. 1993. Evidence for representations of perceptually similar natural categories by 3-month-old and 4-month-old infants. *Perception* 22:463–75

Ramachandran VS. 1990. Perceiving shape from shading. In *The Perceptual World.* ed. I Rock, pp. 127–38. New York: Freeman

Reed SK. 1972. Pattern recognition and categorization. *Cogn. Psychol.* 3:382–407

Richmond BJ, Optican LM, Podell M, Spitzer H. 1987. Temporal encoding of two-dimensional patterns by single units in primate inferior temporal cortex. 1. Response characteristics. *J. Neurophysiol.* 57:132–46

Robinson DL, Goldberg ME, Stanton GB. 1978. Parietal association cortex in the primate: sensory mechanisms and behavioral modulations. *J. Neurophysiol.* 41:910–32

Rock I, DiVita J. 1987. A case of viewer-centered object perception. *Cogn. Psychol.* 19: 280–93

Rock I, DiVita J, Barbeito R. 1981. The effect on form perception of change of orientation in the third dimension. *J. Exp. Psychol.* 7: 719–32

Rock I, Wheeler D, Tudor L. 1989. Can we imagine how objects look from other viewpoints? *Cogn. Psychol.* 21:185–210

Rockland KS, Saleem KS, Tanaka K. 1994. Divergent feedback connections from areas V4 and TEO in the macaque. *Vis. Neurosci.* 11:579–600

Rodman HR, Ò Scalaidhe SP, Gross CG. 1993. Response properties of neurons in temporal cortical visual areas of infant monkeys. *J. Neurophysiol.* 70:1115–36

Rolls ET. 1984. Neurons in the cortex of the temporal lobe and in the amygdala of the monkey with responses selective for faces. *Hum. Neurobiol.* 3:209–22

Rolls ET. 1992a. Neurophysiolgy and functions of the primate amygdala. In *The Amygdala,* ed. JP Aggleton, pp. 143–65. New York: Wiley-Liss

Rolls ET. 1992b. Neurophysiological mechanisms underlying face processing within and beyond the temporal cortical visual areas. *Philos. Trans. R. Soc. London Ser. B* 35:11–21

Rolls ET. 1994. Brain mechanisms for invariant visual recognition and learning. *Behav. Process.* 22:113–38

Rosch E, Mervis CB, Gray WD, Johnson DM, Boyes-Braem P. 1976a. Basic objects in natural categories. *Cogn. Psychol.* 8:382–439

Rosch E, Simpson C, Miller RS. 1976b. Structural bases of typicality effects. *J. Exp. Psychol.: Hum. Percept. Perform.* 2:491–502

Rosenfeld SA, Van Hoesen GW. 1979. Face recognition in the rhesus monkey. *Neuropsychologia* 17:503–9

Ryan CME. 1982. Concept formation and individual recognition in the domestic chicken (*gallus gallus*). *Behav. Anal. Lett.* 2:213–20

Sakata H, Taira M. 1994. Parietal control of hand action. *Curr. Opin. Neurobiol.* 4:847–56

Sanghera MK, Rolls ET, Roper-Hall A. 1979. Visual responses of neurons in the dorsolateral amygdala of the alert monkey. *Exp. Neurol.* 63:610–26

Sáry C, Vogels R, Orban GA. 1993. Cue-invariant shape selectivity of macaque inferior temporal neurons. *Science* 260:995–97

Sato T, Kawamura T, Iwai E. 1980. Responsiveness of inferotemporal single units to visual pattern stimuli in monkeys performing discrimination. *Exp. Brain Res.* 38:313–19

Schwartz EL, Desimone R, Albright TD, Gross CG. 1983. Shape recognition and inferior temporal neurons. *Proc. Natl. Acad. Sci. USA* 80:5776–78

Seltzer B, Pandya DN. 1978. Afferent cortical connections and architectonics of the superior temporal sulcus and surrounding cortex in the rhesus monkey. *Brain Res.* 149:1–24

Seltzer B, Pandya DN. 1994. Parietal, temporal, and occipital projections to cortex of the superior temporal sulcus in the rhesus monkey: a retrograde tracer study. *J. Comp. Neurol.* 343:445–63

Sereno A, Maunsell JHR. 1995. Spatial and shape selective sensory and attentional effects in neurons in the macaque lateral intraparietal cortex (LIP). *Invest. Ophthalmol. Vis. Sci. Suppl.* 36:S692

Sergent J, Ohta S, MacDonald B. 1992. Functional neuroanatomy of face and object processing. A positron emission tomography study. *Brain* 115:15–36

Shinar D, Owen DH. 1973. Effects of form rotation on the speed of classification: the development of shape constancy. *Percept. Psychophys.* 14:149–54

Shiwa T. 1987. Corticocortical projections to the monkey temporal lobe with particular reference to the visual processing pathways. *Arch. Ital. Biol.* 125:139–54

Simion F, Bagnara S, Roncato S, Umilta C. 1982. Transformation processes upon the visual code. *Percept. Psychophys.* 31:13–25

Snowden RJ. 1994. Motion processing in the primate cerebral cortex. In *Visual Detection of Motion,* ed. AT Smith, RJ Snowden, pp. 51–83. New York: Academic

Srinivas K. 1993. Perceptual specificity in nonverbal priming. *J. Exp. Psychol.: Learn. Mem. Cogn.* 19:582–602

Strange W, Kenney T, Kessel F, Jenkins J. 1970. Abstraction over time of prototypes from distortions of random dot patterns. *J. Exp. Psychol.* 83:508–10

Suzuki WA, Amaral DG. 1994. Topographic organization of the reciprocal connections between the monkey entorhinal cortex and the perirhinal and parahippocampal cortices. *J. Neurosci.* 14:1856–77

Taira M, Mine S, Georgopoulos AP, Murata A, Sakata H. 1990. Parietal cortex neurons of the monkey related to the visual guidance of hand movement. *Exp. Brain Res.* 83:29–36

Tanaka JW, Taylor M. 1991. Object categories

and expertise: Is the basic level in the eye of the beholder? *Cogn. Psychol.* 23:457–82

Tanaka K. 1993. Neuronal mechanisms of object recognition. *Science* 262:685–88

Tanaka K. 1996. Inferotemporal cortex and object vision. *Annu. Rev. Neurosci.* 19:109–39

Tanaka K, Saito HA, Fukada Y, Moriya M. 1991. Coding visual images of objects in the inferotemporal cortex of the macaque monkey. *J. Neurophysiol.* 66:170–89

Tarr M, Pinker S. 1989. Mental rotation and orientation-dependence in shape recognition. *Cogn. Psychol.* 21:233–82

Tarr MJ. 1995. Rotating objects to recognize them: a case study of the role of mental transformations in the recognition of three-dimensional objects. *Psychol. Bull. Rev.* 2: 55–82

Tranel D, Damasio AR, Damasio H. 1988. Intact recognition of facial expression, gender, and age in patients with impaired recognition of face identity. *Neurology* 38:690–96

Tulving E, Schacter DL. 1990. Priming and human memory systems. *Science* 247:301–6

Turner BH, Mishkin M, Knapp M. 1980. Organization of the amygdalopetal projections from modality specific cortical association areas in the monkey. *J. Comp. Neurol.* 191: 515–43

Tversky B, Hemenway K. 1984. Objects, parts, and categories. *J. Exp. Psychol.* 113:169–93

Ullman S. 1989. Aligning pictorial descriptions: an approach to object recognition. *Cognition* 32:193–254

Ungerleider LG, Desimone R. 1986. Projections to the superior temporal sulcus from the central and peripheral field representations of VI and V2. *J. Comp. Neurol.* 248:147–63

Ungerleider LG, Mishkin M. 1982. Two cortical visual systems. In *Analysis of Visual Behavior*, ed. DJ Ingle, pp. 549–86. Cambridge, MA: MIT Press

Valentine T. 1988. Upside-down faces: a review of the effect of inversion upon face recognition. *Br. J. Psychol.* 79:471–92

Van Hoesen GW. 1982. The parahippocampal gyrus: new observations regarding its cortical connections in the monkey. *Trends Neurosci.* 52:345–50

Vetter T, Poggio T, Bülthoff HH. 1994. The importance of symmetry and virtual views in three-dimensional object recognition. *Curr. Biol.* 4:18–23

Vishton PM, Cutting JE. 1995. Veridical size perception for action reaching vs. estimating. *Invest. Ophthalmol. Vis. Sci. Suppl.* 36:S358

Von Bonin G, Bailey P. 1947. *The Neocortex of Macaca Mulatta.* Urbana, IL: Univ. Ill. Press. 4th ed.

Von Bonin G, Bailey P. 1950. *The Neocortex of the Chimpanzee.* Urbana, IL: Univ. Ill. Press

Wachsmuth E, Oram MW, Perrett DI. 1994.

Recognition of objects and their component parts: responses of single units in the temporal cortex of the macaque. *Cereb. Cortex* 4: 509–22

Warrington EK, Taylor AM. 1973. The contribution of the right parietal lobe to object recognition. *Cortex* 9:152–64

Watanabe S, Ito Y. 1991. Discrimination of individuals in pigeons. *Bird Behav.* 9:20–29

Watanabe S, Lea SEG, Dittrich WH. 1993. What can we learn from experiments on pigeon concept discrimination? In *Vision, Brain, and Behavior in Birds,* ed. HP Zeigler, HJ Bischof, pp. 351–76. Cambridge, MA: MIT Press

Waxman AM, Seibert M, Bernardon AM, Fay DA. 1993. Neural systems for automatic target learning and recognition. *Lincoln Lab J.* 6:77–116

Webster MJ, Bachevalier J, Ungerleider LG. 1993. Subcortical connections of inferior temporal areas TE and TEO in macaque monkeys. *J. Comp. Neurol.* 335:73–91

Weiskrantz L. 1956. Behavioral changes associated with ablation of the amygdaloid complex in monkeys. *J. Comp. Physiol. Psychol.* 49:381–91

Whiteley AM, Warrington EK. 1977. Prosopagnosia: a clinical, psychological, and anatomical study of three patients. *J. Neurol. Neurosurg. Psychiatry* 40:395–403

Williams GV, Rolls ET, Leonard CM, Stern C. 1993. Neuronal responses in the ventral striatum of the behaving macaque. *Behav. Brain Res.* 55:243–52

Wilson FA, Rolls ET. 1993. The effects of stimulus novelty and familiarity on neuronal activity in the amygdala of monkeys performing recognition memory tasks. *Exp. Brain Res.* 93:367–82

Wilson FAW, Ò Scalaidhe SP, Goldman-Rakic PS. 1994. Functional synergism between putative gamma-aminobutyrate-containing neurons and pyramidal neurons in prefrontal cortex. *Proc. Natl. Acad. Sci. USA* 91:4009–13

Yamane S, Kaji S, Kawano K. 1988. What facial features activate face neurons in the inferotemporal cortex of the monkey? *Exp. Brain Res.* 73:209–14

Yeterian EH, Van Hoesen GW. 1978. Corticostriate projections in the rhesus monkey: the organization of certain cortico-caudate connections. *Brain Res.* 139:43–63

Young AW, Hellawell D, Hay DC. 1987. Configurational information in face perception. *Perception* 16:747–60

Young MP. 1993. Visual cortex: modules for pattern recognition. *Curr. Biol.* 3:44–46

Young MP, Yamane S. 1992. An analysis at the population level of the processing of faces in the inferotemporal cortex. In *Brain Mechanisms of Perception and Memory: From Neuron to Behaviour,* ed. L Squire, T

Ono, M Fukuda, D Perrett, pp. 47–71. New York: Oxford Univ. Press

Yukie M, Iwai E. 1988. Direct projections from the ventral TE area of the inferotemporal cortex to hippocampal field CA1 in the monkey. *Neurosci. Lett.* 88:6–10

Zeki SM. 1973. Comparison of the cortical de-generation in the visual regions of the temporal lobe of the monkey following section of the anterior commissure and the splenium. *J. Comp. Neurol.* 148:167–76

Zola-Morgan S, Squire LR. 1993. Neuro-anatomy of memory. *Annu. Rev. Neurosci.* 16:547–63

# SUBJECT INDEX

## A

Aβ deposition
  Alzheimer's disease and, 54,
    70–73
  isoform-specific binding of
    apolipoprotein E and, 71–
    72
Acetylcholine
  heterosynaptic LTD modula-
    tion and, 441
  ventral tegmental area
    dopamine–containing neu-
    rons and
    modulation of, 418–19
Acyclovir, 274
Addiction, 319–33, 405–6, 423–
  28
  possible dopamine system role
    in, 406, 423–28
  reinforcement effects and,
    423–28
  possible sites of rewarding ac-
    tion, 328–33
  reward circuitry, 322
  self-administration and, 320
  (See also Brain stimulation re-
    ward)
Adeno-associated viruses
  (AAVs), 267
Adenosine 3',5'-cyclic monophos-
  phate (See cAMP)
Adenosine deaminase, 43–45
Adenovirus, 266–67
Adenylyl cyclase, 236, 522–23
  type III
    cAMP olfactory signal
      transduction pathway
      and, 522–23
Adrenal chromaffin cells
  mechanisms of exocytosis in,
    221
n-Aequorin-J, 221
Affected Pedigree Member
  (APM) analysis, 65–67
AIDS, 1–20
  HIV-associated dementia and,
    1–20
  (See also HIV, neurotoxicity)
AIDS-dementia complex (See
  HIV, neurotoxicity)
Alzheimer's disease, 53–73
  Affected Pedigree Member
    analysis of, 65–67
  APOE 4 and late-onset famil-
    ial, 57–66, 72–73

apolipoprotein E metabolism
  and, 53, 55, 67–73
  Aβ deposition and, 54, 70–
    73
  isoform specific effects of,
    67–73
  MAP2 and, 55, 70–73
  tau and, 55, 70–73
  clinical phenotype, 53–55, 64,
    68–73
  neuritic plaques and, 53–55,
    64, 73
  neurofibrillary tangles and,
    53–55, 64, 68–73
  early onset
    identified forms of, 61–62
  genetic heterogeneity of, 55–61
  herpes simplex virus type 1–
    mediated gene transfer
    and, 282
  linkage studies of, 61–64
γ-Aminobutyric acid (GABA)
  ventral tegmental area
    dopamine–containing neu-
    rons and, 417–18, 425–28
α-Amino-3-hydroxy-5-
  methylisoxazole-4-propion-
  ate (See AMPA)
AMPA, 36–37
Amphetamine, 327–33
Amplicons, 277–78, 281
  tyrosine hydroxylase gene
    transfer and, 281
Amygdala
  object vision and, 131–35
Amyloid precursor protein
  (APP), 54, 64
  in neuritic plaques, 54, 64
Amyotrophic lateral sclerosis,
  187, 205–8
  neurofilament abnormalities
    and, 205–8
Anemone toxin, 155, 157
Anterior inferotemporal cortex
  afferents to, 119–24
  columnar organization of, 116–
    17, 119–31
    binding and, 130–31
    continuous mapping and,
      129–30
    functions of, 129–30
    optical imaging of, 124–30
  selectivity of cells in, 110–16,
    122–24
    changeability of 127–30
    position, orientation, and
      size in, 114–15, 122–24

stimulus, 110–14, 116
  visual object recognition in,
    109–36
Anterior middle temporal sulcus
  (AMTS)
  object vision and, 110, 134
α₁-Antichymotrypsin, 54
Apolipoprotein E (ApoE)
  Alzheimer's disease risk and,
    53, 67–73
  intraneural localization of, 67–
    70
  isoform-specific interactions
    of, 67–68, 70–73
    Aβ peptide and, 71–72
    HDL receptor and, 67
    LDL receptor and, 67–68
    VLDL receptor and, 67
  metabolism of, 53, 67–73
  mRNA expression, 68
  neuronal metabolism and, 68–
    70
Apolipoprotein E locus (APOE),
  53–73
  and Alzheimer's disease, 58–73
Apomorphine, 328
Apterous, 547
Arachidonic acid
  metabolites of
    HIV neurotoxicity and, 15–
      16
β-Arrestin (BARR), 526
Astrocytosis
  HIV-associated dementia and,
    6
Attraction
  chemo-, 341–46
  contact-mediated, 341–46
  growth cone guidance and,
    341–46
Axon guidance, 341–71, 556–60
  (See also Growth cone guid-
    ance)
Axonal cytoskeleton
  neurofilaments and, 187–208
Axonal transport, 187, 199–206
  of neuronal intermediate fila-
    ments, 199–206
Axonogenesis
  Drosophila, 545, 554, 556–60

## B

β-arrestin (BARR), 526
Balint's syndrome, 597
BAPTA, 223–24, 450
  LTD induction with

623

# CUMULATIVE INDEXES

## CONTRIBUTING AUTHORS, VOLUMES 11–19

# CHAPTER TITLES, VOLUMES 11–19

637

# ANNUAL REVIEWS INC.
4139 El Camino Way • P.O. Box 10139
Palo Alto, CA 94303-0139 • USA

BB96

**Step 1** *Ordered by:*

Name _____

Address _____

_____ Zip Code _____

**Call from USA or Canada**
 **1.800.523.8635**

FAX orders 24 hours a day
 **1.415.424.0910**

Today's Date _____ Day Phone: ( )_____

Fax ( )_____ e-mail _____

**Step 4** *Payment Method*

☐ Check or money order enclosed. Make checks payable to "Annual Reviews Inc." *or charge*

☐ VISA   ☐ M/C   ☐ AMEX

Account Number _____

Mo __/__ Yr __/__   Print name exactly as it appears on credit card.
Expiration Date _____

Signature _____

| Qty | Annual Review of | Vol. | Place on Standing Order? Save 10% now with payment | Price | Total |
|---|---|---|---|---|---|
| | | | ☐ Yes, save 10%  ☐ No | $ | |
| | | | ☐ Yes, save 10%  ☐ No | $ | |
| | | | ☐ Yes, save 10%  ☐ No | $ | |
| | | | ☐ Yes, save 10%  ☐ No | $ | |
| | | | ☐ Yes, save 10%  ☐ No | $ | |

**Step 2** ☐ Student / Recent Graduate (past three years) discount 30% off. Not applicable to standing orders. Proof of status enclosed.

***Enter Order***
☐ California customers. Add applicable California sales tax for your location.
☐ Canadian customers. Add 7% Canadian GST. **(Reg. # 121449029 RT)**

**Step 3**
✔ Handling Charges. Add $3 per volume. Applies to all orders.

***Shipping and Handling***
☐ Standard shipping, US Mail 4th class bookrate (surface) No extra charge.   **N/C**

☐ Optional UPS Ground service, $3 extra per volume in 48 contiguous states only. UPS not available to PO boxes.

UPS Next Day Air ☐   UPS Second Day Air ☐   US Airmail ☐   Note option at left. We will calculate amount and add to your total.
Optional shipping to anywhere. Charged at actual cost and added to total. Prices vary by weight of volumes.

**Total** _____

 **Call Toll Free 1.800.523.8635 from USA or Canada** 8am–4pm, M-F, Pacific Time. From elsewhere call 1.415.493.4400 ext. 1

✉ Mail Orders, fill in form, send in attached envelope.   @ e-mail **service@annurev.org**

**ANNUAL REVIEWS INC.** on the Web **http://www.annurev.org**

Thank You for Your Order

From_____

ANNUAL REVIEWS INC
4139 EL CAMINO WAY
P O BOX 10139
PALO ALTO   CA    94303-0139

Place
Stamp
Here